T0189192

Lecture Notes in Computer Science 12366

More information about this series at http://www.springer.com/series/7412

Andrea Vedaldi · Horst Bischof ·
Thomas Brox · Jan-Michael Frahm (Eds.)

Computer Vision – ECCV 2020

16th European Conference
Glasgow, UK, August 23–28, 2020
Proceedings, Part XXI

Springer

Editors
Andrea Vedaldi ⓘ
University of Oxford
Oxford, UK

Horst Bischof ⓘ
Graz University of Technology
Graz, Austria

Thomas Brox ⓘ
University of Freiburg
Freiburg im Breisgau, Germany

Jan-Michael Frahm
University of North Carolina at Chapel Hill
Chapel Hill, NC, USA

ISSN 0302-9743   ISSN 1611-3349   (electronic)
Lecture Notes in Computer Science
ISBN 978-3-030-58588-4   ISBN 978-3-030-58589-1   (eBook)
https://doi.org/10.1007/978-3-030-58589-1

LNCS Sublibrary: SL6 – Image Processing, Computer Vision, Pattern Recognition, and Graphics

This Springer imprint is published by the registered company Springer Nature Switzerland AG
The registered company address is: Gewerbestrasse 11, 6330 Cham, Switzerland

Foreword

Hosting the European Conference on Computer Vision (ECCV 2020) was certainly exciting journey. From the 2016 plan to hold it at the Edinburgh Internatior Conference Centre (hosting 1,800 delegates) to the 2018 plan to hold it at Glasgow Scottish Exhibition Centre (up to 6,000 delegates), we finally ended with movir online because of the COVID-19 outbreak. While possibly having fewer delegates th: expected because of the online format, ECCV 2020 still had over 3,100 register participants.

Although online, the conference delivered most of the activities expected at face-to-face conference: peer-reviewed papers, industrial exhibitors, demonstratior and messaging between delegates. In addition to the main technical sessions, tl conference included a strong program of satellite events with 16 tutorials and 4 workshops.

Furthermore, the online conference format enabled new conference features. Evel paper had an associated teaser video and a longer full presentation video. Along wi the papers and slides from the videos, all these materials were available the week befor the conference. This allowed delegates to become familiar with the paper content ar be ready for the live interaction with the authors during the conference week. The liv event consisted of brief presentations by the oral and spotlight authors and industri: sponsors. Question and answer sessions for all papers were timed to occur twice s delegates from around the world had convenient access to the authors.

As with ECCV 2018, authors' draft versions of the papers appeared online wit open access, now on both the Computer Vision Foundation (CVF) and the Europea Computer Vision Association (ECVA) websites. An archival publication arrangemer was put in place with the cooperation of Springer. SpringerLink hosts the final versio of the papers with further improvements, such as activating reference links and sup plementary materials. These two approaches benefit all potential readers: a versio available freely for all researchers, and an authoritative and citable version wit additional benefits for SpringerLink subscribers. We thank Alfred Hofmann an Aliaksandr Birukou from Springer for helping to negotiate this agreement, which w expect will continue for future versions of ECCV.

August 2020

Vittorio Ferrar
Bob Fishe
Cordelia Schmic
Emanuele Trucc

Preface

Welcome to the proceedings of the European Conference on Computer Vision (ECCV 2020). This is a unique edition of ECCV in many ways. Due to the COVID-19 pandemic, this is the first time the conference was held online, in a virtual format. This was also the first time the conference relied exclusively on the Open Review platform to manage the review process. Despite these challenges ECCV is thriving. The conference received 5,150 valid paper submissions, of which 1,360 were accepted for publication (27%) and, of those, 160 were presented as spotlights (3%) and 104 as orals (2%). This amounts to more than twice the number of submissions to ECCV 2018 (2,439). Furthermore, CVPR, the largest conference on computer vision, received 5,850 submissions this year, meaning that ECCV is now 87% the size of CVPR in terms of submissions. By comparison, in 2018 the size of ECCV was only 73% of CVPR.

The review model was similar to previous editions of ECCV; in particular, it was double blind in the sense that the authors did not know the name of the reviewers and vice versa. Furthermore, each conference submission was held confidentially, and was only publicly revealed if and once accepted for publication. Each paper received at least three reviews, totalling more than 15,000 reviews. Handling the review process at this scale was a significant challenge. In order to ensure that each submission received as fair and high-quality reviews as possible, we recruited 2,830 reviewers (a 130% increase with reference to 2018) and 207 area chairs (a 60% increase). The area chairs were selected based on their technical expertise and reputation, largely among people that served as area chair in previous top computer vision and machine learning conferences (ECCV, ICCV, CVPR, NeurIPS, etc.). Reviewers were similarly invited from previous conferences. We also encouraged experienced area chairs to suggest additional chairs and reviewers in the initial phase of recruiting.

Despite doubling the number of submissions, the reviewer load was slightly reduced from 2018, from a maximum of 8 papers down to 7 (with some reviewers offering to handle 6 papers plus an emergency review). The area chair load increased slightly, from 18 papers on average to 22 papers on average.

Conflicts of interest between authors, area chairs, and reviewers were handled largely automatically by the Open Review platform via their curated list of user profiles. Many authors submitting to ECCV already had a profile in Open Review. We set a paper registration deadline one week before the paper submission deadline in order to encourage all missing authors to register and create their Open Review profiles well on time (in practice, we allowed authors to create/change papers arbitrarily until the submission deadline). Except for minor issues with users creating duplicate profiles, this allowed us to easily and quickly identify institutional conflicts, and avoid them, while matching papers to area chairs and reviewers.

Papers were matched to area chairs based on: an affinity score computed by the Open Review platform, which is based on paper titles and abstracts, and an affinity score computed by the Toronto Paper Matching System (TPMS), which is based on the paper's full text, the area chair bids for individual papers, load balancing, and conflict avoidance. Open Review provides the program chairs a convenient web interface to experiment with different configurations of the matching algorithm. The chosen configuration resulted in about 50% of the assigned papers to be highly ranked by the area chair bids, and 50% to be ranked in the middle, with very few low bids assigned.

Assignments to reviewers were similar, with two differences. First, there was a maximum of 7 papers assigned to each reviewer. Second, area chairs recommended up to seven reviewers per paper, providing another highly-weighed term to the affinity scores used for matching.

The assignment of papers to area chairs was smooth. However, it was more difficult to find suitable reviewers for all papers. Having a ratio of 5.6 papers per reviewer with a maximum load of 7 (due to emergency reviewer commitment), which did not allow for much wiggle room in order to also satisfy conflict and expertise constraints. We received some complaints from reviewers who did not feel qualified to review specific papers and we reassigned them wherever possible. However, the large scale of the conference, the many constraints, and the fact that a large fraction of such complaints arrived very late in the review process made this process very difficult and not all complaints could be addressed.

Reviewers had six weeks to complete their assignments. Possibly due to COVID-19 or the fact that the NeurIPS deadline was moved closer to the review deadline, a record 30% of the reviews were still missing after the deadline. By comparison, ECCV 2018 experienced only 10% missing reviews at this stage of the process. In the subsequent week, area chairs chased the missing reviews intensely, found replacement reviewers in their own team, and managed to reach 10% missing reviews. Eventually, we could provide almost all reviews (more than 99.9%) with a delay of only a couple of days on the initial schedule by a significant use of emergency reviews. If this trend is confirmed, it might be a major challenge to run a smooth review process in future editions of ECCV. The community must reconsider prioritization of the time spent on paper writing (the number of submissions increased a lot despite COVID-19) and time spent on paper reviewing (the number of reviews delivered in time decreased a lot presumably due to COVID-19 or NeurIPS deadline). With this imbalance the peer-review system that ensures the quality of our top conferences may break soon.

Reviewers submitted their reviews independently. In the reviews, they had the opportunity to ask questions to the authors to be addressed in the rebuttal. However, reviewers were told not to request any significant new experiment. Using the Open Review interface, authors could provide an answer to each individual review, but were also allowed to cross-reference reviews and responses in their answers. Rather than PDF files, we allowed the use of formatted text for the rebuttal. The rebuttal and initial reviews were then made visible to all reviewers and the primary area chair for a given paper. The area chair encouraged and moderated the reviewer discussion. During the discussions, reviewers were invited to reach a consensus and possibly adjust their ratings as a result of the discussion and of the evidence in the rebuttal.

After the discussion period ended, most reviewers entered a final rating and recommendation, although in many cases this did not differ from their initial recommendation. Based on the updated reviews and discussion, the primary area chair then made a preliminary decision to accept or reject the paper and wrote a justification for it (meta-review). Except for cases where the outcome of this process was absolutely clear (as indicated by the three reviewers and primary area chairs all recommending clear rejection), the decision was then examined and potentially challenged by a secondary area chair. This led to further discussion and overturning a small number of preliminary decisions. Needless to say, there was no in-person area chair meeting, which would have been impossible due to COVID-19.

Area chairs were invited to observe the consensus of the reviewers whenever possible and use extreme caution in overturning a clear consensus to accept or reject a paper. If an area chair still decided to do so, she/he was asked to clearly justify it in the meta-review and to explicitly obtain the agreement of the secondary area chair. In practice, very few papers were rejected after being confidently accepted by the reviewers.

This was the first time Open Review was used as the main platform to run ECCV. In 2018, the program chairs used CMT3 for the user-facing interface and Open Review internally, for matching and conflict resolution. Since it is clearly preferable to only use a single platform, this year we switched to using Open Review in full. The experience was largely positive. The platform is highly-configurable, scalable, and open source. Being written in Python, it is easy to write scripts to extract data programmatically. The paper matching and conflict resolution algorithms and interfaces are top-notch, also due to the excellent author profiles in the platform. Naturally, there were a few kinks along the way due to the fact that the ECCV Open Review configuration was created from scratch for this event and it differs in substantial ways from many other Open Review conferences. However, the Open Review development and support team did a fantastic job in helping us to get the configuration right and to address issues in a timely manner as they unavoidably occurred. We cannot thank them enough for the tremendous effort they put into this project.

Finally, we would like to thank everyone involved in making ECCV 2020 possible in these very strange and difficult times. This starts with our authors, followed by the area chairs and reviewers, who ran the review process at an unprecedented scale. The whole Open Review team (and in particular Melisa Bok, Mohit Unyal, Carlos Mondragon Chapa, and Celeste Martinez Gomez) worked incredibly hard for the entire duration of the process. We would also like to thank René Vidal for contributing to the adoption of Open Review. Our thanks also go to Laurent Charling for TPMS and to the program chairs of ICML, ICLR, and NeurIPS for cross checking double submissions. We thank the website chair, Giovanni Farinella, and the CPI team (in particular Ashley Cook, Miriam Verdon, Nicola McGrane, and Sharon Kerr) for promptly adding material to the website as needed in the various phases of the process. Finally, we thank the publication chairs, Albert Ali Salah, Hamdi Dibeklioglu, Metehan Doyran, Henry Howard-Jenkins, Victor Prisacariu, Siyu Tang, and Gul Varol, who managed to compile these substantial proceedings in an exceedingly compressed schedule. We express our thanks to the ECVA team, in particular Kristina Scherbaum for allowing open access of the proceedings. We thank Alfred Hofmann from Springer who again

serve as the publisher. Finally, we thank the other chairs of ECCV 2020, including in
particular the general chairs for very useful feedback with the handling of the program.

August 2020   Andrea Vedaldi
                                                       Horst Bischof
                                                      Thomas Brox
                                                 Jan-Michael Frahm

Organization

General Chairs

Vittorio Ferrari   Google Research, Switzerland
Bob Fisher   University of Edinburgh, UK
Cordelia Schmid   Google and Inria, France
Emanuele Trucco   University of Dundee, UK

Program Chairs

Andrea Vedaldi   University of Oxford, UK
Horst Bischof   Graz University of Technology, Austria
Thomas Brox   University of Freiburg, Germany
Jan-Michael Frahm   University of North Carolina, USA

Industrial Liaison Chairs

Jim Ashe   University of Edinburgh, UK
Helmut Grabner   Zurich University of Applied Sciences, Switzerland
Diane Larlus   NAVER LABS Europe, France
Cristian Novotny   University of Edinburgh, UK

Local Arrangement Chairs

Yvan Petillot   Heriot-Watt University, UK
Paul Siebert   University of Glasgow, UK

Academic Demonstration Chair

Thomas Mensink   Google Research and University of Amsterdam, The Netherlands

Poster Chair

Stephen Mckenna   University of Dundee, UK

Technology Chair

Gerardo Aragon Camarasa   University of Glasgow, UK

Tutorial Chairs

Carlo Colombo	University of Florence, Italy
Sotirios Tsaftaris	University of Edinburgh, UK

Publication Chairs

Albert Ali Salah	Utrecht University, The Netherlands
Hamdi Dibeklioglu	Bilkent University, Turkey
Metehan Doyran	Utrecht University, The Netherlands
Henry Howard-Jenkins	University of Oxford, UK
Victor Adrian Prisacariu	University of Oxford, UK
Siyu Tang	ETH Zurich, Switzerland
Gul Varol	University of Oxford, UK

Website Chair

Giovanni Maria Farinella	University of Catania, Italy

Workshops Chairs

Adrien Bartoli	University of Clermont Auvergne, France
Andrea Fusiello	University of Udine, Italy

Area Chairs

Lourdes Agapito	University College London, UK
Zeynep Akata	University of Tübingen, Germany
Karteek Alahari	Inria, France
Antonis Argyros	University of Crete, Greece
Hossein Azizpour	KTH Royal Institute of Technology, Sweden
Joao P. Barreto	Universidade de Coimbra, Portugal
Alexander C. Berg	University of North Carolina at Chapel Hill, USA
Matthew B. Blaschko	KU Leuven, Belgium
Lubomir D. Bourdev	WaveOne, Inc., USA
Edmond Boyer	Inria, France
Yuri Boykov	University of Waterloo, Canada
Gabriel Brostow	University College London, UK
Michael S. Brown	National University of Singapore, Singapore
Jianfei Cai	Monash University, Australia
Barbara Caputo	Politecnico di Torino, Italy
Ayan Chakrabarti	Washington University, St. Louis, USA
Tat-Jen Cham	Nanyang Technological University, Singapore
Manmohan Chandraker	University of California, San Diego, USA
Rama Chellappa	Johns Hopkins University, USA
Liang-Chieh Chen	Google, USA

Yung-Yu Chuang	National Taiwan University, Taiwan
Ondrej Chum	Czech Technical University in Prague, Czech Republic
Brian Clipp	Kitware, USA
John Collomosse	University of Surrey and Adobe Research, UK
Jason J. Corso	University of Michigan, USA
David J. Crandall	Indiana University, USA
Daniel Cremers	University of California, Los Angeles, USA
Fabio Cuzzolin	Oxford Brookes University, UK
Jifeng Dai	SenseTime, SAR China
Kostas Daniilidis	University of Pennsylvania, USA
Andrew Davison	Imperial College London, UK
Alessio Del Bue	Fondazione Istituto Italiano di Tecnologia, Italy
Jia Deng	Princeton University, USA
Alexey Dosovitskiy	Google, Germany
Matthijs Douze	Facebook, France
Enrique Dunn	Stevens Institute of Technology, USA
Irfan Essa	Georgia Institute of Technology and Google, USA
Giovanni Maria Farinella	University of Catania, Italy
Ryan Farrell	Brigham Young University, USA
Paolo Favaro	University of Bern, Switzerland
Rogerio Feris	International Business Machines, USA
Cornelia Fermuller	University of Maryland, College Park, USA
David J. Fleet	Vector Institute, Canada
Friedrich Fraundorfer	DLR, Austria
Mario Fritz	CISPA Helmholtz Center for Information Security, Germany
Pascal Fua	EPFL (Swiss Federal Institute of Technology Lausanne), Switzerland
Yasutaka Furukawa	Simon Fraser University, Canada
Li Fuxin	Oregon State University, USA
Efstratios Gavves	University of Amsterdam, The Netherlands
Peter Vincent Gehler	Amazon, USA
Theo Gevers	University of Amsterdam, The Netherlands
Ross Girshick	Facebook AI Research, USA
Boqing Gong	Google, USA
Stephen Gould	Australian National University, Australia
Jinwei Gu	SenseTime Research, USA
Abhinav Gupta	Facebook, USA
Bohyung Han	Seoul National University, South Korea
Bharath Hariharan	Cornell University, USA
Tal Hassner	Facebook AI Research, USA
Xuming He	Australian National University, Australia
Joao F. Henriques	University of Oxford, UK
Adrian Hilton	University of Surrey, UK
Minh Hoai	Stony Brooks, State University of New York, USA
Derek Hoiem	University of Illinois Urbana-Champaign, USA

Timothy Hospedales	University of Edinburgh and Samsung, UK
Gang Hua	Wormpex AI Research, USA
Slobodan Ilic	Siemens AG, Germany
Hiroshi Ishikawa	Waseda University, Japan
Jiaya Jia	The Chinese University of Hong Kong, SAR China
Hailin Jin	Adobe Research, USA
Justin Johnson	University of Michigan, USA
Frederic Jurie	University of Caen Normandie, France
Fredrik Kahl	Chalmers University, Sweden
Sing Bing Kang	Zillow, USA
Gunhee Kim	Seoul National University, South Korea
Junmo Kim	Korea Advanced Institute of Science and Technology, South Korea
Tae-Kyun Kim	Imperial College London, UK
Ron Kimmel	Technion-Israel Institute of Technology, Israel
Alexander Kirillov	Facebook AI Research, USA
Kris Kitani	Carnegie Mellon University, USA
Iasonas Kokkinos	Ariel AI, UK
Vladlen Koltun	Intel Labs, USA
Nikos Komodakis	Ecole des Ponts ParisTech, France
Piotr Koniusz	Australian National University, Australia
M. Pawan Kumar	University of Oxford, UK
Kyros Kutulakos	University of Toronto, Canada
Christoph Lampert	IST Austria, Austria
Ivan Laptev	Inria, France
Diane Larlus	NAVER LABS Europe, France
Laura Leal-Taixe	Technical University Munich, Germany
Honglak Lee	Google and University of Michigan, USA
Joon-Young Lee	Adobe Research, USA
Kyoung Mu Lee	Seoul National University, South Korea
Seungyong Lee	POSTECH, South Korea
Yong Jae Lee	University of California, Davis, USA
Bastian Leibe	RWTH Aachen University, Germany
Victor Lempitsky	Samsung, Russia
Ales Leonardis	University of Birmingham, UK
Marius Leordeanu	Institute of Mathematics of the Romanian Academy, Romania
Vincent Lepetit	ENPC ParisTech, France
Hongdong Li	The Australian National University, Australia
Xi Li	Zhejiang University, China
Yin Li	University of Wisconsin-Madison, USA
Zicheng Liao	Zhejiang University, China
Jongwoo Lim	Hanyang University, South Korea
Stephen Lin	Microsoft Research Asia, China
Yen-Yu Lin	National Chiao Tung University, Taiwan, China
Zhe Lin	Adobe Research, USA

Haibin Ling	Stony Brooks, State University of New York, USA
Jiaying Liu	Peking University, China
Ming-Yu Liu	NVIDIA, USA
Si Liu	Beihang University, China
Xiaoming Liu	Michigan State University, USA
Huchuan Lu	Dalian University of Technology, China
Simon Lucey	Carnegie Mellon University, USA
Jiebo Luo	University of Rochester, USA
Julien Mairal	Inria, France
Michael Maire	University of Chicago, USA
Subhransu Maji	University of Massachusetts, Amherst, USA
Yasushi Makihara	Osaka University, Japan
Jiri Matas	Czech Technical University in Prague, Czech Republic
Yasuyuki Matsushita	Osaka University, Japan
Philippos Mordohai	Stevens Institute of Technology, USA
Vittorio Murino	University of Verona, Italy
Naila Murray	NAVER LABS Europe, France
Hajime Nagahara	Osaka University, Japan
P. J. Narayanan	International Institute of Information Technology (IIIT), Hyderabad, India
Nassir Navab	Technical University of Munich, Germany
Natalia Neverova	Facebook AI Research, France
Matthias Niessner	Technical University of Munich, Germany
Jean-Marc Odobez	Idiap Research Institute and Swiss Federal Institute of Technology Lausanne, Switzerland
Francesca Odone	Università di Genova, Italy
Takeshi Oishi	The University of Tokyo, Tokyo Institute of Technology, Japan
Vicente Ordonez	University of Virginia, USA
Manohar Paluri	Facebook AI Research, USA
Maja Pantic	Imperial College London, UK
In Kyu Park	Inha University, South Korea
Ioannis Patras	Queen Mary University of London, UK
Patrick Perez	Valeo, France
Bryan A. Plummer	Boston University, USA
Thomas Pock	Graz University of Technology, Austria
Marc Pollefeys	ETH Zurich and Microsoft MR & AI Zurich Lab, Switzerland
Jean Ponce	Inria, France
Gerard Pons-Moll	MPII, Saarland Informatics Campus, Germany
Jordi Pont-Tuset	Google, Switzerland
James Matthew Rehg	Georgia Institute of Technology, USA
Ian Reid	University of Adelaide, Australia
Olaf Ronneberger	DeepMind London, UK
Stefan Roth	TU Darmstadt, Germany
Bryan Russell	Adobe Research, USA

Kwang Moo Yi   University of Victoria, Canada
Zhaozheng Yin   Stony Brook, State University of New York, USA
Chang D. Yoo   Korea Advanced Institute of Science and Technology,
                              South Korea
Shaodi You   University of Amsterdam, The Netherlands
Jingyi Yu   ShanghaiTech University, China
Stella Yu   University of California, Berkeley, and ICSI, USA
Stefanos Zafeiriou   Imperial College London, UK
Hongbin Zha   Peking University, China
Tianzhu Zhang   University of Science and Technology of China, China
Liang Zheng   Australian National University, Australia
Todd E. Zickler   Harvard University, USA
Andrew Zisserman   University of Oxford, UK

Technical Program Committee

Sathyanarayanan   Samuel Albanie   Pablo Arbelaez
  N. Aakur   Shadi Albarqouni   Shervin Ardeshir
Wael Abd Almgaeed   Cenek Albl   Sercan O. Arik
Abdelrahman   Hassan Abu Alhaija   Anil Armagan
  Abdelhamed   Daniel Aliaga   Anurag Arnab
Abdullah Abuolaim   Mohammad   Chetan Arora
Supreeth Achar   S. Aliakbarian   Federica Arrigoni
Hanno Ackermann   Rahaf Aljundi   Mathieu Aubry
Ehsan Adeli   Thiemo Alldieck   Shai Avidan
Triantafyllos Afouras   Jon Almazan   Angelica I. Aviles-Rivero
Sameer Agarwal   Jose M. Alvarez   Yannis Avrithis
Aishwarya Agrawal   Senjian An   Ismail Ben Ayed
Harsh Agrawal   Saket Anand   Shekoofeh Azizi
Pulkit Agrawal   Codruta Ancuti   Ioan Andrei Bârsan
Antonio Agudo   Cosmin Ancuti   Artem Babenko
Eirikur Agustsson   Peter Anderson   Deepak Babu Sam
Karim Ahmed   Juan Andrade-Cetto   Seung-Hwan Baek
Byeongjoo Ahn   Alexander Andreopoulos   Seungryul Baek
Unaiza Ahsan   Misha Andriluka   Andrew D. Bagdanov
Thalaiyasingam Ajanthan   Dragomir Anguelov   Shai Bagon
Kenan E. Ak   Rushil Anirudh   Yuval Bahat
Emre Akbas   Michel Antunes   Junjie Bai
Naveed Akhtar   Oisin Mac Aodha   Song Bai
Derya Akkaynak   Srikar Appalaraju   Xiang Bai
Yagiz Aksoy   Relja Arandjelovic   Yalong Bai
Ziad Al-Halah   Nikita Araslanov   Yancheng Bai
Xavier Alameda-Pineda   Andre Araujo   Peter Bajcsy
Jean-Baptiste Alayrac   Helder Araujo   Slawomir Bak

Mahsa Baktashmotlagh
Kavita Bala
Yogesh Balaji
Guha Balakrishnan
V. N. Balasubramanian
Federico Baldassarre
Vassileios Balntas
Shurjo Banerjee
Aayush Bansal
Ankan Bansal
Jianmin Bao
Linchao Bao
Wenbo Bao
Yingze Bao
Akash Bapat
Md Jawadul Hasan Bappy
Fabien Baradel
Lorenzo Baraldi
Daniel Barath
Adrian Barbu
Kobus Barnard
Nick Barnes
Francisco Barranco
Jonathan T. Barron
Arslan Basharat
Chaim Baskin
Anil S. Baslamisli
Jorge Batista
Kayhan Batmanghelich
Konstantinos Batsos
David Bau
Luis Baumela
Christoph Baur
Eduardo
   Bayro-Corrochano
Paul Beardsley
Jan Bednavr'ik
Oscar Beijbom
Philippe Bekaert
Esube Bekele
Vasileios Belagiannis
Ohad Ben-Shahar
Abhijit Bendale
Róger Bermúdez-Chacón
Maxim Berman
Jesus Bermudez-cameo

Florian Bernard
Stefano Berretti
Marcelo Bertalmio
Gedas Bertasius
Cigdem Beyan
Lucas Beyer
Vijayakumar Bhagavatula
Arjun Nitin Bhagoji
Apratim Bhattacharyya
Binod Bhattarai
Sai Bi
Jia-Wang Bian
Simone Bianco
Adel Bibi
Tolga Birdal
Tom Bishop
Soma Biswas
Mårten Björkman
Volker Blanz
Vishnu Boddeti
Navaneeth Bodla
Simion-Vlad Bogolin
Xavier Boix
Piotr Bojanowski
Timo Bolkart
Guido Borghi
Larbi Boubchir
Guillaume Bourmaud
Adrien Bousseau
Thierry Bouwmans
Richard Bowden
Hakan Boyraz
Mathieu Brédif
Samarth Brahmbhatt
Steve Branson
Nikolas Brasch
Biagio Brattoli
Ernesto Brau
Toby P. Breckon
Francois Bremond
Jesus Briales
Sofia Broomé
Marcus A. Brubaker
Luc Brun
Silvia Bucci
Shyamal Buch

Pradeep Buddharaju
Uta Buechler
Mai Bui
Tu Bui
Adrian Bulat
Giedrius T. Burachas
Elena Burceanu
Xavier P. Burgos-Artizzu
Kaylee Burns
Andrei Bursuc
Benjamin Busam
Wonmin Byeon
Zoya Bylinskii
Sergi Caelles
Jianrui Cai
Minjie Cai
Yujun Cai
Zhaowei Cai
Zhipeng Cai
Juan C. Caicedo
Simone Calderara
Necati Cihan Camgoz
Dylan Campbell
Octavia Camps
Jiale Cao
Kaidi Cao
Liangliang Cao
Xiangyong Cao
Xiaochun Cao
Yang Cao
Yu Cao
Yue Cao
Zhangjie Cao
Luca Carlone
Mathilde Caron
Dan Casas
Thomas J. Cashman
Umberto Castellani
Lluis Castrejon
Jacopo Cavazza
Fabio Cermelli
Hakan Cevikalp
Menglei Chai
Ishani Chakraborty
Rudrasis Chakraborty
Antoni B. Chan

Kwok-Ping Chan
Siddhartha Chandra
Sharat Chandran
Arjun Chandrasekaran
Angel X. Chang
Che-Han Chang
Hong Chang
Hyun Sung Chang
Hyung Jin Chang
Jianlong Chang
Ju Yong Chang
Ming-Ching Chang
Simyung Chang
Xiaojun Chang
Yu-Wei Chao
Devendra S. Chaplot
Arslan Chaudhry
Rizwan A. Chaudhry
Can Chen
Chang Chen
Chao Chen
Chen Chen
Chu-Song Chen
Dapeng Chen
Dong Chen
Dongdong Chen
Guanying Chen
Hongge Chen
Hsin-yi Chen
Huaijin Chen
Hwann-Tzong Chen
Jianbo Chen
Jianhui Chen
Jiansheng Chen
Jiaxin Chen
Jie Chen
Jun-Cheng Chen
Kan Chen
Kevin Chen
Lin Chen
Long Chen
Min-Hung Chen
Qifeng Chen
Shi Chen
Shixing Chen
Tianshui Chen

Weifeng Chen
Weikai Chen
Xi Chen
Xiaohan Chen
Xiaozhi Chen
Xilin Chen
Xingyu Chen
Xinlei Chen
Xinyun Chen
Yi-Ting Chen
Yilun Chen
Ying-Cong Chen
Yinpeng Chen
Yiran Chen
Yu Chen
Yu-Sheng Chen
Yuhua Chen
Yun-Chun Chen
Yunpeng Chen
Yuntao Chen
Zhuoyuan Chen
Zitian Chen
Anchieh Cheng
Bowen Cheng
Erkang Cheng
Gong Cheng
Guangliang Cheng
Jingchun Cheng
Jun Cheng
Li Cheng
Ming-Ming Cheng
Yu Cheng
Ziang Cheng
Anoop Cherian
Dmitry Chetverikov
Ngai-man Cheung
William Cheung
Ajad Chhatkuli
Naoki Chiba
Benjamin Chidester
Han-pang Chiu
Mang Tik Chiu
Wei-Chen Chiu
Donghyeon Cho
Hojin Cho
Minsu Cho

Nam Ik Cho
Tim Cho
Tae Eun Choe
Chiho Choi
Edward Choi
Inchang Choi
Jinsoo Choi
Jonghyun Choi
Jongwon Choi
Yukyung Choi
Hisham Cholakkal
Eunji Chong
Jaegul Choo
Christopher Choy
Hang Chu
Peng Chu
Wen-Sheng Chu
Albert Chung
Joon Son Chung
Hai Ci
Safa Cicek
Ramazan G. Cinbis
Arridhana Ciptadi
Javier Civera
James J. Clark
Ronald Clark
Felipe Codevilla
Michael Cogswell
Andrea Cohen
Maxwell D. Collins
Carlo Colombo
Yang Cong
Adria R. Continente
Marcella Cornia
John Richard Corring
Darren Cosker
Dragos Costea
Garrison W. Cottrell
Florent Couzinie-Devy
Marco Cristani
Ioana Croitoru
James L. Crowley
Jiequan Cui
Zhaopeng Cui
Ross Cutler
Antonio D'Innocente

Rozenn Dahyot
Bo Dai
Dengxin Dai
Hang Dai
Longquan Dai
Shuyang Dai
Xiyang Dai
Yuchao Dai
Adrian V. Dalca
Dima Damen
Bharath B. Damodaran
Kristin Dana
Martin Danelljan
Zheng Dang
Zachary Alan Daniels
Donald G. Dansereau
Abhishek Das
Samyak Datta
Achal Dave
Titas De
Rodrigo de Bem
Teo de Campos
Raoul de Charette
Shalini De Mello
Joseph DeGol
Herve Delingette
Haowen Deng
Jiankang Deng
Weijian Deng
Zhiwei Deng
Joachim Denzler
Konstantinos G. Derpanis
Aditya Deshpande
Frederic Devernay
Somdip Dey
Arturo Deza
Abhinav Dhall
Helisa Dhamo
Vikas Dhiman
Fillipe Dias Moreira
    de Souza
Ali Diba
Ferran Diego
Guiguang Ding
Henghui Ding
Jian Ding

Mingyu Ding
Xinghao Ding
Zhengming Ding
Robert DiPietro
Cosimo Distante
Ajay Divakaran
Mandar Dixit
Abdelaziz Djelouah
Thanh-Toan Do
Jose Dolz
Bo Dong
Chao Dong
Jiangxin Dong
Weiming Dong
Weisheng Dong
Xingping Dong
Xuanyi Dong
Yinpeng Dong
Gianfranco Doretto
Hazel Doughty
Hassen Drira
Bertram Drost
Dawei Du
Ye Duan
Yueqi Duan
Abhimanyu Dubey
Anastasia Dubrovina
Stefan Duffner
Chi Nhan Duong
Thibaut Durand
Zoran Duric
Iulia Duta
Debidatta Dwibedi
Benjamin Eckart
Marc Eder
Marzieh Edraki
Alexei A. Efros
Kiana Ehsani
Hazm Kemal Ekenel
James H. Elder
Mohamed Elgharib
Shireen Elhabian
Ehsan Elhamifar
Mohamed Elhoseiny
Ian Endres
N. Benjamin Erichson

Jan Ernst
Sergio Escalera
Francisco Escolano
Victor Escorcia
Carlos Esteves
Francisco J. Estrada
Bin Fan
Chenyou Fan
Deng-Ping Fan
Haoqi Fan
Hehe Fan
Heng Fan
Kai Fan
Lijie Fan
Linxi Fan
Quanfu Fan
Shaojing Fan
Xiaochuan Fan
Xin Fan
Yuchen Fan
Sean Fanello
Hao-Shu Fang
Haoyang Fang
Kuan Fang
Yi Fang
Yuming Fang
Azade Farshad
Alireza Fathi
Raanan Fattal
Joao Fayad
Xiaohan Fei
Christoph Feichtenhofer
Michael Felsberg
Chen Feng
Jiashi Feng
Junyi Feng
Mengyang Feng
Qianli Feng
Zhenhua Feng
Michele Fenzi
Andras Ferencz
Martin Fergie
Basura Fernando
Ethan Fetaya
Michael Firman
John W. Fisher

Matthew Fisher
Boris Flach
Corneliu Florea
Wolfgang Foerstner
David Fofi
Gian Luca Foresti
Per-Erik Forssen
David Fouhey
Katerina Fragkiadaki
Victor Fragoso
Jean-Sébastien Franco
Ohad Fried
Iuri Frosio
Cheng-Yang Fu
Huazhu Fu
Jianlong Fu
Jingjing Fu
Xueyang Fu
Yanwei Fu
Ying Fu
Yun Fu
Olac Fuentes
Kent Fujiwara
Takuya Funatomi
Christopher Funk
Thomas Funkhouser
Antonino Furnari
Ryo Furukawa
Erik Gärtner
Raghudeep Gadde
Matheus Gadelha
Vandit Gajjar
Trevor Gale
Juergen Gall
Mathias Gallardo
Guillermo Gallego
Orazio Gallo
Chuang Gan
Zhe Gan
Madan Ravi Ganesh
Aditya Ganeshan
Siddha Ganju
Bin-Bin Gao
Changxin Gao
Feng Gao
Hongchang Gao

Jin Gao
Jiyang Gao
Junbin Gao
Katelyn Gao
Lin Gao
Mingfei Gao
Ruiqi Gao
Ruohan Gao
Shenghua Gao
Yuan Gao
Yue Gao
Noa Garcia
Alberto Garcia-Garcia
Guillermo
    Garcia-Hernando
Jacob R. Gardner
Animesh Garg
Kshitiz Garg
Rahul Garg
Ravi Garg
Philip N. Garner
Kirill Gavrilyuk
Paul Gay
Shiming Ge
Weifeng Ge
Baris Gecer
Xin Geng
Kyle Genova
Stamatios Georgoulis
Bernard Ghanem
Michael Gharbi
Kamran Ghasedi
Golnaz Ghiasi
Arnab Ghosh
Partha Ghosh
Silvio Giancola
Andrew Gilbert
Rohit Girdhar
Xavier Giro-i-Nieto
Thomas Gittings
Ioannis Gkioulekas
Clement Godard
Vaibhava Goel
Bastian Goldluecke
Lluis Gomez
Nuno Gonçalves

Dong Gong
Ke Gong
Mingming Gong
Abel Gonzalez-Garcia
Ariel Gordon
Daniel Gordon
Paulo Gotardo
Venu Madhav Govindu
Ankit Goyal
Priya Goyal
Raghav Goyal
Benjamin Graham
Douglas Gray
Brent A. Griffin
Etienne Grossmann
David Gu
Jiayuan Gu
Jiuxiang Gu
Lin Gu
Qiao Gu
Shuhang Gu
Jose J. Guerrero
Paul Guerrero
Jie Gui
Jean-Yves Guillemaut
Riza Alp Guler
Erhan Gundogdu
Fatma Guney
Guodong Guo
Kaiwen Guo
Qi Guo
Sheng Guo
Shi Guo
Tiantong Guo
Xiaojie Guo
Yijie Guo
Yiluan Guo
Yuanfang Guo
Yulan Guo
Agrim Gupta
Ankush Gupta
Mohit Gupta
Saurabh Gupta
Tanmay Gupta
Danna Gurari
Abner Guzman-Rivera

JunYoung Gwak
Michael Gygli
Jung-Woo Ha
Simon Hadfield
Isma Hadji
Bjoern Haefner
Taeyoung Hahn
Levente Hajder
Peter Hall
Emanuela Haller
Stefan Haller
Bumsub Ham
Abdullah Hamdi
Dongyoon Han
Hu Han
Jungong Han
Junwei Han
Kai Han
Tian Han
Xiaoguang Han
Xintong Han
Yahong Han
Ankur Handa
Zekun Hao
Albert Haque
Tatsuya Harada
Mehrtash Harandi
Adam W. Harley
Mahmudul Hasan
Atsushi Hashimoto
Ali Hatamizadeh
Munawar Hayat
Dongliang He
Jingrui He
Junfeng He
Kaiming He
Kun He
Lei He
Pan He
Ran He
Shengfeng He
Tong He
Weipeng He
Xuming He
Yang He
Yihui He

Zhihai He
Chinmay Hegde
Janne Heikkila
Mattias P. Heinrich
Stéphane Herbin
Alexander Hermans
Luis Herranz
John R. Hershey
Aaron Hertzmann
Roei Herzig
Anders Heyden
Steven Hickson
Otmar Hilliges
Tomas Hodan
Judy Hoffman
Michael Hofmann
Yannick Hold-Geoffroy
Namdar Homayounfar
Sina Honari
Richang Hong
Seunghoon Hong
Xiaopeng Hong
Yi Hong
Hidekata Hontani
Anthony Hoogs
Yedid Hoshen
Mir Rayat Imtiaz Hossain
Junhui Hou
Le Hou
Lu Hou
Tingbo Hou
Wei-Lin Hsiao
Cheng-Chun Hsu
Gee-Sern Jison Hsu
Kuang-jui Hsu
Changbo Hu
Di Hu
Guosheng Hu
Han Hu
Hao Hu
Hexiang Hu
Hou-Ning Hu
Jie Hu
Junlin Hu
Nan Hu
Ping Hu

Ronghang Hu
Xiaowei Hu
Yinlin Hu
Yuan-Ting Hu
Zhe Hu
Binh-Son Hua
Yang Hua
Bingyao Huang
Di Huang
Dong Huang
Fay Huang
Haibin Huang
Haozhi Huang
Heng Huang
Huaibo Huang
Jia-Bin Huang
Jing Huang
Jingwei Huang
Kaizhu Huang
Lei Huang
Qiangui Huang
Qiaoying Huang
Qingqiu Huang
Qixing Huang
Shaoli Huang
Sheng Huang
Siyuan Huang
Weilin Huang
Wenbing Huang
Xiangru Huang
Xun Huang
Yan Huang
Yifei Huang
Yue Huang
Zhiwu Huang
Zilong Huang
Minyoung Huh
Zhuo Hui
Matthias B. Hullin
Martin Humenberger
Wei-Chih Hung
Zhouyuan Huo
Junhwa Hur
Noureldien Hussein
Jyh-Jing Hwang
Seong Jae Hwang

Sung Ju Hwang
Ichiro Ide
Ivo Ihrke
Daiki Ikami
Satoshi Ikehata
Nazli Ikizler-Cinbis
Sunghoon Im
Yani Ioannou
Radu Tudor Ionescu
Umar Iqbal
Go Irie
Ahmet Iscen
Md Amirul Islam
Vamsi Ithapu
Nathan Jacobs
Arpit Jain
Himalaya Jain
Suyog Jain
Stuart James
Won-Dong Jang
Yunseok Jang
Ronnachai Jaroensri
Dinesh Jayaraman
Sadeep Jayasumana
Suren Jayasuriya
Herve Jegou
Simon Jenni
Hae-Gon Jeon
Yunho Jeon
Koteswar R. Jerripothula
Hueihan Jhuang
I-hong Jhuo
Dinghuang Ji
Hui Ji
Jingwei Ji
Pan Ji
Yanli Ji
Baoxiong Jia
Kui Jia
Xu Jia
Chiyu Max Jiang
Haiyong Jiang
Hao Jiang
Huaizu Jiang
Huajie Jiang
Ke Jiang

Lai Jiang
Li Jiang
Lu Jiang
Ming Jiang
Peng Jiang
Shuqiang Jiang
Wei Jiang
Xudong Jiang
Zhuolin Jiang
Jianbo Jiao
Zequn Jie
Dakai Jin
Kyong Hwan Jin
Lianwen Jin
SouYoung Jin
Xiaojie Jin
Xin Jin
Nebojsa Jojic
Alexis Joly
Michael Jeffrey Jones
Hanbyul Joo
Jungseock Joo
Kyungdon Joo
Ajjen Joshi
Shantanu H. Joshi
Da-Cheng Juan
Marco Körner
Kevin Köser
Asim Kadav
Christine Kaeser-Chen
Kushal Kafle
Dagmar Kainmueller
Ioannis A. Kakadiaris
Zdenek Kalal
Nima Kalantari
Yannis Kalantidis
Mahdi M. Kalayeh
Anmol Kalia
Sinan Kalkan
Vicky Kalogeiton
Ashwin Kalyan
Joni-kristian Kamarainen
Gerda Kamberova
Chandra Kambhamettu
Martin Kampel
Meina Kan

Christopher Kanan
Kenichi Kanatani
Angjoo Kanazawa
Atsushi Kanehira
Takuhiro Kaneko
Asako Kanezaki
Bingyi Kang
Di Kang
Sunghun Kang
Zhao Kang
Vadim Kantorov
Abhishek Kar
Amlan Kar
Theofanis Karaletsos
Leonid Karlinsky
Kevin Karsch
Angelos Katharopoulos
Isinsu Katircioglu
Hiroharu Kato
Zoltan Kato
Dotan Kaufman
Jan Kautz
Rei Kawakami
Qiuhong Ke
Wadim Kehl
Petr Kellnhofer
Aniruddha Kembhavi
Cem Keskin
Margret Keuper
Daniel Keysers
Ashkan Khakzar
Fahad Khan
Naeemullah Khan
Salman Khan
Siddhesh Khandelwal
Rawal Khirodkar
Anna Khoreva
Tejas Khot
Parmeshwar Khurd
Hadi Kiapour
Joe Kileel
Chanho Kim
Dahun Kim
Edward Kim
Eunwoo Kim
Han-ul Kim

Hansung Kim
Heewon Kim
Hyo Jin Kim
Hyunwoo J. Kim
Jinkyu Kim
Jiwon Kim
Jongmin Kim
Junsik Kim
Junyeong Kim
Min H. Kim
Namil Kim
Pyojin Kim
Seon Joo Kim
Seong Tae Kim
Seungryong Kim
Sungwoong Kim
Tae Hyun Kim
Vladimir Kim
Won Hwa Kim
Yonghyun Kim
Benjamin Kimia
Akisato Kimura
Pieter-Jan Kindermans
Zsolt Kira
Itaru Kitahara
Hedvig Kjellstrom
Jan Knopp
Takumi Kobayashi
Erich Kobler
Parker Koch
Reinhard Koch
Elyor Kodirov
Amir Kolaman
Nicholas Kolkin
Dimitrios Kollias
Stefanos Kollias
Soheil Kolouri
Adams Wai-Kin Kong
Naejin Kong
Shu Kong
Tao Kong
Yu Kong
Yoshinori Konishi
Daniil Kononenko
Theodora Kontogianni
Simon Korman

Adam Kortylewski
Jana Kosecka
Jean Kossaifi
Satwik Kottur
Rigas Kouskouridas
Adriana Kovashka
Rama Kovvuri
Adarsh Kowdle
Jedrzej Kozerawski
Mateusz Kozinski
Philipp Kraehenbuehl
Gregory Kramida
Josip Krapac
Dmitry Kravchenko
Ranjay Krishna
Pavel Krsek
Alexander Krull
Jakob Kruse
Hiroyuki Kubo
Hilde Kuehne
Jason Kuen
Andreas Kuhn
Arjan Kuijper
Zuzana Kukelova
Ajay Kumar
Amit Kumar
Avinash Kumar
Suryansh Kumar
Vijay Kumar
Kaustav Kundu
Weicheng Kuo
Nojun Kwak
Suha Kwak
Junseok Kwon
Nikolaos Kyriazis
Zorah Lähner
Ankit Laddha
Florent Lafarge
Jean Lahoud
Kevin Lai
Shang-Hong Lai
Wei-Sheng Lai
Yu-Kun Lai
Iro Laina
Antony Lam
John Wheatley Lambert

Xiangyuan lan
Xu Lan
Charis Lanaras
Georg Langs
Oswald Lanz
Dong Lao
Yizhen Lao
Agata Lapedriza
Gustav Larsson
Viktor Larsson
Katrin Lasinger
Christoph Lassner
Longin Jan Latecki
Stéphane Lathuilière
Rynson Lau
Hei Law
Justin Lazarow
Svetlana Lazebnik
Hieu Le
Huu Le
Ngan Hoang Le
Trung-Nghia Le
Vuong Le
Colin Lea
Erik Learned-Miller
Chen-Yu Lee
Gim Hee Lee
Hsin-Ying Lee
Hyungtae Lee
Jae-Han Lee
Jimmy Addison Lee
Joonseok Lee
Kibok Lee
Kuang-Huei Lee
Kwonjoon Lee
Minsik Lee
Sang-chul Lee
Seungkyu Lee
Soochan Lee
Stefan Lee
Taehee Lee
Andreas Lehrmann
Jie Lei
Peng Lei
Matthew Joseph Leotta
Wee Kheng Leow

Gil Levi
Evgeny Levinkov
Aviad Levis
Jose Lezama
Ang Li
Bin Li
Bing Li
Boyi Li
Changsheng Li
Chao Li
Chen Li
Cheng Li
Chenglong Li
Chi Li
Chun-Guang Li
Chun-Liang Li
Chunyuan Li
Dong Li
Guanbin Li
Hao Li
Haoxiang Li
Hongsheng Li
Hongyang Li
Houqiang Li
Huibin Li
Jia Li
Jianan Li
Jianguo Li
Junnan Li
Junxuan Li
Kai Li
Ke Li
Kejie Li
Kunpeng Li
Lerenhan Li
Li Erran Li
Mengtian Li
Mu Li
Peihua Li
Peiyi Li
Ping Li
Qi Li
Qing Li
Ruiyu Li
Ruoteng Li
Shaozi Li

Sheng Li
Shiwei Li
Shuang Li
Siyang Li
Stan Z. Li
Tianye Li
Wei Li
Weixin Li
Wen Li
Wenbo Li
Xiaomeng Li
Xin Li
Xiu Li
Xuelong Li
Xueting Li
Yan Li
Yandong Li
Yanghao Li
Yehao Li
Yi Li
Yijun Li
Yikang LI
Yining Li
Yongjie Li
Yu Li
Yu-Jhe Li
Yunpeng Li
Yunsheng Li
Yunzhu Li
Zhe Li
Zhen Li
Zhengqi Li
Zhenyang Li
Zhuwen Li
Dongze Lian
Xiaochen Lian
Zhouhui Lian
Chen Liang
Jie Liang
Ming Liang
Paul Pu Liang
Pengpeng Liang
Shu Liang
Wei Liang
Jing Liao
Minghui Liao

Renjie Liao
Shengcai Liao
Shuai Liao
Yiyi Liao
Ser-Nam Lim
Chen-Hsuan Lin
Chung-Ching Lin
Dahua Lin
Ji Lin
Kevin Lin
Tianwei Lin
Tsung-Yi Lin
Tsung-Yu Lin
Wei-An Lin
Weiyao Lin
Yen-Chen Lin
Yuewei Lin
David B. Lindell
Drew Linsley
Krzysztof Lis
Roee Litman
Jim Little
An-An Liu
Bo Liu
Buyu Liu
Chao Liu
Chen Liu
Cheng-lin Liu
Chenxi Liu
Dong Liu
Feng Liu
Guilin Liu
Haomiao Liu
Heshan Liu
Hong Liu
Ji Liu
Jingen Liu
Jun Liu
Lanlan Liu
Li Liu
Liu Liu
Mengyuan Liu
Miaomiao Liu
Nian Liu
Ping Liu
Risheng Liu

Sheng Liu
Shu Liu
Shuaicheng Liu
Sifei Liu
Siqi Liu
Siying Liu
Songtao Liu
Ting Liu
Tongliang Liu
Tyng-Luh Liu
Wanquan Liu
Wei Liu
Weiyang Liu
Weizhe Liu
Wenyu Liu
Wu Liu
Xialei Liu
Xianglong Liu
Xiaodong Liu
Xiaofeng Liu
Xihui Liu
Xingyu Liu
Xinwang Liu
Xuanqing Liu
Xuebo Liu
Yang Liu
Yaojie Liu
Yebin Liu
Yen-Cheng Liu
Yiming Liu
Yu Liu
Yu-Shen Liu
Yufan Liu
Yun Liu
Zheng Liu
Zhijian Liu
Zhuang Liu
Zichuan Liu
Ziwei Liu
Zongyi Liu
Stephan Liwicki
Liliana Lo Presti
Chengjiang Long
Fuchen Long
Mingsheng Long
Xiang Long

Yang Long
Charles T. Loop
Antonio Lopez
Roberto J. Lopez-Sastre
Javier Lorenzo-Navarro
Manolis Lourakis
Boyu Lu
Canyi Lu
Feng Lu
Guoyu Lu
Hongtao Lu
Jiajun Lu
Jiasen Lu
Jiwen Lu
Kaiyue Lu
Le Lu
Shao-Ping Lu
Shijian Lu
Xiankai Lu
Xin Lu
Yao Lu
Yiping Lu
Yongxi Lu
Yongyi Lu
Zhiwu Lu
Fujun Luan
Benjamin E. Lundell
Hao Luo
Jian-Hao Luo
Ruotian Luo
Weixin Luo
Wenhan Luo
Wenjie Luo
Yan Luo
Zelun Luo
Zixin Luo
Khoa Luu
Zhaoyang Lv
Pengyuan Lyu
Thomas Möllenhoff
Matthias Müller
Bingpeng Ma
Chih-Yao Ma
Chongyang Ma
Huimin Ma
Jiayi Ma

K. T. Ma
Ke Ma
Lin Ma
Liqian Ma
Shugao Ma
Wei-Chiu Ma
Xiaojian Ma
Xingjun Ma
Zhanyu Ma
Zheng Ma
Radek Jakob Mackowiak
Ludovic Magerand
Shweta Mahajan
Siddharth Mahendran
Long Mai
Ameesh Makadia
Oscar Mendez Maldonado
Mateusz Malinowski
Yury Malkov
Arun Mallya
Dipu Manandhar
Massimiliano Mancini
Fabian Manhardt
Kevis-kokitsi Maninis
Varun Manjunatha
Junhua Mao
Xudong Mao
Alina Marcu
Edgar Margffoy-Tuay
Dmitrii Marin
Manuel J. Marin-Jimenez
Kenneth Marino
Niki Martinel
Julieta Martinez
Jonathan Masci
Tomohiro Mashita
Iacopo Masi
David Masip
Daniela Massiceti
Stefan Mathe
Yusuke Matsui
Tetsu Matsukawa
Iain A. Matthews
Kevin James Matzen
Bruce Allen Maxwell
Stephen Maybank

Helmut Mayer
Amir Mazaheri
David McAllester
Steven McDonagh
Stephen J. Mckenna
Roey Mechrez
Prakhar Mehrotra
Christopher Mei
Xue Mei
Paulo R. S. Mendonca
Lili Meng
Zibo Meng
Thomas Mensink
Bjoern Menze
Michele Merler
Kourosh Meshgi
Pascal Mettes
Christopher Metzler
Liang Mi
Qiguang Miao
Xin Miao
Tomer Michaeli
Frank Michel
Antoine Miech
Krystian Mikolajczyk
Peyman Milanfar
Ben Mildenhall
Gregor Miller
Fausto Milletari
Dongbo Min
Kyle Min
Pedro Miraldo
Dmytro Mishkin
Anand Mishra
Ashish Mishra
Ishan Misra
Niluthpol C. Mithun
Kaushik Mitra
Niloy Mitra
Anton Mitrokhin
Ikuhisa Mitsugami
Anurag Mittal
Kaichun Mo
Zhipeng Mo
Davide Modolo
Michael Moeller

Pritish Mohapatra
Pavlo Molchanov
Davide Moltisanti
Pascal Monasse
Mathew Monfort
Aron Monszpart
Sean Moran
Vlad I. Morariu
Francesc Moreno-Noguer
Pietro Morerio
Stylianos Moschoglou
Yael Moses
Roozbeh Mottaghi
Pierre Moulon
Arsalan Mousavian
Yadong Mu
Yasuhiro Mukaigawa
Lopamudra Mukherjee
Yusuke Mukuta
Ravi Teja Mullapudi
Mario Enrique Munich
Zachary Murez
Ana C. Murillo
J. Krishna Murthy
Damien Muselet
Armin Mustafa
Siva Karthik Mustikovela
Carlo Dal Mutto
Moin Nabi
Varun K. Nagaraja
Tushar Nagarajan
Arsha Nagrani
Seungjun Nah
Nikhil Naik
Yoshikatsu Nakajima
Yuta Nakashima
Atsushi Nakazawa
Seonghyeon Nam
Vinay P. Namboodiri
Medhini Narasimhan
Srinivasa Narasimhan
Sanath Narayan
Erickson Rangel
    Nascimento
Jacinto Nascimento
Tayyab Naseer

Lakshmanan Nataraj
Neda Nategh
Nelson Isao Nauata
Fernando Navarro
Shah Nawaz
Lukas Neumann
Ram Nevatia
Alejandro Newell
Shawn Newsam
Joe Yue-Hei Ng
Trung Thanh Ngo
Duc Thanh Nguyen
Lam M. Nguyen
Phuc Xuan Nguyen
Thuong Nguyen Canh
Mihalis Nicolaou
Andrei Liviu Nicolicioiu
Xuecheng Nie
Michael Niemeyer
Simon Niklaus
Christophoros Nikou
David Nilsson
Jifeng Ning
Yuval Nirkin
Li Niu
Yuzhen Niu
Zhenxing Niu
Shohei Nobuhara
Nicoletta Noceti
Hyeonwoo Noh
Junhyug Noh
Mehdi Noroozi
Sotiris Nousias
Valsamis Ntouskos
Matthew O'Toole
Peter Ochs
Ferda Ofli
Seong Joon Oh
Seoung Wug Oh
Iason Oikonomidis
Utkarsh Ojha
Takahiro Okabe
Takayuki Okatani
Fumio Okura
Aude Oliva
Kyle Olszewski

Björn Ommer
Mohamed Omran
Elisabeta Oneata
Michael Opitz
Jose Oramas
Tribhuvanesh Orekondy
Shaul Oron
Sergio Orts-Escolano
Ivan Oseledets
Aljosa Osep
Magnus Oskarsson
Anton Osokin
Martin R. Oswald
Wanli Ouyang
Andrew Owens
Mete Ozay
Mustafa Ozuysal
Eduardo Pérez-Pellitero
Gautam Pai
Dipan Kumar Pal
P. H. Pamplona Savarese
Jinshan Pan
Junting Pan
Xingang Pan
Yingwei Pan
Yannis Panagakis
Rameswar Panda
Guan Pang
Jiahao Pang
Jiangmiao Pang
Tianyu Pang
Sharath Pankanti
Nicolas Papadakis
Dim Papadopoulos
George Papandreou
Toufiq Parag
Shaifali Parashar
Sarah Parisot
Eunhyeok Park
Hyun Soo Park
Jaesik Park
Min-Gyu Park
Taesung Park
Alvaro Parra
C. Alejandro Parraga
Despoina Paschalidou

Nikolaos Passalis
Vishal Patel
Viorica Patraucean
Badri Narayana Patro
Danda Pani Paudel
Sujoy Paul
Georgios Pavlakos
Ioannis Pavlidis
Vladimir Pavlovic
Nick Pears
Kim Steenstrup Pedersen
Selen Pehlivan
Shmuel Peleg
Chao Peng
Houwen Peng
Wen-Hsiao Peng
Xi Peng
Xiaojiang Peng
Xingchao Peng
Yuxin Peng
Federico Perazzi
Juan Camilo Perez
Vishwanath Peri
Federico Pernici
Luca Del Pero
Florent Perronnin
Stavros Petridis
Henning Petzka
Patrick Peursum
Michael Pfeiffer
Hanspeter Pfister
Roman Pflugfelder
Minh Tri Pham
Yongri Piao
David Picard
Tomasz Pieciak
A. J. Piergiovanni
Andrea Pilzer
Pedro O. Pinheiro
Silvia Laura Pintea
Lerrel Pinto
Axel Pinz
Robinson Piramuthu
Fiora Pirri
Leonid Pishchulin
Francesco Pittaluga

Daniel Pizarro
Tobias Plötz
Mirco Planamente
Matteo Poggi
Moacir A. Ponti
Parita Pooj
Fatih Porikli
Horst Possegger
Omid Poursaeed
Ameya Prabhu
Viraj Uday Prabhu
Dilip Prasad
Brian L. Price
True Price
Maria Priisalu
Veronique Prinet
Victor Adrian Prisacariu
Jan Prokaj
Sergey Prokudin
Nicolas Pugeault
Xavier Puig
Albert Pumarola
Pulak Purkait
Senthil Purushwalkam
Charles R. Qi
Hang Qi
Haozhi Qi
Lu Qi
Mengshi Qi
Siyuan Qi
Xiaojuan Qi
Yuankai Qi
Shengju Qian
Xuelin Qian
Siyuan Qiao
Yu Qiao
Jie Qin
Qiang Qiu
Weichao Qiu
Zhaofan Qiu
Kha Gia Quach
Yuhui Quan
Yvain Queau
Julian Quiroga
Faisal Qureshi
Mahdi Rad

Filip Radenovic
Petia Radeva
Venkatesh
  B. Radhakrishnan
Ilija Radosavovic
Noha Radwan
Rahul Raguram
Tanzila Rahman
Amit Raj
Ajit Rajwade
Kandan Ramakrishnan
Santhosh
  K. Ramakrishnan
Srikumar Ramalingam
Ravi Ramamoorthi
Vasili Ramanishka
Ramprasaath R. Selvaraju
Francois Rameau
Visvanathan Ramesh
Santu Rana
Rene Ranftl
Anand Rangarajan
Anurag Ranjan
Viresh Ranjan
Yongming Rao
Carolina Raposo
Vivek Rathod
Sathya N. Ravi
Avinash Ravichandran
Tammy Riklin Raviv
Daniel Rebain
Sylvestre-Alvise Rebuffi
N. Dinesh Reddy
Timo Rehfeld
Paolo Remagnino
Konstantinos Rematas
Edoardo Remelli
Dongwei Ren
Haibing Ren
Jian Ren
Jimmy Ren
Mengye Ren
Weihong Ren
Wenqi Ren
Zhile Ren
Zhongzheng Ren

Zhou Ren
Vijay Rengarajan
Md A. Reza
Farzaneh Rezaeianaran
Hamed R. Tavakoli
Nicholas Rhinehart
Helge Rhodin
Elisa Ricci
Alexander Richard
Eitan Richardson
Elad Richardson
Christian Richardt
Stephan Richter
Gernot Riegler
Daniel Ritchie
Tobias Ritschel
Samuel Rivera
Yong Man Ro
Richard Roberts
Joseph Robinson
Ignacio Rocco
Mrigank Rochan
Emanuele Rodolà
Mikel D. Rodriguez
Giorgio Roffo
Grégory Rogez
Gemma Roig
Javier Romero
Xuejian Rong
Yu Rong
Amir Rosenfeld
Bodo Rosenhahn
Guy Rosman
Arun Ross
Paolo Rota
Peter M. Roth
Anastasios Roussos
Anirban Roy
Sebastien Roy
Aruni RoyChowdhury
Artem Rozantsev
Ognjen Rudovic
Daniel Rueckert
Adria Ruiz
Javier Ruiz-del-solar
Christian Rupprecht

Chris Russell
Dan Ruta
Jongbin Ryu
Ömer Sümer
Alexandre Sablayrolles
Faraz Saeedan
Ryusuke Sagawa
Christos Sagonas
Tonmoy Saikia
Hideo Saito
Kuniaki Saito
Shunsuke Saito
Shunta Saito
Ken Sakurada
Joaquin Salas
Fatemeh Sadat Saleh
Mahdi Saleh
Pouya Samangouei
Leo Sampaio
  Ferraz Ribeiro
Artsiom Olegovich
  Sanakoyeu
Enrique Sanchez
Patsorn Sangkloy
Anush Sankaran
Aswin Sankaranarayanan
Swami Sankaranarayanan
Rodrigo Santa Cruz
Amartya Sanyal
Archana Sapkota
Nikolaos Sarafianos
Jun Sato
Shin'ichi Satoh
Hosnieh Sattar
Arman Savran
Manolis Savva
Alexander Sax
Hanno Scharr
Simone Schaub-Meyer
Konrad Schindler
Dmitrij Schlesinger
Uwe Schmidt
Dirk Schnieders
Björn Schuller
Samuel Schulter
Idan Schwartz

William Robson Schwartz
Alex Schwing
Sinisa Segvic
Lorenzo Seidenari
Pradeep Sen
Ozan Sener
Soumyadip Sengupta
Arda Senocak
Mojtaba Seyedhosseini
Shishir Shah
Shital Shah
Sohil Atul Shah
Tamar Rott Shaham
Huasong Shan
Qi Shan
Shiguang Shan
Jing Shao
Roman Shapovalov
Gaurav Sharma
Vivek Sharma
Viktoriia Sharmanska
Dongyu She
Sumit Shekhar
Evan Shelhamer
Chengyao Shen
Chunhua Shen
Falong Shen
Jie Shen
Li Shen
Liyue Shen
Shuhan Shen
Tianwei Shen
Wei Shen
William B. Shen
Yantao Shen
Ying Shen
Yiru Shen
Yujun Shen
Yuming Shen
Zhiqiang Shen
Ziyi Shen
Lu Sheng
Yu Sheng
Rakshith Shetty
Baoguang Shi
Guangming Shi

Hailin Shi
Miaojing Shi
Yemin Shi
Zhenmei Shi
Zhiyuan Shi
Kevin Jonathan Shih
Shiliang Shiliang
Hyunjung Shim
Atsushi Shimada
Nobutaka Shimada
Daeyun Shin
Young Min Shin
Koichi Shinoda
Konstantin Shmelkov
Michael Zheng Shou
Abhinav Shrivastava
Tianmin Shu
Zhixin Shu
Hong-Han Shuai
Pushkar Shukla
Christian Siagian
Mennatullah M. Siam
Kaleem Siddiqi
Karan Sikka
Jae-Young Sim
Christian Simon
Martin Simonovsky
Dheeraj Singaraju
Bharat Singh
Gurkirt Singh
Krishna Kumar Singh
Maneesh Kumar Singh
Richa Singh
Saurabh Singh
Suriya Singh
Vikas Singh
Sudipta N. Sinha
Vincent Sitzmann
Josef Sivic
Gregory Slabaugh
Miroslava Slavcheva
Ron Slossberg
Brandon Smith
Kevin Smith
Vladimir Smutny
Noah Snavely

Roger
    D. Soberanis-Mukul
Kihyuk Sohn
Francesco Solera
Eric Sommerlade
Sanghyun Son
Byung Cheol Song
Chunfeng Song
Dongjin Song
Jiaming Song
Jie Song
Jifei Song
Jingkuan Song
Mingli Song
Shiyu Song
Shuran Song
Xiao Song
Yafei Song
Yale Song
Yang Song
Yi-Zhe Song
Yibing Song
Humberto Sossa
Cesar de Souza
Adrian Spurr
Srinath Sridhar
Suraj Srinivas
Pratul P. Srinivasan
Anuj Srivastava
Tania Stathaki
Christopher Stauffer
Simon Stent
Rainer Stiefelhagen
Pierre Stock
Julian Straub
Jonathan C. Stroud
Joerg Stueckler
Jan Stuehmer
David Stutz
Chi Su
Hang Su
Jong-Chyi Su
Shuochen Su
Yu-Chuan Su
Ramanathan Subramanian
Yusuke Sugano

Masanori Suganuma
Yumin Suh
Mohammed Suhail
Yao Sui
Heung-Il Suk
Josephine Sullivan
Baochen Sun
Chen Sun
Chong Sun
Deqing Sun
Jin Sun
Liang Sun
Lin Sun
Qianru Sun
Shao-Hua Sun
Shuyang Sun
Weiwei Sun
Wenxiu Sun
Xiaoshuai Sun
Xiaoxiao Sun
Xingyuan Sun
Yifan Sun
Zhun Sun
Sabine Susstrunk
David Suter
Supasorn Suwajanakorn
Tomas Svoboda
Eran Swears
Paul Swoboda
Attila Szabo
Richard Szeliski
Duy-Nguyen Ta
Andrea Tagliasacchi
Yuichi Taguchi
Ying Tai
Keita Takahashi
Kouske Takahashi
Jun Takamatsu
Hugues Talbot
Toru Tamaki
Chaowei Tan
Fuwen Tan
Mingkui Tan
Mingxing Tan
Qingyang Tan
Robby T. Tan

Xiaoyang Tan
Kenichiro Tanaka
Masayuki Tanaka
Chang Tang
Chengzhou Tang
Danhang Tang
Ming Tang
Peng Tang
Qingming Tang
Wei Tang
Xu Tang
Yansong Tang
Youbao Tang
Yuxing Tang
Zhiqiang Tang
Tatsunori Taniai
Junli Tao
Xin Tao
Makarand Tapaswi
Jean-Philippe Tarel
Lyne Tchapmi
Zachary Teed
Bugra Tekin
Damien Teney
Ayush Tewari
Christian Theobalt
Christopher Thomas
Diego Thomas
Jim Thomas
Rajat Mani Thomas
Xinmei Tian
Yapeng Tian
Yingli Tian
Yonglong Tian
Zhi Tian
Zhuotao Tian
Kinh Tieu
Joseph Tighe
Massimo Tistarelli
Matthew Toews
Carl Toft
Pavel Tokmakov
Federico Tombari
Chetan Tonde
Yan Tong
Alessio Tonioni

Andrea Torsello
Fabio Tosi
Du Tran
Luan Tran
Ngoc-Trung Tran
Quan Hung Tran
Truyen Tran
Rudolph Triebel
Martin Trimmel
Shashank Tripathi
Subarna Tripathi
Leonardo Trujillo
Eduard Trulls
Tomasz Trzcinski
Sam Tsai
Yi-Hsuan Tsai
Hung-Yu Tseng
Stavros Tsogkas
Aggeliki Tsoli
Devis Tuia
Shubham Tulsiani
Sergey Tulyakov
Frederick Tung
Tony Tung
Daniyar Turmukhambetov
Ambrish Tyagi
Radim Tylecek
Christos Tzelepis
Georgios Tzimiropoulos
Dimitrios Tzionas
Seiichi Uchida
Norimichi Ukita
Dmitry Ulyanov
Martin Urschler
Yoshitaka Ushiku
Ben Usman
Alexander Vakhitov
Julien P. C. Valentin
Jack Valmadre
Ernest Valveny
Joost van de Weijer
Jan van Gemert
Koen Van Leemput
Gul Varol
Sebastiano Vascon
M. Alex O. Vasilescu

Subeesh Vasu
Mayank Vatsa
David Vazquez
Javier Vazquez-Corral
Ashok Veeraraghavan
Erik Velasco-Salido
Raviteja Vemulapalli
Jonathan Ventura
Manisha Verma
Roberto Vezzani
Ruben Villegas
Minh Vo
MinhDuc Vo
Nam Vo
Michele Volpi
Riccardo Volpi
Carl Vondrick
Konstantinos Vougioukas
Tuan-Hung Vu
Sven Wachsmuth
Neal Wadhwa
Catherine Wah
Jacob C. Walker
Thomas S. A. Wallis
Chengde Wan
Jun Wan
Liang Wan
Renjie Wan
Baoyuan Wang
Boyu Wang
Cheng Wang
Chu Wang
Chuan Wang
Chunyu Wang
Dequan Wang
Di Wang
Dilin Wang
Dong Wang
Fang Wang
Guanzhi Wang
Guoyin Wang
Hanzi Wang
Hao Wang
He Wang
Heng Wang
Hongcheng Wang

Hongxing Wang
Hua Wang
Jian Wang
Jingbo Wang
Jinglu Wang
Jingya Wang
Jinjun Wang
Jinqiao Wang
Jue Wang
Ke Wang
Keze Wang
Le Wang
Lei Wang
Lezi Wang
Li Wang
Liang Wang
Lijun Wang
Limin Wang
Linwei Wang
Lizhi Wang
Mengjiao Wang
Mingzhe Wang
Minsi Wang
Naiyan Wang
Nannan Wang
Ning Wang
Oliver Wang
Pei Wang
Peng Wang
Pichao Wang
Qi Wang
Qian Wang
Qiaosong Wang
Qifei Wang
Qilong Wang
Qing Wang
Qingzhong Wang
Quan Wang
Rui Wang
Ruiping Wang
Ruixing Wang
Shangfei Wang
Shenlong Wang
Shiyao Wang
Shuhui Wang
Song Wang

Tao Wang
Tianlu Wang
Tiantian Wang
Ting-chun Wang
Tingwu Wang
Wei Wang
Weiyue Wang
Wenguan Wang
Wenlin Wang
Wenqi Wang
Xiang Wang
Xiaobo Wang
Xiaofang Wang
Xiaoling Wang
Xiaolong Wang
Xiaosong Wang
Xiaoyu Wang
Xin Eric Wang
Xinchao Wang
Xinggang Wang
Xintao Wang
Yali Wang
Yan Wang
Yang Wang
Yangang Wang
Yaxing Wang
Yi Wang
Yida Wang
Yilin Wang
Yiming Wang
Yisen Wang
Yongtao Wang
Yu-Xiong Wang
Yue Wang
Yujiang Wang
Yunbo Wang
Yunhe Wang
Zengmao Wang
Zhangyang Wang
Zhaowen Wang
Zhe Wang
Zhecan Wang
Zheng Wang
Zhixiang Wang
Zilei Wang
Jianqiao Wangni

Anne S. Wannenwetsch
Jan Dirk Wegner
Scott Wehrwein
Donglai Wei
Kaixuan Wei
Longhui Wei
Pengxu Wei
Ping Wei
Qi Wei
Shih-En Wei
Xing Wei
Yunchao Wei
Zijun Wei
Jerod Weinman
Michael Weinmann
Philippe Weinzaepfel
Yair Weiss
Bihan Wen
Longyin Wen
Wei Wen
Junwu Weng
Tsui-Wei Weng
Xinshuo Weng
Eric Wengrowski
Tomas Werner
Gordon Wetzstein
Tobias Weyand
Patrick Wieschollek
Maggie Wigness
Erik Wijmans
Richard Wildes
Olivia Wiles
Chris Williams
Williem Williem
Kyle Wilson
Calden Wloka
Nicolai Wojke
Christian Wolf
Yongkang Wong
Sanghyun Woo
Scott Workman
Baoyuan Wu
Bichen Wu
Chao-Yuan Wu
Huikai Wu
Jiajun Wu

Jialin Wu
Jiaxiang Wu
Jiqing Wu
Jonathan Wu
Lifang Wu
Qi Wu
Qiang Wu
Ruizheng Wu
Shangzhe Wu
Shun-Cheng Wu
Tianfu Wu
Wayne Wu
Wenxuan Wu
Xiao Wu
Xiaohe Wu
Xinxiao Wu
Yang Wu
Yi Wu
Yiming Wu
Ying Nian Wu
Yue Wu
Zheng Wu
Zhenyu Wu
Zhirong Wu
Zuxuan Wu
Stefanie Wuhrer
Jonas Wulff
Changqun Xia
Fangting Xia
Fei Xia
Gui-Song Xia
Lu Xia
Xide Xia
Yin Xia
Yingce Xia
Yongqin Xian
Lei Xiang
Shiming Xiang
Bin Xiao
Fanyi Xiao
Guobao Xiao
Huaxin Xiao
Taihong Xiao
Tete Xiao
Tong Xiao
Wang Xiao

Yang Xiao
Cihang Xie
Guosen Xie
Jianwen Xie
Lingxi Xie
Sirui Xie
Weidi Xie
Wenxuan Xie
Xiaohua Xie
Fuyong Xing
Jun Xing
Junliang Xing
Bo Xiong
Peixi Xiong
Yu Xiong
Yuanjun Xiong
Zhiwei Xiong
Chang Xu
Chenliang Xu
Dan Xu
Danfei Xu
Hang Xu
Hongteng Xu
Huijuan Xu
Jingwei Xu
Jun Xu
Kai Xu
Mengmeng Xu
Mingze Xu
Qianqian Xu
Ran Xu
Weijian Xu
Xiangyu Xu
Xiaogang Xu
Xing Xu
Xun Xu
Yanyu Xu
Yichao Xu
Yong Xu
Yongchao Xu
Yuanlu Xu
Zenglin Xu
Zheng Xu
Chuhui Xue
Jia Xue
Nan Xue

Tianfan Xue
Xiangyang Xue
Abhay Yadav
Yasushi Yagi
I. Zeki Yalniz
Kota Yamaguchi
Toshihiko Yamasaki
Takayoshi Yamashita
Junchi Yan
Ke Yan
Qingan Yan
Sijie Yan
Xinchen Yan
Yan Yan
Yichao Yan
Zhicheng Yan
Keiji Yanai
Bin Yang
Ceyuan Yang
Dawei Yang
Dong Yang
Fan Yang
Guandao Yang
Guorun Yang
Haichuan Yang
Hao Yang
Jianwei Yang
Jiaolong Yang
Jie Yang
Jing Yang
Kaiyu Yang
Linjie Yang
Meng Yang
Michael Ying Yang
Nan Yang
Shuai Yang
Shuo Yang
Tianyu Yang
Tien-Ju Yang
Tsun-Yi Yang
Wei Yang
Wenhan Yang
Xiao Yang
Xiaodong Yang
Xin Yang
Yan Yang

Yanchao Yang
Yee Hong Yang
Yezhou Yang
Zhenheng Yang
Anbang Yao
Angela Yao
Cong Yao
Jian Yao
Li Yao
Ting Yao
Yao Yao
Zhewei Yao
Chengxi Ye
Jianbo Ye
Keren Ye
Linwei Ye
Mang Ye
Mao Ye
Qi Ye
Qixiang Ye
Mei-Chen Yeh
Raymond Yeh
Yu-Ying Yeh
Sai-Kit Yeung
Serena Yeung
Kwang Moo Yi
Li Yi
Renjiao Yi
Alper Yilmaz
Junho Yim
Lijun Yin
Weidong Yin
Xi Yin
Zhichao Yin
Tatsuya Yokota
Ryo Yonetani
Donggeun Yoo
Jae Shin Yoon
Ju Hong Yoon
Sung-eui Yoon
Laurent Younes
Changqian Yu
Fisher Yu
Gang Yu
Jiahui Yu
Kaicheng Yu

Ke Yu
Lequan Yu
Ning Yu
Qian Yu
Ronald Yu
Ruichi Yu
Shoou-I Yu
Tao Yu
Tianshu Yu
Xiang Yu
Xin Yu
Xiyu Yu
Youngjae Yu
Yu Yu
Zhiding Yu
Chunfeng Yuan
Ganzhao Yuan
Jinwei Yuan
Lu Yuan
Quan Yuan
Shanxin Yuan
Tongtong Yuan
Wenjia Yuan
Ye Yuan
Yuan Yuan
Yuhui Yuan
Huanjing Yue
Xiangyu Yue
Ersin Yumer
Sergey Zagoruyko
Egor Zakharov
Amir Zamir
Andrei Zanfir
Mihai Zanfir
Pablo Zegers
Bernhard Zeisl
John S. Zelek
Niclas Zeller
Huayi Zeng
Jiabei Zeng
Wenjun Zeng
Yu Zeng
Xiaohua Zhai
Fangneng Zhan
Huangying Zhan
Kun Zhan

Xiaohang Zhan
Baochang Zhang
Bowen Zhang
Cecilia Zhang
Changqing Zhang
Chao Zhang
Chengquan Zhang
Chi Zhang
Chongyang Zhang
Dingwen Zhang
Dong Zhang
Feihu Zhang
Hang Zhang
Hanwang Zhang
Hao Zhang
He Zhang
Hongguang Zhang
Hua Zhang
Ji Zhang
Jianguo Zhang
Jianming Zhang
Jiawei Zhang
Jie Zhang
Jing Zhang
Juyong Zhang
Kai Zhang
Kaipeng Zhang
Ke Zhang
Le Zhang
Lei Zhang
Li Zhang
Lihe Zhang
Linguang Zhang
Lu Zhang
Mi Zhang
Mingda Zhang
Peng Zhang
Pingping Zhang
Qian Zhang
Qilin Zhang
Quanshi Zhang
Richard Zhang
Rui Zhang
Runze Zhang
Shengping Zhang
Shifeng Zhang

Shuai Zhang
Songyang Zhang
Tao Zhang
Ting Zhang
Tong Zhang
Wayne Zhang
Wei Zhang
Weizhong Zhang
Wenwei Zhang
Xiangyu Zhang
Xiaolin Zhang
Xiaopeng Zhang
Xiaoqin Zhang
Xiuming Zhang
Ya Zhang
Yang Zhang
Yimin Zhang
Yinda Zhang
Ying Zhang
Yongfei Zhang
Yu Zhang
Yulun Zhang
Yunhua Zhang
Yuting Zhang
Zhanpeng Zhang
Zhao Zhang
Zhaoxiang Zhang
Zhen Zhang
Zheng Zhang
Zhifei Zhang
Zhijin Zhang
Zhishuai Zhang
Ziming Zhang
Bo Zhao
Chen Zhao
Fang Zhao
Haiyu Zhao
Han Zhao
Hang Zhao
Hengshuang Zhao
Jian Zhao
Kai Zhao
Liang Zhao
Long Zhao
Qian Zhao
Qibin Zhao

Qijun Zhao
Rui Zhao
Shenglin Zhao
Sicheng Zhao
Tianyi Zhao
Wenda Zhao
Xiangyun Zhao
Xin Zhao
Yang Zhao
Yue Zhao
Zhichen Zhao
Zijing Zhao
Xiantong Zhen
Chuanxia Zheng
Feng Zheng
Haiyong Zheng
Jia Zheng
Kang Zheng
Shuai Kyle Zheng
Wei-Shi Zheng
Yinqiang Zheng
Zerong Zheng
Zhedong Zheng
Zilong Zheng
Bineng Zhong
Fangwei Zhong
Guangyu Zhong
Yiran Zhong
Yujie Zhong
Zhun Zhong
Chunluan Zhou
Huiyu Zhou
Jiahuan Zhou
Jun Zhou
Lei Zhou
Luowei Zhou
Luping Zhou
Mo Zhou
Ning Zhou
Pan Zhou
Peng Zhou
Qianyi Zhou
S. Kevin Zhou
Sanping Zhou
Wengang Zhou
Xingyi Zhou

Yanzhao Zhou
Yi Zhou
Yin Zhou
Yipin Zhou
Yuyin Zhou
Zihan Zhou
Alex Zihao Zhu
Chenchen Zhu
Feng Zhu
Guangming Zhu
Ji Zhu
Jun-Yan Zhu
Lei Zhu
Linchao Zhu
Rui Zhu
Shizhan Zhu
Tyler Lixuan Zhu

Wei Zhu
Xiangyu Zhu
Xinge Zhu
Xizhou Zhu
Yanjun Zhu
Yi Zhu
Yixin Zhu
Yizhe Zhu
Yousong Zhu
Zhe Zhu
Zhen Zhu
Zheng Zhu
Zhenyao Zhu
Zhihui Zhu
Zhuotun Zhu
Bingbing Zhuang
Wei Zhuo

Christian Zimmermann
Karel Zimmermann
Larry Zitnick
Mohammadreza
    Zolfaghari
Maria Zontak
Daniel Zoran
Changqing Zou
Chuhang Zou
Danping Zou
Qi Zou
Yang Zou
Yuliang Zou
Georgios Zoumpourlis
Wangmeng Zuo
Xinxin Zuo

Additional Reviewers

Victoria Fernandez
    Abrevaya
Maya Aghaei
Allam Allam
Christine
    Allen-Blanchette
Nicolas Aziere
Assia Benbihi
Neha Bhargava
Bharat Lal Bhatnagar
Joanna Bitton
Judy Borowski
Amine Bourki
Romain Brégier
Tali Brayer
Sebastian Bujwid
Andrea Burns
Yun-Hao Cao
Yuning Chai
Xiaojun Chang
Bo Chen
Shuo Chen
Zhixiang Chen
Junsuk Choe
Hung-Kuo Chu

Jonathan P. Crall
Kenan Dai
Lucas Deecke
Karan Desai
Prithviraj Dhar
Jing Dong
Wei Dong
Turan Kaan Elgin
Francis Engelmann
Erik Englesson
Fartash Faghri
Zicong Fan
Yang Fu
Risheek Garrepalli
Yifan Ge
Marco Godi
Helmut Grabner
Shuxuan Guo
Jianfeng He
Zhezhi He
Samitha Herath
Chih-Hui Ho
Yicong Hong
Vincent Tao Hu
Julio Hurtado

Jaedong Hwang
Andrey Ignatov
Muhammad
    Abdullah Jamal
Saumya Jetley
Meiguang Jin
Jeff Johnson
Minsoo Kang
Saeed Khorram
Mohammad Rami Koujan
Nilesh Kulkarni
Sudhakar Kumawat
Abdelhak Lemkhenter
Alexander Levine
Jiachen Li
Jing Li
Jun Li
Yi Li
Liang Liao
Ruochen Liao
Tzu-Heng Lin
Phillip Lippe
Bao-di Liu
Bo Liu
Fangchen Liu

Hanxiao Liu
Hongyu Liu
Huidong Liu
Miao Liu
Xinxin Liu
Yongfei Liu
Yu-Lun Liu
Amir Livne
Tiange Luo
Wei Ma
Xiaoxuan Ma
Ioannis Marras
Georg Martius
Effrosyni Mavroudi
Tim Meinhardt
Givi Meishvili
Meng Meng
Zihang Meng
Zhongqi Miao
Gyeongsik Moon
Khoi Nguyen
Yung-Kyun Noh
Antonio Norelli
Jaeyoo Park
Alexander Pashevich
Mandela Patrick
Mary Phuong
Bingqiao Qian
Yu Qiao
Zhen Qiao
Sai Saketh Rambhatla
Aniket Roy
Amelie Royer
Parikshit Vishwas
   Sakurikar
Mark Sandler
Mert Bülent Sarıyıldız
Tanner Schmidt
Anshul B. Shah

Ketul Shah
Rajvi Shah
Hengcan Shi
Xiangxi Shi
Yujiao Shi
William A. P. Smith
Guoxian Song
Robin Strudel
Abby Stylianou
Xinwei Sun
Reuben Tan
Qingyi Tao
Kedar S. Tatwawadi
Anh Tuan Tran
Son Dinh Tran
Eleni Triantafillou
Aristeidis Tsitiridis
Md Zasim Uddin
Andrea Vedaldi
Evangelos Ververas
Vidit Vidit
Paul Voigtlaender
Bo Wan
Huanyu Wang
Huiyu Wang
Junqiu Wang
Pengxiao Wang
Tai Wang
Xinyao Wang
Tomoki Watanabe
Mark Weber
Xi Wei
Botong Wu
James Wu
Jiamin Wu
Rujie Wu
Yu Wu
Rongchang Xie
Wei Xiong

Yunyang Xiong
An Xu
Chi Xu
Yinghao Xu
Fei Xue
Tingyun Yan
Zike Yan
Chao Yang
Heran Yang
Ren Yang
Wenfei Yang
Xu Yang
Rajeev Yasarla
Shaokai Ye
Yufei Ye
Kun Yi
Haichao Yu
Hanchao Yu
Ruixuan Yu
Liangzhe Yuan
Chen-Lin Zhang
Fandong Zhang
Tianyi Zhang
Yang Zhang
Yiyi Zhang
Yongshun Zhang
Yu Zhang
Zhiwei Zhang
Jiaojiao Zhao
Yipu Zhao
Xingjian Zhen
Haizhong Zheng
Tiancheng Zhi
Chengju Zhou
Hao Zhou
Hao Zhu
Alexander Zimin

Contents – Part XXI

DVI: Depth Guided Video Inpainting for Autonomous Driving. 1
  Miao Liao, Feixiang Lu, Dingfu Zhou, Sibo Zhang, Wei Li,
  and Ruigang Yang

Incorporating Reinforced Adversarial Learning in Autoregressive
Image Generation 18
  Kenan E. Ak, Ning Xu, Zhe Lin, and Yilin Wang

APRICOT: A Dataset of Physical Adversarial Attacks
on Object Detection. 35
  A. Braunegg, Amartya Chakraborty, Michael Krumdick, Nicole Lape,
  Sara Leary, Keith Manville, Elizabeth Merkhofer, Laura Strickhart,
  and Matthew Walmer

Visual Question Answering on Image Sets 51
  Ankan Bansal, Yuting Zhang, and Rama Chellappa

Object as Hotspots: An Anchor-Free 3D Object Detection Approach
via Firing of Hotspots 68
  Qi Chen, Lin Sun, Zhixin Wang, Kui Jia, and Alan Yuille

Placepedia: Comprehensive Place Understanding
with Multi-faceted Annotations. 85
  Huaiyi Huang, Yuqi Zhang, Qingqiu Huang, Zhengkui Guo, Ziwei Liu,
  and Dahua Lin

DELTAS: Depth Estimation by Learning Triangulation and Densification
of Sparse Points 104
  Ayan Sinha, Zak Murez, James Bartolozzi, Vijay Badrinarayanan,
  and Andrew Rabinovich

Dynamic Low-Light Imaging with Quanta Image Sensors 122
  Yiheng Chi, Abhiram Gnanasambandam, Vladlen Koltun,
  and Stanley H. Chan

Disambiguating Monocular Depth Estimation with a Single Transient 139
  Mark Nishimura, David B. Lindell, Christopher Metzler,
  and Gordon Wetzstein

DSDNet: Deep Structured Self-driving Network 156
  Wenyuan Zeng, Shenlong Wang, Renjie Liao, Yun Chen, Bin Yang,
  and Raquel Urtasun

QuEST: Quantized Embedding Space for Transferring Knowledge   173
Himalaya Jain, Spyros Gidaris, Nikos Komodakis, Patrick Pérez,
and Matthieu Cord

EGDCL: An Adaptive Curriculum Learning Framework for Unbiased
Glaucoma Diagnosis   190
Rongchang Zhao, Xuanlin Chen, Zailiang Chen, and Shuo Li

Backpropagated Gradient Representations for Anomaly Detection   206
Gukyeong Kwon, Mohit Prabhushankar, Dogancan Temel,
and Ghassan AlRegib

Dense RepPoints: Representing Visual Objects with Dense Point Sets   227
Ze Yang, Yinghao Xu, Han Xue, Zheng Zhang, Raquel Urtasun,
Liwei Wang, Stephen Lin, and Han Hu

On Dropping Clusters to Regularize Graph Convolutional
Neural Networks.   245
Xikun Zhang, Chang Xu, and Dacheng Tao

Adaptive Video Highlight Detection by Learning from User History   261
Mrigank Rochan, Mahesh Kumar Krishna Reddy, Linwei Ye,
and Yang Wang

Improving 3D Object Detection Through Progressive Population
Based Augmentation   279
Shuyang Cheng, Zhaoqi Leng, Ekin Dogus Cubuk, Barret Zoph,
Chunyan Bai, Jiquan Ngiam, Yang Song, Benjamin Caine,
Vijay Vasudevan, Congcong Li, Quoc V. Le, Jonathon Shlens,
and Dragomir Anguelov

DR-KFS: A Differentiable Visual Similarity Metric for 3D
Shape Reconstruction   295
Jiongchao Jin, Akshay Gadi Patil, Zhang Xiong, and Hao Zhang

SPAN: Spatial Pyramid Attention Network for Image
Manipulation Localization   312
Xuefeng Hu, Zhihan Zhang, Zhenye Jiang, Syomantak Chaudhuri,
Zhenheng Yang, and Ram Nevatia

Adversarial Learning for Zero-Shot Domain Adaptation.   329
Jinghua Wang and Jianmin Jiang

YOLO in the Dark - Domain Adaptation Method
for Merging Multiple Models   345
Yukihiro Sasagawa and Hajime Nagahara

Identity-Aware Multi-sentence Video Description   360
  Jae Sung Park, Trevor Darrell, and Anna Rohrbach

VQA-LOL: Visual Question Answering Under the Lens of Logic   379
  Tejas Gokhale, Pratyay Banerjee, Chitta Baral, and Yezhou Yang

Piggyback GAN: Efficient Lifelong Learning for Image
Conditioned Generation   397
  *Mengyao Zhai, Lei Chen, Jiawei He, Megha Nawhal, Frederick Tung,
  and Greg Mori*

TRRNet: Tiered Relation Reasoning for Compositional Visual
Question Answering   414
  Xiaofeng Yang, Guosheng Lin, Fengmao Lv, and Fayao Liu

Mining Inter-Video Proposal Relations for Video Object Detection   431
  Mingfei Han, Yali Wang, Xiaojun Chang, and Yu Qiao

TVR: A Large-Scale Dataset for Video-Subtitle Moment Retrieval   447
  Jie Lei, Licheng Yu, Tamara L. Berg, and Mohit Bansal

Minimum Class Confusion for Versatile Domain Adaptation   464
  Ying Jin, Ximei Wang, Mingsheng Long, and Jianmin Wang

Large Batch Optimization for Object Detection: Training COCO
in 12 minutes   481
  *Tong Wang, Yousong Zhu, Chaoyang Zhao, Wei Zeng, Yaowei Wang,
  Jinqiao Wang, and Ming Tang*

Towards Practical and Efficient High-Resolution HDR Deghosting
with CNN   497
  *K. Ram Prabhakar, Susmit Agrawal, Durgesh Kumar Singh,
  Balraj Ashwath, and R. Venkatesh Babu*

Monocular Differentiable Rendering for Self-supervised 3D
Object Detection   514
  *Deniz Beker, Hiroharu Kato, Mihai Adrian Morariu, Takahiro Ando,
  Toru Matsuoka, Wadim Kehl, and Adrien Gaidon*

Shape Prior Deformation for Categorical 6D Object Pose
and Size Estimation   530
  Meng Tian, Marcelo H. Ang Jr., and Gim Hee Lee

Dynamic and Static Context-Aware LSTM for Multi-agent
Motion Prediction   547
  Chaofan Tao, Qinhong Jiang, Lixin Duan, and Ping Luo

Image-Based Table Recognition: Data, Model, and Evaluation 564
   Xu Zhong, Elaheh ShafieiBavani, and Antonio Jimeno Yepes

Group Activity Prediction with Sequential Relational Anticipation Model. . . . 581
   Junwen Chen, Wentao Bao, and Yu Kong

PiP: Planning-Informed Trajectory Prediction for Autonomous Driving 598
   Haoran Song, Wenchao Ding, Yuxuan Chen, Shaojie Shen,
   Michael Yu Wang, and Qifeng Chen

PSConv: Squeezing Feature Pyramid into One Compact Poly-Scale
Convolutional Layer 615
   Duo Li, Anbang Yao, and Qifeng Chen

Hierarchical Context Embedding for Region-Based Object Detection 633
   Zhao-Min Chen, Xin Jin, Borui Zhao, Xiu-Shen Wei, and Yanwen Guo

Attention-Driven Dynamic Graph Convolutional Network for Multi-label
Image Recognition 649
   Jin Ye, Junjun He, Xiaojiang Peng, Wenhao Wu, and Yu Qiao

Gen-LaneNet: A Generalized and Scalable Approach for 3D
Lane Detection 666
   Yuliang Guo, Guang Chen, Peitao Zhao, Weide Zhang, Jinghao Miao,
   Jingao Wang, and Tae Eun Choe

Sparse-to-Dense Depth Completion Revisited: Sampling Strategy
and Graph Construction 682
   Xin Xiong, Haipeng Xiong, Ke Xian, Chen Zhao, Zhiguo Cao,
   and Xin Li

MEAD: A Large-Scale Audio-Visual Dataset for Emotional
Talking-Face Generation 700
   Kaisiyuan Wang, Qianyi Wu, Linsen Song, Zhuoqian Yang, Wayne Wu,
   Chen Qian, Ran He, Yu Qiao, and Chen Change Loy

Detecting Human-Object Interactions with Action Co-occurrence Priors 718
   Dong-Jin Kim, Xiao Sun, Jinsoo Choi, Stephen Lin, and In So Kweon

Learning Connectivity of Neural Networks from
a Topological Perspective. 737
   Kun Yuan, Quanquan Li, Jing Shao, and Junjie Yan

JSTASR: Joint Size and Transparency-Aware Snow Removal Algorithm
Based on Modified Partial Convolution and Veiling Effect Removal 754
   Wei-Ting Chen, Hao-Yu Fang, Jian-Jiun Ding, Cheng-Che Tsai,
   and Sy-Yen Kuo

Ocean: Object-Aware Anchor-Free Tracking.   771
    Zhipeng Zhang, Houwen Peng, Jianlong Fu, Bing Li, and Weiming Hu

Author Index   789

DVI: Depth Guided Video Inpainting for Autonomous Driving

Miao Liao, Feixiang Lu$^{(\boxtimes)}$, Dingfu Zhou, Sibo Zhang, Wei Li, and Ruigang Yang

Baidu Research, Baidu Inc., Beijing, China
miao.liao@gmail.com, lufeixiang@baidu.com, dingfuzhou@gmail.com, sibozhang1@gmail.com, liweimcc@gmail.com, ryang2@uky.edu

Abstract. To get clear street-view and photo-realistic simulation in autonomous driving, we present an automatic video inpainting algorithm that can remove traffic agents from videos and synthesize missing regions with the guidance of depth/point cloud. By building a dense 3D map from stitched point clouds, frames within a video are geometrically correlated via this common 3D map. In order to fill a target inpainting area in a frame, it is straightforward to transform pixels from other frames into the current one with correct occlusion. Furthermore, we are able to fuse multiple videos through 3D point cloud registration, making it possible to inpaint a target video with multiple source videos. The motivation is to solve the long-time occlusion problem where an occluded area has never been visible in the entire video. To our knowledge, we are the first to fuse multiple videos for video inpainting. To verify the effectiveness of our approach, we build a large inpainting dataset in the real urban road environment with synchronized images and Lidar data including many challenge scenes, e.g., long time occlusion. The experimental results show that the proposed approach outperforms the state-of-the-art approaches for all the criteria, especially the RMSE (Root Mean Squared Error) has been reduced by about **13%**.

Keywords: Video inpainting · Autonomous driving · Depth · Image synthesis · Simulation

1 Introduction

As computational power increases, multi-modality sensing has become more and more popular in recent years. Especially in the area of Autonomous Driving (AD), multiple sensors are combined to overcome the drawbacks of individual ones, which can provide redundancy for safety. Nowadays, most self-driving cars are equipped with lidar and cameras for both perception and mapping. Simulation systems have become essential to the development and validation of AD

Electronic supplementary material The online version of this chapter (https://doi.org/10.1007/978-3-030-58589-1_1) contains supplementary material, which is available to authorized users.

© Springer Nature Switzerland AG 2020
A. Vedaldi et al. (Eds.): ECCV 2020, LNCS 12366, pp. 1–17, 2020.
https://doi.org/10.1007/978-3-030-58589-1_1

technologies. Instead of using computer graphics to create virtual driving scenarios, Li et al. [11] proposed to augment real-world pictures with simulated traffic flow to create photorealistic simulation images and renderings. One key component in their pipeline is to remove those moving agents on the road to generate clean background street images. AutoRemover [27] generated those kinds of data using the augmented platform and proposed a video inpainting method based on the deep learning techniques. Those map service companies, which display street-level panoramic views in their map Apps, also choose to place depth sensors in addition to image sensors on their capture vehicles. Due to privacy protection, those street view images have to be post-processed to blur human faces and vehicle license plates before posted for public access. There is a strong desire to totally remove those agents on the road for better privacy protection and more clear street images.

Significant progress has been made in image inpainting in recent years. The mainstream approaches [4,6,21] adopt the patch-based method to complete missing regions by sampling and pasting similar patches from known regions or other source images. The method has been naturally extended to video inpainting, where not only spatial coherence but also temporal coherence are preserved.

The basic idea behind video inpainting is that the missing regions/pixels within a frame are observed in some other frames of the same video. Under this observation, some prior works [8,23,24] use optical flow as guidance to fill the missing pixels either explicitly or implicitly. They are successfully applied in different scenarios with seamless inpainting results. However, flow computation suffers from textureless areas, no matter it's learning based or not. Furthermore, perspective changes in the video could also degrade the quality of optical flow estimation. These frame-wise flow errors are accumulated when we fill missing pixels from a temporally distant frame, resulting in distorted inpainting results, which will be shown in the experiment section.

The emergence of deep learning, especially Generative Adversarial Networks (GAN), has provided us a powerful tool for inpainting. For images, [9,15,25] formulate inpainting as a conditional image generation problem. Although formulated differently, GAN based inpainting approaches are essentially the same as the patch-based approach, since the spirit is still looking for similar textures in the training data and fill the holes. Therefore, they have to delicately choose their training data to match the domain of the input images. And domain adaptation is not an easy task once the input images come from different scenarios. Moreover, GAN-based approaches share the same problem as the patch-based methods that they are poor at handling perspective changes in images.

As image+depth sensors become standard for AD cars, we propose a method to inpaint street-view videos with the guidance of depth. Depending on the tasks, target objects are either manually labeled or automatically detected in color images, and then removed from their depth counterpart. A 3D map is built by stitching all point clouds together and projected back onto individual frames. Most of the frame pixels are assigned with a depth value via 3D projection and those remaining pixels get their depth by interpolation. With a dense depth map and known extrinsic camera parameters, we are able to sample colors from other frames to fill holes within the current frame. These colors serve as an initial guess for those missing pixels, followed by regularization enforcing spatial and photometric smoothness. After that, we apply color harmonization to make smooth and seamless blending boundaries. In the end, a moving average is applied along the optical flow to make the final inpainted video look smooth temporally.

Unlike learning-based methods, our approach can't inpaint occluded areas if they are never visible in the video. To solve this problem, we propose fusion inpainting, which makes use of multiple video clips to inpainting a target region. Compared to state-of-the-art inpainting approaches, we are able to preserve better details in the missing region with correct perspective distortion. Temporal coherence is implicitly enforced since the 3D map is consistent across all frames. We are even able to inpaint multiple video clips captured at different times by registering all the frames into a common 3D point map. Although our experiments are conducted on datasets captured from a self-driving car, the proposed method is not limited to this scenario only. It can be generalized to both indoor and outdoor scenarios, as long as we have synchronized image+depth data.

In this paper, we propose a novel video inpainting method with the guidance of 3D maps in AD scenarios. We avoid using deep learning-based methods so that our entire pipeline only runs on CPUs. This makes it easy to be generalized to different platforms and different use cases because it doesn't require GPUs and domain adaptation of training data. 3D map guided inpainting is a new direction for the inpainting community to explore, given that more and more videos are accompanied with depth data. The main contributions of this paper are listed as follows:

1. We propose a novel approach of depth guided video inpainting for autonomous driving;
2. We are the first to fuse multiple videos for inpainting, in order to solve the long time occlusion problem;
3. We collect a new dataset in the urban road with synchronized images and Lidar data including many challenge inpainting scenes such as long time occlusion;
4. Furthermore, we designed Candidate Color Sampling Criteria and Color Harmonization for inpainting. Our approach shows smaller RMSE compared with other state-of-art methods.

2 Related Work

The principle of inpainting is essentially filling the target holes by borrowing appearance information from known sources. The sources could be regions other than the hole in the same image, images from the same video or images/videos of similar scenarios. It's critical to reduce the search space for the right pixels. Following different cues, prior works can be categorized into 3 major classes: propagation-based inpainting, patch-based inpainting, and learning-based inpainting.

Propogation-based Inpainting. Propagation-based methods [1,5] extrapolate boundary pixels around the holes for image completion. These approaches are successfully applied to regions of uniform colors. However, it has difficulties to fill large holes with rich texture variations. Thus, Propagation-based approaches usually repair small holes and scratches in an image.

Patch-based Inpainting. Patch-based methods [2,4,6,21] on the other hand, not only look at the boundary pixels but also search in the other regions/images for similar appearance in order to complete missing regions. This kind of approach has been extended to the temporal domain for video inpainting [13,14,20]. Huang et al. [8] jointly estimate optical flow and color in the missing regions to address the temporal consistency problem. In general, patch-based methods can better handle non-stationary visual data. As suggested by its name, Patch-based methods depend on reliable pixel matches to copy and paste image patches to missing regions. When a pixel match can't be robustly obtained, for example in cases of big perspective changes or illumination changes, the inpainting results are problematic.

Learning-Based Inpainting. The success of deep learning techniques inspires recent works on applying it for image inpainting. Ren et al. [19] adds a few feature maps in the new Shepard layers, achieving stronger results than a much deeper network architecture. Generative Adversarial Networks(GAN [7]) was first introduced to generate novel photos. It's straightforward to extend it to inpainting by formulating inpainting as a conditional image generation problem [9,15,25]. Pathak et al. [15] proposed context encoders, which is a convolutional neural network trained to generate the contents of an arbitrary image region conditioned on its surroundings. The context encoders are trained to both understand the content of the entire image, and produce a plausible hypothesis for the missing parts. Iizuka et al. [9] used global and local context discriminators to distinguish real images from fake ones. The global discriminator looks at the entire image to ensure it is coherent as a whole, while the local discriminator looks only at a small area centered at the completed region to ensure the local consistency of the generated patches. More recently, Yu et al. [25] presented a contextual attention mechanism in a generative inpainting framework, which further improves the inpainting quality. For video inpainting, Xu et al. [24] formulated an effective framework that is specially designed to exploit redundant information across video frames. They first synthesize a spatially and temporally coherent optical flow field across video frames, then the synthesized flow field is used to guide the propagation of pixels to fill up the missing regions in the video.

3   Proposed Approach

Figure 1 shows a brief pipeline of our approach. A 3D map is first built by stitching all point clouds together, and projected back onto individual frames. With dense depth map and known extrinsic camera parameters, we are able to sample candidate colors from other frames to fill holes within current frame.

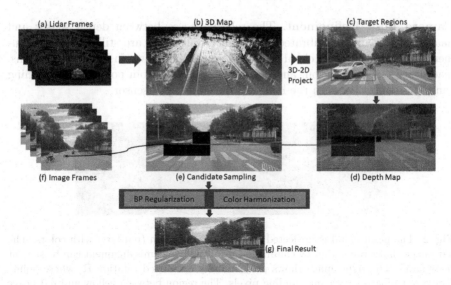

Fig. 1. Frame-wise point clouds (a) are stitched into a 3D map (b) using LOAM. The 3D map is projected onto a frame (c) to generate a depth map. For each pixel in the target region (e), we use its depth (d) as guidance to sample colors from other frames (f). Final pixel values are determined by BP regularization and color harmonization to ensure photometric consistency. (g) shows the final inpainting result.

Then, a belief propagation based regularization is applied to make sure pixel colors within the inpainting region are consistent with each other. It is followed by a color harmonization step which ensures that colors within inpainting region are consistent with outside regions. More details will be described in the following subsections.

3.1 3D Depth Map

Dynamic Object Removal. We first remove the moving objects from the point cloud, only keep the background points in the final 3D map. It is straight-forward to do so once the calibration between the depth sensor and the image sensor is performed. All points that are projected in the bounding box of the image are removed. The bounding boxes can be automatically detected or manually labeled. Alternatively, we can use PointNet++ [18] to detect and remove those typical moving objects directly from the point cloud.

3D Map Stitching. For lidar sensors, LOAM [26] is a quite robust tool to fuse the multiple frames to build the 3D map. It is capable to match and track geometric features even with a sparse 16-beam lidar. For other dense depth sensors, such as Kinect, [10] and [22] proposed real-time solutions to reconstruct a 3D map which can be further down-sampled to generate the final point cloud with a reasonable resolution.

Camera Pose Refinement. The relative poses between depth sensor and image sensor can be calibrated in advance, but there are still some misalignments between the point cloud and image pixels, as shown in Fig. 2. Vibrations, inaccurate synchronization, and accumulative errors from point cloud stitching cause pose offset between the image sensor and depth sensor.

Fig. 2. The point cloud is projected into the target region (red box) with colors. The left image shows projection by calibration result. Obvious misalignment can be seen at boundaries. The right image shows projection by optimized rotation $\mathbf{R}$, where points match much better with surrounding pixels. The region between yellow and red boxes is where we compare colors of projected 3D points and image pixels to optimize camera rotation matrix $\mathbf{R}$.

In order to produce seamless inpainting results, such offset should be compensated even if it's minor in most times. From the initial extrinsic calibration between the image sensor and depth sensor, we optimize their relative rotation R and translation T by minimizing the photometric projection error. The error is defined as:

$$E = \sum_{p \in \Omega} |c(p) - c(q)|^2, \tag{1}$$

where p is a pixel projection from 3D map. Ω is an area surrounding the target inpainting region, which is illustrated in Fig. 2 as the region between red and yellow boxes. q is original pixel in the image overlaid by p. The function $c(\cdot)$ returns the value of a pixel.

Note that the colors and locations of a pixel are discrete values, making the error function E not continuous on $\mathbf{R}$ and $\mathbf{T}$. We can't solve the following equation directly using the standard solvers, such as Levenberg-Marquardt algorithm or Gauss-Newton algorithm. Instead, we search discrete spaces of $\mathbf{R}$ and $\mathbf{T}$ to minimize E. However, $\mathbf{R}$ and $\mathbf{T}$ have 6 degrees of freedom (DoF) in total, making the searching space extremely large. We choose to fix $\mathbf{T}$ and only optimize $\mathbf{R}$ because $\mathbf{R}$ is dominant at determining projection location when the majority of the 3D map are distant points. Moreover, in most cases, we only need to move projection pixels slightly in vertical and horizontal directions in the image space, which are determined by pitch and yaw angles of the camera. We finally reduce our search space to 2 DoF, which significantly speed up the optimization process.

Depth Map. Once the camera pose is refined, we project the 3D map onto each image frame to generate the corresponding depth map. Note that some point clouds are captured far from the current image, which can be occluded and de-occluded during the projection process. Hence, we employ z-buffer to get the nearest depth.

Fig. 3. A color image and its corresponding dense depth map. Note that the depth is only rendered for background points and all moving objects have been removed.

To get a fully dense depth map, we could definitely borrow some of the fancy algorithms (e.g. [3]) that learn to produce dense depth maps from sparse ones. However, we find that the simple linear interpolation is good enough to generate the dense 3D map in our cases. We further apply a 5×5 median filter to remove some individual noise points. The final depth map is shown in Fig. 3.

3.2   Candidate Color Sampling Criteria

As every pixel is assigned a depth value, it is possible to map a pixel from one image to other images. There are multiple choices of colors to fill in the pixels of the target inpainting region, a guideline should be followed to find the best candidate color. We have 2 principles to choose the best color candidate: 1) always choose from the frame that is closer to the current frame temporally and 2) always choose from the frame where the 3D background is closer to the camera. Please refer to Fig. 4 for an example of our candidate selection criteria. The first requirement ensures our inpainting approach suffers less from perspective distortion and occlusion. And the second requirement is because image records more texture details when it's closer to objects, so that we can retain more details in the inpainting regions.

Under this guideline and the fact that sensors only move forwards during capture, our algorithm works by first searching forwards temporally to the end of video and then backwards until beginning. The first valid pixel is chosen as the candidate. And the valid pixel means its location doesn't fall into the target inpainting regions.

3.3   Regularization with Belief Propagation

At this point, every pixel gets color value individually. If the camera pose and depth value are 100% correct, we can generate perfect inpainting results with smooth boundaries and neighbors. However, it's not the case in the real world,

Fig. 4. Color candidate selection criteria. Top row: a pixel finds its candidate colors in 2 later frames where road texture appears clear in both images. In this case, we choose the frame that is temporally close to the current frame, in order to minimize the impact of perspective change and potential occlusion or de-occlusion. Bottom row: a pixel finds its candidate colors in one previous frame and one later frame. In this case, we prefer the later frame over the previous one, since road texture is lost in the previous frame. (Color figure online)

especially, the depth map always carries errors. Therefore, we have to explicitly enforce some smoothness constraints.

We formulate the color selection as a discrete global optimization problem and solve it using belief propagation (BP). Before explaining the formulation, we first define the color space and neighbors of a target pixel. As shown in left image pair in Fig. 5, a target pixel (left red box) finds its candidate pixel (right red box) from a source image, due to depth inaccuracy, the true color might not lie exactly on the candidate pixel, but a small window around. So we collect all pixel colors from this small n by n window to form the color space for the target pixel. The right image pair in Fig. 5 illustrates how to find out the expected colors of neighbors. Because of perspective changes, the 4 neighbors of a target pixel are not necessarily neighbors in the source image. Hence, we warp neighbor pixels into the source image by their depth value to sample the expected colors.

Let P be the set of pixels in the target inpainting region and L be a set of labels. The labels correspond to the indices of potential colors in the color space. A labeling function l assigns a $l_p \in L$ to each pixel $p \in P$. We assume that the labels should vary smoothly almost everywhere but may change dramatically at some places such as object boundaries. The quality of labeling is given by an energy function as

$$E = \sum_{(p,q)\in N} V(l_p, l_q) + \sum_{p\in P} D_p(l_p), \qquad (2)$$

where N are the number of edges in the four-connected image grid graph. $V(l_p, l_q)$ is the cost of assigning labels l_p and l_q to two neighboring pixels, and is normally referred to as the discontinuity cost. $D_p(l_p)$ is the cost of assigning label l_p to pixel p, which is referred to as the data cost. Determining a labeling with minimum energy corresponds to the Maximum A Posteriori (MAP) estimation problem.

Fig. 5. Left image pair: potential color choices for a target pixel. a pixel within inpainting region find its candidate pixel (red box) from a source image. A small window of pixels around this candidate (yellow boxes) are all potential colors to fill the target pixel. Right image pair: the 4 neighbors of a target pixel are not necessarily neighbors in the source image due to perspective change. In order to get neighbor colors, we need to warp neighbor pixels into the source image by their depth value. (Color figure online)

We incorporate boundary smoothness constraint into the data cost as following:

$$D_p(l_p) = \begin{cases} |C_{pl}(l_p) - I(q)|, & \text{if p is left boundary pixel} \\ |C_{pr}(l_p) - I(q)|, & \text{if p is right boundary pixel} \\ |C_{pt}(l_p) - I(q)|, & \text{if p is top boundary pixel} \\ |C_{pb}(l_p) - I(q)|, & \text{if p is bottom boundary pixel} \\ \alpha, & \text{otherwise} \end{cases} \quad , \qquad (3)$$

where $C_{pl}, C_{pr}, C_{pt}, C_{pb}$ return expected colors of pixel p's left, right, top and bottom neighbors respectively. q is the neighbor pixel of p outside of the inpainting region in the target image, so it has known color, which is returned by the function $I(q)$. Here we take the difference of true neighbor color and expected neighbor color as a measure of labeling quality. For those pixels not on the inpainting boundary, we give equal opportunities to all the labels by assigning a constant value of α. The discontinuity cost is defined as

$$V(l_p, l_q) = \begin{cases} |C_{pl}(l_p) - C_q(l_q)| + |C_p(l_p) - C_{qr}(l_q)| & \text{if L} \\ |C_{pr}(l_p) - C_q(l_q)| + |C_p(l_p) - C_{ql}(l_q)| & \text{if R} \\ |C_{pt}(l_p) - C_q(l_q)| + |C_p(l_p) - C_{qb}(l_q)| & \text{if T} \\ |C_{pb}(l_p) - C_q(l_q)| + |C_p(l_p) - C_{qt}(l_q)| & \text{if B} \end{cases} \qquad (4)$$

Here, $C_p(\cdot)$ and $C_q(\cdot)$ fetch colors for p and q at label l_p and l_q. L, R, T, B stand for q is on left, on right, on top and on bottom respectively. For a pair of 2 neighboring pixels p and q, we compute differences between p's color and q's expected color of p and vice versa.

Fig. 6. Left image: input image. Middle image: inpainting result. Note the color discontinuity in yellow box and blank pixels in red box. Right image: result after color harmonization. (Color figure online)

3.4 Color Harmonization

Pixels from different frames may have different colors due to changing camera exposure time and white balance, causing color discontinuities (Fig. 6). We borrow Poisson image editing [17] to solve these problems. Poisson image editing is originally proposed to clone an image patch from source image into a destination image with seamless boundary and original texture. It achieve this by solving the following minimization problem

$$\min_{f} \int \int_{\Omega} |\Delta f - v| \text{ with } f|_{\partial\Omega} = f^*|_{\partial\Omega}. \tag{5}$$

Here all the notations are inherited from [17]. Ω is the inpainting region with boundary $\partial\Omega$. f^* is color function of destination image and f is color function of the target inpainting region within destination image. $\Delta. = [\partial./\partial x, \partial./\partial y]$ is gradient operator. v is the desired color gradient defined over Ω.

In our case, v is computed using the output from the belief propagation step, with one exception. If two neighboring pixels within Ω are from different frames, we set their gradient to 0. This guarantee color consistency within the inpainting regions. The effectiveness of this solution is demonstrated in Fig. 6. Note that the blank-pixel region is also filled up. Since blank pixels have 0 gradient values, solving the Poisson equation on this part is equivalent to smooth color interpolation.

3.5 Video Fusion

Our algorithm has an implicit assumption that the inpainting regions must be visible in some other frames. Otherwise, some pixels will remain blank, as can be seen from Fig. 6. Learning-based methods can hallucinate inpainting colors from their training data. In contrast, our approach can't inpaint occluded areas if they are never visible in the video, leaving blank pixels.

For small areas of blank pixels, a smooth interpolation is sufficient to fill the hole. However, in some cases, a vehicle in front could block a wide field of view for the entire video duration, leaving big blank holes. A simple interpolation will not be capable of handling this problem. A better way to address this issue would be capturing another video of the same scene, where the occluded parts become visible. Fortunately, it is straightforward to register newly captured frames into an existing 3D map using LOAM [26]. Once new frames are registered and merged into the existing 3D map, inpainting is performed exactly the same way. Some of our results of video fusion can be found in the next section as well as in supplemental materials.

3.6 Temporal Smoothing

Finally, we compute both forward and backward optical flows for all the result frames. For every pixel in the target inpainting areas, we trace it into neighboring frames using the optical flow and replace its original color with average of colors sampled from neighbor frames.

4 Experiments and Results

To our best knowledge, all the public datasets (including DAVIS Dataset [16]) for video inpainting don't come with depth, which is a must for our algorithm. Autonomous driving dataset ApolloScape [12] indeed have both camera images and point clouds, but it's not adopted by research community to evaluate video inpainting. Plus, its dataset was captured by a professional mapping Lidar RIEGL, which is not a typical setup for an autonomous driving car. Thus, we captured our own dataset and compare to previous works on our dataset.

Fig. 7. 5 frames from different video clips are demonstrated to compare our results with others.

4.1 Inpainting Dataset

We use an autonomous driving car to collect large-scale datasets in urban streets. The data is generated from a variety of sensors, including Hesai Pandora all in one sensor (40-beam LiDAR, 4 mono cameras covering 360 degrees, 1 forward-facing color camera), and a localization system working at 10 HZ. The LiDAR is synchronized with embedded frontal facing wide-angle color camera. We recorded a total of 5 h length of RGB videos includes 100K synchronized 1280 × 720 images and point cloud. The dataset includes many challenging scenes e.g. background is occluded by large bus, shuttle or truck in the intersection and the front car is blocking the front view all the time. For those long time occlusion scenarios, the background is missing in the whole video sequence. We captured these difficult streets/intersections more than once, providing us the data for video fusion inpainting. Our new dataset will be published with the paper.

4.2 Comparisons

We qualitatively and quantitatively compare our results to three state-of-the-art works: two video inpainting approaches [8,24] and one image inpainting approach [25]. For those two deep learning-based approaches [24] and [25], we retrain their models on our dataset by randomly sampling missing regions on input frames to perform a fair comparison.

Qualitative Comparison. In Fig. 7, we compare our results with three other methods. It is clear that our method produces better results than others. Even though Huang [8] got smooth inpainting results, almost all the texture details are missing in their results. As shown, Yu [25] and Xu [24] sometimes fill totally messy texture in the target regions.

Figure 8 illustrates our capability to handle perspective change between source and target frames. Since our method is based on 3D geometry, perspective changes are inherently handled correctly. However, existing methods have a hard time overcoming this issue. They either fail to recover detailed texture or fail to place the texture in the right place.

Quantitative Comparison. To quantitatively compare our method with other methods, we manually labeled some background areas as the target inpainting regions and use them as the ground truth. We utilize four metrics for the evaluations: Mean Absolute Error (MAE), Root Mean Squared Error (RMSE), Peak Signal to Noise Ratio (PSNR), and Structural Similarity Index (SSIM). Table 1 shows the evaluation results of the baseline methods and our method. Note that our method outperforms others on all four metrics. Our method reduce RMSE by 13% compared to SOTA method.

Fig. 8. Top row: a patch from source image needs to be used to inpaint an occluded region in target image. Although there is significant perspective change from source to target images, our method produces geometrically and visually correct results. While other methods either fail to recover detailed texture or fail to place the texture in the right place.

Table 1. Quantitative comparison with other methods, where the best results are highlighted in bold. To be clear, the values of "MAE" and "RMSE" are the lower the better while the values of "PSNR" and "SSIM" are the higher the better.

Methods	MAE	RMSE	PSNR	SSIM
Yu [25]	10.961	16.848	20.821	0.850
Xu [24]	7.569	12.932	19.220	0.594
Huang [8]	6.924	11.017	20.022	0.762
Ours	**6.135**	**9.633**	**21.631**	**0.895**

4.3   Ablation Study

Poisson Image Blending. Figure 9 shows the effectiveness of applying Poisson image blending. Visible seams are obvious at boundaries of pixels coming from different frames. This is because our capturing camera is working under auto exposure and auto white balance mode. A same object may have different color tones in different frames from the same video, not to mention videos captured on different days. Table 2 shows the quantitative results with and without Poisson color blending. It is clear that color blending indeed improves the results.

Video Fusion. Figure 9 shows fusion of 2 videos. 1st row shows four frames from a video and the 2nd row shows 4 frames from another video captured on a different day at the same traffic intersection. Here, our goal is to inpaint those foreground objects in the 2nd video. 3rd row shows output using video 2 only, where exists large blank regions. That is because front vehicles keep blocking certain areas during the entire capture time. It is clear that Poisson

Table 2. Ablation study on Poisson color blending, where the best results are highlighted in bold. To be clear, the values of "MAE" and "RMSE" are the lower the better while the values of "PSNR" and "SSIM" are the higher the better.

Strategies	MAE	RMSE	PSNR	SSIM
No blending	9.410	17.484	21.783	0.911
blending	**6.497**	**13.009**	**22.312**	**0.917**

Table 3. Ablation study on multiple video fusion, where the best results are highlighted in bold. To be clear, the values of "MAE" and "RMSE" are the lower the better while the values of "PSNR" and "SSIM" are the higher the better.

Strategies	MAE	RMSE	PSNR	SSIM
No fusion	10.427	14.967	20.941	0.879
fusion	**6.059**	**8.333**	**21.195**	**0.882**

Fig. 9. 1st row: frames from video 1; **2nd row**: frames from video 2 captured on a different day; **3rd row**: results after Poisson color blending using video 2 only; **4th row**: direct inpainting results by fusing both videos; **5th row**: results after Poisson color blending using both videos. (Color figure online)

image blending is not capable of completing large blank holes. 4th row shows BP output after we fuse the 1st video into the 2nd one, where the blank holes are all gone. 5th row shows the final results after color blending and optical flow temporal smoothing. Table 3 demonstrates effectiveness of video fusion.

The fusion of multiple videos for inpainting demonstrates another advantage of our proposed approach. For those existing video inpainting works, they haven't address the issue of long-time occlusion, neither did they proposed to fuse multiple videos for inpainting purpose. Please checkout video demos here: https://youtu.be/iOIxdQIzjQs .

5   Conclusion

In this paper, we propose an automatic video inpainting algorithm that removes object from videos and synthesizes missing regions with the guidance of depth. It outperforms existing state-of-the-art inpainting methods on our inpainting dataset by preserving accurate texture details. The experiments indicate that our approach could reconstruct cleaner and better background images, especially in the challenging scenarios with long time occlusion scenes. Furthermore, our method may be generalized to any videos as long as depth exists, in contrast to those deep learning-based approaches whose success heavily depend on comprehensiveness and resemblance of training dataset.

References

1. Ballester, C., Bertalmio, M., Caselles, V., Sapiro, G., Verdera, J.: Filling-in by joint interpolation of vector fields and gray levels. Trans. Img. Proc. **10**(8), 1200–1211 (2001). https://doi.org/10.1109/83.935036
2. Bertalmio, M., Vese, L., Sapiro, G., Osher, S.: Simultaneous structure and texture image inpainting. Trans. Img. Proc. **12**(8), 882–889 (2003). https://doi.org/10.1109/TIP.2003.815261
3. Cheng, X., Wang, P., Yang, R.: Depth estimation via affinity learned with convolutional spatial propagation network. In: Proceedings of the European Conference on Computer Vision (ECCV), pp. 103–119 (2018)
4. Darabi, S., Shechtman, E., Barnes, C., Goldman, D.B., Sen, P.: Image melding: combining inconsistent images using patch-based synthesis. ACM Trans. Graph. (TOG) (Proc. SIGGRAPH 2012), **31**(4), 82:1–82:10 (2012)
5. Ebdelli, M., Le Meur, O., Guillemot, C.: Video inpainting with short-term windows: application to object removal and error concealment. IEEE Trans. Image Process. **24**, 3034–3047 (2015). https://doi.org/10.1109/TIP.2015.2437193
6. Efros, A.A., Freeman, W.T.: Image quilting for texture synthesis and transfer. In: Proceedings of the 28th Annual Conference on Computer Graphics and Interactive Techniques, pp. 341–346. SIGGRAPH 2001, ACM, New York, NY, USA (2001). https://doi.org/10.1145/383259.383296, http://doi.acm.org/10.1145/383259.383296
7. Goodfellow, I., et al.: Generative adversarial nets. In: Ghahramani, Z., Welling, M., Cortes, C., Lawrence, N.D., Weinberger, K.Q. (eds.) Advances in Neural Information Processing Systems, vol. 27, pp. 2672–2680. Curran Associates, Inc. (2014). http://papers.nips.cc/paper/5423-generative-adversarial-nets.pdf
8. Huang, J.B., Kang, S.B., Ahuja, N., Kopf, J.: Temporally coherent completion of dynamic video. ACM Trans. Graph. (TOG) **35**(6), 196 (2016)
9. Iizuka, S., Simo-Serra, E., Ishikawa, H.: Globally and locally consistent image completion. ACM Trans. Graph. (Proc. SIGGRAPH 2017) **36**(4), 107:1–107:14 (2017)
10. Izadi, S., et al.: Kinectfusion: real-time 3d reconstruction and interaction using a moving depth camera. In: Proceedings of the 24th Annual ACM Symposium on User Interface Software and Technology, pp. 559–568. UIST 2011, ACM, New York, NY, USA (2011). DOIurlhttp://doi.org/10.1145/2047196.2047270, http://doi.acm.org/10.1145/2047196.2047270

11. Li, W., et al.: Aads: augmented autonomous driving simulation using data-driven algorithms. Sci. Robot. **4**(28) (2019). https://doi.org/10.1126/scirobotics.aaw0863, https://robotics.sciencemag.org/content/4/28/eaaw0863

12. Ma, Y., Zhu, X., Zhang, S., Yang, R., Wang, W., Manocha, D.: Trafficpredict: trajectory prediction for heterogeneous traffic-agents. In: Proceedings of the AAAI Conference on Artificial Intelligence, vol. 33, pp. 6120–6127 (2019). https://arxiv.org/pdf/1811.02146.pdf

13. Newson, A., Almansa, A., Fradet, M., Gousseau, Y., Pérez, P.: Towards fast, generic video inpainting. In: Proceedings of the 10th European Conference on Visual Media Production, pp. 7:1–7:8. CVMP 2013, ACM, New York, NY, USA (2013). https://doi.org/10.1145/2534008.2534019, http://doi.acm.org/10.1145/2534008.2534019

14. Newson, A., Almansa, A., Fradet, M., Gousseau, Y., Pérez, P.: Video inpainting of complex scenes. SIAM J. Imaging Sci. **7**, 1993–2019 (2014). https://doi.org/10.1137/140954933

15. Pathak, D., Krähenbühl, P., Donahue, J., Darrell, T., Efros, A.: Context encoders: Feature learning by inpainting. In: Computer Vision and Pattern Recognition (CVPR) (2016)

16. Perazzi, F., Pont-Tuset, J., McWilliams, B., Van Gool, L., Gross, M., Sorkine-Hornung, A.: A benchmark dataset and evaluation methodology for video object segmentation. In: Proceedings of the IEEE Conference on Computer Vision and Pattern Recognition, pp. 724–732 (2016)

17. Pérez, P., Gangnet, M., Blake, A.: Poisson image editing. In: ACM SIGGRAPH 2003 Papers, pp. 313–318. SIGGRAPH 2003, ACM, New York, NY, USA (2003). https://doi.org/10.1145/1201775.882269, http://doi.acm.org/10.1145/1201775.882269

18. Qi, C.R., Yi, L., Su, H., Guibas, L.J.: Pointnet++: deep hierarchical feature learning on point sets in a metric space. arXiv preprint arXiv:1706.02413 (2017)

19. Ren, J.S., Xu, L., Yan, Q., Sun, W.: Shepard convolutional neural networks. In: Cortes, C., Lawrence, N.D., Lee, D.D., Sugiyama, M., Garnett, R. (eds.) Advances in Neural Information Processing Systems, vol. 28, pp. 901–909. Curran Associates, Inc. (2015). http://papers.nips.cc/paper/5774-shepard-convolutional-neural-networks.pdf

20. Shih, T.K., Tang, N.C., Hwang, J.N.: Exemplar-based video inpainting without ghost shadow artifacts by maintaining temporal continuity. IEEE Trans. Cir. and Sys. for Video Technol. **19**(3), 347–360 (2009). https://doi.org/10.1109/TCSVT.2009.2013519, http://dx.doi.org/10.1109/TCSVT.2009.2013519

21. Simakov, D., Caspi, Y., Shechtman, E., Irani, M.: Summarizing visual data using bidirectional similarity. In: 2008 IEEE Conference on Computer Vision and Pattern Recognition, pp. 1–8. IEEE (2008)

22. Steinbrücker, F., Sturm, J., Cremers, D.: Real-time visual odometry from dense RGB-D images. In: 2011 IEEE International Conference on Computer Vision Workshops (ICCV Workshops), pp. 719–722, November 2011. https://doi.org/10.1109/ICCVW.2011.6130321

23. Wang, C., Huang, H., Han, X., Wang, J.: Video inpainting by jointly learning temporal structure and spatial details. In: Proceedings of the 33th AAAI Conference on Artificial Intelligence (2019)

24. Xu, R., Li, X., Zhou, B., Loy, C.C.: Deep flow-guided video inpainting. In: The IEEE Conference on Computer Vision and Pattern Recognition (CVPR), June 2019

25. Yu, J., Lin, Z., Yang, J., Shen, X., Lu, X., Huang, T.S.: Generative image inpainting with contextual attention. arXiv preprint arXiv:1801.07892 (2018)
26. Zhang, J., Singh, S.: Loam: lidar odometry and mapping in real-time. In: Robotics: Science and Systems Conference, July 2014
27. Zhang, R., et al.: Autoremover: automatic object removal for autonomous driving videos. arXiv preprint arXiv:1911.12588 (2019)

Incorporating Reinforced Adversarial Learning in Autoregressive Image Generation

Kenan E. Ak[1]([✉]), Ning Xu[2], Zhe Lin[2], and Yilin Wang[2]

[1] Institute for Infocomm Research, A*STAR, Singapore, Singapore
kenanea@i2r.a-star.edu.sg
[2] Adobe Research, San Jose, USA
{nxu,zlin,yilwang}@adobe.com

Abstract. Autoregressive models recently achieved comparable results versus state-of-the-art Generative Adversarial Networks (GANs) with the help of Vector Quantized Variational AutoEncoders (VQ-VAE). However, autoregressive models have several limitations such as exposure bias and their training objective does not guarantee visual fidelity. To address these limitations, we propose to use Reinforced Adversarial Learning (RAL) based on policy gradient optimization for autoregressive models. By applying RAL, we enable a similar process for training and testing to address the exposure bias issue. In addition, visual fidelity has been further optimized with adversarial loss inspired by their strong counterparts: GANs. Due to the slow sampling speed of autoregressive models, we propose to use partial generation for faster training. RAL also empowers the collaboration between different modules of the VQ-VAE framework. To our best knowledge, the proposed method is first to enable adversarial learning in autoregressive models for image generation. Experiments on synthetic and real-world datasets show improvements over the MLE trained models. The proposed method improves both negative log-likelihood (NLL) and Fréchet Inception Distance (FID), which indicates improvements in terms of visual quality and diversity. The proposed method achieves state-of-the-art results on Celeba for 64×64 image resolution, showing promise for large scale image generation.

Keywords: Autoregressive models · Reinforcement learning · Vector quantized variational autoencoders · Generative adversarial networks

1 Introduction

Image generation is a central problem in computer vision and has numerous applications. Nowadays, powerful image generation methods are mostly based

Electronic supplementary material The online version of this chapter (https://doi.org/10.1007/978-3-030-58589-1_2) contains supplementary material, which is available to authorized users.

Fig. 1. Sample comparison of the proposed reinforced adversarial learning with the MLE trained model on CelebA [30] and LSUN-bedroom [52] datasets. Results are randomly sampled.

on Generative Adversarial Networks (GANs), which was first introduced by Goodfellow *et al.* [19]. With the development of advanced network structures and large-scale training [11,55], GANs are able to generate high-quality and high-resolution images. However, it is known that GANs do not capture the complete diversity of the true distribution [7,8]. Additionally, GANs are difficult to evaluate where hand-crafted metrics such as Inception Score [40] and Fréchet Inception Distance (FID) [22] must be used to test their performance (Fig. 1).

Autoregressive models [28,33,34] directly optimize negative log-likelihood (NLL) on training data offer another way for image generation. These models are less likely to face the mode collapse issue due to their objective [38]. Additionally, the objective itself provides a good evaluation metric. PixelCNN [34] is a common choice due to its performance and computational efficiency. The introduction of two recent works [16,38], which make use of PixelCNNs, vector quantization [32] and hierarchical-structure have shown comparable results vs. GANs. These advancements could open up a new avenue for image generation research.

On the other hand, there are several open problems in likelihood-based methods. As pointed out in [47], optimizing NLL does not necessarily lead to generating realistic images as this objective is not a good measure of visual quality. Another issue is the exposure bias. The sampling of PixelCNN is sequential, which is a different procedure from its training stage. This discrepancy can be troublesome as small errors during the sampling can accumulate towards the next steps, which may lead to unrealistic samples [9]. To alleviate these issues, VQ-VAE-2 [38] proposed to use the classifier based rejection sampling. This technique can eliminate low-quality samples based on an ImageNet pre-trained classifier. Classifier based rejection sampling helps VQ-VAE-2 achieve competitive results versus BigGAN [11]. Nevertheless, this method is time exhaustive, especially considering the slow sampling speed of autoregressive models. Additionally, class information might not be always available in most datasets.

In this paper, we aim to leverage both advantages of GANs and likelihood models. Consequently, our objective is to further improve the visual fidelity of autoregressive models while addressing the exposure bias issue. We focus on the VQ-VAE framework [32,38], which relies upon PixelCNN [34] for latent code generation and uses a pre-trained decoder to reconstruct images. However, PixelCNN [34] generates sequences of discrete codes. Therefore, we cannot directly optimize it directly with GAN objectives. To fix this issue, we regard the sampling process of PixelCNN as a sequential decision-making process and optimize it by Reinforced Adversarial Learning (RAL). A discriminator network is trained to distinguish between real & fake images and provides rewards to the Pixel-CNN. We compare our proposed method on two image generation benchmarks to demonstrate that RAL can indeed improve both FID and NLL significantly.

Adversarial training for sequence generation with RL is first explored in Seq-GAN [54] but there is a lot of room for improvement, especially for images. Compared to text or music generation tasks studied in SeqGAN, image generation is more complex due to longer sequences and strong spatial correlation of images. In this paper, in addition to applying RAL work for image generation for the first time, we also propose a partial generation idea to address the issue with long-sequences. Moreover, we include a method that enables the use of intermediate rewards for spatial correlation. The proposed method also enables the incorporation of independently trained modules in VQ-VAE. Our experiments show that RAL can greatly improve one of the most successful image generation methods for autoregressive models, which has a lot of potential for future work.

In summary, our contributions are four-fold:

- We propose to augment the autoregressive image generation with adversarial training by using reinforcement learning to further improve the image quality while incorporating collaboration between independently trained modules.
- We utilize a patch-based discriminator to enable intermediate rewards.
- We propose partial generation for faster training to addresses the sampling issue of long sequences.
- We conduct extensive experiments to show that the proposed method significantly improves over the MLE trained baselines in different settings for both synthetic and real datasets.

The rest of the paper is organized as follows. In Sect. 2, we briefly introduce related works. In Sect. 3, we introduce the basics of VQ-VAE & VQ-VAE-2 and we describe our method in detail in Sect. 4. Implementation details are provided in Sect. 5.1, experimental results are presented in Sect. 5 and finally, we conclude the paper in Sect. 6.

2 Related Work

2.1 Generative Models

GANs [19] and conditional-GANs [31] have shown a huge success for many problems in computer vision [1–5,13,21,51] and speech processing [39,41–44,56]. Advanced architectures such as SaGAN [55], StyleGAN [26] and BigGAN [11] have shown GANs' superiority in terms of image quality compared to Variational Autoencoder (VAE) [28] and autoregressive models, e.g., PixelCNN [34]. Additionally, GANs are much faster in inference compared to autoregressive models. However, it is known that GANs models can not fully capture the data distribution and may sometimes suddenly drop modes [7,8]. Moreover, there are no perfect metrics to evaluate the sampling quality of GANs.

Autoregressive models are less likely to face the mode collapse and can be easily evaluated by measuring log-likelihood [38]. However, these models are expensive to train and extremely slow at the sampling time when trained in the pixel space. Performing vector quantization [14,32,48] and training a PixelCNN prior in discrete latent space is much more efficient than pixel space, which enables the scaling to large resolution images. Following this idea, the recently introduced VQ-VAE-2 framework managed to achieve competitive results compared to state-of-the-art GANs. Concurrently with VQ-VAE-2, Fauw et al. [16] followed a similar approach to combine autoencoders with autoregressive decoders. The advantage of VQ-VAE-2 is the feed-forward design of the decoder, making it faster for sampling. Due to faster sampling time, VQ-VAE-2 is more feasible for our work, which requires sampling of numerous images during the training.

Although these aforementioned autoregressive models achieve decent improvements in terms of image coherence and fidelity than before [12,34], GANs are still preferred due to sampling speed and image quality. Additionally, as mentioned by Theis et al. [47], good performance on log-likelihood training objective does not always guarantee good samples. In order to reduce this gap, VQ-VAE-2 proposes to use classifier-based rejection sampling to trade-off diversity and quality inspired by BigGAN [11]. However, this approach requires the autoregressive prior to sample many latent codes, which is time exhaustive and not ideal for real-world problems. Additionally, the rejection sampling method relies on a ImageNet pretrained classifier and might not be available for another dataset such as CelebA [30] and LSUN-bedroom [52]. Lastly, the rejection sampling method decreases diversity in image samples. We propose to use reinforcement learning to improve the sampling quality of autoregressive models and enable a similar mechanism for training and testing.

2.2 Reinforcement Learning in Sequence Generation

RL for sequence generation has been applied by [37], which uses bilingual evaluation understudy (BLEU) [35] as a reward function to guide the generation process. SeqGAN [54] is the first to train a discriminator through adversarial learning and use reinforcement learning based on policy gradient [46] to provide rewards for seq-to-seq network [45]. Recent works [18,23] also showed that adversarial training can be used to iteratively update images where rewards are estimated by a discriminator network. In this work, we similarly employ a discriminator network to provide rewards for the generated samples and improve sampling quality. We enable intermediate rewards by using a PatchGAN [24] network and use a partial generation procedure to handle long sequences.

3   Background: VQ-VAE & VQ-VAE-2

Our method is built upon VQ-VAE [32] & VQ-VAE-2 [38] frameworks that are trained in two stages as follows:

Stage 1: Learning Hierarchical Latent Codes. In this stage, VQ-VAE learns an *encoder* E that transforms an image into a set of features, a *codebook* C that maps the real-valued features into a set of discrete latent codes and a *decoder* U that reconstructs the image from these latent codes. E, C and U are learned together as in [32]. However, the reconstruction in VQ-VAE is not perfect due to the quantization, therefore VQ-VAE-2 [38] proposed to learn hierarchical latent codes to alleviate this problem. In VQ-VAE-2, the first hierarchy called top-level latent codes captures global image information such as shape and structures while the other hierarchy called bottom-level latent captures fine-grained details. Once stage 1 is learned, an input image can be represented as the hierarchical latent codes for compression and reconstructed by the decoder.

Stage 2: Learning Priors Over Latent Codes. PixelCNN [34] is an autoregressive model and has been previously applied to learn image distributions in the pixel space. In VQ-VAE, PixelCNN is used to learn priors in the latent space, which has much smaller dimensionality than the pixel space. For the two hierarchies in VQ-VAE-2, PixelSNAIL [12] is used to learn the top-level latent codes while a lighter PixelCNN is used to learn the bottom-level as the size of the bottom-level latent codes is larger. These two models are trained separately by MLE. After the training, latent codes sampled from the PixelCNN priors can be reconstructed with the pretrained decoder to an image.

4   Reinforced Adversarial Learning

There are several drawbacks of PixelCNNs priors used in the VQ-VAE frameworks [32,38]. First, they are trained with MLE which is not a good measure of sampling quality [47]. Second, during the training, all latent codes are estimated from the real images while during inference each code is sampled, which could never be observed during the MLE training. Such discrepancy could result in an unrealistic generation. Lastly, each hierarchy and the decoder in the VQ-VAE-2 framework are trained separately, which may not collaborate well when put together for sampling images.

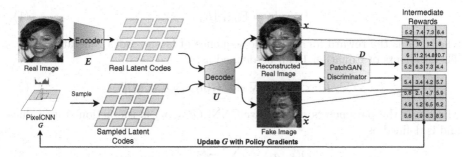

Fig. 2. Overview of the proposed Reinforced Adversarial Learning (RAL) framework. Initially, PixelCNN G is used to sample fake latent codes which are then fed into Decoder to reconstruct the fake image. Similarly, the real image can be reconstructed from the real latent codes. We use PatchGAN Discriminator to provide rewards for overlapping regions in the image. These rewards can then be used to update G with policy gradients.

In this section, we describe how we introduce adversarial learning from GANs into likelihood models to leverage the best from both worlds. Our motivation is to let the PixelCNNs be able to generate samples that can fool the discriminator that is trained to distinguish between real and fake images. In this way, our PixelCNNs are trained to generate realistic sequences and the training process is exactly the same as inference. In addition, the harmony of different PixelCNNs and the decoder are also improved since they are optimized together. One issue is that PixelCNN cannot be directly optimized by the adversarial loss, which we solve by employing reinforcement learning. Parameters of Encoder & Decoder are fixed since including them to the training corrupts the image decoding procedure, which results in having poor quality of images even when using real latent maps.

We illustrate our solution in Sect. 4.1. Our patch-based discriminator and reward definition are presented in Sect. 4.2. We describe the idea of partial generation that is useful for the generation of large images in Sect. 4.3. Training details are given in Sect. 4.4. An overview of the proposed RAL framework is shown in Fig. 2.

4.1 Policy Gradients

For simplicity, let us first consider learning a single PixelCNN G on one hierarchy of latent codes. The extension to multiple PixelCNNs on multiple hierarchies (VQ-VAE-2) is straightforward.

In adversarial training, a generator is directly optimized to maximally confuse a discriminator. However, at each time $t \in [0, T-1]$, our PixelCNN generates a discrete code c_t based on the conditional probability $G(c_t|c_0, ..., c_{t-1})$ given all the previous codes $(c_0, ..., c_{t-1})$. This process is non-differentiable and thus cannot be directly optimized. However it can be regarded as a decision making process where the state $s_t = (c_0, ..., c_{t-1})$ and the action $a_t = c_t$. We use policy gradients [46] to solve this problem where the objective is:

$$J_G = \mathbf{E}_{a \sim G}[R_T] \tag{1}$$

where R_T is the reward for the whole sequence of latent codes generated by G. The gradients of Equation 1 can be defined as:

$$\nabla_\theta J_G \propto \mathbf{E}_{a_t \sim G}\left[\nabla_\theta \log G(a_t|s_t)Q(s_t, a_t)\right] \tag{2}$$

where θ is the parameters of the PixelCNN. $Q(s_t, a_t)$ is the action-state value and is defined as:

$$Q(s_t, a_t) = \sum_{k=t+1}^{T} \gamma^{k-t-1} r_k \tag{3}$$

where r_k is the reward at time step k, which can be obtained by the discriminator and will be discussed later. γ is the discounted factor within the range of $[0, 1]$.

We use the REINFORCE [50] algorithm to roll-out the whole sequence after a_t is sampled by using the same PixelCNN G. In our experiment, we only do one Monte-Carlo roll-out for the training-speed concern. Finally, G can be updated as follows where the α is learning rate:

$$\theta \leftarrow \theta + \alpha \nabla_\theta J_G \tag{4}$$

4.2   Discriminator

After a sequence of latent codes $(c_0, ..., c_{T-1})$ is generated, the decoder U is used to reconstruct the image $\tilde{x}$, as shown in Fig. 2. Then a discriminator D is trained to distinguish between the generated and real images (real images are also their reconstructed version). We use the WGAN loss [6,20] instead of the original GAN loss proposed in [19]. Since it provides smoother gradients which enable the generator to still learn even when the discriminator is performing strong. The loss function of our discriminator is defined as:

$$\begin{aligned} L_D = &-\mathbf{E}_{\mathbf{x} \sim p_d}[D(\mathbf{x})] + \mathbf{E}_{\tilde{\mathbf{x}} \sim G}[D(\tilde{\mathbf{x}})] \\ &+ \lambda_{gp}\mathbf{E}_{\hat{\mathbf{x}}}[(\|\nabla_{\hat{\mathbf{x}}} D(\hat{\mathbf{x}}))\|_2 - 1)^2] \end{aligned} \tag{5}$$

where p_d is the distribution of real images. The final term is the gradient penalty weighted by λ_{gp}. $\hat{\mathbf{x}}$ is an image sampled uniformly along a straight line between real and generated images.

Our discriminator follows a similar structure to the PatchGAN discriminator [24]. In contrast to traditional GAN discriminators that only produce a single scalar output, the PatchGAN discriminator can provide a score map S, each element of which corresponds to the score of a local image patch as shown in Fig. 2. There are several advantages of using this type of discriminator. First, it provides a good measure of the realism of local patches. Second, the scores can be used as intermediate rewards which can alleviate the issue of sparse rewards in RL training. Third, its fully convolutional structure can handle arbitrary image sizes, which is convenient for our partial generation.

The *intermediate reward* at every time step can be computed by upsampling the score map S to the original size of the latent codemap as shown in Fig. 2.

The reward r_{t+1} of the sampled code c_t has a corresponding location at the upsampled score map and can be easily obtained. Since each code is contained in a local region, the action-state value function $Q(s_t, a_t)$ thus more focus on the local realism.

Our *single reward* function is defined as:

$$r_t = \begin{cases} 0, & t < T \\ D(\tilde{\mathbf{x}}), & t = T \end{cases} \tag{6}$$

where $D(\tilde{\mathbf{x}})$ is the average value of the score map S. In this case the action-state value function $Q(s_t, a_t)$ focus more on the long-term reward, i.e. the realism of the whole image.

4.3   Partial Generation

In order to generate large-size images, the length of sampled latent codes also needs to be increased, which is time-exhaustive. This is especially troublesome as our RL training requires many Monte-Carlo roll-outs at each iteration. As a solution, we propose to use partial generation whose central idea is to only sample partial images, which can already provide meaningful rewards while improving sampling efficiency greatly.

Specifically, the partial generation works under two modes: (1) continue generation from real latent codes and (2) partial generation from scratch. During each roll-out in RL training, the algorithm first randomly decides one mode. If mode-1 is selected, then a random number representing the number of rows of real latent codes is chosen. Then following these real codes, G samples the rest of the sequence. If mode-2 is selected, a random number representing the number of rows of fake latent codes is chosen. Then G samples this number of codes from scratch. The two modes are essential and improve the generation of the Pixel-CNN on different image regions. In addition, mode-1 provides real latent codes as context and is useful for the image completion task. Figure 3 shows some sampled examples of both the two modes. It should be noted that although the sampled codes correspond to different image sizes, as our discriminator is fully convolutional and thus has no issue.

Fig. 3. Examples for partial generation from G. The proposed idea, improves the training speed of RL algorithm.

4.4 Training

VQ-VAE is first trained with the two stages as described in Sect. 3 and PixelCNN is pretrained with the MLE objective. Then our RL training starts. We use Adam optimizer [27] with $\beta_1 = 0.5$, $\beta_2 = 0.999$ and set the learning rate of the discriminator and PixelCNN to 1e-4, 4e-6 respectively with the mini-batch size of 16. Initially, the discriminator is trained for 100 iterations to catch up with the pretrained PixelCNN. During the RL training, for each PixelCNN update, the discriminator is updated 5 times and updates are performed iteratively. When there are two hierarchies of latent codes available, both the PixelCNN models are trained at the same time which encourages collaboration between them. Outputs from the discriminator are first normalized to the range of (-1, 1) before updating the PixelCNNs. The normalization is performed with respect to the highest value from a set of fake and real images. Without normalizing the discriminator's outputs, rewards become unreliable as, after each update, the discriminator's outputs may change drastically even with the WGAN penalty. By normalizing rewards based on the maximum value of real and fake images, rewards act as an evaluation score from the current state of D. For the partial generation, the algorithm randomly switches between different modes and samples a random number of rows. We set λ_{gp} in Eq. 5 to 10 and the discounted factor γ to 0.99.

5 Experiments

We first conduct experiments on a synthetic dataset similarly to [49,54] in Sect. 5.2. The synthetic dataset is constructed from a pre-trained PixelCNN [34] prior, which we denote as the oracle model G_{oracle}. The oracle model is used to provide the true data distribution and the generator is trained to fit the oracle distribution. The advantage of having an oracle model is that the generated samples can be evaluated with NLL which is not possible for real data.

Our second experiment in Sect. 5.3 is for real-world images where we train our models in the CelebA dataset [30] with different scales: 64×64, 128×128 and 256×256, in addition to the LSUN-bedroom [52]. As an evaluation metric, we use Fréchet Inception Distance (FID) [22], which uses an Inception network to extract features to compare the closeness of real and fake images statistics. Lower FID means better image quality and diversity. Several experiments with different settings and architecture details are included in the supplementary material.

We also perform an ablation study in Sect. 5.4 followed up by a use-case of the proposed method for image completion task in Sect. 5.5, which also demonstrates the proposed method captures global structure better than MLE trained model as the generated sequences are more correlated with the real ones.

5.1 Implementation Details

This section describes the implementation details of different modules used in RAL. The detailed information on architecture choices is included in our supplementary material.

Encoder. VQ-VAE [32] encoder is used to compress images to latent space. Each layer includes a transposed convolution for downsampling. For hierarchical codes, two latent levels are used i.e., top and bottom. Images are encoded into following latents codes for different resolutions: a) $64{\times}64 \rightarrow 8{\times}8$, b) $128{\times}128 \rightarrow 16{\times}16$, $32{\times}32$, c) $256{\times}256 \rightarrow 32{\times}32$, $64{\times}64$.

Decoder. VQ-VAE decoder is used to reconstruct images from latent codes. For hierarchical codes, the bottom-level code is conditioned on the top level. The network consists of transposed convolutional layers for upsampling.

PixelCNN/PixelSNAIL. Similar to VQ-VAE-2, we use PixelCNN with self-attention layers to model top-priors. For bottom-priors, which is more computationally expensive, we remove self-attention layers and reduce the number of residual channels. To improve the sampling speed from prior networks, we use caching similar to [36] and avoid redundant computation. Note that, we use a smaller VQ-VAE-2 architecture compared to [38] due to computational limitations. The main issue is the model capacity requirement for MLE pre-training. Even with 8×V100 GPUs (8×16GB), which is the best machine we can get, we were barely able to use similar networks designed in the VQ-VAE-2 for $128{\times}128$, but not enough for $256{\times}256$ due to the increase in latent code sizes.

PatchGAN Discriminator. The discriminator has a similar convolutional architecture to the encoder network. while at the patch level to classify whether a patch is real or fake. The discriminator has 5 strided convolutions which output a 4×4 reward map from a 128×128 image. We also experiment with different output scales in our ablation study to find the optimum architecture.

5.2 Synthetic Experiments

We first perform synthetic experiments on the CelebA dataset with the image resolution of 64×64, which are mapped into one level of 8×8 latent codes by the encoder. Instead of directly training on real images, we first train a PixelCNN G_{oracle} on the 8×8 latent codes. Then this model is used as an oracle data generator to provide training data. The advantage is that we can have accurate NLL estimation, which cannot be achieved with real images since the oracle generator for real images is unknown.

Next, a lighter PixelCNN[1] G_{MLE} is trained on training samples generated by G_{oracle} with the MLE objective, which is the VQ-VAE baseline. Our method fine tunes G_{MLE} with our RL training to estimate another model G_{RL}. For this experiment we do not use partial generation since the size of latent codes is already small. The ground truth NLL estimation of a generator G can be computed as:

$$\text{NLL}_{\text{oracle}} = -\mathbb{E}_{c_t \sim G} \left[\sum_{t=0}^{T-1} \log G_{\text{oracle}} \left(c_t | c_{0:t-1} \right) \right] \tag{7}$$

---

[1] We use a lighter PixelCNN so that the MLE model cannot be easily trained towards the oracle model.

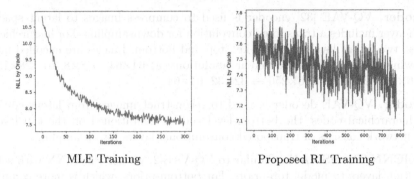

| MLE Training | Proposed RL Training |

Fig. 4. Negative log-likelihood during the training iterations.

Table 1. NLL and FID values for Oracle Experiments.

	NLL	FID on Oracle data	FID on Real data
MLE	7.55	7.49	15.63
RL - Single Reward	7.07	**5.17**	14.28
RL - Intermediate Reward	**7.05**	5.62	**13.69**

Table 2. FID values for real-world experiments. The number in parenthesis is FID on reconstructed real images.

Dataset	CelebA 64 × 64	CelebA 128 × 128	CelebA 256 × 256	LSUN-bedroom 128 × 128
MLE	5.35	58.07 (21.28)	69.37 (55.11)	47.77 (23.14)
Single Reward	**3.24**	**49.24 (16.49)**	66.44 (53.54)	39.86 (19.37)
Intermediate Reward	4.10	51.91 (19.68)	**64.72 (51.38)**	**36.47 (16.06)**
PGGAN [25]	–	7.30	–	8.34 — 256 × 256
COCO-GAN [29]	4.00	5.74	–	5.99 — 256 × 256

We report the NLL results in Table 1 and the learning curves in Fig. 4. It can be seen that our RAL with different rewards improve the MLE-trained model even though our objective is not NLL. In Table 1, we also show that our methods get better FID scores when compared with both fake images produced by G_{oracle} and real images.

5.3   Real World Experiments

Given the success of oracle experiments, we further evaluate our method for real-word image generation and report the FID values [22] in Table 2. Note that the FID scores outside/inside parenthesis are computed between generated images and the real images/reconstructed real images. The large FID scores on 128×128 images are mainly caused by the lossy compression of the VQ-VAE encoder-decoder, which is also mentioned by [38]. In addition, only reconstructed real images are used during our training. Therefore the scores inside the parenthesis are more meaningful for comparison.

Fig. 5. Image samples on CelebA dataset with 128×128 resolution. The ratio of good samples from MLE model is less than ours.

Fig. 6. Image samples on LSUN-bedroom dataset with 128×128 resolution. The ratio of good samples from MLE model is less than ours.

Results show that our models improve the MLE trained model in all settings by a large margin. In addition, our single-reward model achieves the best FID score on CelebA 64×64 resolution even when compared to the state-of-the-art GAN models [25, 29] (the compression loss is relatively small on 64×64 resolution and thus our FID score is not affected much). For 128×128 resolution, the use of intermediate reward performs better on the LSUN-bedroom while the single reward is better for CelebA. We use a much smaller architecture than the one reported in [38] for CelebA 256×256 experiments. The proposed RAL still

Table 3. Ablation experiments on CelebA with 128×128 resolution. The number in parenthesis is FID score on reconstructed real images.

	FID		FID
Partial-Gen. (Proposed)	**49.24 (16.54)**	Single reward, D output: 8×8	53.15 (18.28)
Full-Gen.	51.91 (20.25)	Single reward, D output: 4×4	**49.24 (16.54)**
RAL on Top	54.82 (19.20)	Single reward, D output: 2×2	49.78 (17.68)
RAL on Bottom	55.02 (19.80)	Interm. reward, D output: 8×8	50.22 (18.65)
		Interm. reward, D output: 4×4	54.835 (19.47)
		Interm. reward, D output: 2×2	55.37 (20.99)

improves the MLE trained model which demonstrates the effectiveness of the proposed method for larger latent codes.

We provide qualitative visual comparisons with the MLE trained model for the Celeba dataset in Fig. 5 as well as LSUN-bedroom dataset in Fig. 6. As can be seen, the main problem of the MLE trained model is that it generates visually good and bad images from the same model. This is mostly due to accumulating errors during the sampling and mismatch of training/testing procedures. Also, note that we do not use any classifier-based rejection sampling proposed in [38]. However, as the MLE trained models is additionally optimized via GAN loss, better-quality samples are produced.

5.4   Ablation Study

In Table 3, we investigate different modules of the proposed method using the CelebA dataset with the image resolution of 128×128. In Table 3(a), the use of partial generation outperforms the full-generation. Note that both configurations are trained for the same time interval for a fair comparison. For all settings in Table 3 (a), we used single-reward, estimated from a discriminator, which outputs a 4×4 output. In Table 3(a), we also test the effect of applying RAL on top and bottom priors independently, which achieves worse FID than the proposed joint training. This presents that the proposed RAL indeed enhances the collaboration of independently MLE trained priors.

Secondly, we test single and intermediate reward configurations with different score outputs from the discriminator in Table 3(b). The score map is adjusted by changing the number of convolutional layers. The single reward works best with the score map of 4×4. Intermediate reward works best with an 8×8 score map. We also tried using larger score map outputs but did not observe improvements.

5.5   Image Completion

Image completion/inpainting [10,15] is a process of restoring missing/damaged parts of the image. As the generation of autoregressive models is sequential, i.e., the generation of the current pixel is conditioned on the previous pixels, autoregressive models can be used for image completion. As in most image generation problems, GANs are top-performing models for this task [17,53] but we believe this is a good application for comparisons with the MLE model.

Fig. 7. Image completion experiment for the MLE trained method versus the proposed method. Images in the first row are hidden with masking function in the second. The generation shown in rows 3, 4 and 5 continue from visible regions of the image and the corresponding location can be approximated by the encoder.

In Fig. 7, we perform some qualitative results with the LSUN-bedroom (128 × 128) dataset with different image completion settings. Firstly, a number of rows of input images (first row) are hidden (second). These hidden rows can be simply eliminated by zeroing out their corresponding locations in latent maps. Next, autoregressive models can continue sampling the remaining codes where completed reconstructed images are shown in the last three rows. We illustrate two modes of the proposed model to show the proposed method can generate diverse samples.

In the first column, a glimpse of a room is given where both methods successfully include the window in the generated image. The proposed method tends to produce more realistic samples due to its adversarial learning. Another interesting example can be seen in the final column, where the proposed method can reproduce the remaining regions more realistically compared to the MLE trained model.

6   Conclusion

Our proposed idea has several advantages compared to traditional autoregressive models. By utilizing reinforcement learning, we bridge the gap between training and testing procedures while enabling the autoregressive model to take advantage of GAN training. Additionally, we use partial generation to improve the training and use two different image rewards. Last but not least, we show that the proposed method can improve the collaboration of independently trained hierarchical modules. To our best knowledge, our framework is first to enable the adversarial learning in PixelCNNs and does not solely depend on the traditional objective function of autoregressive models for good sample quality. A possible future direction would be to extend the proposed framework to higher resolution images which would be possible with faster sampling times.

References

1. Ak, K.E.: Deep learning approaches for attribute manipulation and text-to-image synthesis. Ph.D. thesis (2019)
2. Ak, K.E., Lim, J.H., Tham, J.Y., Kassim, A.A.: Attribute manipulation generative adversarial networks for fashion images. In: ICCV. IEEE (2019)
3. Ak, K.E., Lim, J.H., Tham, J.Y., Kassim, A.A.: Semantically consistent hierarchical text to fashion image synthesis with an enhanced-attentional generative adversarial network. In: ICCVW (2019)
4. Ak, K.E., Lim, J.H., Tham, J.Y., Kassim, A.A.: Semantically consistent text to fashion image synthesis with an enhanced attentional generative adversarial network. PRL (2020)
5. Ak, K.E., Ying, S., Lim, J.H.: Learning cross-modal representations for language-based image manipulation. In: ICIP (2020)
6. Arjovsky, M., Chintala, S., Bottou, L.: Wasserstein generative adversarial networks. In: ICML, pp. 214–223 (2017)
7. Arora, S., Ge, R., Liang, Y., Ma, T., Zhang, Y.: Generalization and equilibrium in generative adversarial nets (gans). In: ICML, pp. 224–232 (2017)
8. Arora, S., Risteski, A., Zhang, Y.: Do GANs learn the distribution? some theory and empirics. In: International Conference on Learning Representations (2018)
9. Bengio, S., Vinyals, O., Jaitly, N., Shazeer, N.: Scheduled sampling for sequence prediction with recurrent neural networks. In: Advances in Neural Information Processing Systems, pp. 1171–1179 (2015)
10. Bertalmio, M., Sapiro, G., Caselles, V., Ballester, C.: Image inpainting. In: Proceedings of the 27th annual conference on Computer graphics and interactive techniques, pp. 417–424. ACM Press/Addison-Wesley Publishing Co. (2000)
11. Brock, A., Donahue, J., Simonyan, K.: Large scale gan training for high fidelity natural image synthesis. arXiv preprint arXiv:1809.11096 (2018)
12. Chen, X., Mishra, N., Rohaninejad, M., Abbeel, P.: Pixelsnail: an improved autoregressive generative model. arXiv preprint arXiv:1712.09763 (2017)
13. Choi, Y., Choi, M., Kim, M., Ha, J.W., Kim, S., Choo, J.: Stargan: unified generative adversarial networks for multi-domain image-to-image translation. In: CVPR, June 2018
14. Chorowski, J., Weiss, R.J., Bengio, S., van den Oord, A.: Unsupervised speech representation learning using wavenet autoencoders. IEEE/ACM Trans. Audio Speech Lang. Process. **27**(12), 2041–2053 (2019)
15. Criminisi, A., Perez, P., Toyama, K.: Object removal by exemplar-based inpainting. In: CVPR, vol. 2, pp. II-II (2003)
16. De Fauw, J., Dieleman, S., Simonyan, K.: Hierarchical autoregressive image models with auxiliary decoders. arXiv preprint arXiv:1903.04933 (2019)
17. Fawzi, A., Samulowitz, H., Turaga, D., Frossard, P.: Image inpainting through neural networks hallucinations. In: 2016 IEEE 12th Image, Video, and Multidimensional Signal Processing Workshop (IVMSP), pp. 1–5. IEEE (2016)

18. Ganin, Y., Kulkarni, T., Babuschkin, I., Eslami, S., Vinyals, O.: Synthesizing programs for images using reinforced adversarial learning. arXiv preprint arXiv:1804.01118 (2018)
19. Goodfellow, I., et al.: Generative adversarial nets. In: NeurIPS, pp. 2672–2680 (2014)
20. Gulrajani, I., Ahmed, F., Arjovsky, M., Dumoulin, V., Courville, A.C.: Improved training of wasserstein gans. In: NeurIPS, pp. 5767–5777 (2017)
21. Heqing, Z., Ak, K.E., Kassim, A.A.: Learning cross-modal representations for language-based image manipulation. In: ICIP (2020)
22. Heusel, M., Ramsauer, H., Unterthiner, T., Nessler, B., Hochreiter, S.: Gans trained by a two time-scale update rule converge to a local nash equilibrium. In: NeurIPS, pp. 6626–6637 (2017)
23. Huang, Z., Heng, W., Zhou, S.: Learning to paint with model-based deep reinforcement learning. arXiv preprint arXiv:1903.04411 (2019)
24. Isola, P., Zhu, J.Y., Zhou, T., Efros, A.A.: Image-to-image translation with conditional adversarial networks. In: CVPR (2017)
25. Karras, T., Aila, T., Laine, S., Lehtinen, J.: Progressive growing of gans for improved quality, stability, and variation. arXiv preprint arXiv:1710.10196 (2017)
26. Karras, T., Laine, S., Aila, T.: A style-based generator architecture for generative adversarial networks. In: Proceedings of the IEEE Conference on Computer Vision and Pattern Recognition, pp. 4401–4410 (2019)
27. Kingma, D.P., Ba, J.: Adam: A method for stochastic optimization. arXiv preprint arXiv:1412.6980 (2014)
28. Kingma, D.P., Welling, M.: Auto-encoding variational bayes. arXiv preprint arXiv:1312.6114 (2013)
29. Lin, C.H., Chang, C.C., Chen, Y.S., Juan, D.C., Wei, W., Chen, H.T.: Coco-gan: generation by parts via conditional coordinating. arXiv preprint arXiv:1904.00284 (2019)
30. Liu, Z., Luo, P., Wang, X., Tang, X.: Deep learning face attributes in the wild. In: ICCV, December 2015
31. Mirza, M., Osindero, S.: Conditional generative adversarial nets. arXiv:1411.1784 (2014)
32. van den Oord, A., Vinyals, O., et al.: Neural discrete representation learning. In: Advances in Neural Information Processing Systems, pp. 6306–6315 (2017)
33. Oord, A.v.d., Kalchbrenner, N., Kavukcuoglu, K.: Pixel recurrent neural networks. arXiv preprint arXiv:1601.06759 (2016)
34. Oord, A.v.d., Kalchbrenner, N., Vinyals, O., Espeholt, L., Graves, A., Kavukcuoglu, K.: Conditional image generation with pixelcnn decoders. In: NeurIPS, pp. 4797–4805 (2016)
35. Papineni, K., Roukos, S., Ward, T., Zhu, W.J.: Bleu: a method for automatic evaluation of machine translation. In: Proceedings of the 40th annual meeting on association for computational linguistics, pp. 311–318. Association for Computational Linguistics (2002)
36. Ramachandran, P., et al.: Fast generation for convolutional autoregressive models. arXiv preprint arXiv:1704.06001 (2017)
37. Ranzato, M., Chopra, S., Auli, M., Zaremba, W.: Sequence level training with recurrent neural networks. arXiv preprint arXiv:1511.06732 (2015)
38. Razavi, A., Oord, A.v.d., Vinyals, O.: Generating diverse high-fidelity images with vq-vae-2. arXiv preprint arXiv:1906.00446 (2019)
39. Saito, Y., Takamichi, S., Saruwatari, H.: Statistical parametric speech synthesis incorporating generative adversarial networks. IEEE/ACM **26**(1), 84–96 (2017)

40. Salimans, T., Goodfellow, I., Zaremba, W., Cheung, V., Radford, A., Chen, X.: Improved techniques for training gans. In: NeurIPS, pp. 2234–2242 (2016)
41. Sisman, B., Vijayan, K., Dong, M., Li, H.: Singan: Singing voice conversion with generative adversarial networks. In: APSIPA ASC, pp. 112–118 (2019)
42. Sisman, B., Li, H.: Generative adversarial networks for singing voice conversion with and without parallel data. In: Speaker Odyssey, pp. 238–244 (2020)
43. Sisman, B., Zhang, M., Dong, M., Li, H.: On the study of generative adversarial networks for cross-lingual voice conversion. In: 2019 IEEE Automatic Speech Recognition and Understanding Workshop (ASRU), pp. 144–151. IEEE (2019)
44. Sisman, B., Zhang, M., Sakti, S., Li, H., Nakamura, S.: Adaptive wavenet vocoder for residual compensation in gan-based voice conversion. In: SLT, pp. 282–289 (2018)
45. Sutskever, I., Vinyals, O., Le, Q.V.: Sequence to sequence learning with neural networks. In: NeurIPS, pp. 3104–3112 (2014)
46. Sutton, R.S., McAllester, D.A., Singh, S.P., Mansour, Y.: Policy gradient methods for reinforcement learning with function approximation. In: Advances in neural information processing systems, pp. 1057–1063 (2000)
47. Theis, L., Oord, A.v.d., Bethge, M.: A note on the evaluation of generative models. arXiv preprint arXiv:1511.01844 (2015)
48. Tjandra, A., Sisman, B., Zhang, M., Sakti, S., Li, H., Nakamura, S.: VQVAE unsupervised unit discovery and multi-scale code2spec inverter for zerospeech challenge 2019. arXiv preprint arXiv:1905.11449 (2019)
49. Toyama, J., Iwasawa, Y., Nakayama, K., Matsuo, Y.: Toward learning better metrics for sequence generation training with policy gradient (2018)
50. Williams, R.J.: Simple statistical gradient-following algorithms for connectionist reinforcement learning. Mach. Learn. **8**(3–4), 229–256 (1992)
51. Xu, T., et al.: Attngan: Fine-grained text to image generation with attentional generative adversarial networks. In: CVPR (2018)
52. Yu, F., Zhang, Y., Song, S., Seff, A., Xiao, J.: LSUN: construction of a large-scale image dataset using deep learning with humans in the loop. arXiv preprint arXiv:1506.03365 (2015)
53. Yu, J., Lin, Z., Yang, J., Shen, X., Lu, X., Huang, T.S.: Free-form image inpainting with gated convolution. In: ICCV, pp. 4471–4480 (2019)
54. Yu, L., Zhang, W., Wang, J., Yu, Y.: Seqgan: sequence generative adversarial nets with policy gradient. In: AAAI (2017)
55. Zhang, H., Goodfellow, I., Metaxas, D., Odena, A.: Self-attention generative adversarial networks. arXiv preprint arXiv:1805.08318 (2018)
56. Zhou, K., Sisman, B., Li, H.: Transforming spectrum and prosody for emotional voice conversion with non-parallel training data. arXiv preprint arXiv:2002.00198 (2020)

APRICOT: A Dataset of Physical Adversarial Attacks on Object Detection

A. Braunegg, Amartya Chakraborty, Michael Krumdick, Nicole Lape,
Sara Leary, Keith Manville$^{(\boxtimes)}$, Elizabeth Merkhofer, Laura Strickhart,
and Matthew Walmer

The MITRE Corporation, McLean, VA, USA
{abraunegg,achakraborty,mkrumdick,nflett,sleary,kmanville,emerkhofer,
lstrickhart,mwalmer}@mitre.org

Abstract. Physical adversarial attacks threaten to fool object detection systems, but reproducible research on the real-world effectiveness of physical patches and how to defend against them requires a publicly available benchmark dataset. We present APRICOT, a collection of over 1,000 annotated photographs of printed adversarial patches in public locations. The patches target several object categories for three COCO-trained detection models, and the photos represent natural variation in position, distance, lighting conditions, and viewing angle. Our analysis suggests that maintaining adversarial robustness in uncontrolled settings is highly challenging but that it is still possible to produce targeted detections under white-box and sometimes black-box settings. We establish baselines for defending against adversarial patches via several methods, including using a detector supervised with synthetic data and using unsupervised methods such as kernel density estimation, Bayesian uncertainty, and reconstruction error. Our results suggest that adversarial patches can be effectively flagged, both in a high-knowledge, attack-specific scenario and in an unsupervised setting where patches are detected as anomalies in natural images. This dataset and the described experiments provide a benchmark for future research on the effectiveness of and defenses against physical adversarial objects in the wild. The APRICOT project page and dataset are available at apricot.mitre.org.

Keywords: Adversarial attacks · Adversarial defense · Datasets and evaluation · Object detection

1 Introduction

Image detection and classification models have driven important advances in safety-critical contexts such as medical imaging [13] and autonomous vehicle technology [12]. However, deep models have been demonstrated to be vulnerable

Electronic supplementary material The online version of this chapter (https://doi.org/10.1007/978-3-030-58589-1_3) contains supplementary material, which is available to authorized users.

© Springer Nature Switzerland AG 2020
A. Vedaldi et al. (Eds.): ECCV 2020, LNCS 12366, pp. 35–50, 2020.
https://doi.org/10.1007/978-3-030-58589-1_3

Fig. 1. APRICOT images run through FRCNN; only detections caused by the patch are shown. (a) FRCNN adversarial patch targeting apple category. (b) SSD adversarial patch targeting person category. (c) RetinaNet adversarial patch targeting bottle category. (d) FRCNN adversarial patch targeting suitcase category.

to adversarial attacks [3,32], images designed to fool models into making incorrect predictions. The majority of research on adversarial examples has focused on digital domain attacks, which directly manipulate the pixels of an image [6,15,23,25], but it has been shown that it is possible to make physical domain attacks in which an adversarial object is manufactured (usually printed) and then placed in a real-world scene. When photographed, these adversarial objects can trigger model errors in a variety of systems, including those for classification [1,4,10,17], facial recognition [29], and object detection [7,33]. These attacks are more flexible, as they do not require the attacker to have direct access to a system's input pixels, and they present a clear, pressing danger for systems like autonomous vehicles, in which the consequences of incorrect judgements can be dire. In addition, most adversarial defense research has focused on the task of classification, while for real-world systems, defenses for deep detection networks are arguably more important.

Unlike digital attacks, physical adversarial objects must be designed to be robust against naturally occurring distortions like angle and illumination, such that they retain their adversariality after being printed and photographed under a variety of conditions. The key to achieving robustness is a method called Expectation over Transformation [1], which directly incorporates simulated distortions during the creation of the adversarial pattern. Without this strategy, adversarial examples typically fail in the real world [21]. For this reason, physical adversarial examples face a robustness-vs-blatancy trade-off [7], as robust adversarial textures are typically more distinctive and visible to the human eye. This is in contrast to digital attacks, which only need to alter pixels by a tiny amount. It is a critical unanswered question whether making physical adversarial patches robust also makes them easy to detect using defensive systems.

Securing safety-critical detection systems against adversarial attacks is essential, but conducting research into how to detect and defeat these attacks is difficult without a benchmark dataset of adversarial attacks in real-world situations. We present APRICOT, a dataset of Adversarial Patches Rearranged In COnText. This dataset consists of over 1,000 images of printed adversarial patches photographed in real-world scenes. The patches in APRICOT are crafted to trigger false positive detections of 10 target COCO [19] classes in 3 different COCO-trained object detection models [16]. While prior works in physical adversarial examples have included some real-world experiments in fairly controlled environments [4,7,10,21], our approach with APRICOT was to use a group of 20 volunteer photographers to capture images of adversarial patches in the wild. The dataset incorporates a wide variety of viewing angles, illuminations, distances-to-camera, and scene compositions, which all push the necessary level of patch robustness further than prior works. To the best of our knowledge, APRICOT is the first publicly released dataset of physical adversarial patches in context. APRICOT allows us to perform a comprehensive analysis of the effectiveness of adversarial patches in the wild. We analyze white-box effectiveness, black-box transferability [25], and the effect of factors such as patch size and viewing angle.

Despite the danger that physical adversarial examples pose against object detection networks, there has been limited research into defensive mechanisms against them. In addition to the APRICOT dataset, we present baseline experiments measuring the detectability of adversarial patches in real-world scenes. We adapt several digital domain defenses [11,22,27] to the task of defending a Faster-RCNN model [26] from black-box attacks in the real world. We consider two broad categories of defenses. "Supervised defenses" like [22] assume that the defender is aware of the attacker's method and can therefore generate their own attack samples and train a defensive mechanism with them. We show that, given this prior knowledge, it is possible to use images with synthetically inserted patches to train an adversarial patch detector with high precision. However, we believe that such defenses are insufficient in the long run, as they may not generalize as well to newly emerging attacks. "Unsupervised defenses" like [11,27] are attack-agnostic, operating without any prior knowledge of attack appearance. We test several unsupervised defense strategies, both for network defense and for adversarial patch localization. We hope that APRICOT will provide a testbed to drive future developments in defenses for object detection networks.

To summarize, our main contributions are as follows: (1) APRICOT, the first publicly available dataset of physical adversarial patches in context, (2) an assessment of the robustness of adversarial patches in the wild, as both white-box and black-box attacks, and (3) several supervised and unsupervised strategies for flagging adversarial patches in real-world scenes. In Sect. 2, we summarize related works in adversarial examples in both the digital and physical domains. In Sect. 3, we describe the creation of the APRICOT dataset. In Sect. 4, we measure the effectiveness of the adversarial patches when placed in real-world scenes. Finally, in Sect. 5, we present several defenses designed to flag adversarial patches in the wild.

2   Related Work

Adversarial Examples in Digital Images. The phenomenon of adversarial examples in digital images has been explored extensively in recent years [3,6, 15,23,32]. Attack methods tend to follow a common framework of optimizing a perturbation vector that, when added to a digital image, causes the target classifier to produce an incorrect answer with high confidence. The most effective adversarial methods make perturbations that are imperceptible to humans. It is even possible to create "black-box" attacks without direct access to the target model by using a surrogate [25].

Adversarial Examples in the Real World. Digital domain attacks assume that the attacker is able to manipulate the pixels of an image directly. However, some works have expanded the scope of adversarial examples into the physical domain [1,4,10,17,29]. Adversarial examples in the real world are particularly dangerous for applications that rely on object detection. [33] developed a real-world adversarial patch that can block person detection. [7] developed an optimization framework called ShapeShifter that can create adversarial stop signs that will be detected but incorrectly classified. In creating the APRICOT dataset, we modify ShapeShifter to produce patches that cause detectors to hallucinate objects. Such patches could confuse machine learning systems and lead them to take pointless or dangerous actions that would be incomprehensible to a human observer.

Defending Against Adversarial Examples. Most prior adversarial defense literature focuses on digital attacks. In this work, we adapt several digital domain defenses with a focus on flagging physical adversarial examples. [22] proposed training a binary classifier to detect adversarial examples directly. This approach falls in the category of supervised defenses. They showed that training on one type of attack can transfer to weaker attacks, but not stronger ones. For unsupervised defense strategies, several authors have suggested that adversarial examples do not exist on the natural image manifold. [11] used kernel density estimation (KDE) to determine whether images are in low-density regions of the training distribution and Bayesian uncertain estimation with dropout to identify samples in low-confidence regions. PixelDefend [31] uses a PixelCNN [24] to model the natural image manifold and back-propagation to remove adversarial perturbations. Defense-GAN [27] uses a generative adversarial network (GAN) [14] to model the image distribution and GAN inversion and reconstruction to remove perturbations. Real-world adversarial objects need to be much more blatant than digital attacks to be robust to natural distortions [7]. This suggests that physical attacks might be inherently more detectable, though this question has yet to be answered due to the lack of a dataset like APRICOT.

3   The APRICOT Dataset

APRICOT is a dataset of physical adversarial patches photographed in real-world scenes. APRICOT contains 1,011 images of 60 unique adversarial patches designed to attack 3 different COCO-trained detection networks. While previous works have tested physical adversarial objects under controlled settings, APRI-COT provides data for testing them in diverse scenes captured by a variety of photographers. The dataset includes both indoor and outdoor scenes, taken at different times of day, featuring various objects in context, and with patch placements that vary in position, scale, rotation, and viewing angle.

The attacks in [7] and [33] can be characterized as "false negative attacks," as they cause detectors to miss genuine objects. The attack style used in APRICOT is instead a "false positive attack," making detectors hallucinate objects that do not exist. Such attacks can be just as dangerous. For example, imagine that a malicious agent has designed a patch to trigger hallucination of the "car" class and has placed it on the road. A self-driving car with a camera and detector model would see the patch, mistake it for a car, and suddenly stop or swerve to avoid a hallucinated collision. Furthermore, an untrained human observer would see the patch on the road but would not understand why the vehicle started to behave erratically. This would both threaten user safety and undercut user confidence.

The intended uses of APRICOT are twofold: firstly, to assess the risk posed by adversarial patches in real-world operating conditions, and secondly, to aid in the creation of new defenses. Research is stalled by the lack of physical datasets, which are much more costly to create than digital datasets. We contribute APRICOT as a benchmark for future works to use in developing and comparing defensive strategies for object detection networks.

3.1 Generating Adversarial Patches

To generate adversarial examples that are effective against detectors in the physical domain, we follow the approach from Chen et al. [7], where the authors optimize physical adversarial examples constrained to the shape and color of a stop sign. Since one goal of our dataset is to promote understanding of real-world adversarial performance across a variety of target objects and arbitrary locations, we choose to drop the object-specific mask and color constraints. We also optimize them with random COCO images in the background. This makes our adversarial examples patches [4] that target detectors and are universal in that they can be placed in arbitrary locations in the real world. The key algorithm behind these approaches is Expectation over Transformation [1], which optimizes a given patch over many inputs, X, subject to a variety of transformations, T, making the adversarial nature of the patch robust over the range of transformations. Given a loss function $\mathcal{L}$ and detector f and parameters θ, we optimize to find a patch $\hat{p}$ with target category y' according to:

$$\hat{p} = \underset{p \in \mathbb{R}^{h \times w \times 3}}{\arg \min} \; \mathbb{E}_{x \sim X, t \sim T}[\mathcal{L}(f_\theta(t(x, p)), y')] \tag{1}$$

By choosing transformations that model different location, scale, and rotation, the optimized patch can remain effective when printed and placed in the

Fig. 2. Circular patches targeting the Person category for FRCNN, SSD, and RetinaNet, respectively.

real world, where the adversary does not have knowledge of the photographer's position relative to the patch. Our attacks target 10 object categories from the COCO dataset: apple, backpack, bottle, cat, clock, laptop, microwave, person, potted plant, and suitcase. A patch for each of the 10 categories using both circle and square masks is trained with each of 3 detection models: Faster R-CNN with Resnet-50 [26], SSD with MobileNet [20], and RetinaNet (SSD with ResNet-50-FPN) [18]. In total, we create and print 60 patches for collection. Examples of the trained patches can be seen in Fig. 2. Each model is obtained from the TensorFlow Object Detection Model Zoo [16], and comes pre-trained on COCO. We adapt the code provided by Chen et al. [7] in order to accommodate single shot models. All patches are optimized using Expectation over Transformation (Eq. 1) with the full range of rotations ($0° - 360°$) and with scale ranging from 5 to 50 percent of the image. The Adam optimizer [8] is used with a learning rate of 0.05. The patches are trained for 500 iterations.

Each patch was printed on the high-quality setting of a Canon PRO-100 series printer. The 20 SSD patches are of a different resolution (256x256 pixels) than the Faster R-CNN (FRCNN) and RetinaNet patches (both 600x600 pixels). We scaled all patches to the same size physically, assuming the effects of pixelation are negligible at this size given the collection instruments used. We printed using the Matte Photo Paper setting, printing the entire image on 12 in. × 12 in. non-reflective matte 88lb. 16.5mil (cardstock) paper. The circle patches have a 10 in. diameter, and square patches are 10 in. × 10 in.

3.2   Dataset Description

APRICOT comprises 1,011 photos taken by 20 volunteer photographers on cell phones during Summer 2019. Each photo is approximately 12 megapixels. Each photographer was randomly assigned three patches and asked to photograph them in at least five public locations using a variety of distances, angles, indoor and outdoor settings, and photo compositions. The dataset contains between 10 and 42 photos of each patch, with a median of 15. Each photo in the dataset contains exactly one adversarial patch annotated with a bounding box. The patch generally occupies about 2% of the image pixels.

Previous work finds that viewing angle affects patch adversariality [7]. Because our photos are taken under comparatively uncontrolled settings, we were not able to measure viewing angle at collection time. Instead, we provide three redundant annotations for each photograph that indicate annotators' perception of whether the patch is pictured "Near Head-On" ($<5°$), "Angled" ($5° - 45°$), or "Severely Angled" ($>45°$). Annotators' perceptions are unanimous for approximately 70% of images. Our experiments use the mode as the definitive categorization: respectively, 39.1%, 59.5% and 1.5% of images are assigned to each category. One image with no annotator agreement is excluded from analyses that include angle. In addition, about 1/4 of the images are labeled as containing a warped patch; i.e., the cardstock the patch is printed on is not flat in the image.

Six patches (138 photos) are assigned to the development set, while the other 54 patches (873 photos) are in the test set. These partitions are photographed by disjoint sets of photographers. The COCO 2017 val and test partitions can be paired with APRICOT for experiments requiring images without adversaries, resulting in approximately 2.5% adversarial (APRICOT) images per partition. We do not include a training set for fully-supervised models. Annotations are distributed in a format compatible with the COCO API.

4   Effectiveness of Adversarial Patches

We present the first systematic analysis of physical adversarial patches in the wild and comparison to digital results. We evaluate the efficacy of our digital adversarial patches by running the COCO-trained detectors on COCO images with the patches inserted (Sect. 4.1) and on the APRICOT images of physical patches in the real world (Sect. 4.2). Then we measure the effect of the adversarial patch. We define a 'fooling event' as a detection that overlaps a ground truth adversarial bounding box, where a 'targeted' fooling event is classified as the same object class as the patch's target. Performance is reported as the percentage of ground truth adversarial bounding boxes that produced at least one fooling event.

In order to count only predictions caused by the presence of an adversarial patch in the image and not true predictions or non-adversarial model errors, we calculate the intersection over union (IoU) between predicted bounding boxes and ground truth patch bounding boxes and only include predictions with an IoU of at least 0.10. We use a small IoU for our metric because the patches will sometimes generate many small, overlapping predictions in the region of the attack, as can be seen in Fig. 1. These predictions should be preserved because they represent a valid threat to detectors. Still, we find that performance is only slightly degraded by increasing the minimum IoU to 0.50, as most small fooling events are accompanied by larger detections. Some patches overlap larger objects that also create detections, like a patch sitting on a bench. To avoid attributing these to the adversaries, we discard any predicted bounding boxes that are more than twice as large as the ground truth box for the patch. This is necessary because APRICOT is not annotated with COCO object categories.

We report only fooling events with a confidence greater than 0.30. This threshold was chosen because it is the standard threshold used by models in the TensorFlow object detection API model zoo. However, we find that the confidence of fooling event predictions and non-fooling event predictions follow similar distributions. For this evaluation, patches are considered small if they take up less than 10% of the image, medium if they take up between 10% and 20%, and large if they take up greater than 20%.

4.1 Digital Performance

Digital evaluation confirms the attacks' effectiveness before they are inserted into the unconstrained physical world. They also serve as an upper threshold of performance before degradation occurs from real-world effects. Patches have been digitally inserted into random images from the COCO dataset at varying image locations, scales (5–25% of the image), and angles of rotation (0°–360°). Each patch was inserted into 15 COCO images in this evaluation.

Table 1 shows the performance of the patches when digitally inserted, broken down by the model used to generate the patch and evaluation model. Results are also broken down by targeted and untargeted fooling events. The patches, for the most part, perform much better when the patch model and evaluation model are the same (white-box attack) than when they are different (black-box attack [25]). For targeted fooling events, white-box performance ranges from 23–50%, and black-box performance ranges from almost completely ineffective in the worst pairings to nearing white-box performance in the case of FRCNN patches on RetinaNet. We find producing untargeted fooling events to be much easier and see fooling event rates of over 60% even in the black-box case.

We examine the effect of size and shape of the patches on fooling event rate in Table 2. The efficacy of adversarial patches increases along with their size in the image in both the targeted and untargeted cases. Square patches perform a little better than circle patches, possibly because patch area was computed using bounding boxes, so circles are consistently over estimated. When measuring untargeted performance, we observe that predicted object categories are clearly correlated with the patch's own geometry; square patches frequently produce kite predictions and circular patches produce doughnut predictions. This highlights the importance of patch shape and is a factor in the difficulty of fooling detectors using patches not tailored to the shape of the targeted category.

Table 1. Performance of digital and physical patches @ 0.3 confidence and 0.10 IoU

| | | | Evaluation Models | | | | | |
| | | | Digital | | | Physical | | |
			FRCNN	SSD	RetinaNet	FRCNN	SSD	RetinaNet
Patch Models	FRCNN (344)	Targeted	0.50	0.04	0.37	0.17	0.0	0.05
		Untargeted	0.89	0.64	0.84	0.43	0.15	0.34
	SSD (331)	Targeted	0.08	0.23	0.07	0.14	0.06	0.04
		Untargeted	0.71	0.70	0.67	0.39	0.20	0.30
	RetinaNet (336)	Targeted	0.14	0.01	0.39	0.01	0.0	0.03
		Untargeted	0.72	0.62	0.78	0.33	0.15	0.29
	All (1,011)	Targeted	0.24	0.09	0.27	0.11	0.02	0.04
		Untargeted	0.77	0.65	0.76	0.38	0.17	0.31

Table 2. Black-box and white-box performance of digital and physical patches broken down by size, shape, and angle. Angle is only taken into account for physical patches. Grayed boxes correspond to white-box case. Count columns indicate the number of images used for evaluation.

		Digital FRCNN Targeted	Untargeted	SSD Targeted	Untargeted	RetinaNet Targeted	Untargeted	Count	Physical FRCNN Targeted	Untargeted	SSD Targeted	Untargeted	RetinaNet Targeted	Untargeted	Count
FRCNN	Small	0.07	0.80	0.0	0.31	0.05	0.70	100	0.15	0.41	0.0	0.15	0.04	0.34	333
	Medium	0.64	0.92	0.04	0.83	0.5	0.88	120	1.0	1.0	0.0	0.22	0.22	0.33	9
	Large	0.83	0.98	0.09	0.78	0.59	0.99	80	1.0	1.0	0.0	0.0	1.0	0.0	2
	Circle	0.47	0.84	0.03	0.64	0.37	0.76	150	0.23	0.59	0.0	0.26	0.06	0.48	180
	Square	0.53	0.95	0.05	0.64	0.38	0.93	150	0.12	0.26	0.0	0.04	0.04	0.19	164
	Head-On	-	-	-	-	-	-	-	0.21	0.42	0.0	0.16	0.06	0.37	126
	Angled	-	-	-	-	-	-	-	0.16	0.42	0.0	0.14	0.04	0.31	209
	Severe	-	-	-	-	-	-	-	0.0	0.78	0.0	0.44	0.0	0.56	9
SSD	Small	0.05	0.62	0.03	0.53	0.04	0.66	110	0.11	0.34	0.01	0.14	0.03	0.27	297
	Medium	0.06	0.61	0.22	0.62	0.09	0.62	110	0.32	0.64	0.41	0.63	0.0	0.50	22
	Large	0.14	0.97	0.51	0.92	0.10	0.78	80	0.33	0.83	0.67	1.0	0.33	0.83	12
	Circle	0.08	0.59	0.20	0.70	0.05	0.58	150	0.09	0.35	0.01	0.14	0.03	0.31	156
	Square	0.07	0.82	0.25	0.70	0.10	0.77	150	0.18	0.42	0.10	0.25	0.05	0.30	175
	Head-On	-	-	-	-	-	-	-	0.17	0.43	0.06	0.20	0.04	0.26	141
	Angled	-	-	-	-	-	-	-	0.11	0.35	0.06	0.19	0.03	0.30	188
	Severe	-	-	-	-	-	-	-	0.0	0.0	0.0	1.0	0.0	0.0	1
RetinaNet	Small	0.02	0.77	0.0	0.35	0.05	0.61	100	0.02	0.33	0.0	0.13	0.01	0.27	317
	Medium	0.18	0.65	0.02	0.77	0.48	0.77	120	0.0	0.31	0.0	0.38	0.38	0.51	16
	Large	0.24	0.78	0.03	0.77	0.60	0.94	80	0.0	1.0	0.0	1.0	0.0	1.0	3
	Circle	0.17	0.70	0.01	0.64	0.35	0.70	150	0.03	0.40	0.0	0.21	0.06	0.37	170
	Square	0.11	0.74	0.02	0.61	0.39	0.82	150	0.0	0.27	0.0	0.09	0.0	0.21	166
	Head-On	-	-	-	-	-	-	-	0.03	0.37	0.0	0.16	0.02	0.25	127
	Angled	-	-	-	-	-	-	-	0.01	0.32	0.0	0.15	0.03	0.30	204
	Severe	-	-	-	-	-	-	-	0.0	0.2	0.0	0.0	0.0	0.4	5

4.2 Physical Performance

Table 1 compares physical and digital patch performance in the white-box and black-box cases. In the physical domain, the division of patch size and angle is not evenly distributed, as this is difficult to ensure when people are taking photographs. The efficacy of patches decreases when photographed in the real world. We observe many of the same trends as in the digital results: untargeted fooling events are much easier to produce and certain models are easier to fool in both the white-box and black-box cases. The correlation between digital and physical results indicates that an adversary can develop an attack digitally and have some confidence as to its efficacy in the real world. However, these results also reveal that fooling detectors in the wild is challenging, especially given black-box knowledge.

As in the digital evaluation, the physical performance of the patches is analysed with respect to several patch characteristics: shape, size, and, additionally, angle to the camera. Comparison to the digital results can be found in Table 2. Again, larger patches have a higher fooling rate for all models in targeted and untargeted cases. Circular patches appear to be more effective on FRCNN and RetinaNet, but square patches are more effective on SSD. As in the digital evaluation, patch geometry has an effect on untargeted category predictions. Circular patches favor kite and umbrella, while also producing frisbee, sports ball, and doughnut. In contrast, square patches favor categories such as TV, handbag, book, and kite. We hypothesize kite is a common prediction due to sharing bright colors with the patches. As angle becomes more extreme, targeted performance decreases. Note that shear transformations were not used during patch training. Interestingly, in the untargeted case, head-on and angled patches perform roughly the same, but severely angled patches perform best by a large margin. Untargeted predictions for severely angled patches favor categories such as kite and umbrella, and also include surfboard and laptop.

Some of the patches included in APRICOT contain regions that closely resemble the target object class. For example, apple-like patterns can be seen in the patch in Fig. 1a. This phenomenon was also observed by [7]. This result is not entirely surprising, as the optimization framework of [7] resembles [30] which instead used gradient ascent as a tool for network explanation and object class visualization. Even though some patches resemble natural objects, they may still be considered adversarial if they possess an unnatural hyper-objectness characteristic. The work of [4] was able to produce adversarial patches that influence a classifier's output more strongly than a genuine object instance. This shows that adversarial patterns can manipulate a network's internal representations in an unnatural way to draw more attention than a natural object. This phenomenon could explain why the patches in APRICOT trigger so many untargeted detections even when severely angled.

5   Flagging Adversarial Patches

Detection systems that can accurately flag adversaries could neutralize their risk by identifying and filtering malicious results. There are multiple ways to formulate the task of flagging adversaries. One could train a detector to directly output predictions for patch locations. One could also apply defensive techniques to a detection network's outputs to identify if one of its predictions was caused by an adversary. Or one could ignore the detector and directly assess image regions to predict whether they contain an adversarial patch. We present baseline experiments to address each of these strategies.

We consider both supervised and unsupervised defenses. In this context, "supervised defense" means that we assume the defender has some knowledge of the attacker's algorithm and thus can sample their own patches from the same or similar distribution and use them to train a defensive mechanism. In the digital defense domain, this is analogous to [22], however it has not yet been studied if robust physical adversarial patches can be easily detected this way. Even if such a defense works, it may not be sufficient if it fails to generalize to new attack strategies, as was observed in [22]. In contrast, "unsupervised defenses" have no prior knowledge of the attack strategy, and instead must learn to identify adversarial anomalies after training on only normal data. This is clearly more challenging, and we believe it presents a more realistic defense scenario as new attacks are always being developed. These approaches are completely attack-agnostic but may simply expose anomalies in the data without regard for adversariality. Results are reported on the APRICOT test partition.

5.1   Detecting Adversarial Patches Using Synthetic Supervision

We train a Faster R-CNN [26] object detection model to detect adversarial patches using a synthetic dataset of patches digitally overlaid on the COCO

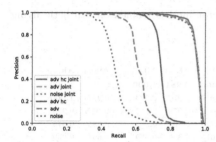

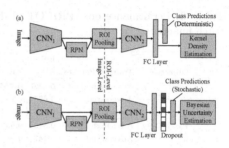

Fig. 3. Precision-Recall curves on APRICOT for models trained on synthetic flying patch images

Fig. 4. Adversarial detection strategies of [11] modified for Faster R-CNN. (a) Shows the layer used for KDE. (b) A dropout layer is added for Bayesian uncertainty estimation.

2017 training set. APRICOT does not contain a training partition, and the cost of physical data collection makes it impractical to collect an appropriately large dataset for fully-supervised adversary detector training. Instead we create several synthetic Flying Patch Datasets (named in homage to [9]) by overlaying patches of several shapes in random positions, sizes, and angles on natural images, similar to the method used in our digital performance analysis. Our adversarial Flying Patch Datasets also use the method of [7] to create patches. Under this framework, patches have a confidence parameter which limits their intensity through L2 regularization. We train models on two datasets, one with a range of confidence settings similar to [7], and one with only *high confidence* patches with the L2 regularization disabled. A dataset of Gaussian noise patches is used as a control to examine whether the models learn features of the attack patterns or more superficial features such as shape and color. For each flying patch collection, two models are trained. The *joint* variant detects both patches and COCO objects, while *patch only* models detect only adversarial patches.

Our patch detector models are fine-tuned from a publicly available COCO detection model, which is the same model used in [7]. This model is different from any used to create patches in APRICOT, so all attacks in APRICOT would be considered black-box attacks. Table 3 reports patch detection performance as Average Precision (AP) and Average Recall (AR). We report results under the default settings of the COCO API [19]: maximum detections 100 and IoU threshold ranging from 0.5 to 0.95. In practice, if an automated system is able to correctly detect a sub-region of an adversarial patch, this detection should be sufficient to alert the system to the presence of the patch. For this reason, we also report AP and AR under a more lenient IoU threshold of 0.5. We also provide a breakdown of recall by target model for the APRICOT patches. Figure 3 visualizes precision and recall with a 0.5 IoU threshold and max 100 detections.

We find that synthetic data is sufficient to train a high-performing adversarial patch detector for exemplars in the physical world. For all training configurations, performance is significantly improved by jointly training to detect both

Table 3. Performance on APRICOT of detector models trained on synthetic flying patch images

		IoU$_{0.5:0.95}$		IoU$_{0.5}$				
Objective	Training Patches	AP	AR	AP	AR	AR$_{frcnn}$	AR$_{ret}$	AR$_{ssd}$
Joint	Adv. High Conf.	**0.686**	**0.757**	**0.943**	**0.981**	**1.000**	0.941	**1.000**
	Adversarial	0.670	0.749	0.937	0.976	0.983	0.944	**1.000**
	Gaussian	0.635	0.714	0.936	0.979	0.987	**0.958**	0.993
Patch Only	Adv. High Conf.	0.540	0.655	0.739	0.885	0.893	0.785	0.979
	Adversarial	0.447	0.588	0.615	0.805	0.759	0.691	0.969
	Gaussian	0.329	0.500	0.478	0.742	0.696	0.562	0.872

patches and normal objects. This suggests that modeling normal objects reduces false positives for adversaries. The best performing model is the one trained on high confidence adversarial patches; however, under a lenient IoU threshold, the performance difference between the jointly trained models becomes much less significant. This suggests that patch shape, color, and texture intensity provide enough cues to detect adversarial patches. Error analysis of high-scoring false positives in APRICOT and COCO reveals mostly colorful, high-texture objects.

While these supervised models achieve high average precision, there is a risk they will fail to generalize to new attacks like in [22]. This motivates the need for attack-agnostic unsupervised defenses like [27,31]. In addition, clever attackers could likely evade detection with white-box access to the patch detector, or by using regularizations that create patches closer to the color and texture distributions of natural scenes.

5.2 Determining if a Detection is Adversarial Using Uncertainty and Density

We adapt the approaches of [11] to determine whether we can discriminate detections produced by adversarial patches from normal predictions made by an object detection model. This method is a self-defense for the object detection model and does not address adversaries that fail to produce false positives. These approaches examine all predictions made by the same model as Sect. 5.1 with >0.3 confidence, resulting in 273 untargeted fooling events out of 4428 detections in APRICOT's test partition. The results are therefore described using retrieval statistics over the 4428 model detections, and these methods' results are tied to the particular detection model's accuracy and internal representations.

Figure 4 illustrates how embeddings and uncertainty are extracted for the Faster R-CNN model. The density approach learns a Kernel Density Estimation (KDE) model for each output class. In this case, we learn a KDE over the final latent representation passed to the detector's object classifier, for all detections in COCO val 2017 with at least 0.3 confidence. The embedding for each detection in APRICOT is scored with respect to the detected class. The Bayesian uncertainty approach of [11] is considered to be more robust to adversarial circumvention [5]. We adapt this approach by applying 0.5 dropout to the same

Faster R-CNN layer at inference time and measuring variance in resulting object category predictions. It would not be possible to add dropout before the region of interest (ROI) pooling step, as each random trial would alter the region proposals themselves. Performing dropout at only the second to last layer is also computationally efficient, as added random trials incur minimal cost. Scores from these two methods are then used as features in a logistic regression learned from the APRICOT dev set, requiring a small amount of supervision. Note that the individual metrics require no supervision.

Figure 5 shows Receiver Operating Characteristics for these three approaches under a low-information scenario where a single threshold is set for all object classes. All three approaches are effective for this task. There is insufficient data to set individual thresholds per detected class, as would be most appropriate for the KDE models; only four object classes have at least 10 fooling events under this model. Many fooling events do not produce the targeted class prediction; this suggests that uncertainty is particularly effective because these detections lie near decision boundaries. The additional supervision required to produce the combined model is only marginally beneficial over uncertainty alone. These results are confounded by non-adversarial false positives or unlikely exemplars of COCO categories. More investigation is needed to determine if there is a difference in how true and false positives are scored in these analyses.

5.3 Localizing Adversarial Regions with Density and Reconstruction

Experiments using autoencoders localize anomalous regions without regard for any attack strategy or target model. These tests are conducted on (32,32) windows of images. We model natural images with the ACAI autoencoder described in [2], trained on windows of the images from COCO's 2017 train partition. Before constructing the windows, all images are resized to (600,600). Unlike the detection models in Sect. 5.1, these localization results label small, uniformly gridded windows of the image that are not necessarily aligned to patch bounding boxes. Therefore, they are presented using receiver operator characteristics over small windows. In this experiment, a (32,32) pixel window is considered adversarial if at least half of the pixels are contained in an APRICOT bounding box.

The density approach is further adapted to learn a one class model over embeddings of the non-adversarial data in the COCO 2017 validation partition. A Gaussian Mixture Model is employed on the intuition that since the embeddings are not separated by class, they will be multimodal. Five components were chosen empirically. Each image window is represented by mean-pooling the spatially invariant dimensions of the autoencoder bottleneck.

A previous work [28] used GAN inversion and reconstruction error as a tool for anomaly detection in medical imagery. In the context of APRICOT, adversarial patches may be detectable as anomalies. Another work [27] used GANs trained on normal data as a tool to remove adversarial perturbations from digital attacks. However, both of these GAN-based reconstruction methods rely on expensive computations to recover the latent space vector and reconstruct the

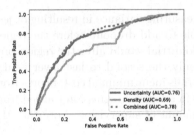
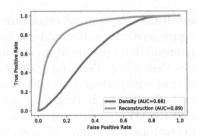

Fig. 5. ROC of KDE and Bayesian uncertainty for finding adversarial detections

Fig. 6. Performance of autoencoder methods for localizing adversary-containing 32 × 32 image regions

image. We reconstruct each window instead using the autoencoder and measure the L2 distance of the reconstruction from the original.

As shown in Fig. 6, the density-based approach has weak predictive power, while reconstruction error seems to effectively discriminate adversarial windows. High-frequency natural textures like foliage generate a high reconstruction error, producing false positives. These differences in performance suggest that the discriminative power of the autoencoder comes not from the information encoded in the latent space but rather from the information lost during encoding.

6 Conclusion

Research into the effectiveness and detectability of adversarial objects in the wild is limited by the absence of labeled datasets of such objects in the real world. We provide the first publicly available dataset of physical adversarial patches in a variety of real-world settings. Annotations of the patches' locations, attacked model, warping, shape, and angle to the camera are included to allow investigation into the effects of these attributes on the patches' adversariality.

Our analysis reveals that photographed physical patches in the wild can fool both white-box and, in some cases, black-box object detection models. Benchmark flagging experiments establish that adversarial patches can be flagged both with and without prior knowledge of the attacks. While our proposed baseline defenses seem to perform well on APRICOT, these efforts should not be considered the answer; in industry, even a small percentage of error can mean the difference between success and catastrophic failure. Moving forward, we want to encourage further experimentation to increase detection accuracy as adversarial attacks continue to become more and more complex.

Acknowledgments. We would like to thank Mikel Rodriguez, David Jacobs, Rama Chellappa, and Abhinav Shrivastava for helpful discussions and feedback on this work. We would also like to thank our MITRE colleagues who participated in collecting and annotating the APRICOT dataset and creating the adversarial patches.

References

1. Athalye, A., Engstrom, L., Ilyas, A., Kwok, K.: Synthesizing robust adversarial examples, pp. 284–293 (2018)
2. Berthelot, D., Raffel, C., Roy, A., Goodfellow, I.: Understanding and improving interpolation in autoencoders via an adversarial regularizer. In: International Conference on Learning Representations (2019). https://openreview.net/forum?id=S1fQSiCcYm
3. Biggio, B., et al.: Evasion attacks against machine learning at test time. In: Blockeel, H., Kersting, K., Nijssen, S., Železný, F. (eds.) ECML PKDD 2013. LNCS (LNAI), vol. 8190, pp. 387–402. Springer, Heidelberg (2013). https://doi.org/10.1007/978-3-642-40994-3_25
4. Brown, T.B., Mané, D., Roy, A., Abadi, M., Gilmer, J.: Adversarial patch (2017)
5. Carlini, N., Wagner, D.: Adversarial examples are not easily detected: bypassing ten detection methods. In: Proceedings of the 10th ACM Workshop on Artificial Intelligence and Security, pp. 3–14. ACM (2017)
6. Carlini, N., Wagner, D.: Towards evaluating the robustness of neural networks. In: 2017 IEEE Symposium on Security and Privacy (SP), pp. 39–57. IEEE (2017)
7. Chen, S.T., Cornelius, C., Martin, J., Chau, D.H.: Shapeshifter: robust physical adversarial attack on faster R-CNN object detector (2018)
8. Diederik P. Kingma, J.B.: Adam: A method for stochastic optimization. In: International Conference on Learning Representations (ICLR) (2015)
9. Dosovitskiy, A., et al.: Flownet: learning optical flow with convolutional networks. In: Proceedings of the IEEE International Conference on Computer Vision, pp. 2758–2766 (2015)
10. Eykholt, K., et al.: Robust physical-world attacks on deep learning visual classification. In: Proceedings of the IEEE Conference on Computer Vision and Pattern Recognition, pp. 1625–1634 (2018)
11. Feinman, R., Curtin, R.R., Shintre, S., Gardner, A.B.: Detecting adversarial samples from artifacts. arXiv preprint arXiv:1703.00410 (2017)
12. Fridman, L., et al.: MIT advanced vehicle technology study: large-scale naturalistic driving study of driver behavior and interaction with automation. IEEE Access **7**, 102021–102038 (2019). https://doi.org/10.1109/access.2019.2926040. http://dx.doi.org/10.1109/ACCESS.2019.2926040
13. Fu, G.S., Levin-Schwartz, Y., Lin, Q.H., Zhang, D.: Machine learning for medical imaging. J. Healthcare Eng. (2019)
14. Goodfellow, I., et al.: Generative adversarial nets. In: Advances in Neural Information Processing Systems, pp. 2672–2680 (2014)
15. Goodfellow, I.J., Shlens, J., Szegedy, C.: Explaining and harnessing adversarial examples (2014)
16. Eykholt, K., et al.: Robust physical-world attacks on deep learning visual classification. In: Proceedings of the IEEE Conference on Computer Vision and Pattern Recognition, pp. 1625–1634 (2018)
17. Kurakin, A., Goodfellow, I., Bengio, S.: Adversarial examples in the physical world. arXiv preprint arXiv:1607.02533 (2016)
18. Lin, T.Y., Goyal, P., Girshick, R., He, K., Dollar, P.: Focal loss for dense object detection. In: The IEEE International Conference on Computer Vision (ICCV), October 2017
19. Lin, T.-Y., et al.: Microsoft COCO: common objects in context. In: Fleet, D., Pajdla, T., Schiele, B., Tuytelaars, T. (eds.) ECCV 2014. LNCS, vol. 8693, pp. 740–755. Springer, Cham (2014). https://doi.org/10.1007/978-3-319-10602-1_48

20. Liu, W., et al.: SSD: Single shot multibox detector. In: European Conference on Computer Vision (2016)
21. Lu, J., Sibai, H., Fabry, E., Forsyth, D.: No need to worry about adversarial examples in object detection in autonomous vehicles. arXiv preprint arXiv:1707.03501 (2017)
22. Metzen, J.H., Genewein, T., Fischer, V., Bischoff, B.: On detecting adversarial perturbations (2017)
23. Moosavi-Dezfooli, S.M., Fawzi, A., Frossard, P.: Deepfool: a simple and accurate method to fool deep neural networks. In: Proceedings of the IEEE Conference on Computer Vision and Pattern Recognition, pp. 2574–2582 (2016)
24. Oord, A.v.d., Kalchbrenner, N., Kavukcuoglu, K.: Pixel recurrent neural networks, pp. 1747–1756 (2016)
25. Papernot, N., McDaniel, P., Goodfellow, I., Jha, S., Celik, Z.B., Swami, A.: Practical black-box attacks against machine learning. In: Proceedings of the 2017 ACM on Asia Conference on Computer and Communications Security, pp. 506–519. ACM (2017)
26. Ren, S., He, K., Girshick, R., Sun, J.: Faster R-cnn: towards real-time object detection with region proposal networks. In: Advances in Neural Information Processing Systems, pp. 91–99 (2015)
27. Samangouei, P., Kabkab, M., Chellappa, R.: Defense-gan: Protecting classifiers against adversarial attacks using generative models (2018)
28. Schlegl, T., Seeböck, P., Waldstein, S.M., Schmidt-Erfurth, U., Langs, G.: Unsupervised anomaly detection with generative adversarial networks to guide marker discovery. In: Niethammer, M., et al. (eds.) IPMI 2017. LNCS, vol. 10265, pp. 146–157. Springer, Cham (2017). https://doi.org/10.1007/978-3-319-59050-9_12
29. Sharif, M., Bhagavatula, S., Bauer, L., Reiter, M.K.: Accessorize to a crime: real and stealthy attacks on state-of-the-art face recognition. In: Proceedings of the 2016 ACM SIGSAC Conference on Computer and Communications Security, pp. 1528–1540. ACM (2016)
30. Simonyan, K., Vedaldi, A., Zisserman, A.: Deep inside convolutional networks: visualising image classification models and saliency maps. arXiv preprint arXiv:1312.6034 (2013)
31. Song, Y., Kim, T., Nowozin, S., Ermon, S., Kushman, N.: Pixeldefend: leveraging generative models to understand and defend against adversarial examples (2018)
32. Szegedy, C., et al.: Intriguing properties of neural networks (2014)
33. Thys, S., Van Ranst, W., Goedemé, T.: Fooling automated surveillance cameras: adversarial patches to attack person detection. In: Proceedings of the IEEE Conference on Computer Vision and Pattern Recognition Workshops (2019)

Visual Question Answering on Image Sets

Ankan Bansal[1]([✉]), Yuting Zhang[2], and Rama Chellappa[1]

[1] University of Maryland, College Park, USA
{ankan,rama}@umd.edu
[2] Amazon Web Services (AWS), Beijing, China
yutingzh@amazon.com

Abstract. We introduce the task of Image-Set Visual Question Answering (ISVQA), which generalizes the commonly studied single-image VQA problem to multi-image settings. Taking a natural language question and a set of images as input, it aims to answer the question based on the content of the images. The questions can be about objects and relationships in one or more images or about the entire scene depicted by the image set. To enable research in this new topic, we introduce two ISVQA datasets – indoor and outdoor scenes. They simulate the real-world scenarios of indoor image collections and multiple car-mounted cameras, respectively. The indoor-scene dataset contains 91,479 human-annotated questions for 48,138 image sets, and the outdoor-scene dataset has 49,617 questions for 12,746 image sets. We analyze the properties of the two datasets, including question-and-answer distributions, types of questions, biases in dataset, and question-image dependencies. We also build new baseline models to investigate new research challenges in ISVQA.

1 Introduction

Answering natural-language questions about images requires understanding both linguistic and visual data. Since its introduction [4], Visual Question Answering (VQA) has attracted significant attention. Several related datasets [14,22,33] and methods [9,12,20] have been proposed.

In this paper, we introduce the new task of Image Set Visual Question Answering (ISVQA)[1]. It aims to answer a free-form natural-language question based on a set of images. The proposed ISVQA task requires reasoning over objects and concepts in different images to predict the correct answer. For example, for Fig. 1 (Left), a model has to find the relationship between the bed in the top-left image and the mirror in the top-right, via pillows which are common to both the images. This example shows the unique challenges associated with

---

[1] Project page: https://ankanbansal.com/isvqa.html

This work was done when Ankan Bansal was an intern at AWS.

Electronic supplementary material The online version of this chapter (https://doi.org/10.1007/978-3-030-58589-1_4) contains supplementary material, which is available to authorized users.

© Springer Nature Switzerland AG 2020
A. Vedaldi et al. (Eds.): ECCV 2020, LNCS 12366, pp. 51–67, 2020.
https://doi.org/10.1007/978-3-030-58589-1_4

image-set VQA. A model for solving this type of problems has to understand the question, find the connections between the images, and use those connections to relate objects across images. Similarly, in Fig. 1 (Right), the model has to avoid double-counting recurring objects in multiple images. These challenges associated with scene understanding have not been explored in existing single-image VQA settings but frequently happen in the real world.

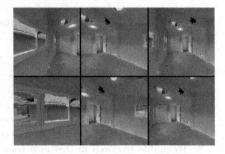

Fig. 1. (Left) Given the set of images above, and the question "What is hanging above the bed?", it is necessary to connect the bed in the top-left image to the mirror in the top-right image. To answer this question a model needs to understand the concepts of "bed", "mirror", "above", "hanging", etc. and be able to relate the bed in the first image with the headrest and pillows in the third image. (Right) When asked the question "How many rectangles are on the interior doors?", the model should be able to provide the answer ("four") and avoid counting the same rectangles multiple times.

ISVQA reflects information retrieval from multiple images of relevance but with no obvious continuous correspondences. Such image sets can be any albums and images captured by multiple devices, *e.g.*, images under the same story on Facebook/Instagram, images of the same product on Craigslist and Amazon, pictures of the same house on real estate websites, and images from different car-mounted cameras. Other instances of the ISVQA task include answering questions about images taken at different times (*e.g.* like in camera trap photography), at different locations (*e.g.* multiple cameras from an indoor or outdoor location), or from different viewpoints (*e.g.* live sports coverage). Some of these settings contain images taken from the same scene, while others might involve images of a larger span. While ISVQA can be generally applied to any type of images, in this paper, we focus on images from multiple views of an environment, especially street and indoor scenes.

ISVQA may require finding the same objects in different images or determining the relationships between different objects within or across images. It can also entail determining which images are the most relevant for the question and then answering based only on them, ignoring the other images. ISVQA leads to new research challenges, including: a) How to use natural language to guide scene understanding across multiple views/images; and b) how to fuse information from relevant images to reason about relationships among entities.

To enable research into these problems, we built two datasets for ISVQA - one for indoor scenes and the other for outdoor scenes. The indoor scenes dataset comes from Gibson Environment [31] and contains 91,479 human-generated questions, each for a set of images - for a total of 48,138 image sets. Similarly, the outdoor scenes dataset comprises of 49,617 questions for 12,746 image sets. The images in the outdoor scenes dataset come from the nuScenes dataset [5]. We explain the data collection methodology and statistics in Sect. 3.

The indoor ISVQA dataset contains two parts: 1.) Gibson-Room - containing images from the same room; and 2.) Gibson-Building - containing images from different places in the same building. This is to facilitate spatial and semantic reasoning both in localized and extended regions in the same scene. The outdoor dataset contains image sets taken from mostly urban environments.

We propose two extensions of single-image VQA methods as baseline approaches to investigate the ISVQA task and the datasets. In addition, we also use an existing Video VQA approach as a simple baseline. Finally, we also propose to use use a transformer-based approach which can specifically target ISVQA. Such baselines meet significant difficulties in solving the ISVQA problem, and they reflect the particular challenges of the ISVQA task. We also present the statistics of the datasets, by analyzing the types of question, distributions of answers for different types of questions, and biases present in the dataset.

In summary, we make the following major contributions in this work.

- We propose ISVQA as a new setting for scene understanding via Q&A;
- We introduce two large-scale datasets for targeting the ISVQA problem. In total, these datasets contain 141,096 questions for 60,884 sets of images.
- We establish baseline methods on ISVQA tasks to recognize the challenges and encourage future research.

2 Related Works

VQA Settings. The basic VQA setting [4] involves answering natural language questions about images. The VisualGenome dataset [17] also contains annotations for visual question-answer pairs at both image and region levels. Visual7W [33] built upon the basic VQA setting and introduced visual grounding to VQA. Several other VQA settings target specific problems or applications. For example, VizWiz [13] was designed to help answer questions asked by blind people. RecipeQA [32] is targeted for answering questions about recipes from multi-modal cues. TallyQA [1], and HowMany-QA [30] specifically target counting questions for single images. Unlike these, the CLEVR [16] benchmark and dataset uses synthetically generated images of rendered 3D shapes and is aimed towards understanding the geometric relationships between objects. IQA [11] is also a synthetic setting where an agent is required to navigate a scene and reach the desired location in order to answer the question.

Unlike existing work, ISVQA targets scene understanding by answering questions which might require multiple images to answer. This important setting has not been studied before and necessitates a specialized dataset. Additionally, answering most of the questions requires a model to ignore some of the images in the set. This capability is absent from many state-of-the-art VQA models.

We also distinguish our work from video VQA. Unlike many such datasets (*e.g.* TVQA+ [18], MovieQA [29]), our datasets do not contain any textual cues like scripts or subtitles. Also, videos are temporally continuous and are usually taken from a stationary view-point. This makes finding associations between objects across frames easy, even if datasets do not provide textual cues (*e.g.* tGIF-QA [15]). The image sets in ISVQA dataset are not akin to video frames. Also, unlike embodied QA [6], ISVQA does not have an agent interacting with the environment. ISVQA algorithms can use only the few given images, which resembles real-world applications. Embodied QA does not require sophisticated inference of the correspondence between images, as the frames that an agent sees are continuous. The agent can reach the desired location, and answer the question using only the final frame. In contrast, ISVQA often needs reasoning across images and an implicit understanding of a scene.

VQA Methods. Most of the recent VQA methods use attention mechanisms to focus on the most relevant image regions. For example, [3] proposed a bottom-up and top-down attention mechanism for answering visual questions. In addition, several methods which use co-attention (or bi-directional) attention over questions and images have been proposed. Such methods include [9,21], all of which use the information from one modality (text or image) to attend to the other. Somewhat different from these is the work from Gao *et al.* [10] which proposed the multi-modality latent interaction module which can model the relationships between visual and language summaries in the form of latent vectors.

Unlike these, [7] used reasoning modules over detected objects to answer questions about geometric relationships between objects. Similarly, Santoro *et al.* [24] proposed using Relation Networks to solve specific relational reasoning problems. Neither of these approaches used attention mechanisms. In this paper, we mostly focus on attention-based mechanisms to design the baseline models.

3   Dataset

The main goal of our data collection effort is to aid multi-image scene understanding via question answering. We use two publicly available datasets (Gibson [31] and nuScenes [5]) as the source of images to build our datasets. Gibson provides navigable 3D-indoor scenes. We use the Habitat API [25] to extract images from multiple locations and viewpoints from it. nuScences contains sets of images taken simultaneously from multiple cameras on a car.

3.1   Annotation Collection

Indoor Scenes. Gibson is a collection of 3D scans of indoor spaces, particularly houses and offices. It provides virtualized scans of real indoor spaces like houses and offices. Using the Habitat platform, we place an agent at different locations and orientations in a scene and store the views visible to the agent. We generate a set of images by obtaining several views from the same scene. Therefore, together, each image set can be considered to represent the scene.

We collect two types of indoor scenes: 1.) Gibson-Building; and 2.) Gibson-Room. Gibson-Building contains multiple images taken from the same building by placing the agent at random locations and recording its viewpoint while Gibson-Room is collected by obtaining several views from the same room.

For Gibson-Building, we sample image sets by placing an agent at random locations in the scene. We show images from Gibson-Building sets to annotators and request them to ask questions about the scene.

We obtain question-answer annotations for a scene from several annotators using Amazon Mechanical Turk. We let each annotator ask a question about the scene and also provide the corresponding answer. We request that the annotators should ensure that their question can be answered using only the scene shown and no additional knowledge should be required.

From a pilot study, we observed that it is easier for humans to frame questions if they are shown the full 3D view of a scene, simulating the situation of them being present in the scene. Humans are able to frame better questions about locations of objects, and their relationships in such a setting. For Gibson-Room, we simulate such immersion by creating a 360° video from a room. We show these videos (see supplementary material for examples of how these videos are created) to the annotators and ask them to provide questions and answers about the scenes. This process helped annotators understand the entire scene more easily and enabled us to collect more questions requiring across-image reasoning. Videos are not used for annotating nuScene and Gibson-Building.

Note that the ISVQA problem and datasets do not have videos. The images for Gibson-Room still came from random views as previously described. It is possible that the image set has less coverage of the scene than the video. Just using the image set, it might not be possible to answer the questions collected on the video. We prune out those cases by by asking other human-annotators to verify if the question can be answered using the provided image set.

Outdoor Scenes. We collect annotations for the nuScenes dataset similar to the Gibson-Building setting. We show the annotators images from an image set. These represent a 360° view of a scene. We, again, ask them to write questions and answers about the scene as before.

Refining Annotations. We showed all the image sets in our datasets and the associated questions obtained from the previous step to up to three other annotators. We asked them to provide an answer to the question based only on the image set shown. We also asked them to say "Not possible to answer" if the question cannot be answered. This step increases confidence about an answer if there is a consensus among the annotators. This step has the added benefit of ensuring that the question can be answered using the image set.

In addition, we also asked the annotators at this stage to mark the images which are required to answer the given question. This provides us information about which images are the most salient for answering a question.

Train and Test Splits. After refinement, we divided the datasets into train and test splits. The statistics of these splits are given in Table 1. For test splits, we have select samples for which at least two annotators agreed on the answer. We also ensured that the train and test sets have the same set of answers.

Table 1. Statistics of train and test splits of the datasets.

Dataset	#Train sets	#Test sets	#Unique answers
Indoor - Gibson (Room + Building)	69,207	22,272	961
Outdoor - nuScenes	33,973	15,644	650

3.2 Dataset Analysis

 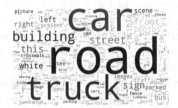

Fig. 2. Question wordclouds for Gibson (left) and nuScenes (right) datasets.

Question Word Distributions. The question word clouds for datasets are shown in Fig. 2. We have removed the first few words from each question before plotting these. This gives us a better picture of which objects people are interested in. Clearly, for outdoor scenes, people are most interested in objects commonly found on the streets and their properties (types, colors, numbers). On the other hand, for indoor scenes, the most frequent questions are about objects hanging on walls and kept on beds, and the room layouts in general.

Types of Questions. Figure 3a shows the distribution of question lengths for the dataset. We observe that a large chunk of the questions contain between 5 and 10 words. Further, Fig. 3b shows the numbers of the most frequent types of questions for the dataset. We observe that the most frequent questions are about properties of objects, and spatial relationships between different entities.

To understand the types of questions in the dataset, we plot the distribution of the most frequent first five words of the questions in the whole dataset in Fig. 3c. Note that a large portion of the questions are about the numbers of different kinds of objects. Another major subset of the questions are about geometric relationships between objects in a scene. A third big part of the dataset contains questions about colors of objects in scenes. Answering questions about the colors of things in a scene requires localization of the object of interest. Depending on the question, this might require reasoning about the relationships between objects in different images. Similarly, counting the number of a particular type of object requires keeping track of previously counted objects to avoid double counting if the same object appears in different images.

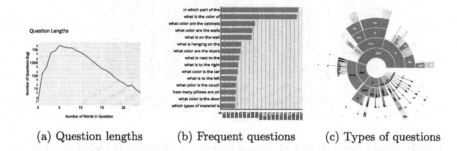

(a) Question lengths   (b) Frequent questions   (c) Types of questions

Fig. 3. (Left) Distribution of questions over no. of words. (Middle) Most frequent types of questions in the dataset. (Right) First five words of the questions.

Answer Distributions. Figure 4 shows the distribution of answers in the dataset for frequently occurring questions types. On one hand, due to human bias in asking questions, dominant answers exist for a few types of questions, such as "can you see the" (usually for an object in the image) and "what is this" (usually a large object). In ISVQA and other VQA datasets humans' tendency to only ask questions about objects that they can see leads to a higher frequency of "yes" answers. On the other hand, most question types do not have a dominant answer. Of particular note are the questions about relative locations and orientations of objects, *e.g.* "What is next to", and questions about the numbers of objects *e.g.* "How many chairs are" etc. This means that it is difficult for a model to perform well by lazily exploiting the statistics of question types.

Number of Images Required. While refining the annotations, we also collect annotations for which images are required to answer the given question. In Fig. 5, we plot the number of images required to answer each question for all datasets. For the plot in Fig. 5, we only consider those image sets for which at least 2 annotators agree about the images which are needed. We observe that one-third of the samples (about 7,000/21,000) in Gibson-Room require at least two images to answer the question. As expected, this ratio is lower for Gibson-Building dataset. However, for all three datasets, we have a large number of questions which require more than one image to answer. The large number of samples in both cases enable the study of both across-image reasoning and image-level focusing. In particular, the latter case also involves rejecting most of the images in the image set and focusing only on one image. In theory, such questions can potentially be answered by using existing single-image VQA models. However, this would require the single-image VQA model to say "Not possible to answer"

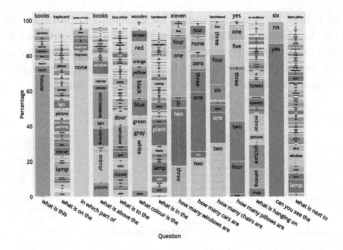

Fig. 4. Answer distributions for several types of questions in the whole dataset.

for all the irrelevant images and finding only the most relevant one. Current VQA models do not have the ability to do this in many cases. (see supplementary material for examples).

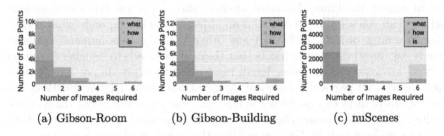

(a) Gibson-Room   (b) Gibson-Building   (c) nuScenes

Fig. 5. Number of images required to answer different types of questions.

4   ISVQA Problem Formulation and Baselines

4.1   Problem Definition

Refer to Fig. 6 for some examples of the ISVQA setting. Given a set of images, $S = \{I_1, I_2, \ldots, I_n\}$, and a natural language question, $Q = \{v_1, v_2, \ldots, v_T\}$, where v_i is the ith word in the question, the task is to provide an answer, $a = f(S, Q)$, which is true for the given question and image set. The function f can either output a probability distribution over a pre-defined set of possible answers, $\mathcal{A}$, or select the best answer from several choices which are input along with the question, i.e., $a = f(S, Q, C_Q)$, where C_Q is the list of choices associated with Q. The former is usually called open-ended QA and the latter is called multiple-choice QA. In this work, we mainly deal with the open-ended setting. Another possible setting is to actually generate the answer using a text generation method similar to image-captioning. But, most existing VQA works focus on either of the first two settings and therefore, we also consider the open-ended setting in this work. We leave the harder problem of generating answers to future work.

what the largest object in the room?   what is above the toilet wall?   what kind of car is in front of the white car?

Fig. 6. Some examples from our dataset which demonstrate the ISVQA problem setting. In each case, input is a set of images and a question.

4.2 Model Definitions

Now, we describe some baselines for the ISVQA problem. These baselines directly adapt single image VQA models. The first of these processes each image separately and concatenates the features obtained from each image to predict the answer. The second baseline directly adapts VQA methods by simply stitching the images and using single image VQA methods to predict the answer.

We also propose an approach to address the special challenges in ISVQA. A fundamental direction to solve ISVQA problem is to enable finer-grained and across-image interactions in a VQA model, where self-attention-based transformers can fit well. In particular, we adapt LXMERT [28], which is designed for cross-modality learning, to both cross-modality and cross-image scenarios.

Concatenate-Feature. Starting from a given set of n images $S = \{I_1, I_2, \ldots, I_n\}$, we use a region proposal network (RPN) to extract region proposals $R_i, i = 1, 2, \ldots, n$ and the corresponding RoI-pooled features (fc6). With some abuse of notation, we denote the region features obtained from each image as $R_i \in \mathbb{R}^{p \times d}, i = 1, 2, \ldots, n$, where p is the number of region features obtained from each image and d is the dimension of the features. We are also given a natural language question $Q = \{v_1, v_2, \ldots, v_T\}$, where v_i is the ith word, encoded as a one-hot vector over a fixed vocabulary V of size d_V. For all the models, we first obtain question token embeddings $E = \{W_w^T v_i\}_{i=1}^{T}$, where $W_w \in \mathbb{R}^{d_V \times d_q}$ is a continuous word-vector embedding matrix. We obtain the question embedding feature using an LSTM-attention module, i.e., $q = \text{AttentionPool}(\text{LSTM}(E)) \in \mathbb{R}^{d_q}$.

Figure 7 shows an outline of the model. For each image, I_i, we obtain the image embedding, x_i by attending over the corresponding region features R_i using the question embedding q.

$$x_i = \text{AttentionPool}(\text{Combine}(R_i, q)) \tag{1}$$

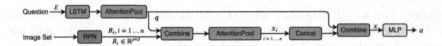

Fig. 7. Concatenate-Feature Baseline. This method adapts a single-image VQA model to an image set $S = \{I_1 \ldots I_n\}$. We first extract region proposals, R_i from each image I_i. The model attends over the regions in each image separately using the question embedding q. Pooling the region features gives a representation of an image as x_i. These are concatenated and combined (element-wise multiplied) by the question embedding to give the joint scene representation x. We use fully-connected layers to predict the final answer a.

where, we use element-wise multiplication (after projecting to suitable dimensions) as the Combine layer and AttentionPool is a combination of an Attention module over the region features which is calculated through a softmax operation and a Pool operation. The region features are multiplied by the attention and added to obtain the pooled image representation. For a single image, this model is an adaptation of the recent Pythia model [27] without its OCR functionality. We concatenate the image features x_i and element-wise multiply by the question embedding to obtain the joint embedding

$$x = \text{Combine}(\text{Concat}(x_1, x_2, \ldots, x_n), q) \qquad (2)$$

where the Combine layer is again an element-wise multiplication. This is passed through a small MLP to obtain the distribution over answers, $P_A = \text{MLP}(x)$.

Stitched Image. Our next baseline is also an adaptation of existing single-image VQA methods. We start by stitching all the images in an image set into a mosaic, similar to the ones shown in Fig. 6. Note that the ISVQA setting does not require the images in an image set to follow an order. Therefore, the stitched image obtained need not be panoramic. We train the recent Pythia [27] model on the stitched images and report performance in Table 2.

Video VQA. To highlight the differences between Video VQA and ISVQA, we adapt the recent state-of-the-art method HME-VideoQA [8]. This model consists of heterogeneous memory module which can potentially learn global context information. We consider images in the image set as frames of a video. Note that, the images in an image set in ISVQA do not necessarily constitute the frames of a video. Therefore, it is reasonable to expect such Video VQA methods to not provide any advantages over our baselines.

Using these baselines, we show that ISVQA is not a trivial extension of VQA. Solving ISVQA requires development of specialized methods.

Transformer-Based Method. We utilize the power of transformers and adapt the LXMERT model [28] to both cross-modality and cross-image scenarios. The transformer can summarize the relevant information within an image set and also model the across-image finer-grained dependencies. Here, we briefly described the original LXMERT model and then describe our modifications.

LXMERT learns cross-modality representations between regions in an image and sentences. It first uses separate visual and language encoders to obtain visual and semantic embeddings. The visual encoder consists of several self-attention sub-layers which help in encoding the relationships between objects. Similarly, the language encoder consists of multiple self-attention sub-layers and feed-forward sub-layers which provide a semantic embedding for the sentence or question. The visual and semantic embeddings are then used to attend to each other via cross-attention sub-layers. This helps the LXMERT model learn final visual and language embeddings which can tightly couple the information from visual and semantic domains. These coupled embeddings can be seen as the joint representations of the image and sentence and are used for inference.

Instead of using features from only a single image as input to the object-relationship encoder, we propose to use the region features from each image in our image-set. As described above, we start by extracting p region proposals and the corresponding features from each of the n images in the image set. We pass the $p \times n$ region features as inputs to the object-relationship encoder in LXMERT. We note that this enables the our model to encode relationships between objects across different images.

Let us denote the image features as $R = [R_1; R_2; \ldots; R_n] \in \mathbb{R}^{pn \times d}$, where R_i are the region features obtained from I_i. We also have the corresponding position encodings of each region in the images $P = [P_1; P_2; \ldots; P_n] \in \mathbb{R}^{pn \times 4}$, where P_i contains the bounding box co-ordinates of the regions in I_i. We combine the region features and position encodings to obtain position-aware embeddings, $S \in \mathbb{R}^{pn \times d'}$, where $S = \text{LayerNorm}(\text{FC}(R)) + \text{LayerNorm}(\text{FC}(P))$. Within- and across-image object relationships are encoded by applying N_R layers of the object relationship encoder. The l-th layer can be represented as

$$x_l = \text{FC}(\text{FC}(\text{SelfAttention}(x_{l-1}))) \tag{3}$$

where, $x_0 = S$, and $X\,(=x_{N_R})$ is the final visual embedding of the object-relationship encoder.

Similarly, given the word embeddings of the question, E, and the index embeddings of each word in the question, $E' = \{\text{IdxEmbed}(1), \ldots, \text{IdxEmbed}(T)\}$, the index-aware word embedding of the i-th word is obtained as $H_i = \text{LayerNorm}(E_i + E'_i)$. Note that the index embedding, IdxEmbed, is just a projection of the position of the word to a vector using fully-connected layers. We apply a similar operation as Eq. 3 N_L times to the word embeddings $H = [H_1; H_2; \ldots; H_T] \in \mathbb{R}^{T \times d_q}$ to give the question embedding, L.

Finally, LXMERT consists of N_X cross-modality encoders stacked one-after-another. Each encoder consists of two operations: 1.) language to vision cross attention, $X = \text{FC}(\text{SelfAttention}(\text{CrossAttention}_{LV}(X, L)))$; and 2.) vision to language cross attention, $L = \text{FC}(\text{SelfAttention}(\text{CrossAttention}_{VL}(L, X)))$. The final output of the N_X encoders are used to predict the answer.

Evaluating Biases in the Datasets. We also evaluate the following prior-based baselines to reveal and understand the biases present in the datasets.
Naïve Baseline. The model always predicts the most frequent answer from the training set. For nuScenes, it always predicts "yes", while for Gibson it predicts "white". Ideally, this should set a minimum performance bar.
Hasty-Student Baseline. In this baseline, a model simply finds the most frequent answer for each type of question. In this case, we define a "question type" as the first two words of a question. For example, a hasty-student might always answer "one" for all "How many" questions. This is similar to the hasty-student baseline used in [19] (MovieQA).
Question-Only Baseline. In this model, we ignore the visual information and only use question text to train a model. Our implementation takes as input only the question embedding, q which is passed through several fully-connected layers to predict the answer distribution. This baseline is meant to reveal the language-bias present in the dataset.

5   Experiments

5.1   Human Performance

An ideal image-set question answering system should be able to reach at least the accuracy achieved by humans. We evaluate the human performance using the annotations with the standard VQA-accuracy metric described below. For the outdoor scenes dataset, humans obtain a VQA-accuracy of 91.88% and for the indoor scenes they obtain 88.80%. Comparing this with Table 2 shows that ISVQA is extremely challenging and requires specialized methods. The reason for the human performance being lower than 100% is that, in many cases an annotator has given an answer which is not exactly similar to the other two but is still semantically similar. For example, the majority answer might have been "black and white" but the third annotator answered "white and black".

5.2   Implementation Details

We start by using Faster R-CNN in Detectron to extract the region proposals and features R_i for each image. Each region feature is 2048-D and we use the top 100 region proposals from each image. To obtain the word-vector embeddings we use 300-D GloVe [23] vectors. The joint visual-question embedding, x is taken to be 5000-D. For evaluation, we use the VQA-Accuracy metric [4]. A predicted answer is given a score of one if it matches at least two out of the three annotations. If it matches only one annotations, it is given a score of 0.5. All of our VQA models are implemented in the Pythia framework [26] and are trained on two NVIDIA V100 GPUs for 22,000 iterations with a batch size of 32. The initial learning rate is warmed up to 0.01 in the first 1,000 iterations. The learning rate is dropped by a factor of 10 at iterations 12,000 and 18,000. For the HME-VideoQA model, we use the implementation provided by the authors. We train the model for 22,000 iterations with a batch size of 32 and a starting learning rate of 0.001. For the transformer-based model, we use $N_L = 9, N_R = 5, N_X = 5$. We use a batch-size of 32, learning rate of 0.00005, and we train the model for 20 epochs. All feature dimensions are kept the same as LXMERT.

5.3 Results

We report the VQA-accuracy for all methods in Table 2. The accuracy achieved by both of the VQA-based baselines is only around 50−54% and the Video VQA model achieves only 39.88% for the indoor dataset and 52.14% for the outdoor dataset. This highlights the need for advanced models for ISVQA.

Comparison Between Baselines. Table 2 shows that the naïve baseline reaches a VQA-Accuracy of only 8.6% for the indoor scenes dataset compared to 47.57% given by the Concatenate-Feature baseline and 50.53% given by the Stitched-Image baseline model. This shows that single-image VQA methods are not enough to overcome the challenges presented by ISVQA. On the other hand, our proposed transformer-based model performs the best for both indoor and outdoor scenes out-performing other models by over 10%.

Language Biases. Recent works (*e.g.* [2]) show that high performance in VQA could be achieved using only the language components. Deep networks can easily exploit biases in the datasets to find short-cuts for answering questions. We observe that most VQA-based baselines perform much better than the question-only baseline. This shows that the ISVQA datasets are less biased compared to many existing VQA datasets [2] and validates the utility of developing ISVQA models that can utilize both the visual and language components simultaneously.

Table 2. Results for both indoor and outdoor datasets.

	Method	VQA-accuracy (%)	
		Gibson	nuScenes
Prior-based baselines	Naïve	8.61	22.46
	Hasty-student	27.22	41.65
	Question-Only	40.26	46.06
Approaches	Video-VQA	39.88	52.14
	Concatenate-feature	47.57	53.66
	Stitched-image	50.53	54.32
	Transformer-based	61.58	64.91
Human performance		88.80	91.88

Performance by Question Type. Figure 8 shows the accuracy bar-chart of our single-image VQA-based baselines for various types of questions. Using this chart, we have the following observations and hypotheses:

Single-Image VQA Baselines can Predict Single-Object Attributes.
Both baseline models can answer questions about colors of single objects well
(black and gray bars). This is expected because no cross-image dependency is
needed.

General Cases May Need Cross-image Inference. A large portion of
questions involve multiple objects, which may appear in different images. The
two VQA baselines using simple attention do not perform well on such questions.
The across-image transformer-based approach performs much better.

Stitched-Image Captures Cross-image Dependency Better. The
Stitched-Image baseline allows direct pooling from regions in all images,
which may capture across-image dependency better. It also outperforms the
Concatenate-Feature baseline for most question types, except for the counting
questions. The Stitched-Image cannot avoid double counting. The transformer-
based approach has all the advantages of the Stitched-Image and can do more
sophisticated inference.

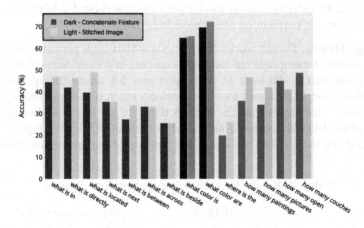

Fig. 8. Performance of the two VQA-based baselines for different types of questions for
the combined Gibson test set. Dark colors represent the performance for Concatenate-
Feature baseline and light colors for Stitched Image baseline. **Blue** is used for geometric
relationship questions, green for counting questions, red for location, and **black** for
color questions. We notice that the VQA-based baselines are able to answer simple
questions like those about colors of single objects very well. However, questions involving spatial reasoning between objects in one image or across images are extremely
challenging for such methods. (Color figure online)

6   Conclusion and Discussion

We proposed the new task of image-set visual question answering (ISVQA).
This task can lead to new research challenges, such as language-guided cross-
image attentions and reasoning. To establish the ISVQA problem and enable its

research, we introduced two ISVQA datasets for indoor and outdoor scenes. Large-scale annotations were collected for questions and answers with novel ways to present the scene to the annotators. We performed bias analysis of the datasets to set up performance lower bounds. We also extended a single-image VQA method to two simple attention-based baseline models and showed the performance of state-of-the-art Video VQA model. Their limited performance reflects the unique challenges of ISVQA, which cannot be solved trivially by the capabilities of existing models. Approaches for solving the ISVQA problem may need to pass information across images in a sophisticated way, understand the scene behind the image set, and attend the relevant images. Another potential direction could be to create explicit maps of the scenes. However, depending on the complexity of the scene, different techniques might be required to explicitly construct a coherent map. Where such maps can be obtained accurately, reconstruction-based ISVQA solutions can be more accurate than the baselines. Meanwhile, humans do not have to do exact scene reconstruction to answer questions. So, in this paper, we have focused on methods that can model across-image dependencies implicitly.

References

1. Acharya, M., Kafle, K., Kanan, C.: TallyQA: answering complex counting questions. In: AAAI Conference on Artificial Intelligence (2019)
2. Agrawal, A., Batra, D., Parikh, D.: Analyzing the behavior of visual question answering models. In: Empirical Methods in Natural Language Processing (2016)
3. Anderson, P., et al.: Bottom-up and top-down attention for image captioning and visual question answering. In: Conference on Computer Vision and Pattern Recognition (2018)
4. Antol, S., et al.: VQA: Visual question answering. In: International Conference on Computer Vision (2015)
5. Caesar, H., et al.: nuScenes: a multimodal dataset for autonomous driving. arXiv preprint arXiv:1903.11027 (2019)
6. Das, A., Datta, S., Gkioxari, G., Lee, S., Parikh, D., Batra, D.: Embodied question answering. In: Conference on Computer Vision and Pattern Recognition (2018)
7. Desta, M.T., Chen, L., Kornuta, T.: Object-based reasoning in VQA. In: Winter Conference on Applications of Computer Vision (2018)
8. Fan, C., Zhang, X., Zhang, S., Wang, W., Zhang, C., Huang, H.: Heterogeneous memory enhanced multimodal attention model for video question answering. In: Conference on Computer Vision and Pattern Recognition (2019)
9. Gao, P., et al.: Dynamic fusion with intra- and inter-modality attention flow for visual question answering. In: Conference on Computer Vision and Pattern Recognition (2019)
10. Gao, P., You, H., Zhang, Z., Wang, X., Li, H.: Multi-modality latent interaction network for visual question answering. In: International Conference on Computer Vision (2019)
11. Gordon, D., Kembhavi, A., Rastegari, M., Redmon, J., Fox, D., Farhadi, A.: IQA: visual question answering in interactive environments. In: Conference on Computer Vision and Pattern Recognition (2018)

12. Goyal, Y., Khot, T., Agrawal, A., Summers-Stay, D., Batra, D., Parikh, D.: Making the V in VQA matter: elevating the role of image understanding in visual question answering. In: International Journal of Computer Vision (2019)
13. Gurari, D., et al.: Vizwiz grand challenge: answering visual questions from blind people. In: Conference on Computer Vision and Pattern Recognition (2018)
14. Hudson, D.A., Manning, C.D.: GQA: a new dataset for real-world visual reasoning and compositional question answering. In: Conference on Computer Vision and Pattern Recognition (2019)
15. Jang, Y., Song, Y., Yu, Y., Kim, Y., Kim, G.: TGIF-QA: Toward spatio-temporal reasoning in visual question answering. In: Conference on Computer Vision and Pattern Recognition. Institute of Electrical and Electronics Engineers Inc. (2017)
16. Johnson, J., Hariharan, B., van der Maaten, L., Fei-Fei, L., Lawrence Zitnick, C., Girshick, R.: Clevr: a diagnostic dataset for compositional language and elementary visual reasoning. In: Conference on Computer Vision and Pattern Recognition (2017)
17. Krishna, R., et al.: Visual genome: connecting language and vision using crowd-sourced dense image annotations. International Journal of Computer Vision **123**(1), 32–73 (2017)
18. Lei, J., Yu, L., Berg, T.L., Bansal, M.: TVQA+: spatio-temporal grounding for video question answering. arXiv preprint arXiv:1904.11574 (2019)
19. Liang, J., Jiang, L., Cao, L., Li, L.J., Hauptmann, A.: Focal visual-text attention for visual question answering. In: Conference on Computer Vision and Pattern Recognition (2018)
20. Lin, X., Parikh, D.: Don't just listen, use your imagination: leveraging visual common sense for non-visual tasks. In: Conference on Computer Vision and Pattern Recognition (2015)
21. Lu, J., Yang, J., Batra, D., Parikh, D.: Hierarchical question-image co-attention for visual question answering. In: Advances In Neural Information Processing Systems (2016)
22. Marino, K., Rastegari, M., Farhadi, A., Mottaghi, R.: OK-VQA : a visual question answering benchmark requiring external knowledge. In: Conference on Computer Vision and Pattern Recognition (2019)
23. Pennington, J., Socher, R., Manning, C.: Glove: global vectors for word representation. In: Conference on Empirical Methods in Natural Language Processing (2014)
24. Santoro, A., et al.: A simple neural network module for relational reasoning. Advances in Neural Information Processing Systems (2017)
25. Savva, M., et al.: Habitat: a platform for embodied AI research. arXiv preprint arXiv:1904.01201 (2019)
26. Singh, A., et al.: Pythia-a platform for vision & language research. In: SysML Workshop, NeurIPS (2018)
27. Singh, A., et al.: Towards VQA models that can read. In: Conference on Computer Vision and Pattern Recognition (2019)
28. Tan, H., Bansal, M.: Lxmert: learning cross-modality encoder representations from transformers. In: Conference on Empirical Methods in Natural Language Processing and the 9th International Joint Conference on Natural Language Processing (2019)
29. Tapaswi, M., Zhu, Y., Stiefelhagen, R., Torralba, A., Urtasun, R., Fidler, S.: MovieQA: understanding stories in movies through question-answering. In: Conference on Computer Vision and Pattern Recognition (2016)

30. Trott, A., Xiong, C., Socher, R.: Interpretable counting for visual question answering. In: International Conference on Learning Representations (2018)
31. Xia, F., Zamir, A.R., He, Z., Sax, A., Malik, J., Savarese, S.: Gibson ENV: real-world perception for embodied agents. In: Conference on Computer Vision and Pattern Recognition (2018)
32. Yagcioglu, S., Erdem, A., Erdem, E., Ikizler-Cinbis, N.: Recipeqa: a challenge dataset for multimodal comprehension of cooking recipes. arXiv preprint arXiv:1809.00812 (2018)
33. Zhu, Y., Groth, O., Bernstein, M., Fei-Fei, L.: Visual7W: grounded question answering in images. In: Conference on Computer Vision and Pattern Recognition (2016)

Object as Hotspots: An Anchor-Free 3D Object Detection Approach via Firing of Hotspots

Qi Chen[1,2], Lin Sun[1(✉)], Zhixin Wang[3], Kui Jia[3,4], and Alan Yuille[2]

[1] Samsung Strategy and Innovation Center, San Jose, CA 95134, USA
lin1.sun@samsung.com
[2] The Johns Hopkins University, Baltimore, MD 21218, USA
{qchen42,ayuille1}@jhu.edu
[3] South China University of Technology, Guangzhou, China
wang.zhixin@mail.scut.edu.cn, kuijia@scut.edu.cn
[4] Pazhou Lab, Guangzhou 510335, China

Abstract. Accurate 3D object detection in LiDAR based point clouds suffers from the challenges of data sparsity and irregularities. Existing methods strive to organize the points regularly, e.g. voxelize, pass them through a designed 2D/3D neural network, and then define object-level anchors that predict offsets of 3D bounding boxes using collective evidences from all the points on the objects of interest. Contrary to the state-of-the-art anchor-based methods, based on the very nature of data sparsity, we observe that even points on an individual object part are informative about semantic information of the object. We thus argue in this paper for an approach opposite to existing methods using object-level anchors. Inspired by compositional models, which represent an object as parts and their spatial relations, we propose to represent an object as composition of its interior non-empty voxels, termed hotspots, and the spatial relations of hotspots. This gives rise to the representation of Object as Hotspots (OHS). Based on OHS, we further propose an anchor-free detection head with a novel ground truth assignment strategy that deals with inter-object point-sparsity imbalance to prevent the network from biasing towards objects with more points. Experimental results show that our proposed method works remarkably well on objects with a small number of points. Notably, our approach ranked 1st on KITTI 3D Detection Benchmark for cyclist and pedestrian detection, and achieved state-of-the-art performance on NuScenes 3D Detection Benchmark.

Keywords: Point clouds · 3D detection · Inter-object point-sparsity imbalance

1 Introduction

Great success has been witnessed in 2D detection recently thanks to the evolution of CNNs. However, extending 2D detection methods to LiDAR based 3D detection is not trivial because point clouds have very different properties from

© Springer Nature Switzerland AG 2020
A. Vedaldi et al. (Eds.): ECCV 2020, LNCS 12366, pp. 68–84, 2020.
https://doi.org/10.1007/978-3-030-58589-1_5

those of RGB images. Point clouds are irregular, so [25,38,47] have converted the point clouds to regular grids by subdividing points into voxels and process them using 2D/3D CNNs. Another unique property and challenge of LiDAR point clouds is the sparseness. LiDAR points lie on the objects' surfaces and meanwhile due to occlusion, self-occlusion, reflection or bad weather conditions, very limited quantity of points can be captured by LiDAR.

Inspired by compositional part-based models [4,6,12,15,50], which have shown robustness when classifying partially occluded 2D objects and for detecting partially occluded object parts [43], we propose to detect objects in LiDAR point clouds by representing them as composition of their interior non-empty voxels. We define the non-empty voxels which contain points within the objects as **spots**. Furthermore, to encourage the most discriminative features to be learned, we select a small subset of spots in each object as **hotspots**, thus introducing the concept of hotspots. The selection criteria are elaborated in Sect. 3.2. Technically, during training, hotspots are spots assigned with positive labels; during inference hotspots are activated by the network with high confidences.

Compositional models represent objects in terms of object parts and their corresponding spatial relations. For example, it can not be an actual dog if a dog's tail is found on the head of the dog. We observe the ground truth box implicitly provides relative spatial information between hotspots and therefore propose a **spatial relation encoding** to reinforce the inherent spatial relations between hotspots.

We further realize that our hotspot selection can address an **inter-object point-sparsity imbalance** issue caused by different object sizes, different distances to the sensor, different occlusion/truncation levels, and reflective surfaces etc. A large number of points are captured on large objects or nearby objects to the sensor while much fewer points are collected for small objects and occluded ones. In the KITTI training dataset, the number of points in annotated bounding boxes ranges from 4874 to 1. We categorize this issue as feature imbalance: objects with more points tend to have rich and redundant features for predicting semantic classes and localization while those with few points have few features to learn from.

The concept of hotspots along with their spatial relations gives rise to a novel representation of **Object as Hotspots (OHS)**. Based on OHS, we design an OHS detection head with a hotspot assignment strategy that deals with inter-object point-sparsity imbalance by selecting a limited number of hotspots and balancing positive examples in different objects. This strategy encourages the network to learn from limited but the most discriminative features from each object and prevents a bias towards objects with more points.

Our concept of OHS is more compatible with anchor-free detectors. Anchor-based detectors assign ground truth to anchors which match the ground truth bounding boxes with IoUs above certain thresholds. This strategy is object-holistic and cannot discriminate different parts of the objects while anchor-free detectors usually predict heatmaps and assign ground truth to individual points inside objects. However, it's nontrivial to design an anchor-free detector. Without the help of human-defined anchor sizes, bounding box regression becomes difficult. We identify the challenge as **regression target imbalance** due to scale variance and therefore adopt soft *argmin* from stereo vision [13] to regress bounding boxes. We show the effectiveness of soft *argmin* in handling regression target imbalance in our algorithm.

The main contributions of proposed method can be summarized as follows:

- We propose a novel representation, termed Object as HotSpots (OHS) to compositionally model objects from LiDAR point clouds as hotspots with spatial relations between them.
- We propose a unique hotspot assignment strategy to address inter-object point-sparsity imbalance and adopt soft *argmin* to address the regression target imbalance in anchor-free detectors.
- Our approach shows robust performance for objects with very few points. The proposed method sets the new state-of-the-art on Nuscene dataset and KITTI test dataset for cyclist and pedestrian detection. Our approach achieves real-time speed with 25 FPS on KITTI dataset.

2 Related Work

Anchor-Free Detectors for RGB Images. Anchor-free detectors for RGB images represent objects as points. Our concept of object as hotspots is closely related to this spirit. ExtremeNet [46] generates the bounding boxes by detecting top-most, left-most, bottom-most, right-most, and center points of the objects. CornerNet [18] detects a pair of corners as keypoints to form the bounding boxes. Zhou et al. [45] focuses on box centers, while CenterNet [5] regards both box centers and corners as keypoints. FCOS [33] and FSAF[49] detect objects by dense points inside the bounding boxes. The difference between these detectors and our OHS is, ours also takes advantage of the unique property of LiDAR point clouds. We adaptively assign hotspots according to different point-sparsity within each bounding box, which can be obtained from annotations. Whereas in RGB images CNNs tend to learn from texture information [8], from which it is hard to measure how rich the features are in each object.

Anchor-Free Detectors for Point Clouds. Some algorithms without anchors are proposed for indoors scenes. SGPN [35] segments instances by semantic segmentation and learning a similarity matrix to group points together. This method is not scalable since the size of similarity matrix grows quadratically with the number of points. 3D-BoNet [39] learns bounding boxes to provide a boundary for points from different instances. Unfortunately, both methods will fail when only partial point clouds have been observed, which is common in LiDAR point clouds. PIXOR [38] and LaserNet [26] project LiDAR points into bird's eye view (BEV) or range view and use standard 2D CNNs to produce bounding boxes in BEV. Note that we do not count VoteNet [27] and Point-RCNN [30] as anchor-free methods due to usage of anchor sizes.

Efforts Addressing Regression Target Imbalance. The bounding box centers and sizes appear in different scales. Some objects have relatively large sizes while others do not. The scale variances in target values give rise to the scale variances in gradients. Small values tend to have smaller gradients and have less impact during training. Regression target imbalance is a great challenge for anchor-free detectors. Anchor-free detectors [5,14,18,33,45,49] became popular after Feature Pyramid Networks (FPN) [21] was proposed to handle objects of different sizes.

Complimentary to FPNs, anchor-based detectors [9,22,23,29] rely on anchor locations and sizes to serve as normalization factors to guarantee that regression targets are mostly small values around zero. Multiple sizes and aspect ratios are hand-designed to capture the multi-modal distribution of bounding box sizes. Anchor-free detectors can be regarded as anchor-based detectors with one anchor of unit size at each location and thus anchor-free detectors don't enjoy the normalizing effect of different anchor sizes.

3   Object as Hotspots

3.1   Hotspot Definition

We represent an object as composition of hotspots. **Spots** are defined as non-empty voxels which have points and overlap with objects. Only a subset of spots are assigned as **hotspots** and used for training, to mitigate the imbalance of number of points and the effect of missing or occluded part of objects. Hotspots are responsible for aggregating minimal and the most discriminative features of an object for background/foreground or inter-class classification. In training, hotspots are assigned by ground truth; in inference, hotspots are predicted by the network.

Intuitively the hotspots should satisfy three properties: 1) they should compose distinguishable parts of the objects in order to capture discriminative features; 2) they should be shared among objects of the same category so that common features can be learned from the same category; 3) they should be minimal so that when only a small number of LiDAR points are scanned in an object, hotspots still contain essential information to predict semantic information and localization, i.e. hotspots should be robust to objects with a small number of points.

3.2   Hotspot Selection and Assignment

Hotspot selection & assignment is illustrated in Fig. 2(a). Unlike previous anchor-free detectors [38,45], which densely assign positive samples inside objects, we only select a subset of spots on objects as hotspots. We assign hotspots to the output feature map of the backbone network. After passing through the backbone network, a neuron on the feature map can be mapped to a super voxel in input point cloud space. We denote a voxel corresponding to a neuron on the output feature map as V_n, where n indexes a neuron.

The annotations do not tell which parts are distinguishable, but we can infer them from the ground truth bounding boxes B_{gt}. We assume V_n is an interior voxel of the object if inside B_{gt}. Then we consider V_n as a spot if it's both non-empty and inside B_{gt}. We choose hotspots as nearest spots to the object center based on two motivations: 1) Points away from the object center are less reliable compared to those near the object centers, i.e., they are more vulnerable to the change of view angle. 2) As stated in FCOS [33], locations closer to object centers tend to provide more accurate localization.

We choose at most M nearest spots as hotspots in each object. M is an adaptive number determined by $M = \frac{C}{Vol}$, where C is a hyperparameter we choose and Vol is the volume of the bounding box. Because relatively large objects tend to have more points and richer features, we use M to further suppress the number of hotspots in these objects. If the number of spots in an object is less than M, we assign all spots as hotspots.

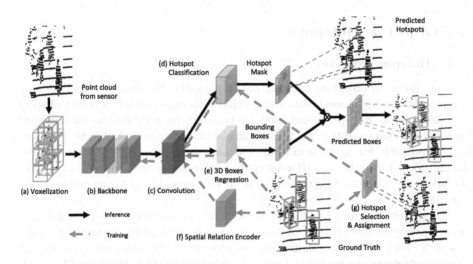

Fig. 1. Outline of HotSpotNet. The point cloud is (a) voxelized, and passed through the (b) backbone network to produce 3D feature maps. These feature maps go through (c) a shared convolution layer, pass into three modules to perform (d) Hotspot Classification and (e) 3D Bounding Box regression (f) Spatial Relation Encoder to train the network, and (g) selected hotspots are assigned as positive labels to (d) Hotspot Classification. During inference only (d) Hotspot Classification and (e) 3D Bounding Box Regression are performed to obtain hotspots and bounding boxes respectively.

4 HotSpot Network

Based on OHS, we architect the Hotspot Network (HotSpotNet) for LiDAR point clouds. HotSpotNet consists of a 3D feature extractor and Object-as-Hotspots (OHS) head. OHS head has three subnets for hotspot classification, box regression and spatial relation encoder.

The overall architecture of our proposed HotSpotNet is shown in Fig. 1. The input LiDAR point clouds are voxelized into cuboid-shape voxels. The input voxels pass through the 3D CNN to generate the feature maps. The three subnets will guide the supervision and generate the predicted 3D bounding boxes. Hotspot assignment happens at the last convolutional feature maps of the backbone. The details of network architecture and the three subnets for supervision are described below.

4.1 Object-as-Hotspots Head

Our OHS head network consists of three subnets: 1) a hotspot classification subnet that predicts the likelihood of class categories; 2) a box regression subnet that regresses the center locations, dimensions and orientations of the 3D boxes. 3) a spatial relation encoder for hotspots.

Hotspot Classification. The classification module is a convolutional layer with K heatmaps each corresponding to one category. The hotspots are labeled as ones. The targets for all the non-hotspots are zeros. We apply a gradient mask so that gradients for *non-hotspots inside the ground truth bounding boxes* are set to zero. That means they are ignored during training and do not contribute to back-propagation. Binary classification is applied to hotspots and non-hotspots. Focal loss [22] is applied at the end,

$$\mathcal{L}_{cls} = \sum_{k=1}^{K} \alpha(1 - p_k)^{\gamma} log(p_k) \tag{1}$$

where,

$$p_k = \left\{ \begin{array}{ll} p & ,\text{hotspots} \\ (1 - p) & ,\text{non-hotspots} \end{array} \right.$$

p is the output probability, and K is the number of categories. The total classification loss is averaged over the total number of hotspots and non-hotspots, excluding the non-hotspots within ground truth bounding boxes.

Box Regression. The bounding box regression only happens on hotspots. For each hotspot, an eight-dimensional vector $[d_x, d_y, z, \log(l), \log(w), \log(h), \cos(r), \sin(r)]$ is regressed to represent the object in LiDAR point clouds. d_x, d_y

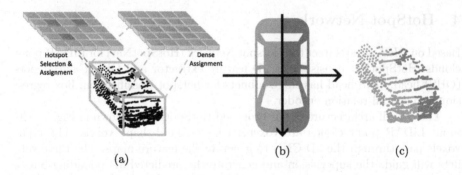

(a)   (b)   (c)

Fig. 2. (a) Illustration of hotspot selection & assignment. Only selected non-empty voxels on objects are assigned as hotspots. Previous anchor-free detectors [33,49] densely assign locations inside objects as positive samples. (b) Spatial relation encoding: we divide the object bounding box in BEV into quadrants by the orientation (front-facing direction) and its perpendicular direction. Quadrants I, II, III, and IV are color-coded with green, blue, purple and orange respectively in the illustration. (c) Illustration of how points of a vehicle are classified into different quadrants, with the same set of color-coding as (b). (Color figure online)

are the axis-aligned deviations from the hotspot to the object centroid. The hotspot centroid in BEV can be obtained by:

$$[x_h, y_h] = (\frac{j + 0.5}{L}(x_{max} - x_{min}) + x_{min}, \frac{i + 0.5}{W}(y_{max} - y_{min}) + y_{min}), \quad (2)$$

where i, j is the spatial index of its corresponding neuron on the feature map with size $W \times L$, and $[x_{min}, x_{max}]$, $[y_{min}, y_{max}]$ are the ranges for x, y when we voxelize all the points.

As discussed in Sect. 2, anchor-free detectors suffer from regression target imbalance. Instead of introducing FPN, i.e. extra layers and computational overhead to our network, we tackle regression target imbalance by carefully designing the targets: 1) We regress $\log(l)$, $\log(w)$, $\log(h)$ instead of their original values because log scales down the absolute values; 2) We regress $\cos(r)$, $\sin(r)$ instead of r directly because they are constrained in $[-1, 1]$ instead of the original angle value in $[-\pi, \pi]$; 3) We use soft $argmin$ [13] to help regress d_x, d_y and z. To regress a point location in a segment ranging from a to b by soft $argmin$, we divide the segment into N bins, each bin accounting for a length of $\frac{b-a}{N}$. The target location can be represented as $t = \Sigma_i^N (S_i C_i)$, where S_i represents the softmax score of the i_{th} bin and C_i is the center location of the i_{th} bin. Soft $argmin$ is widely used in stereo vision to predict disparity in sub-pixel resolution. We notice soft $argmin$ can address regression target imbalance by turning the regression into classification problem and avoiding regressing absolute values.

Smooth L1 loss [9] is adopted for regressing these bounding box targets and the regression loss is only computed over hotspots.

$$\mathcal{L}_{loc}(x) = \begin{cases} 0.5x^2 & , |x| < 1 \\ |x| - 0.5 & , \text{otherwise} \end{cases} \quad (3)$$

Spatial Relation Encoder. Inspired by compositional models, we incorporate hotspot spatial relations to our HotSpotNet. Since convolution is translation-invariant, it's hard for a CNN to learn spatial relations without any supervision. Therefore, we explore the implicit spatial relation from annotations. We observe that most target objects for autonomous driving can be considered as rigid objects (e.g. cars), so the relative locations of hotspots to object centers do not change, which can be determined with the help of bounding box centers and orientations. We thus categorize the relative hotspot location to the object center on BEV into a one-hot vector representing quadrants, as shown in Fig. 2 (b)&(c). We train hotspot spatial relation encoder as quadrant classification with binary cross-entropy loss and we compute the loss only for hotspots.

$$\mathcal{L}_q = \sum_{i=0}^{3} -[q_i \log(p_i) + (1 - q_i) \log(1 - p_i)] \tag{4}$$

where i indexes the quadrant, q_i is the target and p_i the predicted likelihood falling into the specific quadrant.

4.2 Learning and Inference

The final loss for our proposed HotSpotNet is the weighted sum of losses from three branches:

$$\mathcal{L} = \delta \mathcal{L}_{cls} + \beta \mathcal{L}_{loc} + \zeta \mathcal{L}_q \tag{5}$$

Where, δ, β and ζ are the weights to balance the classification, box regression and spatial relation encoder loss.

During inference, if the corresponding largest entry value of the K-dimensional vector of the classification heatmaps is above the threshold, we consider the location as hotspot firing for the corresponding object. Since one instance might have multiple hotspots, we further use Non-Maximum Supression (NMS) with the Intersection Over Union (IOU) threshold to pick the most confident hotspot for each object. The spatial relation encoder does not contribute to inference.

5 Experiments

In this section, we summarize the dataset in Sect. 5.1 and present the implementation details of our proposed HotSpotNet in 5.2. We evaluate our method on KITTI 3D detection Benchmark [7] in Sect. 5.3 and NuScenes 3D detection dataset [1] in Sect. 5.4. We also analyze the advantages of HotSpotNet in Sect. 5.5 and present ablation studies in Sect. 5.6.

5.1 Datasets and Evaluation

KITTI Dataset. KITTI has 7,481 annotated LiDAR point clouds for training with 3D bounding boxes for classes such as cars, pedestrians and cyclists. It also provides 7,518 LiDAR point clouds for testing. In the rest of paper, all the ablation studies are conducted on the common train/val split, i.e. 3712 LiDAR point clouds for training and 3769 LiDAR point clouds for validation. To further compare the results with other approaches on KITTI 3D detection benchmark, we randomly split the KITTI annotated data into 4 : 1 for training and validation and report the performance on KITTI test dataset. Following the official KITTI evaluation protocol, average precision (AP) based on 40 points is applied for evaluation. The IoU threshold is 0.7 for cars and 0.5 for pedestrians and cyclists.

NuScenes Dataset. The dataset contains $1,000$ scenes, including 700 scenes for training, 150 scenes for validation and 150 scenes for test. $40,000$ frames are annotated in total, including 10 object categories. The mean average precision (mAP) is calculated based on the distance threshold (i.e. 0.5 m, 1.0 m, 2.0 m and 4.0 m). Additionally, a new metric, nuScenes detection score (NDS) [1], is introduced as a weighted sum of mAP and precision on box location, scale, orientation, velocity and attributes.

5.2 Implementation Details

Backbone Network. In experiments on KITTI, we adopt the same backbone as used by SECOND [37]. We set point cloud range as $[0, 70, 4]$, $[-40, 40]$, $[-3, 1]$ and voxel size as $(0.025, 0.025, 0.05)$ m along x, y, z axis. A maximum of five points are randomly sampled from each voxel and voxel features are obtained by averaging corresponding point features.

As for NuScenes, we choose the state-of-the-art method CBGS [48] as our baseline. Input point cloud range is set to $[-50.4, 50.4]$, $[-50.4, 50.4]$, $[-5, 3]$ along x, y, z, respectively. We implement our method with ResNet [10] and PointPillars (PP) [17] backbones and report each performance. We set voxel size as $(0.1, 0.1, 0.16)$ m for ResNet backbone and $(0.2, 0.2)$ m for PP backbone. For each hotspot, we also set $(\log l, \log w, \log h)$ as outputs of soft *argmin* to handle the size variances for 10 object categories.

Object-as-Hotspots Head. Since the output feature map of the backbone network is consolidated to BEV, in this paper we assign hotspots in BEV as well. Our OHS head consists of a shared 3×3 convolution layer with stride 1. We use a 1×1 convolution layer followed by sigmoid to predict confidence for hotspots. For regression, we apply several 1×1 convolution layers to different regressed values. Two 1×1 convolution layers are stacked to predict soft *argmin* for (d_x, d_y) and z. Additional two 1×1 convolution layers to predict the dimensions and rotation. We set the range $[-4, 4]$ with 16 bins for d_x, d_y and 16 bins for z, with the same vertical range as the input point cloud. We set $C = 64$ to assign hotspots. For hotspot spatial relation encoder, we use another a 1×1 convolution layer with softmax for cross-entropy classification. We set $\gamma = 2.0$

and $\alpha = 0.25$ for focal loss. For KITTI, the loss weights are set as $\delta = \beta = \zeta = 1$. For NuScenes we set $\delta = 1$ and $\beta = \zeta = 0.25$.

Training and Inference. For KITTI, we train the entire network end-to-end with adamW [24] optimizer and one-cycle policy [32] with LR max $2.25e^{-3}$, division factor 10, momentum ranges from 0.95 to 0.85 and weight decay 0.01. We train the network with batch size 8 for 150 epochs. During testing, we keep 100 proposals after filtering the confidence lower than 0.3, and then apply the rotated NMS with IOU threshold 0.01 to remove redundant boxes.

For NuScenes, we set LR max as 0.001. We train the network with batch size 48 for 20 epochs. During testing, we keep 80 proposals after filtering the confidence lower than 0.1, and IOU threshold for rotated NMS is 0.02.

Data Augmentation. Following SECOND[37], for KITTI, we apply random flipping, global scaling, global rotation, rotation and translation on individual objects, and GT database sampling. For NuScenes, we adopt same augmentation strategies as in CBGS [48] except we add random flipping along x axis and attach GT objects from the annotated frames. Half of points from GT database are randomly dropped and GT boxes containing fewer than five points are abandoned.

5.3 Experiment Results on KITTI Benchmark

As shown in Table 1, we evaluate our method on the KITTI test dataset. For fair comparison, we also show the performance of our implemented SECOND [37] with same voxel size as ours, represented by HR-SECOND in the table. For the 3D object detection benchmark, solely LiDAR-based, our proposed HotSpotNet outperforms all published LiDAR-based, one-stage detectors on cars, cyclists and pedestrians of all difficulty levels. In particular, by the time of submission our method ranks $1st$ among all published methods on KITTI test set for cyclist and pedestrian detection. HotSpotNet shows its advantages on objects with a small number of points. The results demonstrate the success of representing objects as hotspots. Our one-stage approach also beats some classic 3D two-stage detectors for car detection, including those fusing LiDAR and RGB images information. Still, our proposed OHS detection head is complimentary to architecture design in terms of better feature extractors.

Inference Speed. The inference speed of HotSpotNet is 25FPS, tested on KITTI dataset with a Titan V100. We compare inference speed with other approaches in Table 1. We achieve significant performance gain while maintaining the speed as our baseline SECOND [37].

5.4 Experiment Results on NuScenes Dataset

We present results on NuScenes validation set (Table 2) and test set (Table 3). We reproduced the baseline CBGS [48] based on implementation from Center-Point [42] without double-flip testing. Our reproduced mAPs are much higher

Table 1. Performance of 3D object detection on KITTI test set. "L", "I" and "L+I" indicates the method uses LiDAR point clouds, RGB images and fusion of two modalities, respectively. FPS stands for frame per second. Bold numbers denotes the best results for single-modal one-stage detectors. Blue numbers are results for best-performing detectors.

Method	Input	Stage	FPS	3D Detection (Car)			3D Detection (Cyclist)			3D Detection (Pedestrian)		
				Mod	Easy	Hard	Mod	Easy	Hard	Mod	Easy	Hard
ComplexYOLO [31]	L	One	17	47.34	55.93	42.60	18.53	24.27	17.31	13.96	17.60	12.70
VoxelNet [47]	L	One	4	65.11	77.47	57.73	48.36	61.22	44.37	39.48	33.69	31.51
SECOND-V1.5 [37]	L	One	33	75.96	84.65	68.71	–	–	–	–	–	–
HR-SECOND [37]	L	One	25	75.32	84.78	68.70	60.82	75.83	53.67	35.52	45.31	33.14
PointPillars [17]	L	One	62	74.31	82.58	68.99	58.65	77.10	51.92	41.92	51.45	38.89
3D IoU Loss [44]	L	One	13	76.50	86.16	71.39	–	–	–	–	–	–
HRI-VoxelFPN [34]	L	One	50	76.70	85.64	69.44	–	–	–	–	–	–
ContFuse [20]	I + L	One	17	68.78	83.68	61.67	–	–	–	–	–	–
MV3D [2]	I + L	Two	3	63.63	74.97	54.00	–	–	–	–	–	–
AVOD-FPN [16]	I + L	Two	10	71.76	83.07	65.73	50.55	63.76	44.93	42.27	50.46	39.04
F-PointNet [28]	I + L	Two	6	69.79	82.19	60.59	56.12	72.27	49.01	42.15	50.53	38.08
F-ConvNet [36]	I + L	Two	2	76.39	87.36	66.69	65.07	81.89	56.64	43.38	52.16	38.8
MMF [19]	I + L	Two	13	77.43	88.40	70.22	–	–	–	–	–	–
PointRCNN [30]	L	Two	10	75.64	86.96	70.70	58.82	74.96	52.53	39.37	47.98	36.01
FastPointRCNN [3]	L	Two	17	77.40	85.29	70.24	–	–	–	–	–	–
STD [40]	L	Two	13	79.71	87.95	75.09	61.59	78.69	55.30	42.47	53.29	38.35
HotSpotNet	L	One	25	78.31	87.60	73.34	65.95	82.59	59.00	45.37	53.10	41.47

than the results presented in the original CBGS paper. As shown in Table 2, our HotSpotNet outperforms CBGS by 1.8 and 3.2 in mAP for the PointPillars and ResNet backbone respectively. In Table 3, our approach outperforms all detectors on the NuScenes 3D Detection benchmark using a single model.

Table 2. 3D object detection mAP on NuScenes val set.

Method	Car	Truck	Bus	Trailer	Construction vehicle	Pedestrian	Motorcycle	Bike	Traffic cone	Barrier	mAP	NDS
CBGS-PP	81.3	49.7	59.0	32.1	13.4	73.1	51.5	23.5	51.3	52.6	48.8	59.2
HotSpotNet-PP	83.3	52.7	63.7	35.3	15.3	74.8	53.7	25.5	50.3	52.0	50.6	59.8
CBGS-ResNet	82.9	52.9	64.6	37.5	18.3	80.3	60.1	39.4	64.8	61.8	56.3	62.8
HotSpotNet-ResNet	84.0	56.2	67.4	38.0	20.7	82.6	66.2	49.7	65.8	64.3	59.5	66.0

5.5   Analysis

We argue that our approach advances in preventing the network from biasing towards objects with more points without compromising performance on these objects. We analyze the effect of different number of hotspots and performance on objects with different number of points.

Table 3. 3D detection mAP on the NuScenes test set

Method	Car	Truck	Bus	Trailer	Construction vehicle	Pedestrian	Motorcycle	Bike	Traffic cone	Barrier	mAP	NDS
SARPNET [41]	59.9	18.7	19.4	18.0	11.6	69.4	29.8	14.2	44.6	38.3	31.6	49.7
PointPillars [17]	68.4	23.0	28.2	23.4	4.1	59.7	27.4	1.1	30.8	38.9	30.5	45.3
WYSIWYG [11]	79.1	30.4	46.6	40.1	7.1	65.0	18.2	0.1	28.8	34.7	35.0	41.9
CBGS [48]	81.1	48.5	54.9	42.9	10.5	80.1	51.5	22.3	70.9	65.7	52.8	63.3
HotSpotNet-ResNet (Ours)	**83.1**	**50.9**	**56.4**	**53.3**	**23.0**	**81.3**	**63.5**	**36.6**	**73.0**	**71.6**	**59.3**	**66.0**

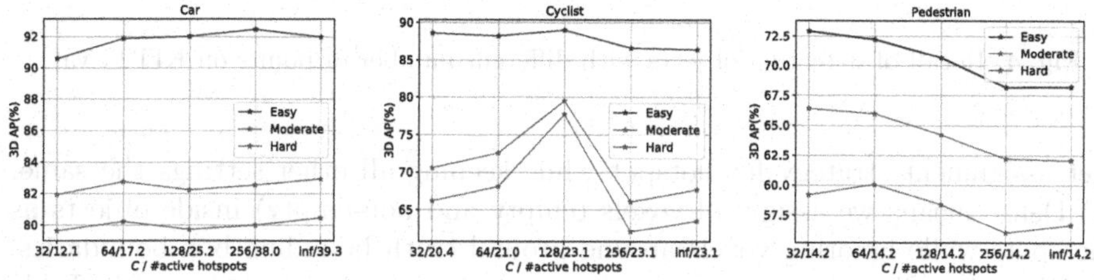

Fig. 3. Performances with different C values on KITTI val. The horizontal axis also shows the number of active hotspots on average with different C values.

Different Number of Hotspots. In Sect. 3.2, we set $M = \frac{C}{Vol}$ as the maximum number of hotspots in each object during training. Here we present the performances with different C values: $32, 64, 128, 256, Inf$, where Inf means we assign all spots as hotspots. The results are shown in Fig. 3. We can see that generally the larger C is, the higher performance in detecting cars. We only perceive a significant drop when $C = 32$ and the overall performance in detecting cars is not sensitive to different values of C. The performance in detecting cyclists reaches its peak when $C = 128$. The lower the C value, the better performance in detecting pedestrians. The performance of detecting pedestrians does not change much when $C \leq 64$. To balance the performance on all classes and prevent over-fitting on one class, we choose $C = 64$ in our paper.

Performance on Objects with Different Number of Points. Comparison between SECOND [37] and our approach for objects with different number of points is shown in Fig. 4. Our approach is consistently better to detect objects with different number of points and less likely to miss objects even with a small number of points. Notably, the relative gain of our approach compared to SECOND increases as the number of points decreases, showing our approach is more robust to sparse objects.

5.6 Ablation Studies

Effect of Different Target Assignment Strategies. We show the effect of our hotspot assignment strategy in Table 4. We present three types of tar-

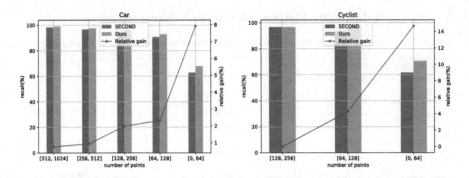

Fig. 4. Recall of detecting objects with different number of points on KITTI val.

get assignment strategy for hotspot while keeping all other settings the same. 1) Dense means we assign all voxels (empty and non-empty) inside objects as hotspots while ignoring voxels around ground truth bounding box boundaries. 2) We assign all non-empty spots as hotspots, corresponding to $C = \inf$ in Table 4. The maximum number of hotspots in each object is $M = \frac{C}{Vol}$ as explained in Sect. 3.2. 3) We set $C = 64$ in our approach to adaptively limit the number of hotspots in each objects. For reference, we also include our baseline, SECOND [37]. The results show that ours (Dense) and ours ($C = \inf$) have similar performances. When considering pedestrian detection ours ($C = \inf$) is slightly better than ours (Dense). Compared to SECOND, they are both better in car and cyclist detection, especially in the hard cases, but worse in pedestrian detection. The inter-object point-sparsity imbalance makes the pedestrian category hard to train. After balancing the number of hotspots over all objects, ours ($C = 64$) outperforms all other target assignment strategies by a large margin in both cyclist and pedestrian detection, while the performance for cars barely changes. This justifies our motivation to force the network to learn the minimal and most discriminative features for each objects.

Table 4. Effect of different target assignment strategy. **Dense**: assigning both empty and non-empty voxels inside objects as hotspots; **C = inf**: assigning all spots as hotspots; **C = 64**: assigning limited number of spots as hotspots.

Method	3D Detection on Car			3D Detection on Cyclist			3D Detection on Pedestrian		
	Mod	Easy	Hard	Mod	Easy	Hard	Mod	Easy	Hard
SECOND [37]	81.96	90.95	77.24	61.62	80.13	57.77	64.19	69.14	57.99
Ours (Dense)	82.2	91.09	79.69	66.45	85.85	62.16	62.82	68.88	55.78
Ours ($C = \inf$)	**82.93**	**91.98**	**80.46**	67.66	86.41	63.5	62.08	68.22	56.64
Ours ($C = 64$)	82.75	91.87	80.22	**72.55**	**88.22**	**68.08**	**65.9**	**72.23**	**60.06**

Effect of Spatial Relation Encoder. To prove the effectiveness of our hotspot spatial encoder, we show the results of our HotSpotNet with and without spatial relation encoder on KITTI validation split for cars in Table 5. We can see that when our algorithm is trained with the spatial relation encoder, the overall performance is boosted. Especially, the great improvement can be observed in hard cases for cyclists and pedestrians.

Table 5. Effect of quadrants as spatial relation encoding.

Method	3D Detection on Car			3D Detection on Cyclist			3D Detection on Pedestrian		
	Mod	Easy	Hard	Mod	Easy	Hard	Mod	Easy	Hard
Ours w/o quadrant	82.27	91.75	79.96	69.31	**89.48**	65.04	65.45	**72.77**	58.36
Ours w quadrant	**82.75**	**91.87**	**80.22**	**72.55**	88.22	**68.08**	**65.9**	72.23	**60.06**
Diff	↑ 0.48	↑ 0.12	↑ 0.26	↑ 3.24	↓ −1.24	↑ 3.04	↑ 0.45	↓ −0.54	↑ 1.7

Effect of Soft *argmin*. We show the importance of soft *argmin* in Table 6. We perceive improvements by using soft *argmin* instead of the raw values. Particularly on small objects, e.g. cyclists and pedestrians, soft *argmin* considerably improves the performance by avoiding regression on absolute values with different scales.

Table 6. Performance of soft *argmin* on (x, y, z) coordination.

Method	3D Detection on Car			3D Detection on Cyclist			3D Detection on Pedestrian		
	Mod	Easy	Hard	Mod	Easy	Hard	Mod	Easy	Hard
Ours w/o soft *argmin*	82.31	91.53	79.88	68.65	88.11	64.36	63.7	67.62	57.15
Ours w/ soft *argmin*	**82.75**	**91.87**	**80.22**	**72.55**	**88.22**	**68.08**	**65.9**	**72.23**	**60.06**
Diff	↑ 0.44	↑ 0.34	↑ 0.34	↑ 3.9	↑ 0.11	↑ 3.72	↑ 2.9	↑ 4.59	↑ 2.91

6 Conclusion

We propose a novel representation, Object-as-Hotspots and an anchor-free detection head with its unique target assignment strategy to tackle inter-object point-sparsity imbalance. Spatial relation encoding as quadrants strengthens features of hotspots and further boosts accurate 3D localization. Extensive experiments show that our approach is effective and robust to sparse point clouds. Meanwhile we address regression target imbalance by carefully designing regression targets, among which soft *argmin* is applied. We believe our work sheds insights on rethinking 3D object representations and understanding characteristics of point clouds and corresponding challenges.

References

1. Caesar, H., et al.: Nuscenes: a multimodal dataset for autonomous driving. arXiv preprint arXiv:1903.11027 (2019)
2. Chen, X., Ma, H., Wan, J., Li, B., Xia, T.: Multi-view 3d object detection network for autonomous driving. In: CVPR (2017)
3. Chen, Y., Liu, S., Shen, X., Jia, J.: Fast point R-CNN. In: ICCV, October 2019
4. Dai, J., Hong, Y., Hu, W., Zhu, S.C., Nian Wu, Y.: Unsupervised learning of dictionaries of hierarchical compositional models. In: CVPR (2014)
5. Duan, K., Bai, S., Xie, L., Qi, H., Huang, Q., Tian, Q.: Centernet: keypoint triplets for object detection. In: ICCV, pp. 6569–6578 (2019)
6. Fidler, S., Boben, M., Leonardis, A.: Learning a hierarchical compositional shape vocabulary for multi-class object representation. arXiv preprint arXiv:1408.5516 (2014)
7. Geiger, A., Lenz, P., Urtasun, R.: Are we ready for autonomous driving? the kitti vision benchmark suite. In: CVPR (2012)
8. Geirhos, R., Rubisch, P., Michaelis, C., Bethge, M., Wichmann, F.A., Brendel, W.: Imagenet-trained CNNs are biased towards texture; increasing shape bias improves accuracy and robustness. In: International Conference on Learning Representations (2019). https://openreview.net/forum?id=Bygh9j09KX
9. Girshick, R.: Fast R-CNN. In: ICCV (2015)
10. He, K., Zhang, X., Ren, S., Sun, J.: Deep residual learning for image recognition. In: CVPR, pp. 770–778 (2016)
11. Hu, P., Ziglar, J., Held, D., Ramanan, D.: What you see is what you get: exploiting visibility for 3d object detection. arXiv preprint arXiv:1912.04986 (2019)
12. Jin, Y., Geman, S.: Context and hierarchy in a probabilistic image model. In: CVPR (2006)
13. Kendall, A., et al.: End-to-end learning of geometry and context for deep stereo regression. In: ICCV (2017)
14. Kong, T., Sun, F., Liu, H., Jiang, Y., Shi, J.: Foveabox: beyond anchor-based object detector. arXiv preprint arXiv:1904.03797 (2019)
15. Kortylewski, A., et al.: Greedy structure learning of hierarchical compositional models. arXiv preprint arXiv:1701.06171 (2017)
16. Ku, J., Mozifian, M., Lee, J., Harakeh, A., Waslander, S.L.: Joint 3d proposal generation and object detection from view aggregation. In: IEEE/RSJ International Conference on Intelligent Robots and Systems (IROS) (2018)
17. Lang, A.H., Vora, S., Caesar, H., Zhou, L., Yang, J., Beijbom, O.: Pointpillars: fast encoders for object detection from point clouds. In: CVPR (2019)
18. Law, H., Deng, J.: Cornernet: Detecting objects as paired keypoints. In: ECCV, pp. 734–750 (2018)
19. Liang, M., Yang, B., Chen, Y., Hu, R., Urtasun, R.: Multi-task multi-sensor fusion for 3d object detection. In: CVPR (2019)
20. Liang, M., Yang, B., Wang, S., Urtasun, R.: Deep continuous fusion for multi-sensor 3d object detection. In: ECCV (2018)
21. Lin, T.Y., Dollár, P., Girshick, R., He, K., Hariharan, B., Belongie, S.: Feature pyramid networks for object detection. In: CVPR (2017)
22. Lin, T.Y., Goyal, P., Girshick, R., He, K., Dollár, P.: Focal loss for dense object detection. In: ICCV (2017)

23. Liu, W., Anguelov, D., Erhan, D., Szegedy, C., Reed, S., Fu, C.-Y., Berg, A.C.: SSD: single shot multibox detector. In: Leibe, B., Matas, J., Sebe, N., Welling, M. (eds.) ECCV 2016. LNCS, vol. 9905, pp. 21–37. Springer, Cham (2016). https:// doi.org/10.1007/978-3-319-46448-0_2

24. Loshchilov, I., Hutter, F.: Fixing weight decay regularization in adam. arXiv preprint arXiv:1711.05101 (2017)

25. Maturana, D., Scherer, S.: Voxnet: a 3d convolutional neural network for real-time object recognition. In: IROS (2015)

26. Meyer, G.P., Laddha, A., Kee, E., Vallespi-Gonzalez, C., Wellington, C.K.: Laser-net: an efficient probabilistic 3d object detector for autonomous driving. In: CVPR (2019)

27. Qi, C.R., Litany, O., He, K., Guibas, L.J.: Deep hough voting for 3d object detection in point clouds. In: ICCV (2019)

28. Qi, C.R., Liu, W., Wu, C., Su, H., Guibas, L.J.: Frustum pointnets for 3d object detection from RGB-d data. In: CVPR (2018)

29. Ren, S., He, K., Girshick, R., Sun, J.: Faster R-CNN: towards real-time object detection with region proposal networks. In: Neural Information Processing Systems (2015)

30. Shi, S., Wang, X., Li, H.: Pointrcnn: 3d object proposal generation and detection from point cloud. In: CVPR (2019)

31. Simon, M., Milz, S., Amende, K., Gross, H.M.: Complex-yolo: an euler-region-proposal for real-time 3d object detection on point clouds. In: ECCV (2018)

32. Smith, L.N., Topin, N.: Super-convergence: very fast training of neural networks using large learning rates. In: Artificial Intelligence and Machine Learning for Multi-Domain Operations Applications, vol. 11006, p. 1100612. International Society for Optics and Photonics (2019)

33. Tian, Z., Shen, C., Chen, H., He, T.: FCOS: fully convolutional one-stage object detection. arXiv preprint arXiv:1904.01355 (2019)

34. Wang, B., An, J., Cao, J.: Voxel-FPN: multi-scale voxel feature aggregation in 3d object detection from point clouds. arXiv preprint arXiv:1907.05286 (2019)

35. Wang, W., Yu, R., Huang, Q., Neumann, U.: SGPN: Similarity group proposal network for 3d point cloud instance segmentation. In: CVPR (2018)

36. Wang, Z., Jia, K.: Frustum convnet: sliding frustums to aggregate local point-wise features for amodal 3d object detection. In: IROS (2019)

37. Yan, Y., Mao, Y., Li, B.: Second: sparsely embedded convolutional detection. Sensors **18**(10), 3337 (2018)

38. Yang, B., Luo, W., Urtasun, R.: Pixor: real-time 3d object detection from point clouds. In: CVPR (2018)

39. Yang, B., et al: Learning object bounding boxes for 3d instance segmentation on point clouds. arXiv preprint arXiv:1906.01140 (2019)

40. Yang, Z., Sun, Y., Liu, S., Shen, X., Jia, J.: STD: sparse-to-dense 3d object detector for point cloud. arXiv preprint arXiv:1907.10471 (2019)

41. Ye, Y., Chen, H., Zhang, C., Hao, X., Zhang, Z.: Sarpnet: shape attention regional proposal network for lidar-based 3d object detection. Neurocomputing **379**, 53–63 (2020)

42. Yin, T., Zhou, X., Krähenbühl, P.: Center-based 3d object detection and tracking. arXiv:2006.11275 (2020)

43. Zhang, Z., Xie, C., Wang, J., Xie, L., Yuille, A.L.: Deepvoting: a robust and explainable deep network for semantic part detection under partial occlusion. In: CVPR (2018)

44. Zhou, D., et al.: IOU loss for 2d/3d object detection. arXiv preprint arXiv:1908.03851 (2019)
45. Zhou, X., Wang, D., Krähenbühl, P.: Objects as points. arXiv preprint arXiv:1904.07850 (2019)
46. Zhou, X., Zhuo, J., Krahenbuhl, P.: Bottom-up object detection by grouping extreme and center points. In: CVPR, pp. 850–859 (2019)
47. Zhou, Y., Tuzel, O.: Voxelnet: end-to-end learning for point cloud based 3d object detection. In: CVPR (2018)
48. Zhu, B., Jiang, Z., Zhou, X., Li, Z., Yu, G.: Class-balanced grouping and sampling for point cloud 3d object detection. arXiv preprint arXiv:1908.09492 (2019)
49. Zhu, C., He, Y., Savvides, M.: Feature selective anchor-free module for single-shot object detection. arXiv preprint arXiv:1903.00621 (2019)
50. Zhu, L.L., Lin, C., Huang, H., Chen, Y., Yuille, A.: Unsupervised structure learning: hierarchical recursive composition, suspicious coincidence and competitive exclusion. In: ECCV (2008)

Placepedia: Comprehensive Place Understanding with Multi-faceted Annotations

Huaiyi Huang[(✉)][iD], Yuqi Zhang[iD], Qingqiu Huang[iD], Zhengkui Guo[iD], Ziwei Liu[iD], and Dahua Lin[iD]

The Chinese University of Hong Kong, Hong Kong, China
{hh016,zy016,hq016,gz019,dhlin}@ie.cuhk.edu.hk, zwliu.hust@gmail.com

Abstract. Place is an important element in visual understanding. Given a photo of a building, people can often tell its functionality, *e.g.* a restaurant or a shop, its cultural style, *e.g.* Asian or European, as well as its economic type, *e.g.* industry oriented or tourism oriented. While place recognition has been widely studied in previous work, there remains a long way towards comprehensive place understanding, which is far beyond categorizing a place with an image and requires information of multiple aspects. In this work, we contribute Placepedia[1], a large-scale place dataset with more than $35M$ photos from $240K$ unique places. Besides the photos, each place also comes with massive multi-faceted information, *e.g.* GDP, population, *etc.*, and labels at multiple levels, including function, city, country, *etc.* This dataset, with its large amount of data and rich annotations, allows various studies to be conducted. Particularly, in our studies, we develop 1) PlaceNet, a unified framework for multi-level place recognition, and 2) a method for city embedding, which can produce a vector representation for a city that captures both visual and multi-faceted side information. Such studies not only reveal key challenges in place understanding, but also establish connections between visual observations and underlying socioeconomic/cultural implications. ([1]The dataset is available at: https://hahehi.github.io/placepedia.html).

1 Introduction

Imagine that you are visiting a new country, and you are traveling among different cities. In each city, you will encounter countless places, and you may see a fancy building, experience some natural wild beauty, or enjoy the unique culture, *etc.* All of these experiences impress you and lead you to a deeper understanding of the place. As we browse through a city, based on certain common visual elements therein, we implicitly establish connections between visual characteristics with other multi-faceted information, such as its function, socioeconomic

Electronic supplementary material The online version of this chapter (https://doi.org/10.1007/978-3-030-58589-1_6) contains supplementary material, which is available to authorized users.

A. Vedaldi et al. (Eds.): ECCV 2020, LNCS 12366, pp. 85–103, 2020.
https://doi.org/10.1007/978-3-030-58589-1_6

Fig. 1. Hierarchical structure of Placepedia with places from all over the world. Each place is associated with its *district, city/town/village, state/province, country, continent,* and a large amount of diverse photos. Both administrative areas and places have rich side information, *e.g. description, population, category, function,* which allows various large-scale studies to be conducted on top of it

status and culture. We therefore believe that it would be a rewarding adventure to move beyond conventional place categorization and explore the connections among different aspects of a place. It indicates that multi-dimension labels are essential for comprehensive place understanding. To support this exploration, a large-scale dataset that cover a diverse set of places with both images and comprehensive multi-faceted information is needed.

However, existing datasets for place understanding [40,43,69], as shown in Table 1 are subject to at least one of the following drawbacks: 1) *Limited Scale.* Some of them [42,43] contain only several thousand images from one particular city. 2) *Restrictive Scope.* Most datasets are constructed for only one task, *e.g.* place retrieval [40] or scene recognition [24,69]. 3) *Lack of Attributes.* These datasets often contain just a very limited set of attributes. For example, [40] contains just photographers and titles. Clearly, these datasets, due to their limitations in scale, diversity, and richness, are not able to support the development of comprehensive place understanding.

In this work, we develop *Placepedia*, a comprehensive place dataset that contains images for places of interest from all over the world with massive attributes, as shown in Fig. 1. *Placepedia* is distinguished in several aspects: **1)** *Large Scale.* It contains over $35M$ images from $240K$ places, several times larger than previous ones. **2)** *Comprehensive Annotations.* The places in Placepedia are tagged with categories, functions, administrative divisions at different levels, *e.g.* city and country, as well as lots of multi-faceted side information, *e.g.* descriptions and coordinates. **3)** *Public Availability.* Placepedia will be made public to the research community. We believe that it will greatly benefit the research on comprehensive place understanding and beyond.

Meanwhile, Placepedia also enables us to rigorously benchmark the performance of existing and future algorithms for place recognition. We create four benchmarks based on Placepedia in this paper, namely *place retrieval, place categorization, function categorization,* and *city/country recognition.* By comparing

Table 1. Comparing Placepedia with other existing datasets. Placepedia offers the largest number of images and the richest information

	# places	# images	# categories	Meta data
Google Landmarks [40]	203,094	5,012,248	N/A	Authors, titles
Places365 [69]	N/A	10,624,928	434	N/A
Holidays [24]	N/A	1,491	500	N/A
Oxford 5k [42]	11	5,062	N/A	N/A
Paris 6k [43]	11	6,412	N/A	N/A
SFLD [6]	N/A	1,700,000	N/A	Coordinates
Pitts 250k [59]	N/A	254,064	N/A	Coordinates
Google Street View [12]	10,343	62,058	N/A	Coordinates, addresses, *etc.*.
Tokyo 24/7 [58]	125	74,000	N/A	Coordinates
Cambridge Landmarks [9]	6	>12,000	N/A	N/A
Vietnam Landscape*a*	103	118,000	N/A	N/A
Placepedia	**>240,000**	**> 35,000,000**	**>3,000**	divisions, descriptions, city info, *etc.*.

a https://blog.facebit.net/2018/09/07/zalo-ai-challenge-problems-and-solutions

different methods and modeling choices on these benchmarks, we gain insights into their pros and cons, which we believe would inspire more effective techniques for place recognition. Furthermore, to provide a trigger for comprehensive place understanding, we develop *PlaceNet*, a unified deep network for multi-level place recognition. It simultaneously predicts place item, category, function, city, and country. Experiments show that by leveraging the multi-level annotations in Placepedia, PlaceNet can learn better representation of a place than previous works. We also leverage both visual and side information from Placepedia to learn city embeddings, which demonstrate strong expressive power as well as the insights on *what distinguish a city*.

From the empirical studies on Placepedia, we see lots of challenges in performing place recognition. 1) The visual appearance can vary significantly due to the changes of angle, illumination, and other environmental factors. 2) A place may look completely different when viewed from inside and outside. 3) A big place, *e.g.* a university, usually consists of a number of small places that have nothing in common in appearance. All these problems remain open. We hope that Placepedia, with its large scale, high diversity and massive annotations, would provide a gold mine for the community to develop more expressive models to meet the aforementioned challenges.

Our contributions in this work can be summarized as below. **1)** We build Placepedia, a large-scale place dataset with comprehensive annotations in multiple aspects. To the best of our knowledge, Placepedia is the largest and the most comprehensive dataset for place understanding. **2)** We design four task-specific benchmarks on Placepedia *w.r.t.* the multi-faceted information. **3)** We conduct systematic studies on place recognition and city embedding, which demonstrate important challenges in place understanding as well as the connections between the visual characteristics and the underlying socioeconomic or cultural implications.

2   Related Work

Place Understanding Datasets. Datasets play an important role for various research topics in computer vision [4,19–23,28,31,32,34,44,45,63,65–68]. During the past decade, lots of place datasets were constructed to facilitate place-related studies. There are mainly three kinds of datasets. The first kind [6,12,40,58,59] focuses on the tasks of place recognition/retrieval, where images are labeled as particular place items, e.g. White House or Eiffel Tower. The second kind [69] targets place categorization or scene recognition. In these datasets, each image is attached with a place type, e.g. parks or museums. The third kind [24,42,43] is for object/image retrieval. The statistics is summarized in Tabler 1. Compared with these datasets, our Placepedia has much larger amount of image and context data, containing over $240K$ places with 35 million images labeled with $3K$ categories. Hence, Placepedia can be used for all these tasks. Besides, the provision of hierarchical administrative divisions for places allows us to study place recognition in different scales, *e.g.* city recognition or country recognition. Also, the function information (*See*, *Do*, *Sleep*, *Eat*, *Buy*, etc.) of places may lead to a new task, namely place function recognition.

Place Understanding Tasks. Lots of work aims at 2D place recognition [1,2, 5,6,12,13,17,25,27,33,37,38,41,46,48,52,54,58,59,71] or place retrieval [10,11, 25,40,47,55,57,61]. Given an image, the goal is to recognize what the particular place is or to retrieve images representing the same place. Scene recognition [29,60,62,64,69,70], on the other hand, defines a diverse list of environment types as labels, *e.g.* nature, classroom, bar. And their job is to assign each image a scene type. [8,50] collects images from several different cities, studies on distinguishing images of one city from others, and discovers what elements are characteristics of a given city. There also exist some other humanities-related studies. [53] classifies keywords of city description into *Economic*, *Cultural*, or *Political*, and then counts the occurrences of these three types to represent city branch. [35] uses satellite data from both daytime and nighttime to discover "ghost cities" which consist of abandoned buildings or housing structures which may hurt urbanization process. [26] collects place images from social media to extract human emotions at different places, to find out the relationship between human emotions and environment factors. [39] uses neural networks trained with map imagery to understand the connection among cities, transportation, and human health. Placepedia, with its large amount of data in both visual and textual domains, allows various studies to be conducted on top in large scale.

3   The Placepedia Dataset

We contribute Placepedia, a large-scale place dataset, to the community. Some example images along with labels are shown in Fig. 2. In this section, we introduce the procedure of building Placepedia. First, a hierarchical structure is organized to store places and their multi-level administrative divisions, where each place/division is associated with rich side information. With this structure,

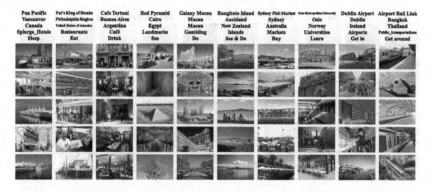

Fig. 2. The text in red, orange, green, blue, purple represents place names, cities, countries/territories, categories, functions, respectively. We see that the appearance of a place can vary from: 1) daytime to nighttime, 2) different angles, 3) inside and outside (Color figure online)

Fig. 3. (a) The number of places for top-10 functions. (b) The number of places for top-50 categories

global places are connected and classified on different levels, which allows us to investigate numerous place-related issues, *e.g.* city recognition [8,50], country recognition, and city embedding. With types (*e.g. park, airport*) provided we can explore tasks such as place categorization [29,69]. With functions (*e.g. See, Sleep*) provided we are able to model the functionality of places. Second, we download place images from Google Image, which are cleaned automatically and manually.

3.1 Hierarchical Administrative Areas and Places

Place Collection. We collect place items with side information from Wikivoyage[1], a free worldwide travel guide website, through the public channel. Pages of Wikivoyage are organized in a hierarchical way, *i.e.*, each destination is obtained by walking through its *continent, country, state/province, city/town/village, district, etc.* As illustrated in Fig. 1, these administrative areas serve as non-leaf nodes in Placepedia, and leaf nodes represent all the places. This process results in a list of 361,524 places together with 24,333 administrative areas.

---

[1] https://en.wikivoyage.org/wiki/Destinations.

Fig. 4. This figure shows ten *function* labels with five example places for each

Meta Data Collection. In Wikivoyage, all destinations are associated with some of the attributes below: *function, category, description, GPS coordinates, address, phone number, opening hour, price, homepage, wikipedia link*. The place number of top-10 functions and top-50 categories are shown in Fig. 3. Table 2 shows the definition for ten functions in Placepedia and Fig. 4 shows several examples of these ten *functions*. Function labels are the section names of places from Wikivoyage. Place functions serve as a good indicator for travelers to choose where to go. For example, some people love to go shopping when traveling; Some prefer to enjoy various flavors of food; Some people are addicted to distinctive landscapes. For administrative areas, Wikivoyage often lack meta data. Hence, we acquire the missing information by parsing their Wikipedia page. At last, the following attributes are extracted: *description, GDP, population density, population, elevation, time zone, area, GPS coordinates, establish time, etc.*

Place Cleaning. To refine the place list, we only keep places satisfying at least one of the two following criteria: 1) It has the attribute *GPS coordinates* or *address*; 2) It is identified as a *location* by Google Entity Recognition [2] or Stanford Entity Recognition[3]. After the removal, 44, 997 items are deleted, and 316, 527 valid place entities remain.

3.2 Place Images

Image Collection. We collect all place images from Google Image engine in the public domain. For each location, its name plus its country is used as the keyword for searching. To increase the probability that images are relevant to a particular location, we only download those whose stem words of image titles contain all stem words of the location name. By this process, a total of over $30M$ images are collected from Google Image.

---

[2] https://cloud.google.com/natural-language.
[3] https://nlp.stanford.edu/software/CRF-NER.html.

Table 2. The description and examples for the 10 *function* labels of Places-Fine and Places-Coarse, which are collected from Wikivoyage

Label	Description	Category examples	Place examples
See	People can enjoy beautiful scenes, arts, And architectures therefrom	Park, Tower, Museum, Gallery, Historical_site	Bergen Aquarium, Lake Parramatta
Do	People can do significant things such as Reading books, watching movies, going On vacation, playing sports	Library, Theater, Resort, Sport, Beach	Apollo Theater, Moray Golf Club
See & Do	People can *see* and *do* in these places For instance, people can not only enjoy Mountain landscape but also climb them	Land_nature, Theater, Resort, Park, Island	Treasure Island, Sydney Opera House
Sleep	People can have a sleep there.	Splurge_hotel, Mid-range_hotel	Othon Palace Rio, Kviknes Hotel
Eat	People can eat food there	Restaurant, Street	Louis' Lunch, Taste on Willis
Drink	People can drink something there	Café, Pub, Street	The Oxford Bar, Cafe Brasilero
Get in	Places for intercity or intercountry Transportation	Airport, Train_station, Public_transportation	Treviso Airport, Kaohsiung Station
Get around	People can travel from one place to Another inside a city or a town, such as Bus stations, metros, and some port	Train_station, Public_transportation	Lujiazui Station, Rathen Ferry
Buy	People usually go shopping there	Mall, Market, Street, Shop, Town, Square	Labrador Mall, Alexa Center
Learn	People usually learn new knowledge or Skills there	University, Sport	Kyoto University, St. Clair College

Image Cleaning. There are 28, 154 places containing Wikipedia links with 8, 125, 108 images. We use this subset to further study place-related tasks. Image set is refined by two stages. Firstly, we use Image Hashing to remove duplicate images. Secondly, we ask human annotators to remove irrelevant images for each place, including those: 1) whose main body represents another place; 2) that are selfie images with faces occupying a large proportion; 3) that are maps indicating the geolocation of the place. In total, $26K$ places with $5M$ images are kept to form this subset. For those places without category labels, we manually annotate the labels for them. And after merging some similar labels, we obtain 50 categories.

Placepedia also helps solve some problems on place understanding, like label confusion and label noise. On one hand, all the labels are collected automatically from the Wikivoyage website. Since it is a popular website that provides worldwide travel guidance and the labels in Wikivoyage have been well organized, there are less label confusion in Placepedia dataset. On the other hand, we have manually checked the labels in Placepedia, which would significantly reduce label noise.

From the examples shown in Fig. 2, we observe that: 1) Images of places may look changeable from daytime to nighttime or during different seasons; 2) It can be significantly different viewed from multiple angles; 3) The appearances from inside and outside usually have little in common; 4) Some places such as universities span very large area and consist of different types of small places. These factors make place-related tasks very challenging. In the rest of this paper, we conduct a series of experiments to demonstrate important challenges in place understanding as well as strong expressive power of city embeddings.

4   Study on Comprehensive Place Understanding

This section introduces our exploration on comprehensive place understanding. Firstly, we carefully design benchmarks, and we evaluate the dataset with a lot of state-of-the-art models with different backbones for different tasks. Secondly, we develop a multi-task model, PlaceNet, which is trained to simultaneously predict place items, categories, functions, cities, and countries. This unified framework for place recognition can serve as a reasonable baseline for further studies in our Placepedia dataset. From the experimental results we also demonstrate the challenges of place recognition on multiple aspects.

4.1   Benchmarks

We build the following benchmarks based on the well-cleaned Placepedia subset, for evaluating different methods.

Datasets

- *Places-Coarse.* We select 200 places for validation and 400 places for testing, from 50 famous cities of 34 countries. The remained $25K$ places are used for training. For validation/testing set, we double checked the annotation results. Places without category labels are manually annotated. After merging similar items of labels, we obtain 50 categories and 10 functions. The training/validation/testing set have $5M/60K/120K$ images respectively, from $7K$ cities of more than 200 countries.
- *Places-Fine.* Places-Fine shares the same validation/testing set with Places-Coarse. For training set, we selected 400 places from the 50 cities of validation/testing places. Different from Places-Coarse, we also double checked the annotation of training data. The training/validation/testing set have $110K/60K/120K$ images respectively, which are tagged with 50 categories, 10 functions, 50 cities, and 34 countries.

Tasks

- *Place Retrieval.* This task is to determine if two images belong to the same place. It is important when people want to find more photos of places they adore. For validation and testing set, 20 images for each place are selected as queries and the rest images form the gallery. Top-k retrieval accuracy is adopted to measure the performance of place retrieval, such that a successful retrieval is counted if at least one image of the same place has been found in the top-k retrieved results.

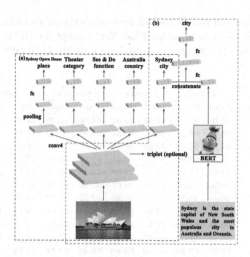

Fig. 5. (a) Pipeline of PlaceNet, which learns five tasks simultaneously. (b) Pipeline of city embedding, which learns city representations considering both vision and text information

- *Place Categorization.* This task is to classify places into 50 place categories, e.g. *museums, parks, churches, temples.* For place categorization, we employ the standard top-k classification accuracy as evaluation metric.
- *Function Categorization.* This task is to classify places into 10 place functions: *See, Do, Sleep, Eat, Drink, See & Do, Get In, Get Around, Buy, Learn.* Again, we employ the standard top-k classification accuracy as evaluation metric.
- *City/Country Recognition.* This task is to classify places into 50 cities or 34 countries. The goal is to determine what city/country an image belongs to. Also, the standard top-k classification accuracy is applied as evaluation metric.

4.2   PlaceNet

We construct a CNN-based model to predict all tasks simultaneously. The training procedure performs in an iterative manner and the system is learned end-to-end.

Network Structures. The network structure of PlaceNet is similar to ResNet50 [16], which has been demonstrated powerful in various vision tasks. As illustrated in Fig. 5 (a), the structures of PlaceNet below the last convolution layer are the same as ResNet50. The last convolution/pooling/fc layers are duplicated to five branches, namely, *place, category, function, city,* and *country,* which is carefully designed for places. Each branch contains two FC layers. Different loss functions and pooling methods are studied in this work.

Loss Functions. We study three losses for PlaceNet, namely, softmax loss, focal loss, and triplet loss. *Softmax loss* or *Focal loss* [30] is adopted to classify place,

Table 3. The experimental results for different methods on all tasks. We vary different pooling methods and loss functions for PlaceNet. Except for the last line, models are trained on Places-Fine. The figures in bold/blue indicate optimal/sub-optimal performance, respectively

		Place		Category		Function		City		Country	
		Top-1	Top-5	Top-1	Top-5	Top-1	Top-5	Top-1	Top-5	Top-1	Top-5
Backbone	AlexNet	33.78	48.19	24.16	53.03	64.97	96.70	12.47	32.52	17.97	43.30
	GoogLeNet	53.48	66.23	26.01	54.81	65.69	97.20	16.34	37.19	20.98	46.43
	VGG16	43.84	59.03	26.89	55.68	65.97	97.11	18.65	41.13	24.86	51.35
	ResNet50	54.53	67.01	25.22	53.62	68.25	96.89	17.15	38.55	19.72	45.51
Pooling	Average	54.33	67.66	25.95	55.07	67.35	97.34	18.73	40.30	24.80	51.03
	Max	49.66	63.26	25.11	54.07	65.45	97.12	16.93	38.18	22.83	48.61
	SPP	28.18	45.55	27.21	53.86	67.02	96.37	15.36	34.48	21.00	43.08
Loss	Softmax	54.31	67.66	25.95	55.07	67.35	97.34	18.73	40.30	24.80	51.03
	Triplet	50.33	64.06	21.15	48.92	64.84	95.61	14.73	36.56	20.43	46.66
	Focal	55.03	67.38	25.27	55.48	67.62	97.53	18.67	40.87	24.73	51.46
PlaceNet on Places-Coarse		**67.85**	**79.35**	**40.42**	**68.98**	**75.48**	**97.58**	**29.25**	**53.47**	**35.83**	**63.78**

category, function, city, and country. To learn the metric described by place pairs, we employ *Triplet loss* [49], which enforces distance constraints among positive and negative samples. When using triplet loss, the network is optimized by a combination of $L_{softmax}$ and $L_{triplet}$.

Pooling Methods. We also study different pooling methods for PlaceNet, namely, average pooling, max pooling, spatial pyramid pooling [15]. Spatial pyramid pooling (SPP) is used to learn multi-scale pooling, which is robust to object deformations and can augment data to confront overfitting.

4.3   Experimental Settings

Data. We use Places-Fine and Places-Coarse defined in Sect. 4.1 as our experimental datasets. Note that Places-Fine and Places-Coarse share the same validation data and testing data, while training data size of the latter is much larger.

Backbone Methods. Deep Convolutional Neural Networks (CNNs) [14,18,28] have shown the impressive power for classification and retrieval tasks. Here we choose four popular CNN architectures, **AlexNet** [28], **GoogLeNet** [56], **VGG16** [51], and **ResNet50** [16], then train them on Places-Fine to create backbone models.

Training Details. We train each model for 90 epochs. For all tasks and all methods, the initial learning rate is set to be 0.5. And the learning rate is multiplied by 0.1 at epoch 63 and epoch 81. The weight decay is $1e^{-4}$. For the optimizer, we use stochastic gradient descent with 0.9 momentum. We also augment the data following the operation on ImageNet, including randomly cropping and horizontally flipping the images. All images are resized to 224×224 and normalized with mean [123, 117, 109] and standard deviation [58, 56, 58]. Each model is

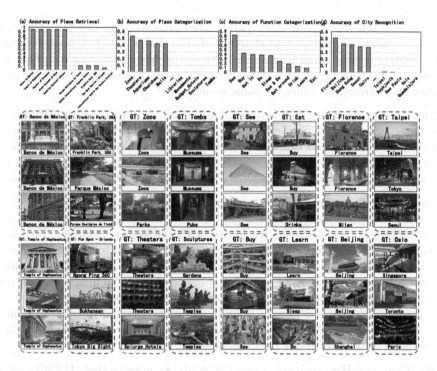

Fig. 6. The 4 tables show the performance of 4 tasks, where each presents the most and the least accurate 5 classes. Below each table are 4 sets of examples, including 2 green/red dash boxes representing sample classes with high/low accuracies. Inside each dash box is the ground truth at the top and three images associated with predicted labels. Green/red solid boxes of images mean right/wrong predictions (Color figure online)

pre-trained on ImageNet and then trained with our Placepedia Dataset in an end-to-end manner.

All experiments are conducted on Places-Fine. And we also train our PlaceNet on Places-Coarse to see if larger scale of datasets can further benefit the recognition performance.

4.4   Analysis on Recognition Results

Quantitative evaluations of different methods on the four benchmarks are provided. Table 3 summarizes the performance of different methods on all tasks. We first analyze the results on Places-Fine for all benchmark tasks.

Place Retrieval. PlaceNet with focal loss achieves the best retrieval results when evaluated using the top-1 accuracy. Some sample places with high/low accuracies are shown in Fig. 6 (a). We observe that: 1) Places with distinctive architectures can be easily recognized, *e.g. Banco de México* and *Temple of Hephaestus.* 2) For some parks, *e.g. Franklin Park*, there is usually no clear evidence to tell them from other parks. The same scenario can take place in categories such as gardens and churches. 3) Big places like *Fun Spot Orlando* may contain several small places, where their appearance may have nothing in common, which makes it very difficult to recognize. Places like resorts, towns, parks, and universities suffer the same issue.

Place Categorization. The best result is yielded by PlaceNet plus SPP. Some sample categories with high/low accuracies are shown in Fig. 6 (b). We observe that: 1) *Zoos* are the most distinctive. Intuitively, if animals are seen in a place, that is probably a zoo. However, photos in zoos may be mistaken for taking in parks. 2) *Tombs* can be confused with *Pubs*, due to bad illumination condition.

Function Categorization. The best setting for learning the function of a place is to use ResNet models. Some sample functions with high/low accuracies are shown in Fig. 6 (c). 1) *See* is recognized with the highest accuracy. 2) Some examples of *Buy* are very difficult to identify, *e.g.* the third image in *Buy*. Even human cannot tell what a street is mainly used for. Is it for shopping, eating, or just for transportation? Same logic applies to shops. The images of *Eat* are often categorized as shops for *buy*ing or *drink*ing. One possible way to recognize the function of a shop is to extract and analyze its name, or to recognize and classify the food type therein. 3) Universities are often unrecognized either, due to its large area with various buildings/scenes.

City/Country Recognition. From Fig. 6 (d), we observe that: 1) Cities with long history (*e.g. Florence, Beijing, Cairo*) are more likely to distinguish from others, because they often preserve the oldest arts and architectures. 2) Travelers often conclude that Taiwan and Japan look quite alike. The results do show that places of *Taipei* may be regarded as in *Tokyo*. 3) Although places can be wrongly classified to another city, the prediction often belongs to the same country with the ground truth city. For instance, Florence and Milan are both in Italy; Beijing and Shanghai are both in China. The results of country recognition are not presented here. They demonstrate similar findings to city recognition.

To conclude, we see that place-related tasks are often very challenging: 1) Places of parks, gardens, churches, *etc.* are easy to classify; However, it is difficult to distinguish one park/garden/church from another. 2) Under bad environmental condition, photos can be extremely difficult to categorize; 3) To recognize the function of a street or a shop is non-trivial, *i.e.* it is hard to determine their use for people to have a dinner, take a drink, or go shopping. 4) Cities of long history such as Beijing and Florence are often recognized with a high accuracy. While images of others are more likely to be misclassified as similar ones inside and outside their countries. We hope that Placepedia with its well-defined benchmarks can foster more effective studies and thus benefit place recognition and retrieval. The last line of Table 3 shows that, to train PlaceNet on larger amount of data, we can further obtain performance gain, by 7 to 16% on different tasks.

5 Study on Multi-faceted City Embedding

We embed each city in an expressive vector to understand places on a city level. Also, the connections between the visual characteristics and the underlying economic or cultural implications are studied therefrom.

5.1 City Embedding

City embedding is to use a vector to represent a city, the items of which indicate different aspects, *e.g.* the economy level, the cultural deposits, the politics atmosphere, *etc.* In this study, cities are embedded from both visual and textual domains. 1) Visual representations of cities are obtained by extracting features from models supervised by city items. 2) Leading paragraphs collected from Wikipedia are used as the description of each city. [7] provided a pre-trained model on language understanding to embed the content of texts into numeric space. We use this model to extract the textual representations for all cities.

Network Structure. The model for city embedding is illustrated in Fig. 5 (b). The input is constructed by concatenating visual and textual vectors. Two fully connected layers are then applied to learn city embedding representations. The corresponding activation functions are ReLU. At last, a classifier and cross entropy loss are used to supervise the learning procedure.

Representative Vectors. We train the network iteratively. The well-trained network is then used to extract the embedding vectors for all images. City embeddings are then acquired by averaging image embeddings city-wise.

5.2 Experimental Results

We analyze city embedding results from two aspects. Firstly, we compare the expressive power of embeddings using different information, namely vision, text, and vision & text, in order to see if learning from both can yield a better city representation. Secondly, we investigate the embedding results neuron-wise to explore what kinds of images can express economic/cultural/political levels of cities the most.

Visual and Text Embedding. In Fig. 7, we demonstrate three embedding results using t-SNE [36]. 1) The left graph shows embeddings using only visual features. We observe that it tends to cluster cities that are visually similar. For example, Tokyo looks like Taipei; Beijing, Shanghai, and Shenzhen are all from China, and Seoul shares lots of similar architectures with them; Florence, Venice, Milan, and Rome are all Italy cities. 2) The graph on the middle shows embeddings using only textual features. We can see that textual features usually express the functionality and geolocation of a city. For example, Tokyo and Oslo are both capitals; London and New York are both financial centers. However, they are not visually alike. Also, cities from the same continent are clustered. 3) The right graph shows embeddings learned from both visual and textual domains.

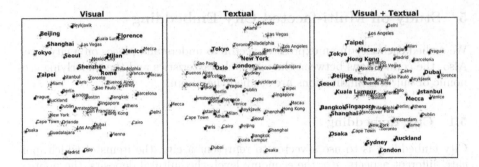

Fig. 7. These three figures show t-SNE representation for city embeddings using vision, text, and vision & text info, respectively. Points with the same color belong to the same continent. We can see that learning from both generates the best embedding results

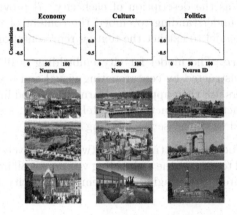

Fig. 8. Three graphs rank Pearson correlation based on neurons in terms of *economy*, *culture*, and *politics*. Below each presents top-3 places activating the neuron of the largest correlation value

They can express the resemblance visually and functionally. For example, cities from east/west-Asia are all clustered together, and cities from Commonwealth of Nations like Sydney, Auckland, and London, are also close to each other on the graph. From the comparison of these graphs, we conclude that learning embeddings from both vision and text content produces the most expressive power of cities.

Economic, Cultural, or Political. We follow the work in [53] to represent each city in three dimensions, namely *economy*, *culture*, and *politics*. In [53], word lists indicating *economy*, *culture*, *politics* are predefined. In this work, leading paragraphs of Wikipedia pages are adopted as our city description. For each city, we calculate the weights of *economic*, *cultural*, and *political* therefrom as in [53]. And we match each neuron to them using Pearson correlation [3], in order to quantify the connection between each neuron and them. Quinnipiac University[4] concludes that a correlation above 0.4 or below −0.4 can be viewed as a strong correlation. From Fig. 8, we see that neurons can express *culture* most confidently, with the highest correlation score larger than 0.6. This is consistent with our knowledge, *i.e.* culture usually is expressed from distinctive architectures or some unique human activities. Looking at the top-3 places that activate the most relevant neuron, we observe that: 1) Economy level is usually conveyed by a cluster of buildings or the crowd on streets, indicating a prosperous place; 2) Cultural atmosphere can be expressed by distinguished architecture styles and human activities; (3) Political elements are often related to temples, churches, and some historical sites, which usually indicate religious activities and politics-related historical movements.

6   Conclusion

In this work, we construct a large-scale place dataset which is comprehensively annotated with multiple aspects. To our knowledge, it is the largest place-related dataset available. To explore place understanding, we carefully build several benchmarks and study contemporary models. The experimental results show that there still remains lots of challenges in place recognition. To learn city embedding representations, we demonstrate that learning from both visual and textual domains can better characterize a city. The learned embeddings also demonstrate that *economic*, *cultural*, and *political* elements can be represented in different types of images. We hope that, with comprehensively annotated Placepedia contributed to the community, more powerful and robust systems will be developed to foster future place-related studies.

Acknowledgment. This work is partially supported by the SenseTime Collaborative Grant on Large-scale Multi-modality Analysis (CUHK Agreement No. TS1610626 & No. TS1712093), the General Research Fund (GRF) of Hong Kong (No. 14203518 & No. 14205719). My gratitude also goes to Yuqi Zhang. As an equal contributor, she spent tons of time in collecting and organizing data.

References

1. Arandjelovic, R., Gronat, P., Torii, A., Pajdla, T., Sivic, J.: Netvlad: CNN architecture for weakly supervised place recognition. In: CVPR, pp. 5297–5307 (2016)
2. Arandjelović, R., Zisserman, A.: DisLocation: scalable descriptor distinctiveness for location recognition. In: Cremers, D., Reid, I., Saito, H., Yang, M.-H. (eds.) ACCV 2014. LNCS, vol. 9006, pp. 188–204. Springer, Cham (2015). https://doi.org/10.1007/978-3-319-16817-3_13
3. Benesty, J., Chen, J., Huang, Y., Cohen, I.: Pearson correlation coefficient. In: Noise reduction in speech processing, vol. 2, pp. 1–4. Springer, Heidelberg (2009). https://doi.org/10.1007/978-3-642-00296-0_5

---

[4] http://faculty.quinnipiac.edu/libarts/polsci/Statistics.html.

4. Caba Heilbron, F., Escorcia, V., Ghanem, B., Carlos Niebles, J.: Activitynet: a large-scale video benchmark for human activity understanding. In: Proceedings of the IEEE conference on computer vision and pattern recognition, pp. 961–970 (2015)
5. Cao, S., Snavely, N.: Graph-based discriminative learning for location recognition. In: CVPR, pp. 700–707 (2013)
6. Chen, D.M., et al.: City-scale landmark identification on mobile devices. In: CVPR 2011, pp. 737–744. IEEE (2011)
7. Devlin, J., Chang, M.W., Lee, K., Toutanova, K.: Bert: pre-training of deep bidirectional transformers for language understanding. arXiv preprint arXiv:1810.04805 (2018)
8. Doersch, C., Singh, S., Gupta, A., Sivic, J., Efros, A.: What makes paris look like paris? (2012)
9. En, S., Lechervy, A., Jurie, F.: RPnet: an end-to-end network for relative camera pose estimation. In: ECCV (2018)
10. Gavves, E., Snoek, C.G.: Landmark image retrieval using visual synonyms. In: Proceedings of the 18th ACM International Conference on Multimedia, pp. 1123–1126. ACM (2010)
11. Gavves, E., Snoek, C.G., Smeulders, A.W.: Visual synonyms for landmark image retrieval. Comput. Vis. Image Underst. **116**(2), 238–249 (2012)
12. Gronat, P., Havlena, M., Sivic, J., Pajdla, T.: Building streetview datasets for place recognition and city reconstruction. Research Reports of CMP, Czech Technical University in Prague (2011)
13. Gronat, P., Obozinski, G., Sivic, J., Pajdla, T.: Learning and calibrating per-location classifiers for visual place recognition. In: CVPR, pp. 907–914 (2013)
14. He, K., Zhang, X., Ren, S., Sun, J.: Delving deep into rectifiers: surpassing human-level performance on imagenet classification. In: CVPR, pp. 1026–1034 (2015)
15. He, K., Zhang, X., Ren, S., Sun, J.: Spatial pyramid pooling in deep convolutional networks for visual recognition. IEEE Trans. Pattern Anal. Mach. Intell. **37**(9), 1904–1916 (2015)
16. He, K., Zhang, X., Ren, S., Sun, J.: Deep residual learning for image recognition. In: CVPR, pp. 770–778 (2016)
17. Hong, Z., Petillot, Y., Lane, D., Miao, Y., Wang, S.: Textplace: visual place recognition and topological localization through reading scene texts. In: ICCV 2019 (2019)
18. Hu, J., Shen, L., Sun, G.: Squeeze-and-excitation networks. In: CVPR, pp. 7132–7141 (2018)
19. Huang, Q., Liu, W., Lin, D.: Person search in videos with one portrait through visual and temporal links. In: Proceedings of the European Conference on Computer Vision (ECCV), pp. 425–441 (2018)
20. Huang, Q., Xiong, Y., Lin, D.: Unifying identification and context learning for person recognition. In: The IEEE Conference on Computer Vision and Pattern Recognition (CVPR), June 2018
21. Huang, Q., Xiong, Y., Rao, A., Wang, J., Lin, D.: Movienet: a holistic dataset for movie understanding. In: Proceedings of the European Conference on Computer Vision (ECCV) (2020)
22. Huang, Q., Xiong, Y., Xiong, Y., Zhang, Y., Lin, D.: From trailers to storylines: An efficient way to learn from movies. arXiv preprint arXiv:1806.05341 (2018)
23. Huang, Q., Yang, L., Huang, H., Wu, T., Lin, D.: Caption-supervised face recognition: Training a state-of-the-art face model without manual annotation. In: Proceedings of the European Conference on Computer Vision (ECCV) (2020)

24. Jegou, H., Douze, M., Schmid, C.: Hamming embedding and weak geometric consistency for large scale image search. In: Forsyth, D., Torr, P., Zisserman, A. (eds.) ECCV 2008. LNCS, vol. 5302, pp. 304–317. Springer, Heidelberg (2008). https://doi.org/10.1007/978-3-540-88682-2_24
25. Johns, E., Yang, G.Z.: Ransac with 2d geometric cliques for image retrieval and place recognition. In: CVPR Workshop, pp. 4321–4329 (2015)
26. Kang, Y., et al.: Extracting human emotions at different places based on facial expressions and spatial clustering analysis. Transactions in GIS (2019)
27. Knopp, J., Sivic, J., Pajdla, T.: Avoiding confusing features in place recognition. In: Daniilidis, K., Maragos, P., Paragios, N. (eds.) ECCV 2010. LNCS, vol. 6311, pp. 748–761. Springer, Heidelberg (2010). https://doi.org/10.1007/978-3-642-15549-9_54
28. Krizhevsky, A., Sutskever, I., Hinton, G.E.: Imagenet classification with deep convolutional neural networks. In: Advances in neural information processing systems, pp. 1097–1105 (2012)
29. Li, Y., Crandall, D.J., Huttenlocher, D.P.: Landmark classification in large-scale image collections. In: 2009 IEEE 12th international conference on computer vision, pp. 1957–1964. IEEE (2009)
30. Lin, T.Y., Goyal, P., Girshick, R., He, K., Dollár, P.: Focal loss for dense object detection. In: CVPR, pp. 2980–2988 (2017)
31. Liu, Z., Luo, P., Qiu, S., Wang, X., Tang, X.: Deepfashion: powering robust clothes recognition and retrieval with rich annotations. In: Proceedings of the IEEE Conference on Computer Vision and Pattern Recognition, pp. 1096–1104 (2016)
32. Liu, Z., Luo, P., Wang, X., Tang, X.: Deep learning face attributes in the wild. In: Proceedings of the IEEE International Conference on Computer Vision, pp. 3730–3738 (2015)
33. Lopez-Antequera, M., Gomez-Ojeda, R., Petkov, N., Gonzalez-Jimenez, J.: Appearance-invariant place recognition by discriminatively training a convolutional neural network. Pattern Recogn. Lett. **92**, 89–95 (2017)
34. Loy, C.C., et al.: Wider face and pedestrian challenge 2018: Methods and results. arXiv preprint arXiv:1902.06854 (2019)
35. Lu, H., Zhang, C., Liu, G., Ye, X., Miao, C.: Mapping china's ghost cities through the combination of nighttime satellite data and daytime satellite data. Remote Sensing **10**(7), 1037 (2018)
36. Maaten, L.V.D., Hinton, G.: Visualizing data using T-SNE. J. Mach. Learn. Res. **9**(8), 2579–2605 (2008)
37. Milford, M., et al.: Sequence searching with deep-learnt depth for condition-and viewpoint-invariant route-based place recognition. In: CVPR Workshops, pp. 18–25 (2015)
38. Mishkin, D., Perdoch, M., Matas, J.: Place recognition with WXBS retrieval. In: CVPR 2015 Workshop on Visual Place Recognition in Changing Environments, vol. 30 (2015)
39. Nice, K.A., Thompson, J., Wijnands, J.S., Aschwanden, G.D., Stevenson, M.: The 'paris-end'of town? urban typology through machine learning. arXiv preprint arXiv:1910.03220 (2019)
40. Noh, H., Araujo, A., Sim, J., Weyand, T., Han, B.: Large-scale image retrieval with attentive deep local features. In: CVPR, pp. 3456–3465 (2017)
41. Panphattarasap, P., Calway, A.: Visual place recognition using landmark distribution descriptors. In: Lai, S.-H., Lepetit, V., Nishino, K., Sato, Y. (eds.) ACCV 2016. LNCS, vol. 10114, pp. 487–502. Springer, Cham (2017). https://doi.org/10.1007/978-3-319-54190-7_30

42. Philbin, J., Chum, O., Isard, M., Sivic, J., Zisserman, A.: Object retrieval with large vocabularies and fast spatial matching. In: CVPR 2007, pp. 1–8. IEEE (2007)
43. Philbin, J., Chum, O., Isard, M., Sivic, J., Zisserman, A.: Lost in quantization: improving particular object retrieval in large scale image databases. In: CVPR 2008, pp. 1–8. IEEE (2008)
44. Rao, A., et al.: A unified framework for shot type classification based on subject centric lens. In: Proceedings of the European Conference on Computer Vision (ECCV) (2020)
45. Rao, A., et al.: A local-to-global approach to multi-modal movie scene segmentation. In: Proceedings of the IEEE/CVF Conference on Computer Vision and Pattern Recognition. pp. 10146–10155 (2020)
46. Sattler, T., Havlena, M., Radenovic, F., Schindler, K., Pollefeys, M.: Hyperpoints and fine vocabularies for large-scale location recognition. In: CVPR, pp. 2102–2110 (2015)
47. Sattler, T., Weyand, T., Leibe, B., Kobbelt, L.: Image retrieval for image-based localization revisited. In: BMVC, vol. 1, p. 4 (2012)
48. Schindler, G., Brown, M., Szeliski, R.: City-scale location recognition. In: CVPR 2007, pp. 1–7. Citeseer (2007)
49. Schroff, F., Kalenichenko, D., Philbin, J.: Facenet: a unified embedding for face recognition and clustering. In: CVPR, pp. 815–823 (2015)
50. Shi, X., Khademi, S., van Gemert, J.: Deep visual city recognition visualization. arXiv preprint arXiv:1905.01932 (2019)
51. Simonyan, K., Zisserman, A.: Very deep convolutional networks for large-scale image recognition. arXiv preprint arXiv:1409.1556 (2014)
52. Sizikova, E., Singh, V.K., Georgescu, B., Halber, M., Ma, K., Chen, T.: Enhancing place recognition using joint intensity - depth analysis and synthetic data. In: Hua, G., Jégou, H. (eds.) ECCV 2016. LNCS, vol. 9915, pp. 901–908. Springer, Cham (2016). https://doi.org/10.1007/978-3-319-49409-8_74
53. Son, J.S., Thill, J.-C.: Is your city economic, cultural, or political? recognition of city image based on multidimensional scaling of quantified web pages. In: Thill, J.-C. (ed.) Spatial Analysis and Location Modeling in Urban and Regional Systems. AGIS, pp. 63–95. Springer, Heidelberg (2018). https://doi.org/10.1007/978-3-642-37896-6_4
54. Stumm, E., Mei, C., Lacroix, S., Nieto, J., Hutter, M., Siegwart, R.: Robust visual place recognition with graph kernels. In: CVPR, pp. 4535–4544 (2016)
55. Sun, X., Ji, R., Yao, H., Xu, P., Liu, T., Liu, X.: Place retrieval with graph-based place-view model. In: Proceedings of the 1st ACM international conference on Multimedia information retrieval, pp. 268–275. ACM (2008)
56. Szegedy, C., et al.: Going deeper with convolutions. In: CVPR, pp. 1–9 (2015)
57. Teichmann, M., Araujo, A., Zhu, M., Sim, J.: Detect-to-retrieve: efficient regional aggregation for image search. In: CVPR, pp. 5109–5118 (2019)
58. Torii, A., Arandjelovic, R., Sivic, J., Okutomi, M., Pajdla, T.: 24/7 place recognition by view synthesis. In: CVPR, pp. 1808–1817 (2015)
59. Torii, A., Sivic, J., Pajdla, T., Okutomi, M.: Visual place recognition with repetitive structures. In: CVPR, pp. 883–890 (2013)
60. Torralba, A., Fergus, R., Freeman, W.T.: 80 million tiny images: a large data set for nonparametric object and scene recognition. IEEE Trans. Pattern Anal. Mach. Intell. 30(11), 1958–1970 (2008)
61. Wang, Y., Lin, X., Wu, L., Zhang, W.: Effective multi-query expansions: robust landmark retrieval. In: Proceedings of the 23rd ACM International Conference on Multimedia, pp. 79–88. ACM (2015)

62. Xiao, J., Hays, J., Ehinger, K.A., Oliva, A., Torralba, A.: Sun database: large-scale scene recognition from abbey to zoo. In: 2010 IEEE Computer Society Conference on Computer Vision and Pattern Recognition, pp. 3485–3492. IEEE (2010)
63. Xiong, Y., Huang, Q., Guo, L., Zhou, H., Zhou, B., Lin, D.: A graph-based framework to bridge movies and synopses. In: The IEEE International Conference on Computer Vision (ICCV), October 2019
64. Yang, J., Zhang, S., Wang, G., Li, M.: Scene and place recognition using a hierarchical latent topic model. Neurocomputing **148**, 578–586 (2015)
65. Yang, L., Chen, D., Zhan, X., Zhao, R., Loy, C.C., Lin, D.: Learning to cluster faces via confidence and connectivity estimation. In: Proceedings of the IEEE Conference on Computer Vision and Pattern Recognition (2020)
66. Yang, L., Huang, Q., Huang, H., Xu, L., Lin, D.: Learn to propagate reliably on noisy affinity graphs. In: Proceedings of the European Conference on Computer Vision (ECCV) (2020)
67. Yang, L., Zhan, X., Chen, D., Yan, J., Loy, C.C., Lin, D.: Learning to cluster faces on an affinity graph. In: Proceedings of the IEEE Conference on Computer Vision and Pattern Recognition (CVPR) (2019)
68. Zhang, X., Yang, L., Yan, J., Lin, D.: Accelerated training for massive classification via dynamic class selection. In: AAAI (2018)
69. Zhou, B., Lapedriza, A., Khosla, A., Oliva, A., Torralba, A.: Places: a 10 million image database for scene recognition. IEEE Trans. Pattern Anal. Mach. Intell. **40**(6), 1452–1464 (2017)
70. Zhou, B., Lapedriza, A., Xiao, J., Torralba, A., Oliva, A.: Learning deep features for scene recognition using places database. In: Advances in Neural Information Processing Systems, pp. 487–495 (2014)
71. Zhu, Y., Wang, J., Xie, L., Zheng, L.: Attention-based pyramid aggregation network for visual place recognition. In: 2018 ACM Multimedia Conference on Multimedia Conference, pp. 99–107. ACM (2018)

DELTAS: Depth Estimation by Learning Triangulation and Densification of Sparse Points

Ayan Sinha[1]([⊠]), Zak Murez[1], James Bartolozzi[1], Vijay Badrinarayanan[2], and Andrew Rabinovich[3]

[1] Magic Leap Inc., Sunnyvale, CA, USA
{asinha,zmurez}@magicleap.com
bartolozzij@gmail.com
[2] Wayve.ai, London, UK
vijay@wayve.ai
[3] InsideIQ Inc., San Francisco, CA, USA
andrew@insideiq.team

Abstract. Multi-view stereo (MVS) is the golden mean between the accuracy of active depth sensing and the practicality of monocular depth estimation. Cost volume based approaches employing 3D convolutional neural networks (CNNs) have considerably improved the accuracy of MVS systems. However, this accuracy comes at a high computational cost which impedes practical adoption. Distinct from cost volume approaches, we propose an efficient depth estimation approach by first (a) detecting and evaluating descriptors for interest points, then (b) learning to match and triangulate a small set of interest points, and finally (c) densifying this sparse set of 3D points using CNNs. An end-to-end network efficiently performs all three steps within a deep learning framework and trained with intermediate 2D image and 3D geometric supervision, along with depth supervision. Crucially, our first step complements pose estimation using interest point detection and descriptor learning. We demonstrate state-of-the-art results on depth estimation with lower compute for different scene lengths. Furthermore, our method generalizes to newer environments and the descriptors output by our network compare favorably to strong baselines.

Keywords: 3D from multi-view and sensors · Stereo depth estimation · Multi-task learning

J. Bartolozzi, V. Badrinarayanan and A. Rabinovich—Work done at Magic Leap.

Electronic supplementary material The online version of this chapter (https://doi.org/10.1007/978-3-030-58589-1_7) contains supplementary material, which is available to authorized users.

1   Motivation

Depth sensing is crucial for a wide range of applications ranging from Augmented Reality (AR)/Virtual Reality (VR) to autonomous driving. Estimating depth can be broadly divided into classes: active and passive sensing. Active sensing techniques include LiDAR, structured-light and time-of-flight (ToF) cameras, whereas depth estimation using a monocular camera or stereopsis of an array of cameras is termed passive sensing. Active sensors are currently the de-facto standard of applications requiring depth sensing due to good accuracy and low latency in varied environments [48]. However, active sensors have their own of limitation. LiDARs are prohibitively expensive and provide sparse measurements. Structured-light and ToF depth cameras have limited range and completeness due to the physics of light transport. Furthermore, they are power hungry and inhibit mobility critical for AR/VR applications on wearables. Consequently, computer vision researchers have pursued passive sensing techniques as a ubiquitous, cost-effective and energy-efficient alternative to active sensors [31].

Passive depth sensing using a stereo cameras requires a large baseline and careful calibration for accurate depth estimation [3]. A large baseline is infeasible for mobile devices like phones and wearables. An alternative is to use MVS techniques for a moving monocular camera to estimate depth. MVS generally refers to the problem of reconstructing 3D scene structure from multiple images with known camera poses and intrinsics [14]. The unconstrained nature of camera motion alleviates the baseline limitation of stereo-rigs, and the algorithm benefits from multiple observations of the same scene from continuously varying viewpoints [17]. However, camera motion also makes depth estimation more challenging relative to rigid stereo-rigs due to pose uncertainty and added complexity of motion artifacts. Most MVS approaches involve building a 3D cost volume, usually with a plane sweep stereo approach [18,45]. Accurate depth estimation using MVS rely on 3D convolutions on the cost volume, which is both memory as well as computationally expensive, scaling cubically with the resolution. Furthermore, redundant compute is added by ignoring useful image-level properties such as interest points and their descriptors, which are a necessary precursor to camera pose estimation, and hence, any MVS technique. This increases the overall cost and energy requirements for passive sensing.

Passive sensing using a single image is fundamentally unreliable due to scale ambiguity in 2D images. Deep learning based monocular depth estimation approaches formulate the problem as depth regression [10,11] and have reduced the performance gap to those of active sensors [24,26], but still far from being practical. Recently, sparse-to-dense depth estimation approaches have been proposed to remove the scale ambiguity and improve robustness of monocular depth estimation [31]. Indeed, recent sparse-to-dense approaches with less than 0.5% depth samples have accuracy comparable to active sensors, with higher range and completeness [6]. However, these approaches assume accurate or seed depth samples from an active sensor which is limiting. The alternative is to use the sparse 3D landmarks output from the best performing algorithms for

Simultaneous Localization and Mapping (SLAM) [32] or Visual Inertial Odometry (VIO) [34]. However, using depth evaluated from these sparse landmarks in lieu of depth from active sensors, significantly degrades performance [47]. This is not surprising as the learnt sparse-to-dense network ignores potentially useful cues, structured noise and biases present in SLAM or VIO algorithm.

Here we propose to learn the sparse 3D landmarks in conjunction with the sparse to dense formulation in an end-to-end manner so as to (a) remove dependence on a cost volume in the MVS technique, thus, significantly reducing compute, (b) complement camera pose estimation using sparse VIO or SLAM by reusing detected interest points and descriptors, (c) utilize geometry-based MVS concepts to guide the algorithm and improve the interpretability, and (d) benefit from the accuracy and efficiency of sparse-to-dense techniques. Our network is a multitask model [22], comprised of an encoder-decoder structure composed on two encoders, one for RGB image and one for sparse depth image, and three decoders: one for interest point detection, one for descriptors and one for the dense depth prediction. We also contribute a differentiable module that efficiently triangulates points using geometric priors and forms the critical link between the interest point decoder, descriptor decoder, and the sparse depth encoder enabling end-to-end training.

The rest of the paper is organized as follows. Section 2 discussed related work and Sect. 3 describes our approach. We perform experimental evaluation in Sect. 4, and finally conclusions and future work are presented in Sect. 5.

2   Related Work

Interest Point Detection and Description: Sparse feature based methods are standard for SLAM or VIO techniques due to their high speed and accuracy. The detect-then-describe approach is the most common approach to sparse feature extraction, wherein, interest points are detected and then described for a patch around the point. The descriptor encapsulates higher level information, which are missed by typical low-level interest points such as corners, blobs, etc. Prior to the deep learning revolution, classical systems like SIFT [28] and ORB [38] were ubiquitously used as descriptors for feature matching for low level vision tasks. Deep neural networks directly optimizing for the objective at hand have now replaced these hand engineered features across a wide array of applications. However, such an end-to-end network has remained elusive for SLAM [33] due to the components being non-differentiable. General purpose descriptors learnt by methods such as SuperPoint [9], LIFT [46], GIFT [27] aim to bridge the gap towards differentiable SLAM.

MVS: MVS approaches either directly reconstruct a 3D volume or output a depth map which can be flexibly used for 3D reconstruction or other applications. Methods reconstructing 3D volumes [5,45] are restricted to small spaces or isolated objects either due to the high memory load of operating in a 3D voxelized space [36,40], or due to the difficulty of learning point representations in complex environments [35]. Here, we use multi-view images captured in indoor environments for depth estimation due to the versatility of depth map representation. This area has lately seen a lot of progress starting with DeepMVS [18] which proposed a learnt patch matching approach. MVDepthNet [44], and DPSNet [19] build a cost volume for depth estimation. GP-MVSNet [17] built upon MVDepthNet to coherently fuse temporal information using gaussian processes. All these methods utilize the plane sweep algorithm during some stage of depth estimation, resulting in an accuracy vs efficiency trade-off.

Sparse to Dense Depth Prediction: Sparse-to-dense depth estimation has recently emerged as a way to supplement active depth sensors due to their range limitations when operating on a power budget, and to fill in depth in hard to detect regions such as dark or reflective objects. The first such approach was proposed by Ma et al. [31], and following work by Chen et al. [6] and [47] introduced innovations in the representation and network architecture. A convolutional spatial propagation module is proposed in [7] to in-fill the missing depth values. Self-supervised approaches [12,13] have concurrently been explored for the sparse-to-dense problem [30]. Recently, a learnable triangulation technique was proposed to learn human pose key-points [21]. We leverage their algebraic triangulation module for the purpose of sparse reconstruction of 3D points.

3   Method

Our method can be broadly sub-divided into three steps as illustrated in Fig. 1 for a prototypical target image and two view-points. In the first step, the target or anchor image and the multi-view images are passed through a shared RGB encoder and descriptor decoder to output a descriptor field for each image. Interest points are also detected for the target or the anchor image. In the second step, the interest points in the anchor image in conjunction with the relative poses are used to determine the search space in the reference or auxiliary images from alternate view-points. Descriptors are sampled in the search space and are matched with descriptors for the interest points. Then, the matched key-points are triangulated using SVD and the output 3D points are used to create a sparse depth image. In the third and final step, the output feature maps for the sparse depth encoder and intermediate feature maps from the RGB encoder are collectively used to inform the depth decoder and output a dense depth image. Each of the three steps are described in greater detail below.

3.1   Interest Point Detector and Descriptor

We adopt SuperPoint-like [9] formulation of a fully-convolutional neural network architecture which operates on a full-resolution image and produces interest point detection accompanied by fixed length descriptors. The model has a single, shared encoder to process and reduce the input image dimensionality. The feature maps from the encoder feed into two task- specific decoder "heads",

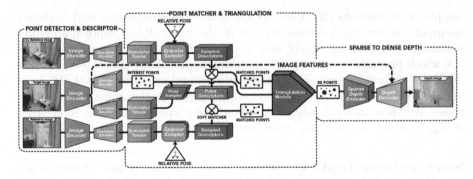

Fig. 1. End-to-end network for detection and description of interest points, matching and triangulation of the points and densification of 3D points for depth estimation.

which learn weights for interest point detection and interest point description. This joint formulation of interest point detection and description in SuperPoint enables sharing compute for the detection and description tasks, as well as the down stream task of depth estimation. However, SuperPoint was trained on grayscale images with focus on interest point detection and description for continuous pose estimation on high frame rate video streams, and hence, has a relatively shallow encoder. On the contrary, we are interested in image sequences with sufficient baseline, and consequently longer intervals between subsequent frames. Furthermore, SuperPoint's shallow backbone suitable for sparse point analysis has limited capacity for our downstream task of dense depth estimation. Hence, we replace the shallow backbone with a ResNet-50 [16] encoder which balances efficiency and performance. The output resolution of the interest point detector decoder is identical to that of SuperPoint. In order to fuse fine and coarse level image information critical for point matching, we use a U-Net [37] like architecture for the descriptor decoder. This decoder outputs an N-dimensional descriptor tensor at $1/8^{th}$ the image resolution, similar to SuperPoint. The architecture is illustrated in Fig. 2. We train the interest point detector network by distilling the output of the original SuperPoint network and the descriptors are trained by the matching formulation described below.

3.2 Point Matching and Triangulation

The previous step provides interest points for the anchor image and descriptors for all images, i.e., the anchor image and full set of auxiliary images. A naive approach will be to match descriptors of the interest points sampled from the descriptor field of the anchor image to all possible positions in each auxiliary image. However, this is computationally prohibitive. Hence, we invoke geometrical constraints to restrict the search space and improve efficiency. Using concepts from multi-view geometry, we only search along the epipolar line in the auxiliary images [14]. The epipolar line is determined using the fundamental matrix, F, using the relation $xFx^T = 0$, where x is the set of points in the image.

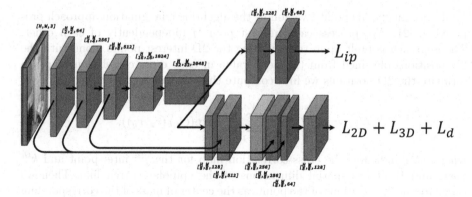

Fig. 2. SuperPoint-like network with detector and descriptor heads.

The matched point is guaranteed to lie on the epipolar line in an ideal scenario as illustrated in Fig. 3 (Left). However, practical limitations to obtain perfect pose lead us to search along the epipolar line with a small fixed offset on either side; Fig. 3 (Middle). Furthermore, the epipolar line stretches for depth values from $-\infty$ to ∞. We clamp the epipolar line to lie within feasible depth sensing range, and vary the sampling rate within this restricted range in order to obtain descriptor fields with the same output shape for implementation purposes, shown in Fig. 3 (Right). We use bilinear sampling to obtain the descriptors at the desired points in the descriptor field. The descriptor of each interest point is convolved with the descriptor field along its corresponding epipolar line for each image view-point:

$$C_{j,k} = \hat{D}_j * D_j^k, \forall x \in \mathcal{E}, \tag{1}$$

where $\hat{D}$ is the descriptor field of the anchor image, D^k is the descriptor field of the kth auxiliary image, and convolved over all sampled points x along the clamped epipolar line $\mathcal{E}$ for point j. This effectively provides a cross-correlation map [2] between the descriptor field and interest point descriptors. High values in this map indicate potential key-point matches in the auxiliary images to the interest points in the anchor image. In practice, we add batch normalization [20] and ReLU non-linearity [23] to output $C_{j,k}$ in order to ease training.

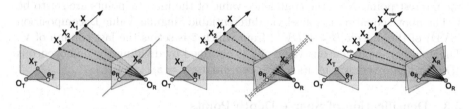

Fig. 3. Left: Epipolar sampling; Middle: Offset sampling due to relative pose error; Right: Constrained depth range sampling

To obtain the 3D points, we follow the algebraic triangulation approach proposed in [21]. We process each interest point j independently of each other. The approach is built upon triangulating the 2D interest points along with the 2D positions obtained from the peak value in each cross correlation map. To estimate the 2D positions we first compute the softmax across the spatial axes:

$$C'_{j,k} = \exp(C_{j,k})/(\sum_{r_x=1}^{W} \sum_{r_y=1}^{H} \exp(C_{j,k}(r_x, r_y))), \tag{2}$$

where, $C_{j,k}$ indicates the cross-correlation map for the jth inter-point and k^{th} view, and W, H are spatial dimensions of the epipolar search line. Then we calculate the 2D positions of the points as the center of mass of the corresponding cross-correlation maps, also termed soft-argmax operation:

$$x_{j,k} = \sum_{r_x=1}^{W} \sum_{r_y=1}^{H} r(x, y)(C'_{j,k}(r(x, y))). \tag{3}$$

The soft-argmax operation enables differentiable routing between the 2D position of the matched points $x_{j,k}$ and the cross-correlation maps $C_{j,k}$. We use the linear algebraic triangulation approach proposed in [21] to estimate the 3D points from the matched 2D points $x_{j,k}$. Their method reduces the finding of the 3D coordinates of a point z_j to solving the over-determined system of equations on homogeneous 3D coordinate vector of the point $\bar{z}$:

$$A_j \bar{z}_j = 0, \tag{4}$$

where $A_j \in \mathcal{R}^{2k,4}$ is a matrix composed of the components from the full projection matrices and $x_{j,k}$. Different view-points may contribute unequally to the triangulation of a point due to occlusions and motion artifacts. Weighing the contributions equally leads to sub-optimal performance. The problem is solved in a differentiable way by adding weights w_k to the coefficients of the matrix corresponding to different views:

$$(w_j A_j)\bar{z}_j = 0. \tag{5}$$

The weights w are set to be the max value in each cross-correlation map. This allows the contribution of the each camera view to be controlled by the quality of match, and low-confidence matches to be weighted less while triangulating the interest point. Note the confidence value of the interest points are set to be 1. The above equation is solved via differentiable Singular Value Decomposition (SVD) of the matrix $B = UDV^T$, from which $\bar{z}$ is set as the last column of V. The final non-homogeneous value of z is obtained by dividing the homogeneous 3D coordinate vector $\bar{z}$ by its fourth coordinate: $z = \bar{z}/(\bar{z})_4$ [21].

3.3   Densification of Sparse Depth Points

The interest-point detector network provides the 2D position of the points. The z coordinate of the triangulated points provides the depth. We impute a sparse

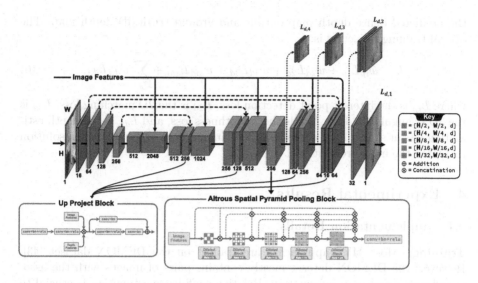

Fig. 4. Proposed sparse-to-dense network architecture showing the concatenation of image and sparse depth features. We use deep supervision over 4 image scales. The blocks below illustrate the upsampling and the altrous spatial pyramid pooling (ASPP) block.

depth image of the same resolution as the input image with depth of these sparse points. Note that the gradients can propagate from the sparse depth image back to the 3D key-points all the way to the input image. This is akin to switch unpooling in SegNet [1]. We pass the sparse depth image through an encoder network which is a narrower version of the image encoder network. Specifically, we use a ResNet-50 encoder with the channel widths after each layer to be $1/4^{th}$ of the image encoder. We concatenate these features with the features obtained from the image encoder. We use a U-net style decoder with intermediate feature maps from both the image as well as sparse depth encoder concatenated with the intermediate feature maps of the same resolution in the decoder, similar to [6]. We provide deep supervision over 4 scales [25]. We also include a spatial pyramid pooling block to encourage feature mixing at different receptive field sizes [4,15]. The details of the architecture are shown in the Fig. 4.

3.4 Overall Training Objective

The entire network is trained with a combination of (a) cross entropy loss between the output tensor of the interest point detector decoder and ground truth interest point locations obtained from SuperPoint, (b) a smooth-L1 loss between the 2D points output after soft argmax and ground truth 2D point matches, (c) a smooth-L1 loss between the 3D points output after SVD triangulation and ground truth 3D points, (d) an edge aware smoothness loss on the output dense depth map, and (e) a smooth-L1 loss over multiple scales between the predicted dense depth map output and ground truth 3D depth map. The overall training objective is:

$$L = w_{ip}L_{ip} + w_{2d}L_{2d} + w_{3d}L_{3d} + w_{sm}L_{sm} + \sum_i w_{d,i}L_{d,i}, \tag{6}$$

where L_{ip} is the interest point detection loss, L_{2d} is the 2D matching loss, L_{3d} is the 3D triangulation loss, L_{sm} is the smoothness loss, and $L_{d,i}$ is the depth estimation loss at scale i for 4 different scales ranging from original image resolution to $1/16^{th}$ the image resolution.

4   Experimental Results

4.1   Implementation Details

Training: Most MVS approaches are trained on the DEMON dataset [43]. However, the DEMON dataset mostly contains pairs of images with the associated depth and pose information. Relative confidence estimation is crucial to accurate triangulation in our algorithm, and needs sequences of length three or greater in order to estimate the confidence accurately and holistically triangulate an interest point. Hence, we diverge from traditional datasets for MVS depth estimation, and instead use ScanNet [8]. ScanNet is an RGB-D video dataset containing 2.5 million views in more than 1500 scans, annotated with 3D camera poses, surface reconstructions, and instance-level semantic segmentations. Three views from a scan at a fixed interval of 20 frames along with the pose and depth information forms a training data point in our method. The target frame is passed through SuperPoint in order to detect interest points, which are then distilled using the loss L_{ip} while training our network. We use the depth images to determine ground truth 2D matches, and unproject the depth to determine the ground truth 3D points. We train our model for 100K iterations using PyTorch framework with batch-size of 24 and ADAM optimizer with learning rate 0.0001 ($\beta_1 = 0.9$, $\beta_2 = 0.999$), which takes about 3 days across 4 Nvidia Titan RTX GPUs. We fix the resolution of the image to be qVGA (240 × 320) and number of interest points to be 512 in each image with at most half the interest points chosen from the interest point detector thresholded at 0.0005, and the rest of the points chosen randomly from the image. Choosing random points ensures uniform distribution of sparse points in the image and helps the densification process. We set the length of the sampled descriptors along the epipolar line to be 100, albeit, we found that the matching is robust even for lengths as small as 25. We set the range of depth estimation to be between 0.5 and 10 m, as common for indoor environments. We empirically set the weights to be [0.1,1.0,2.0,1.0,2.0] for $w_{ip}, w_{2d}, w_{3d}, w_{sm}, w_{d,1}$, respectively. We damp $w_{d,1}$ by a factor of 0.7 for each subsequent scale.

Evaluation: The ScanNet test set consists of 100 scans of unique scenes different for the 707 scenes in the training dataset. We first evaluate the performance of our detector and descriptor decoder for the purpose of pose estimation on Scan-Net. We use the evaluation protocol and metrics proposed in SuperPoint, namely the mean localization error (MLE), the matching score (MScore), repeatability (Rep) and the fraction of correct pose estimated using descriptor matches and PnP algorithm at 5° threshold for rotation and 5 cm for translation. We compare against SuperPoint, SIFT, ORB and SURF at a NMS threshold of 3 pixels for Rep, MLE, and MScore as suggested in the SuperPoint paper. Next, we use standard metrics to quantitatively measure the quality of our estimated depth: : absolute relative error (Abs Rel), absolute difference error (Abs diff), square relative error (Sq Rel), root mean square error and its log scale (RMSE and RMSE log) and inlier ratios ($\delta < 1.25^i$ where $i \in 1, 2, 3$). Note higher values for inlier ratios are desirable, whereas all other metrics warrant lower values.

We compare our method to recent deep learning approaches for MVS: (a) DPSNet: Deep plane sweep approach, (b) MVDepthNet: Multi-view depth net, and (c) GPMVSNet temporal non-parametric fusion approach using Gaussian processes. Note that these methods perform much better than traditional geometry-based stereo algorithms. Our primary results are on sequences of length 3, but we also report numbers on sequences of length 2,4,5 and 7 in order to understand the performance as a function of scene length. We evaluate the methods on Sun3D dataset, in order to understand the generalization of our approach to other indoor scenes. We also discuss the multiply-accumuate operations (MACs) for the different methods to understand the operating efficiency at run-time.

4.2 Detector and Descriptor Quality

Table 1 shows the results of the our detector and descriptor evaluation. Note that MLE and repeatability are detector metrics, MScore is a descriptor metric, and rotation@5° and translation@5cm are combined metrics. We set the threshold for our detector at 0.0005, the same as that used during training. This results in a large number of interest points being detected (Num) which artificially inflates the repeatability score (Rep) in our favour, but has poor localization performance as indicated by MLE metric. However, our MScore is comparable to SuperPoint although we trained our network to only match along the epipolar line, and not for the full image. Furthermore, we have the best rotation@5° and translation@5cm metric indicating that the matches found using our descriptors help accurately determine rotation and translation, i.e., pose. These results are indicative that our training procedure can complement the homographic adaptation technique of SuperPoint and boost the overall performance. Incorporation of evaluated pose using ideas discussed in [39], in lieu of ground truth pose to train our network is left for future work.

Table 1. Performance of different descriptors on ScanNet.

	MLE	MScore	Num	Rep	rot@5°	trans@5cm
ORB	2.584	0.194	401	0.613	0.142	0.064
SIFT	**2.327**	0.201	203	0.496	0.311	0.148
SURF	2.577	0.198	268	0.460	0.303	0.134
SuperPoint	2.545	**0.375**	129	0.519	0.489	0.244
Ours	3.101	0.329	1511	**0.738**	**0.518**	**0.254**

4.3 Depth Results

We set the same hyper-parameters for evaluating our network for all scenarios and across all datasets, i.e., fix the number of points detected to be 512, length of the sampled descriptors to be 100, and the detector threshold to be 5e-4. In order to ensure uniform distribution of the interest points and avoid clusters, we set a high NMS value of 9 as suggested in [9]. The supplement has analysis of the sparse depth output from our network and ablation study over different choices of hyper parameters. Table 2 shows the performance of depth estimation on sequences of length 3 and gap 20 as used in the training set. For fair comparison, we evaluate two versions of the competing approaches (1) The author provided open source trained model, (2) The trained model fine-tuned on Scan-Net for 100K iterations with the default training parameters as suggested in the manuscript or made available by the authors. We use a gap of 20 frames to train each network, similar to ours. The fine-tuned models are indicated by the suffix *FT* in the table. Unsurprisingly, the fine-tuned models fare much better than the original models on ScanNet evaluation. MVDepthNet has least improvement after fine-tuning, which can be attributed to the heavy geometric and photometric augmentation used during training, hence making it generalize well. DPSNet benefits maximally from fine-tuning with over 25% drop in absolute error. However, our network outperforms all methods across all metrics. Figure 6 shows qualitative comparison between the different methods and Fig. 5 show sample 3D reconstructions of the scene from the estimated depth maps. In Fig. 6, we see that MVDepthNet has gridding artifacts, which are removed by GPMVS-Net. However, GPMVSNet has poor metric performance. DPSNet washes away finer details and also suffers from gridding artifacts. Our method preserves finer details while maintaining global coherence compared to all other methods. As we use geometry to estimate sparse depth, and the network in-fills the missing values, we retain metric performance while leveraging the generative ability of CNNs with sparse priors. In Fig. 5 we see our method consistently output less noisy scene reconstructions compared to MVDepthNet and DPSNet. Moreover, we see planes and corners being respected better than the other methods.

An important feature of any multiview stereo method is the ability to improve with more views. Table 3 shows the performance for different number of images. We set the frame gap to be 20, 15, 12 and 10 for 2,4,5 and 7 frames respectively.

Table 2. Performance of depth estimation on ScanNet. We use sequences of length 3 and sample every 20 frames. FT indicates fine-tuned on ScanNet.

	Abs Rel	Abs	Sq Rel	RMSE	RMSE log	$\delta < 1.25$	$\delta < 1.25^2$	$\delta < 1.25^3$
GPMVS	0.1306	0.2600	0.0944	0.3451	0.1881	0.8481	0.9462	0.9753
GPMVS-FT	0.1079	0.2255	0.0960	0.4659	0.1998	0.8905	0.9591	0.9789
MVDepth	0.1191	0.2096	0.0910	0.3048	0.1597	0.8690	0.9599	0.9851
MVDepth-FT	0.1054	0.1911	0.0970	0.3053	0.1553	0.8952	0.9707	0.9895
DPS	0.1470	0.2248	0.1035	0.3468	0.1952	0.8486	0.9474	0.9761
DPS-FT	0.1025	0.1675	0.0574	0.2679	0.1531	0.9102	0.9708	0.9872
Ours	**0.0932**	**0.1540**	**0.0506**	**0.2505**	**0.1426**	**0.9287**	**0.9767**	**0.9893**

These gaps ensure that each set approximately span similar volumes in 3D space, and any performance improvement emerges from the network better using the available information as opposed to acquiring new information. We again see that our method outperforms all other methods on all three metrics for different sequence lengths. Closer inspection of the values indicate that the DPSNet and GPMVSNet do not benefit from additional views, whereas, MVDepthNet benefits from a small number of additional views but stagnates for more than 4 frames. On the contrary, we show steady improvement in all three metrics with additional views. This can be attributed to our point matcher and triangulation module which naturally benefits from additional views.

Table 3. Performance of depth estimation on ScanNet. Results on sequences of various lengths are presented. GPN: GPMVSNet, MVN: MVDepthNet, DPS: DPSNet. AbR: Absolute Relative, Abs: Absolute difference, SqR: Square Relative.

Method	2 frames			4 frames			5 frames			7 frames		
	AbR	Abs	SqR	AbR	Abs	SqR	AbR	Abs	SqR	AbR	Abs	SqR
GPN	0.112	0.233	0.101	0.109	0.226	0.100	0.107	0.226	0.112	0.109	0.230	0.116
MVN	0.126	0.238	0.471	0.105	0.191	0.078	0.106	0.192	0.071	0.108	0.195	0.067
DPS	**0.099**	0.181	0.062	0.102	0.168	0.057	0.102	0.168	0.057	0.102	0.167	0.057
Ours	0.106	**0.173**	**0.057**	**0.090**	**0.150**	**0.049**	**0.088**	**0.147**	**0.048**	**0.087**	**0.144**	**0.043**

As a final experiment, we test our network on Sun3D test dataset consisting of 80 pairs of images. Sun3D also captures indoor environments, albeit at a much smaller scale compared to ScanNet. Table 4 shows the performance for the two versions of DPSNet and MVDepthNet discussed previously, and our network. Note DPSNet and MVDepthNet were originally trained on the Sun3D training database. The fine-tuned version of DPSNet performs better than the original network on the Sun3D test set owing to the greater diversity in ScanNet training database. MVDepthNet on the contrary performs worse, indicating that it overfit to ScanNet and the original network was sufficiently trained and generalized well. Remarkably, we again outperform both methods although our trained network has never seen any image from the Sun3D database. This indicates that our principled way of determining sparse depth, and then densifying has good generalizability. The supplement shows additional qualitative results.

Table 4. Performance of depth estimation on Sun3D. We use sequences of length 2.

	Abs Rel	Abs	Sq Rel	RMSE	RMSE log	$\delta < 1.25$	$\delta < 1.25^2$	$\delta < 1.25^3$
MVDepth	0.1377	0.3199	0.1564	0.4523	0.1853	0.8245	0.9601	0.9851
MVDepth-FT	0.3092	0.7209	4.4899	1.718	0.319	0.7873	0.9117	0.9387
DPS	0.1590	0.3341	0.1564	0.4516	0.1958	0.8087	0.9363	0.9787
DPS-FT	0.1274	0.2858	0.0855	0.3815	0.1768	0.8396	0.9459	0.9866
Ours	**0.1245**	**0.2662**	**0.0741**	**0.3602**	**0.1666**	**0.8551**	**0.9728**	**0.9902**

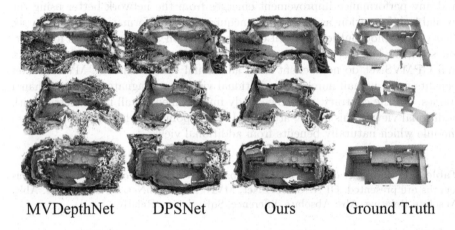

MVDepthNet   DPSNet   Ours   Ground Truth

Fig. 5. 3D scene reconstruction using predicted depth over the full sequence.

We evaluate the total number of multiply-accumulate operations (MACs) needed for our approach. For a 2 image sequence, we perform 16.57 Giga Macs (GMacs) for the point detector and descriptor module, less than 0.002 GMacs for the matcher and triangulation module, and 67.90 GMacs for the sparse-to-dense module. A large fraction of this is due to the U-Net style feature tensors connecting the image and sparse depth encoder to the decoder. We perform a total of 84.48 GMacs to estimate the depth for a 2 image sequence. This is considerably lower than DPSNet which performs 295.63 GMacs for a 2 image sequence, and also less than the real-time MVDepthNet which performs 134.8 GMacs for a pair of images to estimate depth. It takes 90 milliseconds to estimate depth on Nvidia Titan RTX GPU, which we evaluated to be 2.5 times faster than DPSNet. Inference time for MVDepthNet and GPMVSNet is ≈60 ms. We believe our method can be further sped up by replacing Pytorch's native SVD with a custom implementation for triangulation. Furthermore, as we do not depend on a cost volume, compound scaling laws as those derived for image [41] and object [42] recognition can be straightforwardly extended to our method.

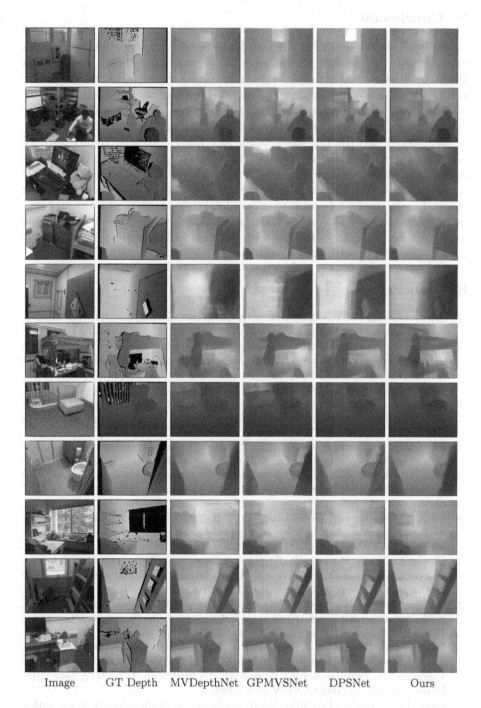

Image   GT Depth   MVDepthNet   GPMVSNet   DPSNet   Ours

Fig. 6. Qualitative Performance of our networks on sampled images from ScanNet.

5   Conclusion

In this work we developed an efficient depth estimation algorithm by learning to triangulate and densify sparse points in a multi-view stereo scenario. On all of the existing benchmarks, we have exceeded the state-of-the-art results, and demonstrated computation efficiency over competitive methods. In future work, we will expand on incorporating more effective attention mechanisms for interest point matching, and more anchor supporting view selection. Jointly learning depth and the full scene holistically using truncated signed distance function (TSDF) or similar representations is another promising direction. Video depth estimation approaches such as [29] are closely related to MVS, and our approach can be readily extended to predict consistent and efficient depth for videos. Finally, we look forward to deeper integration with the SLAM problem, as depth estimation and SLAM are duals of each other. Overall, we believe that our approach of coupling geometry with the power of conventional 2D CNNs is a promising direction for learning 3D Vision.

References

1. Badrinarayanan, V., Kendall, A., Cipolla, R.: Segnet: a deep convolutional encoder-decoder architecture for image segmentation (2015)
2. Bertinetto, L., Valmadre, J., Henriques, J.F., Vedaldi, A., Torr, P.H.S.: Fully-convolutional siamese networks for object tracking. In: Hua, G., Jégou, H. (eds.) ECCV 2016. LNCS, vol. 9914, pp. 850–865. Springer, Cham (2016). https://doi.org/10.1007/978-3-319-48881-3_56
3. Chang, J.R., Chen, Y.S.: Pyramid stereo matching network. In: Proceedings of the IEEE Conference on Computer Vision and Pattern Recognition, pp. 5410–5418 (2018)
4. Chen, L.C., Papandreou, G., Schroff, F., Adam, H.: Rethinking atrous convolution for semantic image segmentation. arXiv preprint arXiv:1706.05587 (2017)
5. Chen, R., Han, S., Xu, J., Su, H.: Point-based multi-view stereo network. In: Proceedings of the IEEE International Conference on Computer Vision, pp. 1538–1547 (2019)
6. Chen, Z., Badrinarayanan, V., Drozdov, G., Rabinovich, A.: Estimating depth from RGB and sparse sensing. In: Proceedings of the European Conference on Computer Vision (ECCV), pp. 167–182 (2018)
7. Cheng, X., Wang, P., Yang, R.: Depth estimation via affinity learned with convolutional spatial propagation network. In: Proceedings of the European Conference on Computer Vision (ECCV), pp. 103–119 (2018)
8. Dai, A., Chang, A.X., Savva, M., Halber, M., Funkhouser, T., Nießner, M.: Scannet: richly-annotated 3D reconstructions of indoor scenes. In: Proceedings Computer Vision and Pattern Recognition (CVPR). IEEE (2017)
9. DeTone, D., Malisiewicz, T., Rabinovich, A.: Superpoint: self-supervised interest point detection and description. In: 2018 IEEE/CVF Conference on Computer Vision and Pattern Recognition Workshops (CVPRW), pp. 337–33712, June 2018. https://doi.org/10.1109/CVPRW.2018.00060
10. Eigen, D., Fergus, R.: Predicting depth, surface normals and semantic labels with a common multi-scale convolutional architecture. In: Proceedings of the IEEE International Conference on Computer Vision, pp. 2650–2658 (2015)

11. Fu, H., Gong, M., Wang, C., Batmanghelich, K., Tao, D.: Deep ordinal regression network for monocular depth estimation In: Proceedings of the IEEE Conference on Computer Vision and Pattern Recognition, pp. 2002–2011 (2018)

12. Garg, R., B.G., V.K., Carneiro, G., Reid, I.: Unsupervised CNN for single view depth estimation: geometry to the rescue. In: Leibe, B., Matas, J., Sebe, N., Welling, M. (eds.) ECCV 2016. LNCS, vol. 9912, pp. 740–756. Springer, Cham (2016). https://doi.org/10.1007/978-3-319-46484-8_45

13. Godard, C., Mac Aodha, O., Brostow, G.J.: Unsupervised monocular depth estimation with left-right consistency. In: Proceedings of the IEEE Conference on Computer Vision and Pattern Recognition, pp. 270–279 (2017)

14. Hartley, R., Zisserman, A.: Multiple View Geometry in Computer Vision. Cambridge University Press, Cambridge (2003)

15. He, K., Zhang, X., Ren, S., Sun, J.: Spatial pyramid pooling in deep convolutional networks for visual recognition. IEEE Trans. Pattern Anal. Mach. Intell. **37**(9), 1904–1916 (2015)

16. He, K., Zhang, X., Ren, S., Sun, J.: Deep residual learning for image recognition. In: Proceedings of the IEEE Conference on Computer Vision and Pattern Recognition, pp. 770–778 (2016)

17. Hou, Y., Kannala, J., Solin, A.: Multi-view stereo by temporal nonparametric fusion. In: Proceedings of the IEEE International Conference on Computer Vision, pp. 2651–2660 (2019)

18. Huang, P.H., Matzen, K., Kopf, J., Ahuja, N., Huang, J.B.: Deepmvs: learning multi-view stereopsis. In: Proceedings of the IEEE Conference on Computer Vision and Pattern Recognition, pp. 2821–2830 (2018)

19. Im, S., Jeon, H.G., Lin, S., Kweon, I.S.: DPSNET: End-to-end deep plane sweep stereo. In: 7th International Conference on Learning Representations, ICLR 2019. International Conference on Learning Representations, ICLR (2019)

20. Ioffe, S., Szegedy, C.: Batch normalization: accelerating deep network training by reducing internal covariate shift. arXiv preprint arXiv:1502.03167 (2015)

21. Iskakov, K., Burkov, E., Lempitsky, V., Malkov, Y.: Learnable triangulation of human pose. In: Proceedings of the IEEE International Conference on Computer Vision, pp. 7718–7727 (2019)

22. Kendall, A., Gal, Y., Cipolla, R.: Multi-task learning using uncertainty to weigh losses for scene geometry and semantics. In: Proceedings of the IEEE Conference on Computer Vision and Pattern Recognition, pp. 7482–7491 (2018)

23. Kendall, A., Gal, Y., Cipolla, R.: Multi-task learning using uncertainty to weigh losses for scene geometry and semantics. In: Proceedings of the IEEE Conference on Computer Vision and Pattern Recognition, pp. 7482–7491 (2018)

24. Lasinger, K., Ranftl, R., Schindler, K., Koltun, V.: Towards robust monocular depth estimation: mixing datasets for zero-shot cross-dataset transfer. arXiv preprint arXiv:1907.01341 (2019)

25. Lee, C.Y., Xie, S., Gallagher, P., Zhang, Z., Tu, Z.: Deeply-supervised nets. In: Artificial Intelligence and Statistics, pp. 562–570 (2015)

26. Lee, J.H., Han, M.K., Ko, D.W., Suh, I.H.: From big to small: multi-scale local planar guidance for monocular depth estimation. arXiv preprint arXiv:1907.10326 (2019)

27. Liu, Y., Shen, Z., Lin, Z., Peng, S., Bao, H., Zhou, X.: Gift: learning transformation-invariant dense visual descriptors via group CNNs. In: Advances in Neural Information Processing Systems, pp. 6990–7001 (2019)

28. Lowe, D.G.: Distinctive image features from scale-invariant keypoints. Int. J. Comput. Vision **60**(2), 91–110 (2004). https://doi.org/10.1023/B:VISI.0000029664.99615.94

29. Luo, X., Huang, J., Szeliski, R., Matzen, K., Kopf, J.: Consistent video depth estimation, vol. 39, p. 4 (2020)

30. Ma, F., Cavalheiro, G.V., Karaman, S.: Self-supervised sparse-to-dense: Self-supervised depth completion from lidar and monocular camera. In: 2019 International Conference on Robotics and Automation (ICRA), pp. 3288–3295. IEEE (2019)

31. Ma, F., Karaman, S.: Sparse-to-dense: Depth prediction from sparse depth samples and a single image (2018)

32. Mur-Artal, R., Montiel, J.M.M., Tardos, J.D.: ORB-SLAM: a versatile and accurate monocular SLAM system. IEEE Trans. Rob. **31**(5), 1147–1163 (2015)

33. Murthy Jatavallabhula, K., Iyer, G., Paull, L.: gradSLAM: dense SLAM meets automatic differentiation. arXiv preprint arXiv:1910.10672 (2019)

34. Nistér, D., Naroditsky, O., Bergen, J.: Visual odometry. In: Proceedings of the 2004 IEEE Computer Society Conference on Computer Vision and Pattern Recognition, 2004. CVPR 2004, vol. 1, p. I. IEEE (2004)

35. Qi, C.R., Su, H., Mo, K., Guibas, L.J.: Pointnet: deep learning on point sets for 3D classification and segmentation. In: Proceedings of the IEEE Conference on Computer Vision and Pattern Recognition, pp. 652–660 (2017)

36. Riegler, G., Osman Ulusoy, A., Geiger, A.: Octnet: learning deep 3D representations at high resolutions. In: Proceedings of the IEEE Conference on Computer Vision and Pattern Recognition, pp. 3577–3586 (2017)

37. Ronneberger, O., Fischer, P., Brox, T.: U-Net: convolutional networks for biomedical image segmentation. In: Navab, N., Hornegger, J., Wells, W.M., Frangi, A.F. (eds.) MICCAI 2015. LNCS, vol. 9351, pp. 234–241. Springer, Cham (2015). https://doi.org/10.1007/978-3-319-24574-4_28

38. Rublee, E., Rabaud, V., Konolige, K., Bradski, G.: ORB: an efficient alternative to SIFT or SURF. In: 2011 International Conference on Computer Vision, pp. 2564–2571. IEEE (2011)

39. Sarlin, P.E., DeTone, D., Malisiewicz, T., Rabinovich, A.: Superglue: learning feature matching with graph neural networks. arXiv preprint arXiv:1911.11763 (2019)

40. Sinha, A., Unmesh, A., Huang, Q., Ramani, K.: Surfnet: generating 3D shape surfaces using deep residual networks. In: Proceedings of the IEEE Conference on Computer Vision and Pattern Recognition, pp. 6040–6049 (2017)

41. Tan, M., Le, Q.V.: Efficientnet: rethinking model scaling for convolutional neural networks. arXiv preprint arXiv:1905.11946 (2019)

42. Tan, M., Pang, R., Le, Q.V.: Efficientdet: scalable and efficient object detection. arXiv preprint arXiv:1911.09070 (2019)

43. Ummenhofer, B., et al.: Demon: depth and motion network for learning monocular stereo. In: Proceedings of the IEEE Conference on Computer Vision and Pattern Recognition, pp. 5038–5047 (2017)

44. Wang, K., Shen, S.: Mvdepthnet: real-time multiview depth estimation neural network. In: 2018 International Conference on 3D Vision (3DV), pp. 248–257. IEEE (2018)

45. Yao, Y., Luo, Z., Li, S., Fang, T., Quan, L.: Mvsnet: depth inference for unstructured multi-view stereo. In: Proceedings of the European Conference on Computer Vision (ECCV), pp. 767–783 (2018)

46. Yi, K.M., Trulls, E., Lepetit, V., Fua, P.: LIFT: learned invariant feature transform. In: Leibe, B., Matas, J., Sebe, N., Welling, M. (eds.) ECCV 2016. LNCS, vol. 9910, pp. 467–483. Springer, Cham (2016). https://doi.org/10.1007/978-3-319-46466-4_28
47. Zhang, Y., Funkhouser, T.: Deep depth completion of a single RGB-D image. In: Proceedings of the IEEE Conference on Computer Vision and Pattern Recognition, pp. 175–185 (2018)
48. Zhang, Z.: Microsoft kinect sensor and its effect. IEEE Multimedia 19(2), 4–10 (2012)

Dynamic Low-Light Imaging with Quanta Image Sensors

Yiheng Chi[1(✉)], Abhiram Gnanasambandam[1], Vladlen Koltun[2],
and Stanley H. Chan[1]

[1] School of ECE, Purdue University, West Lafayette, IN 47907, USA
chi14@purdue.edu
[2] Intel Labs, Santa Clara, CA 95054, USA

Abstract. Imaging in low light is difficult because the number of photons arriving at the sensor is low. Imaging dynamic scenes in low-light environments is even more difficult because as the scene moves, pixels in adjacent frames need to be aligned before they can be denoised. Conventional CMOS image sensors (CIS) are at a particular disadvantage in dynamic low-light settings because the exposure cannot be too short lest the read noise overwhelms the signal. We propose a solution using Quanta Image Sensors (QIS) and present a new image reconstruction algorithm. QIS are single-photon image sensors with photon counting capabilities. Studies over the past decade have confirmed the effectiveness of QIS for low-light imaging but reconstruction algorithms for dynamic scenes in low light remain an open problem. We fill the gap by proposing a student-teacher training protocol that transfers knowledge from a motion teacher and a denoising teacher to a student network. We show that dynamic scenes can be reconstructed from a burst of frames at a photon level of 1 photon per pixel per frame. Experimental results confirm the advantages of the proposed method compared to existing methods.

Keywords: Quanta Image Sensors · Low light · Burst photography

1 Introduction

Imaging in photon-starved situations is one of the biggest technological challenges for applications such as security, robotics, autonomous cars, and health care. However, the growing demand for higher resolution, smaller pixels, and smaller form factors have limited the photon sensing area of the sensors. This, in turn, puts a fundamental limit on the signal-to-noise ratio that the sensors can achieve. Over the past few years, there is an increasing amount of effort in developing alternative sensors that have photon-counting ability. Quanta Image Sensors (QIS) are one of these new types of image sensors that can count individual photons at a very high frame rate and have a high spatial resolution [44,45]. Various prototype QIS have been reported, and numerous studies have confirmed their capability for high speed imaging [5], high dynamic range imaging [19,28], color imaging [20,27], and tracking [33].

© Springer Nature Switzerland AG 2020
A. Vedaldi et al. (Eds.): ECCV 2020, LNCS 12366, pp. 122–138, 2020.
https://doi.org/10.1007/978-3-030-58589-1_8

 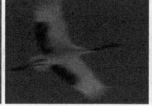 

(a) Real image by CIS,   (b) Real image by QIS,   (c) Our reconstruction
avg of 8 frames, 0.5 ppp   avg of 8 frames, 0.5 ppp   using 8 QIS frames

Fig. 1. Goal of this paper. The images above are the *real* captures by a CMOS
Image Sensor (CIS) and a QIS prototype [27] at the same photon level of 0.5 photons
per pixel (ppp) per frame. Our goal is to reconstruct images with dynamic content
from a burst of QIS frames.

Despite the increasing literature on QIS sensor development [24,44,45] and
signal processing algorithms [10,61], one of the most difficult problems in QIS
is image reconstruction for *dynamic* scenes. Image reconstruction for dynamic
scenes is important for broad adoption of QIS: solving the problem can open
the door to a wide range of low-light applications such as videography, moving
object detection, non-stationary facial recognition, etc. However, motion in low
light is difficult because it must deal with two types of distortions: low light
causes shot noise which is random and affects the entire image, whereas motion
causes geometric warping which is often local. In this paper, we address this
problem with a new algorithm.

Figure 1 summarizes our objective. Figure 1(a) shows real data captured by
a conventional CMOS image sensor (CIS). The photon level is 0.5 photons per
pixel (ppp). Figure 1(b) shows the data captured by a QIS at the same photon
level. To illustrate the effect of motion, we show the average of 8 consecutive
frames. Figure 1(c) shows the result of the proposed image reconstruction algorithm applied to the 8 QIS frames. Although the scene is in motion, the presented
approach recovers most of the image details. This brings out the two contributions of this paper:

(i) We demonstrate low-light image reconstruction of dynamic scenes at a photon level of 1 photon per pixel (ppp) per frame. This is lower than most of
the results reported in the computational photography literature.
(ii) We propose a student-teacher framework and show that this training
method is effective in handling noise and motion simultaneously.

2 Background

2.1 Quanta Image Sensors

Quanta Image Sensors (QIS) were originally proposed in 2005 as a candidate solution for the shrinking pixel problem [22,23]. The idea is to partition a CIS pixel

into many tiny cells called "jots" where each jot is a single-photon detector. By oversampling the scene in space and time, the underlying image can be recovered using a carefully designed image reconstruction algorithm. Numerous studies have analyzed the theoretical properties of these sensors, including their performance limit [62], photon statistics [24], threshold analysis [19], dynamic range [28], and color filter array [20]. On the hardware side, a number of prototypes have become available [17,18,44]. The prototype QIS we use in this paper is based on [45].

As photon counting devices, QIS share many similarities with single-photon avalanche diodes (SPAD) [17]. However, SPAD amplify signals using avalanche multiplication. This requires a high electrical voltage (typically higher than 20V) to accelerate the photoelectron. Because avalanche multiplication requires space for electrons to multiply, SPAD have high dark current ($> 10e^-/\mathrm{pix/s}$), large pitch ($> 5\mu m$), low fill-factor ($< 70\%$), and low quantum efficiency ($< 50\%$). In contrast, QIS do not require avalanche multiplication. They have significantly better fill-factor, quantum efficiency, dark current, and read noise. SPAD are excellent candidates for resolving time-stamps, e.g., time-of-flight applications [6,26,32,40,52], although new studies have shown other applications [46]. QIS have higher resolution which makes them suitable for low-light photography. Recent literature provides a more detailed comparison [27].

2.2 How Dark Is One Photon per Pixel?

All photon levels in this paper are measured in terms of photons per pixel (ppp). "Photons per pixel" is the average number of photons a pixel detects during the exposure period. We use photons per pixel as the metric because the amount of photons detected by a sensor depends on the exposure time and sensor size. A large sensor can collect more photons, and longer exposure time would allow more photons to arrive at the sensor. Therefore, even for the same scene with the same illuminance (measured in lux), the number of photons per pixel seen by two sensors can be different. To give readers an idea of the amount of noise we are dealing with in this paper, Fig. 2(a,b) shows a pair of real images captured by CIS and QIS at 0.25 ppp. Note that the signal at this photon level is significantly worse than what is commonly considered "heavy noise" in the denoising literature, illustrated in Fig. 2(d). We should also highlight that while QIS is a better sensor, at low light the signal-to-noise ratio is upper bounded by the fundamental limit of the Poisson process. As shown in Fig. 2(c), an ideal sensor with zero read noise and zero dark current will still produce an image contaminated by shot noise. Therefore, reconstruction algorithms are needed to recover the images even though QIS have higher photon sensitivity than CIS.

2.3 Related Work

QIS Image Reconstruction. Image reconstruction for QIS is challenging because of the unique Poisson-Gaussian statistics of the sensor. Early reconstruction techniques are based on solving maximum-likelihoods using gradient descent [61], dynamic programming [63], and convex optimization techniques

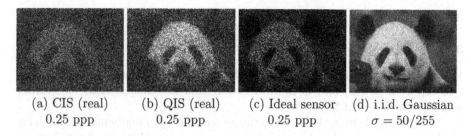

(a) CIS (real)   (b) QIS (real)   (c) Ideal sensor   (d) i.i.d. Gaussian
    0.25 ppp   0.25 ppp   0.25 ppp $\sigma = 50/255$

Fig. 2. Photon level and sensor limitations. (a) and (b) show a pair of real images captured by CIS and QIS at 0.25 ppp. (c) shows a simulated image acquired by an "ideal sensor" which is free of read noise and dark current. The random shot noise in this ideal image suggests that although QIS has higher sensitivity than CIS, image reconstruction algorithms still play a critical role because there is a fundamental limit due to the Poisson statistics. (d) shows an image distorted by i.i.d. Gaussian noise of a strength $\sigma = 50/255$, considered high in the denoising literature.

[8,9]. The first non-iterative algorithm for QIS image reconstruction was proposed by Chan et al. [7]. It was shown that if one assumes spatial independence, then the truncated Poisson likelihood can be simplified to Binomial. Consequently, the Anscombe binomial transform can be used to stabilize the variance, and off-the-shelf denoising (e.g., BM3D [15]) can be used to denoise the image. Choi et al. [13] followed the idea by replacing the denoiser with a deep neural network. Alternative solutions using end-to-end deep neural networks have also been proposed for QIS [55] and SPAD [10]. To the best of our knowledge, ours is the first dynamic scene reconstruction for QIS.

Low-light Denoising. The majority of existing denoising algorithms are designed for CIS. Single-frame image denoising methods are abundant, e.g., non-local means [4], BM3D [15], Poisson denoising [48], and many others [25,31,36,49]. On the deep neural network side, there are numerous networks dedicated to single-image denoising [43,56,64,65]. However, recent benchmark experiments found that BM3D is often better than deep learning methods for real sensor data [53,60]. Specific to low-light imaging, Chen et al. [11,12] observed that by modeling the entire image and signal processing pipeline using an end-to-end network, better reconstruction results can be obtained from the raw sensor data. However, since the images are still captured by CIS, the photon levels are much higher than what we study in this paper.

For dynamic scenes, extensions of the static methods to videos are available, e.g., based on non-local means [3,16,57], optical flow [41,42,58], and sparse representation [37,54]. The most relevant approach for this paper is the burst photography technique [34], which can be traced back to earlier methods based on optical flow [38,41,42]. Recent reports on burst photography have focused on using deep neural networks [2,29,39,59]. Among these, the kernel prediction network (KPN) by Mildenhall et al. [51] is the most relevant work for us. However, as we will demonstrate later in the paper, the performance of KPN is not as satisfactory in the extreme noise conditions we deal with.

3   Method

3.1   QIS Imaging Model

We first present the image formation model. Our model is based on the prototype QIS reported in [45] and is more detailed than the models used in existing literature such as [7,62].

As light travels from the scene to the sensor, the main mathematical model is the Poisson process which describes how photons arrive. However, due to various sources of distortions, the measured QIS signal, x_{QIS}, is given by

$$\underbrace{x_{\mathrm{QIS}}}_{\text{observed}} = \mathrm{ADC}\bigg\{ \underbrace{\mathrm{Poisson}}_{\text{photon arrival}} \bigg(\underbrace{\alpha}_{\text{sensor gain}} \cdot \big(\underbrace{x_{\mathrm{true}}}_{\text{scene}} + \underbrace{\eta_{\mathrm{dc}}}_{\text{dark current}} \big) \bigg) + \underbrace{\eta_{\mathrm{r}}}_{\text{read noise}} \bigg\}. \quad (1)$$

Here we assume that the sensor is monochromatic because the real data reported in this paper are based on a monochromatic prototype QIS. To simulate color data we need to include a sub-sampling step to model the color filter array. η_{dc} denotes the dark current and η_{r} denotes the read noise arising from the read-out circuit. The analog-to-digital converter (ADC) describes the sensor output. In single-bit QIS, the output is a binary signal obtained by thresholding the Poisson count [19]. In multi-bit QIS, the output is the Poisson count clipped to the maximum number of bits. To image a dynamic scene, we use QIS to collect a stack of short-exposure frames. Akin to previous work [7,62], we assume that noise is independent over time.

For the prototype sensor we use in this paper, the dark current η_{dc} in Eq. (1) has an average value of $0.0068e^-$/pix/s and the read noise η_{r} takes the value of $0.25e^-$/pix [45]. The sensor gain α controls the exposure time and the dynamic range, which changes from scene to scene. For all experiments we conduct in this paper, the analog-to-digital conversion is 3-bit. The spatial resolution of the sensor is 1024×1024, although we typically crop regions of the image for analysis.

3.2   The Dilemma of Noise and Motion

At the heart of dynamic image reconstruction is the coexistence of noise and motion. The dilemma here is that they are intertwined. To remove noise in a dynamic scene, we often need to either align the frames or construct a steerable kernel over the space-time volume. The alignment step is roughly equivalent to estimating optical flow [35], whereas constructing the steerable kernel is equivalent to non-local means [3,57] or kernel prediction [51]. However, if the images are contaminated by noise, then both optical flow and kernel prediction will fail. When this step fails, denoising will be difficult because we will not be able to easily find neighboring patches for filtering.

Existing algorithms in the denoising literature can usually only handle one of the two situations. For example, the kernel prediction network (KPN) [51] can extract motion information from a dynamic scene but its performance drops when noise becomes heavy. Similarly, the residual encoder-decoder networks

(a) QIS raw data   (b) KPN [51]   (c) sRED [50]   (d) Ours
8-frame avg   23.09 dB   17.74 dB   26.74 dB

Fig. 3. The dilemma of noise and motion. (a) A simulated QIS sequence at 2 ppp, averaged over 8 frames. (b) Result of Kernel Prediction Network (KPN) [51], a burst photography method that handles motion. (c) Result of a single-frame image denoiser sRED [50] applied to the 8-frame avg. (d) Result of our proposed method.

REDNet [50] and DnCNN [64] are designed for static scenes. In Fig. 3, we show the results of a synthetic experiment. The results illustrate the limitations of the motion-based KPN [51] and the single-frame REDNet (sRED) [50]. Our goal is to leverage the strengths of both.

3.3 Student-Teacher Learning

If a kernel prediction network can handle clean image sequences well and a denoising network can handle static image sequences well, is there a way we can leverage their strengths to address the dynamic low-light setting? Our solution is to develop a training scheme using the concept of student-teacher learning.

Figure 4 describes our method. There are three players in this training protocol: a teacher for motion (based on kernel prediction), a teacher for denoising (based on image denoiser networks), and a student which is the network we are going to use eventually. The two teachers are individually pretrained using their respective imaging conditions. For example, the motion teacher is trained using sequences of clean and dynamic contents, whereas the denoising teacher is trained using sequences of noisy but static contents. During the training step, the teachers will transfer their knowledge to the student. During testing, only the student is used.

To transfer knowledge from the two teachers to the student, the student is first designed to have two branches, one branch duplicating the architecture of the motion teacher and another branch duplicating the architecture of the denoising teacher. When training the student, we generate three versions of the training samples. The motion teacher sees training samples that are clean and only contain motion, x_{motion}. The denoising teacher sees a training sample containing no motion but corrupted by noise, x_{noise}. The student sees the noisy dynamic sequence x_{QIS}.

Because the student has identical branches to the teachers, we can compare the features extracted by the teachers and the student. Specifically, if we denote

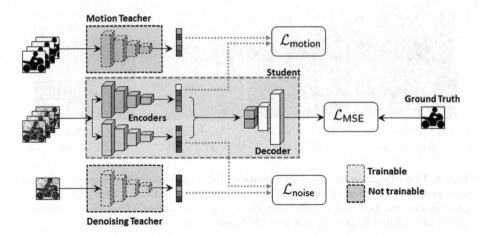

Fig. 4. Overview of the proposed method. The proposed student-teacher setup consists of two teachers and a student. The motion teacher shares motion features, whereas the denoising teacher shares denoising features. To compare the respective feature differences, perceptual losses $\mathcal{L}_{\text{noise}}$ and $\mathcal{L}_{\text{motion}}$ are defined. The student network has two encoders and one decoder. The final estimates are compared with the ground truth using the MSE loss $\mathcal{L}_{\text{MSE}}$.

$\phi(\cdot)$ as the feature extraction performed by the motion teacher, $\widehat{\phi}(\cdot)$ the student motion branch, $\varphi(\cdot)$ the denoising teacher, and $\widehat{\varphi}(\cdot)$ the student denoising branch, then we can define a pair of *perceptual similarities*: the motion similarity

$$\mathcal{L}_{\text{motion}} = \|\ \underbrace{\widehat{\phi}(\boldsymbol{x}_{\text{QIS}})}_{\text{motion student}} - \underbrace{\phi(\boldsymbol{x}_{\text{motion}})}_{\text{motion teacher}}\ \|^2 \tag{2}$$

and the denoising similarity

$$\mathcal{L}_{\text{noise}} = \|\ \underbrace{\widehat{\varphi}(\boldsymbol{x}_{\text{QIS}})}_{\text{denoising student}} - \underbrace{\varphi(\boldsymbol{x}_{\text{noise}})}_{\text{denoising teacher}}\ \|^2. \tag{3}$$

Intuitively, what this pair of equations does is ensure that the features extracted by the student branches are similar to those extracted by the respective teachers, which are features that can be extracted in good conditions. If this can be achieved, then we will have a good representation of the noisy dynamic sample and hence we can do a better reconstruction.

The two student branches can be considered as two autoencoders which convert the input images to codewords. As shown on the right side of Fig. 4, we have a "decoder" which translates the concatenated codewords back to an image. The loss function of the decoder is given by the standard mean squared error (MSE) loss:

$$\mathcal{L}_{\text{MSE}} = \|f(\boldsymbol{x}_{\text{QIS}}) - \boldsymbol{x}_{\text{true}}\|^2, \tag{4}$$

where f is the student network and so $f(x_{QIS})$ denotes the estimated image. The overall loss function is the sum of these losses:

$$\mathcal{L}_{\text{overall}} = \mathcal{L}_{\text{MSE}} + \lambda_1 \mathcal{L}_{\text{motion}} + \lambda_2 \mathcal{L}_{\text{noise}}, \tag{5}$$

where λ_1 and λ_2 are tunable parameters. Training the network is equivalent to finding the encoders $\widehat{\phi}$ and $\widehat{\varphi}$, and the decoder f.

3.4   Choice of Teacher and Student Networks

The proposed student-teacher framework is quite general. Specific to this paper, the two teachers and the student are chosen as follows.

The motion teacher is the kernel prediction network (KPN) [51]. We modify it by removing the skip connections to maintain the information kept by the encoder. In addition, we remove the pooling layers and the bilinear upsampling layers to maximize the amount of information being fed to the feature layer. With these changes, the KPN becomes a fully convolutional-deconvolutional network.

The denoising teacher we use is a modified version of REDNet [50], which is also used in another QIS reconstruction method [13]. To differentiate this single-frame REDNet and another modified version (to be discussed in the experiment section), we refer to this single-frame REDNet denoising teacher as sRED. Like the motion teacher, we remove the residual connections since they have a negative impact on the feature transfer in student-teacher learning.

The student network has two encoders and a decoder. The encoders have exactly the same architectures as the teachers. The decoder is a stack of 15 layers where each layer is a 128-channel up-convolution. The entrance layer is used to concatenate the motion and denoising features.

4   Experiments

4.1   Setting

Training Data. The training data consists of two parts. The first part is for *global motion*. We use the Pascal VOC 2008 dataset [21] which contains 2000 training images. The second part is for *local motion*. We use the Stanford Background Dataset [30] which contains 715 images with segmentation. For both datasets, we randomly crop patches of size 64×64 from the images to serve as ground truth. An additional 500 images are used for validation. To create global motion, we shift the patches according to a random continuous camera motion where the number of pixels traveled by the camera range from 7 to 35 across 8 consecutive frames. This is approximately 1 m/s. For local motion, we fix the background and shift the foreground using translations and rotations. The implementation of the translation is the same as that of the global motion but applied to foreground objects. The rotation is implemented by rotating the object with angle ranging from 0 to $15°$.

Training the Teachers. The motion teacher is trained using a set of noise-free and dynamic sequences. The loss function is the mean squared error (MSE) loss suggested by [51]. The network is trained for 200 epochs using the dataset described above. The denoising teacher is trained using a set of noisy but static images. Therefore, for every ground-truth sequence we generate a triplet of sequences: A noise-free dynamic sequence for the motion teacher, a noisy static image for the denoising teacher, and a noisy dynamic sequence for the student. We remark that such a data synthesis approach works for our problem because the simulated QIS data matches the statistics of real measurements.

Baselines. We compare the proposed methods with three existing dynamic scene reconstruction methods: (i) BM4D [47], (ii) Kernel Prediction Network (KPN) [51], and (iii) a modified version of REDNet [50]. Our modification generalizes REDNet to multi-frame inputs, by introducing a 3D convolution at the input layer to pool the features. We refer to the modified version as multi-frame RED (mRED). Note that mRED has residual connections while sRED (denoising teacher) does not. We consider mRED a more fair baseline since it takes an input of 8 consecutive frames rather than a single frame. For KPN, the original method [51] suggested using a fixed kernel size of $K = 5$; we modify the setting by defining K as the maximum number of pixels traveled by the motion.

Implementation. All networks are implemented using Keras [14] and Tensor-Flow [1]. The student-teacher training is done using a semi-annealing process. Specifically, the regularization parameters λ_1 and λ_2 are updated once every 25 epochs such that λ_1 and λ_2 decay exponentially for the first 100 epochs. For the next 100 epochs, λ_1 and λ_2 are set to 0 and the overall loss function becomes $\mathcal{L}_{\text{overall}} = \mathcal{L}_{\text{MSE}}$.

4.2   Synthetic Experiments

We begin by conducting synthetic experiments. We first visually compare the reconstructed images of the proposed method and the competing methods. Figure 5 shows some results using global translation. The motion magnitude is 28 pixels across 8 frames, at 2 ppp. Figure 6 shows some results using arbitrary global motion, at 4 ppp. The motion trajectory is shown in the inset in the figure. Figure 7 shows some results of local motion. We simulate QIS data with a real motion video of 30 fps. The photon level is 1.5 ppp. The average inference time of KPN on a 512×512 patch is 0.0886 s using an NVIDIA GeForce RTX 2080 Ti graphics card. For the same testing setting, mRED takes 0.0653 s, and the proposed method takes 0.1943 s. The average time for BM4D (MATLAB version) is 23.6985 s.

To quantitatively analyze the performance, we use the linear global motion to plot two sets of curves as shown in Fig. 8. In the first plot, we show PSNR as a function of the motion magnitude. The magnitude of the motion is defined as the number of pixels traveled along the dominant direction, over 8 consecutive frames. As shown in Fig. 8(a), the proposed method has a consistently higher PSNR compared to the three competing methods, ranging from 1.5 dB to 3 dB. This suggests that the presence of both teachers has provided a positive impact

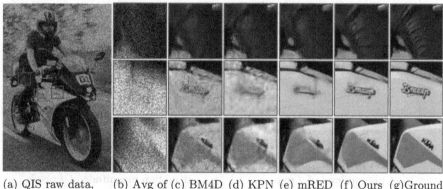

(a) QIS raw data,   (b) Avg of (c) BM4D (d) KPN (e) mRED (f) Ours (g)Ground
1 frame   8 frames 23.04 dB 25.45 dB 26.42 dB 29.39 dB   Truth

Fig. 5. Simulated QIS data with linear global motion. (a) The raw QIS image is simulated at 2 ppp, with a global motion of 28 pixels uniformly spaced across 8 frames. (b) An average 8 QIS raw frames. (c) BM4D [47]. (d) KPN [51]. (e) mRED, a modification of REDNet [50]. (f) Proposed method. (g) Ground truth.

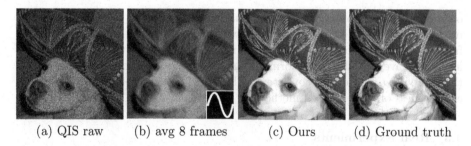

(a) QIS raw   (b) avg 8 frames   (c) Ours   (d) Ground truth

Fig. 6. Simulated QIS data with arbitrary global motion. (a) QIS raw data simulated at 4 ppp. The motion trajectory is shown in the inset. (b) Average of 8 frames. (c) Proposed method. (d) Ground truth.

(a) QIS raw   (b) avg 8 frames   (c) Ours   (d) Ground truth

Fig. 7. Simulated QIS data with local motion. In this example, only the car moves. The background is static. (a) Raw QIS frame assuming 1.5 ppp. (b) The average of 8 QIS frames. (c) Proposed method. (d) Ground truth.

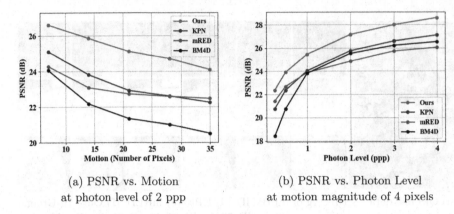

(a) PSNR vs. Motion
at photon level of 2 ppp

(b) PSNR vs. Photon Level
at motion magnitude of 4 pixels

Fig. 8. Quantitative analysis using synthetic data. (a) PSNR as a function of the motion magnitude, at a photon level of 2 ppp. The magnitude of the motion is defined as the number of pixels traveled along the dominant direction, over 8 consecutive frames. (b) PSNR as a function of photon level. The motion magnitude is fixed at 4 pixels, but the photon level changes. Our method consistently outperforms BM4D [47], KPN [51], and mRED (a modified version of [50]).

on solving the motion and noise dilemma, which is difficult for both KPN and mRED. The second set of curves is shown in Fig. 8(b) and reports PSNR as a function of the photon level. The curves in Fig. 8(b) suggest that for the photon levels we have tested, the performance gap between the proposed method and the competing methods is consistent. This provides additional evidence of the effectiveness of the proposed method.

4.3   Real Experiments

We verify the results using real QIS data. The real data is collected using a prototype Gigajot PathFinder camera [45]. The camera has a spatial resolution of 1024×1024. The integration time of each frame is 75 μs. Each reconstruction is based on 8 consecutive QIS frames. At the time this experiment is conducted, the readout circuit of this camera is still a prototype that is not optimized for speed. Thus, instead of demonstrating a real high-speed video, we capture a slowly moving real dynamic scene where the motion is continuous but slow. We make the exposure period short so that it is equivalent to a high-speed video. We expect that the problem will be solved in the next generation of QIS.

The physical setup of the experiment is shown in Fig. 9(a). We put the camera approximately 1 m away from the objects. The photon level is controlled by a light source. To create motion, the objects are mounted on an Ashanks SmoothONE C300S motorized camera slider, which allows us to control the location of the objects remotely. The "ground truth" (reference images) in this experiment is obtained by capturing a static scene via 8 consecutive QIS frames. Since these static images are noisy (due to photon shot noise), we apply mRED to denoise the images before using them as the references.

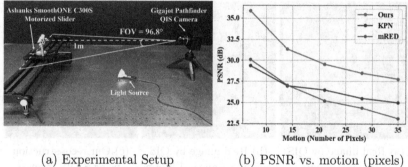

(a) Experimental Setup   (b) PSNR vs. motion (pixels)

Fig. 9. (a) Setup of QIS data collection. The QIS camera is placed 1 m from the object which is attached to a motorized slider. The horizontal field of view (FOV) of the lens is 96.8°. The motion is continuous but slow. **(b) Quantitative analysis on real data.** The plot shows the PSNR values as a function of the motion magnitude, under a photon level of 0.5 ppp. The "reference" in this experiment is determined by reconstructing an image using a stack of static frames of the same scene. The reconstruction method is based on [13].

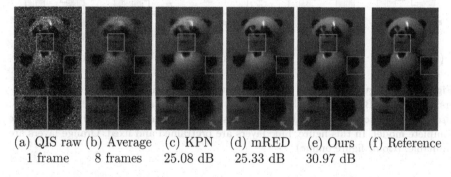

(a) QIS raw (b) Average   (c) KPN   (d) mRED   (e) Ours   (f) Reference
  1 frame   8 frames   25.08 dB   25.33 dB   30.97 dB

Fig. 10. Real QIS data. (a) A snapshot of a real QIS frame captured at 2 ppp per frame. The number of pixels traveled by the object over the 8 frames is 28 pixels. (b) The average of 8 QIS frames. Note the blur in the image. (c) Reconstruction result of KPN [51]. (d) Reconstruction result of mRED, a modification of [50]. (e) Our proposed method. (f) Reference image is a static scene denoised using mRED.

A visual comparison for this experiment is shown in Fig. 10. The quantitative analysis is shown in Fig. 9(b), where we plot the PSNR curves as functions of the number of pixels traveled by the object. As we can see, the performance of the proposed method and the competing methods are similar to those reported in the synthetic experiments. The gap appears to be consistent with the synthetic experiments. An additional real data experiment is shown in Fig. 11, where we use QIS to capture a rotating fan scene.

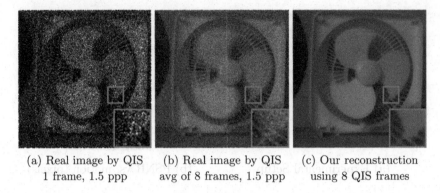

(a) Real image by QIS   (b) Real image by QIS   (c) Our reconstruction
1 frame, 1.5 ppp   avg of 8 frames, 1.5 ppp   using 8 QIS frames

Fig. 11. Real QIS data with rotational motion. The image is captured at 1.5 ppp. Note the motion blur in the 8-frame average.

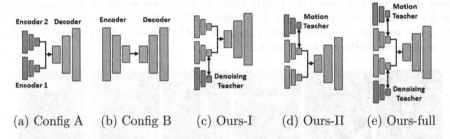

(a) Config A   (b) Config B   (c) Ours-I   (d) Ours-II   (e) Ours-full

Fig. 12. Configurations for ablation study. (a) Config A: Uses pretrained teachers. (b) Config B: Uses a single encoder instead of two smaller encoders. (c) Ours-I: Uses denoising teacher only. (d) Ours-II: Uses motion teacher only. (e) Our-full: The complete model. In this figure, blue layers are pretrained and fixed. Orange layers are trainable. (Color figure online)

4.4   Ablation Study

We conduct an ablation study to evaluate the significance of the proposed student-teacher training protocol. Figure 12 summarizes the 5 configurations we study. Config A is a vanilla baseline where the denoising and motion teachers are pretrained. Config B uses a single encoder instead of two encoders. Ours-I uses a student-teacher setup to train the denoising encoder. Ours-II is similar to Ours-I, but we use the motion teacher in lieu of the denoising teacher. Ours-full uses both teachers. All networks are trained using the same set of noisy and dynamic sequences. The experiments are conducted using synthetic data, at a photon level of 1 ppp and motion of 28 pixels across 8 frames. The results are summarized in Table 1.

Is student-teacher training necessary? Configurations A and B do not use any teacher. Comparing with Ours-full, the PSNR values of Config A and Config B are worse by more than 1 dB. Even if we compare with a single teacher, e.g.,

Table 1. Ablation Study Results. This table summarizes the influence of different teachers on the proposed method. The experiments are conducted using synthetic data, at a photon level of 1 ppp and a motion of 28 pixels along the dominant direction.

Configuration	# of Encoders	Which Teacher?	Test PSNR
A	2	None	21.51 dB
B	1	None	22.74 dB
Ours-I	2	Denoising	23.53 dB
Ours-II	2	Motion	23.65 dB
Ours-full	2	Both	23.87 dB

Ours-I, it is still 0.8 dB ahead of Config B. Therefore, the student-teacher training protocol has a positive impact on performance.

Do teacher encoders extract meaningful information? Config A uses two pretrained encoders and a trainable decoder. The network achieves 21.51 dB, which means that some features are useful for reconstruction. However, when comparing with Ours-full, it is substantially worse (23.87 dB compared to 21.51 dB). Since the network architectures are identical, the performance gap is likely caused by the training protocol. This indicates that the student-teacher setup is a better way to transfer knowledge from teachers to a student network.

Which teacher to use? Configurations Ours-I and Ours-II both use one teacher. The results suggest that if we only use one teacher, the motion teacher has a small gain (0.1 dB) over the denoising teacher. However, if we use both teachers as in the proposed method, we observe another 0.2 dB improvement. Thus, the presence of both teachers is helpful.

5   Conclusion

Dynamic low-light imaging is an important capability in application such as autonomous driving, security, and health care. CMOS image sensors (CIS) have fundamental limitations due to their inability to count photons. This paper considers Quanta Image Sensors (QIS) as an alternative solution. By developing a deep neural network using a new student-teacher training protocol, we demonstrated the effectiveness of transferring knowledge from a motion teacher and a denoising teacher to the student network. Experimental results indicate that the proposed method outperforms existing solutions trained under the same conditions. The proposed student-teacher protocol can also be applied to CIS problems. However, at a photon level of 1 photon per pixel or lower, QIS are necessary. Future work will focus on generalizing the reconstruction to more complex motions.

Acknowledgement. This work is supported in part by the US National Science Foundation under grant CCF-1718007.

References

1. Abadi, M., et al.: TensorFlow: large-scale machine learning on heterogeneous systems (2015). https://www.tensorflow.org/
2. Aittala, M., Durand, F.: Burst image deblurring using permutation invariant convolutional neural networks. In: ECCV (2018)
3. Buades, A., Coll, B., Morel, J.M.: Denoising image sequences does not require motion estimation. In: IEEE Conference Advanced Video and Signal Based Surveillance, pp. 70–74 (2005)
4. Buades, A., Coll, B., Morel, J.M.: A review of image denoising algorithms, with a new one. SIAM Multiscale Modeling Simul. 4(2), 490–530 (2005)
5. Burri, S., Maruyama, Y., Michalet, X., Regazzoni, F., Bruschini, C., Charbon, E.: Architecture and applications of a high resolution gated SPAD image sensor. Optics Express 22(14), 17573–17589 (2014)
6. Callenberg, C., Lyons, A., den Brok, D., Henderson, R., Hullin, M.B., Faccio, D.: EMCCD-SPAD camera data fusion for high spatial resolution time-of-flight imaging. In: Computational Optical Sensing and Imaging. Optical Society of America (2019)
7. Chan, S.H., Elgendy, O.A., Wang, X.: Images from bits: non-iterative imagereconstruction for Quanta Image Sensors. Sensors 16(11) (2016)
8. Chan, S.H., Lu, Y.M.: Efficient image reconstruction for gigapixel Quantum Image Sensors. In: IEEE Global Conference Signal and Information Processing (2014)
9. Chan, S.H., Wang, X., Elgendy, O.A.: Plug-and-play ADMM for image restoration: fixed-point convergence and applications. IEEE Trans. Comput. Imaging 3(1), 84–98 (2017)
10. Chandramouli, P., Burri, S., Bruschini, C., Charbon, E., Kolb, A.: A bit too much?. In: ICCP, High Speed Imaging from Sparse Photon Counts (2019)
11. Chen, C., Chen, Q., Do, M.N., Koltun, V.: Seeing motion in the dark. In: ICCV (2019)
12. Chen, C., Chen, Q., Xu, J., Koltun, V.: Learning to see in the dark. In: CVPR (2018)
13. Choi, J.H., Elgendy, O.A., Chan, S.H.: Image reconstruction for Quanta Image Sensors using deep neural networks. In: ICASSP (2018)
14. Chollet, F., et al.: Keras (2015). https://www.keras.io
15. Dabov, K., Foi, A., Katkovnik, V., Egiazarian, K.: Image denoising by sparse 3-D transform-domain collaborative filtering. IEEE Trans. Image Process. 16(8), 2080–2095 (2007)
16. Davy, A., et al.: A non-local CNN for video denoising. In: ICIP (2019)
17. Dutton, N.A., et al.: A SPAD-based QVGA image sensor for single-photon counting and quanta imaging. IEEE Trans. Electron Devices 63(1), 189–196 (2015)
18. Dutton, N.A., Parmesan, L., Holmes, A.J., Grant, L.A., Henderson, R.K.: 320×240 oversampled digital single photon counting image sensor. In: Symposium on VLSI Circuits Digest of Technical Papers (2014)
19. Elgendy, O.A., Chan, S.H.: Optimal threshold design for Quanta Image Sensor. IEEE Trans. Comput. Imaging 4(1), 99–111 (2017)
20. Elgendy, O.A., Chan, S.H.: Color Filter Arrays for Quanta Image Sensors. arXiv preprint arXiv:1903.09823 (2019)
21. Everingham, M., Van Gool, L., Williams, C.K.I., Winn, J., Zisserman, A.: The pascal visual object classes (VOC) challenge. Int. J. Comput. Vis. 88(2), 303–338 (2010)

22. Fossum, E.R.: Gigapixel digital film Sensor (DFS) proposal. In: Nanospace Manipulation of Photons and Electrons for Nanovision Systems (2005)
23. Fossum, E.R.: Some thoughts on future digital still cameras. In: Image Sensors and Signal Processing for Digital Still Cameras (2006)
24. Fossum, E.R.: Modeling the performance of single-bit and multi-bit quanta image sensors. IEEE J. Electron Devices Soc. 1(9), 166–174 (2013)
25. Fu, Q., Jung, C., Xu, K.: Retinex-based perceptual contrast enhancement in images using luminance adaptation. IEEE Access 6, 61277–61286 (2018)
26. Gariepy, G., et al.: Single-photon sensitive light-in-fight imaging. Nat. Commun. 6(1), 1–7 (2015)
27. Gnanasambandam, A., Elgendy, O., Ma, J., Chan, S.H.: Megapixel photon-counting color imaging using Quanta Image Sensor. Optics Express 27(12), 17298–17310 (2019)
28. Gnanasambandam, A., Ma, J., Chan, S.H.: High dynamic range imaging using Quanta Image Sensors. In: International Image Sensors Workshop (2019)
29. Godard, C., Matzen, K., Uyttendaele, M.: Deep burst denoising. In: ECCV (2018)
30. Gould, S., Fulton, R., Koller, D.: Decomposing a scene into geometric and semantically consistent regions. In: ICCV (2009)
31. Guo, X., Li, Y., Ling, H.: LIME: low-light image enhancement via illumination map estimation. IEEE Trans. Image Process. 26(2), 982–993 (2016)
32. Gupta, A., Ingle, A., Gupta, M.: Asynchronous single-photon 3D imaging. In: ICCV (2019)
33. Gyongy, I., Dutton, N., Henderson, R.: Single-photon tracking for high-speed vision. Sensors 18(2), 323 (2018)
34. Hasinoff, S.W., et al.: Burst photography for high dynamic range and low-light imaging on mobile cameras. ACM Trans. Graph. 35(6) (2016)
35. Horn, B.K., Schunck, B.G.: Determining optical flow. In: Techniques and Applications of Image Understanding, vol. 281. International Society Optics and Photonics (1981)
36. Hu, Z., Cho, S., Wang, J., Yang, M.H.: Deblurring low-light images with light streaks. In: CVPR (2014)
37. Ji, H., Liu, C., Shen, Z., Xu, Y.: Robust video denoising using low rank matrix completion. In: CVPR (2010)
38. Joshi, N., Cohen, M.: Seeing Mt. Rainier: Lucky imaging for multi-image denoising, sharpening, and haze removal. In: ICCP (2010)
39. Kokkinos, F., Lefkimmiatis, S.: Iterative residual CNNs for burst photography applications. In: CVPR (2019)
40. Lindell, D.B., O'Toole, M., Wetzstein, G.: Single-photon 3D imaging with deepsensor fusion. ACM Trans. Graph. 37(4) (2018)
41. Liu, C., Freeman, W.: A high-quality video denoising algorithm based on reliable motion estimation. In: ECCV (2010)
42. Liu, Z., Yuan, L., Tang, X., Uyttendaele, M., Sun, J.: Fast burst images denoising. ACM Trans. Graph., 33(6) (2014)
43. Lore, K.G., Akintayo, A., Sarkar, S.: LLNet: a deep autoencoder approach to natural low-light image enhancement. Pattern Recogn. 61, 650–662 (2017)
44. Ma, J., Fossum, E.: A pump-gate jot device with high conversion gain for a Quanta Image Sensor. IEEE J. Electron Devices Soc. 3(2), 73–77 (2015)
45. Ma, J., Masoodian, S., Starkey, D., Fossum, E.R.: Photon-number-resolving megapixel image sensor at room temperature without avalanche gain. Optica 4(12), 1474–1481 (2017)

46. Ma, S., Gupta, S., Ulku, A.C., Brushini, C., Charbon, E., Gupta, M.: Quanta burst photography. ACM Trans. Graph. (TOG), **39**(4) (2020)
47. Maggioni, M., Katkovnik, V., Egiazarian, K., Foi, A.: Nonlocal transform-domain filter for volumetric data denoising and reconstruction. IEEE Trans. Image Process. **22**(1), 119–133 (2012)
48. Makitalo, M., Foi, A.: Optimal inversion of the Anscombe transformation in low-count Poisson image denoising. IEEE Trans. Image Process. **20**(1), 99–109 (2010)
49. Malm, H., Oskarsson, M., Warrant, E., et al.: Adaptive enhancement and noise reduction in very low light-level video. In: ICCV (2007)
50. Mao, X.J., Shen, C., Yang, Y.B.: Image restoration using convolutional auto-encoders with symmetric skip connections. arXiv preprint arXiv:1606.08921 (2016)
51. Mildenhall, B., et al.: Burst denoising with kernel prediction networks. In: CVPR (2018)
52. O'Toole, M., Heide, F., Lindell, D.B., Zang, K., Diamond, S., Wetzstein, G.: Reconstructing transient images from single-photon sensors. In: CVPR (2017)
53. Plotz, T., Roth, S.: Benchmarking denoising algorithms with real photographs. In: CVPR (2017)
54. Protter, M., Elad, M.: Image sequence denoising via sparse and redundant representations. IEEE Trans. Image Process. **18**(1), 27–35 (2008)
55. Remez, T., Litany, O., Bronstein, A.: A picture is worth a billion bits: real-time image reconstruction from dense binary threshold pixels. In: ICCP (2016)
56. Remez, T., Litany, O., Giryes, R., Bronstein, A.: Deep convolutional denoising of low-light images. arXiv preprint arXiv:1701.01687 (2017)
57. Sutour, C., Deledalle, C.A., Aujol, J.F.: Adaptive regularization of the NL-means: application to image and video denoising. IEEE Trans. Image Process. **23**(8), 3506–3521 (2014)
58. Werlberger, M., Pock, T., Unger, M., Bischof, H.: Optical flow guided TV-L 1 video interpolation and restoration. In: International Workshop on Energy Minimization Methods in Computer Vision and Pattern Recognition (2011)
59. Xia, Z., Perazzi, F., Gharbi, M., Sunkavalli, K., Chakrabarti, A.: Basis prediction networks for effective burst denoising with large kernels. arXiv preprint arXiv:1912.04421 (2019)
60. Xu, J., Li, H., Liang, Z., et al.: Real-world noisy image denoising: a new benchmark. arXiv preprint arXiv:1804.02603 (2018)
61. Yang, F., Lu, Y.M., Sbaiz, L., Vetterli, M.: An optimal algorithm for reconstructing images from binary measurements. In: Proceedings SPIE, vol. 7533 (2010)
62. Yang, F., Lu, Y.M., Sbaiz, L., Vetterli, M.: Bits from photons: oversampled image acquisition using binary poisson statistics. IEEE Trans. Image Process. **21**(4), 1421–1436 (2011)
63. Yang, F., Sbaiz, L., Charbon, E., Süsstrunk, S., Vetterli, M.: Image reconstruction in the gigavision camera. In: ICCV Workshops (2009)
64. Zhang, K., Zuo, W., Chen, Y., et al.: Beyond a Gaussian denoiser: residual learning of deep CNN for image denoising. IEEE Trans. Image Process. **26**(7) (2017)
65. Zhang, K., Zuo, W., Zhang, L.: FFDNet: toward a fast and flexible solution for CNN-based image denoising. IEEE Trans. Image Process. **27**(9) (2018)

Disambiguating Monocular Depth Estimation with a Single Transient

Mark Nishimura$^{(\boxtimes)}$ (ID), David B. Lindell (ID), Christopher Metzler (ID), and Gordon Wetzstein (ID)

Stanford University, Stanford, CA, USA
{markn1,lindell,cmetzler,gordon.wetzstein}@stanford.edu

Abstract. Monocular depth estimation algorithms successfully predict the relative depth order of objects in a scene. However, because of the fundamental scale ambiguity associated with monocular images, these algorithms fail at correctly predicting true metric depth. In this work, we demonstrate how a depth histogram of the scene, which can be readily captured using a single-pixel time-resolved detector, can be fused with the output of existing monocular depth estimation algorithms to resolve the depth ambiguity problem. We validate this novel sensor fusion technique experimentally and in extensive simulation. We show that it significantly improves the performance of several state-of-the-art monocular depth estimation algorithms.

Keywords: Depth estimation · Time-of-flight imaging

1 Introduction

Estimating dense 3D geometry from 2D images is an important problem with applications to robotics, autonomous driving, and medical imaging. Depth maps are a common representation of scene geometry and are useful precursors to higher-level scene understanding tasks such as pose estimation and object detection. Additionally, many computer vision tasks rely on depth sensing, including navigation [11], semantic segmentation [16,43,49], 3D object detection [17,27,48,50,51], and 3D object classification [32,41,58].

Traditional depth sensing techniques include those based on stereo or multiview, active illumination, camera motion, or focus cues [55]. However, each of these techniques has aspects that may make their deployment challenging in certain scenarios. For example, stereo or multiview techniques require multiple cameras, active illumination techniques may have limited resolution or require time-consuming scanning procedures, and other techniques require camera motion or multiple exposures at different focus distances.

Electronic supplementary material The online version of this chapter (https://doi.org/10.1007/978-3-030-58589-1_9) contains supplementary material, which is available to authorized users.

A. Vedaldi et al. (Eds.): ECCV 2020, LNCS 12366, pp. 139–155, 2020.
https://doi.org/10.1007/978-3-030-58589-1_9

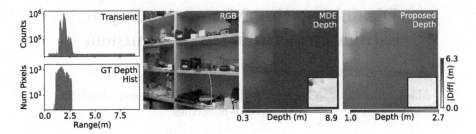

Fig. 1. Monocular depth estimation predicts a depth map (second from right) from a single RGB image (second from left). The ill-posedness of the problem prevents reliable absolute depth estimation, resulting in large errors (inset images). The proposed method uses a single transient measurement aggregating the time-of-flight information of the entire scene (leftmost) to correct the output of the depth estimation and optimize the quality of the estimated absolute depth (rightmost).

One of the most promising approaches to overcoming these challenges is monocular depth estimation (MDE), which requires only a single RGB image from a conventional camera to recover a dense depth map [2,7,9,23,45]. Recent approaches to MDE employ neural networks that learn to predict depth by exploiting pictorial depth cues such as perspective, occlusion, shading, and relative object size. While such models have significantly improved over recent years, MDE approaches to date are incapable of reliably estimating absolute distances in a scene due to the inherent scale ambiguities of monocular image cues. Instead, these models excel in predicting ordinal depth, or the relative ordering of objects in a scene [7,9]. Interestingly, Alhashim and Wonka [2] recently showed that if the median ground truth depth of the scene is known, the initial output of a MDE network can be corrected to produce accurate absolute depth.

Although access to the median ground truth depth is impossible in a realistic scenario, low-cost sensors capable of capturing aggregated depth information from a scene are readily available. For example, the proximity sensor on recent generation Apple iPhones uses a low-power pulsed light source and a single-pixel time-resolved detector to sense distance to an object directly in front of the phone. Time-resolved detectors, such as avalanche photon diodes (APDs) or single-photon avalanche diodes (SPADs), can measure the full waveform of time-resolved incident radiance at each pixel (Fig. 1). These detectors form the backbone of modern LiDAR systems [22,26,40]. However, single-photon sensor arrays have not yet been used for 3D imaging on consumer electronics, primarily because the requirement for ultra-fast timing electronics makes it difficult to produce high-resolution arrays at low cost and because the scanning requirement for single-pixel systems introduces a point of mechanical failure and complicates high-resolution, high-framerate imaging.

Here, we propose to use a single-pixel time-resolved detector and pulsed light source in an unconventional way: rather than optically focusing them to record the distance to a single scene point, we diffuse the emitted light and aggregate the reflected light over the entire scene with the detector. The resulting transient measurement resembles a histogram of the scene's depth and can be used to achieve accurate absolute depth in conjunction with a monocular depth estimate (Fig. 1).

To this end, we develop a sensor fusion strategy that processes the ordinal depth computed by a monocular depth estimator to be consistent with the measurements captured by the aggregated time-resolved detector. We demonstrate in extensive simulations that our approach achieves substantial improvements in the quality of the estimated depth maps, regardless of which specific depth estimator is used. Moreover, we build a camera prototype that combines an RGB camera and a single-pixel time-resolved detector and use it to validate the proposed depth estimation technique.

In summary, we make the following contributions:

- We propose augmenting an RGB camera with a global depth transient aggregated by a time-resolved detector to address scale ambiguity in MDE.
- We introduce a depth reconstruction algorithm that uses the detector's image formation model in conjunction with a modified version of histogram matching, to produce a depth map from a single RGB image and transient. The algorithm can be applied instantly to any existing and future MDE algorithms.
- We analyze this approach on indoor scenes using the NYU Depth v2 dataset and demonstrate that our approach is able to resolve scale ambiguity while being fast and easy to implement.
- We build a prototype camera and evaluate its efficacy on captured data, assessing both the quality and the ability of our method to improve generalization of monocular depth estimators across scene types.

2   Related Work

Monocular Depth Estimation. Estimating a depth map from a single RGB image has been approached using Markov Random Fields [45], geometric approaches [20], and non-parametric, SIFT-based methods [21]. More recently, deep neural networks have been applied to this problem, for example using a multi-scale neural network to predict depth maps [7], using an unsupervised approach that trains a network using stereo pairs [12], and using a logarithmic depth discretization scheme combined with an ordinal regression loss function [9]. Various experiments using different types of encoder networks (*e.g.*, ResNet, DenseNet) [2,23] have also been employed with some success, as have approaches mixing deep learning with conditional random fields [60], and attention-based approaches [18,61]. Recently, Lasinger et al. [25] improved the robustness of monocular depth estimation using cross-dataset transfer.

Despite achieving remarkable success on estimating ordinal depth from a single image, none of these methods is able to resolve inherent scale ambiguity in a principled manner. We introduce a new approach that leverages existing monocular depth estimation networks and disambiguates the output using depth histogram–like measurements obtained from a single time-resolved detector. Other approaches to disambiguating monocular depth estimation use optimized freeform lenses [6,57] or dual-pixel sensors [10], but these approaches require custom lenses or sensors and specialized image reconstruction methods. In contrast, our approach adds minimal additional hardware to a single RGB camera, and may leverage sensors currently deployed in consumer electronics.

Depth Imaging and Sensor Fusion with Time-Resolved Detectors. Emerging LiDAR systems use avalanche photon diodes (APDs) or single-photon avalanche diodes (SPADs) to record the time of flight of individual photons. These time-resolved detectors can be fabricated using standard CMOS processes, but the required time-stamping electronics are challenging to miniaturize and fabricate at low cost. For this reason, many LiDAR systems, especially those using SPADs, use a single or a few detectors combined with a scanning mechanism [15,22,24,26,40]. Unfortunately, this makes it challenging to scan dynamic scenes at high resolution and scanners can also be expensive, difficult to calibrate, and prone to mechanical failure. To reduce the scanning complexity to one dimension, 1D detector arrays have been developed [3,4,38], and 2D SPAD arrays are also an active area of research [35,52,56,62]. Yet, single-pixel time-resolved detectors remain the only viable option for low-cost consumer devices today.

The proposed method uses a single-pixel APD or SPAD and pulsed light source that are diffused across the entire scene instead of aimed at a single point, as with proximity sensors. This unique configuration captures a measurement that closely resembles the depth histogram of the scene. Our sensor fusion algorithm achieves reliable absolute depth estimation by combining the transient measurement with the output of a monocular depth estimator using a histogram matching technique. While other recent work also explored RGB-SPAD sensor fusion [1,28,53], the RGB image was primarily used to guide the denoising and upsampling of measurements from a SPAD array.

Histogram Matching and Global Hints. Histogram matching is a well-known image processing technique for adjusting an image so that its histogram matches some pre-specified histogram (often derived from another image) [13,14]. Nikolova et al. [36] use optimization to recover a strict ordering of the image pixels, yielding an exact histogram match. Morovic et al. [34] provide an efficient and precise method for fast histogram matching which supports weighted pixel values. In the image reconstruction space, Swoboda and Schnörr [54] use a histogram to form an image prior based on the Wasserstein distance for image denoising and inpainting. Rother et al. [44] use a histogram prior to create an energy function that penalizes foreground segmentations with dissimilar histograms. In the area of non-line-of-sight imaging [8,29–31,39], Caramazza et al. [5] use a single non-line-of-sight transient to recover the identity of a person hidden from view. In a slightly different application area, Zhang et al. [63] train a neural network to produce realistically colorized images given only a black-and-white image and a histogram of global color information.

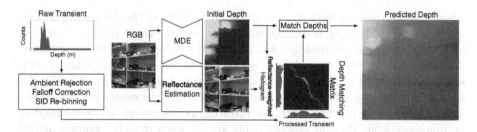

Fig. 2. Overview of processing pipeline. The processing pipeline uses the input transient measurement and an RGB image to produce an accurate depth map. The transient is preprocessed to adjust for ambient photon detections, radiometric falloff factors, and to calibrate the bin widths. From the RGB image, an MDE estimates an initial depth map and the scene reflectance is estimated. A reflectance-weighted depth histogram is compared to the processed transient to calculate a histogram matching matrix which is used to output the corrected depth.

In our procedure, the transient measurements closely resemble a histogram of the depth map where the histogram values are weighted by spatially varying scene reflectances and inverse-square falloff effects. We therefore adapt the algorithm in Morovic et al. [34] in order to accommodate general per-pixel weights during histogram matching.

3   Method

In this section, we describe the image formation of a diffused pulsed laser and time-resolved detector. Although our model is derived for the specific case of imaging with a single-photon avalanche diode (SPAD), the resulting image formation model equally applies to other time-resolved detectors. We also describe an approach for correcting a depth map generated with a monocular depth estimator to match the global scene information captured by the transient.

3.1   Image Formation Model of a Diffused SPAD

Consider a diffused laser that emits a pulse at time $t = 0$ with time-varying intensity $g(t)$ illuminating some 3D scene. We parameterize the geometry of the scene as a distance map $z(x, y)$, where each of the 3D points has also some unknown reflectivity α at the wavelength of the laser. Ignoring interreflections of the emitted light within the scene, a single-pixel diffused SPAD integrates light scattered back from the scene onto the detector as

$$s(t) = \int_{\Omega_x} \int_{\Omega_y} \frac{\alpha(x, y)}{z(x, y)^2} \cdot g\left(t - \frac{2z(x, y)}{c}\right) dx dy, \tag{1}$$

where c is the speed of light, $\Omega_{x,y}$ is the spatial extent of the diffused light, and we assume that the light is diffused uniformly over the scene points. Each time such a light pulse is emitted into the scene and scattered back to the detector, the single-pixel SPAD time-stamps up to one of the returning photons with some probability. The process is repeated millions of times per second with the specific number of emitted pulses being controlled by the repetition rate of the laser. As derived in previous work, the resulting measurement can be modeled as an inhomogeneous Poisson process $\mathcal{P}$ [22,46,47]. Each detected photon arrival event is discretized into a histogram h of the form

$$h[n] \sim \mathcal{P} \left(\eta \int_{n\Delta t}^{(n+1)\Delta t} (f * s)(t)\,dt + b \right), \tag{2}$$

where $[n\Delta t, (n+1)\Delta t)$ models the nth time interval or bin of the temporal histogram, η is the photon detection probability of the SPAD, f is a function that models the temporal uncertainty in the detector, and b represents background detections from ambient light and false positive detections known as *dark count*. Like previous work, we neglect scene interreflections and confine ourselves to the low-flux condition (where the number of photon detections is controlled to be much smaller than the number of emitted pulses) to avoid pileup [47]. Finally, we adopt the term *transient* for the histogram $h[n]$ [59].

3.2   Ambient Rejection and Falloff Correction

Before performing histogram matching, we apply three preprocessing steps to (1) remove background counts from the transient, (2) compensate for distance falloff effects, and (3) re-bin the transient to improve relative accuracy with increasing distance. An overview of the processing pipeline, including these preprocessing steps and the histogram matching procedure is depicted in Fig. 2.

Background Subtraction. In the first step, we remove the background counts from the transient by initially estimating the average amount of background counts in each time bin. For nearly all natural scenes, the closest objects to the camera are a finite distance away, and so the first bins of the SPAD measurement contain only background counts without any backscattered signal. We can therefore estimate the average number of background and noise counts $\hat{b}$ as

$$\hat{b} = \frac{1}{N} \sum_{n=0}^{N} h[n]. \tag{3}$$

where we choose the number of bins N to correspond to time values before the backscattered signal arrives.

While simply subtracting $\hat{b}$ from the measurements would remove many of the background counts, a large number of bins containing only background counts would still have non-zero values, resulting in a skewed estimate after applying histogram matching. Instead, we estimate the temporal support of transient bins containing signal photons (*i.e.*, the range of depths in the scene) and only subtract $\hat{b}$ from these bins (clipping negative bin values to 0). We assume that other transient bins contain only background counts that can be discarded.

Specifically, we identify the first and last bins that record backscattered signal photons by locating discontinuities in the recorded counts [59]. An initial spike in the measurements at bin n_{first} results from the onset of backscattered signal from the closest object, and a steep dropoff occurs after bin n_{last} after backscattered photons from the furthest object are recorded. We estimate n_{first} and n_{last} by calculating first order differences of the transient $d[n] = |h[n] - h[n+1]|$. For a moderate number of background counts, each background bin $h[n]$ can be approximated as a Gaussian with mean and variance b, and thus $h[n] - h[n+1]$ can be approximated as a Gaussian with mean 0 and variance $2b$. We identify candidate discontinuities $\mathcal{E}$ with a threshold on the measured differences:

$$\mathcal{E} = \left\{ n : d[n] > \beta\sqrt{2\hat{b}} \right\}. \tag{4}$$

We find that $\beta = 5$ yields good results across both simulated and captured data.

Initial estimates n'_{first} and n'_{last} are set to the minimum value in $\mathcal{E}$ and the maximum value, incremented by one bin. Then, we refine these estimates by selecting the closest bins that remain above a threshold τ such that

$$\hat{n}_{first} = \min\{n : h[n] > \tau, h[n+1] > \tau, \cdots, h[n'_{first}] > \tau\}$$
$$\hat{n}_{last} = \max\{n : h[n'_{last}] > \tau, \cdots, h[n-1] > \tau, h[n] > \tau\}. \tag{5}$$

The remaining ambient counts are discarded by setting the recorded counts to zero for all bins where $n < \hat{n}_{first}$ and $n > \hat{n}_{last}$. We use $\tau = \hat{b} + \sqrt{\hat{b}}$ in all of our experiments.

Falloff Compensation. In the second step, we compensate for distance falloff effects by multiplying the transient by the distance-dependent scaling factor,

$$h'[n] = h[n] \cdot z_n^2. \tag{6}$$

Here, $z_n = \left(n + \frac{1}{2}\right)\left(\frac{c\Delta t}{2}\right)$ is the distance corresponding to bin n, and this radiometric falloff model is consistent with measurements captured with our prototype.

Transient Re-binning. Last, we re-bin the transient so that the bin widths increase for increasingly distant objects. We select the Spacing-Increasing Discretization (SID) method of [9], which changes the bin widths according to an exponential function, allocating more bins to closer distances and fewer bins to farther distances for a fixed number of bins. The bin edges t_i are given by the following equation, parameterized by the number of bins K and the range of distances $[\ell, u]$:

$$t_i = e^{\log(\ell) + \frac{\log(u/\ell) \cdot i}{K}} \qquad \text{for} \qquad i = 0, \ldots, K. \tag{7}$$

This rebinning procedure allows us to use a reduced number of bins in the histogram matching, reducing computation time while maintaining accuracy.

For the simulated results we use $K = 140$ bins with (ℓ, u) corresponding to the depth values of bins $\hat{n}_{\text{first}}$ and $\hat{n}_{\text{last}}$ respectively. The output of the rebinning procedure is the target histogram h_{target} which we use for histogram matching.

3.3 Histogram Matching

Histogram matching is a procedure that adjusts pixel values from an input image so that the image histogram matches a target histogram. We apply this procedure to match the histogram of an input depth map, obtained from a monocular depth estimator, to the post-processed target histogram h_{target} from the SPAD. This initialize-then-refine approach allows us to swap out the monocular depth estimator to deal with different scene types without requiring end-to-end retraining.

The input depth map cannot be directly histogram-matched to the target histogram because the target histogram incorporates the spatially varying reflectance of the scene. To account for reflectance in the histogram matching procedure, we use the normalized image color channel closest to the laser wavelength as an estimate of the reflectance and compute a reflectance-weighted depth histogram h_{source}; instead of incrementing a bin in the depth histogram by one for every pixel in the MDE at the corresponding depth, we add the estimated reflectance value of the pixel to the histogram bin. We also re-bin this histogram, following Fu et al. and using $K = 140$ with $(\ell, u) = (0.657, 9.972)$ [9].

We match the re-binned histogram h_{source} to h_{target} using the method of Morovic et al. [34]. The method involves computing a pixel movement matrix T such that $T[m, n]$ is the fraction of $h_{\text{source}}[m]$ that should be moved to $h_{\text{target}}[n]$. We refer the reader to the supplement for pseudocode. Intuitively, the procedure starts from the first bin of the source histogram and distributes its contents to the first bins of the target histogram, with successive source histogram bins being shifted to successive target bins in sequence.

Finally, we use the movement matrix T to shift the pixels of the input depth map to match the global depth of the target histogram. For a depth map pixel with depth bin k, we select the corrected bin by sampling from the distribution $T[k, :]/\sum_{n=1}^{N} T[k, n]$. This sampling procedure handles the case where a single input depth bin of the MDE is mapped to multiple output bins [34].

Pseudo-code for this procedure is included in the supplement; we will make source code and data available.

4   Evaluation and Assessment

4.1   Implementation Details

We use the NYU Depth v2 dataset to evaluate our method. This dataset consists of 249 training and 215 testing RGB-D images captured with a Kinect.

To simulate a transient, we take the provided depth map and calculate a weighted depth histogram by weighting the pixel contributions to each depth bin

Table 1. Quantitative evaluation using NYU Depth v2. Bold indicates best performance for that metric, while underline indicates second best. The proposed scheme outperforms DenseDepth and DORN on all metrics, and it closely matches or even outperforms the median rescaling scheme and histogram matching with the exact depth map histogram, even though those methods have access to ground truth. Metric definitions can be found in [7].

	$\delta^1 \uparrow$	$\delta^2 \uparrow$	$\delta^3 \uparrow$	$rel \downarrow$	$rmse \downarrow$	$log10 \downarrow$
DORN	0.846	0.954	0.983	0.120	0.501	0.053
DORN + median rescaling	0.871	0.964	0.988	0.111	0.473	0.048
DORN + GT histogram matching	<u>0.906</u>	**0.972**	**0.990**	0.095	0.419	<u>0.040</u>
Proposed (SBR = 5)	0.902	0.970	<u>0.989</u>	0.092	0.423	<u>0.040</u>
Proposed (SBR = 10)	0.905	<u>0.971</u>	**0.990**	<u>0.090</u>	<u>0.413</u>	**0.039**
Proposed (SBR = 50)	<u>0.906</u>	<u>0.971</u>	**0.990**	0.089	**0.408**	**0.039**
Proposed (SBR = 100)	**0.907**	<u>0.971</u>	**0.990**	0.089	**0.408**	**0.039**
DenseDepth	0.847	0.973	<u>0.994</u>	0.123	0.461	0.053
DenseDepth + median rescaling	0.888	0.978	**0.995**	0.106	0.409	0.045
DenseDepth + GT histogram matching	**0.930**	**0.984**	**0.995**	**0.079**	**0.338**	**0.034**
Proposed (SBR = 5)	0.922	0.981	<u>0.994</u>	0.083	0.361	0.036
Proposed (SBR = 10)	0.924	0.982	**0.995**	0.082	0.352	<u>0.035</u>
Proposed (SBR = 50)	0.925	<u>0.983</u>	**0.995**	<u>0.081</u>	0.347	<u>0.035</u>
Proposed (SBR = 100)	<u>0.926</u>	<u>0.983</u>	**0.995**	<u>0.081</u>	<u>0.346</u>	<u>0.035</u>
MiDaS + GT histogram matching	**0.801**	**0.943**	**0.982**	**0.149**	**0.558**	**0.062**
Proposed (SBR = 5)	0.792	0.937	0.978	0.153	0.579	0.064
Proposed (SBR = 10)	0.793	0.937	<u>0.979</u>	0.152	0.572	0.064
Proposed (SBR = 50)	<u>0.794</u>	<u>0.938</u>	<u>0.979</u>	<u>0.151</u>	<u>0.570</u>	<u>0.063</u>
Proposed (SBR = 100)	<u>0.794</u>	<u>0.938</u>	<u>0.979</u>	<u>0.151</u>	<u>0.570</u>	0.064

by the luminance of each pixel. To model radiometric falloff, we multiply each bin by $1/z^2$, and convolve with a modeled system temporal response, which we approximate as a Gaussian with a full-width at half-maximum of 70 ps. We scale the histogram by the total number of observed signal photon counts (set to 10^6) and add a fixed number of background photons $b \in \{2 \times 10^5, 10^5, 2 \times 10^4, 10^4\}$. The background counts are evenly distributed across all bins to simulate the ambient and dark count detections, and the different background levels correspond to signal-to-background ratios (SBR) of $5, 10, 50$ and 100 respectively. Finally, each bin is Poisson sampled to produce the final simulated transient.

4.2   Simulated Results

We show an extensive quantitative evaluation in Table 1. Here, we evaluate three recent monocular depth estimation CNNs: DORN [9], DenseDepth [2], and MiDaS [25]. To evaluate the quality of DORN and DenseDepth, we report various standard error metrics [7]. Moreover, we show a simple post-processing step that rescales their outputs to match the median ground truth depth [2].

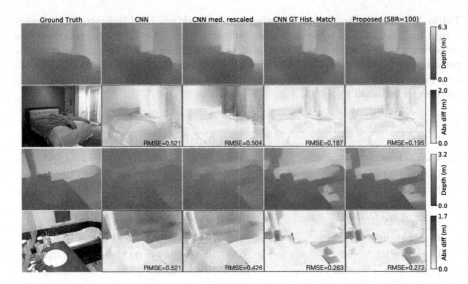

Fig. 3. Simulated results from NYU Depth v2 computed with the DenseDepth CNN [2]. The depth maps estimated by the CNN are reasonable, but contain systematic error. Oracle access to the ground truth depth maps, either through the median depth or the depth histogram, can remove this error and correct the depth maps. The proposed method uses a single transient measurement and does not rely on ground truth depth, but it achieves a quality that closely matches the best-performing oracle.

We also show the results of histogram matching the output of the CNNs with the ground truth depth map histogram. Note that we do not report the quality of the direct output of MiDaS as this algorithm does not output metric depth. However, we do show its output histogram matched with the ground truth depth map histogram. In all cases, post-processing the estimated depth maps either with the median depth or depth histogram significantly improves the absolute depth estimation, often by a large margin compared to the raw output of the CNNs. Unfortunately, ground truth depth is typically not accessible so neither of these two post-processing methods are viable in practical application scenarios.

Instead, our method uses the simulated measurements from a single aggregated transient to correct the depth map. In Table 1, results are shown for several different signal-to-background ratios (SBRs). We see that the proposed method achieves high-quality results for correcting the raw depth map estimated by the respective CNNs for all cases. The quality of the resulting depth maps is almost as good as that achieved with the oracle ground truth histogram, which can be interpreted as an approximate upper bound on the performance, despite a relatively high amount of noise and background signal. These results demonstrate that the proposed method is agnostic to the specific depth estimation CNN applied to get the initial depth map and that it generally achieves significant improvements in the estimated depth maps, clearly surpassing the variation in performance between depth estimation CNNs.

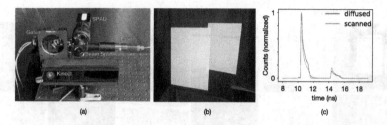

Fig. 4. (a) Prototype scanning setup. The pulsed light from the laser travels through a beam splitter before being guided by the galvo to the scene. Returning light is measured by the single-pixel SPAD. The Kinect v2 RGB camera is used to capture the image used to generate the monocular depth estimate (the depth camera is not used). (b) Scene and (c) measurements for diffused and summed scanned mode (10 s capture, aligned peaks). The observed counts in the diffuse mode match closely with the sum of the raster-scanned measurements.

In Fig. 3, we also show qualitative results of our simulations. For each of these scenes, we show the RGB reference image, the ground truth depth map, the raw output of the DenseDepth CNN, the result of rescaling the CNN output with the median ground truth depth, the result of histogram-matching the CNN output by the ground truth depth map histogram, and the result achieved by the proposed method for an SBR of 100. Error maps for all the depth estimation methods are shown. As expected, the CNN outputs depth maps that look reasonable but that have an average root mean squared error (RMSE) of about 50–60 cm. Rescaling this depth map to match the median ground truth depth value slightly improves the quality and histogram-matching with the ground truth depth histogram shows a large amount of improvement. The quality of the proposed method is close to using the oracle histogram, despite relying on noisy transient measurements. Additional simulations using DenseDepth and other depth estimation CNNs for a variety of scenes are shown in the supplement.

5   Experimental Demonstration

5.1   Prototype RGB-SPAD Camera Hardware

As shown in Fig. 4, our prototype comprises a color camera (Microsoft Kinect v2), a single-pixel SPAD (Micro Photon Devices 100 μm PDM series, free-running), a laser (ALPHALAS PICOPOWER-LD-450-50), and a two-axis galvanometer mirror system (Thorlabs GVS012). The laser operates at 670 nm with a pulse repetition rate of 10 MHz with a peak power of 450 mW and average power of 0.5 mW. The ground truth depth map is raster-scanned at a resolution of 512×512 pixels, and the single transient is generated by summing all of these measurements for a specific scene. This allows us to validate the accuracy of the proposed histogram matching algorithm, which only uses the integrated single histogram, by comparing it with the captured depth. To verify that our digitally aggregated scanned SPAD measurements match measurements produced by an

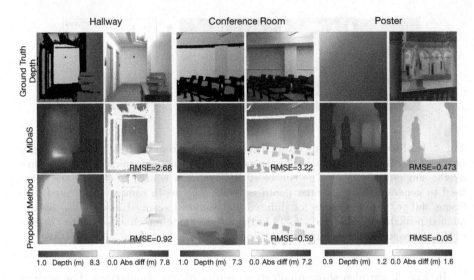

Fig. 5. Experimental results. For each scene, we record a ground truth depth map that is raster-scanned with the SPAD (upper left subimages), and an RGB image (upper right subimages). A monocular depth CNN predicts an initial depth map (top middle left subimages), which is corrected with the digitally aggregated SPAD histogram using the proposed method (bottom left subimages), as shown by the error maps and root mean squared error (RMSE) for each example (middle left, bottom subimages). The CNN is confused when we show it a photograph of a poster (rightmost scene); it incorrectly predicts the depth of the scene depicted on the flat print. Our method is able to correct this error. (Color figure online)

optically diffused SPAD (see Fig. 4(b, c)), we set up a slightly modified version of our prototype consisting of both scanned and optically diffused SPADs side-by-side. Additional details about the hardware prototype can be found in the supplement.

We determined camera intrinsics and extrinsics for the Kinect's RGB camera and the scanning system using MATLAB's camera calibration toolbox. The SPAD histogram and RGB image were captured from slightly different view-points; we account for this in the SPAD histogram by shifting the 1D transient according to the SPAD's offset from the RGB camera. We re-bin the captured 1D transient for the indoor captured results using Eq. 7 with $K = 600$ bins, and $(\ell, u) = (0.4, 9.)$. For the outdoor captured result, we use $K = 600$ and $(\ell, u) = (0.4, 11)$.

5.2 Experimental Results

Using the hardware prototype, we captured a number of scenes as shown in Figs. 5, 6, and in the supplement. We crop the RGB image to have dimensions that are multiples of 32. For DORN only, we further downsample the image to a resolution of 353×257. We then feed this RGB image into the monocular depth

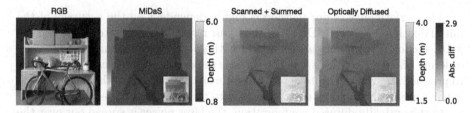

Fig. 6. Captured result comparing the direct output of the MiDaS MDE and depth maps corrected by our method using the digitally aggregated transient of the scanned SPAD (center right) and a single transient captured by an optically diffused SPAD and laser (right). Both of these approaches result in very similar results and both are significantly better than the output of the MDE, as shown by the error maps in the insets. The diffused SPAD results are captured at ~25 mW laser power indoors.

estimation algorithm. In Fig. 5 we show a subset of the scenes we captured and processed with MiDaS [25], which achieved the best results among the depth estimators we tested. Additional scenes, also processed with other MDE approaches, including DenseDepth [2] and DORN [9], are included in the supplement. The ground truth depth is captured with the scanned SPAD, as described above, and regions with low signal-to-noise ratio are masked out (shown in black).

In the first two examples, the "Hallway" and "Conference Room" scenes, we see that the monocular depth CNN estimates the ordinal depth of the scene reasonably well. However, the root mean squared error (RMSE) for these two scenes is relatively high ranging from 2.6–3.2 m (see red/white error maps in Fig. 5). The proposed method using a single diffused SPAD measurement corrects this systematic depth estimation error and brings the RMSE down to 0.6–0.9 m. The "Poster" scene is meant to confuse the CNN—it shows a flat poster with a printed scene. As expected, the CNN predicts that the statue is closer than the arches in the background, which is incorrect in this case. The proposed method uses the SPAD histogram to correctly flatten the estimated depth map.

Figure 6 shows the RGB image of a scene along with the monocular depth estimate computed by MiDaS and depth maps corrected by our method using both the digitally aggregated transients from the scanned SPAD and the single optically diffused measurement, which are very similar.

6   Discussion

In summary, we demonstrate a method to greatly improve depth estimates from monocular depth estimators by correcting the scale ambiguity errors inherent with such techniques. Our approach produces depth maps with accurate absolute depth, and helps MDE neural networks generalize across scene types, including on data captured with our hardware prototype. Moreover, we require only minimal additional sensing hardware; we show that a single measurement histogram from a diffused SPAD sensor contains enough information about global scene geometry to correct errors in monocular depth estimates.

The performance of our method is highly dependent on the accuracy of the initial depth map of the MDE algorithm. Our results demonstrate that when the MDE technique produces a depth map with good ordinal accuracy, where the ordering of object depths is roughly correct, the depth estimate can be corrected to produce accurate absolute depth. However, if the ordering of the initial depths is not correct, these errors may propagate to the final output depth map.

In the optically diffused configuration, the laser power is spread out over the entire scene. Accordingly, for distant scene points very little light may return to the SPAD, making reconstruction difficult (an analogous problem occurs with dark objects). Thus, our method is best suited to short- to medium-range scenes. On the other hand, in bright environments, pileup will ultimately limit the range of our method. However, this can be mitigated with optical elements to reduce the amount of incident light, with pileup correction [19,42], or even by taking two transient measurements, one with and one without laser illumination, and using their difference to approximate the background-free transient. Finally, under normal indoor conditions, it is theoretically possible to achieve an SBR of 5 at a range of 3 m with a laser of only 21 mW while remaining in the low-flux regime. We confirm this empirically with our diffused setup, which operates without significant pileup effects while using approximately 25 mW of laser power (see Fig. 6 and the supplement for details).

Future Work. Future work could implement our algorithm or similar sensor fusion algorithms on smaller platforms such as existing cell phones with single-pixel SPAD proximity sensors and RGB cameras. Necessary adjustments, such as pairing near-infrared (NIR) SPADs with NIR sensors, could be made. The small baseline of such sensors would also mitigate the effects of shading and complex BRDFs on the reflectance estimation step. More sophisticated intrinsic imaging techniques could also be employed.

Conclusions. Since their introduction, monocular depth estimation algorithms have improved tremendously. However, recent advances, which have generally relied on new network architectures or revised training procedures, have produced only modest performance improvements. In this work we dramatically improve the performance of several monocular depth estimation algorithms by fusing their estimates with transient measurements. Such histograms are easy to capture using time-resolved single-photon detectors and are poised to become an important component of future low-cost imaging systems.

Acknowledgments. D.L. was supported by a Stanford Graduate Fellowship. C.M. was supported by an ORISE Intelligence Community Postdoctoral Fellowship. G.W. was supported by an NSF CAREER Award (IIS 1553333), a Sloan Fellowship, by the KAUST Office of Sponsored Research through the Visual Computing Center CCF grant, and a PECASE by the ARL.

References

1. Ahmad Siddiqui, T., Madhok, R., O'Toole, M.: An extensible multi-sensor fusion framework for 3D imaging. In: Proceedings of the IEEE/CVF Conference on Computer Vision and Pattern Recognition Workshops, pp. 1008–1009 (2020)
2. Alhashim, I., Wonka, P.: High quality monocular depth estimation via transfer learning. arXiv:1812.11941v2 (2018)
3. Burri, S., Bruschini, C., Charbon, E.: Linospad: a compact linear SPAD camera system with 64 FPGA-based TDC modules for versatile 50 ps resolution time-resolved imaging. Instruments 1(1), 6 (2017)
4. Burri, S., Homulle, H., Bruschini, C., Charbon, E.: Linospad: a time-resolved 256×1 CMOS SPAD line sensor system featuring 64 FPGA-based TDC channels running at up to 8.5 giga-events per second. In: Optical Sensing and Detection IV, vol. 9899, p. 98990D. International Society for Optics and Photonics (2016)
5. Caramazza, P., et al.: Neural network identification of people hidden from view with a single-pixel, single-photon detector. Sci. Rep. 8(1), 11945 (2018)
6. Chang, J., Wetzstein, G.: Deep optics for monocular depth estimation and 3D object detection. In: Proceedings of ICCV (2019)
7. Eigen, D., Puhrsch, C., Fergus, R.: Depth map prediction from a single image using a multi-scale deep network. In: Proceedings of NeurIPS (2014)
8. Faccio, D., Velten, A., Wetzstein, G.: Non-line-of-sight imaging. Nat. Rev. Phys. 1–10 (2020)
9. Fu, H., Gong, M., Wang, C., Batmanghelich, K., Tao, D.: Deep ordinal regression network for monocular depth estimation. In: Proceedings of CVPR (2018)
10. Garg, R., Wadhwa, N., Ansari, S., Barron, J.T.: Learning single camera depth estimation using dual-pixels. In: Proceedings of ICCV (2019)
11. Geiger, A., Lenz, P., Stiller, C., Urtasun, R.: Vision meets robotics: the KITTI dataset. Int. J. Robot. Res. 32(11), 1231–1237 (2013)
12. Godard, C., Mac Aodha, O., Brostow, G.J.: Unsupervised monocular depth estimation with left-right consistency. In: Proceedings of CVPR (2017)
13. Gonzales, R., Fittes, B.: Gray-level transformations for interactive image enhancement. Mech. Mach. Theory 12(1), 111–122 (1977)
14. Gonzalez, R.C., Woods, R.E.: Digital Image Processing. Prentice-Hall Inc, Upper Saddle River (2008)
15. Gupta, A., Ingle, A., Velten, A., Gupta, M.: Photon-flooded single-photon 3D cameras. In: Proceedings of CVPR. IEEE (2019)
16. Gupta, S., Arbelaez, P., Malik, J.: Perceptual organization and recognition of indoor scenes from RGB-D images. In: Proceedings of CVPR (2013)
17. Gupta, S., Girshick, R., Arbeláez, P., Malik, J.: Learning rich features from RGB-D images for object detection and segmentation. In: Proceedings of ECCV (2014)
18. Hao, Z., Li, Y., You, S., Lu, F.: Detail preserving depth estimation from a single image using attention guided networks. In: Proceedings of 3DV (2018)
19. Heide, F., Diamond, S., Lindell, D.B., Wetzstein, G.: Sub-picosecond photon-efficient 3D imaging using single-photon sensors. Sci. Rep. 8(17726), 1–8 (2018)
20. Hoiem, D., Efros, A.A., Hebert, M.: Automatic photo pop-up. ACM Trans. Graph. 24(3), 577–584 (2005)
21. Karsch, K., Liu, C., Kang, S.: Depth transfer: depth extraction from video using non-parametric sampling. IEEE Trans. Pattern Anal. Mach. Intell. 36(11), 2144–2158 (2014)

22. Kirmani, A., Venkatraman, D., Shin, D., Colaço, A., Wong, F.N., Shapiro, J.H., Goyal, V.K.: First-photon imaging. Science **343**(6166), 58–61 (2014)

23. Laina, I., Rupprecht, C., Belagiannis, V., Tombari, F., Navab, N.: Deeper depth prediction with fully convolutional residual networks. In: Proceedings of 3DV. IEEE (2016)

24. Lamb, R., Buller, G.: Single-pixel imaging using 3D scanning time-of-flight photon counting. SPIE Newsroom (2010)

25. Lasinger, K., Ranftl, R., Schindler, K., Koltun, V.: Towards robust monocular depth estimation: Mixing datasets for zero-shot cross-dataset transfer. arXiv:1907.01341 (2019)

26. Li, Z.P., et al.: Single-photon computational 3D imaging at 45 km. arXiv preprint arXiv:1904.10341 (2019)

27. Lin, D., Fidler, S., Urtasun, R.: Holistic scene understanding for 3D object detection with RGBD cameras. In: Proceedings of ICCV (2013)

28. Lindell, D.B., O'Toole, M., Wetzstein, G.: Single-photon 3D imaging with deep sensor fusion. ACM Trans. Graph. (SIGGRAPH) **37**(4), 113 (2018)

29. Lindell, D.B., Wetzstein, G., O'Toole, M.: Wave-based non-line-of-sight imaging using fast F-K migration. ACM Trans. Graph. **38**(4), 1–13 (2019)

30. Liu, X., Bauer, S., Velten, A.: Phasor field diffraction based reconstruction for fast non-line-of-sight imaging systems. Nat. Commun. **11**(1), 1–13 (2020)

31. Liu, X., et al.: Non-line-of-sight imaging using phasor-field virtual wave optics. Nature **572**(7771), 620–623 (2019)

32. Maturana, D., Scherer, S.: Voxnet: a 3D convolutional neural network for real-time object recognition. In: Proceedings of IROS (2015)

33. McManamon, P.: Review of ladar: a historic, yet emerging, sensor technology with rich phenomenology. Opt. Eng. **51**(6), 060901 (2012)

34. Morovic, J., Shaw, J., Sun, P.L.: A fast, non-iterative and exact histogram matching algorithm. Pattern Recognit. Lett. **23**(1–3), 127–135 (2002)

35. Niclass, C., Rochas, A., Besse, P.A., Charbon, E.: Design and characterization of a CMOS 3-D image sensor based on single photon avalanche diodes. IEEE J. Solid-State Circuits **40**(9), 1847–1854 (2005)

36. Nikolova, M., Wen, Y.W., Chan, R.: Exact histogram specification for digital images using a variational approach. J. Math. Imaging. Vis. **46**(3), 309–325 (2013)

37. O'Connor, D.V., Phillips, D.: Time-Correlated Single Photon Counting. Academic Press, London (1984)

38. O'Toole, M., Heide, F., Lindell, D.B., Zang, K., Diamond, S., Wetzstein, G.: Reconstructing transient images from single-photon sensors. In: Proceedings of CVPR (2017)

39. O'Toole, M., Lindell, D.B., Wetzstein, G.: Confocal non-line-of-sight imaging based on the light-cone transform. Nature **555**(7696), 338–341 (2018)

40. Pawlikowska, A.M., Halimi, A., Lamb, R.A., Buller, G.S.: Single-photon three-dimensional imaging at up to 10 kilometers range. Opt. Express **25**(10), 11919–11931 (2017)

41. Qi, C.R., Su, H., Nießner, M., Dai, A., Yan, M., Guibas, L.J.: Volumetric and multi-view CNNs for object classification on 3D data. In: Proceedings of CVPR (2016)

42. Rapp, J., Ma, Y., Dawson, R.M.A., Goyal, V.K.: Dead time compensation for high-flux depth imaging. In: Proceedings of ICASSP (2019)

43. Ren, X., Bo, L., Fox, D.: RGB-(D) scene labeling: features and algorithms. In: Proceedings of CVPR (2012)

44. Rother, C., Minka, T., Blake, A., Kolmogorov, V.: Cosegmentation of image pairs by histogram matching-incorporating a global constraint into MRFs. In: Proceedings of CVPR (2006)
45. Saxena, A., Chung, S.H., Ng, A.Y.: Learning depth from single monocular images. In: Proceedings of NeurIPS (2006)
46. Shin, D., Kirmani, A., Goyal, V.K., Shapiro, J.H.: Photon-efficient computational 3-D and reflectivity imaging with single-photon detectors. IEEE Trans. Computat. Imag. 1(2), 112–125 (2015)
47. Shin, D., et al.: Photon-efficient imaging with a single-photon camera. Nat. Commun. 7, 12046 (2016)
48. Shrivastava, A., Gupta, A.: Building part-based object detectors via 3D geometry. In: Proceedings of ICCV (2013)
49. Silberman, N., Hoiem, D., Kohli, P., Fergus, R.: Indoor segmentation and support inference from RGBD images. In: Proceedings of ECCV (2012)
50. Song, S., Xiao, J.: Sliding shapes for 3D object detection in depth images. In: Proceedings of ECCV (2014)
51. Song, S., Xiao, J.: Deep sliding shapes for amodal 3D object detection in RGB-D images. In: Proceedings of CVPR (2016)
52. Stoppa, D., Pancheri, L., Scandiuzzo, M., Gonzo, L., Dalla Betta, G.F., Simoni, A.: A CMOS 3-D imager based on single photon avalanche diode. IEEE Trans. Circuits Syst. I Reg. Papers 54(1), 4–12 (2007)
53. Sun, Z., Lindell, D.B., Solgaard, O., Wetzstein, G.: Spadnet: deep RGB-SPAD sensor fusion assisted by monocular depth estimation. Opt. Express 28(10), 14948–14962 (2020)
54. Swoboda, P., Schnörr, C.: Convex variational image restoration with histogram priors. SIAM J. Imaging Sci. 6(3), 1719–1735 (2013)
55. Szeliski, R.: Computer Vision: Algorithms and Applications. Springer, Heidelberg (2010)
56. Veerappan, C., et al.: A 160 × 128 single-photon image sensor with on-pixel 55ps 10b time-to-digital converter. In: Proceedings of ISSCC (2011)
57. Wu, Y., Boominathan, V., Chen, H., Sankaranarayanan, A., Veeraraghavan, A.: PhaseCam3D–learning phase masks for passive single view depth estimation. In: Proceedings of ICCP (2019)
58. Wu, Z., et al.: 3D shapenets: a deep representation for volumetric shapes. In: Proceedings of CVPR (2015)
59. Xin, S., Nousias, S., Kutulakos, K.N., Sankaranarayanan, A.C., Narasimhan, S.G., Gkioulekas, I.: A theory of Fermat paths for non-line-of-sight shape reconstruction. In: Proceedings of CVPR (2019)
60. Xu, D., Ricci, E., Ouyang, W., Wang, X., Sebe, N.: Multi-scale continuous CRFs as sequential deep networks for monocular depth estimation. In: Proceedings of CVPR (2017)
61. Xu, D., Wang, W., Tang, H., Liu, H., Sebe, N., Ricci, E.: Structured attention guided convolutional neural fields for monocular depth estimation. In: Proceedings of CVPR (2018)
62. Zhang, C., Lindner, S., Antolovic, I., Wolf, M., Charbon, E.: A CMOS SPAD imager with collision detection and 128 dynamically reallocating TDCs for single-photon counting and 3D time-of-flight imaging. Sensors 18(11), 4016 (2018)
63. Zhang, R., et al.: Real-time user-guided image colorization with learned deep priors. ACM Trans. Graph. 9(4) (2017)

DSDNet: Deep Structured Self-driving Network

Wenyuan Zeng[1,2]($\boxtimes$), Shenlong Wang[1,2], Renjie Liao[1,2], Yun Chen[1], Bin Yang[1,2], and Raquel Urtasun[1,2]

[1] Uber ATG, Pittsburgh, USA
{wenyuan,slwang,rjliao,yun.chen,byang10,urtasun}@uber.com
[2] University of Toronto, Toronto, Canada

Abstract. In this paper, we propose the Deep Structured self-Driving Network (DSDNet), which performs object detection, motion prediction, and motion planning with a single neural network. Towards this goal, we develop a deep structured energy based model which considers the interactions between actors and produces socially consistent multimodal future predictions. Furthermore, DSDNet explicitly exploits the predicted future distributions of actors to plan a safe maneuver by using a structured planning cost. Our sample-based formulation allows us to overcome the difficulty in probabilistic inference of continuous random variables. Experiments on a number of large-scale self driving datasets demonstrate that our model significantly outperforms the state-of-the-art.

Keywords: Autonomous driving · Motion prediction · Motion planning

1 Introduction

The self-driving problem can be described as safely, comfortably and efficiently maneuvering a vehicle from point A to point B. This task is very complex; Even the most intelligent agents to date (*i.e.*, humans) are very frequently involved in traffic accidents. Despite the development of Advanced Driver-Assistance Systems (ADAS), 1.3 million people die every year on the road, and 20 to 50 million are severely injured.

Avoiding collisions in complicated traffic scenarios is not easy, primarily due to the fact that there are other traffic participants, whose future behaviors are unknown and very hard to predict. A vehicle that is next to our lane and blocked by its leading vehicle might decide to stay in its lane or cut in front of us. A pedestrian waiting on the edge of the road might decide to cross the road at any time. Moreover, the behavior of each actor depends on the actions taken

Electronic supplementary material The online version of this chapter (https:// doi.org/10.1007/978-3-030-58589-1_10) contains supplementary material, which is available to authorized users.

© Springer Nature Switzerland AG 2020
A. Vedaldi et al. (Eds.): ECCV 2020, LNCS 12366, pp. 156–172, 2020.
https://doi.org/10.1007/978-3-030-58589-1_10

by other actors, making the prediction task even harder. Thus, it is extremely important to model the future motions of actors with multi-modal distributions that also consider the interactions between actors.

To safely drive on the road, a self-driving vehicle (SDV) needs to detect surrounding actors, predict their future behaviors, and plan safe maneuvers. Despite the recent success of deep learning for perception, the prediction task, due to the aforementioned challenges, remains an open problem. Furthermore, there is also a need to develop motion planners that can take the uncertainty of the predictions into account. Previous works have utilized parametric distributions to model multimodality of motion prediction. Mixture of Gaussians [11,20] are a natural approach due to their close-form inference. However, it is hard to decide the number of modes in advance. Furthermore, these approaches suffer from mode collapse during training [20,22,38]. An alternative is to learn a model distribution from data using, *e.g.*, neural networks. As shown in [25,39,46], a CVAE [44] can be applied to capture multi-modality, and the interactions between actors can be modeled through latent variables. However, it is typically hard/slow to do probabilistic inference and the interaction mechanism does not explicitly model collision which humans want to avoid at all causes. Besides, none of these works have shown the effects upon planning on real-world datasets.

In this paper we propose the **D**eep **S**tructured self-**D**riving **Net**work (DSD-Net), a single neural network that takes raw sensor data as input to jointly detect actors in the scene, predict a multimodal distribution over their future behaviors, and produce safe plans for the SDV. This paper has three key contributions:

- Our prediction module uses an energy-based formulation to explicitly capture the interactions among actors and predict multiple future outcomes with calibrated uncertainty.
- Our planning module considers multiple possibilities of how the future might unroll, and outputs a safe trajectory for the self-driving car that respects the laws of traffic and is compliant with other actors.
- We address the costly probabilistic inference with a sample-based framework.

DSDNet conducts efficient inference based on message passing over a sampled set of continuous trajectories to obtain the future motion predictions. It then employs a structured motion planning cost function, which combines a cost learned in a data-driven manner and a cost inspired by human prior knowledge on driving (*e.g.*, traffic rules, collision avoidance) to ensure that the SDVs planned path is safe. We refer the reader to Fig. 1 for an overview of our full model.

We demonstrate the effectiveness of our model on two large-scale real-world datasets: **nuScenes** [6] and **ATG4D**, as well as one simulated dataset **CARLA-Precog** [15,39]. Our method significantly outperforms previous state-of-the-art results on both prediction and planning tasks.

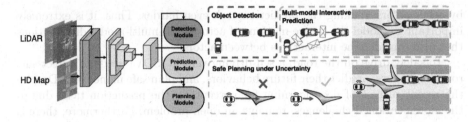

Fig. 1. DSDNet overview: The model takes LiDAR and map as inputs, processes them with a CNN backbone, and jointly performs object detection, multi-modal socially-consistent prediction, and safe planning under uncertainty. *Rainbow* patterns mean highly likely actors' future positions predicted by our model.

2   Related Work

Motion Prediction: Two of the main challenges of prediction are modeling interactions among actors and making accurate multi-modal predictions. To address these, [1,7,18,25,26,39,46,57] learn per-actor latent representations and model interactions by communicating those latent representations among actors. These methods can naturally work with VAE [23] and produce multi-modal predictions. However, they typically lack interpretability and it is hard to encode prior knowledge, such as the traffic participants' desire to avoid collisions. Different from building implicit distributions with VAE, [11,27] build explicit distributions using mixture of modes (e.g., GMM) where it is easier to perform efficient probabilistic inference. In this work, we further enhance the model capacity with a non-parametric explicit distribution constructed over a dense set of trajectory samples. In concurrent work [36] use similar representation to ours, but they do not model social interactions and do not demonstrate how such a representation can benefit planning.

Recently, a new prediction paradigm of performing joint detection and prediction has been proposed [8–10,28,29,47,54], in which actors' location information is not known a-priori, and needs to be inferred from the sensors. In this work, we will demonstrate our approach in both settings: using sensor data or history of actors' locations as input.

Motion Planning: Provided with perception and prediction results, planing is usually formulated as a cost minimization problem over trajectories. The cost function can be either manually engineered to guarantee certain properties [5,16,33,61], or learned from data through imitation learning or inverse reinforcement learning [41,49,54,60]. However, most of these planners assume detection and prediction to be accurate and certain, which is not true in practice. Thus, [2,19,40,56] consider uncertainties in other actors' behaviors, and formulate collision avoidance in a probabilistic manner. Following this line of work, we also conduct uncertainty-aware motion planning.

End-to-end self-driving methods try to fully utilize the power of data-driven approaches and enjoy simple inference. They typically use a neural network

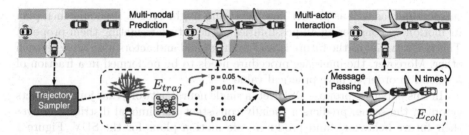

Fig. 2. Details of the multimodal social prediction module: For each actor, we sample a set of physically valid trajectories, and use a neural network E_{traj} to assign energies (probabilities) to them. To make different actors' behaviors socially consistent, we employ message passing steps which explicitly model interactions and can encode human prior knowledge (collision avoidance). The final predicted socially-consistent distribution is shown on top right.

to directly map from raw sensor data to planning outputs, and are learned through imitation learning [4,37], or reinforcement learning [13,34] when a simulator [15,30] is available. However, most of them lack interpretability and do not explicitly ensure safety. While our method also benefits from the power of deep learning, in contrast to the aforementioned approaches, we explicitly model interactions between the SDV and the other dynamic agents, achieving safer planning. Furthermore, safety is explicitly accounted for in our planning cost functions (Fig. 2).

Structured models and Belief Propagation: To encode prior knowledge, there is a recent surge of deep structured models [3,12,17,43], which use deep neural networks (DNNs) to provide the energy terms of a probabilistic graphical models (PGMs). Combining the powerful learning capacity of DNNs and the task-specific structure imposed by PGMs, deep structured models have been successfully applied to various computer vision problems, *e.g.*, semantic segmentation [43], anomaly detection [55], contour segmentation [31]. However, for continuous random variables, inference is very challenging. Sample-based belief propagation (BP) [21,45,48,50,51,53], address this issue by first constructing the approximation of the continuous distribution via Markov Chain Monte Carlo (MCMC) samples and then performing inference via BP. Inspired by these works, we design a deep structured model that can learn complex human behaviors from large data while incorporating our prior knowledge. We also bypass the difficulty in continuous variable inference using a physically valid sampling procedure.

3   Deep Structured Self-driving Network

Given sensor measurements and a map of the environment, the objective of a self-driving vehicle (SDV) is to select a trajectory to execute (amongst all feasible ones) that is safe, comfortable, and allows the SDV to reach its destination. In order to plan a safe maneuver, a self-driving vehicle has to first understand its surroundings as well as predict how the future might evolve. It should then plan its motion by considering all possibilities of the future weighting them properly. This is not trivial as the future is very multi-modal and actors interact with each other. Moreover, the inference procedure needs to be performed in a fraction of a second in order to have practical value.

In this paper we propose DSDNet, a single neural network that jointly detects actors in the scene, predicts a socially consistent multimodal distribution over their future behaviors, and produces safe motion plans for the SDV. Figure 1 gives an overview of our proposed approach. We first utilize a backbone network to compute the intermediate feature-maps, which are then used for detection, prediction and planning. After detecting actors with a detection header, a deep structured probabilistic inference module computes the distributions of actors' future trajectories, taking into account the interactions between them. Finally, our planning module outputs the planned trajectory by considering both the contextual information encoded in the feature-maps as well as possible futures predicted from the model.

In the following, we first briefly explain the input representation, backbone network and detection module. We then introduce our novel probabilistic prediction and motion planning framework in Sects. 3.2 and 3.3 respectively. Finally, we illustrate how to train our model end-to-end in Sect. 3.4.

3.1   Backbone Feature Network and Object Detection

Let $\mathbf{X}$ be the LiDAR point clouds and the HD map given as input to our system. Since LiDAR point clouds can be very sparse and the actors' motion is an important cue for detection and prediction, we use the past 10 LiDAR sweeps (e.g., 1 s of measurements) and voxelize them into a 3D tensor [29,52,54,58]. We utilize HD maps as they provide a strong prior about the scene. Following [54], we rasterize the lanes with different semantics (*e.g.*, straight, turning, blocked by traffic light) into different channels and concatenate them with the 3D LiDAR tensor to form our input representation. We then process this 3D tensor with a deep convolutional network backbone and compute a backbone feature map $\mathbf{F} \in \mathbb{R}^{H \times W \times C}$, where H, W correspond to the spatial resolution after downsampling (backbone) and C is the channel number. We then employ a single-shot detection header on this feature map to output detection bounding boxes for the actors in the scene. We apply two Conv2D layers separately on $\mathbf{F}$, one for classifying if a location is occupied by an actor, the other for regressing the position offset, size, orientation and speed of each actor. Our prediction and planning modules will then take these detections and the feature map as input to produce both a distribution over the actors' behaviors and a safe planning maneuver. For more details on our detector and backbone network please refer to the supplementary material.

3.2   Probabilistic Multimodal Social Prediction

In order to plan a safe maneuver, we need to predict how other actors could potentially behave in the next few seconds. As actors move on the ground, we represent their possible future behavior using a trajectory defined as a sequence of 2D waypoints on birds eye view (BEV) sampled at T discrete timestamps. Note that T is the same duration as our planning horizon, and we compute the motion prediction distribution and a motion plan each time a new sensor measurement arises (i.e., every 100 ms).

Output Parameterization: Let $\mathbf{s}_i \in \mathbb{R}^{T \times 2}$ be the future trajectory of the i-th actor. We are interested in modeling the joint distribution of all actors condition on the input, that is $p(\mathbf{s}_1, \cdots, \mathbf{s}_N | \mathbf{X})$. Modeling this joint distribution and performing efficient inference is challenging, as each actor has a high-dimensional continuous action space. Here, we propose to approximate this high-dimensional continuous space with a finite number of samples, and construct a non-parametric distribution over the sampled space. Specifically, for each actor, we randomly sample K possible future trajectory $\{\hat{\mathbf{s}}_i^1, \cdots, \hat{\mathbf{s}}_i^K\}$ from the original continuous trajectory space $\mathbb{R}^{T \times 2}$. We then constrain the possible future state of each actor to be one of those K samples. To ensure samples are always diverse, dense[1] and physically plausible, we follow the **Neural Motion Planner** (NMP) [54] and use a combination of straight, circle, and clothoid curves. More details and analysis of the sampler can be found in the supplementary material.

Modeling Future Behavior of All Actors: We employ an energy formulation to measure the probability of each possible future configuration of all actors in the scene: a configuration $(\mathbf{s}_1, \cdots, \mathbf{s}_N)$ has low energy if it is likely to happen. We can then compute the joint distribution of all actors' future behaviors as

$$p(\mathbf{s}_1, \cdots, \mathbf{s}_N | \mathbf{X}, \mathbf{w}) = \frac{1}{Z} \exp\left(-E(\mathbf{s}_1, \cdots, \mathbf{s}_N | \mathbf{X}, \mathbf{w})\right), \quad (1)$$

where $\mathbf{w}$ are learnable parameters, $\mathbf{X}$ is the raw sensor data and Z is the partition function $Z = \sum \exp(-E(\hat{\mathbf{s}}_1^{k_1}, \cdots, \hat{\mathbf{s}}_N^{k_N}))$ summing over all actors' possible states.

We construct the energy $E(\mathbf{s}_1, \cdots, \mathbf{s}_N | \mathbf{X}, \mathbf{w})$ inspired by how humans drive, e.g., following common sense as well as traffic rules. For example, humans drive smoothly along the road and avoid collisions with each other. Therefore, we decompose the energy E into two terms. The first term encodes the goodness of a future trajectory (independent of other actors) while the second term explicitly encodes the fact that pairs of actors should not collide in the future.

$$E(\mathbf{s}_1, \cdots, \mathbf{s}_N | \mathbf{X}, \mathbf{w}) = \sum_{i=1}^{N} E_{traj}(\mathbf{s}_i | \mathbf{X}, \mathbf{w}_{traj}) + \sum_{i=1}^{N} \sum_{i \neq j}^{N} E_{coll}(\mathbf{s}_i, \mathbf{s}_j | \mathbf{X}, \mathbf{w}_{coll}) \quad (2)$$

where N is the number of detected actors and $\mathbf{w}_{traj}$ and $\mathbf{w}_{coll}$ are the parameters.

---

[1] We would like the samples to cover the original continuous space and have high recall wrt the ground-truth future trajectories.

Since the goodness $E_{traj}(\mathbf{s}_i|\mathbf{X}, \mathbf{w}_{traj})$ is hard to define manually, we use a neural network to learn it from data (see Fig. 3). Given the sensor data $\mathbf{X}$ and a proposed trajectory $\mathbf{s}_i$, the network will output a scalar value. Towards this goal, we first use the detected bounding box of the i-th actor and apply ROIAlign to the backbone feature map, followed by several convolution layers to compute the actor's feature. Note that the backbone feature map is expected to encode rich information about both the environment and the actor.

We then index (bilinear interpolation) T features on the backbone feature map at the positions of trajectory's waypoints, and concatenate them together with $(x_t, y_t, cos(\theta_t), sin(\theta_t), distance_t)$ to form the trajectory feature of $\mathbf{s}_i$. Finally, we feed both actor and trajectory features into an MLP which outputs $E_{traj}(\mathbf{s}_i|\mathbf{X}, \mathbf{w}_{traj})$. Figure 3 shows an illustration of the architecture.

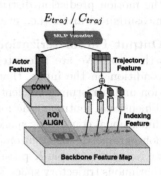

We use a simple yet effective collision energy: $E(\mathbf{s}_i, \mathbf{s}_j) = \gamma$ if $\mathbf{s}_i$ collides with $\mathbf{s}_j$, and $E(\mathbf{s}_i, \mathbf{s}_j) = 0$ otherwise, to explicitly model the collision avoidance interaction between actors as explained in the next paragraph. We found this simple pairwise energy empirically works well, the exploration of other learnable pairwise energy is thus left as future work.

Fig. 3. Neural header for evaluating E_{traj} and C_{traj}.

Message Passing Inference: For safety, our motion planner needs to take all possible actor's future into consideration. Therefore, motion forecasting needs to infer the probability of each actor taking a particular future trajectory: $p(\mathbf{s}_i = \hat{\mathbf{s}}_i^k|\mathbf{X}, \mathbf{w})$. We thus conduct marginal inference over the joint distribution. Note that the joint probability defined in Eq. (2) represents a deep structured model (i.e., a Markov random field with potentials computed with deep neural networks). We utilize sum-product message passing [53] to estimate the marginal distribution per actor, taking into account the effects of all other actors by marginalization. The marginal $p(\mathbf{s}_i|\mathbf{X}, \mathbf{w})$ reflects the uncertainty and multi-modality in an actor's future behavior and will be leveraged by our planner. We use the following update rule in an iterative manner for each actor $(\mathbf{s}_i)$:

$$m_{ij}(\mathbf{s}_j) \propto \sum_{\mathbf{s}_i \in \{\mathbf{s}_i^k\}} e^{-E_{traj}(\mathbf{s}_i) - E_{coll}(\mathbf{s}_i, \mathbf{s}_j)} \prod_{n \neq i, j} m_{ni}(\mathbf{s}_i) \qquad (3)$$

where m_{ij} is the message sent from actor i to actor j and $\propto$ means equal up to a normalization constant. Through this message passing procedure, actors communicate with each others their future intentions $\mathbf{s}_i$ and how probable those intentions are $E_{traj}(\mathbf{s}_i)$. The collision energy E_{coll} helps to coordinate intentions from different actors such that the behaviors are compliant and do not result in collision. After messages have been passed for a fixed number of iterations, we compute the approximated marginal as

$$p(\mathbf{s}_i = \hat{\mathbf{s}}_i^k | \mathbf{X}, \mathbf{w}) \propto e^{-Etraj(\hat{\mathbf{s}}_i^k)} \prod_{j \neq i} m_{ji}(\hat{\mathbf{s}}_i^k). \qquad (4)$$

Since we typically have a small graph (less than 100 actors in a scene) and each $\mathbf{s}_i$ only has K possible values $\{\hat{\mathbf{s}}_i^1, \cdots, \hat{\mathbf{s}}_i^K\}$, we can efficiently evaluate Eq. (3) and Eq. (4) via matrix multiplication on GPUs. In practice we find that our energy in Eq. (2) usually results in sparse graphs: most actor will only interact with nearby actors, especially the actors in the front and in the back. As a result, the message passing converges within 5 iterations [2]. With our non-highly-optimized implementation, the prediction module takes less than 100 ms on average, and thus it satisfies our real-time requirements.

3.3 Safe Motion Planning Under Uncertain Future

The motion planning module fulfills our final goal, that is, navigating towards a destination while avoiding collision and obeying traffic rules. Towards this goal, we build a cost function C, which assigns lower cost values to "good" trajectory proposals and higher values to "bad" ones. Planning is then performed by finding the optimal trajectory with the minimum cost

$$\boldsymbol{\tau}^* = \arg\min_{\boldsymbol{\tau} \in \mathcal{P}} C\left(\boldsymbol{\tau} | p(\mathbf{s}_1, \cdots, \mathbf{s}_N), \mathbf{X}, \mathbf{w}\right), \qquad (5)$$

with $\boldsymbol{\tau}^*$ the planned optimal trajectory and $\mathcal{P}$ the set of physically realizable trajectories that do not violate the SDV's dynamics. In practice, we sample a large number of future trajectories for the SDV conditioned on its current dynamic state (*e.g.*, velocity and acceleration) to form $\mathcal{P}$, which gives us a finite set of feasible trajectories $\mathcal{P} = \{\hat{\boldsymbol{\tau}}^1, \cdots, \hat{\boldsymbol{\tau}}^K\}$. We use the same sampler as described in Sect. 3.2 to ensure we get a wide variety of physically possible trajectories.

Planning Cost: Given a SDV trajectory $\boldsymbol{\tau}$, we compute the cost based on how good $\boldsymbol{\tau}$ is 1) conditioned on the scene, (*e.g.*, traffic lights and road topology); 2) considering all other actors' future behaviors (*i.e.*, marginal distribution estimated from the prediction module). We thus define our cost as

$$C(\boldsymbol{\tau} | p(\mathbf{s}_1, \cdots, \mathbf{s}_N), \mathbf{X}, \mathbf{w}) = C_{traj}(\boldsymbol{\tau} | \mathbf{X}, \mathbf{w}) + \sum_{i=1}^{N} \mathbb{E}_{p(\mathbf{s}_i | \mathbf{X}, \mathbf{w})} \left[C_{coll}(\boldsymbol{\tau}, \mathbf{s}_i | \mathbf{X}, \mathbf{w}) \right],$$

$$(6)$$

where C_{traj} models the goodness of a SDV trajectory using a neural network. Similar to E_{traj} in the prediction module, we use the trajectory feature and ROIAlign extracted from the backbone feature map to compute this scalar cost value. The collision cost C_{coll} is designed for guaranteeing safety and avoid collision: *i.e.*, $C_{coll}(\boldsymbol{\tau}, \mathbf{s}_i) = \lambda$ if $\boldsymbol{\tau}$ and $\mathbf{s}_i$ colide, and 0 otherwise. This ensures a dangerous trajectory $\boldsymbol{\tau}$ will incur a very high cost and will be rejected by our cost

---

[2] Although the sum-product algorithm is only exact for tree structures, it is shown to work well in practice for graphs with cycles [35,48].

minimization inference process. Furthermore, Eq. (6) evaluates the expected collision cost $\mathbb{E}_{p(\mathbf{s}_i|\mathbf{X},\mathbf{w})}[C_{coll}]$. Such a formulation is helpful for safe motion planning since the future is uncertain and we need to consider all possibilities, properly weighted by how likely they are to happen.

Inference: We conduct exact minimization over $\mathcal{P}$. C_{traj} is a neural network based cost and we can evaluate all K possible trajectories with a single batch forward pass. C_{coll} can be computed with a GPU based collision checker. As a consequence, we can efficiently evaluate $C(\boldsymbol{\tau})$ for all K samples and select the trajectory with minimum-cost as our final planning result.

3.4   Learning

We train the full model (backbone, detection, prediction and planning) jointly with a multi-class loss defined as follows

$$\mathcal{L} = \mathcal{L}_{\text{planning}} + \alpha \mathcal{L}_{\text{prediction}} + \beta \mathcal{L}_{\text{detection}}. \tag{7}$$

where α, β are constant hyper-parameters. Such a multi-task loss can fully exploit the supervision for each task and help the training[3].

Detection Loss: We employ a standard detection loss $\mathcal{L}_{detection}$, which is a sum of classification and regression loss. We use a cross-entropy classification loss and assign an anchor's label based on its IoU with any actor. The regression loss is a smooth ℓ_1 between our model regression outputs and the ground-truth targets. Those targets include position, size, orientation and velocity. We refer the reader to the supplementary material for more details.

Prediction Loss: As our prediction module outputs a discrete distribution for each actor's behavior, we employ cross-entropy between our discrete distribution and the true target. as our prediction loss. Once we sampled K trajectories per actor, this loss can be regarded as a standard classification loss over K classes (one for each trajectory sample). The target class is set to be the closest trajectory sample to the ground-truth future trajectory (in ℓ_2 distance).

Planning Loss: We expect our model to assign lower planning costs to better trajectories (*e.g.*, collision free, towards the goal), and higher costs to bad ones. However, we do not have direct supervision over the cost. Instead, we utilize a max-margin loss, using expert behavior as positive examples and randomly sampled trajectories as negative ones. We set large margins for dangerous behaviors such as trajectories with collisions. This allows our model to penalize dangerous behaviors more severely. More formally, our planning loss is defined as

$$\mathcal{L}_{\text{planning}} = \sum_{data} \max_k \left(\left[C\left(\boldsymbol{\tau}^{gt}|\mathbf{X}\right) - C\left(\hat{\boldsymbol{\tau}}^k|\mathbf{X}\right) + d^k + \gamma^k \right]_+ \right),$$

---

[3] We find that using only $\mathcal{L}_{planning}$ without the other two terms prevents the model from learning reasonable detection and prediction.

where $[\cdot]_+$ is a ReLU function, τ^{gt} is the expert trajectory and $\hat{\tau}^k$ is the k-th trajectory sample. We also define d^k as the ℓ_2 distance between $\hat{\tau}^k$ and τ^{gt}, and γ^k is a constant positive penalty if $\hat{\tau}^k$ behave dangerously, *e.g.*, $\gamma_{\text{collision}}$ if τ^k collides with another actor and 0 otherwise.

4 Experimental Evaluation

We evaluate our model on all three tasks: detection, prediction, and planning. We show results on two large scale real-world self-driving datasets: **nuScenes** [6] and our in-house dataset **ATG4D**, as well as the **CARLA** simulated dataset [15,39]. We show that **1)** our prediction module largely outperforms the state-of-the-art on public benchmarks and we demonstrate the benefits of explicitly modeling the interactions between actors. **2)** Our planning module achieves the safest planning results and largely decreases the collision and lane violation rate, compared to competing methods. **3)** Although sharing a single backbone to speedup inference, our model does not sacrifice detection performance compared to the state-of-the-art. We provide datasets' details and implementation details in the supplementary material.

Table 1. Prediction performance on nuScenes, ATG4D and CARLA

nuScenes	L2 (m)			Col (‰)			ATG4D	L2 (m)			Col (‰)		
Method	1s	2s	3s	1s	2s	3s	Method	1s	2s	3s	1s	2s	3s
Social-LSTM [1]	0.71	-	1.85	0.8	-	9.6	FaF [29]	0.60	1.11	1.82	-	-	-
CSP [14]	0.70	-	1.74	0.4	-	5.8	IntentNet [10]	0.51	0.93	1.52	-	-	-
CAR-Net [42]	0.61	-	1.58	0.4	-	4.9	NMP [54]	0.45	0.80	1.31	0.2	1.1	5.9
NMP [54]	0.43	0.83	1.40	0.0	1.4	6.5							
DSDNet	**0.40**	**0.76**	**1.27**	**0.0**	**0.0**	**0.2**	**DSDNet**	**0.43**	**0.75**	**1.22**	**0.1**	**0.1**	**0.2**

(a) Prediction from raw sensor data: ℓ_2 and Col (collision rate), lower the better.

CARLA	DESIRE	SocialGAN	R2P2-MA	MultiPath	ESP (Flex)	MFP-4	**DSDNet**
Town 1	2.422	1.141	0.770	0.68	0.447	0.279	**0.195**
Town 2	1.697	0.979	0.632	0.69	0.435	0.290	**0.213**

(b) Prediction from ground-truth perception: minMSD with K = 12, lower the better.

4.1 Multi-modal Interactive Prediction

Baselines: On CARLA, we compare with the state-of-the-art reproduced and reported from [11,39,46]. On nuScenes[4], we compare our method against several powerful multi-agent prediction approaches reproduced and reported from [7][5]:

---

[4] Numbers are reported on official validation split, since there is no joint detection and prediction benchmark.

[5] [7] replaced the original encoder (taking the ground-truth detection and tracking as input) with a learned CNN that takes LiDAR as input for a fair comparison.

Social-LSTM [1], **Convolutional Social Pooling (CSP)** [14] and **CAR-Net** [42]. On ATG4D, we compare with LiDAR-based joint detection and prediction models, including **FaF** [29] and **IntentNet** [10]. We also compare with **NMP** [54] on both datasets.

Metrics: Following previous works [1,18,25,29,54], we report **L2 Displacement Error** between our prediction (most likely) and the ground-truth at different future timestamps. We also report **Collision Rate**, defined as the percentage of actors that will collide with others if they follow the predictions. We argue that a socially consistent prediction model should achieve low collision rate, as avoiding collision is always one of the highest priorities for a human driver. On **CARLA**, we follow [39] and use **minMSD** as our metric, which is the minimal mean squared distance between the top 12 predictions and the ground-truth.

Quantitative Results: As shown in Table 1a, our method achieves the best results on both datasets. This is impressive as most baselines use ℓ_2 as training objective, and thus are directly favored by the ℓ_2 error metric, while our approaches uses cross-entropy loss to learn proper distributions and capture multi-modality. Note that multimodal techniques are thought to score worst in this metric (see e.g., [11]). Here, we show that it is possible to model multimodality while achieving lower ℓ_2 error, as the model can better understand actors' behavior. Our approach also significantly reduces the collisions between the actors' predicted trajectories, which justifies the benefit of our multi-agent interaction modeling. We further evaluate our prediction performance when assuming ground-truth perception/history are known, instead of predicting using noisy detections from the model. We conduct this evaluation on CARLA where all previous methods use this settings. As shown in Table 1b, our method again significantly outperforms previous best results.

Table 2. Motion planning performance on ATG4D. All metrics are computed in a cumulative manner across time, lower the better.

Model	Collision Rate (%)			Lane Violation Rate (%)			L2 (m)		
	1s	2s	3s	1s	2s	3s	1s	2s	3s
Ego-motion	0.01	0.54	1.81	0.51	2.72	6.73	0.28	0.90	2.02
IL	0.01	0.55	1.72	0.44	2.63	5.38	**0.23**	**0.84**	**1.92**
Manual Cost	0.02	0.22	2.21	0.39	2.73	5.02	0.40	1.43	2.99
Learnable-PLT [41]	0.00	0.13	0.83	-	-	-	-	-	-
NMP [54]	0.01	0.09	0.78	0.35	0.77	2.99	0.31	1.09	2.35
DSDNet	**0.00**	**0.05**	**0.26**	**0.11**	**0.57**	**1.55**	0.29	0.97	2.02

4.2 Motion Planning

Baselines: We implement multiple baselines for comparison, including both neural-based and classical motion planners: **Ego-motion** takes past 1 s posi-

Table 3. Detection performance: higher is better. Note that although our method uses single backbone for multiple challenging tasks, our detection module can achieve on-par performance with the state-of-the-art.

nuScenes	Det AP @ meter				ATG4D	Det AP @ IoU			
Method	0.5	1.0	2.0	4.0	Method	0.5	0.6	0.7	0.8
Mapillary[32]	10.2	36.2	64.9	80.1	FaF [29]	89.8	82.5	68.1	35.8
PointPillars [24]	55.5	71.8	76.1	78.6	IntentNet [10]	94.4	89.4	78.8	43.5
NMP [54]	71.7	82.5	85.5	87.0	Pixor [52]	93.4	89.4	78.8	52.2
Megvii [59]	**72.9**	82.5	85.9	**87.7**	NMP [54]	**94.2**	**90.8**	**81.1**	53.7
DSDNet	72.1	**83.2**	**86.2**	87.4	**DSDNet**	92.5	90.2	**81.1**	**55.4**

tions of the ego-car and use a 4-layer MLP to predict the future locations, as the ego-motion usually providse strong cues of how the SDV will move in the future. **Imitation Learning (IL)** uses the same backbone network as our model but directly regresses an output trajectory for the SDV. We train such a regression model with ℓ_2 loss w.r.t. the ground-truth planning trajectory. **Manual Cost** is a classical sampling based motion planner based on a manually designed cost function encoding collision avoidance and route following. The planned trajectory is chosen by finding the trajectory sample with minimal cost. We also include previously published learnable motion planning methods: **Learnable-PLT** [41] and **Neural Motion Planner (NMP)** [54]. These two method utilize a similar max-margin planning loss as ours. However, **Learnable-PLT** only consider the most probable future prediction, while **NMP** assumes planning is independent of prediction given the features.

Metrics: We exploit three metrics to evaluate motion planning performance. **Collision Rate** and **Lane Violation rate** are the ratios of frames at which our planned trajectory either collides with other actors' ground-truth future behaviors, or touches/crosses a lane boundary, up to a specific future timestamp. Those are important safety metrics (lower is better). **L2 to expert path** is the average ℓ_2 distance between the planning trajectory and the expert driving path. Note that the expert driving path is just one among many possibilities, thus rendering this metric not perfect.

Quantitative Results: The planning results are shown in Table 2. We can observe that: **1)** our proposed method provides the safest plans, as we achieve much lower collision and lane violation rates compared to all other methods. **2)** Ego-motion and IL achieves the best ℓ_2 metric, as they employ the power of neural networks and directly optimize the ℓ_2 loss. However, they have high collision rate, which indicates directly mimicking expert demonstrations is still insufficient to learn a safety-aware self-driving stack. In contrast, by learning interpretable intermediate results (detection and prediction) and by incorporating prior knowledge (collision cost), our model can achieve much better results. The later point is further validated by comparing to NMP, which, despite learning detection and prediction, does not explicitly condition on them during planning. **3)** Manual-Cost and Learnable-PLT explicitly consider collision avoidance and traffic rules. However, unlike our approach, they only take the most likely motion forecast into consideration. Consequently, these methods have a higher collision rate than our approach.

Table 4. Ablation Study for prediction and planning modules

Multi-modal	Interaction	Pred Col (‰)	Pred L2 (m)
		4.9	1.28
✓		2.4	**1.22**
✓	✓	**0.2**	**1.22**

(a) Prediction ablation studies on multi-modality and pairwise interaction modeling.

w/ Prediction (Type)	Plan Col (%)	Lane Vio (%)
N/A	0.60	1.64
most likely	0.45	1.60
multi-modal	**0.26**	**1.55**

(b) Motion planning ablation studies on incorporating different prediction results.

4.3 Object Detection Results

We show our object detection results on **nuScenes** and **ATG4D**. Although we use a single backbone for all three challenging tasks, we show in Table 3 that our model can achieve similar or better results than state-of-the-art LiDAR detectors on both datasets. On **nuScenes**[6], we follow the official benchmark and use detection average precision (AP) at different distance (in meters) thresholds as our metric. Here Megvii [59] is the leading method on the leaderboard at the time of our submission. We use a smaller resolution (Megvii has 1000 pixels on each side while ours is 500) for faster online inference, yet achieve on-par performance. We also conduct experiments on ATG4D. Since our model uses the same backbone as Pixor [52], which only focuses on detection, we demonstrate that a multi-task formulation does not sacrifice detection performance.

4.4 Ablation Study and Qualitative Results

Table 4a compares different prediction modules with the same backbone network. We can see that explicitly modeling a multi-modal future significantly boosts the prediction performance in terms of both collision rate and ℓ_2 error, comparing to a deterministic (unimodal) prediction. The performance is further boosted if the prediction module explicitly models the future interaction between multiple actors, particularly in collision rate. Table 4b compares motion planners that consider different prediction results. We can see that explicitly incorporating

---

[6] We conduct the comparison on the official validation split, as our model currently only focuses on vehicles while the testing benchmark is built for multi-class detection.

future prediction, even only the most likely prediction, will boost the motion planning performance, especially the collision rate. Furthermore, if the motion planner takes multi-modal futures into consideration, it achieves the best performance among the three. This further justifies our model design.

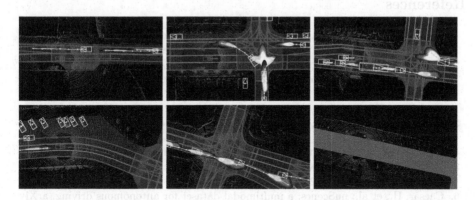

Fig. 4. Qualitative results on ATG4D. Our prediction module can capture multi-modalities when vehicles approach intersections, while being certain and accurate when vehicles drive along a single lane (top). Our model can produce smooth trajectories which follow the lane and are compliant with other actors (bottom). Cyan boxes: detection. Cyan trajectory: prediction. Red box: ego-car. Red trajectory: our planning. We overlay the predicted marginal distribution for different timestamps with different colors and only show high-probability regions. (Color figure online)

We show qualitative results in Fig. 4, where we visualize our detections, predictions, motion planning trajectories, and the predicted uncertainties. We use different colors for different future timestamps to visualize high-probability actors' future positions estimated from our prediction module. Thus larger 'rainbow' areas mean more uncertain. On the first row, we can see the predictions are certain when vehicles drive along the lanes (left), while we see multi-modal predictions when vehicles approach an intersection (middle, right). On the second row, we can see our planning can nicely follow the lane (left), make a smooth left turn (middle), and take a nudge when an obstacle is blocking our path (right).

5   Conclusion

In this paper, we propose DSDNet, which is built on top of a single backbone network that takes as input raw sensor data and an HD map and performs perception, prediction, and planning under a unified framework. In particular, we build a deep energy based model to parameterize the joint distribution of future trajectories of all traffic participants. We resort to a sample based formulation, which enables efficient inference of the marginal distribution via belief propagation. We design a structured planning cost which encourages traffic-rule following and collision avoidance for the SDV. We show that our model outperforms the state-of-the-art on several challenging datasets. In the future, we plan to explore more types of structured energies and cost functions.

References

1. Alahi, A., Goel, K., Ramanathan, V., Robicquet, A., Fei-Fei, L., Savarese, S.: Social LSTM: human trajectory prediction in crowded spaces. In: CVPR (2016)
2. Bandyopadhyay, T., Won, K.S., Frazzoli, E., Hsu, D., Lee, W.S., Rus, D.: Intention-aware motion planning. In: Frazzoli, E., Lozano-Perez, T., Roy, N., Rus, D. (eds.) Algorithmic Foundations of Robotics X. STAR, vol. 86, pp. 475–491. Springer, Heidelberg (2013). https://doi.org/10.1007/978-3-642-36279-8_29
3. Belanger, D., McCallum, A.: Structured prediction energy networks. In: ICML (2016)
4. Bojarski, M., et al.: End to end learning for self-driving cars. arXiv (2016)
5. Buehler, M., Iagnemma, K., Singh, S.: The DARPA Urban Challenge: Autonomous Vehicles in City Traffic (2009)
6. Caesar, H., et al.: nuScenes: a multimodal dataset for autonomous driving. arXiv (2019)
7. Casas, S., Gulino, C., Liao, R., Urtasun, R.: Spatially-aware graph neural networks for relational behavior forecasting from sensor data. arXiv (2019)
8. Casas, S., Gulino, C., Suo, S., Luo, K., Liao, R., Urtasun, R.: Implicit latent variable model for scene-consistent motion forecasting. In: ECCV (2020)
9. Casas, S., Gulino, C., Suo, S., Urtasun, R.: The importance of prior knowledge in precise multimodal prediction. In: IROS (2020)
10. Casas, S., Luo, W., Urtasun, R.: IntentNet: learning to predict intention from raw sensor data. In: Proceedings of The 2nd Conference on Robot Learning (2018)
11. Chai, Y., Sapp, B., Bansal, M., Anguelov, D.: Multipath: multiple probabilistic anchor trajectory hypotheses for behavior prediction. arXiv (2019)
12. Chen, L.C., Schwing, A., Yuille, A., Urtasun, R.: Learning deep structured models. In: ICML (2015)
13. Codevilla, F., Miiller, M., López, A., Koltun, V., Dosovitskiy, A.: End-to-end driving via conditional imitation learning. In: ICRA (2018)
14. Deo, N., Trivedi, M.M.: Convolutional social pooling for vehicle trajectory prediction. In: CVPR (2018)
15. Dosovitskiy, A., Ros, G., Codevilla, F., Lopez, A., Koltun, V.: Carla: an open urban driving simulator. arXiv (2017)
16. Fan, H., et al.: Baidu apollo em motion planner. arXiv (2018)
17. Graber, C., Meshi, O., Schwing, A.: Deep structured prediction with nonlinear output transformations. In: NeurIPS (2018)
18. Gupta, A., Johnson, J., Fei-Fei, L., Savarese, S., Alahi, A.: Social GAN: socially acceptable trajectories with generative adversarial networks. In: CVPR (2018)
19. Hardy, J., Campbell, M.: Contingency planning over probabilistic obstacle predictions for autonomous road vehicles. IEEE Trans. Robot. D (2013)
20. Hong, J., Sapp, B., Philbin, J.: Rules of the road: predicting driving behavior with a convolutional model of semantic interactions. In: CVPR (2019)
21. Ihler, A., McAllester, D.: Particle belief propagation. In: Artificial Intelligence and Statistics (2009)

22. Jain, A., et al.: Discrete residual flow for probabilistic pedestrian behavior prediction. arXiv (2019)
23. Kingma, D.P., Welling, M.: Auto-encoding variational Bayes. arXiv (2013)
24. Lang, A.H., Vora, S., Caesar, H., Zhou, L., Yang, J., Beijbom, O.: PointPillars: fast encoders for object detection from point clouds. In: CVPR (2019)
25. Lee, N., Choi, W., Vernaza, P., Choy, C.B., Torr, P.H., Chandraker, M.: Desire: distant future prediction in dynamic scenes with interacting agents. In: CVPR (2017)
26. Li, L., et al.: End-to-end contextual perception and prediction with interaction transformer. In: IROS (2020)
27. Liang, M., et al.: Learning lane graph representations for motion forecasting. In: ECCV (2020)
28. Liang, M., et al.: PnPNet: end-to-end perception and prediction with tracking in the loop. In: CVPR (2020)
29. Luo, W., Yang, B., Urtasun, R.: Fast and furious: real time end-to-end 3d detection, tracking and motion forecasting with a single convolutional net
30. Manivasagam, S., et al.: LiDARsim: realistic lidar simulation by leveraging the real world. In: CVPR (2020)
31. Marcos, D., et al.: Learning deep structured active contours end-to-end. In: CVPR (2018)
32. Min Choi, H., Kang, H., Hyun, Y.: Multi-view reprojection architecture for orientation estimation. In: ICCV (2019)
33. Montemerlo, M., et al.: Junior: the stanford entry in the urban challenge. J. Field Robot. (2008)
34. Müller, M., Dosovitskiy, A., Ghanem, B., Koltun, V.: Driving policy transfer via modularity and abstraction. arXiv (2018)
35. Murphy, K.P., Weiss, Y., Jordan, M.I.: Loopy belief propagation for approximate inference: an empirical study. In: Proceedings of the Fifteenth Conference on Uncertainty in Artificial Intelligence (1999)
36. Phan-Minh, T., Grigore, E.C., Boulton, F.A., Beijbom, O., Wolff, E.M.: CoverNet: multimodal behavior prediction using trajectory sets. In: CVPR (2020)
37. Pomerleau, D.A.: ALVINN: an autonomous land vehicle in a neural network. In: NeurIPS (1989)
38. Rhinehart, N., Kitani, K.M., Vernaza, P.: R2P2: a reparameterized pushforward policy for diverse, precise generative path forecasting. In: Ferrari, V., Hebert, M., Sminchisescu, C., Weiss, Y. (eds.) ECCV 2018. LNCS, vol. 11217, pp. 794–811. Springer, Cham (2018). https://doi.org/10.1007/978-3-030-01261-8_47
39. Rhinehart, N., McAllister, R., Kitani, K., Levine, S.: PRECOG: prediction conditioned on goals in visual multi-agent settings. arXiv (2019)
40. Sadat, A., Casas, S., Ren, M., Wu, X., Dhawan, P., Urtasun, R.: Perceive, predict, and plan: Safe motion planning through interpretable semantic representations. In: ECCV (2020)
41. Sadat, A., Ren, M., Pokrovsky, A., Lin, Y.C., Yumer, E., Urtasun, R.: Jointly learnable behavior and trajectory planning for self-driving vehicles. arXiv (2019)
42. Sadeghian, A., Legros, F., Voisin, M., Vesel, R., Alahi, A., Savarese, S.: CAR-Net: clairvoyant attentive recurrent network. In: Ferrari, V., Hebert, M., Sminchisescu, C., Weiss, Y. (eds.) ECCV 2018. LNCS, vol. 11215, pp. 162–180. Springer, Cham (2018). https://doi.org/10.1007/978-3-030-01252-6_10
43. Schwing, A.G., Urtasun, R.: Fully connected deep structured networks. arXiv (2015)

44. Sohn, K., Lee, H., Yan, X.: Learning structured output representation using deep conditional generative models. In: NeurIPS (2015)
45. Sudderth, E.B., Ihler, A.T., Isard, M., Freeman, W.T., Willsky, A.S.: Nonparametric belief propagation. Commun. ACM (2010)
46. Tang, Y.C., Salakhutdinov, R.: Multiple futures prediction. arXiv (2019)
47. Wang, T.H., Manivasagam, S., Liang, M., Yang, B., Zeng, W., Raquel, U.: V2VNET: vehicle-to-vehicle communication for joint perception and prediction. In: ECCV (2020)
48. Weiss, Y., Pearl, J.: Belief propagation: technical perspective. Commun. ACM (2010)
49. Wulfmeier, M., Ondruska, P., Posner, I.: Maximum entropy deep inverse reinforcement learning. arXiv (2015)
50. Yamaguchi, K., Hazan, T., McAllester, D., Urtasun, R.: Continuous Markov random fields for robust stereo estimation. In: Fitzgibbon, A., Lazebnik, S., Perona, P., Sato, Y., Schmid, C. (eds.) ECCV 2012. LNCS, vol. 7576, pp. 45–58. Springer, Heidelberg (2012). https://doi.org/10.1007/978-3-642-33715-4_4
51. Yamaguchi, K., McAllester, D., Urtasun, R.: Efficient joint segmentation, occlusion labeling, stereo and flow estimation. In: Fleet, D., Pajdla, T., Schiele, B., Tuytelaars, T. (eds.) ECCV 2014. LNCS, vol. 8693, pp. 756–771. Springer, Cham (2014). https://doi.org/10.1007/978-3-319-10602-1_49
52. Yang, B., Luo, W., Urtasun, R.: Pixor: Real-time 3D object detection from point clouds
53. Yedidia, J.S., Freeman, W.T., Weiss, Y.: Understanding belief propagation and its generalizations. In: Exploring Artificial Intelligence in the New Millennium (2003)
54. Zeng, W., Luo, W., Suo, S., Sadat, A., Yang, B., Casas, S., Urtasun, R.: End-to-end interpretable neural motion planner. In: CVPR (2019)
55. Zhai, S., Cheng, Y., Lu, W., Zhang, Z.: Deep structured energy based models for anomaly detection. In: ICML (2016)
56. Zhan, W., Liu, C., Chan, C.Y., Tomizuka, M.: A non-conservatively defensive strategy for urban autonomous driving. In: 2016 IEEE 19th International Conference on Intelligent Transportation Systems (ITSC) (2016)
57. Zhao, T., et al.: Multi-agent tensor fusion for contextual trajectory prediction. In: CVPR (2019)
58. Zhou, Y., Tuzel, O.: VoxelNet: end-to-end learning for point cloud based 3D object detection. In: CVPR (2018)
59. Zhu, B., Jiang, Z., Zhou, X., Li, Z., Yu, G.: Class-balanced grouping and sampling for point cloud 3D object detection. arXiv (2019)
60. Ziebart, B.D., Maas, A.L., Bagnell, J.A., Dey, A.K.: Maximum entropy inverse reinforcement learning. In: AAAI (2008)
61. Ziegler, J., Bender, P., Dang, T., Stiller, C.: Trajectory planning for bertha–a local, continuous method. In: Intelligent Vehicles Symposium Proceedings, 2014 IEEE (2014)

QuEST: Quantized Embedding Space for Transferring Knowledge

Himalaya Jain[1]([✉]), Spyros Gidaris[1], Nikos Komodakis[2,3], Patrick Pérez[1], and Matthieu Cord[1,4]

[1] Valeo.ai, Paris, France
himalaya.jain@valeo.com
[2] University of Crete, Heraklion, Greece
[3] LIGM, Paris, France
[4] Sorbonne University, Paris, France

Abstract. Knowledge distillation refers to the process of training a student network to achieve better accuracy by learning from a pre-trained teacher network. Most of the existing knowledge distillation methods direct the student to follow the teacher by matching the teacher's output, feature maps or their distribution. In this work, we propose a novel way to achieve this goal: by distilling the knowledge through a *quantized visual words* space. According to our method, the teacher's feature maps are first quantized to represent the main visual concepts (i.e., visual words) encompassed in these maps and then the student is asked to predict those visual word representations. Despite its simplicity, we show that our approach is able to yield results that improve the state of the art on knowledge distillation for model compression and transfer learning scenarios. To that end, we provide an extensive evaluation across several network architectures and most commonly used benchmark datasets.

Keywords: Knowledge distillation · Transfer learning · Model compression

1 Introduction

Knowledge distillation is an interesting learning problem with many practical applications. It was initially introduced by Hinton et al. [13] (KD method) as a means to achieve model compression. The main idea is to train a network using the output of another pre-trained network. Specifically, KD trains a low capacity "student network" (i.e., compressed model) to mimic the *softened classification predictions* of a higher capacity pre-trained "teacher network" (i.e., original uncompressed model). Adding such an auxiliary objective to the standard training loss of the student network, leads to learning a more accurate model. Apart from model compression, knowledge distillation has also been shown to be beneficial to semi-supervised learning [19,31] and transfer learning [20] problems.

Electronic supplementary material The online version of this chapter (https://doi.org/10.1007/978-3-030-58589-1_11) contains supplementary material, which is available to authorized users.

A. Vedaldi et al. (Eds.): ECCV 2020, LNCS 12366, pp. 173–189, 2020.
https://doi.org/10.1007/978-3-030-58589-1_11

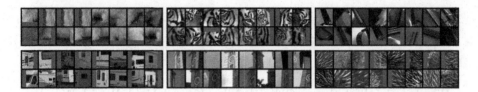

Fig. 1. Teacher-word clusters from layer4 of ResNet34. For each cluster we depict the 16 patch members with the smallest Euclidean distance to the cluster centroid. We see that they encode localized mid-to-high level image patterns

Since KD, several other methods have been proposed for knowledge distillation [4,7,12,17,21,24,28,35,39]. A popular paradigm is that of transferring knowledge through intermediate features of the two networks [28,39]. For instance, in FitNet [28] the student is trained to regress the raw feature maps of the teacher. The main intuition is that these features tend to encode useful semantic knowledge for transfer learning. However, empirical evidence [1,11,12,36,39] suggests that naively regressing the teacher features might be a difficult task which over-regularizes the student, thus leading to a suboptimal knowledge transfer solution. This can be because (a) the exact feature magnitudes (i.e., feature "details") of the feature maps are not semantically important aspects of the teacher network in the context of knowledge distillation, and/or (b) features behaviour largely depends on the architecture design (e.g., ResNet versus MobileNet) and capacity (i.e., width and depth) of a network. For instance, AT [39] shows that transferring attention maps (feature maps aggregated into a single channel) might be a better option in many cases; FSP [36] opts to remove the spatial information by considering only the correlations among feature channels; AB [12] transfers binarized versions of feature maps; and VID [1] proposes a regression loss (inspired from information theory) that is meant to "ignore" difficult to regress channels.

In our work we propose to address the knowledge transfer task in the context of *a spatially dense quantization of feature maps into visual codes*. Specifically, first we learn with k-means a vocabulary of deep features from the teacher network, which we call visual teacher-words. Then, given an image, we use this vocabulary to quantize/assign each location of the teacher's feature maps into the closest visual words. As distillation task, the student has to predict these visual teacher-word assignments maps using an auxiliary convolutional module. Hence, our method departs from prior work and essentially *transfers discretized (into words) versions of the feature maps*. Besides being remarkably simple, this strategy has the advantage of aligning the student's behavior with that of the teacher by focusing only on the main local visual concepts (i.e., teacher-words) that the teacher has learnt to detect in an image (see Fig. 1). As a result, it ignores unimportant feature "details" that are difficult to regress, thus avoiding over-regularizing the student. Compared to prior work (e.g., [12,36,39]), we argue that our strategy of transferring visual word representations is more effective as it does not withhold important aspects/cues of the teacher's feature maps.

We extensively evaluate our method on two knowledge distillation scenarios (i.e., model compression and transfer learning on small-sized datasets), with various pairs of teacher-student architectures, and across a variety of datasets. In all cases our simple but effective approach achieves state-of-the-art results.

2   Related Work

The aim of knowledge distillation is to transfer knowledge from a trained teacher network to a student network. To carry out the knowledge transfer, the student is trained to imitate some facets of the teacher network. For instance, Buciluǎ et al. [3] proposed to train the student as an approximator of a large ensemble of teacher networks by predicting the output of this network ensemble and Hinton et al. [13] proposed to train the student to output the same softened predictions (i.e., classification probabilities) as the trained teacher network.

The teacher and student are deep networks, having a sequential structure that is known to learn hierarchical representations of increasing abstraction. Inspired by this, some methods propose to match the intermediate activation or feature maps of the teacher as the distillation task for the student. This encourages the student to follow the intermediate solutions produced by the teacher. FitNet [28] proposes to train the student with an additional layer to regress the feature maps of the teacher. While AT [39] builds "attention maps" by aggregating feature maps, and ℓ_2 error between normalized attention maps of student and teacher is used as distillation loss. In FSP [36], the difference between Gram matrices of feature maps of the two networks is minimized for distillation.

Another line of works focuses on matching distributions of feature maps rather than feature maps themselves. In NST [15], the maximum mean discrepancy between the distributions of feature maps of teacher and student is minimized. VID [1] and CRD [32] consider maximizing mutual information between the two networks as the knowledge distillation task. Mutual information is maximized by maximizing variational lower bound (VID) or contrastive loss (CRD).

SP [34] departs from matching feature maps or their distributions. This relieves the student network from the demanding task of copying the teacher network, which could be too ambitious given the difference in their capacity and architecture. SP proposes instead a similarity preserving constraint by using difference in pairwise similarity, computed on a mini-batch, as distillation loss. Similar methods are proposed in other recent works RKD [23] and CC [25].

In our work, similar to SP, we propose a different space for distillation than feature space. We propose to use a quantized space where the teacher's feature maps are encoded by quantization with learned visual teacher-words. These visual words are learned by k-means clustering on the feature maps of the teacher thus, they represent useful semantic concepts. With the proposed quantized space we concentrate more on the important semantic concepts and their spatial correlation for knowledge distillation.

We note that quantizing features into visual words is key a ingredient of the bag-of-visual-word techniques that were extensively used in the past [5,16,26, 30,33]. Although these visual-word techniques have now been used with deep

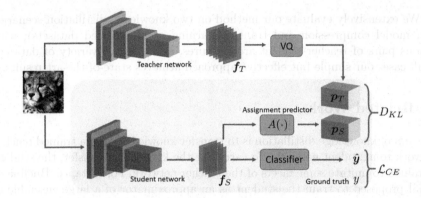

Fig. 2. Overview of the proposed method for knowledge transfer based on visual word assignments. The feature maps $\boldsymbol{f}_T$ of the teacher are converted into a visual words representation $\boldsymbol{p}_T$ via vector quantization (VQ) with soft-assignment. Then, as distillation task, the student has to predict the teacher's words assignments $\boldsymbol{p}_T$ using its own feature maps $\boldsymbol{f}_S$ and an auxiliary assignment prediction module A. The distillation loss is the KL-divergence D_{KL} of the predicted word assignments $\boldsymbol{p}_S$ with $\boldsymbol{p}_T$. The visual word vocabulary that is used in VQ is learned off-line by applying the k-means clustering algorithm to teacher feature maps extracted from the training dataset

learning [2,8,9], our work is the first to leverage a visual-word strategy with deep learning for knowledge distillation.

3   Approach

3.1   Preliminaries

Here we briefly explain the learning setting of knowledge distillation methods. Let S be a student network that we want to train using a dataset of $(\boldsymbol{x}, y)$ examples, where $\boldsymbol{x}$ is the image and y is the label. In the standard supervised learning setting the teacher S is trained to minimise for each example $(\boldsymbol{x}, y)$ the classification loss $\mathcal{L}_{CLS}(S) = \mathcal{L}_{CE}(y, \sigma(\boldsymbol{z}_S))$, where $\mathcal{L}_{CE}$ is the cross-entropy loss, $\boldsymbol{z}_S = S(\boldsymbol{x})$ are the classification logits predicted from S for the image $\boldsymbol{x}$, and $\sigma(\cdot)$ is the softmax function. Knowledge distillation methods assume that there is available a pre-trained teacher network T, which has higher capacity than that of the student S. The goal is to exploit the pre-trained knowledge of teacher T for training a better (i.e., more accurate) student S than the standard supervised learning setting. To that end, they define for each training example $(\boldsymbol{x}, y)$ an additional loss term $\mathcal{L}_{DIST}(S; T)$ based on the pre-trained teacher T. For instance, in the seminal work of Hinton et al.[13], the additional distillation loss is the cross entropy of the two network output softmax distributions for the same training example $(\boldsymbol{x}, y)$:

$$\mathcal{L}_{DIST}(S; T) = \rho^2 \mathcal{L}_{CE}\left(\sigma(\frac{\boldsymbol{z}_T}{\rho}), \sigma(\frac{\boldsymbol{z}_S}{\rho})\right), \tag{1}$$

where $z_T = T(x)$ are the logits produced from T for the image x, and $\rho > 0$ is a temperature that softens (i.e., lowers the peakiness) of the two softmax distributions. Through this loss, the student learns from the teacher by mimicking the teacher's output on the training data. Therefore, the final objective that the student S has to minimize per training example (x, y) is:

$$\mathcal{L} = \alpha \mathcal{L}_{CLS}(S) + \beta \mathcal{L}_{DIST}(S; T), \qquad (2)$$

where α and β are the weights of the two loss terms.

In the next subsection we explain the distillation loss $\mathcal{L}_{DIST}(S; T)$ that we propose in our work.

3.2 Distilling Visual Teacher-Word Assignments

An overview of our approach is provided in Fig. 2. The proposed distillation task first converts the teacher feature maps into visual word assignments and then trains the student to predict those assignments from its own feature maps.

Converting the Teacher Feature Maps into Visual Word Assignments. Given an image x, let $f_T \in \mathbb{R}^{C_T \times H_T \times W_T}$ be the feature map (with C_T channels and $H_T \times W_T$ spatial dimensions) that T generates at one of its hidden layers, and $f_T^{(h,w)} \in \mathbb{R}^{C_T}$ be the feature vector at the location (h, w) of this feature map. We quantize f_T in a spatially dense way using a predefined vocabulary $V = \{v_k\}_{k=1...K}$ of K C_T-dimensional visual teacher-word embeddings. Specifically, we first compute for each location (w, h) the squared Euclidean distances between the feature vector $f_T^{(h,w)}$ and the K teacher-words:

$$d^{(h,w)} = \left[\left\| v_k - f_T^{(h,w)} \right\|_2^2 \right]_{k=1...K}. \qquad (3)$$

Then, using $d^{(h,w)}$ we compute the K-dimensional soft-assignment vector $p_T^{(h,w)}$, which lies on a K-dimensional probability simplex, as:

$$p_T^{(h,w)} = \sigma(-d^{(h,w)}/\tau), \qquad (4)$$

where $\tau > 0$ is a temperature value used for controlling the softness of the assignment. The visual teacher-word vocabulary V is learned off-line by k-means clustering feature vectors $f_T^{(h,w)}$ extracted from the available training dataset. As we see in Fig. 1, the vocabulary V includes the main local image patterns that are captured by the teacher's feature maps (e.g., a dog's leg on grass, the eye of a tiger, or a clock tower). Hence, $p_T^{(h,w)}$ essentially encodes what semantic concepts (defined by V) the teacher network detects at location (h, w) of its feature map.

Soft-Assignment vs. Hard-Assignment. In our experiments we noticed that the optimal τ value produces very peaky soft-assignments, e.g., the softmax probability for the closest visual teacher-words is on average around 0.996, which is very close to hard assignment. However, very peaky soft-assignment still leads to slightly better students, which can be attributed to either **(a)** the fact that, as in KD [13], the probability mass for the remaining visual teacher-words carries useful teacher knowledge, or **(b)** soft-assignments regularize the visual teacher-words prediction task that the student has to perform [37].

Predicting the Teacher's Visual Word Assignments. The distillation task that we propose is to train the student S to predict the teacher soft-assignment map $\boldsymbol{p}_T$ (of visual words) based on its own feature map $\boldsymbol{f}_S \in \mathbb{R}^{C_S \times H_S \times W_S}$. As $\boldsymbol{p}_T$ encodes semantic information with spatial structure, we posit that, to predict $\boldsymbol{p}_T$ the student must build an understanding of semantics similar to the teacher. We assume for simplicity that the student's feature map $\boldsymbol{f}_S$ has the same spatial size as that of the teacher[1], i.e., $H_S = H_T$ and $W_S = W_T$. We also note that we impose *no constraint on the number of student feature channels C_S*. Therefore, to predict $\boldsymbol{p}_T$, we use an *assignment predictor A*, which consists of one cosine-similarity-based convolutional layer with 1×1 kernel size followed by a learnable scaling factor and a softmax function. Specifically, at each location (h, w), the student predicts the K-dimensional soft-assignment vector $\boldsymbol{p}_S^{(h,w)}$ from its feature vector $\boldsymbol{f}_S^{(h,w)}$, as:

$$\boldsymbol{p}_S^{(h,w)} = A(\boldsymbol{f}_S^{(h,w)}) = \sigma\left(\left[\gamma \frac{\langle \boldsymbol{W}_i, \boldsymbol{f}_S^{(h,w)} \rangle}{\|\boldsymbol{W}_i\| \|\boldsymbol{f}_S^{(h,w)}\|}\right]_{i=1\ldots K}\right), \tag{5}$$

where γ is the learnable scaling factor, $\boldsymbol{W} \in \mathbb{R}^{C_S \times K}$ are the parameters of the convolutional layer, and $\boldsymbol{W}_i$ is the i-th column of $\boldsymbol{W}$. Hence, $\boldsymbol{W}$ plays essentially the role of a new student vocabulary, and soft assignment is done according to cosine similarity (instead of Euclidean distance).

Finally, we define our distillation loss $\mathcal{L}_{DIST}(S;T)$ (for a training example) as the summation over all the locations (h, w) of the KL-divergence D_{KL} of the predicted soft-assignment distribution $\boldsymbol{p}_S^{(h,w)}$ from the target distribution $\boldsymbol{p}_T^{(h,w)}$:

$$\mathcal{L}_{DIST}(S;T) = \sum_{h,w} D_{KL}(\boldsymbol{p}_T^{(h,w)} \| \boldsymbol{p}_S^{(h,w)}) , \tag{6}$$

where $D_{KL}(\boldsymbol{p}_T^{(h,w)} \| \boldsymbol{p}_S^{(h,w)})$ consists of the cross-entropy of $\boldsymbol{p}_T^{(h,w)}$ and $\boldsymbol{p}_S^{(h,w)}$, and the entropy of $\boldsymbol{p}_T^{(h,w)}$. The latter is independent to the student and thus does not actually affect the training. We minimize the total loss $\mathcal{L} = \alpha \mathcal{L}_{CLS}(S) + \beta \mathcal{L}_{DIST}(S;T)$ of the student network over W, γ, and the parameters of S. Note that it is possible to apply the proposed distillation loss to more than one feature levels. In this case the final distillation loss is the sum of all the per-level losses.

---

[1] Otherwise, we simply down-sample the biggest feature map to the size of the smaller one with an (adaptive) average pooling layer.

3.3 Discussion

Transferring "Discretized" Representations. In our method, the teacher's feature maps are essentially discretized into visual words (i.e., as already described, the softmax probability for the closest visual teacher-words is on average around 0.996). Therefore, by minimizing the KL-divergence between the predicted visual word assignments p_S and the teacher assignments p_T, we essentially "discretize" the student feature maps according to the "discretization" at the teacher side. The difference is that in the student case, the visual vocabulary is implicitly defined by the parameters W of the assignment predictor A, and the assignment is done via cosine similarity instead of Euclidean distance.[2] Hence, the student feature maps must learn to discriminate (i.e., to cluster together) the same local image patterns (represented by the visual words) as the teacher feature maps. This way, the student learns to align its feature maps with those of the teacher network both spatially and semantically, but without being forced to regress the exact feature "details" of its teacher (which can over-regularize the student).

Posing Feature Transfer as a Classification Problem. In a sense, our method converts the direct feature regression problem into a classification one where the teacher-words are the classification prototypes. As has been shown in prior works (e.g., see methods based on contrastive learning [22,32]), this leads to a training task that is easier to solve/optimize and, as a result, constitutes a far more effective learning strategy.

Relation to KD Method [13]. Our method relates to KD in the sense that it also tries to "mimic" softened classification predictions. However, in our case those classification predictions are spatially dense and over the visual teacher-words, which represent visual concepts that are not only more localized but also much more numerous than the available image classes. Due to this fact, our distillation loss leads to a more efficient (i.e., richer) knowledge transfer. We should note at this point that many prior methods combine their proposed distillation loss with the KD loss. In our case, however, we empirically observed that adding the KD loss to our method is not required as it does not offer any significant performance improvement, which further verifies the effectiveness of our approach.

4 Experiments

In this section, we experimentally evaluate our proposed approach and compare it with several state-of-the art knowledge distillation methods. In the remainder of this section, we first compare our approach against prior art with extensive experiments on knowledge distillation in Sect. 4.1 followed by results on transfer learning in Sect. 4.2. Finally, in Sect. 4.3, we analyse and discuss the impact of various hyper-parameters and design choices of our approach.

---

[2] In Sect. A of supplementary, we discuss the use of cosine similarity on student side.

4.1   Comparison with Prior Work

We extensively evaluate our approach on three different datasets: ImageNet [6], CIFAR-100 [18] and CIFAR-10. We start with ImageNet, as this is the most challenging and interesting dataset among the three. Then, on CIFAR-100 we conduct an extensive evaluation on many network architectures and compare against several state-of-the-art methods. Finally, we conclude with CIFAR-10.

Table 1. Evaluation on ImageNet. Top-1 and Top-5 error rate of student network on ImageNet validation set, for two teacher-student combinations, ResNet34 (21.8M) to ResNet18 (11.7M) and ResNet50 (25.6M) to MobileNet (4.2M). The results for other methods are taken from [32] and [11]. FT refers to [17]

	ResNet34	ResNet18	KD	AT+KD	SP	CC	CRD	CRD+KD	Ours
Top-1	26.69	30.25	29.34	29.30	29.38	30.04	28.83	28.62	**28.33**
Top-5	8.58	10.93	10.12	10.00	10.20	10.83	9.87	9.51	**9.33**
	ResNet50	MobileNet	KD	AT+KD	FT	AB+KD	OFD	Ours	
Top-1	23.87	31.13	31.42	30.44	30.12	31.11	28.75	**27.46**	
Top-5	7.14	11.24	11.02	10.67	10.50	11.29	9.66	**8.87**	

Table 2. Experiments on CIFAR-100. Top-1 test accuracy of student networks for various student-teacher combinations and knowledge distillation methods. The pre-trained teacher models are taken from [32] and for all methods except ours we use the results reported in [32] for a fair comparison. We report average over 5 runs as in [32]

Model		Knowledge distillation methods										Teacher	Student
Teacher	Student	KD	FitNet	AT	AB	FSP	SP	VID	CRD	CRD+KD	Ours		
WRN-40-2	WRN-16-2	74.92	73.58	74.08	72.50	72.91	73.83	74.11	75.48	75.64	**76.10**	75.61	73.26
WRN-40-2	WRN-40-1	73.54	72.24	72.77	72.38	–	72.43	73.30	74.14	74.38	**74.58**	75.61	71.98
ResNet56	ResNet20	70.66	69.21	70.55	69.47	69.95	69.67	70.38	71.16	71.63	**71.84**	72.34	69.06
ResNet110	ResNet20	70.67	68.99	70.22	69.53	70.11	70.04	70.16	71.46	71.56	**71.89**	74.31	69.06
ResNet110	ResNet32	73.08	71.06	72.31	70.98	71.89	72.69	72.61	73.48	73.75	**74.08**	74.31	71.14
ResNet32x4	ResNet8x4	73.33	73.50	73.44	73.17	72.62	72.94	73.09	75.51	75.46	**75.88**	79.42	72.50
VGG13	VGG8	72.98	71.02	71.43	70.94	70.23	72.68	71.23	73.94	**74.29**		73.81	74.64	70.36

For all the experiments we use $\alpha = 1$, $\beta = 1$, and for ImageNet and CIFAR-100 $K = 4096$, $\tau = 0.2$ while for CIFAR-10 $K = 256$, $\tau = 0.005$. We apply our distillation loss on the feature maps of the last convolutional layer. For complete implementation details, see Sections B and C of the supplementary material.

ImageNet Results. In Table 1 we provide results for the ImageNet dataset, which contains 1.28M training images and 50K test images, over 1000 semantic classes. It is much more challenging than CIFAR-100 and CIRAR-10, which makes it the ultimate benchmark for evaluating distillation methods.

We evaluate our approach on it with two teacher-student combinations: ResNet34 to ResNet18 (same architecture design) and ResNet50 to MobileNet

Table 3. Distillation between different architectures. Top-1 test accuracy on CIFAR100 dataset of the student networks. The student models are learned with knowledge distillation from a teacher with different architecture. The table compares various distillation methods. Similar to Table 2, we use the pre-trained teacher networks provided by [32] and the results for other methods are also taken from [32]. Following the protocol of [32], we report average over 3 runs

Model		Knowledge distillation methods									Teacher	Student
Teacher	Student	KD	FitNet	AT	AB	SP	VID	CRD	CRD+KD	Ours		
VGG13	MobileNetV2	67.37	64.14	59.4	66.06	66.3	65.56	69.73	**69.94**	68.79	74.64	64.6
ResNet50	MobileNetV2	67.35	63.16	58.58	67.20	68.08	67.57	69.11	69.54	**69.81**	79.34	64.6
ResNet50	VGG8	73.81	70.69	71.84	70.65	73.34	70.30	74.30	74.58	**75.17**	79.34	70.36
ResNet32x4	ShuffleNetV1	74.07	73.59	71.73	73.55	73.48	73.38	75.11	75.12	**76.28**	79.42	70.50
ResNet32x4	ShuffleNetV2	74.45	73.54	72.73	74.31	74.56	73.4	75.65	76.05	**77.09**	79.42	71.82
WRN-40-2	ShuffleNetV1	74.83	73.73	73.32	73.34	74.52	73.61	76.05	76.27	**76.75**	75.61	70.50

[14] (different architecture design). We observe that our distillation method reduces the Top-1 error of ResNet18 from 30.25% to 28.33% and MobileNet from 31.13% to 27.46%, which sets the new state-of-the-art on this challenging dataset.

For ResNet34 to ResNet18, we use the results reported in CRD [32] for all other methods. Our approach outperforms all including CRD+KD which is second to ours with 0.30 higher error rate. For ResNet50 to MobileNet, we follow OFD [11]. We outperform OFD with 1.29% lower error rate, which is a very significant improvement. Note that OFD extensively studied distillation with feature regression and carefully picks the location for feature regression, uses a modified ReLU (margin-ReLU) activation function, and a more robust (partial L_2) distance loss. Nevertheless, our simple approach outperforms this well-engineered direct feature regression method.

Note that, due to the time-consuming nature of the ImageNet experiments, we did not try to tune the hyper-parameters of our method in this case (instead we reused the ones chosen for CIFAR-100). As a result, a further reduction of the error rates of the student might very well be possible by a more proper adjustment of the hyper-parameters.

CIFAR-100 Results. The CIFAR-100 dataset is one of the most commonly used dataset for evaluating knowledge distillation methods. It consists of small 32×32 resolution images and 100 semantic classes. It has 50K images in the training set and 10K in the test set which are evenly distributed across the semantic classes.

In Tables 2 and 3 we provide an exhaustive evaluation of our method under many different network architectures. Also, we compare against several prior methods, i.e., KD [13], FitNet [28], AT [39], AB [12], FSP [36], SP [34], VID [1], and CRD [32]. For network sizes and compression rates see §D of supplementary.

Table 4. CIFAR-10 experiments. Top-1 error of student networks on CIFAR-10. We use the results reported in [34] and following it, use median of 5 runs

Model		Knowledge Distillation methods				Teacher	Student
Teacher (#params)	Student (#params)	KD	AT+KD	SP	Ours		
WRN-40-1 (0.56M)	WRN-16-1 (0.17M)	8.48	8.30	8.13	**8.02**	6.51	8.74
WRN-16-2 (0.69M)	WRN-16-1 (0.17M)	7.94	8.28	**7.52**	7.55	6.07	8.74
WRN-40-2 (2.24M)	WRN-16-2 (0.69M)	6.00	5.89	**5.52**	5.56	5.18	6.07
WRN-16-8 (11M)	WRN-16-2 (0.69M)	5.62	5.47	5.34	**5.06**	4.24	6.07
WRN-16-8 (11M)	WRN-40-2 (2.24M)	4.86	**4.47**	4.55	4.48	4.24	5.18

Knowledge Transfer Between Networks with the Same Architecture Design. In Table 2, we compare knowledge distillation methods with teacher and student networks that have the same architecture design but different depth or width (width refers to the number of channels per layer). For the network architectures, we evaluate knowledge distillation between WideResNet [38], ResNet [10] and VGG [29].

Our method outperforms all the other methods on all the different teacher-student combinations, except for the *VGG13 to VGG8* case where we are second only to CRD [32] a contemporary method to ours. On average we improve by an absolute 2.97% over students without distillation, *i.e.* relatively 16.9% more than CRD (2.54% over student) and 7.6% more than CRD+KD (2.76%). Moreover, in some cases our students achieve accuracy either very close to that of the teacher (*ResNet56 to ResNet20* and *ResNet110 to ResNet32*), or even exceeds it (*WRN-40-2 to WRN-40-1*), which further confirms the potency of our method.

Knowledge Transfer Between Networks with Different Architectures. In Table 3, we evaluate the merit of our approach for distillation between different network designs. All of them but one have different spatial dimensions between feature maps of student and teacher. For example, MobileNetV2 at the penultimate layer has 2×2 feature maps while the teachers, VGG13 and ResNet50, have 4×4. Similarly, ShuffleNetV1/V2 have 4×4 while ResNet32x4 and WRN-40-2 have 8×8. As the soft-assignment maps p_T and p_S should have the same spatial dimension to apply our distillation loss, we do average pooling on the feature maps of teacher before quantization, as explained in Sect. 3.2.

We observe that our approach outperforms every other method on all but one experiment. In the cases of the ResNet32x4 teacher, we notice an improvement of more than 1% against the other methods. Overall, the proposed approach improves by an average of 5.25% on the student without distillation. While the most competitive CRD and CRD+KD bring respectively an improvement of 4.59% and 4.85%. Note that, in terms of average gain over the student without distillation, we get a relative improvement of 14.29% and 8.21% compared to CRD and CRD+KD respectively.

Table 5. Results for transfer learning. The teacher is pre-trained on ImageNet while the student is trained from scratch on a target dataset with cross-entropy loss and distillation loss. The results for other methods are reported from [1]

(a) MIT-67, ResNet34 to ResNet18

M	80	50	25	10
student	48.13	37.69	27.01	14.25
fine-tuning	70.97	66.04	58.13	47.91
LwF	63.43	51.79	41.04	22.76
FitNet	71.34	60.45	54.78	36.94
AT	58.21	48.66	43.66	27.01
NST	55.52	46.34	33.21	20.82
VID-LP	67.91	58.51	47.09	31.94
VID-I	71.34	63.66	60.07	**50.97**
LwF+FitNet	70.97	60.37	54.48	38.73
VID-LP+VID-I	71.87	65.75	61.79	50.37
Ours	**73.18**	**69.40**	**62.71**	50.92

(b) MIT-67, ResNet34 to VGG-9

M	80	50	25	10
student	53.58	43.96	29.70	15.97
fine-tuning	65.97	58.51	51.72	39.63
LwF	60.90	52.01	41.57	27.76
FitNet	70.90	64.70	54.48	40.82
AT	60.90	52.16	42.76	25.60
NST	55.60	46.04	35.22	21.64
VID-LP	68.88	61.64	50.22	39.25
VID-I	**72.01**	67.01	59.33	45.90
LwF+FitNet	70.52	64.10	54.63	40.15
VID-LP+VID-I	71.72	66.49	58.96	45.89
Ours	71.92	**67.79**	**60.10**	**47.99**

CIFAR-10 Results. The CIFAR-10 dataset is similar to CIFAR-100 with the only difference that there are now 10 semantic classes instead of 100.

Comparison with Prior Work. In Table 4, we compare our approach with KD, AT and SP in terms of error rate on CIFAR-10 test set. We consider distillation between WideResNet student and teacher with different depth and/or width. Again, our method achieves state-of-the-art results on CIFAR-10. Specifically, we outperform the other methods on two settings (*WRN-40-1* to *WRN-16-1* and *WRN-16-8* to *WRN-16-2*), while we achieve almost the same results on the other three settings with statistically negligible difference of less than 0.04%. Our approach achieves an average reduction in error rate of 0.82% compared to the student without distillation, while the most competitive SP method gets 0.75% followed by AT with 0.48%.

4.2 Transfer Learning to Small-Sized Datasets

In this section, we evaluate our method for transfer learning. In transfer learning, the objective is to train a student on a small target dataset with the aid of a teacher which is pre-trained on a large dataset. For our experimental evaluation, we use ResNet34 pre-trained on ImageNet as the teacher and, ResNet18 and VGG-9 [29] as the student network. We use MIT-67 [27] as the target dataset. It contains $15,620$ indoor scene images, classified into 67 classes. Following the evaluation protocol of [1], we sub-sample the training set with $M = \{80, 50, 25, 10\}$ images per classes. This is to assess the performance at various levels of availability of the training data. The student is trained from scratch on target data with cross-entropy loss and distillation loss. For all the transfer learning experiments we use $K = 4096$ visual words which is learned on ImageNet, where the teacher network was pre-trained.

In case of transfer learning we found that it is better to apply the proposed loss at the last two layers (layer3 and layer4) of ResNet34. For student network ResNet18 we use the same, layer3 and layer4, and for VGG-9 we use the last two max-pool layers. Exactly the same two feature levels are being used by the competing methods in this section.

The results for transfer learning experiments are given in Table 5. The table compares our method to several distillation approaches including FitNet, AT, NST VID. In the table, LwF refers to learning without forgetting [20], VID-LP and VID-I are VID [1] loss on logits and on intermediate features respectively, while VID-LP+VID-I uses both. Our proposed method outperforms all the methods on 3 out of 4 configurations of M for both the students. While being second only to VID-I with a statistically insignificant difference of less than 0.1%.

We also give results with the fine-tuning case, i.e., pre-training on ImageNet and then fine-tuning on MIT-67 without any distillation loss. Fine-tuning is a very strong baseline and standard method for transfer learning. However, it is not very practical since it requires pre-training the student on a large-size dataset (which is computationally expensive and needs access and storage of the dataset). In our experiments, the proposed method performs better than even fine-tuning.

4.3   Further Analysis

Here we analyse several aspects of our method for the model compression scenario.

Impact of Temperature Value τ. In Fig. 3a we plot how the temperature value τ of the soft-quantization of the teacher feature maps in Eq. (4) affects the performance of our method. We provide results for two teacher-student configurations, WRN-40-2 to WRN-16-2 and WRN-40-2 to ShuffleNetV1, and 5 different τ values, 0.1, 0.2, 0.5, 1.0, and 2.0. We also provide results for the hard-assignment case, which we denote with the $\tau = 0$ in the plot. We observe that choosing a small τ value, which means a more peaky soft-assignment (i.e., closer to hard-assignment), leads to better distillation performance. However, going to the extreme case of hard-assignment (i.e., $\tau = 0$) drops the distillation performance. Hence, very peaky soft-assignment achieves better knowledge transfer than the hard-assignment case.

Impact of Vocabulary Size K. In Fig. 3b we plot the distillation performance of our method as a function of the vocabulary size K on CIFAR-100. We observe a relatively stable performance for $K \geq 2048$. In our CIFAR-10 experiments we noticed that K between 128 to 256 was leading to better performance[3]. Therefore, it seems that the number of visual words K depends on the number of classes, preferably it should be sufficiently larger than the number of classes.

---

[3] See Fig. 1 in Sect. E of supplementary for performance-vs.-K plot on CIFAR-10.

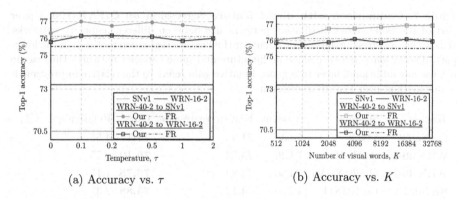

(a) Accuracy vs. τ (b) Accuracy vs. K

Fig. 3. Effect of varying temperature τ and visual words K. Top-1 accuracy on CIFAR-100 of students trained with proposed distillation loss with varying hyper-parameters τ and K. The graphs show performance of two students, WRN-16-2 and ShuffleNetv1 (SNv1) with WRN-40-2 as teacher. (a) Effect of temperature τ. (b) Effect of varying number of visual words. The straight lines refer to the student trained without distillation (solid line) or with feature regression (FR, dash-dotted line)

Table 6. Distillation at different layers. Comparing student networks with the proposed approach when applied at different layers. Block L corresponds to the penultimate layer (before classification layer) while Block $L-1$ refers to the layer at a block before Block L. We report the Top-1 accuracy

CIFAR-100		CIFAR-10	
WRN-40-2→WRN-16-2		WRN-40-1→WRN-16-1	
Block $L-1$	74.16	Block $L-1$	91.64
Block L	**76.10**	Block L	**91.98**

At Which Feature Level to Apply the Distillation Loss? In Table 6 we measure the performance of our method when the distillation loss is applied (1) to the last feature level of the teacher-student networks (which is what we have used for all other experiments on the model compression scenario), and (2) to the feature level of the previous down-sampling stage (i.e., the output of the 2nd residual block in WRN-40-2/WRN-16-2 or WRN-40-1/WRN-16-1). We report results on CIFAR-100 and CIFAR-10. In all cases, switching to the feature maps of the previous down-sampling stage leads to a drop in performance. Therefore, our proposed distillation loss appears to perform best with the last feature level.

Comparison with Direct Feature Regression. In Table 7, we explicitly compare our approach against the baseline method of directly regressing raw feature maps (Feature regression). To that end, we apply the Feature regression distillation loss on the same feature layer as in our method, i.e., the last convolutional layer. Also, we train the student network under the same conditions as in our

Table 7. Comparison with direct feature regression on CIFAR-100. Upper part of the table shows distillation performance between different capacity networks, thus we use a regression layer on student for the Feature regression method. In the lower part, distillation is between same architecture and same capacity networks thus we do not use any additional layer to regress. Relative gain refers to the relative improvement of our method over Feature regression

Teacher	Student	Student	Feature regression	Ours	Relative gain (%)
ResNet50	VGG8	70.36	71.9	**75.17**	4.55
WRN-40-2	WRN-16-2	73.26	75.53	**76.10**	0.75
WRN-40-2	ShuffleNetV1	70.50	75.89	**76.75**	1.13
ResNet32x4	ResNet8x4	72.50	74.12	**75.88**	2.37
ResNet32x4	ShuffleNetV2	71.82	75.58	**77.09**	2.00
ResNet56	ResNet20	69.06	71.56	**71.84**	0.39
ResNet50	ResNet50	79.34	79.12	**80.58**	1.85
WRN40-2	WRN40-2	75.61	**78.26**	77.96	−0.38
ResNet110	ResNet110	74.31	75.52	**76.15**	0.83
ResNet32x4	ResNet32x4	79.42	**80.57**	80.53	−0.05
ResNet56	ResNet56	72.34	73.64	**74.77**	1.53

method. In the upper part of the table, we evaluate distillation performance with different capacity networks. In this case, we add a 1×1 convolutional regression layer on top of the chosen layer for distillation of the student network. This is needed to match feature dimension of the teacher. Not surprisingly, our approach outperforms Feature regression for all the teacher-student pairs. Further, we also evaluate on distillation between same architecture and same capacity networks. This is an ideal case for Feature regression as there is no burden on student to mimic a higher capacity network and also no need for additional regression layer. Even in this ideal setup for Feature regression we outperform it for three architectures while being marginally lower on the other two; overall, our method achieves an average relative gain of 0.76% over Feature regression.

5   Conclusions

Our work deals with the important learning problem of knowledge distillation. The goal of knowledge distillation is to improve the accuracy of a student network by exploiting the learned knowledge of a pre-trained teacher network. To that end, we follow the common paradigm of transferring the knowledge encoded on the learned teacher features. However, instead of performing the distillation task in the context of the initial feature space of the teacher network, we transform it to a new quantized space.

Specifically, our distillation method first densely quantizes the teacher feature maps into visual words and then trains the student to predict this quantization based on its own feature maps. By solving this task the student is forced to align its feature maps with those of the teacher network while ignoring unimportant "feature" details, thus facilitating efficient knowledge transfer between the two networks. To demonstrate the effectiveness of our distillation method, we exhaustively evaluate it on two very common knowledge distillation scenarios, model compression (i.e., on ImageNet, CIFAR-100 and CIFAR-10 datasets) and transfer learning to small-sized datasets (i.e., ImageNet to MIT-67 datasets), and across a variety of deep network architectures. Despite its simplicity, our method manages to surpass prior work, achieving new state-of-the-art results on a variety of benchmarks.

References

1. Ahn, S., Hu, S.X., Damianou, A., Lawrence, N.D., Dai, Z.: Variational information distillation for knowledge transfer. In: Proceedings of the IEEE Conference on Computer Vision and Pattern Recognition (2019)
2. Arandjelovic, R., Gronat, P., Torii, A., Pajdla, T., Sivic, J.: NetVLAD: CNN architecture for weakly supervised place recognition. In: Proceedings of the IEEE Conference on Computer Vision and Pattern Recognition (2016)
3. Buciluǎ, C., Caruana, R., Niculescu-Mizil, A.: Model compression. In: Proceedings of the 12th ACM SIGKDD International Conference on Knowledge Discovery and Data Mining, pp. 535–541. ACM (2006)
4. Cho, J.H., Hariharan, B.: On the efficacy of knowledge distillation. In: Proceedings of the IEEE International Conference on Computer Vision, October 2019
5. Csurka, G., Dance, C., Fan, L., Willamowski, J., Bray, C.: Visual categorization with bags of keypoints. In: ECCV Workshops (2004)
6. Deng, J., Dong, W., Socher, R., Li, L.J., Li, K., Fei-Fei, L.: ImageNet: a large-scale hierarchical image database. In: Proceedings of the IEEE Conference on Computer Vision and Pattern Recognition, pp. 248–255. IEEE (2009)
7. Furlanello, T., Lipton, Z., Tschannen, M., Itti, L., Anandkumar, A.: Born-again neural networks. In: International Conference on Machine Learning, pp. 1602–1611 (2018)
8. Gidaris, S., Bursuc, A., Komodakis, N., Pérez, P., Cord, M.: Learning representations by predicting bags of visual words. In: Proceedings of the IEEE Conference on Computer Vision and Pattern Recognition, pp. 6928–6938 (2020)
9. Girdhar, R., Ramanan, D., Gupta, A., Sivic, J., Russell, B.: ActionvLAD: earning spatio-temporal aggregation for action classification. In: Proceedings of the IEEE Conference on Computer Vision and Pattern Recognition, pp. 971–980 (2017)
10. He, K., Zhang, X., Ren, S., Sun, J.: Deep residual learning for image recognition. In: Proceedings of the IEEE Conference on Computer Vision and Pattern Recognition, pp. 770–778 (2016)
11. Heo, B., Kim, J., Yun, S., Park, H., Kwak, N., Choi, J.Y.: A comprehensive overhaul of feature distillation. In: Proceedings of the IEEE International Conference on Computer Vision (2019)
12. Heo, B., Lee, M., Yun, S., Choi, J.Y.: Knowledge transfer via distillation of activation boundaries formed by hidden neurons. In: Proceedings of the AAAI Conference on Artificial Intelligence, vol. 33, pp. 3779–3787 (2019)
13. Hinton, G., Vinyals, O., Dean, J.: Distilling the knowledge in a neural network. arXiv preprint arXiv:1503.02531 (2015)

14. Howard, A.G., et al.: MobileNets: efficient convolutional neural networks for mobile vision applications. arXiv preprint arXiv:1704.04861 (2017)
15. Huang, Z., Wang, N.: Like what you like: knowledge distill via neuron selectivity transfer. arXiv preprint arXiv:1707.01219 (2017)
16. Jégou, H., Douze, M., Schmid, C., Pérez, P.: Aggregating local descriptors into a compact image representation. In: Proceedings of the IEEE Conference on Computer Vision and Pattern Recognition (2010)
17. Kim, J., Park, S., Kwak, N.: Paraphrasing complex network: network compression via factor transfer. In: Advances in Neural Information Processing Systems, pp. 2760–2769 (2018)
18. Krizhevsky, A.: Learning multiple layers of features from tiny images (2009)
19. Laine, S., Aila, T.: Temporal ensembling for semi-supervised learning. In: International Conference on Learning Representations (2017)
20. Li, Z., Hoiem, D.: Learning without forgetting. IEEE Trans. Pattern Anal. Mach. Intell. **40**(12), 2935–2947 (2017)
21. Liu, Y., et al.: Knowledge distillation via instance relationship graph. In: Proceedings of the IEEE Conference on Computer Vision and Pattern Recognition, June 2019
22. Oord, A.V.D., Li, Y., Vinyals, O.: Representation learning with contrastive predictive coding. arXiv preprint arXiv:1807.03748 (2018)
23. Park, W., Kim, D., Lu, Y., Cho, M.: Relational knowledge distillation. In: Proceedings of the IEEE Conference on Computer Vision and Pattern Recognition, June 2019
24. Passalis, N., Tefas, A.: Learning deep representations with probabilistic knowledge transfer. In: Ferrari, V., Hebert, M., Sminchisescu, C., Weiss, Y. (eds.) ECCV 2018. LNCS, vol. 11215, pp. 283–299. Springer, Cham (2018). https://doi.org/10.1007/978-3-030-01252-6_17
25. Peng, B., et al.: Correlation congruence for knowledge distillation. In: Proceedings of the IEEE International Conference on Computer Vision, October 2019
26. Perronnin, F., Dance, C.: Fisher kernels on visual vocabularies for image categorization. In: Proceedings of the IEEE Conference on Computer Vision and Pattern Recognition (2007)
27. Quattoni, A., Torralba, A.: Recognizing indoor scenes. In: Proceedings of the IEEE Conference on Computer Vision and Pattern Recognition, pp. 413–420. IEEE (2009)
28. Romero, A., Ballas, N., Kahou, S.E., Chassang, A., Gatta, C., Bengio, Y.: FitNets: hints for thin deep nets. In: International Conference on Learning Representations (2015). https://arxiv.org/abs/1412.6550
29. Simonyan, K., Zisserman, A.: Very deep convolutional networks for large-scale image recognition. arXiv preprint arXiv:1409.1556 (2014)
30. Sivic, J., Zisserman, A.: Video google: efficient visual search of videos. In: Ponce, J., Hebert, M., Schmid, C., Zisserman, A. (eds.) Toward Category-Level Object Recognition. LNCS, vol. 4170, pp. 127–144. Springer, Heidelberg (2006). https://doi.org/10.1007/11957959_7
31. Tarvainen, A., Valpola, H.: Mean teachers are better role models: weight-averaged consistency targets improve semi-supervised deep learning results. In: Advances in Neural Information Processing Systems, pp. 1195–1204 (2017)
32. Tian, Y., Krishnan, D., Isola, P.: Contrastive representation distillation. In: International Conference on Learning Representations (2020)

33. Tolias, G., Avrithis, Y., Jégou, H.: To aggregate or not to aggregate: selective match kernels for image search. In: Proceedings of the IEEE International Conference on Computer Vision (2013)
34. Tung, F., Mori, G.: Similarity-preserving knowledge distillation. In: Proceedings of the IEEE International Conference on Computer Vision (2019)
35. Yang, C., Xie, L., Qiao, S., Yuille, A.: Knowledge distillation in generations: more tolerant teachers educate better students. arXiv preprint arXiv:1805.05551 (2018)
36. Yim, J., Joo, D., Bae, J., Kim, J.: A gift from knowledge distillation: fast optimization, network minimization and transfer learning. In: Proceedings of the IEEE Conference on Computer Vision and Pattern Recognition, pp. 4133–4141 (2017)
37. Yuan, L., Tay, F.E., Li, G., Wang, T., Feng, J.: Revisit knowledge distillation: a teacher-free framework. In: Proceedings of the IEEE Conference on Computer Vision and Pattern Recognition (2020)
38. Zagoruyko, S., Komodakis, N.: Wide residual networks. In: Proceedings of the British Machine Vision Conference (2016)
39. Zagoruyko, S., Komodakis, N.: Paying more attention to attention: improving the performance of convolutional neural networks via attention transfer. In: International Conference on Learning Representations (2017). https://arxiv.org/abs/1612.03928

EGDCL: An Adaptive Curriculum Learning Framework for Unbiased Glaucoma Diagnosis

Rongchang Zhao[1], Xuanlin Chen[1], Zailiang Chen[1], and Shuo Li[2]($^{(\boxtimes)}$)

[1] School of Computer Science, Central South University, Changsha, China
zhaorc@csu.edu.cn
[2] Western University, London, ON, Canada
slishuo@gmail.com

Abstract. Today's computer-aided diagnosis (CAD) model is still far from the clinical practice of glaucoma detection, mainly due to the training bias originating from 1) the normal-abnormal class imbalance and 2) the rare but significant hard samples in fundus images. However, debiasing in CAD is not trivial because existing methods cannot cure the two types of bias to categorize fundus images. In this paper, we propose a novel curriculum learning paradigm (EGDCL) to train an unbiased glaucoma diagnosis model with the adaptive dual-curriculum. Innovatively, the dual-curriculum is designed with the guidance of evidence maps to build a training criterion, which gradually cures the bias in training data. In particular, the dual-curriculum emphasizes unbiased training contributions of data from easy to hard, normal to abnormal, and the dual-curriculum is optimized jointly with model parameters to obtain the optimal solution. In comparison to baselines, EGDCL significantly improves the convergence speed of the training process and obtains the top performance in the test procedure. Experimental results on challenging glaucoma datasets show that our EGDCL delivers unbiased diagnosis (0.9721 of Sensitivity, 0.9707 of Specificity, 0.993 of AUC, 0.966 of F2-score) and outperform the other methods. It endows our EGDCL a great advantage to handle the unbiased CAD in clinical application.

Keywords: Curriculum learning · Unbiased diagnosis · Sample imbalance · Hard sample · Computer-aided diagnosis

1 Introduction

Ophthalmic disease seriously affects the visual health of people. For example, as the common irreversible blinding ophthalmopathy, glaucoma will attack about 76 million people in the world by 2020 [27]. Computer-aided diagnosis (CAD) plays a significant role in early detection to prevent vision loss of patients with glaucoma [21,28]. Currently, the success of machine learning model has benefited ophthalmic disease diagnosis with automated algorithm [9,12], in particular,

© Springer Nature Switzerland AG 2020
A. Vedaldi et al. (Eds.): ECCV 2020, LNCS 12366, pp. 190–205, 2020.
https://doi.org/10.1007/978-3-030-58589-1_12

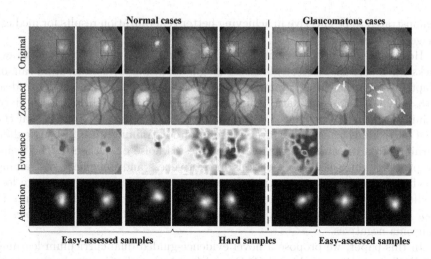

Fig. 1. Training bias is an essential yet challenging problem which seriously impedes the clinical application of CAD algorithms. The challenges originate from extreme normal-abnormal class imbalance and rare hard samples. We observe that, from both sides to the middle, glaucoma identification becomes more and more hard, whereas training samples become rarer.

automatically detecting of glaucoma in fundus images [6,12,28–31]. Through a sequence of advances, those automated diagnosis methods achieves compared accuracy with less time consuming on the challenging benchmarks.

However, training bias seriously impedes clinical applications of existing models due to the introduced false positives. In practice, there are two properties of bias encountered during training a CAD model: 1) the normal-abnormal class imbalance is suffered during collecting the training dataset in the clinic because healthy cases account for the vast majority of populations; 2) A rare of hard samples exists in abnormal cases that are clinically significant for population screening and diagnosis. Therefore, the CAD model is confronted with a great challenge, where the overwhelming majority of training data is composed of normal cases, but the trained models need to robustly recognize the hard abnormal cases, e.g., patients in the early stage of glaucoma (Fig. 1). Obviously, the biased models will misdiagnose those hard abnormal and cannot be absorbed by current healthcare infrastructures because of its limitations on the reliable assessment of hard samples [16,18] and unacceptable sensitivity.

Training bias can be potentially addressed by curriculum learning with the idea of data reweighting that assigns a weight to each sample and minimizes the weighted loss [22]. Curriculum learning [1] benefits to start with easier samples and gradually takes more complex samples into consideration. Curriculum learning highly organizes the training process by introducing different concepts at different times in curriculum to exploit previously learned concepts to ease the learning of complex one. This learning paradigm has been empirically demonstrated to be effective in achieving better generalization results for medical image analysis [10,11,13].

However, existing curriculum learning methods suffer from two crucial drawbacks when used in unbiased glaucoma diagnosis: 1) the fixed curriculum cannot adaptively represent the training criteria of the developing CAD model to deal with the biased training data, which result in inconsistency between the fixed training criteria and the biased data distribution. There is no guarantee that the fixed curriculum leads to a converged solution for training bias. 2) Curriculum learning often discards the rare hard samples as noise or outliers in the training process, which leads to a serious ineffectiveness and imbalance of training benefits. In gradient optimization, frequent easy samples contribute more loss gradients during training while hard samples are not focused. This results in poor sensitivity and biased models that cannot deal with the hard samples in glaucoma negatives.

In this paper, we propose a novel evidence-guided dual-curriculum learning (EGDCL) to train an unbiased CAD model with the adaptive dual-curriculum. The adaptive dual-curriculum is innovatively developed with the guidance of evidence maps to gradually cure the bias in training data. Therefore, the dual-curriculum can be considered as a novel adaptive training criterion to balance the training benefits of biased dataset from easy to hard, from normal to abnormal. In our EGDCL, evidence maps quantitatively provide the discriminative local features and diagnosis difficulty of each sample as the prior knowledge to identify the bias of training data. The dual-curriculum not only inherits the advantages of curriculum learning that select gradually training samples for effective training, but also adaptively learns the effective weights to balance training benefits by feature reweighting and loss reweighting.

Our EGDCL is a teacher-student framework where the student model provides prior knowledge for dual-curriculum generation by identifying the bias of the decision procedure, while the teacher model learns the CAD model for unbiased glaucoma diagnosis by resampling the data distribution with the newly-designed dual-curriculum. EGDCL is capable of achieving effective unbiased glaucoma diagnosis due to two advantages: **1)** The proposed dual-curriculum adaptively encodes training criteria of sample reweighting as sample weights and feature weights to deal with the training bias. **2)** The proposed teacher-student framework jointly optimizes the dual-curriculum designing and glaucoma classifying in a unified model to obtain the optimal solution of curriculum learning.

Our proposed EGDCL achieves top performance on two most competitive glaucoma diagnosis dataset, i.e., LAG [16] and RIM-ONE [8]. The proposed dual-curriculum learning paradigm can benefit both unbiased classification and effective training in other areas. The main contributions of this work are summarized as follows:

- A novel dual-curriculum learning paradigm (EGDCL) is proposed to tackle the issue of training bias for unbiased glaucoma diagnosis consisting of class imbalance and hard sample mining.

- An effective learning method is proposed to jointly optimize dual-curriculum designing and glaucoma classifying for the optimal solution of training bias, which provides a new learning paradigm for deep embedding learning.
- Our EGDCL achieves top performance on various competitive glaucoma datasets, demonstrating its classification effectiveness and optimal convergence speed on unbiased glaucoma diagnosis.

2   Related Work

Computer-Aided Glaucoma Diagnosis: The success of machine learning has benefited CAD applications [20], especially glaucomatous disease classification [6,24,30,31]. Prior works on glaucoma diagnosis devoted to classifier designing with hand-crafted features like texture, higher-order spectra, wavelet-based features. Those methods consider feature representation and classifier design individually, thus leads to lower classification accuracy. Along with the development of deep learning, [3,4] reports their work on automated glaucoma detection based on deep learning models. This type of diagnosis methods employs CNNs and GANs in optic disc segmentation [6,12], medical indices estimation [28–30] or ONH assessment [16,18] to promote the performance of glaucoma diagnosis.

Unbiased Classification: Both image classification [22,23] and object detection [14,19] face a large training bias. Training bias refers to a disproportionate ratio of observations among the different class, which leads to inefficient training because large redundancy of training samples exist in the biased dataset that have no contributions to model training. There have two types of methods developed to tackle the training bias: data resampling [2], which choosing the suitable proportion of data to train a network, and data reweighting that assigns a weight to each sample and minimizing the weighted loss function [22]. Curriculum learning [1] and self-paced learning [15] represents a learning regime inspired by the learning proceeds of humans that gradually proceeds from easy to more complex or hard to deal with the samples imbalance. Besides, hard negative mining samples hard samples during training [25]. Recently, the focal loss is proposed to address the class imbalance in one-stage objection detection [19]. Unfortunately, to our best knowledge, no work has been reported to tackle the special issue of training bias in disease diagnosis originating from both class imbalance and rare hard sample.

3   Methodology

As shown in Fig. 2, our EGDCL consists of three tightly integrated parts: **1)** A self-attention student network $S(\Theta)$ is proposed with an evidence identification algorithm to learn evidence maps E for the representation of training bias, *e.g.*, diagnosis difficulty and discriminative features. **2)** A curriculum generation module is innovatively designed with the help of evidence maps to learn two adaptive sequences of training criteria ($C1$ and $C2$) for training benefits balancing. **3)** A reweighted loss function is constructed for teacher network $T(\Psi)$ according to the dual-curriculum outputs (α and β) to train the unbiased diagnosis model.

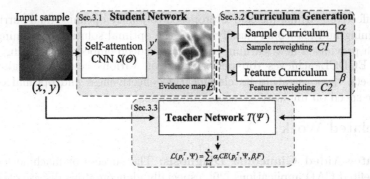

Fig. 2. The proposed evidence-guided dual-curriculum learning (EGDCL) consists of: **Student Network** for evidence identification, **Curriculum Generation** for adaptive training criteria to balance training benefits of biased data, and **Teacher Network** for unbiased glaucoma diagnosis with the regulation of dual-curriculum.

3.1   Student Network for Spatial Evidence Identification

Student network $S(\Theta)$ is constructed with two self-attention modules and an evidence identification algorithm to quantitatively identify evidences E of the decision procedure. The student network discovers evidence maps to represent the diagnosis difficulty of samples and highlight the discriminative local features supporting the disease classification, which provides prior knowledge of training bias for the dual-curriculum generation.

Student Network. The student network is a self-attention deep nets with an evidence identification algorithm for the generation of evidence maps. The self-attention structure develops two separate attention pathways, which not only learns the rich contextual features by inferring the feature interdependencies along two separate attention pathways, but also learns to focus on specific structures and contexts of the varying shapes and appearance to capture reliable biomarkers.

Given the input feature map $\mathbf{F} \in \mathbb{R}^{C \times H \times W}$, the self-attention modules infer a 3D attention map $m(\mathbf{F}) \in \mathbb{R}^{C \times H \times W}$. The refined feature $\tilde{\mathbf{F}}$ can be computed as

$$\tilde{\mathbf{F}} = \mathbf{F} \otimes (1 + m(\mathbf{F})) = \mathbf{F} \otimes (1 + m_s(\mathbf{F}) \cdot m_c(\mathbf{F})) \tag{1}$$

where $\otimes$ denotes element-wise multiplication. We adopt a residual learning scheme along with the two separate attention pathways to facilitate the gradient flow. To apply the attention modules in classification network, we first compute the spatial attention $m_s(\mathbf{F}) \in \mathbb{R}^{H} \times W$ and channel attention $m_c(\mathbf{F}) \in \mathbb{R}^{C}$ at two separate pathways, then integrate them into attention map $m(\mathbf{F}) \in \mathbb{R}^{C \times H \times W}$ by a bilinear operator

$$m(\mathbf{F}) = \|sqrt(m_s(\mathbf{F}) \otimes m_c(\mathbf{F}))\|_2 \tag{2}$$

where $\otimes$ is the cross production (Fig. 3).

Student network for spatial evidence identification

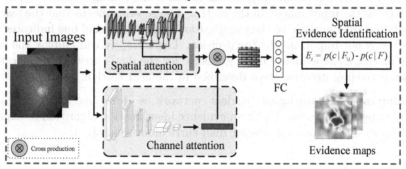

Fig. 3. Student network learns the feature independencies and then provides the quantitative evidence maps with two self-attention modules and an evidence identification algorithm. The evidence maps provide prior knowledge of training bias about diagnosis difficulty and local features for dual-curriculum generation.

Spatial Evidence Identification. Once the student network captures rich contextual features, the prediction difference analysis [32] can be adopted to estimate the spatial evidence maps by producing a relevance matrix $\mathbf{E}$, which reflects the relative importance of all features.

The relevance of a feature $\mathbf{F}_i$ can be estimated by measuring the difference between $p(c|\mathbf{F})$ and $p(c|\mathbf{F}_{\backslash i})$, where $\mathbf{F}_{\backslash i}$ denotes the set of all features except F_i. Here, i indicates the location index of the feature map or evidence map, and $c \in [0, 1]$ represents the class label where 0 indicates normal and 1 is glaucoma. The difference represents how the prediction changes if the feature is unknown.

$$E_i = p(c|\mathbf{F}_{\backslash i}) - p(c|\mathbf{F}) \tag{3}$$

The prediction $p(c|\mathbf{F}_{\backslash i})$ if feature $\mathbf{F}_i$ unknown can be simulated by marginalizing

$$p(c|\mathbf{F}_{\backslash i}) = \sum_{F_i} p(F_i|\mathbf{F}_{\backslash i}) p(c|\mathbf{F}) \tag{4}$$

In Eq. (4), the conditional probability $p(\mathbf{F}_i|\mathbf{F}_{\backslash i})$ of feature $\mathbf{F}_i$ is infeasible to be modeled because pixel value is highly dependent on other pixels in medical image. However, there exits an underlying assumption that the conditional of a pixel given its neighborhood does not depend on the position of the pixel in the image, even though a pixel often depends strongly on its small neighborhood. Therefore, the conditional probability $p(\mathbf{F}_i|\mathbf{F}_{\backslash i})$ can be approximated by assuming that feature $\mathbf{F}_i$ is independent of others $\mathbf{F}_{\backslash i}$ by finding a patch that contains $\mathbf{F}_i$. The prediction can be computed as

$$p(c|\mathbf{F}_{\backslash i}) \approx \sum_{F_i} p(\mathbf{F}_i) p(c|\mathbf{F}) \tag{5}$$

Based on the Eq. (3) and (5), we can estimate the relevance matrix E_i of the same size as the input image. In the matrix, a large value means that the feature contributed substantially to the classification, whereas a small one indicates the feature was not important for the decision. Therefore, we can employ the relevance matrix E_i as evidence maps guidance for the dual-curriculum generation in the succeeding iterative steps described in the next section.

Summarized Advantages: Student network is developed with two self-attention pathways coupled with an evidence identification algorithm to explore prior knowledge of training bias for dual-curriculum generation.

3.2   Curriculum Generation

Innovatively, the dual-curriculum is designed to exploit a novel training criteria to gradually tackle training bias. The dual-curriculum not only to adaptively balance training benefits of biased samples, but also to emphasize the training contribution of rare hard samples (Fig. 5), with the help of two types of weights α and β. The weights are updated along with the training procedure of the diagnosis model according to what knowledge the model has already learned in each iteration as described in Sect. 3.1.

Sample Curriculum $(C1)$. The sample curriculum (Fig. 4(a)) is designed to dynamically encode a set of weights on the loss function to balance the training contributions. Initially, the weights favor easily diagnosed samples, and then gradually involve an adaptive change of weights to increase the training focus of rare hard samples. In EGDCL, we propose to reshape the loss function with a weighting factor α not only to adjust the training benefits of each sample from easy to hard, but also to focus training on rare hard negatives.

Formally, the weighting factor α of samples is defined as

$$\alpha_i = \gamma(\frac{1}{1 - p_i^E})bool_i + (1 - \gamma)(\frac{1}{p_i^T}) \tag{6}$$

where γ is a hyperparameter, and p_i^T and p_i^E denote the model's estimated probability for the class with label $y = 1$ based on teacher network and evidence maps. The weighting factor consists of two parts: the former represents the contribution from evidence maps, whereas the latter denotes contribution from the training model.

For the former, a compact classifier *correct sub-network* is adopted to assess the sample x_i based on the evidence maps E_i, and give the classification probability p_i^E for the class with label $y = 1$ and recognition results y', then a *bool* function is defined to validate the effectiveness of the evidence maps for disease diagnosis and then determine its contribution to sample reweighting. So the *bool* function is defined as

$$bool_i = \begin{cases} 0 & y'_i == y_i \\ 1 & y'_i \neq y_i \end{cases} \tag{7}$$

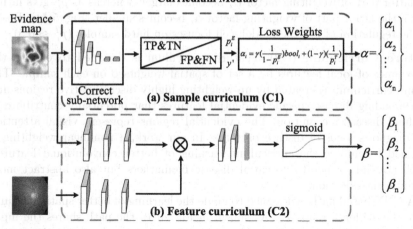

Fig. 4. The dual-curriculum adaptively provides two types of weights α and β along with the training procedure as the sample curriculum (a) and feature curriculum (b), respectively. The weights α and β are adopted in teacher network to balance the training benefits for unbiased glaucoma diagnosis.

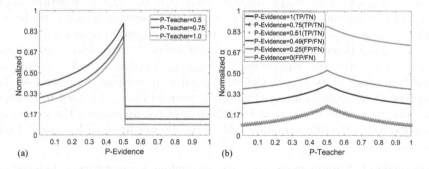

Fig. 5. A weighting factor α is proposed to balance training benefits of samples from normal to abnormal, form easy to hard. Specifically, the weighting factor enables model to focus on rare hard samples by reshaping the loss function. The factor up-weights loss contribution of rare hard samples with the greatest value when it is misclassified and its classification probability p is near to 0.5.

It should be noted that there are three properties of the weighting factor α_i in Eq. (6): **1)** When a sample is misclassified based on the evidence maps $(p_i^E < 0.5)$, $bool_i = 1$ and the weighting factor α_i is regulated by p_i^E, whereas when the sample is classified correctly, the weighting factor α_i is unaffected by p_i^E (Fig. 5(a)). This setting balances the training contribution of samples from both positives and negatives. **2)** When a sample is misclassified based on the evidence maps $(p_i^E < 0.5)$, as p_i^E gets closer to 0.5, the weighting factor α_i becomes larger and the loss is up-weighted (Fig. 5(a)). **3)** The weighting factor α_i is also regulated by p_i^T to focus on the hard samples. As p_i^T gets closer to 0.5, the latter part of weighting factor α_i becomes larger, whereas as p_i^T gets father to 0.5, the latter part of weighting factor α_i becomes smaller (Fig. 5(b)). Based on the regulation of p_i^T, the model well focuses on hard samples ($p_i^T \approx 0.5$)

Feature Curriculum (C2). Feature curriculum is designed to encode the importance of local features by a set of spatial weights β on each sample. The feature curriculum is created by up-weighting highly discriminative regions and corresponding disease-specific evidential features that potentially contribute to the final disease recognition. The evidential regions represent visual attention and diagnosis focus of disease patterns. In our work, a nonlinear weighting is designed to enforce the curriculum learning of better convolutional features, which not only generate potential disease biomarkers but also abstract more semantic classification.

A CNN-based path is designed to guide the learning of better spatial features using the evidence maps E_i. As shown in Fig. 4(b), the path shares the input image and evidence maps from the student network, and models the feature curriculum as a set of weights β of convolutional features in spatial position

$$\beta_i = UpConv(\sigma(MLP(\mathbf{E}_i) \otimes MLP(\mathbf{F}_i))) \tag{8}$$

where $\otimes$ denotes element-wise multiplication, σ is the sigmoid function. MLP and $UpConv$ indicate the operator of multi-layer perceptron and convolution with up-sampling, respectively. $\mathbf{F}_i$ is feature map outputted from MLP.

The convolutional layer with 1×1 kernel is designed to transform the multiple dimensional matrix into single channel. Sigmoid function is used to shape the value to a range of $[0, 1]$ and $UpConv$ operator up-samples the matrix as the same size of the original image (Fig. 4(b)). Sigmoid function is used to reshape the value to a range of $[0, 1]$ and $UpConv$ operator up-samples the matrix as the same size of the original image and exerts one weight on each feature of the position.

Summarized Advantages: The dual-curriculum is innovatively designed to encode two sequences of training criteria C1 and C2 with weighting factors α and β to balance the training benefits of biased data distribution.

3.3   Teacher Network for Glaucoma Diagnosis

Teacher network is a CNN-based classification model with two distinguished characteristics: **1)** an effective training objective is defined with the help of sample weights α, which is updated in each iteration towards a uniform distribution; **2)** a sophisticated feature attention is designed with the renovation of feature curriculum β to guide the teacher network capture the discriminative feature which is meaningful for glaucoma diagnosis in an iterative training process.

In standard training, the model often minimizes the expected loss for the training set and equally weight each input sample in the loss function. However, training contributions from samples with a different disease severity and distribution are unequal because of difference of gradient flow from the biased dataset.

To balance the contributions, here, the proposed EGDCL learns samples weights α_i and feature weights β for the input sample x_i. Therefore, we minimize the newly-designed loss function as

$$\Theta^*, \Psi^* = \arg\min_{\Theta, \Psi} \sum_i^N \alpha_i CE(p_i^T, \Theta, \Psi, \beta_i \mathbf{F}) \tag{9}$$

where α_i, β_i are the loss and feature weights of the ith sample, respectively. $CE(p_i^T, \Theta, \Psi, \beta_i \mathbf{F})$ denotes the standard cross-entropy loss on the sample i with the reweighted feature maps $\beta_i \mathbf{F}$. Note that $\{\alpha_i\}_i^N$ and $\{\beta_i\}_i^N$ are encoded in the dual-curriculum and adaptively assign importance weights to samples and its features in each iteration. The loss function is defined not only on the learning contribution of each sample, but also on the feature aggregation at each position.

Summarized Advantages: A novel reweighted loss function and local feature aggregation are proposed to train the unbiased diagnosis model with the debiasing training criteria (dual-curriculum).

4 Experiments and Results

To demonstrate the superiority of the proposed EGDCL, we conduct some experiments on the unbiased glaucoma diagnosis problem and compare the results with baselines and the state-of-the-art methods.

4.1 Dataset and Evaluation

Dataset. Our EGDCL is validated with the challenging dataset LAG [16], which makes public 4854 fundus images labeled with either positive glaucoma (1711) or negative glaucoma (3143). The dataset is randomly divided into training (2427) and testing (2427) sets. Furthermore, the EGDCL is also validated on other challenging dataset RIM-ONE [8] with 51 glaucomatous and 118 normal eyes. To compare with the baselines, fundus images are all resized to 224×224 before inputting to EGDCL.

Evaluation Metrics. Given the model trained with our method, the results are evaluated in terms of five different metrics: $Accuracy = \frac{TP+TN}{TP+TN+FP+FN}$, $Sensitivity = \frac{TP}{TP+FN}$, $Specificity = \frac{TN}{TN+FP}$, $F2 - score = \frac{5TP}{5TP+4FN+FP}$, and AUC. Here, TP, TN, FP, and FN are the numbers of true positive, true negative, false positive and false negative, respectively. It should be noted that the sensitivity measures the performance at detecting the positives, which is significant to evaluate how good a model is at classifying disease cases, especially hard samples. F2-score is adopted to emphasize the significance of sensitivity because a high sensitivity indicates rare overlooks of the actual positive.

In addition, the receiver operating characteristic curve (ROC) and area under ROC (AUC) are adopted in our experiments. We indicate the teacher network without dual-curriculum learning as experimental *Baseline*, sample curriculum as *C1* and feature curriculum as *C2*.

Table 1. Performance of our EGDCL on LAG under different configurations for glaucoma diagnosis with five evaluation criterion. Each cell contains the corresponding value and its improvement versus baseline.

Method	Accuracy	Sensitivity	Specificity	AUC	F2-score
Baseline	0.9604	0.9467	0.9675	0.9908	0.9448
Baseline+C1	0.9662 (↑ 0.58%)	0.9709 (↑ 0.96%)	0.9638 (↓ 0.37%)	**0.9945** (↑ **0.37%**)	0.9630 (↑ 1.84%)
Baseline+C2	0.9571 (↓ 0.33%)	0.9345 (↓ 1.22%)	0.9688 (↑ 0.13%)	0.9907 (↓ 0.01%)	0.9355 (↓ 0.93%)
Baseline+C1+C2	**0.9712** (↑ **1.08%**)	**0.9721** (↑ **2.54%**)	**0.9707** (↑ **0.32%**)	0.9931 (↑ 0.23%)	**0.9665** (↑ **2.17%**)

4.2 Training and Inference

EGDCL is configured under the teacher-student framework where student network is adopted only in training stage, whereas teacher network is implemented in both training and inference stages. When training EGDCL, the supervision of diagnosis label and attention maps are simultaneously employed for student network to obtain evidence maps. The loss function of Eq. (9) is minimized through the SGD algorithm with Adam optimizer and 0.9 momentum. The initial learning rate is set to 4×10^{-4}. The initial values of α are set as 1. $\gamma = 0.5$ in Eq. (6) and batch size is set to be 8 in our experiments. Inference involves simply forwarding an image through the trained teacher network. The predictions from teacher network are applied to final evaluations directly.

4.3 Performance of Unbiased Glaucoma Diagnosis

As shown in Fig. 6 and Table 1, EGDCL delivers unbiased glaucoma diagnosis and hard sample mining on LAG dataset [16] with the top performance on all the evaluation metrics with 0.9712 of *Accuracy*, 0.9721 of *Sensitivity*, 0.9707 of *Specificity*, 0.9665 of *F2-score* and 0.9931 of *AUC*. The results indicate that our EGDCL well handles the training bias and obtains the accuracy of glaucoma diagnosis with the help of the dual-curriculum. In particular, we need to emphasize the improvement of *Sensitivity* benefited from the accurate assessment of hard samples. Compared with the baselines, our EGDCL obtains the highest scores with *Sensitivity* of 0.9721 given the *Specificity* of 0.9707, which indicates that more cases with glaucoma are correctly identified by our method, even though the cases with heavy diagnosis difficulty. Besides, the highest *F2-score* of 0.9665 indicates our proposed EGDCL not only ensures the specificity by identifying the true negatives, but also obtains excellent sensitivity by correctly finding the true positives. This means our method can help clinicians find more of hard glaucomatous cases.

Figure 7 shows the success of our EGDCL on glaucoma diagnosis with the ROC curves and AUC values. Evidenced by ROC curves and AUC value (0.9931), the glaucoma diagnosis results indicate that our EGDCL achieves a competitive performance by mining the hard positives and negatives cases.

In addition, we conduct extensive experiments on other glaucoma dataset (RIM-ONE [8]) to demonstrate the effectiveness of the EGDCL. This experiment adopts 169 cases for training and testing, and the results show promised performance with 0.951 of *Accuracy*, 0.916 of *Sensitivity*, 0.979 of *Specificity*, 0.976 of *F2-score* and 0.927 of *AUC*.

4.4   Effectiveness of Dual-Curriculum Learning

Convergence Speed. Our EGDCL achieves the optimal convergence speed due to the help of dual-curriculum design. Compared with other configurations, EGDCL saves more than half of the training time to get the minimum of the loss. From Fig. 6 we can observe that our proposed learning paradigm can stably convergence to the minimum after about the 15th epoch. It should be noted that the feature curriculum *C2* introduces improvement of diagnosis effectiveness despite there exists a weak disturbance of the convergence curve between *Baseline+C1* and *Baseline+C1+C2*.

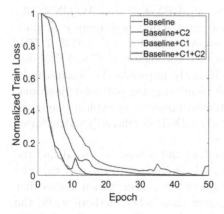
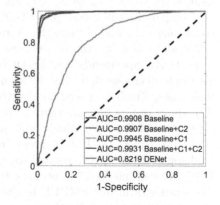

Fig. 6. The curve of train loss along with epoch demonstrates significant improvements of training convergence of diagnosis model.

Fig. 7. The ROC curves with AUC scores for glaucoma diagnosis based on the different configurations of the proposed EGDCL.

The outstanding convergence speed benefits from: **1)** the dual-curriculum gradually selects training samples from easy to hard by sample reweighting. This strategy helps to find better local minimal of a non-convex training criterion by loss reweighting, hence, our EGDCL achieves the global optimization by knowledge accumulation from easy samples. **2)** The dual-curriculum learning emphasizes the training contributions of different samples with different diagnosis difficulty, which gives rise to improved generalization and faster convergence.

Effectiveness for Hard Sample Mining. The effectiveness of the dual curriculum learning on hard sample mining can be proven by Table 1. For all the evaluation metrics, EGDCL outperforms the baseline models with an average of 1.27%, where no dual-curriculum learning is explored during the training. It should be noted that the improvement of sensitivity is 2.54% up to 0.972 whereas the improvement of specificity is 0.32% up to 0.971, which means our EGDCL can accurately discover more patients who have the condition in the early stage. These significant improvements attribute the success to hard negatives mining with the dual-curriculum learning. We can also observe from Table 1 that the integration of two types of curriculum (sample curriculum and feature curriculum) provides the optimal advance for unbiased glaucoma diagnosis.

4.5   Performance Comparison

Comparisons reveal the great advantages of our EGDCL for unbiased glaucoma diagnosis over existing methods [7,16,17,30], as shown in Table 2. We compare our EGDCL with all the methods that have been tested on the LAG dataset, namely the GON [17], DCNN [3], MCL-Net [30], DENet [7] and AG-CNN [16].

Compared results show that EGDCL achieves the best performance on glaucoma diagnosis, and obtains the average improvement of 1.08%, 1.81%, 1.77%, 1.81% and 1.55% in terms of accuracy, sensitivity, specificity, AUC and F2-score, respectively. Specifically, the EGDCL significantly improves the sensitivity of glaucoma diagnosis to 97.21% by accurately identifying the potential glaucomatous cases, which is crucial to identify potential positives in clinical diagnosis. The above results indicate that the proposed EGDCL significantly outperforms other state-of-the-art methods in all metrics.

Figure 7 plots the ROC curves of our method and others, for visualizing the trade-off between sensitivity and specificity. It is easily seen in the plot that the ROC curve of our EGDCL is closer to the upper-left corner, which means that the sensitivity of our EGDCL is always higher than other methods given the same specificity value. Further quantification evaluation is reported in Table 2, which shows the great advantages of our method in terms of AUC value.

To demonstrate the advantaged performance, our EGDCL is compared with other state-of-the-art methods on RIM-ONE dataset, which suffers more serous class balance between positives and negatives. The evaluation metrics in Table 3 indicates that our method obtains a significant improvement of 8.8% compared with the state-of-the-art, which outperforms other methods in all metrics. It is worth noted that our EGDCL performs significantly better than other methods in terms of sensitivity.

Table 2. Comparison with state-of-the-art methods for glaucoma diagnosis on LAG dataset. EGDCL achieves the best performance with the average improvement of 1.08%, 1.81%, 1.77%, 1.81% and 1.55% in terms of accuracy, sensitivity, specificity, AUC and F2-score, respectively, comparing with AG-CNN [16].

Method	Accuracy	Sensitivity	Specificity	AUC	F2-score
GON [17]	0.897	0.914	0.884	0.960	0.901
DCNN [3]	0.892	0.906	0.882	0.956	0.894
MCL-Net [30]	0.962	0.964	0.957	0.979	0.958
DENet [7]	0.756	0.631	0.843	0.822	0.650
AG-CNN [16]	0.953	0.954	0.952	0.975	0.951
Focal loss [19]	0.951	0.908	0.973	0.986	0.915
Class-balance [5]	0.949	0.915	0.968	0.986	0.919
Hard mining [26]	0.958	0.937	0.969	0.991	0.938
Our EGDCL	**0.9712** (↑ **1.08%**)	**0.9721** (↑ **2.54%**)	**0.9707** (↑ **0.32%**)	**0.9931** (↑ **0.23%**)	**0.9665** (↑ **2.17%**)

Table 3. Comparison with state-of-the-art methods for glaucoma diagnosis on RIM-ONE dataset. The proposed EGDCL outperforms others with a significant improvement of 8.8%, comparing with AG-CNN [16].

Method	Accuracy	Sensitivity	Specificity	AUC	F2-score
GON [17]	0.661	0.717	0.623	0.681	0.679
DCNN [3]	0.800	0.696	0.870	0.831	0.711
MCL-Net [30]	0.824	0.786	0.823	0.803	0.721
DENet [7]	0.558	0.492	0.569	0.574	0.338
AG-CNN [16]	0.852	0.848	0.855	0.916	0.837
Our EGDCL	**0.951** (↑ **9.85%**)	**0.916** (↑ **6.77%**)	**0.979** (↑ **12.48%**)	**0.976** (↑ **6.01%**)	**0.927** (↑ **8.98%**)

5   Conclusions

The proposed novel curriculum learning paradigm (EGDCL) performs unbiased glaucoma diagnosis by designing an adaptive dual-curriculum. Innovatively, the dual-curriculum is designed with the guidance of evidence maps to build a training criterion, which gradually cures the bias in training data. The dual-curriculum balances training benefits of biased data and gradually cures the training bias from easy to hard, from normal to abnormal. Generally, the dual-curriculum is designed to represent the training criteria of sample reweighting, which simultaneously encodes the feature and sample weights with the guidance of evidence maps. Experimental results indicate that our EGDCL outperforms the baselines and the state-of-the-art methods. The proposed EGDCL not only gives rise to improved faster convergence, but also obtains the top performance on unbiased glaucoma diagnosis. It endows our EGDCL a great advantage to handle the special issue of training bias in clinical applications.

References

1. Bengio, Y., Louradour, J., Collobert, R., Weston, J.: Curriculum learning. In: ICML, pp. 41–48. ACM (2009)
2. Chawla, N.V., Bowyer, K.W., Hall, L.O., Kegelmeyer, W.P.: SMOTE: synthetic minority over-sampling technique. J. Artif. Intell. Res. **16**, 321–357 (2002)
3. Chen, X., Xu, Y., Wong, D.W.K., Wong, T.Y., Liu, J.: Glaucoma detection based on deep convolutional neural network. In: EMBC, pp. 715–718. IEEE (2015)
4. Chen, X., Xu, Y., Yan, S., Wong, D.W.K., Wong, T.Y., Liu, J.: Automatic feature learning for glaucoma detection based on deep learning. In: Navab, N., Hornegger, J., Wells, W.M., Frangi, A.F. (eds.) MICCAI 2015. LNCS, vol. 9351, pp. 669–677. Springer, Cham (2015). https://doi.org/10.1007/978-3-319-24574-4_80
5. Cui, Y., Jia, M., Lin, T.Y., Song, Y., Belongie, S.: Class-balanced loss based on effective number of samples. In: Proceedings of the IEEE Conference on Computer Vision and Pattern Recognition, pp. 9268–9277 (2019)
6. Fu, H., Cheng, J., Xu, Y., Wong, D.W.K., Liu, J., Cao, X.: Joint optic disc and cup segmentation based on multi-label deep network and polar transformation. IEEE TMI **37**(7), 1597–1605 (2018)
7. Fu, H., et al.: Disc-aware ensemble network for glaucoma screening from fundus image. IEEE TMI **37**(11), 2493–2501 (2018)
8. Fumero, F., Alayón, S., Sanchez, J.L., Sigut, J., Gonzalez-Hernandez, M.: Rimone: an open retinal image database for optic nerve evaluation. In: 2011 24th International Symposium on Computer-based Medical Systems (CBMS), pp. 1–6. IEEE (2011)
9. Gargeya, R., Leng, T.: Automated identification of diabetic retinopathy using deep learning. Ophthalmology **124**(7), 962–969 (2017)
10. Guo, S., et al.: CurriculumNet: weakly supervised learning from large-scale web images. In: Ferrari, V., Hebert, M., Sminchisescu, C., Weiss, Y. (eds.) ECCV 2018. LNCS, vol. 11214, pp. 139–154. Springer, Cham (2018). https://doi.org/10.1007/978-3-030-01249-6_9
11. Haarburger, C., et al.: Multi scale curriculum CNN for context-aware breast MRI malignancy classification (2019)
12. Haleem, M.S., Han, L., Van Hemert, J., Li, B.: Automatic extraction of retinal features from colour retinal images for glaucoma diagnosis: a review. CMIG **37**(7–8), 581–596 (2013)
13. Jiménez-Sánchez, A., et al.: Medical-based deep curriculum learning for improved fracture classification. In: Shen, D., et al. (eds.) MICCAI 2019. LNCS, vol. 11769, pp. 694–702. Springer, Cham (2019). https://doi.org/10.1007/978-3-030-32226-7_77
14. Jin, S., et al.: Unsupervised hard example mining from videos for improved object detection. In: Ferrari, V., Hebert, M., Sminchisescu, C., Weiss, Y. (eds.) ECCV 2018. LNCS, vol. 11217, pp. 316–333. Springer, Cham (2018). https://doi.org/10.1007/978-3-030-01261-8_19
15. Kumar, M.P., Packer, B., Koller, D.: Self-paced learning for latent variable models. In: NeurIPS, pp. 1189–1197 (2010)

16. Li, L., Xu, M., Wang, X., Jiang, L., Liu, H.: Attention based glaucoma detection: a large-scale database and CNN model. In: CVPR, pp. 10571–10580 (2019)
17. Li, Z., He, Y., Keel, S., Meng, W., Chang, R.T., He, M.: Efficacy of a deep learning system for detecting glaucomatous optic neuropathy based on color fundus photographs. Ophthalmology **125**(8), 1199–1206 (2018)
18. Liao, W., Zou, B., Zhao, R., Chen, Y., He, Z., Zhou, M.: Clinical interpretable deep learning model for glaucoma diagnosis. IEEE JBHI (2019)
19. Lin, T.Y., Goyal, P., Girshick, R., He, K., Dollár, P.: Focal loss for dense object detection. In: ICCV, pp. 2980–2988 (2017)
20. Mookiah, M.R.K., Acharya, U.R., Chua, C.K., Lim, C.M., Ng, E., Laude, A.: Computer-aided diagnosis of diabetic retinopathy: a review. Comput. Biol. Med. **43**(12), 2136–2155 (2013)
21. Orlando, J., et al.: Refuge challenge: a unified framework for evaluating automated methods for glaucoma assessment from fundus photographs. Med. Image Anal. **59**, 101570 (2020)
22. Ren, M., Zeng, W., Yang, B., Urtasun, R.: Learning to reweight examples for robust deep learning. In: ICML, pp. 4331–4340 (2018)
23. Sarafianos, N., Xu, X., Kakadiaris, I.A.: Deep imbalanced attribute classification using visual attention aggregation. In: Ferrari, V., Hebert, M., Sminchisescu, C., Weiss, Y. (eds.) ECCV 2018. LNCS, vol. 11215, pp. 708–725. Springer, Cham (2018). https://doi.org/10.1007/978-3-030-01252-6_42
24. Schacknow, P.N., Samples, J.R.: The Glaucoma Book: A Practical, Evidence-based Approach to Patient Care. Springer, New York (2010). https://doi.org/10.1007/978-0-387-76700-0
25. Shrivastava, A., Gupta, A., Girshick, R.: Training region-based object detectors with online hard example mining. In: CVPR, pp. 761–769 (2016)
26. Smirnov, E., Melnikov, A., Oleinik, A., Ivanova, E., Kalinovskiy, I., Luckyanets, E.: Hard example mining with auxiliary embeddings. In: Proceedings of the IEEE Conference on Computer Vision and Pattern Recognition Workshops, pp. 37–46 (2018)
27. Tham, Y.C., Li, X., Wong, T.Y., Quigley, H.A., Cheng, C.Y.: Global prevalence of glaucoma and projections of glaucoma burden through 2040 a systematic review and meta-analysis. Ophthalmology **121**(11), 2081–2090 (2014)
28. Zhao, R., Chen, X., Xiyao, L., Zailiang, C., Guo, F., Li, S.: Direct cup-to-disc ratio estimation for glaucoma screening via semi-supervised learning. IEEE JBHI (2019)
29. Zhao, R., Chen, Z., Liu, X., Zou, B., Li, S.: Multi-index optic disc quantification via multitask ensemble learning. In: Shen, D., et al. (eds.) MICCAI 2019. LNCS, vol. 11764, pp. 21–29. Springer, Cham (2019). https://doi.org/10.1007/978-3-030-32239-7_3
30. Zhao, R., Li, S.: Multi-indices quantification of optic nerve head in fundus image via multitask collaborative learning. Med. Image Anal. **60**, 101593 (2020)
31. Zhao, R., Liao, W., Zou, B., Chen, Z., Li, S.: Weakly-supervised simultaneous evidence identification and segmentation for automated glaucoma diagnosis. In: AAAI, vol. 33, pp. 809–816. AAAI (2019)
32. Zintgraf, L.M., Cohen, T.S., Adel, T., Welling, M.: Visualizing deep neural network decisions: prediction difference analysis. arXiv preprint arXiv:1702.04595 (2017)

Backpropagated Gradient Representations for Anomaly Detection

Gukyeong Kwon$^{(\boxtimes)}$, Mohit Prabhushankar, Dogancan Temel,
and Ghassan AlRegib

Georgia Institute of Technology, Atlanta, GA 30332, USA
{gukyeong.kwon,mohit.p,cantemel,alregib}@gatech.edu

Abstract. Learning representations that clearly distinguish between normal and abnormal data is key to the success of anomaly detection. Most of existing anomaly detection algorithms use activation representations from forward propagation while not exploiting gradients from backpropagation to characterize data. Gradients capture model updates required to represent data. Anomalies require more drastic model updates to fully represent them compared to normal data. Hence, we propose the utilization of backpropagated gradients as representations to characterize model behavior on anomalies and, consequently, detect such anomalies. We show that the proposed method using gradient-based representations achieves state-of-the-art anomaly detection performance in benchmark image recognition datasets. Also, we highlight the computational efficiency and the simplicity of the proposed method in comparison with other state-of-the-art methods relying on adversarial networks or autoregressive models, which require at least 27 times more model parameters than the proposed method.

Keywords: Gradient-based representations · Anomaly detection · Novelty detection · Image recognition

1 Introduction

Recent advancements in deep learning enable algorithms to achieve state-of-the-art performance in diverse applications such as image classification, image segmentation, and object detection. However, the performance of such learning algorithms still suffers when abnormal data is given to the algorithms. Abnormal data encompasses data whose classes or attributes differ from training samples. Recent studies have revealed the vulnerability of deep neural networks against abnormal data [32,42]. This becomes particularly problematic when trained models are deployed in critical real-world scenarios. The neural networks can make wrong prediction for anomalies with high confidence and lead to vital

Electronic supplementary material The online version of this chapter (https://doi.org/10.1007/978-3-030-58589-1_13) contains supplementary material, which is available to authorized users.

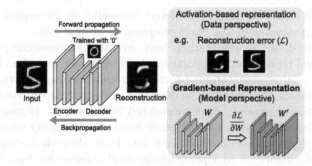

Fig. 1. Activation and gradient-based representation for anomaly detection. While activation characterizes how much of input correspond to learned information, gradients focus on model updates required by the input.

consequences. Therefore, understanding and detecting abnormal data are significantly important research topics.

Representation from neural networks plays a key role in anomaly detection. The representation is expected to clearly differentiate normal data from abnormal data. To achieve the separation, most of existing anomaly detection algorithms deploy a representation obtained in a form of activation. The activation-based representation is constrained during training. During inference, deviation of activation from the constrained representation is formulated as an anomaly score. In Fig. 1, we demonstrate an example of a widely used activation-based representation from an autoencoder. Assume that the autoencoder is trained with digit '0' and learns to accurately reconstruct curved edges. When an abnormal image, digit '5', is given to the network, the top and bottom curved edges are correctly reconstructed but the relatively complicated structure of straight edges in the middle cannot be reconstructed. Reconstruction error measures the difference between the target and the reconstructed image and it can be used to detect anomalies [1,41]. The reconstructed image, which is the activation-based representation from the autoencoder, characterizes what the network knows about input. Thus, abnormality is characterized by measuring *how much of the input does not correspond to the learned information of the network.*

In this paper, we propose using gradient-based representations to detect anomalies by characterizing model updates caused by data. Gradients are generated through backpropagation to train neural networks by minimizing designed loss functions [28]. During training, the gradients with respect to the weights provide directional information to update the neural network and learn knowledge that it has not learned. The gradients from normal data do not guide a significant change of the current weight. However, the gradients from abnormal data guide more drastic updates on the network to fully represent data. In the example given in Fig. 1, the autoencoder needs larger updates to accurately reconstruct the abnormal image, digit '5', than the normal image, digit '0'. Therefore, the gradients can be utilized as representations to characterize abnormality of data. We propose to detect anomalies by measuring *how much model update is required by the input compared to normal data*.

The gradient-based representations have several advantages compared to the activation-based representations, particularly for anomaly detection. First of all, the gradient-based representations provide abnormality characterization at different levels of data abstraction. The deviation of the activation-based representations from the constraint, often formulated as a loss ($\mathcal{L}$), is measured from the output of specific layers. On the other hand, the gradients with respect to the weights ($\frac{\partial \mathcal{L}}{\partial \mathcal{W}}$) can be obtained from any layer through backpropagation. This enables the algorithm to capture fine-grained abnormality both in low-level characteristics such as edge or color and high-level class semantics. In addition, the gradient-based representations provide directional information to characterize anomalies. The loss in the activation-based representation often measures the distance between representations of normal and abnormal data. However, by utilizing a loss defined in the gradient-based representations, we can use vectors to analyze direction in which the representation of abnormal data deviates from that of normal data. Considering that the gradients are obtained in parallel with the activation, the directional information of the gradients provides complementary features for anomaly detection along with the activation.

The gradients as representations have not been actively explored for anomaly detection. The gradients have been utilized in diverse applications such as adversarial attack generation and visualization [8,40]. However, to the best of our knowledge, this paper is the first attempt to explore the representation capability of backpropagated gradients for anomalies. We provide a theoretical explanation for using gradient-based representations to detect anomalies based on the theory of information geometry, particularly using Fisher kernel principal [10]. In addition, through comprehensive experiments with activation-based representations, we validate the effectiveness of gradient-based representations in abnormal class and condition detection, which aims at detecting data from unseen classes and abnormal conditions. We show that the proposed anomaly detection algorithm using the gradient-based representations achieves state-of-the-art performance. The main contributions of this paper are three folds:

i) We propose utilizing backpropagated gradients as representations to characterize anomalies.
ii) We validate the representation capability of gradients for anomaly detection in comparison with activation through comprehensive baseline experiments.
iii) We propose an anomaly detection algorithm using gradient-based representations and show that it outperforms state-of-the-art algorithms using activation-based representations.

2   Related Works

2.1   Anomaly Detection

Most of the existing anomaly detection algorithms are focused on learning constrained activation-based representations during training. Several works propose to directly learn hyperplane or hypersphere in hidden representation space to detect anomalies. One-Class support vector machine (OC-SVM) learns a maximum margin hyperplane which separates data from the origin in the feature space [33]. Abnormal data is expected to lie on the other side of normal data and separated by the hyperplane. The authors in [37] extend the idea of OC-SVM and propose to learn a smallest hypersphere that encloses the most of training data in the feature space. In [26], a deep neural network is trained to constrain the activation-based representations of data into the minimum volume of hypersphere. For a given test sample, an anomaly score is defined by the distance between the sample and the center of hypersphere.

An autoencoder has been a dominant learning framework for anomaly detection. The autoencoder generates two well-constrained representations, which are latent representation and reconstructed image representation. Based on these constrained representations, latent loss or reconstruction error have been widely used as anomaly scores. In [30,41], the authors argue that anomalies cannot be accurately projected in the latent space and are poorly reconstructed. Therefore, they propose to use the reconstruction error to detect anomalies. The authors in [42] fit Gaussian mixture models (GMM) to reconstruction error features and latent variables and estimate the likelihood of inputs to detect anomalies. In [1], the authors develop an autoregressive density estimation model to learn the probability distribution of the latent representation. The likelihood of the latent representation and the reconstruction error are used to detect abnormal data.

Adversarial training is also actively explored to differentiate the representation of abnormal data. In general, a generator learns to generate realistic data similar to training data and a discriminator is trained to discriminate whether the data is generated from the generator (fake) or from training data (real) [7]. The discriminator learns a decision boundary around training data and is utilized as an abnormality detector during testing. In [29], the authors adversarilally train a discriminator with an autoencoder to classify reconstructed images from original images and distorted images. The discriminator is utilized as an anomaly detector during testing. In [32], the mapping from a query image to a latent variable in a generative adversarial network (GAN) [7] is estimated. The loss which measures visual similarity and feature matching for the mapping is utilized as an anomaly score. The authors in [24] use an adversarial autoencoder [18] to learn the parameterized manifold in the latent space and estimate probability distributions for anomaly detection.

Aforementioned works exclusively focus on distinguishing between normal and abnormal data in the activation-based representations. In particular, most of the algorithms use adversarial networks or likelihood estimation networks to further constrain activation-based representations. These networks often require a large amount of training parameters and computations. We show that a directional constraint imposed on the gradient-based representations enables to achieve the state-of-the-art anomaly detection performance using only a backbone autoencoder with significantly less number of model parameters.

2.2 Backpropagated Gradients

The backpropagated gradients have been utilized in diverse applications including but not limited to visualization, adversarial attacks, and image classification. The backpropagated gradients have been widely used for the visualization of deep networks. In [36,40], information that networks have learned for a specific target class is mapped back to the pixel space through the backpropagation and visualized. The authors in [34] utilize the gradients with respect to the activation to weight the activation and visualize the reasoning for prediction that neural networks have made. An adversarial attack is another application of gradients. In [8,14], the authors show that adversarial attacks can be generated by adding an imperceptibly small vector which is the signum of input gradients. Several works have incorporated gradients with respect to the input in a form of regularization during the training of neural networks to improve the robustness [5,25,35]. Although existing works have shown that the gradients with respect to the input or the activation can be useful for diverse applications, the gradients with respect to the weights of neural networks have not been actively explored aside from its role in training deep networks.

A few works have explored the gradients with respect to the model parameters as features for data. The authors in [23] propose to use Fisher kernels which are based on the normalized gradient vectors of the generative model for image categorization. The authors in [2,3] characterize information encoded in the neural network and utilize Fisher information to represent tasks. In [15], the gradients of the neural network are utilized to classify distorted images and objectively estimate the quality of them. The gradients have been also studied as a local liner approximation to a neural network [19]. Our approach differs from other existing works in two main aspects. First, we generalize the Fisher kernel principal using the backpropagated gradients from the neural networks. Since we use the backpropagated gradients to estimate the Fisher score of normal data distribution, the data does not need to be modeled by known probabilistic distributions such as a GMM. Second, we use the gradients to represent information that the networks have not learned. In particular, we provide our interpretation of gradients which characterize abnormal information for the neural networks and validate their effectiveness in anomaly detection.

3   Gradient-Based Representations

In this section, the intuition to using gradient-based representation for anomaly detection is detailed. In particular, we present our interpretation of gradients from a geometric and a theoretical perspective. Geometric interpretation of gradients highlights the advantages of the gradients over activation from a data manifold perspective. Also, theory of information geometry further supports the characterization of anomalies using the gradients.

3.1   Geometric Interpretation of Gradients

We use an autoencoder, which is an unsupervised representation learning framework to explain the geometric interpretation of gradients. An autoencoder consists of an encoder, f_θ, and a decoder, g_ϕ. From an input image, x, a latent variable, z, is generated as $z = f_\theta(x)$ and a reconstructed image is obtained by feeding the latent variable into the decoder, $g_\phi(f_\theta(x))$. The training is performed by minimizing a loss function, $J(x; \theta, \phi)$, defined as follows:

$$J(x; \theta, \phi) = \mathcal{L}(x, g_\phi(f_\theta(x))) + \Omega(z; \theta, \phi), \tag{1}$$

where $\mathcal{L}$ is a reconstruction error, which measures the dissimilarity between the input and the reconstructed image and Ω is a regularization term for the latent variable.

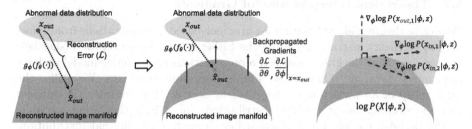

Fig. 2. Geometric interpretation of gradients.   **Fig. 3.**   Gradient constraint on the manifold.

We visualize the geometric interpretation of backpropagated gradients in Fig. 2. The autoencoder is trained to accurately reconstruct training images and the reconstructed training images form a manifold. We assume that the structure of the manifold is a linear plane as shown in the figure for the simplicity of explanation. During testing, any given input to the autoencoder is projected onto the reconstructed image manifold through the projection, $g_\phi(f_\theta(\cdot))$. Ideally, perfect reconstruction is achieved when the reconstructed image manifold includes the input image. Assume that abnormal data distribution is outside of the reconstructed image manifold. When an abnormal image, x_{out}, sampled from the distribution is input to the autoencoder, it will be reconstructed as $\hat{x}_{out}$ through the projection, $g_\phi(f_\theta(x_{out}))$. Since the abnormal image has not been utilized for training, it will be poorly reconstructed. The distance between x_{out} and $\hat{x}_{out}$ is formulated as the reconstruction error and characterizes the abnormality of the data as shown in the left side of Fig. 2. The gradients with respect to the weights, $\frac{\partial \mathcal{L}}{\partial \theta}, \frac{\partial \mathcal{L}}{\partial \phi}$, can be calculated through the backpropagation of the reconstruction error. These gradients represent required changes in the reconstructed image manifold to incorporate the abnormal image and reconstruct it accurately as shown in the right side of Fig. 2. In other words, these gradients characterize orthogonal variations of the abnormal data distribution with respect to the reconstructed image manifold.

The interpretation of gradients from the data manifold perspective highlights the advantages of gradients in anomaly detection. In activation-based representations, the abnormality is characterized by distance information measured using a designed loss function. On the other hand, the gradients provide directional information, which indicates the movement of manifold in which data representations reside. This movement characterizes, in particular, in which direction the abnormal data distribution deviates from the representations of normal data. Furthermore, the gradients obtained from different layers provide a comprehensive perspective to represent anomalies with respect to the current representations of normal data. Therefore, the directional information from gradients can be utilized as complementary information to the distance information from the activation.

3.2   Theoretical Interpretation of Gradients

We derive theoretical explanation for gradient-based representations from information geometry, particularly using the Fisher kernel. Based on the Fisher kernel, we show that the gradient-based representations characterize model updates from query data and differentiate normal from abnormal data. We utilize the same setup of an autoencoder described in Sect. 3.1 but consider the encoder and the decoder as probability distributions [6]. Given the latent variable, z, the decoder models input distribution through a conditional distribution, $P_\phi(x|z)$. The autoencoder is trained to minimize the negative log-likelihood, $-\log P_\phi(x|z)$. When x is a real value and $P_\phi(x|z)$ is assumed to be a Gaussian distribution, the decoder estimates the mean of the Gaussian. Also, the minimization of the negative log-likelihood corresponds to using a mean squared error as the reconstruction error. When x is a binary value, the decoder is assumed to be a Bernoulli distribution. The negative log-likelihood is formulated as a binary cross entropy loss. Considering the decoder as the conditional probability enables to interpret gradients using the Fisher kernel.

The Fisher kernel defines a metric between samples using the gradients of generative probability distribution [10]. Let X be a set of samples and $P(X|\theta)$ is a probability density function of the samples parameterized by $\theta = [\theta_1, \theta_2, ..., \theta_N]^\mathsf{T} \in \mathbb{R}^N$. This probability distribution models a Riemannian manifold with a local metric defined by Fisher information matrix, $F \in \mathbb{R}^{N \times N}$, as follows:

$$F = \mathop{\mathbb{E}}_{x \in X}[U_\theta^X U_\theta^{X\mathsf{T}}] \quad \text{where} \quad U_\theta^X = \nabla_\theta \log P(X|\theta). \tag{2}$$

U_θ^X is called the Fisher score which describes the contribution of the parameters in modeling the data distribution. In [10], the authors propose the Fisher kernel to measure the difference between two samples based on the Fisher score. The Fisher kernel, K_{FK}, is defined as

$$K_{FK}(X_i, X_j) = U_\theta^{X_i \mathsf{T}} F^{-1} U_\theta^{X_j}, \tag{3}$$

where X_i and X_j are two data samples. The Fisher kernels enable to extract discriminant features from the generative model and they have been actively used in diverse applications such as image categorization, image classification, and action recognition [21,23,31].

We use the Fisher kernel estimated from the autoencoder for anomaly detection. The distribution of the decoder is parameterized by the weights, ϕ, and the Fisher score from the decoder is defined as $U_{\phi,z}^X = \nabla_\phi \log P(X|\phi, z)$. Also, since the distribution is learned to be generalizable to the test data, we can use the Fisher kernel to measure the distance between training data and normal test data, and between training data and abnormal test data. The Fisher kernel for normal data (inliers), K_{FK}^{in}, and abnormal data (outliers), K_{FK}^{out}, are derived as follows, respectively:

$$K_{FK}^{in}(X_{tr}, X_{te,in}) = U_\phi^{X_{tr}\mathsf{T}} F^{-1} U_{\phi,z}^{X_{te,in}} \tag{4}$$

$$K_{FK}^{out}(X_{tr}, X_{te,out}) = U_\phi^{X_{tr}\mathsf{T}} F^{-1} U_{\phi,z}^{X_{te,out}}, \tag{5}$$

where $X_{tr}, X_{te,in}, X_{te,out}$ are training data, normal test data, and abnormal test data, respectively. For ideal anomaly detection, K_{FK}^{out} should be larger than K_{FK}^{in} to clearly differentiate normal and abnormal data. The difference between K_{FK}^{in} and K_{FK}^{out} is characterized by the Fisher scores $U_{\phi,z}^{X_{te,in}}$ and $U_{\phi,z}^{X_{te,out}}$. Therefore, the Fisher scores from query data are discriminant features for detecting anomalies. We propose to estimate the Fisher scores using the backpropagated gradients with respect to the weights of the decoder. Since the autoencoder is trained to minimize the negative log-likelihood loss, $\mathcal{L} = -\log P_\phi(x|z)$, the backpropagated gradients, $\frac{\partial \mathcal{L}}{\partial \phi}$, obtained from normal and abnormal data estimate $U_{\phi,z}^{X_{te,in}}$ and $U_{\phi,z}^{X_{te,out}}$ when the autoencoder is trained with a sufficiently large amount of data to model the data distribution. Therefore, we can interpret the gradient-based representations as discriminant representations obtained from the conditional probabilistic modeling of data for anomaly detection.

We visualize the gradients with respect to the weights of the decoder obtained by backpropagating the reconstruction error, $\mathcal{L}$, from normal data, $x_{in,1}, x_{in,2}$, and abnormal data, $x_{out,1}$, in Fig. 3. These gradients estimate the Fisher scores for inliers and outliers, which need to be clearly separated for anomaly detection. Given the definition of the Fisher scores, the gradients from normal data should contribute less to the change of the manifold compared to those from abnormal data. Therefore, the gradients from normal data should reside in the tangent space of the manifold but abnormal data results in the gradients orthogonal to the tangent space. We achieve this separation in gradient-based representations through directional constraint described in the following section.

4   Method: Gradient Constraint

The separation between inliers and outliers in the representation space is often achieved by modeling the normality of data. The deviation from the normality model captures the abnormality. The normality is often modeled through constraints imposed during training. The constraint allows normal data to be easily constrained but makes abnormal data deviates. For example, the autoencoders constrain the output to be similar to the input and the reconstruction error measures the deviation. A variational autoencoder (VAE) [12] and an adversarial autoencoder (AAE) often constrain the latent representation to follow the Gaussian distribution and the deviation from the Gaussian distribution characterizes anomalies. In the gradient-based representations, we also impose a constraint during training to model the normality of data and further differentiate $U_{\phi,z}^{X_{te,in}}$ from $U_{\phi,z}^{X_{te,out}}$ defined in Sect. 3.2.

We propose to train an autoencoder with a directional gradient constraint to model the normality. In particular, based on the interpretation of gradients from the Fisher kernel perspective, we enforce the alignment between gradients. This constraint makes the gradients from normal data aligned with each other and result in small changes to the manifold. On the other hand, the gradients from abnormal data will not be aligned with others and guide abrupt changes to the manifold. We utilize a gradient loss, $\mathcal{L}_{grad}$, as a regularization term in the entire loss function, J. We calculate the cosine similarity between the gradients of a certain layer i in the decoder at the kth iteration of training, $\frac{\partial \mathcal{L}}{\partial \phi_i}^k$, and the average of the training gradients of the same layer i obtained until the $(k-1)^{th}$ iteration, $\frac{\partial J}{\partial \phi_i}^{k-1}_{avg}$. The gradient loss at the kth iteration of training is obtained by averaging the cosine similarity over all the layers in the decoder as follows:

$$\mathcal{L}_{grad} = -\mathop{\mathbb{E}}_{i}\left[\text{cosSIM}\left(\frac{\partial J}{\partial \phi_i}^{k-1}_{avg}, \frac{\partial \mathcal{L}}{\partial \phi_i}^k\right)\right], \quad \frac{\partial J}{\partial \phi_i}^{k-1}_{avg} = \frac{1}{(k-1)}\sum_{t=1}^{k-1}\frac{\partial J}{\partial \phi_i}^t, \quad (6)$$

where J is defined as $J = \mathcal{L} + \Omega + \alpha \mathcal{L}_{grad}$. The first and the second terms are the reconstruction error and the latent loss, respectively and they are defined by different types of autoencoders. α is a weight for the gradient loss. We set sufficiently small α value to ensure that gradients actively explore the optimal weights until the reconstruction error and the latent loss become small enough. Based on the interpretation of the gradients described in Sect. 3.2, we only constrain the gradients of the decoder layers and the encoder layers remain unconstrained.

During training, $\mathcal{L}$ is first calculated from the forward propagation. Through the backpropagation, $\frac{\partial \mathcal{L}}{\partial \phi_i}^k$ is obtained without updating the weights. Based on the obtained gradient, the entire loss J is calculated and finally the weights are updated using backpropagated gradients from the loss J. An anomaly score is defined by the combination of the reconstruction error and the gradient loss as $\mathcal{L} + \beta \mathcal{L}_{grad}$. Although we use α to weight the gradient loss during training, we found that the gradient loss is often more effective than the reconstruction error for anomaly detection. To better balance the two losses, we use $\beta = 4\alpha$ for all the experiments and show that the weighted combination of two losses improve the performance. The proposed anomaly detection algorithm using **Grad**ient **Con**straint is called GradCon.

5   Experiments

5.1   Experimental Setup

We conduct anomaly detection experiments to both qualitatively and quantitatively evaluate the performance of the gradient-based representations. In particular, we perform abnormal class detection and abnormal condition detection using the gradient constraint and compare GradCon with other state-of-the-art activation-based anomaly detection algorithms. In abnormal class detection, images from one class of a dataset are considered as inliers and used for the training. Images from other classes are considered as outliers. In abnormal condition detection, images without any effect are utilized as inliers and images captured under challenging conditions such as distortions or environmental effects are considered as outliers. Both inliers and outliers are given to the network during testing. The anomaly detection algorithms are expected to correctly classify data of which class and condition differ from those of the training data.

Datasets. We utilize four benchmark datasets, which are CIFAR-10 [13], MNIST [16], fashion MNIST (fMNIST) [39], and CURE-TSR [38] to evaluate the performance of the proposed algorithm. We use CIFAR-10, MNIST, fMNIST for abnormal class detection and CURE-TSR for abnormal condition detection. CIFAR-10 dataset consists of 60,000 color images with 10 classes. MNIST dataset contains 70,000 handwritten digit images from 0 to 9 and fMNIST dataset also has 10 classes of fashion products and there are 7,000 images per class. CURE-TSR dataset has 637,560 color traffic sign images which consist of 14 traffic sign types under 5 levels of 12 different challenging conditions. For CIFAR-10, CURE-TSR, and MNIST, we follow the protocol described in [22] to create splits. To be specific, we utilize the original training and the test split of each dataset for training and testing. 10% of training images are held out for validation. For fMNIST, we follow the protocol described in [24]. The dataset is split into 5 folds and 60% of each class is used for training, 20% is used for validation, the remaining 20% is used for testing. In the experiments with CIFAR-10, MNIST, and fMNIST, we use images from one class as inliers for training. During testing, inlier images and the same number of oulier images randomly sampled from other classes are utilized. For CURE-TSR, challenge-free images are utilized as inliers for training. During testing, challenge-free images are utilized as inliers and the same images with challenging conditions are utilized as outliers. We particularly use 5 challenge levels with 8 challenging conditions which are Decolorization, Lens blur, Dirty lens, Exposure, Gaussian blur, Rain, Snow, and Haze. All the results are obtained using area under receiver operation characteristic curve

(AUROC) and we also report F1 score in fMNIST dataset for the fair comparison with the state-of-the-art method [24].

Implementation Details. We train a convolutional autoencoder (CAE) for GradCon. The encoder and the decoder consist of 4 convolutional layers and the dimension of the latent variable is $3 \times 3 \times 64$. The number of convolutional filters for each layer in the encoder is 32, 32, 64, 64 and the kernel size is 4×4 for all the layers. The architecture of the decoder is symmetric to the encoder. Adam optimizer [11] with the learning rate of 0.001 is used for training. We use mean square error as the reconstruction error and do not use any latent loss for the CAE ($\Omega = 0$). $\alpha = 0.03$ is used to weight the gradient loss.

5.2   Baseline Comparison

We compare the performance of the gradient-based representations in characterizing abnormal data with the activation-based representations. Furthermore, we show that the gradient-based representations can complement the activation-based representations and improve the performance of anomaly detection. We train four different autoencoders, which are CAE, CAE with the gradient constraint (CAE + Grad), VAE, VAE with the gradient constraint (VAE + Grad) for the baseline experiments. VAEs are trained using binary cross entropy as the reconstruction error and Kullback Leibler (KL) divergence as the latent loss. Implementation details for VAEs are same as those for the CAE described in Sect. 5.1. We train the autoencoders using images from each class of CIFAR-10. Two losses defined by the activation-based representations, which are the reconstruction error (Recon) and the latent loss (Latent), and the gradient loss (Grad) defined by the gradient-based representations are separately used as anomaly scores for detection. AUROC results are reported in Table 1 and the highest AUROC for each class is highlighted in bold.

Effectiveness of the Gradient Constraint (CAE vs. CAE+Grad). We first compare the performance of CAE and CAE + Grad to analyze the effectiveness of the gradient-based representation with constraint. The reconstruction error from CAE and CAE + Grad achieves comparable average AUROC scores. The gradient loss from CAE + Grad achieves the best performance with an average AUROC of 0.661. This shows that the gradient constraint marginally sacrifices the performance from the activation-based representation and achieve the superior performance from the gradient-based representation.

Performance Sacrifice from the Latent Constraint (CAE vs. VAE). We evaluate the effect of the latent constraint by comparing CAE and VAE. The latent loss of VAE achieves the improved performance compared to the reconstruction error of CAE by an average AUROC of 0.019. However, the performance of the reconstruction error from VAE is lower than that from CAE by 0.038. This shows that the latent constraint sacrifices the performance from another activation-based representation which is the reconstructed image. Since both latent representation and reconstructed image are obtained from forward

Table 1. Baseline anomaly detection results on CIFAR-10. The reconstruction error (Recon) and the latent loss (Latent) are obtained from the activation-based representations and the gradient loss (Grad) is obtained from the gradient-based representations.Baseline anomaly detection results on CIFAR-10. The reconstruction error (Recon) and the latent loss (Latent) are obtained from the activation-based representations and the gradient loss (Grad) is obtained from the gradient-based representations.

Model	Loss	Plane	Car	Bird	Cat	Deer	Dog	Frog	Horse	Ship	Truck	Average
CAE	Recon	0.682	0.353	0.638	0.587	0.669	**0.613**	0.495	0.498	0.711	0.390	0.564
CAE + Grad	Recon	0.659	0.356	**0.640**	0.555	0.695	0.554	0.549	0.478	0.695	0.357	0.554
	Grad	**0.752**	0.619	0.622	0.580	0.705	0.591	0.683	**0.576**	**0.774**	**0.709**	**0.661**
VAE	Recon	0.553	0.608	0.437	0.546	0.393	0.531	0.489	0.515	0.552	0.631	0.526
	Latent	0.634	0.442	**0.640**	0.497	**0.743**	0.515	**0.745**	0.527	0.674	0.416	0.583
VAE + Grad	Recon	0.556	0.606	0.438	0.548	0.392	0.543	0.496	0.518	0.552	0.631	0.528
	Latent	0.586	0.396	0.618	0.476	0.719	0.474	0.698	0.537	0.586	0.413	0.550
	Grad	0.736	**0.625**	0.591	**0.596**	0.707	0.570	0.740	0.543	0.738	0.629	0.647

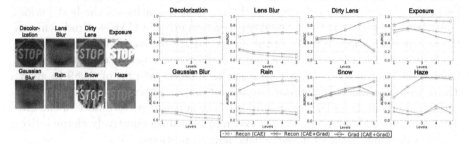

Fig. 4. Baseline anomaly detection results on CURE-TSR.

propagation, the constraint imposed in the latent space affects the reconstruction performance. Therefore, using a combination of multiple activation-based representations faces limitations in improving the performance.

Complementary Features from the Gradient Constraint (VAE vs. VAE + Grad). Comparison between VAE and VAE + Grad shows the effectiveness of using the gradient constraint with the activation constraint. The gradient loss in VAE + Grad achieves the second best average AUROC and outperforms the latent loss in the VAE by 0.064. The performance from the reconstruction error is comparable between VAE and VAE + Grad. The average AUROC of the latent loss from VAE + Grad is marginally sacrificed by 0.033 compared to that from VAE. In both CAE + Grad and VAE + Grad, the performance gain from the gradient loss is always greater than the sacrifice in other activation-based representations. This is contrary to the CAE and VAE comparison where the performance gain is smaller than the sacrifice from the reconstruction error. Since gradients are obtained in parallel with the activation, constraining gradients less affects the anomaly detection performance from the activation-based representations. Thus, the gradient-based representations can provide complementary features to the activation-based representations for anomaly detection.

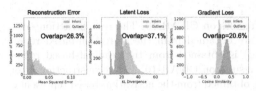

Fig. 5. Histogram analysis on activation losses and gradient loss in MNIST.

Table 2. Anomaly detection results from the gradients of each layer in the decoder.

Layer		1st	2nd	3rd	4th	All
CIFAR-10		0.648	0.649	0.628	0.605	0.661
CURE-TSR	DL	0.688	0.640	0.649	0.681	0.708
	EX	0.859	0.811	0.781	0.833	0.891
	SN	0.677	0.612	0.628	0.693	0.702

Abnormal Condition Detection. We further analyze the discriminant capability of the gradient-based representations for diverse challenging conditions and levels. We compare the performance of CAE and CAE + Grad using the reconstruction error (Recon) and the gradient loss (Grad). Samples with challenging conditions and the AUROC performance are visualized in Fig. 4. For all challenging conditions and levels, CAE + Grad achieves the best performance. In particular, except for snow level 1–3, the gradient loss achieves the best performance and for snow level 1–3, the reconstruction error of CAE + Grad achieves the best performance. In terms of the average AUROC over challenge levels, the gradient loss of CAE + Grad outperforms the reconstruction error of CAE by the largest margin of 0.612 in rain and the smallest margin of 0.089 in snow. These test conditions encompass acquisition imperfection, processing artifact, and environmental challenging conditions. The superior performance of the gradient loss shows that the gradient-based representation effectively characterizes diverse types and levels of unseen challenging conditions.

Decomposition of the Gradient Loss. We decompose the gradient loss and analyze the contribution of gradients from each layer on anomaly detection. Instead of the gradient loss obtained by averaging the cosine similarity over all the layers as (6), we use the cosine similarity from each layer as an anomaly score. The average AUROC results obtained by the gradients from the first to the fourth layer of the decoder are reported in Table 2. Also, results obtained by averaging the cosine similarity over all layers are reported. We use CIFAR-10 and Dirty Lens (DL), Exposure (EX), Snow (SN) challenge types of CURE-TSR. In CIFAR-10, inlier class and outlier classes share most of low-level features such as edges or colors. Also, semantic information mostly differentiate classes. Since the layers close to the latent space focus more on high-level characteristics of data, the gradient loss from the first and the second layer show the largest contribution on anomaly detection. In CURE-TSR, challenging conditions alter low-level characteristics of images such as edges or colors. Therefore, the last layer of the decoder also contributes more than middle layers for abnormal condition detection. This shows that gradients extracted from different layers characterize abnormality at different levels of data abstraction. In both datasets, results obtained by combining all the layers (All) show the best performance. Given that losses defined by activation-based representations can be calculated only from the output of specific layers, using gradients from all the layers enable to capture abnormality in both low-level and high-level characteristics of data.

Table 3. Anomaly detection AUROC results on CIFAR-10.

	Plane	Car	Bird	Cat	Deer	Dog	Frog	Horse	Ship	Truck	Average
OCSVM [33]	0.630	0.440	0.649	0.487	0.735	0.500	0.725	0.533	0.649	0.508	0.586
KDE [4]	0.658	0.520	**0.657**	0.497	0.727	0.496	**0.758**	0.564	0.680	0.540	0.610
DAE [9]	0.411	0.478	0.616	0.562	0.728	0.513	0.688	0.497	0.487	0.378	0.536
VAE [12]	0.634	0.442	0.640	0.497	**0.743**	0.515	0.745	0.527	0.674	0.416	0.583
PixelCNN [20]	**0.788**	0.428	0.617	0.574	0.511	0.571	0.422	0.454	0.715	0.426	0.551
LSA [1]	0.735	0.580	**0.690**	0.542	**0.761**	0.546	**0.751**	0.535	0.717	0.548	0.641
AnoGAN [32]	0.671	0.547	0.529	0.545	0.651	0.603	0.585	**0.625**	0.758	0.665	0.618
DSVDD [27]	0.617	**0.659**	0.508	**0.591**	0.609	**0.657**	0.677	**0.673**	0.759	**0.731**	0.648
OCGAN [22]	0.757	0.531	0.640	**0.620**	0.723	**0.620**	0.723	0.575	**0.820**	0.554	**0.657**
GradCon	**0.760**	**0.598**	0.648	0.586	0.733	0.603	0.684	0.567	**0.784**	**0.678**	**0.664**

Table 4. Anomaly detection AUROC results on MNIST.

	0	1	2	3	4	5	6	7	8	9	Average
OCSVM [33]	0.988	**0.999**	0.902	0.950	0.955	0.968	0.978	0.965	0.853	0.955	0.951
KDE [4]	0.885	0.996	0.710	0.693	0.844	0.776	0.861	0.884	0.669	0.825	0.814
DAE [9]	0.894	**0.999**	0.792	0.851	0.888	0.819	0.944	0.922	0.740	0.917	0.877
VAE [12]	**0.997**	**0.999**	0.936	0.959	**0.973**	0.964	**0.993**	0.976	0.923	**0.976**	0.970
PixelCNN [20]	0.531	0.995	0.476	0.517	0.739	0.542	0.592	0.789	0.340	0.662	0.618
LSA [1]	0.993	**0.999**	**0.959**	**0.966**	0.956	0.964	**0.994**	**0.980**	**0.953**	**0.981**	**0.975**
AnoGAN [32]	0.966	0.992	0.850	0.887	0.894	0.883	0.947	0.935	0.849	0.924	0.913
DSVDD [27]	0.980	**0.997**	0.917	0.919	0.949	0.885	0.983	0.946	**0.939**	0.965	0.948
OCGAN [22]	**0.998**	**0.999**	0.942	0.963	**0.975**	**0.980**	0.991	**0.981**	**0.939**	**0.981**	**0.975**
GradCon	0.995	**0.999**	**0.952**	**0.973**	0.969	**0.977**	**0.994**	0.979	0.919	0.973	**0.973**

5.3 Comparison with State-of-The-Art Algorithms

We evaluate the performance of GradCon which uses the combination of the reconstruction error and the gradient loss as an anomaly score. We compare GradCon with other benchmarking and state-of-the-art algorithms. The AUROC results on CIFAR-10 and MNIST are reported in Table 3 and Table 4, respectively. Top two AUROC scores for each class are highlighted in bold. GradCon achieves the best average AUROC performance in CIFAR-10 while achieving the second best performance in MNIST by the gap of 0.002. In Fig. 5, we visualize the histogram of the reconstruction error, the latent loss, and the gradient loss for inliers and outliers to further analyze the state-of-the-art performance of the proposed method. We calculate each loss for all the inliers and the outliers in MNIST. Also, we provide the percentage of overlap calculated by dividing the number of samples in the overlapped region of the histograms by the total number of samples. Ideally, measured errors on each representation should separate the histograms of inliers and outliers as much as possible for effective anomaly detection. The gradient loss achieves the least number of samples overlapped

Table 5. Anomaly detection results on fMNIST.

% of outlier		10	20	30	40	50
F1	GPND	**0.968**	**0.945**	0.917	0.891	0.864
	Grad	0.964	0.939	0.917	0.899	0.870
	GradCon	0.967	**0.945**	**0.924**	**0.905**	**0.871**
AUC	GPND	0.928	0.932	0.933	0.933	0.933
	Grad	0.931	0.925	0.926	0.928	0.926
	GradCon	**0.938**	**0.933**	**0.935**	**0.936**	**0.934**

Table 6. Number of model parameters.

Method	# of parameters
AnoGAN	6,338,176
GPND	6,766,243
LSA	13,690,160
GradCon	**230,721**

which explains the state-of-the art performance achieved by GradCon. We also evaluate the performance of GradCon in comparison with another state-of-the-art algorithm denoted as GPND [24] in fMNIST. In this fMNIST experiment, we change the ratio of outliers in the test set from 10% to 50% and evaluate the performance in terms of AUROC and F1 score. We report the results from the gradient loss (Grad) and GradCon in Table 5. GradCon outperforms GPND in all outlier ratios in terms of AUROC. Except for the 10% of outlier ratio, GradCon achieves higher F1 scores than GPND. The results of the gradient loss and GradCon show that the combination of the gradient loss and the reconstruction error improves the performance for all the outlier ratios in terms of AUROC and F1 score.

Computational Efficiency of GradCon. GradCon requires significantly less computational resources compared to other state-of-the-art algorithms. To show the computational efficiency of GradCon, we measure the average inference time per image using a machine with two GTX Titan X GPUs and compare computation time. While the average inference time per image for GPND on fMNIST is 5.72 ms, GradCon takes only 3.08 ms which is around 1.9 time faster. Also, we compare the number of model parameters for GradCon with that for the state-of-the-art algorithms in Table 6. AnoGAN, GPND, and LSA are based on a GAN [7], an AAD [18], and an autoregressive model [17], respectively but GradCon is solely based on a CAE. Hence, the number of model parameters for GradCon is approximately 27, 29, 59 times less than that for AnoGAN, GPND, and LSA, respectively. Most of the state-of-the-art algorithms require additional training of adversarial networks or probabilistic modeling on top of the activation-based representations from the encoder and the decoder. Since GradCon is only based on the reconstruction error and the gradient loss of the CAE, it is computationally efficient even while achieving the state-of-the-art performance.

6   Conclusion

We propose using a gradient-based representation for anomaly detection by characterizing model behavior on anomalies. We introduce the geometric interpretation of gradients and derive an anomaly score based on the deviation of gradients from the directional constraint. From thorough baseline analysis, we show the effectiveness of gradient-based representations for anomaly detection in comparison with the activation-based representations. Also, the proposed anomaly detection algorithm, GradCon, which is the combination of the reconstruction error and the gradient loss achieves the state-of-the-art performance in benchmarking image recognition datasets. In terms of the computational efficiency, GradCon has significantly less number of model parameters and shows faster inference time compared to other state-of-the-art anomaly detection algorithms. Given that most of anomaly detection algorithms adopt adversarial training frameworks or probabilistic modelings on activation-based representations, using more sophisticated training frameworks on gradient-based representations remains for future work.

A   Appendix

In Sect. A.1, we compare the performance of GradCon with other benchmarking and state-of-the-art algorithms on fMNIST. In Sect. A.2, we perform statistical analysis and highlight the separation between inliers and outliers achieved by using the gradient-based representations in CIFAR-10. In Sect. A.3, we analyze different parameter settings for GradCon. Finally, we provide additional details on CURE-TSR dataset in Sect. A.4.

A.1   Additional Results on fMNIST

We compared the performance of GradCon with other benchmarking and state-of-the-art algorithms using CIFAR-10 and MNIST in Table 3 and 4. In Table 5 of the paper, we mainly focused on rigorous comparison between GradCon and GPND which shows the second best performacne in terms of the average AUROC on fMNIST. In this section, we report the average AUROC performance of Grad-Con in comparison with that of additional benchmarking and state-of-the-art algorithms using fMNIST in Table 7. The same experimental setup for fMNIST described in Sect. 5.1 is utilized and the test set contains the same number of inliers and outliers. GradCon outperforms all the compared algorithms including GPND. Given that ALOCC, OCGAN, and GPND are all based on adversarial training to further constrain the activation-based representations, GradCon achieves the best performance in fMNIST only based on a CAE and requires significantly less computations.

Table 7. Average AUROC result of GradCon compared with benchmarking and state-of-the-art anomaly detection algorithms on fMNIST.

Method	ALOCC DR [29]	ALOCC D [29]	DCAE [30]	OCGAN [22]	GPND [24]	**GradCon**
AUROC	0.753	0.601	0.908	0.924	0.933	**0.934**

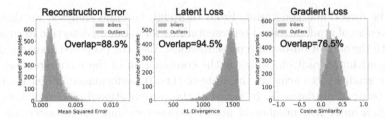

Fig. 6. Histogram analysis on activation losses and gradient loss in CIFAR-10. For each class, we calculate the activation losses and the gradient loss from inliers and outliers. The losses from all 10 classes are visualized using histograms. The percentage of overlap is calculated by dividing the number of samples in the overlapped region of the histograms by the total number of samples.

A.2   Histogram Analysis on CIFAR-10

We presented histogram analysis using gray scale digit images in MNIST to explain the state-of-the-art performance achieved by GradCon in Fig. 5. In this section, we perform the same histogram analysis using color images of general objects in CIFAR-10 to further highlight the separation between inliers and outliers achieved by the gradient-based representations. We obtain histograms for CIFAR-10 through the same procedures that are used to generate histograms for MNIST visualized in Fig. 5. In Fig. 6, we visualize the histograms of the reconstruction error, the latent loss, and the gradient loss in CIFAR-10. Also, we provide the percentage of overlap between histograms from inliers and outliers. The measured error on each representation is expected to differentiate inliers from outliers and achieve as small as possible overlap between histograms. The gradient loss shows the smallest overlap compared to other two losses defined in activation-based representations. This statistical analysis also supports the superior performance of GradCon compared to other reconstruction error or latent loss-based algorithms reported in Table 3.

Comparison between histograms from MNIST visualized in Fig. 5 and those from CIFAR-10 shows that the gradient loss is more effective when data becomes complicated and challenging for anomaly detection. In MNIST, simple low-level features such as curved edges or straight edges can be class discriminant features for anomaly detection. On the other hand, CIFAR-10 contains images with richer structure and features than MNIST. Therefore, normal and abnormal data are not easily separable and the overlap between histograms is significantly larger in CIFAR-10 than MNIST. In CIFAR-10, the overlap of the gradient loss is smaller than the second smallest overlap of the reconstruction error by 12.4%. In MNIST, the overlap of the gradient loss is smaller than the second smallest overlap by 5.7%. GradCon also outperforms other state-of-the-art methods by a larger margin of AUROC in CIFAR-10 compared to MNIST. The overlap and performance differences show that the contribution of the gradient loss becomes more significant when data is complicated and challenging for anomaly detection.

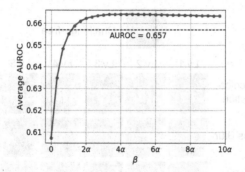

Fig. 7. Average AUROC results with different β parameters in CIFAR-10. $\alpha = 0.03$ is utilized to train the CAE. The dotted line (average AUROC = 0.657) indicates the performance of OCGAN which achieves the second best performance in CIFAR-10.

A.3   Parameter Setting for the Gradient Loss

We analyze the impact of different parameter settings on the performance of GradCon. The final anomaly score of GradCon is given as $\mathcal{L} + \beta\mathcal{L}_{grad}$, where $\mathcal{L}$ is the reconstruction error and $\mathcal{L}_{grad}$ is the gradient loss. While we use α parameter to weight the gradient loss and constrain the gradients during training, we observe that the gradient loss generally shows better performance as an anomaly score than the reconstruction error. Hence, we use $\beta = n\alpha$, where n is constant, to weight the gradient loss more for the anomaly score. We evaluate the average AUROC performance of GradCon with different β parameters using CIFAR-10 in Fig. 7. In particular, we change the scaling constant, n, to change β in the x-axis of the plot. The performance of GradCon improves as we increase β in the range of $\beta = [0, 2\alpha]$. Also, GradCon consistently achieves state-of-the-art performance across a wide range of β parameter settings when $\beta \geq 1.67\alpha$. To be specific, GradCon always outperforms OCGAN which achieves the second best average AUROC performance of 0.657 in CIFAR-10 when $\beta \geq 1.67\alpha$. This analysis shows that GradCon achieves the best performance in CIFAR-10 across a wide range of β.

A.4   Additional Details on CURE-TSR Dataset

We visualize traffic sign images with 8 different challenge types and 5 different levels in Fig. 8. Level 5 images contain the most severe challenge effect and level 1 images are least affected by the challenging conditions. Since level 1 images are perceptually most similar to the challenge-free image, it is more challenging for anomaly detection algorithms to classify level 1 images as outliers. The gradient loss from CAE + Grad outperforms the reconstruction error from CAE in all level 1 challenge types. This result shows that the gradient loss consistently outperforms the reconstruction error even when inliers and outliers become relatively similar under mild challenging conditions.

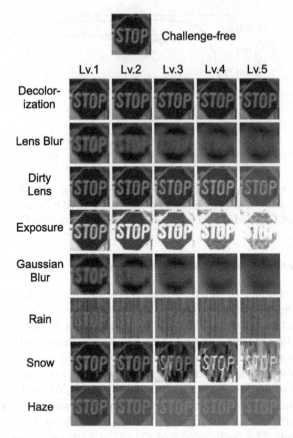

Fig. 8. A challenge-free stop sign and stop signs with 8 different challenge types and 5 different challenge levels. Challenging conditions become more severe as the level becomes higher.

References

1. Abati, D., Porrello, A., Calderara, S., Cucchiara, R.: Latent space autoregression for novelty detection. In: Proceedings of the IEEE Conference on Computer Vision and Pattern Recognition, pp. 481–490 (2019)
2. Achille, A.: Task2vec: task embedding for meta-learning. In: Proceedings of the IEEE International Conference on Computer Vision, pp. 6430–6439 (2019)
3. Achille, A., Paolini, G., Soatto, S.: Where is the information in a deep neural network? arXiv preprint arXiv:1905.12213 (2019)
4. Bishop, C.M.: Pattern Recognition and Machine Learning. Springer, Boston (2006). https://doi.org/10.1007/978-1-4615-7566-5
5. Drucker, H., Le Cun, Y.: Double backpropagation increasing generalization performance. In: IJCNN 1991-Seattle International Joint Conference on Neural Networks, vol. 2, pp. 145–150. IEEE (1991)
6. Goodfellow, I., Bengio, Y., Courville, A., Bengio, Y.: Deep Learning, vol. 1. MIT Press, Cambridge (2016)

7. Goodfellow, I., et al.: Generative adversarial nets. In: Advances in Neural Information Processing Systems, pp. 2672–2680 (2014)
8. Goodfellow, I.J., Shlens, J., Szegedy, C.: Explaining and harnessing adversarial examples. arXiv preprint arXiv:1412.6572 (2014)
9. Hadsell, R., Chopra, S., LeCun, Y.: Dimensionality reduction by learning an invariant mapping. In: Proceedings of the IEEE Conference on Computer Vision and Pattern Recognition, vol. 2, pp. 1735–1742. IEEE (2006)
10. Jaakkola, T., Haussler, D.: Exploiting generative models in discriminative classifiers. In: Advances in Neural Information Processing Systems, pp. 487–493 (1999)
11. Kingma, D.P., Ba, J.: Adam: a method for stochastic optimization. arXiv preprint arXiv:1412.6980 (2014)
12. Kingma, D.P., Welling, M.: Auto-encoding variational bayes. arXiv preprint arXiv:1312.6114 (2013)
13. Krizhevsky, A., Hinton, G.: Learning multiple layers of features from tiny images. Technical report, Citeseer (2009)
14. Kurakin, A., Goodfellow, I., Bengio, S.: Adversarial machine learning at scale. arXiv preprint arXiv:1611.01236 (2016)
15. Kwon, G., Prabhushankar, M., Temel, D., AlRegib, G.: Distorted representation space characterization through backpropagated gradients. In: 2019 26th IEEE International Conference on Image Processing (ICIP). IEEE (2019)
16. LeCun, Y., Bottou, L., Bengio, Y., Haffner, P., et al.: Gradient-based learning applied to document recognition. Proc. IEEE 86(11), 2278–2324 (1998)
17. Luc, P., Neverova, N., Couprie, C., Verbeek, J., LeCun, Y.: Predicting deeper into the future of semantic segmentation. In: Proceedings of the IEEE International Conference on Computer Vision, pp. 648–657 (2017)
18. Makhzani, A., Shlens, J., Jaitly, N., Goodfellow, I., Frey, B.: Adversarial autoencoders. arXiv preprint arXiv:1511.05644 (2015)
19. Mu, F., Liang, Y., Li, Y.: Gradients as features for deep representation learning. In: International Conference on Learning Representations (2020)
20. Van den Oord, A., Kalchbrenner, N., Espeholt, L., Vinyals, O., Graves, A., et al.: Conditional image generation with pixel CNN decoders. In: Advances in Neural Information Processing Systems, pp. 4790–4798 (2016)
21. Peng, X., Zou, C., Qiao, Yu., Peng, Q.: Action recognition with stacked fisher vectors. In: Fleet, D., Pajdla, T., Schiele, B., Tuytelaars, T. (eds.) ECCV 2014. LNCS, vol. 8693, pp. 581–595. Springer, Cham (2014). https://doi.org/10.1007/978-3-319-10602-1_38
22. Perera, P., Nallapati, R., Xiang, B.: OcGAN: one-class novelty detection using GANs with constrained latent representations. In: Proceedings of the IEEE Conference on Computer Vision and Pattern Recognition, pp. 2898–2906 (2019)
23. Perronnin, F., Dance, C.: Fisher kernels on visual vocabularies for image categorization. In: Proceedings of the IEEE Conference on Computer Vision and Pattern Recognition, pp. 1–8. IEEE (2007)
24. Pidhorskyi, S., Almohsen, R., Doretto, G.: Generative probabilistic novelty detection with adversarial autoencoders. In: Advances in Neural Information Processing Systems, pp. 6822–6833 (2018)
25. Ross, A.S., Doshi-Velez, F.: Improving the adversarial robustness and interpretability of deep neural networks by regularizing their input gradients. In: Thirty-second AAAI Conference on Artificial Intelligence (2018)
26. Ruff, L., et al.: Deep one-class classification. In: International Conference on Machine Learning, pp. 4390–4399 (2018)

27. Ruff, L., et al.: Deep one-class classification. In: International Conference on Machine Learning, pp. 4393–4402 (2018)
28. Rumelhart, D.E., Hinton, G.E., Williams, R.J.: Learning representations by back-propagating errors. Nature **323**(6088), 533 (1986)
29. Sabokrou, M., Khalooei, M., Fathy, M., Adeli, E.: Adversarially learned one-class classifier for novelty detection. In: Proceedings of the IEEE Conference on Computer Vision and Pattern Recognition, pp. 3379–3388 (2018)
30. Sakurada, M., Yairi, T.: Anomaly detection using autoencoders with nonlinear dimensionality reduction. In: Proceedings of the MLSDA 2014 2nd Workshop on Machine Learning for Sensory Data Analysis, p. 4. ACM (2014)
31. Sánchez, J., Perronnin, F., Mensink, T., Verbeek, J.: Image classification with the fisher vector: theory and practice. Int. J. Comput. Vis. **105**(3), 222–245 (2013)
32. Schlegl, T., Seeböck, P., Waldstein, S.M., Schmidt-Erfurth, U., Langs, G.: Unsupervised anomaly detection with generative adversarial networks to guide marker discovery. In: Niethammer, M., et al. (eds.) IPMI 2017. LNCS, vol. 10265, pp. 146–157. Springer, Cham (2017). https://doi.org/10.1007/978-3-319-59050-9_12
33. Schölkopf, B., Platt, J.C., Shawe-Taylor, J., Smola, A.J., Williamson, R.C.: Estimating the support of a high-dimensional distribution. Neural Comput. **13**(7), 1443–1471 (2001)
34. Selvaraju, R.R., Cogswell, M., Das, A., Vedantam, R., Parikh, D., Batra, D.: Grad-CAM: visual explanations from deep networks via gradient-based localization. In: Proceedings of the IEEE International Conference on Computer Vision, pp. 618–626 (2017)
35. Sokolić, J., Giryes, R., Sapiro, G., Rodrigues, M.R.: Robust large margin deep neural networks. IEEE Trans. Signal Process. **65**(16), 4265–4280 (2017)
36. Springenberg, J.T., Dosovitskiy, A., Brox, T., Riedmiller, M.: Striving for simplicity: the all convolutional net. arXiv preprint arXiv:1412.6806 (2014)
37. Tax, D.M., Duin, R.P.: Support vector data description. Mach. Learn. **54**(1), 45–66 (2004)
38. Temel, D., Kwon, G., Prabhushankar, M., AlRegib, G.: Cure-TSR: challenging unreal and real environments for traffic sign recognition. arXiv preprint arXiv:1712.02463 (2017)
39. Xiao, H., Rasul, K., Vollgraf, R.: Fashion-MNIST: a novel image dataset for benchmarking machine learning algorithms. arXiv preprint arXiv:1708.07747 (2017)
40. Zeiler, M.D., Fergus, R.: Visualizing and understanding convolutional networks. In: Fleet, D., Pajdla, T., Schiele, B., Tuytelaars, T. (eds.) ECCV 2014. LNCS, vol. 8689, pp. 818–833. Springer, Cham (2014). https://doi.org/10.1007/978-3-319-10590-1_53
41. Zhou, C., Paffenroth, R.C.: Anomaly detection with robust deep autoencoders. In: Proceedings of the 23rd ACM SIGKDD International Conference on Knowledge Discovery and Data Mining, pp. 665–674. ACM (2017)
42. Zong, B., et al.: Deep autoencoding gaussian mixture model for unsupervised anomaly detection. In: International Conference on Learning Representations (2018)

Dense RepPoints: Representing Visual Objects with Dense Point Sets

Ze Yang[1,2], Yinghao Xu[3,4], Han Xue[5], Zheng Zhang[7(✉)], Raquel Urtasun[6], Liwei Wang[1], Stephen Lin[7], and Han Hu[7]

[1] Peking University, Beijing, China
yangze@pku.edu.cn, wanglw@cis.pku.edu.cn
[2] Zhejiang Lab, Hangzhou, China
[3] Zhejiang University, Hangzhou, China
justimyhxu@gmail.com
[4] The Chinese University of Hong Kong, Hong Kong, China
[5] Shanghai Jiao Tong University, Shanghai, China
xiaoxiaoxh@sjtu.edu.cn
[6] University of Toronto, Toronto, Canada
urtasun@cs.toronto.edu
[7] Microsoft Research Asia, Beijing, China
{zhez,stevelin,hanhu}@microsoft.com

Abstract. We present a new object representation, called Dense Rep-Points, that utilizes a large set of points to describe an object at multiple levels, including both box level and pixel level. Techniques are proposed to efficiently process these dense points, maintaining near-constant complexity with increasing point numbers. Dense RepPoints is shown to represent and learn object segments well, with the use of a novel distance transform sampling method combined with set-to-set supervision. The distance transform sampling combines the strengths of contour and grid representations, leading to performance that surpasses counterparts based on contours or grids. Code is available at https://github.com/justimyhxu/Dense-RepPoints.

1 Introduction

Representation matters. While significant advances in visual understanding algorithms have been witnessed in recent years, they all rely on proper representation of visual elements for convenient and effective processing. For example, a single image feature, a rectangular box, and a mask are usually adopted

---

Z. Yang, Y. Xu and H. Xue—Equal contribution.
This work was done when Ze Yang, Yinghao Xu and Han Xue were interns at Microsoft Research Asia.

Electronic supplementary material The online version of this chapter (https://doi.org/10.1007/978-3-030-58589-1_14) contains supplementary material, which is available to authorized users.

© Springer Nature Switzerland AG 2020
A. Vedaldi et al. (Eds.): ECCV 2020, LNCS 12366, pp. 227–244, 2020.
https://doi.org/10.1007/978-3-030-58589-1_14

to represent input for recognition tasks of different granularity, i.e. image classification [19,23,38], object detection [16,28,36] and pixel-level segmentation [7,18,26], respectively. In addition, the representation at one level of granularity may help the recognition task at another granularity, e.g. an additional mask representation may aid the learning of a coarser recognition task such as object detection [18].

We thus consider the question of whether a unified representation for recognition tasks can be devised over various levels of granularity.

Recently, RepPoints [47] was proposed to represent an object by a small set of adaptive points, simultaneously providing a geometric and semantic description of an object. It demonstrates good performance for the coarse localization task of object detection, and also shows potential to conform to more sophisticated object structures such as semantic keypoints. However, the small number of points (9 by default) limits its ability to reveal more detailed structure of an

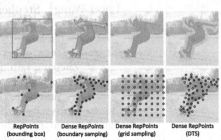

Fig. 1. Visual object in different geometric forms (top row from left to right): bounding box, boudary sampling (Contour), Grid sampling (binary mask), Distance transform sampling. These various object forms can be unified represented by a dense point set, called *Dense RepPoints* (bottom row).

object, such as pixel-level instance segmentation. In addition, the supervision for recognition and coarse localization also may hinder learning of more fine-grained geometric descriptions.

This paper presents *Dense RepPoints*, which utilizes a large number of points along with optional attributes to represent objects in detail, e.g. for instance segmentation. Because of its high representation flexibility, *Dense RepPoints* can effectively model common object segment descriptors, including contours (polygon) [21,29,45] and grid masks [7,18], as illustrated in columns 2 and 3 of Fig. 1. *Dense RepPoints* can also model a *binary boundary mask*, a new geometric descriptor for object segments that combines the description efficiency of contours and the reduced dependence on exact point localization of grid masks, as illustrated in column 4 of Fig. 1.

To learn and represent binary boundary masks by *Dense RepPoints*, three techniques are proposed. The first is a distance transform sampling (DTS) method, which converts a ground-truth boundary mask into a point set by probabilistic sampling based on the distance transform map of the object contour. With this conversion, the *Dense RepPoints* prediction and ground truth are both point sets and are thus comparable. The second is a set-to-set supervision loss, in contrast to the commonly used point-to-point supervision loss, e.g. [33,45]. The set-to-set supervision loss avoids assigning exact geometric meaning for every point, which is usually difficult and semantically inaccurate for instance segmentation but is required by point-to-point methods. The third is a novel conversion method from the learnt non-grid *Dense RepPoints* to an instance mask of any resolution, based on Delaunay triangulation.

With these three novel techniques, *Dense RepPoints* are learnt to well represent the binary boundary map of objects. It also yields better performance than methods based on a contour or grid mask representation. The method achieves 39.1 mask mAP and 45.6 box mAP on the COCO test-dev set using a ResNet-101 backbone network.

In addition to greater representation ability and better accuracy, Dense RepPoints can also be efficiently processed with our proposed techniques. The complexity of vanilla RepPoints increases linearly with the number of points, making it impractical for large point sets, e.g. hundreds of points. To resolve this issue, we propose two techniques, *group pooling* and *shared offset/attribute field*, for object classification and offset/attribute prediction, respectively. These techniques enable near-constant complexity with increasing numbers of points, while maintaining the same accuracy.

The contributions of this work are summarized as follows:

- We propose a new object representation, called *Dense RepPoints*, that models objects by a large number of adaptive points. The new representation shows great flexibility in representing detailed geometric structure of objects. It also provides a unified object representation over different levels of granularity, such as at the box level and pixel level. This allows for coarse detection tasks to benefit from finer segment annotations as well as enable instance segmentation, in contrast to training through separate branches built on top of base features as popularized in [10,18].
- We adapt the general *Dense RepPoints* representation model to the instance segmentation task, where three novel techniques of *distance transform sampling* (DTS), *set-to-set supervision loss* and *Delaunay triangulation based conversion* are proposed. *Dense RepPoints* is found to be superior to previous methods built on a contour or grid mask representation.
- We propose two techniques, of *group pooling* and *shared offset/attribute fields*, to efficiently process the large point set of *Dense RepPoints*, yielding near constant complexity with similar accuracy.

2   Related Work

Bounding Box Representation. Most existing high-level object recognition benchmarks [14,24,29] employ bounding box annotations for object detection. The current top-performing two-stage object detectors [11,16,17,36] use bounding boxes as anchors, proposals and final predictions throughout their pipelines. Some early works have proposed to use rotated boxes [20] to improve upon axis-aligned boxes, but the representation remains in a rectangular form. For other high-level recognition tasks such as instance segmentation and human pose estimation, the intermediate proposals in top-down solutions [10,18] are all based on bounding boxes. However, the bounding box is a coarse geometric representation which only encodes a rough spatial extent of an object.

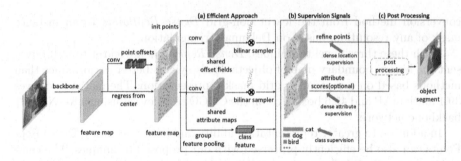

Fig. 2. Overview of *Dense RepPoints*. First, the initial representative points are generated by regressing from the center point as in RepPoints [47]. Then, these initial points are refined by the proposed efficient approaches to obtain refined, attributed representative points. Finally, post-processing is applied to generate the instance segment.

Non-box Object Representations. For instance segmentation, the annotation for objects is either as a binary mask [14] or as a set of polygons [29]. While most current top-performing approaches [6,9,18] use a binary mask as final predictions, recent approaches also exploit contours for efficient interactive annotation [1,4,30] and segmentation [8,45]. This contour representation, which was popular earlier in computer vision [5,21,39–41], is believed to be more compatible with the semantic concepts of objects [32,40]. Some works also use edges and superpixels [22,46] as object representations. Our proposed *Dense RepPoints* has the versatility to model objects in several of these non-box forms, providing a more generalized representation.

Point Set Rrepresentation. There is much research focused on representing point clouds in 3D space [34,35]. A direct instantiation of ordered point sets in 2D perception is 2D pose [2,3,43], which directly addresses the semantic correspondence problem. Recently, there has been increasing interest in the field of object detection on using specific point locations, including corner points [25], extreme points [50], and the center point [13,49]. These point representations are actually designed to recover a bounding box, which is coarse and lacks semantic information. RepPoints [47] proposes a learnable point set representation trained from localization and recognition feedback. However, it uses only a small number ($n = 9$) of points to represent objects, limiting its ability to represent finer geometry. In this work, we extend RepPoints [47] to a denser and finer geometric representation, enabling usage of dense supervision and taking a step towards dense semantic geometric representation.

3   Methodology

In this section, we first review *RepPoints* [47] for object detection in Sect. 3.1. Then, we introduce *Dense RepPoints* in Sect. 3.2 for strengthening the representation ability of *RepPoints* from object detection to fine-grained geometric localization and recognition tasks, such as extracting an instance mask, by associating an attribute vector with each representative point. In addition, these fine-grained tasks usually require higher resolution and many more representative points than object detection, which makes the computational complexity of vanilla *RepPoints* infeasible. We discuss how to reduce the computational complexity of vanilla *RepPoints* in Sect. 3.3 for representing an instance mask. In Sect. 3.4, we describe how to use *Dense RepPoints* to model instance masks with different sampling strategies, and then design appropriate supervision signals in Sect. 3.5. Since representative points are usually sparse and non-grid while an instance segment is dense and grid-aligned, we discuss how to transform representative points into an instance segment in Sect. 3.6. An overview of our method is exhibited in Fig. 2.

3.1 Review of RepPoints for Object Detection

We first review how *RepPoints* [47] detects objects. A set of adaptive representative points $\mathcal{R}$ is used to represent an object in RepPoints:

$$\mathcal{R} = \{p_i\}_{i=1}^{n} \tag{1}$$

where $p_i = (x_i + \Delta x_i, y_i + \Delta y_i)$ is the i-th representative point, x_i and y_i denote an initialized location, Δx_i and Δy_i are learnable offsets, and n is the number of points. The feature of a point $\mathcal{F}(p)$ is extracted from the feature map $\mathcal{F}$ through bilinear interpolation, and the feature of a point set $\mathcal{F}(\mathcal{R})$ is defined as the concatenation of all representative points of $\mathcal{R}$:

$$\mathcal{F}(\mathcal{R}) = \mathrm{concat}(\mathcal{F}(p_1), ..., \mathcal{F}(p_n)) \tag{2}$$

which is used to recognize the class of the point set. The bounding box of a point set can be obtained by conversion functions [47]. In the training phase, explicit supervision and annotation for representative points is not required. Instead, representative points are driven to move to appropriate locations by the classification loss and box localization loss:

$$L_{\mathrm{det}} = L_{\mathrm{cls}}^{b} + L_{\mathrm{loc}}^{b} \tag{3}$$

Both bilinear interpolation used in feature extraction and the conversion function used in bounding box transformation are differentiable with respect to the point locations. These representative points are suitable for representing the object category and accurate position at the same time.

3.2 Dense RepPoints

In vanilla *RepPoints*, the number of representative points is relatively small ($n = 9$). It is sufficient for object detection, since the category and bounding box of an object can be determined with few points. Different from object detection, fine-grained geometric localization tasks such as instance segmentation usually provide pixel-level annotations that require precise estimation. Therefore, the representation capacity of a small number of points is insufficient, and a significantly larger set of points is necessary together with an attribute vector associated with each representative point:

$$\mathcal{R} = \{(x_i + \Delta x_i, y_i + \Delta y_i, \mathbf{a}_i)\}_{i=1}^{n}, \tag{4}$$

where $\mathbf{a}_i$ is the attribute vector associated with the i-th point.

In instance segmentation, the attribute can be a scalar, defined as the foreground score of each point. In addition to the box-level classification and localization terms, L_{cls}^{b} and L_{loc}^{b}, we introduce a point-level classification loss L_{cls}^{p} and a point-level localization loss L_{loc}^{p}. The objective function of Eq. 3 becomes:

$$L = \underbrace{L_{cls}^{b} + L_{loc}^{b}}_{L_{det}} + \underbrace{L_{cls}^{p} + L_{loc}^{p}}_{L_{mask}} \tag{5}$$

where L_{cls}^{p} is responsible for predicting the point foreground score and L_{loc}^{p} is for learning point localization. This new representation is named *Dense RepPoints*.

3.3   Efficient Computation

Intuitively, denser points will improve the capacity of the representation. However, the feature of an object in vanilla *RepPoints* is formed by concatenating the features of all points, so the FLOPs will rapidly increase as the number of points increases. Therefore, directly using a large number of points in *RepPoints* is impractical. To address this issue, we introduce group pooling and shared offset fields to reduce the computational complexity, thereby significantly reducing the extra FLOPs while maintaining performance. In addition, we further introduce a shared attribute map to efficiently predict whether a point is in the foreground.

Group Pooling. Group pooling is designed to effectively extract object features and is used in the box classification branch (see Fig. 3 top). Given n representative points, we equally divide the points into k groups, with each group having n/k points (if k is not divisible by n, the last group will have fewer points than the others to ensure a total of n points). Then, we aggregate the feature of each point within a group by max-pooling to extract a group feature. Finally, a 1×1 convolution is computed over the concatenated group features from all groups. In this way, the object features are represent by groups instead of points, reducing the computational complexity from $O(n)$ to $O(k)$. The groups are driven by the classification target and will learn different semantics for different groups (though no clear geometric separation) to enhance classification power. We empirically find that the number of groups do not need to be increased when the points become denser, thus the computational complexity is not affected by using a larger set of points. In our implementation, we set k to 9 by default, which works relatively well for classification.

Shared Offset Fields. The computational complexity of predicting the offsets for the points is $O(n^2)$ in *RepPoints*, making the dense point set representation unsuitable for real applications. Unlike in the classification branch, we need the information of individual points for point location refinement. Hence, we cannot directly apply the grouped features used in classification. Instead, we empirically find that local point features provide enough information for point refinement, in the same spirit as Curve-GCN [30] which uses local features for contour refinement. To share feature computation among points, we propose to first compute n shared offset field maps based on the image feature map. And then for the i-th representative point, its position is directly predicted via bilinear interpolation at the corresponding location of the i-th offset field (see Fig. 3 middle). This reduces the computational complexity of the regression from $O(n^2)$ to $O(n)$. By using group pooling and shared offset fields, even if a large number of points are used, the added FLOPs is still very small compared to that of the entire network (see Sect. 4.3).

Shared Attribute Map. Predicting the foreground score of each point can be implemented in manner similar to *shared offset fields* by using a shared position-sensitive attribute map, first introduced in R-FCN [11]. In the position-sensitive attribute map, each channel has an explicit positional meaning. Therefore, the foreground score of each representative point can be interpolated on the channel corresponding to its location (see Fig. 3 bottom).

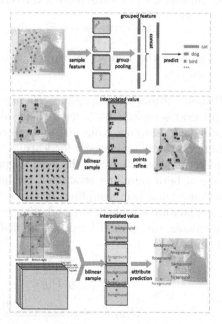

Fig. 3. Illustration of efficient feature extraction for *Dense RepPoints*. **Top:** group pooling operation. **Middle:** shared offset fields for each point index. **Bottom:** shared attribute maps for each relative position.

3.4 Different Sampling Strategies

How to represent object segments effectively is a core problem in visual perception. Contours and binary masks are two typical representations widely used in previous works [7,18,33,45]. In *Dense RepPoints*, these representations can be simulated by different sampling strategies: a binary mask by uniformly sampling grid points over the bounding box of an object, and a contour as all sampling points along the object boundary. We call these two sampling strategies *grid sampling* and *boundary sampling*, respectively, and discuss them in this section. In addition, we introduce a new sampling strategy, named *dis-*

tance transform sampling, which combines the advantages of both grid sampling and boundary sampling.

Boundary Sampling (Contour). An instance segment can be defined as the inner region of a closed object contour. Contour points is a compact object description because of its 1-D nature (defined by a sequence of points). In our method, the contour representation can be simulated through supervising the offsets of representative points along the object boundary, with the score of points set to 1 by default.

Grid Sampling (Binary Mask). A binary mask can be represented as a set of uniformly sampled grid points over the bounding box of an object, and each sampled point has a binary score to represent its category, i.e. foreground or background. This sampling strategy (representation) is widely used, such as in Mask R-CNN [18] and Tensor Mask [7]. In *Dense RepPoints*, grid sampling can be implemented by constraining the offsets of representative points as:

$$\Delta x_i = \alpha(\frac{i}{\sqrt{n}} - 0.5), \quad \Delta y_i = \beta(\frac{i}{\sqrt{n}} - 0.5), \quad i \in \{1...n\} \tag{6}$$

where n is the number of sampling points, and α and β are two learnable parameters.

Distance Transform Sampling (Binary Boundary Mask). Boundary sampling and grid sampling both have their advantages and applications. In general, boundary sampling (contour) is more compact for object segment description, and grid sampling is easier for learning, mainly because its additional attribute (foreground score) avoids the need for precise point localization. To take advantage of both sampling strategies, we introduce a new sampling method called distance transform sampling. In this sampling strategy, points near the object boundary are sampled more and other regions are sampled less. During the training phase, the ground truth is sampled according to distance from the object boundary:

$$P(p) = \frac{g(\mathcal{D}(p))}{\sum_q g(\mathcal{D}(q))} \tag{7}$$

$$\mathcal{D}(p) = \frac{\min_{e \in \mathcal{E}} \|p - e\|_2}{\sqrt{\max_{e,e' \in \mathcal{E}} |e_x - e'_x| \cdot \max_{e,e' \in \mathcal{E}} |e_y - e'_y|}} \tag{8}$$

where $P(p)$ is the sampling probability of point p, $D(p)$ is the normalized distance from the object boundary of point p, $\mathcal{E}$ is the boundary point set, and g is a decreasing function. In our work, we use the Heaviside step function for g:

$$g(x) = \begin{cases} 1 & x \leq \delta \\ 0 & x > \delta \end{cases} \tag{9}$$

Here, we use $\delta = 0.04$ by default. Intuitively, points with a distance less than δ (close to the contour) have a uniform sampling probability, and points with a distance greater than δ (away from the contour) are not sampled.

3.5   Sampling Supervision

The point classification loss L_{cls}^p and the point localization loss L_{loc}^p are used to supervise the different segment representations during training. In our method, L_{cls}^p is defined as a standard cross entropy loss function with softmax activation, where a point located in the foreground is labeled as positive and otherwise its label is negative.

For localization supervision, a point-to-point approach could be taken, where each ground truth point is assigned an exact geometric meaning, e.g. using the polar assignment method in PolarMask [45]. Each ground truth with exact geometric meaning also corresponds to a fixed indexed representative point in *Dense RepPoints*, and the L2 distance is used as the point localization loss L_{loc}^p:

$$L_{point}(\mathcal{R}, \mathcal{R}') = \frac{1}{n} \sum_{k=1}^{n} \|(x_i, y_i) - (x_i', y_i')\|_2 \qquad (10)$$

where $(x_i, y_i) \in \mathcal{R}$ and $(x_i', y_i') \in \mathcal{R}'$ represent the point in the predicted point set and ground-truth point set, respectively.

However, assigning exact geometric meaning to each point is difficult and may be semantically inaccurate for instance segmentation. Therefore, we propose set-to-set supervision, rather than supervise each individual point. The point localization loss is measured by *Chamfer distance* [15,37] between the supervision point set and the learned point set:

$$L_{set}(\mathcal{R}, \mathcal{R}') = \frac{1}{2n} \left(\sum_{i=1}^{n} \min_j \|(x_i, y_i) - (x_j', y_j')\|_2 + \sum_{j=1}^{n} \min_i \|(x_i, y_i) - (x_j', y_j')\|_2 \right)$$

where $(x_i, y_i) \in \mathcal{R}$ and $(x_j', y_j') \in \mathcal{R}'$. We evaluate these two forms of supervision in Sect. 4.3.

3.6   Representative Points to Object Segment

Dense RepPoints represents an object segment in a sparse and non-grid form, and thus an extra post-processing step is required to transform the non-grid points into a binary mask. In this section, we propose two approaches, Concave Hull [31] and Triangulation, for this purpose.

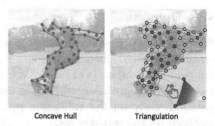

Concave Hull   Triangulation

Fig. 4. Post-Processing. Generating image segments by Concave Hull and Triangulation.

Concave Hull. An instance mask can be defined as a concave hull of a set of foreground points (see Fig. 4 left), which is used by many contour-based methods. In *Dense RepPoints*, boundary sampling naturally uses this post-processing. We first use a threshold to binarize the predicted points by their foreground scores, and then compute their concave hull to obtain the binary mask. In our approach, we empirically set a threshold of 0.5 by default.

Triangulation. Triangulation is commonly used in computer graphics to obtain a mesh from a point set representation, and we introduce it to generate an object segment. Specifically, we first apply Delaunay triangulation to partition the space into triangles with vertices defined by the learned point set. Then, each pixel in the space will fall inside a triangle and its point score is obtained by linearly interpolating from the triangle vertices in the Barycentric coordinates (Fig. 4 right). Finally, a threshold is used to binarize the interpolated score map to obtain the binary mask (Fig. 5).

4 Experiments

Fig. 5. Visualization of points and instance masks by DTS. **Top**: The learned points (225 points) is mainly distributed around the mask boundary. **Bottom**: The foreground masks generated by triangulation post-processing on COCO `test-dev` images with ResNet-50 backbone under '3x' training schedule.

4.1 Datasets

We present experimental results for instance segmentation and object detection on the COCO2017 benchmark [29], which contains 118k images for training, 5k images for validation (`val`) and 20k images for testing (`test-dev`). The standard mean average precision (mAP) is used to measure accuracy. We conduct an ablation study on the validation set, and compare with other state-of-the-art methods on the test-dev set.

4.2 Implementation Details

We follow the training settings of *RepPoints* [47]. Horizontal image flipping augmentation, group normalization [44] and focal loss [28] are used during training. If not specified, ResNet-50 [19] with FPN [27] is used as the default backbone in the ablation study, and weights are initialized from the ImageNet [12] pre-trained model. Distance transform sampling with set-to-set supervision is used

Table 1. Validating the proposed components for greater efficiency. With group pooling (GP) and shared offset fields (SOF), the mAP constantly improve as the number of points increase, while the FLOPS is nearly unaffected.

n	G FLOPS			mAP		
	Base	+ GP	+ SOF	Base	+ GP	+ SOF
9	211.04	208.03	202.05	38.1	37.9	37.9
25	255.14	237.80	205.93	37.7	37.8	37.7
49	321.28	278.86	209.18	37.7	37.6	37.5
81	409.46	331.03	212.60	37.5	37.5	37.5

Table 2. Comparison of different mask representations.

Number of points	9	25	81	225	729
Contour	**19.7**	23.9	26.0	25.2	24.1
Grid points	5.0	17.6	29.7	31.6	32.8
DTS	13.9	**24.5**	**31.5**	**32.8**	**33.8**

as the default training setting, and triangulation is chosen as the default post-processing. For predicting attribute scores, we follow SOLO [42] by using seven 3×3 convs in the attribute score head, and the feature map of P3 is used to fuse with feature maps of other levels through addition operation, which is inspired by the conversion FPN used in TensorMask [7],

The models are trained on 8 GPUs with 2 images per GPU for 12 epochs (1× settings). In SGD training, the learning rate is initialized to 0.01 and then divided by 10 at epochs 8 and 11. The weight decay and momentum parameters are set to 10^{-4} and 0.9, respectively. In inference, we follow SOLO [42] to refine the classification score by using the mask prediction, and we use NMS with IoU threshold of 0.5, following RetinaNet [28].

4.3 Ablation Study

Components for Greater Efficiency. We validate group pooling (GP) and shared offset fields (SOF) by adding them to vanilla *RepPoints* [47] and evaluating the performance on object detection. Results are shown in Table 1. We present the results under different numbers of points: $n = 9, 25, 49, 81$. By using group pooling, FLOPs significantly decreases with increasing number of points compared to vanilla *RepPoints* with similar mAP. By introducing shared offset fields, while mAP is not affected, FLOPs is further reduced and nearly constant with respect to n. Specifically, for $n = 81$, our efficient approach saves 197G FLOPs in total. This demonstrates the effectiveness of our efficient approach representation and makes the use of more representative points in instance segmentation possible.

Table 3. Comparison of triangulation and concave hull.

# of points	9	25	49	81	225
Concave-Hull	9.7	21.0	21.3	20.6	23.4
Triangulation	13.9	24.5	29.6	31.5	**32.8**

Table 4. Comparison of point-to-point and set-to-set supervision.

Number of points	9	25	81	225	729
Point-to-point	10.7	20.7	27.8	31.3	32.6
Set-to-set	13.9	24.5	31.5	32.8	**33.8**

Different Sampling Strategies. We compare different strategies for sampling object points. Since different sampling strategies perform differently under different post-processing, we compare them with the their best-performing post-processing method for fair comparison. Therefore, we use triangulation (Fig. 4 right) for distance transform sampling, bilinear interpolation (`imresize`) for grid sampling, and concave hull (Fig. 4 left) for boundary sampling. Please see the Appendix for more details on the post-processing. Results are shown in Table 2. Boundary sampling has the best performance with few points. When $n = 9$, boundary sampling obtains 19.7 mAP, and grid sampling has only 5.0 mAP. Distance transform sampling has 13.9 mAP, which lies in the middle. The reason is that boundary sampling only samples points on the boundary, which is the most efficient way to represent object masks, so relatively good performance can be achieved with fewer points. Both grid sampling and distance transform sampling need to sample non-boundary points, so their efficiency is lower than boundary sampling, but distance transform sampling samples more points around the boundary than in other regions, thus it performs much better than grid sampling.

When using more points, grid sampling and distance transform sampling perform better than boundary sampling. For $n = 729$, grid sampling and distance transform sampling achieve 32.8 mAP and 33.8 mAP, respectively, while boundary sampling only obtains 24.1 mAP. This is due to the limited representation capability of boundary sampling since it only takes boundary points into consideration. In addition, distance transform sampling outperforms grid sampling in all cases, which indicates that distance transform sampling is more efficient than grid sampling while maintaining the same representation capability.

Concave Hull vs. Triangulation. Concave hull and triangulation both can transform a point set to a binary mask. Here, we compare them using distance transform sampling. Results are shown in Table 3. Triangulation outperforms concave hull consistently with different numbers of points, indicating that triangulation is more suitable for distance transform sampling (DTS). It is noted that concave hull with DTS is worse than contour sampling, because DTS does not

Table 5. Results of *Dense RepPoints* on different numbers of points.

Number of points	81	225	441	729
AP	31.5	32.8	33.3	**33.8**
AP@50	54.2	54.2	54.5	**54.8**
AP@75	32.7	34.4	35.2	**35.9**

Table 6. Effects of dense supervision on detection.

	Dense RepPoints				Mask R-CNN
	n = 9	n = 25	n = 49	n = 81	
w.o. Inst	37.9	37.7	37.5	37.5	36.4
w. Inst	38.1	38.7	39.2	39.4	37.3
Improve	+0.2	+1.0	+1.7	**+1.9**	+0.9

Table 7. Performance of instance segmentation on COCO [29] `test-dev`. Our method significantly surpasses all other state-of-the-arts. '*' indicates training without ATSS [48] assigner and 'jitter' indicates using scale-jitter during training.

Method	Backbone	Epochs	Jitter	AP	AP_{50}	AP_{75}	AP_S	AP_M	AP_L
Mask R-CNN [18]	ResNet-101	12		35.7	58.0	37.8	15.5	38.1	52.4
Mask R-CNN [18]	ResNeXt-101	12		37.1	60.0	39.4	16.9	39.9	53.5
TensorMask [7]	ResNet-101	72	✓	37.1	59.3	39.4	17.4	39.1	51.6
SOLO [42]	ResNet-101	72	✓	37.8	59.5	40.4	16.4	40.6	54.2
ExtremeNet [50]	HG-104	100	✓	18.9	–	–	10.4	20.4	28.3
PolarMask [45]	ResNet-101	24	✓	32.1	53.7	33.1	14.7	33.8	45.3
Ours*	ResNet-50	12		33.9	55.3	36.0	17.5	37.1	44.6
Ours	ResNet-50	12		34.1	56.0	36.1	17.7	36.6	44.9
Ours	ResNet-50	36	✓	37.6	60.4	40.2	20.9	40.5	48.6
Ours	ResNet-101	12		35.8	58.2	38.0	18.7	38.8	47.1
Ours	ResNet-101	36	✓	39.1	62.2	42.1	21.8	42.5	50.8
Ours	ResNeXt-101	36	✓	40.2	63.8	43.1	23.1	43.6	52.0
Ours	ResNeXt-101-DCN	36	✓	41.8	65.7	45.0	24.0	45.2	54.6

strictly sample on the boundary but usually samples points near the boundary. Besides, it also samples points farther from the boundary.

Different Supervision Strategies. Point-to-point is a common and intuitive supervision strategy and it is widely used by other methods [33,45,50]. However, this kind of supervision may prevent *Dense RepPoints* from learning better sampling strategies, since it is restrictive and ignores the relationships among points. This motivates the proposed set-to-set supervision in Sect. 3.5. We compare the two forms of supervision using distance transform sampling. Results are shown in Table 4. Set-to-set supervision consistently outperforms point-to-point supervision, especially for a small number of points.

Table 8. Object detection on COCO [29] `test-dev`. Our method significantly surpasses all other state-of-the-arts. '*' indicates training without ATSS [48] assigner and 'jitter' indicates using scale-jitter during training.

Method	Backbone	Epochs	Jitter	AP	AP_{50}	AP_{75}	AP_S	AP_M	AP_L
Faster R-CNN [27]	ResNet-101	12		36.2	59.1	39.0	18.2	39.0	48.2
Mask R-CNN [18]	ResNet-101	12		38.2	60.3	41.7	20.1	41.1	50.2
Mask R-CNN [18]	ResNeXt-101	12		39.8	62.3	43.4	22.1	43.2	51.2
RetinaNet [28]	ResNet-101	12		39.1	59.1	42.3	21.8	42.7	50.2
RepPoints [47]	ResNet-101	12		41.0	62.9	44.3	23.6	44.1	51.7
ATSS [48]	ResNeXt-101-DCN	24	✓	47.7	66.5	51.9	29.7	50.8	59.4
CornerNet [25]	HG-104	100	✓	40.5	56.5	43.1	19.4	42.7	53.9
ExtremeNet [50]	HG-104	100	✓	40.1	55.3	43.2	20.3	43.2	53.1
CenterNet [49]	HG-104	100	✓	42.1	61.1	45.9	24.1	45.5	52.8
Ours*	ResNet-50	12		39.4	58.9	42.6	22.2	43.0	49.6
Ours	ResNet-50	12		40.1	59.7	43.3	22.8	42.8	50.4
Ours	ResNet-50	36	✓	43.9	64.0	47.6	26.7	46.7	54.1
Ours	ResNet-101	12		42.1	62.0	45.6	24.0	45.1	52.9
Ours	ResNet-101	36	✓	45.6	65.7	49.7	27.7	48.9	56.6
Ours	ResNeXt-101	36	✓	47.0	67.3	51.1	29.3	50.1	58.0
Ours	ResNeXt-101+DCN	36	✓	48.9	69.2	53.4	30.5	51.9	61.2

More Representative Points. *Dense RepPoints* can take advantage of more points than vanilla *RepPoint* [47], and its computation cost does not change as the number of points increases. Table 5 shows the performance of *Dense RepPoints* on different numbers of points using distance transform sampling and triangulation inference. In general, more points bring better performance, but as the number of points increases, the improvement saturates.

Benefit of Dense RepPoints on Detection. Instance segmentation benefits object detection via multi-task learning as reported in Mask R-CNN [18]. Here, we examine whether *Dense RepPoints* can improve object detection performance as well. Results are shown in Table 6. Surprisingly, *Dense RepPoints* not only takes advantage of instance segmentation to strengthen object detection, but also brings greater improvement when more points are used. Specifically, when $n = 81$, *Dense RepPoints* improves detection mAP by 1.9 points. As a comparison, Mask R-CNN improves by 0.9 points compared to Faster R-CNN. This indicates that multi-task learning benefits more from better representation. This suggests that *Dense RepPoints* models a finer geometric representation. This novel application of explicit multi-task learning also verifies the necessity of using a denser point set, and it demonstrates the effectiveness of our unified representation.

4.4 Comparison with Other SOTA Methods

A comparison is conducted with other state-of-the-arts methods in object detection and instance segmentation on the COCO `test-dev` set. We use 729 representative points by default, and trained by distance transform sampling and set-to-set supervision. ATSS [48] is used as the label assignment strategy if not specified. In the inference stage, the instance mask is generated by adopting triangulation as post-processing.

We first compare with other state-of-the-art instance segmentation methods. Results are shown in Table 7. With the same ResNet-101 backbone, our method achieves 39.1 mAP with the 1x setting, outperforming all other methods. By further integrating ResNeXt-101-DCN as a stronger backbone, our method reaches 41.8 mAP.

We then compare with other state-of-the-arts object detection methods. Results are shown in Table 8. With ResNet-101 as the backbone, our method achieves 42.1 mAP with the 1x setting, outperforming RepPoints [47] and Mask R-CNN by 1.1 mAP and 3.9 mAP, respectively. With ResNeXt-101-DCN as a stronger backbone, our method achieves 48.9 mAP, surpassing all other anchor-free SOTA methods.

5 Conclusion

In this paper, we present *Dense RepPoints*, a dense attributed point set representation for 2D objects. By introducing efficient feature extraction and employing dense supervision, this work takes a step towards learning a unified representation for top-down object recognition pipelines, enabling explicit modeling between different visual entities, *e.g.* coarse bounding boxes and fine instance masks. Besides, we also propose a new point sampling method to describe masks, shown to be effective in our experiments. Experimental results show that this new dense 2D representation is not only applicable for predicting dense masks, but also can help improve other tasks such as object detection via its novel multi-granular object representation. We also analyze the upper bound for our representation and plan to explore better score head designs and system-level performance improvements particularly on large objects.

Acknowledgement. We thank Jifeng Dai and Bolei Zhou for discussion and comments about this work. Jifeng Dai was involved in early discussions of the work and gave up authorship after he joined another company.

References

1. Acuna, D., Ling, H., Kar, A., Fidler, S.: Efficient interactive annotation of segmentation datasets with polygon-RNN++ (2018)
2. Alp Güler, R., Neverova, N., Kokkinos, I.: DensePose: dense human pose estimation in the wild. In: CVPR, pp. 7297–7306 (2018)
3. Cao, Z., Simon, T., Wei, S.E., Sheikh, Y.: Realtime multi-person 2D pose estimation using part affinity fields. In: CVPR, pp. 7291–7299 (2017)
4. Castrejon, L., Kundu, K., Urtasun, R., Fidler, S.: Annotating object instances with a polygon-RNN. In: CVPR, pp. 5230–5238 (2017)
5. Chan, T.F., Vese, L.A.: Active contours without edges. IEEE Trans. Image Process. 10(2), 266–277 (2001)
6. Chen, L.C., Hermans, A., Papandreou, G., Schroff, F., Wang, P., Adam, H.: MaskLAB: instance segmentation by refining object detection with semantic and direction features. In: CVPR, pp. 4013–4022 (2018)
7. Chen, X., Girshick, R.B., He, K., Dollár, P.: TensorMask: a foundation for dense object segmentation. In: ICCV (2019)
8. Cheng, D., Liao, R., Fidler, S., Urtasun, R.: Darnet: Deep active ray network for building segmentatio. arXiv preprint arXiv:1905.05889 (2019)
9. Dai, J., He, K., Li, Y., Ren, S., Sun, J.: Instance-sensitive fully convolutional networks. In: Leibe, B., Matas, J., Sebe, N., Welling, M. (eds.) ECCV 2016. LNCS, vol. 9910, pp. 534–549. Springer, Cham (2016). https://doi.org/10.1007/978-3-319-46466-4_32
10. Dai, J., He, K., Sun, J.: Instance-aware semantic segmentation via multi-task network cascades. In: CVPR, pp. 3150–3158 (2016)
11. Dai, J., Li, Y., He, K., Sun, J.: R-FCN: object detection via region-based fully convolutional networks. In: NeurIPS, pp. 379–387 (2016)
12. Deng, J., Dong, W., Socher, R., Li, L.J., Li, K., Fei-Fei, L.: ImageNet: a large-scale hierarchical image database (2009)
13. Duan, K., Bai, S., Xie, L., Qi, H., Huang, Q., Tian, Q.: CenterNet: object detection with keypoint triplets. arXiv preprint arXiv:1904.08189 (2019)
14. Everingham, M., Van Gool, L., Williams, C.K., Winn, J., Zisserman, A.: The pascal visual object classes (VOC) challenge. IJCV 88(2), 303–338 (2010)
15. Fan, H., Su, H., Guibas, L.J.: A point set generation network for 3D object reconstruction from a single image. In: Proceedings of the IEEE Conference on Computer Vision and Pattern Recognition, pp. 605–613 (2017)
16. Girshick, R.: Fast R-CNN. In: ICCV, pp. 1440–1448 (2015)
17. Girshick, R., Donahue, J., Darrell, T., Malik, J.: Rich feature hierarchies for accurate object detection and semantic segmentation. In: CVPR, pp. 580–587 (2014)
18. He, K., Gkioxari, G., Dollár, P., Girshick, R.: Mask R-CNN. In: ICCV, pp. 2961–2969 (2017)
19. He, K., Zhang, X., Ren, S., Sun, J.: Deep residual learning for image recognition. In: CVPR, pp. 770–778 (2016)
20. Huang, C., Ai, H., Li, Y., Lao, S.: High-performance rotation invariant multiview face detection. PAMI 29(4), 671–686 (2007)
21. Kass, M., Witkin, A., Terzopoulos, D.: Snakes: active contour models. IJCV 1(4), 321–331 (1988)
22. Kirillov, A., Levinkov, E., Andres, B., Savchynskyy, B., Rother, C.: InstanceCut: from edges to instances with multicut. In: CVPR, pp. 5008–5017 (2017)

23. Krizhevsky, A., Sutskever, I., Hinton, G.E.: ImageNet classification with deep convolutional neural networks. In: NeurIPS, pp. 1097–1105 (2012)
24. Kuznetsova, A., et al.: The open images dataset V4: unified image classification, object detection, and visual relationship detection at scale. arXiv preprint arXiv:1811.00982 (2018)
25. Law, H., Deng, J.: CornerNet: detecting objects as paired keypoints. In: Ferrari, V., Hebert, M., Sminchisescu, C., Weiss, Y. (eds.) Computer Vision – ECCV 2018. LNCS, vol. 11218, pp. 765–781. Springer, Cham (2018). https://doi.org/10.1007/978-3-030-01264-9_45
26. Li, Y., Qi, H., Dai, J., Ji, X., Wei, Y.: Fully convolutional instance-aware semantic segmentation. In: CVPR, pp. 2359–2367 (2017)
27. Lin, T.Y., Dollár, P., Girshick, R., He, K., Hariharan, B., Belongie, S.: Feature pyramid networks for object detection. In: ICCV, pp. 2117–2125 (2017)
28. Lin, T.Y., Goyal, P., Girshick, R., He, K., Dollár, P.: Focal loss for dense object detection. In: ICCV, pp. 2980–2988 (2017)
29. Lin, T.Y., et al.: Microsoft COCO: common objects in context. In: Fleet, D., Pajdla, T., Schiele, B., Tuytelaars, T. (eds.) ECCV 2014. LNCS, vol. 8693, pp. 740–755. Springer, Cham (2014). https://doi.org/10.1007/978-3-319-10602-1_48
30. Ling, H., Gao, J., Kar, A., Chen, W., Fidler, S.: Fast interactive object annotation with curve-GCN. In: CVPR (2019)
31. Moreira, A., Santos, M.Y.: Concave hull: A k-nearest neighbours approach for the computation of the region occupied by a set of points (2007)
32. Palmer, S.E.: Vision Science: Photons to Phenomenology. MIT Press, Cambridge (1999)
33. Peng, S., Jiang, W., Pi, H., Bao, H., Zhou, X.: Deep snake for real-time instance segmentation. arXiv preprint arXiv:2001.01629 (2020)
34. Qi, C.R., Su, H., Mo, K., Guibas, L.J.: PointNet: deep learning on point sets for 3D classification and segmentation. In: CVPR (2017)
35. Qi, C.R., Yi, L., Su, H., Guibas, L.J.: Pointnet++: deep hierarchical feature learning on point sets in a metric space. In: NeurIPS, pp. 5099–5108 (2017)
36. Ren, S., He, K., Girshick, R., Sun, J.: Faster R-CNN: towards real-time object detection with region proposal networks. In: NeurIPS, pp. 91–99 (2015)
37. Rubner, Y., Tomasi, C., Guibas, L.J.: The earth mover's distance as a metric for image retrieval. IJCV **40**(2), 99–121 (2000)
38. Simonyan, K., Zisserman, A.: Very deep convolutional networks for large-scale image recognition. In: ICLR (2015)
39. Srinivasan, P., Zhu, Q., Shi, J.: Many-to-one contour matching for describing and discriminating object shape. In: CVPR (2010)
40. Toshev, A., Taskar, B., Daniilidis, K.: Shape-based object detection via boundary structure segmentation. IJCV **99**(2), 123–146 (2012)
41. Wang, X., Bai, X., Ma, T., Liu, W., Latecki, L.J.: Fan shape model for object detection. In: CVPR, pp. 151–158. IEEE (2012)
42. Wang, X., Kong, T., Shen, C., Jiang, Y., Li, L.: Solo: segmenting objects by locations. arXiv preprint arXiv:1912.04488 (2019)
43. Wei, S.E., Ramakrishna, V., Kanade, T., Sheikh, Y.: Convolutional pose machines. In: CVPR (2016)
44. Wu, Y., He, K.: Group normalization. In: Ferrari, V., Hebert, M., Sminchisescu, C., Weiss, Y. (eds.) ECCV 2018. LNCS, vol. 11217, pp. 3–19. Springer, Cham (2018). https://doi.org/10.1007/978-3-030-01261-8_1
45. Xie, E., et al.: PolarMask: single shot instance segmentation with polar representation. arXiv preprint arXiv:1909.13226 (2019)

46. Yang, J., Price, B., Cohen, S., Lee, H., Yang, M.H.: Object contour detection with a fully convolutional encoder-decoder network. In: CVPR, pp. 193–202 (2016)
47. Yang, Z., Liu, S., Hu, H., Wang, L., Lin, S.: RepPoints: point set representation for object detection. In: CVPR (2019)
48. Zhang, S., Chi, C., Yao, Y., Lei, Z., Li, S.Z.: Bridging the gap between anchor-based and anchor-free detection via adaptive training sample selection. arXiv preprint arXiv:1912.02424 (2019)
49. Zhou, X., Wang, D., Krähenbühl, P.: Objects as points. arXiv preprint arXiv:1904.07850 (2019)
50. Zhou, X., Zhuo, J., Krähenbühl, P.: Bottom-up object detection by grouping extreme and center points. In: CVPR (2019)

On Dropping Clusters to Regularize Graph Convolutional Neural Networks

Xikun Zhang, Chang Xu, and Dacheng Tao[✉]

UBTECH Sydney AI Centre, School of Computer Science, Faculty of Engineering,
The University of Sydney, Darlington, NSW 2008, Australia
xzha0505@uni.sydney.edu.au, {c.xu,dacheng.tao}sydney.edu.au

Abstract. Dropout has been widely adopted to regularize graph convolutional networks (GCNs) by randomly zeroing entries of the node feature vectors and obtains promising performance on various tasks. However, the information of individually zeroed entries could still present in other correlated entries by propagating (1) spatially between entries of different node feature vectors and (2) depth-wisely between different entries of each node feature vector, which essentially weakens the effectiveness of dropout. This is mainly because in a GCN, neighboring node feature vectors after linear transformations are aggregated to produce new node feature vectors in the subsequent layer. To effectively regularize GCNs, we devise DropCluster which first randomly zeros some seed entries and then zeros entries that are spatially or depth-wisely correlated to those seed entries. In this way, the information of the seed entries is thoroughly removed and cannot flow to subsequent layers via the correlated entries. We validate the effectiveness of the proposed DropCluster by comprehensively comparing it with dropout and its representative variants, such as SpatialDropout, Gaussian dropout and DropEdge, on skeleton-based action recognition.

Keywords: Regularization · Dropout · Graph convolutional networks

1 Introduction

GCNs have gained state-of-the-art results on various tasks including node classification [14,27], graph generation [33,35], skeleton-based action recognition [22,32], tracking [8,31] and so on. GCNs are designed for extracting features from graph structured data [14], and its general working mechanism is iteratively applying linear transformation to each node feature vector and aggregation on neighboring node feature vectors as the node features in the subsequent layer. Despite the successes, regularization of GCNs is not yet sufficiently studied and only started to receive attention recently [20].

Dropout [11] is currently a most widely adopted regularization in GCNs. However, as dropout is not originally designed for GCNs, it lacks an in-depth

Electronic supplementary material The online version of this chapter (https://doi.org/10.1007/978-3-030-58589-1_15) contains supplementary material, which is available to authorized users.

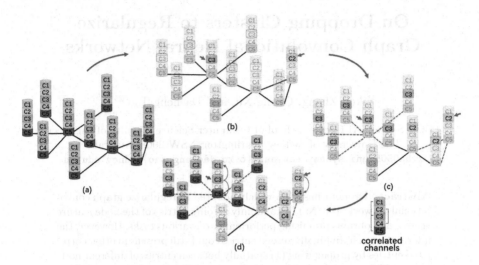

Fig. 1. (a) is a feature map of 10 nodes and 5 channels. In (b), two random seed entries are selected and marked by red arrows. (c) selects the entries spatially correlated to the seed entries. (d) further selects the entries depth-wisely correlated to the selected entries in (c). We assume that channel 2 and 4 are correlated, channel 1, 3, 5 are correlated, and only show 1-hop spatial correlation (Color figure online)

investigation on the information flow in GCNs, and thus its performance could be limited in practice. Specifically, dropout regularizes a network by randomly zeroing entries in a given feature map to remove part of the information. But in GCNs, as linear transformation on each node feature vector and local aggregation on neighboring node feature vectors are iteratively applied, the information is propagated both depth-wisely between different entries of each node feature vector and spatially between different node feature vectors. Thus the information of the individually zeroed entries by dropout could still present in other correlated entries. Although different variants of dropout were proposed to improve the regularization effectiveness, they mostly failed to consider the feature correlation in GCNs. The improvements are made in other aspects like using adaptive dropping rate [2], generalizing the random noise distribution from Bernoulli to Gaussian [23], randomly deleting graph edges to regularize GCNs [20], and so on. The only methods that consider feature correlation are several modified dropout techniques for CNNs, but due to the tremendous difference in data structure, these methods on Euclidean data are incapable of handling graph data. For example, DropBlock [9] considered the spatial correlation in the CNN feature maps and proposed to drop blocks of contiguous entries to remove information more effectively. However, its defined 'block' cannot be straightforwardly applied to graph data. Above all, there are no existing methods that could effectively regularize GCNs by taking the feature correlation into consideration.

In this paper, we propose DropCluster to better regularize GCNs by considering the feature correlations. We first randomly sample seed entries from the given node feature vectors, and then concurrently drop the seed entries as well as other entries with spatial or depth-wise correlation. As the information of a node is propagated to its l-hop neighbors after the l-th GCN layer, we consider nodes within l-hop neighborhood of a node as spatially correlated. The depth-wise correlation between different feature channels is measured by the linear correlation coefficient. By concurrently removing the spatially and depth-wisely correlated entries, the dropped information is eliminated more effectively, and the network gets more properly regularized. Also, as we select to drop entries within l-hop neighborhood of the seed entries, the selected entries form clusters around the seed entries, which is the reason our method is named DropCluster (Fig. 1). To demonstrate the effectiveness of DropCluster, we comprehensively compare it with dropout and other representative variants in experiments. Besides, our method is implemented into different GCNs to demonstrate its generalization capability. Moreover, we also implement it into networks with extended depths to further show its effectiveness in regularizing deep GCNs. The experiments are conducted on skeleton-based action recognition task on Northwestern-UCLA and NTU-RGB+D datasets.

2 Related Work

GCNs. GCNs are designed for extracting features from graph data. Two main streams of GCNs include the spectral- and the spatial-based GCNs. The spectral-based GCNs [3,6,14,16] apply convolutional filters to graph spectrum with good theoretical foundations, but the spatial-based GCNs are more preferred due to efficiency, flexibility and generalization issues. Thus, we only focus on the spatial GCNs in our work. The first spatial GCN was proposed in [17] and the main operation is aggregating the neighboring information of each node to obtain gradually refined node representations. Later, different variants began to emerge [1,10,18,27]. These variants mainly focus on two aspects, i.e. selecting which nodes to include in convolution and how to aggregate the selected nodes. Original GCN [14] selected 1-hop neighbors for the convolution and the receptive field of each convolution is restricted to 1-hop neighborhood. [25] utilized polynomials of functions of adjacency matrix as convolutional kernel thus the multi-hop neighborhood is captured by high order polynomials. Besides selecting distant nodes to enlarge receptive field, [4,10,37] applied different sampling strategies to neighboring nodes to reduce the computation burden. Works on designing aggregation mainly focus on different ways to decide the aggregation weights. [27] proposed to compute the weight between two nodes by inputting the node features into a feed forward network. [24] designed a graph agreement model to predict the probability of each edge correctly connecting two nodes, which then helps aggregating nodes more properly.

Dropout and Variants. Dropout was proposed in [11] to regularize fully connected networks by randomly zeroing entries in the feature maps. Later on, different variants began to emerge, including Gaussian dropout [23] that generalized dropout by extending the Bernoulli distribution to Gaussian, DropEdge [20] that regularizes GCNs by randomly deleting edges, and so on [2,12,13,28,34]. However, although these methods successfully improved dropout, they do not consider feature correlations thus their effectiveness is limited in GCNs, which is shown in our experiments. Besides the techniques dropping information randomly, there are also variants considering the spatial feature correlation, which were specially adapted to regularize CNNs. [26] proposed SpatialDropout to drop entire channels of a feature map, so that spatially correlated entries in the dropped channels are removed together. [7] proposed Cutout to randomly mask out square regions of the input. Inspired by Cutout, [9] proposed DropBlock to drop contiguous regions of a feature map and gained significant improvement. Among them, SpatialDropout [26] can be applied in GCNs and is shown in our experiments. While the other two cannot be applied to graph data as the 'block' on Euclidean data cannot be straightforwardly defined on graph. Moreover, depth-wise correlation is not considered in these methods. In this work, we propose a better regularization by considering feature correlations. Different from the methods above, we do not only consider spatial correlation, but also consider depth-wise correlation between channels. Moreover, our method is customized for GCNs on graph structured data.

3   DropCluster

Our proposed DropCluster aims to better regularize GCNs by considering spatial and depth-wise correlations when dropping entries. In this section, we first present the preliminaries, and then describe our model from two aspects, including addressing the spatial and depth-wise correlations. Finally, we discuss some related methods.

3.1   Preliminaries

We give a general formulation for directed graphs, and generalization to undirected graphs is straightforward with each undirected edge viewed as two directed edges. Each graph is represented as $G = (V, E)$, with $V = \{v_i | i = 1...N\}$ consisting of all the N nodes and $E = \{e_{ij}\}$ consisting of all the edges. Each node v_i has a feature vector $x_i \in \mathbb{R}^d$ with d channels (depth of d), and the feature map $X \in \mathbb{R}^{N \times d}$ is the concatenation of all feature vectors. Each edge e_{ij} denotes a directed edge pointing from v_i to v_j. The edges are also represented by an adjacency matrix $A \in \mathbb{R}^{N \times N}$, with $A_{ij} \in \{0, 1\}$ denoting whether the edge e_{ij} exists. Moreover, we d enote the adjacency matrix with added self-connections as $\tilde{A} = A + I_N$, where $I_N \in \mathbb{R}^{N \times N}$ is an identity matrix.

We denote a channel of a node feature vector as an entry. The j-th channel in x_i is then denoted as entry x_i^j. Given G with N nodes and d-dimensional feature vectors, its feature map $X \in \mathbb{R}^{N \times d}$ has $N \cdot d$ entries. In the following, depth-wise correlation denotes the correlation between different channels of the feature vectors. The neighboring nodes of a node v_i is defined as $\mathcal{N}_G = \{v_k | e_{ik} \in E\}$. The l-hop neighbors of v_i consist of the nodes with a distance of l from v_i, and are denoted as $\mathcal{N}_G^l(v_i)$. The neighboring entries of an entry x_i^j are entries in the same channels of v_i's neighboring nodes, i.e. $\mathcal{N}_e(x_i^j) = \{x_k^j | v_k \in \mathcal{N}_G(v_i)\}$. And the l-hop neighboring entries are similarly defined as $\mathcal{N}_e^l(x_i^j) = \{x_k^j | v_k \in \mathcal{N}_G^l(v_i)\}$.

DropCluster regularizes a network by multiplying the input feature map by a dropping mask M element-wisely. A graph convolution operation with a dropping mask M could be formulated as:

$$X^{(l+1)} = \sigma(\tilde{D}^{-\frac{1}{2}} \tilde{A} \tilde{D}^{-\frac{1}{2}} (M \odot X^{(l)}) W^{(l)}). \tag{1}$$

Here, $X^{(l)}$ is the output feature map from the l-th graph convolutional layer. $\tilde{D} \in \mathbb{R}^{N \times N}$ is a diagonal matrix denoting the degree of each node with $D_{ii} = \sum_j \tilde{A}_{ij}$. The term $W^{(l)}$ is a trainable matrix that conduct linear transformation on the node features at the l-th layer. σ is the activation function.

To generate the dropping mask M, the first step of DropCluster is selecting random seed entries. Specifically, we draw a matrix $M_{seed} \in \mathbb{R}^{N \times d}$ from a Bernoulli distribution parameterized by p_{seed}, i.e. $M_{ij} \sim \text{Bernoulli}(p_{seed})$, where the p_{seed} will be explained in Sect. 3.4. Then we will find other entries that are spatially or depth-wisely correlated to the seed entries to drop them concurrently.

3.2 Spatial Correlation

Spatial correlation comes from graph convolution. The information of each node is propagated to its l-hop neighbors after the l-th convolutional layer, therefore dropping individual entries cannot stop information from flowing to the subsequent layer. Thus, after the l-th layer, we propose to drop the seed entries together with the entries within their l-hop neighborhood. These entries form clusters centered at the seed entries with a radius of l, which is the reason our method is called DropCluster. The initial mask M_{seed} denotes seed entries with 1 and the others with 0, and we will update M_{seed} into mask M_s so that both the seed entries and their neighbors within l-hop are denoted with 1. As multiplying M_{seed} with $\tilde{A}$ propagates the value of seed entries to their 1-hop neighbors, multiplying M_{seed} with $\tilde{A}^l$ propagates the values to all entries within l-hop neighborhood. Thus we construct M_s as:

$$M_s = H(\tilde{A}^l \cdot M_{seed}), \tag{2}$$

where the Heaviside step function $H(\cdot)$ returns 1 for positive input and 0 otherwise, and is applied to binarize the obtained mask.

3.3 Depth-Wise Correlation

In GCNs, a linear transformation ($W^{(l)}$ in Eq. 1) on node feature vectors always follows the convolution operation. Thus, there also exists depth-wise correlation between entries of different channels. In this part, we introduce how to drop the depth-wisely correlated entries. We adopt the linear correlation coefficient as the measurement of depth-wise correlation. Given the input feature map $X \in \mathbb{R}^{N \times d}$

with d channels, the i-th column $X^i \in \mathbb{R}^N$ corresponds to the i-th channel of the feature map. We first generate a correlation matrix $M_{corr} \in \mathbb{R}^{d \times d1}$ to store the correlation coefficients between each pair of channels:

$$M_{corr}^{ij} = corr(X^i, X^j), \tag{3}$$

where M_{corr}^{ij} is at row i and column j of M_{corr}, denoting the correlation coefficient between X^i and X^j. The correlation coefficient $corr(a, b)$ between two variables is a normalized version of their covariance, which is defined as:

$$corr(a, b) = \frac{Cov(a, b)}{\sigma(a) \cdot \sigma(b)}, \tag{4}$$

where $Cov(a, b)$ is the covariance of two variables:

$$Cov(a, b) = E((a - \bar{a}) \cdot (b - \bar{b})), \tag{5}$$

and $\sigma(\cdot)$ is the standard variance of a variable:

$$\sigma(a) = \sqrt{E[(a - \bar{a})^2]}. \tag{6}$$

M_{corr} stores the correlation coefficients ranging from 0 to 1. For the following usage, if the absolute value of the correlation coefficient between two channels exceeds a threshold t_c, we regard the two channels as correlated. Formally, we derive another binary matrix M_c:

$$M_c = H(|M_{corr}| - t_c). \tag{7}$$

The mask M_s stores the already selected entries to drop. With M_c, we update M_s to further include entries depth-wisely correlated to the already selected ones. The i-th row of M_s ($M_{s,i}$) indicates the selected channels of x_i, and the j-th column of M_c (M_c^j) indicates all the channels correlated to channel j. Thus the inner product between $M_{s,i}$ and M_c^j takes a positive value if entry x_i^j is depth-wisely correlated to the already selected entries of x_i and takes zero otherwise. And the matrix multiplication between M_s and M_c is a parallelization of this computation, which turns all the other entries depth-wisely correlated to the already selected ones in M_s into positive. Above all, the update of mask M_s with binarization is formulated as:

$$M_s^c = H(M_s \cdot M_c). \tag{8}$$

M_s^c denotes all the entries to drop with 1, thus the final mask M mentioned in Eq. 1 would be $M = 1 - M_s^c$. The rescaling rate for training is omitted for simplicity and is included in Algorithm 1.

Linear correlation coefficient measures the linear correlation strength between channels. With strong linear correlation, two channels are mutually predictable, i.e. the presence of one feature indicates with high confidence the presence of the other one. Thus, individually dropping an entry in one channel cannot efficiently remove the semantic information, but concurrently dropping entries in other strongly correlated channels can remove the information more completely.

---

[1] Algorithm for parallelized computation of the correlation matrix is included in the supplementary material.

Algorithm 1: Dropluster

Input: $X^{(l)} \in \mathbb{R}^{N \times d}$, $A \in \mathbb{R}^{N \times N}$, r_d, t_c, layer index l

1 Compute correlation matrix $M_{corr} \in \mathbb{R}^{d \times d}$;

2 Apply threshold to M_{corr}: $M_c = H(|M_{corr}| - t_c)$;

3 Average number of correlated channels: $n_c = \frac{SUM(M_c)}{d}$;

4 Average number of edges: $n_e = \frac{SUM(A)}{N}$;

5 Number of seed entries:

6 **if** $l = 1$ **then**

7 $\quad \Big| \quad n_{seed} = \frac{N \times d \times r_d}{(1 + n_e) \times n_c}$;

8 **else**

9 $\quad \Big| \quad n_{seed} = \frac{N \times d \times r_d}{(1 + n_e + \sum_{i=1}^{l-1} n_e \times (n_e - 1)^{i-1}) \times n_c}$;

10 **end**

11 $p_{seed} = \frac{n_{seed}}{N \times d}$;

12 Initialize mask M_{seed} : $M_{seed}^{ij} \sim \text{Bernoulli}(p_{seed})$;

13 $\tilde{A} = A + I_N$;

14 Spatial correlation: $M_s = H(\tilde{A}^l \cdot M_{seed})$;

15 Channel correlation: $M_s^c = H(M_s \cdot M_c)$;

16 Actual dropping rate: $r_d^a = \frac{SUM(M_s^c)}{N \times d}$;

17 Rescaling rate: $r_c = \frac{1}{1 - r_d^a}$;

18 **return** $(1 - M_s^c) \cdot X^{(l)} \cdot r_c$

3.4 Number of Seed Entries

When handling the spatial and depth-wise correlation, we propagate the values of seed entries to neighboring and depth-wisely correlated entries. As the number of neighboring and depth-wisely correlated entries varies from node to node, the final number of chosen entries also varies. Thus a challenge is to choose a proper number of seed entries (p_{seed}) so that the final actual dropping rate is close to what we set. Our solution is as follows: Given the dropping rate r_d, the average number of edges denoted as n_e, the average number of correlated channels of each channel is n_c, the number of seed entries in the first layer should be:

$$n_{seed} = \frac{N \times d \times r_d}{(1 + n_e) \times n_c}, \tag{9}$$

while in the l-th ($l > 1$) layer they should be:

$$n_{seed} = \frac{N \times d \times r_d}{(1 + n_e + \sum_{i=2}^{l} n_e \times (n_e - 1)^{i-1}) \times n_c}, \tag{10}$$

and the p_{seed} is derived as:

$$p_{seed} = \frac{n_{seed}}{N \times d}. \tag{11}$$

Due to the page limit, detailed computations of Eqs. 9 and 10 are included in the supplementary materials.

3.5   Discussions

This part analyzes the difference between DropCluster and several related methods including SpatialDropout [26], DropBlock [9], and DropEdge [20].

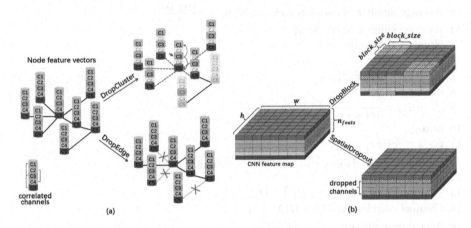

Fig. 2. (a): DropCluster and DropEdge on the given node features. For DropCluster, we highlight the seed entry (marked by red arrow) and other spatially or depth-wisely correlated entries. We assume the feature map is in the first GCN layer, thus only 1-hop neighboring entries are spatially correlated. For DropEdge, randomly deleted edges are marked with red crosses. (b): DropBlock and SpatialDropout on the given CNN feature map. The dropped entries are marked by red slashes (Color figure online)

DropCluster vs. SpatialDropout. Due to the convolution operation and spatial correlation in natural images, entries in CNN feature maps are also spatially correlated, which hampers the effectiveness of dropout. Thus, SpatialDropout [26] proposed to randomly drop entire feature channels to concurrently remove spatially correlated entries (Fig. 2 (b)). Given a feature map $F \in \mathbb{R}^{n_{feats} \times w \times h}$, SpatialDropout samples n_{feats} binary values, indicating whether to drop each channel. Differently, we consider the l-hop neighboring entries as spatially correlated in the l-th layer. Then the l-hop neighboring entries are removed together with the seed entries (Fig. 2 (a)). Moreover, considering depth-wise correlation is another great difference. Instead of randomly selecting channels, we drop entries in channels depth-wisely correlated to the selected entries.

DropCluster vs. DropBlock. DropBlock [9] focused on the spatial correlation in CNN feature maps and drops blocks of contiguous entries. Given a feature map, DropBlock drops entries in several randomly selected square regions of size $block\_size \times block\_size$ (Fig. 2 (b)). DropBlock cannot regularize GCNs as the square region cannot be defined on graphs. DropCluster deals with spatial correlation by concurrently removing entries within l-hop neighborhood of the random seed entries in the l-th layer. Compared to DropBlock, this not only differs in that it could be used in GCNs, but is also more delicate in that the size of dropped region is increased with the number of layer, corresponding to the increased receptive field in deeper layers. Moreover, DropCluster also removes depth-wisely correlated entries, which is not considered in DropBlock.

DropCluster vs. DropEdge. Similar to our work, DropEdge [20] is also specially for GCNs, but the approach is largely different. We aim to regularize GCNs by removing part of the information contained in the node features, while DropEdge proposed to randomly delete edges to sparsify the graph (Fig. 2 (a)). Given an adjacency matrix A denoting $|E|$ edges and the dropping rate as p, the adjacency matrix after dropping is $A_{drop} = A - A'$, where A' contains $|E| \cdot p$ randomly picked edges. In our experiments, DropEdge is also a baseline.

4   Experiments

In this section, we first test different hyperparameters to investigate their influence. Then we conduct ablation studies to verify the effectiveness of different parts of our model. After that, we compare our performance with other state-of-the-art techniques. To further demonstrate the effectiveness of our method in deep GCNs, we implement it to networks with increasing depths and compare the performance with dropout [11]. Finally, we implement our method to different network structures to show its generalization capability.

4.1   Datasets

Northwestern-UCLA [29]. The Multiview 3D event dataset contains RGB, depth and human skeleton data captured simultaneously by three Kinect cameras. This dataset include 10 action categories: pick up with one hand, pick up with two hands, drop trash, walk around, sit down, stand up, donning, doffing, throw, carry. Each action is performed by 10 actors. This dataset contains data taken from a variety of viewpoints. Following the setting in [15], samples obtained from the first two cameras are used for training and samples from the third camera are used for testing.

NTU-RGB+D [21]. NTU-RGB+D contains 56,880 samples from 60 classes, with modality of RGB videos, depth map sequences, 3D skeletal data, and infrared videos. Each human skeleton graph has 25 joints denoted by 3D coordinates (X, Y, Z). Following the recommendation of [21], we split the dataset by cross-subject (x-sub) and cross-view (x-view). In x-sub setting, 40,320 samples generated by one group of people serve as training set, and the other 16,560 samples by the other group of people are used for testing. In x-view setting, 37,920 samples captured by one set of cameras are used for training, and 18,960 samples captured by the other set of cameras are used for testing.

4.2   Implementation Details

On Northwestern-UCLA, we adopt ST-GCN [32] with modification as backbone and denote it as ST-GCN-U. ST-GCN-U has 9 layers. The first 3 layers have 32 channels for output, and the following 2 layers have 64 channels for output. The 6-th and 7-th have 128 channels for output, and the final 2 layers have 256 channels for output. DropCluster is implemented after each convolutional layer. The optimizer is SGD, with starting learning rate of 0.01. The model is trained for 100 epoch, and we decay the learning rate by 0.1 at epoch 20, 50, and 80.

On NTU-RGB+D, we adopt the original ST-GCN [32] with 9 layers as backbone. The first 3 layers have 64 channels for output, the following 3 layers have 128 channels for output, and the final 3 layers have 256 channels for output. DropCluster follows every convolutional layer. The optimizer is SGD with starting learning rate of 0.1. We train the model for 80 epochs, and decay the learning rate by 0.1 after epoch 10 and 50.

In Sect. 4.6, the 2-layer net has 64 input and 128 output channels for the 1-st layer, and the 2-nd layer has 128 channels for input and 256 for output. The 3-layer net has same first two layer structure as the 2-layer net, and the 3-rd layer has 256 channels for both input and output. The 4-layer net is built by inserting a layer with 128 channels for both input and output between the first two layers of the 3-layer net. For the 6-layer net, two identical layers with 64 channels for both input and output are inserted before the 1-st layer of the 4-layer net. Finally, the 11-layer net is based on the 9-layer net. The additional 10-th layer has 256 channels and 512 channels for input and output, and the final layer has 512 channels for both input and output.

4.3   Hyperparameter Analysis

In this part, we implement DropCluster to ST-GCN-U on Northwestern-UCLA with different dropping rate r_d and correlation threshold t_c. First we fix $t_c = 0.5$ and implement different dropping rates with the results shown in Fig. 3.

The red curve in Fig. 3 shows DropCluster performance with different dropping rates. We see that DropCluster works well with smaller rates and degrades when the rate increases. To further investigate this phenomenon, we simultaneously show performance of dropout. The behaviour of DropCluster and dropout are similar in that the performance first increases with dropping rate and degrades after reaching a peak with an optimal rate. As both DropCluster and dropout regularize a network by eliminating part of the information, for any dropping method, it is reasonable to expect an optimal dropping rate corresponding to a proper amount of information eliminated. Thus, as DropCluster eliminates semantic information more effectively, enough information is removed with a small dropping rate. On the contrast, by zeroing entries in randomly, dropout needs a higher rate to remove enough information. Thus we observe a smaller optimal dropping rate for DropCluster and a higher one for dropout. Moreover, even if a higher dropping rate could remove enough information, it

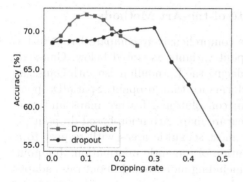

Fig. 3. Performance of DropCluster and dropout with different dropping rates on Northwestern-UCLA dataset (Color figure online)

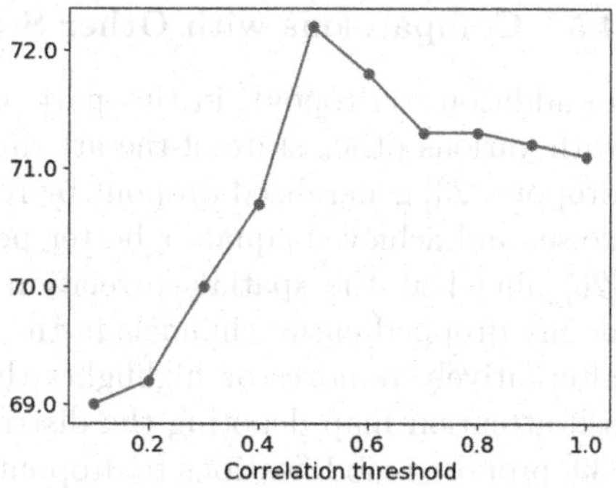

Fig. 4. Performance of DropCluster with different correlation threshold on Northwestern-UCLA dataset

may hurt the data quality as there are too many hollows everywhere in the feature map. Overall, DropCluster can eliminate enough information by zeroing a small part of the data and thus protect the data quality elsewhere. While in contrast, dropout needs higher dropping rate to achieve the same amount of information removal but also hurt data quality, resulting in a lower peak accuracy even with optimal dropping rate. Above, we set 0.1 as dropping rate for all experiments.

Then we vary t_c from 0 to 1, as shown in Fig. 4. According to Fig. 4, neither too small or too large threshold is promising, and 0.5 is approximately optimal. The correlation threshold determines the strength of co-dropping entries depthwisely. Too high threshold causes some correlated channels unable to be also concurrently dropped, while too small ones render weakly correlated channels dropped concurrently. Above all, we set the threshold as 0.5 for all experiments.

4.4 Ablation Study

Spatial and depth-wise correlations are major concerns of our method. In this part, we implement two models that consider only spatial or depth-wise correlation to separately study them. The experiments are conducted on Northwestern-UCLA with ST-GCN-U as backbone.

From Table 1, either considering sole spatial or depth-wise correlation improves

Table 1. Ablation study on Northwestern-UCLA dataset

Method	Top-1	Top-5
Without dropping	68.5%	97.2%
Dropout	70.5%	97.3%
Spatial Correlation	71.1%	99.1%
Depth-wise Correlation	70.8%	97.5%
Full DropCluster	72.2%	99.6%

the performance, but does not show strong superiority over dropout. The full DropCluster considering both correlation yields much better results, implying that both spatial and depth-wise correlations are significant in the GCN feature map and should be concurrently considered.

4.5   Comparisons with Other State-of-the-Art Methods

In addition to dropout, in this part, we comprehensively compare DropCluster with various other state-of-the-art dropout variants as stated below. Gaussian dropout [23] generalized dropout by replacing the Bernoulli noise with Gaussian noise, and achieved equal or better performance than dropout. SpatialDropout [26] aimed at the spatial correlation in convolutional feature maps and randomly dropped entire channels in the feature map. Attention-based dropout [5] alternatively removes or highlights the most semantic areas with respect to a self-attention map denoting the distribution of semantic information. Jumpout [30] proposed modifications to dropout including monotone dropout rate, adapting dropout rate to number of activated neurons, and rescaling the output to work with batch normalization. None of these methods are specially for GCNs. DropEdge [20] randomly removes edges at each layer, aiming to alleviate the over-fitting and over-smoothing problems by sparsifying the graph. We carefully adjusted the hyperparameters to get the most out of them, and the results are listed in Table 2 and Table 3.

Table 2. Comparisons with state-of-the-art regularization methods on Northwestern-UCLA

Methods	Top-1	Top-5
Without dropping	68.5%	97.2%
Jumpout [30]	69.2%	97.0%
Attention-based dropout [5]	69.5%	97.1%
Gaussian dropout [23]	70.0%	97.0%
DropEdge [20]	70.3%	97.2%
Dropout [11]	70.5%	97.3%
SpatialDropout [26]	70.8%	97.4%
DropCluster	72.2%	99.6%

Table 3. Comparisons with state-of-the-art regularization methods on NTU-RGB+D dataset

Methods	X-Sub	X-View
Without dropping	80.6%	88.0%
Jumpout [30]	80.7%	87.9%
Attention-based dropout [5]	81.0%	88.1%
SpatialDropout [26]	81.2%	88.1%
DropEdge [20]	81.3%	88.4%
Dropout [11]	81.5%	88.3%
Gaussian dropout [23]	82.2%	88.4%
DropCluster	83.3%	88.9%

From Table 2 and 3, although most baselines are not specially designed for GCNs, they show promising performance. But DropCluster still significantly outperforms them. On Northwestern-UCLA, the performance is boosted by 3.7% with DropCluster, while the highest improvement obtained by comparison methods is 2.3% by SpatialDropout. On NTU-RGB+D, our method gets 2.3% and 0.9% improvement compared to the model without any dropping on two protocols, and the second highest improvement is 1.6% and 0.4% obtained by Gaussian dropout. The performance of DropCluster on NTU-RGB+D is lower than that on Northwestern-UCLA, which is mainly caused by volume difference of the datasets. NTU-RGB+D is 30 times larger than Northwestern-UCLA, thus the model is more prone to over-fitting on Northwestern-UCLA and would benefit more from regularization. Among all baselines, SpatialDropout is related to our method as it also considers spatial relationship between entries. It is similar to DropCluster without considering depth-wise correlation, corresponding to 'Spatial Correlation' shown in Table 1. The performances of SpatialDropout and 'Spatial Correlation' are also similar with only 0.3% difference. The reason that SpatialDropout performs slightly worse could be that it drops the entire channels of the feature map, which causes severe information loss. On the contrast, for each channel, we drop clusters of nodes within the l-hop neighborhood of seed entries in the l-th layer, which flexibly removes only part of the information.

4.6   Implementation on Networks with Extended Depths

To further show DropCluster performs well in deep GCNs, we apply it to networks with increasing depths. We conduct experiments on NTU-RGB+D dataset with cross-view protocol. The results are shown in Table 4.

Table 4. Implementations of our DropCluster to networks with increasing depths

#layer	#para	w/o dropping	Dropout	Ours
2	0.37×10^6	81.1%	81.3%	81.4%
3	0.96×10^6	83.2%	83.3%	84.0%
4	1.11×10^6	86.4%	86.9%	87.7%
6	1.19×10^6	88.1%	88.4%	88.7%
9	1.96×10^6	88.1%	88.3%	88.9%
11	5.50×10^6	88.0%	88.6%	**90.0%**

Table 5. Implementation to GECNN and SLHM

Model	Regularizer	X-Sub	X-View
GECNN	-	83.6%	89.1%
GECNN	Dropout	84.0%	89.4%
GECNN	DropCluster	**84.7%**	**90.4%**
SLHM	-	84.3%	89.2%
SLHM	Dropout	84.7%	89.7%
SLHM	DropCluster	**85.3%**	**90.7%**

According to Table 4, the performance of the model without dropping first increases rapidly with depth then slows down, and stops at depth of 6, Moreover, the performance even gets lower when depth further increases. This phenomenon is reasonable. At the first stage wherein the depth increases from 2 to 6, the performance is boosted by the higher capacity of deeper networks. However, higher capacity also renders the network more prone to over-fitting. From Table 4, we see the parameters of 11-layer net is more than twice of the amount of 9-layer net, causing much higher probability of over-fitting. With dropout, this is alleviated but a downwards trend still presents when depth increases from 6 to 9. With DropCluster, the improvement is amazing. It not only better regularizes the shallower networks, but also outperforms dropout by 1.3% when the depth is increased to 11. Above all, it is obvious that DropCluster is more effective at regularizing deep networks to better exploit their expressive potential.

4.7   Further Implementations

In this part, we implement DropCluster to more networks to further demonstrate its generalization capability. Specifically, we apply DropCluster to GECNN, SLHM and GCN-NAS with two independent branch models and a full model.

The experiments above are on GCNs that apply convolution on graph nodes. In this part, we first apply our method to GECNN and SLHM [36], which are different from the node-based GCN models in that the graph edges also participate in convolution. In GECNN, only edges are included in convolution, while in SLHM, both edges and nodes participate. Despite the difference in the convolution computation, the correlation between features are similar. Thus our method could be directly implemented, and the results are in Table 5.

From Table 5, DropCluster significantly improves both models. As we directly adopt the hyperparameters of DropCluster from the previous sections, the generalization ability of DropCluster is strongly demonstrated.

GCN-NAS [19] is a recent model that adopted neural architecture search to design GCN for skeleton-based action recognition. The obtained model has a joint- and a bone-stream. We implement DropCluster to both streams as well as the full model, and the results are shown in Table 6.

Table 6. Implementation to GCN-NAS on NTU-RGB+D dataset with cross-subject protocol

Backbone model	Stream	Regularization	Top-1	Top-5
GCN-NAS	Joint	-	87.5%	97.7%
GCN-NAS	Joint	DropCluster	88.0%	97.7%
GCN-NAS	Bone	-	87.5%	97.8%
GCN-NAS	Bone	DropCluster	88.1%	97.7%
GCN-NAS	Joint & Bone	-	89.4%	98.0%
GCN-NAS	Joint & Bone	DropCluster	**89.9%**	**98.2%**

From Table 6, we could see that the performance of GCN-NAS is significantly improved with the regularization of our proposed DropCluster.

Above all, DropCluster demonstrates strong generalization capability and could be easily implemented in different GCNs as an effective regularization without hyperparameter adjustments.

5   Conclusion

In this paper, we propose DropCluster, an effective method to regularize GCNs by considering spatial and depth-wise correlation between features. We implement it to deep GCNs with different structures on different datasets with comparisons to dropout and other variants. Moreover, we apply it to networks with increasing depths to further demonstrate its capability to regularize deep GCNs.

Acknowledgement. This work was supported by Australian Research Council Projects FL-170100117, DP-180103424, and DE-180101438.

References

1. Atwood, J., Towsley, D.: Diffusion-convolutional neural networks. In: Advances in Neural Information Processing Systems, pp. 1993–2001 (2016)
2. Ba, J., Frey, B.: Adaptive dropout for training deep neural networks. In: Advances in Neural Information Processing Systems, pp. 3084–3092 (2013)
3. Bruna, J., Zaremba, W., Szlam, A., LeCun, Y.: Spectral networks and locally connected networks on graphs. arXiv preprint arXiv:1312.6203 (2013)
4. Chen, J., Ma, T., Xiao, C.: Fastgcn: fast learning with graph convolutional networks via importance sampling. arXiv preprint arXiv:1801.10247 (2018)
5. Choe, J., Shim, H.: Attention-based dropout layer for weakly supervised object localization. In: Proceedings of the IEEE Conference on Computer Vision and Pattern Recognition, pp. 2219–2228 (2019)
6. Defferrard, M., Bresson, X., Vandergheynst, P.: Convolutional neural networks on graphs with fast localized spectral filtering. In: Advances in Neural Information Processing Systems, pp. 3844–3852 (2016)
7. DeVries, T., Taylor, G.W.: Improved regularization of convolutional neural networks with cutout. arXiv preprint arXiv:1708.04552 (2017)
8. Gao, J., Zhang, T., Xu, C.: Graph convolutional tracking. In: Proceedings of the IEEE Conference on Computer Vision and Pattern Recognition, pp. 4649–4659 (2019)
9. Ghiasi, G., Lin, T.Y., Le, Q.V.: Dropblock: a regularization method for convolutional networks. In: Advances in Neural Information Processing Systems, pp. 10727–10737 (2018)
10. Hamilton, W., Ying, Z., Leskovec, J.: Inductive representation learning on large graphs. In: Advances in Neural Information Processing Systems, pp. 1024–1034 (2017)
11. Hinton, G.E., Srivastava, N., Krizhevsky, A., Sutskever, I., Salakhutdinov, R.R.: Improving neural networks by preventing co-adaptation of feature detectors. arXiv preprint arXiv:1207.0580 (2012)
12. Kang, G., Li, J., Tao, D.: Shakeout: A new regularized deep neural network training scheme. In: Thirtieth AAAI Conference on Artificial Intelligence (2016)
13. Kingma, D.P., Salimans, T., Welling, M.: Variational dropout and the local reparameterization trick. In: Advances in Neural Information Processing Systems, pp. 2575–2583 (2015)
14. Kipf, T.N., Welling, M.: Semi-supervised classification with graph convolutional networks. arXiv preprint arXiv:1609.02907 (2016)
15. Lee, I., Kim, D., Kang, S., Lee, S.: Ensemble deep learning for skeleton-based action recognition using temporal sliding LSTM networks. In: Proceedings of the IEEE International Conference on Computer Vision, pp. 1012–1020 (2017)
16. Levie, R., Monti, F., Bresson, X., Bronstein, M.M.: Cayleynets: graph convolutional neural networks with complex rational spectral filters. IEEE Trans. Signal Process. **67**(1), 97–109 (2018)
17. Micheli, A.: Neural network for graphs: a contextual constructive approach. IEEE Trans. Neural Netw. **20**(3), 498–511 (2009)
18. Niepert, M., Ahmed, M., Kutzkov, K.: Learning convolutional neural networks for graphs. In: International Conference on Machine Learning, pp. 2014–2023 (2016)
19. Peng, W., Hong, X., Chen, H., Zhao, G.: Learning graph convolutional network for skeleton-based human action recognition by neural searching. arXiv preprint arXiv:1911.04131 (2019)

20. Rong, Y., Huang, W., Xu, T., Huang, J.: Dropedge: towards deep graph convolutional networks on node classification. In: International Conference on Learning Representations (2019)
21. Shahroudy, A., Liu, J., Ng, T.T., Wang, G.: NTU RGB+D: a large scale dataset for 3D human activity analysis. arXiv preprint arXiv:1604.02808 (2016)
22. Shi, L., Zhang, Y., Cheng, J., Lu, H.: Skeleton-based action recognition with directed graph neural networks. In: Proceedings of the IEEE Conference on Computer Vision and Pattern Recognition, pp. 7912–7921 (2019)
23. Srivastava, N., Hinton, G., Krizhevsky, A., Sutskever, I., Salakhutdinov, R.: Dropout: a simple way to prevent neural networks from overfitting. J. Mach. Learn. Res. **15**(1), 1929–1958 (2014)
24. Stretcu, O., Viswanathan, K., Movshovitz-Attias, D., Platanios, E., Ravi, S., Tomkins, A.: Graph agreement models for semi-supervised learning. In: Advances in Neural Information Processing Systems, pp. 8710–8720 (2019)
25. Such, F.P., et al.: Robust spatial filtering with graph convolutional neural networks. IEEE J. Sel. Top. Signal Process. **11**(6), 884–896 (2017)
26. Tompson, J., Goroshin, R., Jain, A., LeCun, Y., Bregler, C.: Efficient object localization using convolutional networks. In: Proceedings of the IEEE Conference on Computer Vision and Pattern Recognition, pp. 648–656 (2015)
27. Veličković, P., Cucurull, G., Casanova, A., Romero, A., Lio, P., Bengio, Y.: Graph attention networks. arXiv preprint arXiv:1710.10903 (2017)
28. Wan, L., Zeiler, M., Zhang, S., Le Cun, Y., Fergus, R.: Regularization of neural networks using dropconnect. In: International Conference on Machine Learning, pp. 1058–1066 (2013)
29. Wang, J., Nie, X., Xia, Y., Wu, Y., Zhu, S.C.: Cross-view action modeling, learning and recognition. In: Proceedings of the IEEE Conference on Computer Vision and Pattern Recognition, pp. 2649–2656 (2014)
30. Wang, S., Zhou, T., Bilmes, J.: Jumpout: improved dropout for deep neural networks with ReLUs. In: International Conference on Machine Learning, pp. 6668–6676 (2019)
31. Wang, X., Türetken, E., Fleuret, F., Fua, P.: Tracking interacting objects optimally using integer programming. In: Fleet, D., Pajdla, T., Schiele, B., Tuytelaars, T. (eds.) ECCV 2014. LNCS, vol. 8689, pp. 17–32. Springer, Cham (2014). https://doi.org/10.1007/978-3-319-10590-1_2
32. Yan, S., Xiong, Y., Lin, D.: Spatial temporal graph convolutional networks for skeleton-based action recognition. In: Thirty-Second AAAI Conference on Artificial Intelligence (2018)
33. You, J., Liu, B., Ying, Z., Pande, V., Leskovec, J.: Graph convolutional policy network for goal-directed molecular graph generation. In: Advances in Neural Information Processing Systems, pp. 6410–6421 (2018)
34. Zhai, K., Wang, H.: Adaptive Dropout With Rademacher Complexity Regularization (2018)
35. Zhang, M., Jiang, S., Cui, Z., Garnett, R., Chen, Y.: D-VAE: a variational autoencoder for directed acyclic graphs. In: Advances in Neural Information Processing Systems, pp. 1586–1598 (2019)
36. Zhang, X., Xu, C., Tian, X., Tao, D.: Graph edge convolutional neural networks for skeleton-based action recognition. IEEE Trans. Neural Netw. Learn. Syst. **31**, 3047–3060 (2019)
37. Zou, D., Hu, Z., Wang, Y., Jiang, S., Sun, Y., Gu, Q.: Layer-dependent importance sampling for training deep and large graph convolutional networks. In: Advances in Neural Information Processing Systems, pp. 11247–11256 (2019)

Adaptive Video Highlight Detection
by Learning from User History

Mrigank Rochan[1]([✉])[iD], Mahesh Kumar Krishna Reddy[1][iD], Linwei Ye[1][iD],
and Yang Wang[1,2][iD]

[1] University of Manitoba, Winnipeg, MB R3T 2N2, Canada
{mrochan,kumarkm,yel3,ywang}@cs.umanitoba.ca
[2] Huawei Technologies, Markham, Canada

Abstract. Recently, there is an increasing interest in highlight detection research where the goal is to create a short duration video from a longer video by extracting its interesting moments. However, most existing methods ignore the fact that the definition of video highlight is highly subjective. Different users may have different preferences of highlight for the same input video. In this paper, we propose a simple yet effective framework that learns to adapt highlight detection to a user by exploiting the user's history in the form of highlights that the user has previously created. Our framework consists of two sub-networks: a fully temporal convolutional highlight detection network H that predicts highlight for an input video and a history encoder network M for user history. We introduce a newly designed temporal-adaptive instance normalization (T-AIN) layer to H where the two sub-networks interact with each other. T-AIN has affine parameters that are predicted from M based on the user history and is responsible for the user-adaptive signal to H. Extensive experiments on a large-scale dataset show that our framework can make more accurate and user-specific highlight predictions.

Keywords: Video highlighght detection · User-adaptive learning

1 Introduction

There is a proliferation in the amount of video data captured and shared everyday. It has given rise to multifaceted challenges, including editing, indexing and browsing of this massive amount of video data. This has drawn attention of the research community to build automated video highlight detection tools. The goal of highlight detection is to reduce an unedited video to its interesting visual moments and events. A robust highlight detection solution can enhance video browsing experience by providing quick video preview, facilitating video sharing on social media and assisting video recommendation systems.

Even though we have made significant progress in highlight detection, the existing methods are missing the ability to adapt its predictions to users. The

© Springer Nature Switzerland AG 2020
A. Vedaldi et al. (Eds.): ECCV 2020, LNCS 12366, pp. 261–278, 2020.
https://doi.org/10.1007/978-3-030-58589-1_16

main thrust of research in highlight detection has been on building generic models. However, different users have different preferences in term of detected highlights [25,34]. Generic highlight detection models ignore the fact that the definition of a video highlight is inherently subjective and depends on each individual user's preference. This can greatly limit the adoption of these models in real-world applications. In Fig. 1, we illustrate the subjective nature of highlights. The input video contains events such as cycling, cooking, and eating. A generic highlight detection model mainly predicts the cycling event as the highlight. But if we examine the user's history (previously created highlights by the user), we can infer that this user is interested in cooking scenes. Therefore, a highlight detection model should predict the cooking event as the highlight instead. Motivated by this observation, we propose *an adaptive highlight detection model* that explicitly takes user's history into consideration while generating highlights.

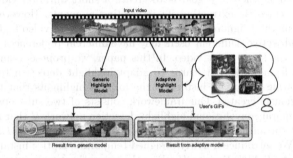

Fig. 1. The definition of highlight of a video is inherently subjective and depends on each user's preference. In contrast to a generic highlight detection model, an adaptive highlight detection model (like ours) incorporates a user's previously created highlights (e.g., GIFs from multiple videos) when predicting highlights of an input video. This allows the model to make more accurate and user-specific highlight predictions.

Although a user's visual highlight history can provide a stronger and more reliable cue of their interests than non-visual meta-data [25], there is very limited research on adapting highlight detection using this visual information. To the best of our knowledge, the recent work by Molino and Gygli [25] is the only prior work on this topic. Their method considers a user's previously created highlights (available as GIFs[1]) from multiple videos when generating new highlights for that user. They propose a ranking model that predicts a higher score for interesting video segments as opposed to non-interesting ones while conditioning on the user's past highlights (i.e., user's history). However, their method has some limitations. Firstly, it operates at the segment level and samples a fixed number of positive and negative segments from a video for learning. This means that the method does not process an entire video which is essential to capture temporal dependencies shown to be vital in many video understanding tasks

---

[1] GIF is an image format with multiple frames played in a loop without sound [11].

[19,24,32,50]. Moreover, it is sensitive to the number of positive and negative samples used in the learning. Secondly, it requires to precompute shot boundaries using a shot detection algorithm [8] for sampling a set of positive/negative video segments. This makes their pipeline computationally complex and expensive. Lastly, their model directly combines user's history features with the features of sampled video segments to predict user-specific highlights. We demonstrate in our experiments that this is not as effective as our proposed model.

In this paper, we introduce a novel user-adaptive highlight detection framework that is simple and powerful. Given an input video for highlight detection and the user's history (highlight GIFs from multiple videos that the user has previously created), our model seamlessly combines the user's history information with the input video to modulate the highlight detection so as to make user-specific and more precise highlight prediction. Our framework consists of two sub-networks: a fully temporal convolutional *highlight detection network* H which produces the highlight for an input video, and a *history encoder network* M that encodes the user's history information. We propose *temporal-adaptive instance normalization* (T-AIN), a conditional temporal instance normalization layer for videos. We introduce T-AIN layers to H where the interaction between the two sub-networks H and M takes place. T-AIN layers have affine transformation parameters that are predicted from M based on the user's history. In other words, M acts a guiding network to H. Through the adjustable affine parameters in T-AIN layers, H can adapt highlight predictions to different users based on their preferences as expressed in their histories. Note that our method does not require expensive shot detection. Moreover, it can utilize an entire video for learning instead of a few sampled video segments.

To summarize, the main contributions of our paper are the following. (1) We study user-adaptive highlight detection using user history in the form of previously created highlights by the user. This problem has many commercial applications, but is not well studied in the literature. (2) Different from ranking models [11,25,45,46] commonly used in highlight detection, we are first to employ a fully temporal convolutional model in highlight detection. Our model does not require expensive shot detection algorithm and can process an entire video at once. This makes our proposed model simpler operationally. (3) We propose temporal-adaptive instance normalization layer (T-AIN) for videos. T-AIN layers have adjustable affine parameters which allow the highlight detection network to adapt to different users. (4) We experiment on a large-scale dataset and demonstrate the effectiveness of our approach. (5) We further explore the application of our model as a pre-trained model for video summarization, a task closely related to highlight detection.

2   Related Work

Highlight detection aims to identify key events in a video that a user is likely to find interesting. Towards this, we exploit the highlights that the user has created in the past. In this section, we discuss the several lines of related work.

Our work is closely related to existing highlight detection methods where the goal is to extract the interesting moments from a long duration video [25,39]. Many prior methods [11,14,25,36,46,47] employ a ranking formulation. They learn to score interesting segments of a video higher than non-interesting segments. Our work is particularly related to [11,25]. Gygli *et al.* [11] propose a GIF creation technique using a highlight detection method. The limitation of this method is that it is a generic highlight detector, whereas we propose a model that is capable of making user-specific predictions. Molino and Gygli [25] propose a model that takes a user's history as an input to make personalized predictions. Their method is a ranking model that operates on a few sampled positive and negative segments of a video combined with the user's highlight history to learn a personalized model. Different from them, our proposed highlight detection model is convolutional in nature. It can accept an entire video (of any length) and the user's history of different sizes as an input to produce a personalized highlight. Unlike [25], we do not require shot detection as pre-processing. Lastly, we develop a more effective way to leverage user history representation in the network instead of directly concatenating with the input video features [25].

This paper has also connection with the research in video summarization. Different from highlight detection, video summarization aims to generate a concise overview of a video [25,39]. Early research [16,17,20,23,24,26,28,32,35,51,54] on summarization mainly develop unsupervised methods. These methods design hand-crafted heuristics to satisfy properties such as diversity and representativeness in the generated summary. Weakly supervised methods also exist that exploit cues from web images or videos [4,16,17,31,35] and video category details [27,30] for summarization. Supervised methods [7,9,10,32,49,50,53] have shown to achieve superior performance due to the fact that they learn directly from annotated training data with human-generated summaries. However, these methods do not consider each user's preference. In theory, it is possible to make user-adaptive predictions by performing user-specific training. But it is expensive in practice as it would result in per-user model [25]. However, we learn a single model that can be adapted to different users by incorporating their histories. Moreover, we do not need to retrain the model for a new user. The adaption to a new user only involves some simple feed-forward computation. Lastly, in video summarization, there is some work that focus on personalization but they either use meta-data [1,2,13,37] or consider user textual query [22,33,41,52] to personalize video summary. Different from them, our method operates on visual features from user video and user's history to make user-specific prediction.

Finally, our approach is partly inspired by recent research in style transfer in images using conditional normalization layers, e.g., adaptive instance normalization [12] and conditional batch normalization [6]. Apart from style transfer, both layers have been applied in many other computer vision problems [3,5,15,21,29,43]. These layers firstly normalize the activations to zero mean and unit variance, then adjust the activations via affine transformation parameters inferred from external data [29]. Since the layers are designed to uniformly modulate the activations spatially, their applications are limited to images. In contrast, in this paper, we propose a normalization layer that uniformly modulates the activations temporally, making it appropriate for video understanding tasks.

3 Our Approach

Given a video $\mathbf{v}$, we represent each frame in the video as a D-dimensional feature vector. The video can be represented as a tensor of dimension $1 \times T \times D$, where T and D are the number of frames and dimension of frame feature vector of each frame in the video, respectively. T varies for different videos. For an user U, let $\mathcal{H} = \{h_1, ..., h_n\}$ denotes the *user's history* which is a collection of visual highlights that the user has created in the past.

Given $\mathbf{v}$ and $\mathcal{H}$, our goal is to learn a mapping function G that predicts two scores for each frame in $\mathbf{v}$ indicating it being non-highlight and highlight. Thus, the final output S is of dimension $1 \times T \times 2$ for the input video $\mathbf{v}$:

$$S = G(\mathbf{v}, \mathcal{H}) = G(\mathbf{v}, \{h_1, ..., h_n\}). \tag{1}$$

We refer to G as the adaptive highlight detector.

3.1 Background: Temporal Convolution Networks

In recent years, temporal convolution networks (e.g., FCSN [32]) have shown to achieve impressive performance on video understanding tasks. These networks mainly perform 1D operations (e.g., 1D convolution, 1D pooling) over the temporal dimension (e.g., over frames in a video). This is analogous to the 2D operations (e.g., 2D convolution, 2D pooling) commonly used in CNN models for image-related tasks. For example, the work in [32] uses temporal convolutional networks for video summarization, where the task is formulated as predicting a binary label for each frame in a video. Our proposed model is based on the temporal convolution network proposed in [32]. But we extend the model to perform user-specific video highlight detection.

3.2 Temporal-Adaptive Instance Normalization

Let $\mathbf{o}^i$ indicate the activations of i-th layer in the temporal convolution neural network (f_T) for the input video $\mathbf{v}$. We use C^i and T^i to denote the number of channels and temporal length of activation in that layer, respectively. We define the *Temporal-Adaptive Instance Normalization* (T-AIN), a conditional normalization layer for videos. T-AIN is inspired by the basic principles of Instance Normalization [40]. The activation is firstly normalized channel-wise along *temporal* dimension (obtaining $\mathbf{o}^i_{norm}$), followed by a uniform modulation with affine transformation. Different from InstanceNorm [40], the affine parameters, scale and bias, in T-AIN are *not learnable* but inferred using external data (i.e., a user's history ($\mathcal{H}$) in our case) which is encoded to a vector m using another temporal convolution network (g_T). Thus, T-AIN is also *conditional* (on $\mathcal{H}$) in nature. The activation value from T-AIN at location $c \in C^i$ and $t \in T^i$ is

$$\left(\frac{o^i_{c,t} - \mathrm{E}[o^i_c]}{\sqrt{\mathrm{Var}[o^i_c] + \epsilon}} \right) \gamma^i_c + \delta^i_c, \tag{2}$$

where $o^i_{c,t}$, $\mathrm{E}[o^i_c]$ and $\mathrm{Var}[o^i_c]$ are the activation before normalization, expectation and variance of the activation $\mathbf{o}^i$ in channel c, respectively. T-AIN computes the $\mathrm{E}[o^i_c]$ and $\mathrm{Var}[o^i_c]$ along temporal dimension independently for each channel and every input sample (video) as:

$$\mathrm{E}[o^i_c] = \mu^i_c = \frac{1}{T^i} \left(\sum_t o^i_{c,t} \right), \tag{3}$$

$$\mathrm{Var}[o^i_c] = \mathrm{E}[(o^i_c - \mathrm{E}[o^i_c])^2] = \frac{1}{T^i} \sum_t \left(o^i_{c,t} - \mu^i_c \right)^2. \tag{4}$$

In Eq. 2, γ^i_c and δ^i_c are the modulation parameters in the T-AIN layer. We obtain γ^i_c and δ^i_c from the encoded vector m generated using the external data. T-AIN firstly scales each value along the temporal length in channel c of temporally normalized activations $\mathbf{o}^i_{norm}$ by γ^i_c and then shifts it by δ^i_c. Similar to InstanceNorm, these statistics vary across channel C^i. We provide details on how we compute these parameters when using user's history $\mathcal{H}$ as an external data in Sect. 3.3. In Fig. 2, we visualize the operations in a T-AIN layer.

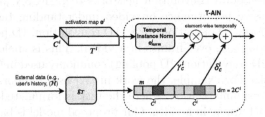

Fig. 2. Overview of a temporal-adaptive instance normalization layer (T-AIN). For an input video $\mathbf{v}$, let $\mathbf{o}^i$ be the activation map with channel dimension C^i and temporal length T^i in the i-th layer of a temporal convolutional network f_T. Let g_T be another temporal convolutional network that encodes external data (e.g., user's history $\mathcal{H}$) into a vector representation m of dimension $2C^i$. T-AIN firstly temporally normalizes $\mathbf{o}^i$ in each channel to obtain $\mathbf{o}^i_{norm}$. It then uniformly scales and shifts $\mathbf{o}^i_{norm}$ in channel c (where $c \in C^i$) over time by γ^i_c and δ^i_c, respectively. The values of γ^i_c and δ^i_c are obtained from m. As can be seen, the main characteristics of T-AIN include temporal operation, no learnable parameters, and conditional on external data.

T-AIN is related to the conditional batch normalization (CBN) [6] and the adaptive instance normalization (AIN) [12]. The main difference is that CBN and AIN work spatially, so they are appropriate for image-related tasks like style transfer. In contrast, T-AIN is designed to operate along time, which makes it suitable for video highlight detection and video understanding tasks in general.

3.3   Adaptive Highlight Detector

The adaptive highlight detector G consists of two sub-networks: a highlight detection network H and a history encoder network M. H is responsible for scoring each frame in an input video to indicate whether or not it should be included in the highlight. The role of M is to firstly encode the user's history information and then guide H in a manner that the generated highlight is adapted to the user's history. Next, we discuss the sub-networks design and learning in detail.

Highlight Detection Network. The highlight detection network H is based on FCSN [32]. It is an encoder-decoder style architecture which is fully convolutional in nature. One advantage of this network is that it is not restricted to fixed-length input videos. It can handle videos with arbitrary lengths. Another advantage is that it is effective in capturing the long-term temporal dynamics among the frames of a video beneficial for video understanding tasks such as highlight detection.

H accepts a video $\mathbf{v}$ with feature representation of dimension $1 \times T \times D$, where T is number of frames in $\mathbf{v}$ and D is the dimension of each frame feature vector. It produces an output of dimension $1 \times T \times 2$ indicating non-highlight and highlight scores for the T frames.

The encoder (F_v) of H has a stack of seven convolutional blocks. The first five blocks (i.e., *conv-blk*1 to *conv-blk*5) consist of several temporal convolution followed by a ReLU and a temporal max pooling operations. The last two blocks (*conv*6 to *conv*7) have a temporal convolution, followed by a ReLU and a dropout layer. The encoder F_v gives two outputs: a feature map from its last layer and a skip connection from block *conv-blk*4.

The output of encoder F_v is fed to the decoder network (D_v). We introduce T-AIN layers at sites where these two outputs enter D_v. We obtain a feature map by applying a 1×1 convolution and a temporal fractionally-strided convolution operation (*deconv*1) to the output of first T-AIN, which is added with the feature map from a 1×1 convolution to the output of second T-AIN layer. Finally, we apply another fractionally-strided temporal convolution (*deconv*2) to obtain a final prediction of shape $1 \times T \times 2$ denoting two scores (non-highlight or highlight) for each frame in the video. Figure 3 (top) visualizes the architecture of H.

History Encoder Network. The history encoder network M is an integral piece of our framework. It acts as a guiding network that guides the highlight network H by adaptively computing the affine transformation parameters of its T-AIN layers. Using these affine parameters, M modulates the activations in H to produce adaptive highlight predictions.

The configuration of this network is same as the encoder F_v in H with few changes towards the end. It performs an average temporal pooling to the output of convolution blocks, which is combined with a skip connection from the input that is then fed to a fully-connected layer. The skip connection involves an average pooling and a fully-connected operation to match dimensions.

The network accepts user's history collection $\mathcal{H}$ of shape $1 \times n \times D$ as an input, where n is the number of elements/highlights in the user's history and

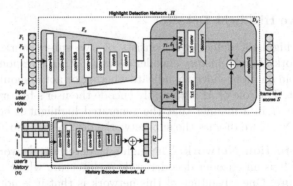

Fig. 3. Overview of our proposed model, `Adaptive-H-FCSN`. The model consists of two sub-networks, a *highlight detection network* H and a *history encoder network* M. H is an encoder-decoder architecture that takes a frame-level vector feature representation of a user input video with T frames. It then generates scores (highlight vs. non-highlight) for each frame in the video while taking information from M. M takes vector feature representation of each element (i.e., highlights the user has previously created) in the user's history as an input and encodes it to a vector $\mathbf{z}_h$. This vector $\mathbf{z}_h$ is then simply fed to a fully connected layer FC to produce the affine parameters γ_j and δ_j in the j-th T-AIN layer of decoder D_v where $j = 1, 2$. This way the highlight detection for the input video is adapted to the user.

D is the dimension of the vector representation of each element. In our implementation, we obtain a D-dimensional vector from a highlight using C3D [38]. Note that n varies for different users. After the combining stage, the network generates a latent code $\mathbf{z}_h$ which is a fixed-length user-specific vector.

Next, we forward $\mathbf{z}_h$ to a fully connected layer (FC) to decode the latent code $\mathbf{z}_h$ into a set of vectors (γ_j, δ_j) where $j = 1, 2$. The parameters γ_j and δ_j denote the scaling and bias parameters of j-th T-AIN layer in the decoder D_v of H, respectively. These affine parameters are uniformly applied at every temporal location in the feature map. This allows to incorporate the user's history information in H and adjust it in a way that the predicted highlight is adapted to the user's history. This way we obtain a user-specific highlight prediction for the input user video. Figure 3 (bottom) presents the architecture of M.

With F_v, D_v and M, we can rewrite Eq. 1 as:

$$S = D_v(F_v(\mathbf{v}), M(\{h_1, ..., h_n\})). \tag{5}$$

By applying this design, we learn a generic video representation using F_v and extract a user-specific latent representation with M. Finally, by injecting user-specific latent information to D_v through T-AIN layers, we allow the model to adapt the highlight detection to the user's history.

Note that we use temporal convolutions over the highlights in the user history. In addition, we also investigate a non-local model [42,48] with self-attention instead of temporal convolutions for M. Here, the output of self-attention is firstly average pooled to produce a single vector and then fed to a fully-connected layer. We find that the non-local model performs slightly inferior to the temporal convolutions model. This is probably because the highlights in the history are ordered based on their time of creation in the dataset, so the temporal convolutions allow the history encoder M to capture some implicit temporal correlations among them.

3.4 Learning and Optimization

We train our adaptive highlight detector G using a cross-entropy loss. For an input video $\mathbf{v}$ with T frames and corresponding binary indicator ground truth label vector (indicating whether a frame is non-highlight or highlight), we define a highlight detection loss $\mathcal{L}_{highlight}$ as:

$$\mathcal{L}_{highlight} = -\frac{1}{T} \sum_{t=1}^{T} \log \left(\frac{\exp(\lambda_{t,l_c})}{\sum_{c=1}^{2} \exp(\lambda_{t,c})} \right), \quad (6)$$

where $\lambda_{t,c}$ is the predicted score of t-th frame to be the c-th class (non-highlight or highlight) and λ_{t,l_c} is the score predicted for the ground truth class.

The goal of our learning is to find optimal parameters $\Theta_{F_v}^*$, $\Theta_{D_v}^*$ and Θ_M^* in the encoder F_v, decoder D_v of the highlight detection network H, and the history encoder network M, respectively. The learning objective can be expressed as:

$$\Theta_{F_v}^*, \Theta_{D_v}^*, \Theta_M^* = \underset{\Theta_{F_v}, \Theta_{D_v}, \Theta_M}{\arg\min} \mathcal{L}_{highlight}(F_v, D_v, M). \quad (7)$$

For brevity, we use `Adaptive-H-FCSN` to denote our adaptive highlight detection model learned using Eq. 7.

4 Experiments

4.1 Dataset

We conduct experiments on the largest publicly available highlight detection dataset, PHD-GIFs [25]. It is also the only large-scale dataset that has user history information for highlight detection. The released dataset consists of $119,938$ videos, $13,822$ users and $222,015$ annotations. The dataset has $11,972$ users in training, $1,000$ users in validation, and 850 users in testing. There is no overlap among users in these three subsets.

Apart from being large-scale, this dataset is also interesting because it contains user-specific highlight examples indicating what exactly a user is interested in when creating highlights. The ground truth segment-level annotation comes from GIFs that a user creates (by extracting key moments) from YouTube videos. In this dataset, a user has GIFs from multiple videos where the last video of the user is considered for highlight prediction and the remaining ones are treated as examples in the user's history.

The dataset only provides YouTube video ID for the videos and not the original videos. So we need to download the original videos from YouTube to carry out the experiments. We were able to download 104, 828 videos and miss the remaining videos of the dataset since they are no longer available on YouTube. In the end, we are able to successfully process 7, 036 users for training, 782 users for validation and 727 users for testing. Note that code of previous methods on this dataset are not available (except pre-trained Video2GIF [11]), so we implement several strong baselines (see Sect. 4.3) in the paper.

4.2   Setup and Implementation Details

Evaluation Metrics: We use the mean Average Precision (mAP) as our evaluation metric. It measures the mean of the average precision of highlight detection calculated for every video in the testing dataset. Different from object detection where all the detections are accumulated from images to compute the average precision, highlight detection treats videos separately because it is not necessary a highlighted moment in a particular video is more interesting than non-highlighted moments in a different video [36]. This metric is commonly used to measure the quality of predictions in highlight detection [11,25,36,45].

Feature Representation: Following prior work [25], we extract C3D [38] (conv5) layer features and use it as feature representation in the model for the input videos and user's history. For an input video, we extract C3D-features at frame-level. For a highlight video in the user's history, we prepare a single vector representation by averaging its frame-level C3D features.

Training Details: We train our models from scratch. All the models are trained with a constant learning rate of 0.0001. We use Adam [18] optimization algorithm for training the models. Note that we apply this training strategy in all our experiments including the other analysis (Sect. 4.5).

Since the dataset has users that create multiple GIFs for a video, we follow [25] to prepare a single ground truth for the video by taking their union.

Testing Details: Given a new test user video and the user's history, we use our trained model to predict highlight score for each frame which is then sent to the evaluation metrics to measure the quality of predicted highlight. We follow the evaluation protocol of previous work [11,25] for fair comparison. Note that our model can handle variable length input videos and variable number of history elements. We consider full user's history while predicting highlights.

4.3   Baselines

We compare the proposed Adaptive-H-FCSN with the following strong baselines:

FCSN [32]: This network is the state of the art in video summarization which we adapt as our highlight detection network. FCSN has no instance normalization layers. We train and evaluate FCSN on the PHD-GIFs dataset.

Video2GIF [11]: This baseline is a state-of-the-art highlight detection model. We evaluate the publicly available pre-trained model.

FCSN-Aggregate: In this baseline, we train FCSN [32] by directly combining the user history with the input video. More specifically, we firstly obtain a vector representation for the user history by averaging the features of elements in the history. Next, we add this aggregated history with the feature representation of each frame in the input video.

H-FCSN: This baseline is a variant of highlight detection network H we presented in Sect. 3.3, where we replace the T-AIN layers in the decoder of H with the unconditional temporal instance normalization layers. We do not have the history encoder network M. This results in `Adaptive-H-FCSN` transformed to a generic highlight detection model with no adaptation to users.

H-FCSN-Aggregate: Similar to FCSN-aggregate, we directly combine the user's history features with an input video features and learn H-FCSN. Different from H-FCSN, this is not a generic highlight detection model but a user-adaptive highlight detection model as we allow the model to leverage the user's history information in the training and inference.

4.4 Results and Comparison

In Table 1, we provide the experiment results (in terms of mAP %) of our final model `Adaptive-H-FCSN` along with the baselines and other alternative methods.

Table 1. Performance (mAP%) comparison between `Adaptive-H-FCSN` and other approaches. We compare with both non-adaptive and adaptive highlight detection methods. Our method `Adaptive-H-FCSN` outperforms the other alternative methods. We also compare with `Adaptive-H-FCSN-attn` that uses self-attention in the history encoder (see Sect. 3.3). Note that all the listed methods use C3D feature representation

Method	mAP (%)	User-adaptive
Random	12.27	✗
FCSN [32]	15.22	✗
Video2GIF [11]	14.75	✗
H-FCSN	15.04	✗
FCSN-aggregate	15.62	✓
H-FCSN-aggregate	15.73	✓
Adaptive-H-FCSN-attn	16.37	✓
Adaptive-H-FCSN	**16.73**	✓

`Adaptive-H-FCSN` outperforms all the baselines. Results show that directly combining (i.e., FCSN-aggregate and H-FCSN-aggregate) history information with input video only slightly improves the highlight detection results in comparison to FCSN and H-FCSN that do not leverage users history information. However, we notice a significant performance gain for Adaptive-H-FCSN model. This result validates that directly combining user history information with the input video is a sub-optimal solution for user-adaptive highlight detection. Additionally, this result reveals that proposed T-AIN layer plays a critical role in producing more accurate and user-specific highlight detection. It is also noteworthy that we obtain a lower performance (nearly 1%) for Video2GIF [11] than reported in PHD-GIFs [25] which implies that our test set is more challenging.

Figure 4 shows some qualitative examples for the generic baseline model (H-FCSN) and our proposed adaptive highlight detection model (Adaptive-H-FCSN). We can see that our model successfully captures the information in user's history and produces highlights that are adapted to the user's preference.

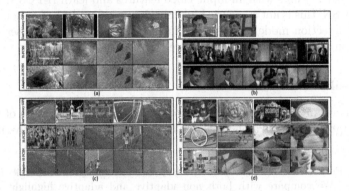

Fig. 4. Qualitative examples for different methods. We show examples of the generic highlight detection model (H-FCSN) and our user-adaptive model (Adaptive-H-FCSN) on four videos. For each video, we show the user's history (multiple GIFs) and few sampled frames from the highlight predictions of the two models. Based on the user's history, we find that in (a) the user has interest in animals; (b) the user is interested in faces that dominate a scene; (c) the user is inclined to highlight goal scoring scenes; and (d) the user focuses on cooking. These visualizations indicate that adaptation to the user's interest is important for a meaningful and accurate highlights. Compared with H-FCSN, the prediction of Adaptive-H-FCSN is more consistent with the user history.

4.5   Analysis

Effect of Affine Parameters. We analyze the importance of affine parameters γ_c^i and δ_c^i (Eq. 2) for adaptive highlight detection. In Table 2, we report the highlight detection performance for different possible choices of these parameters. We find that when these parameters are adaptively computed ($\gamma_c^i = \gamma_c^h$, $\delta_c^i = \delta_c^h$) from another network capturing the user history information (i.e., Adaptive-H-FCSN) significantly boosts the highlight detection performance as opposed to cases when it is directly learned ($\gamma_c^i = \gamma_c^*$, $\delta_c^i = \delta_c^*$) and set to a fixed value ($\gamma_c^i = 1$, $\delta_c^i = 0$) in the main highlight detection network. Thus, the proposed T-AIN layer is key to obtain user-adaptive highlights.

Table 2. Impact of affine parameters on highlight detection. Here we show the performance (mAP%) for different choices of affine parameters γ_c^i and δ_c^i in Eq. 2

Method	$\gamma_c^i = 1, \delta_c^i = 0$	$\gamma_c^i = \gamma_c^*, \delta_c^i = \delta_c^*$	$\gamma_c^i = \gamma_c^h, \delta_c^i = \delta_c^h$
H-FCSN	14.64	15.04	–
H-FCSN-aggregate	14.87	15.73	–
Adaptive-H-FCSN	–	–	**16.73**

Effect of User's History Size. We perform additional study to analyze how sensitive our model is to the length of a user's history (i.e., numbers of highlights previously created). We restrict the number of history elements for users in the training. That is, we consider only h highlight videos from the user's history in training. During testing, we consider the user's full history.

Table 3 shows the results of various methods as a function of number of elements ($h = 0, 1, 5, n$) in user's history. We observe that Adaptive-H-FCSN outperforms generic highlight model (H-FCSN) even when there is a single highlight in the user's history. We notice the performance of Adaptive-H-FCSN gradually improves when we increase the number of history elements, whereas H-FCSN-aggregate doesn't show similar performance trend. It achieves the best results when we utilize a user's full history (i.e., $h = n$).

Table 3. Impact of an user's history size (i.e., number of history elements/highlights) on different methods. Here we vary the history size h as 0 (no history), 1, 5, and n (full history). The performance of our model improves with the increase in history size

History size ($\mathcal{H}$)	$h = 0$	$h = 1$	$h = 5$	$h = n$
H-FCSN	15.04	–	–	–
H-FCSN-aggregate	–	15.62	15.04	15.73
Adaptive-H-FCSN	–	15.57	15.69	**16.73**

4.6 Application to Video Summarization

Video summarization is closely related to highlight detection. Highlight detection aims at extracting interesting moments and events of a video, while video summarization aims to generate a concise synopsis of a video. Popular datasets in summarization are very small [31], making learning and optimization challenging. We argue that pretraining using a large-scale video data from a related task, such as PHD-GIFs [25] in highlight detection, could tremendously help video summarization models. In video summarization, this idea remains unexplored. In order to validate our notion and compare with recent state-of-the-art in [44], we select the SumMe dataset [9] which has only 25 videos.

We evaluate our trained H-FCSN (i.e., the generic highlight detection model we proposed in Sect. 4.3) directly on SumMe. In Table 4, we compare the performance of our H-FCSN (trained on the PHD-GIFs [25] dataset) on SumMe with state-of-the-art supervised video summarization methods. Following prior work [44], we randomly select 20% of data in SumMe for testing. We repeat this experiment five times (as in [44]) and report the average performance. Surprisingly, even though we *do not* train on SumMe, our model achieves state-of-the-art summarization performance, outperforming contemporary supervised models. Therefore, we believe that the future research in video summarization should consider pretraining their model on a large-scale video data from a related task such as highlight detection. We envision that this way we can simultaneously make progress in both highlight detection and video summarization.

Table 4. Performance comparison in term of F-score (%) on SumMe. Note that unlike other methods, we do not train on SumMe rather directly test our trained (using PHD-GIFs) model for summarization. Results of other methods are taken from [44]

Method	F-score (%)
Interesting [9]	39.4
Submodularity [10]	39.7
DPP-LSTM [50]	38.6
GAN$_{sup}$ [24]	41.7
DR-DSN$_{sup}$ [54]	42.1
S^{2}N [44]	43.3
Ours (H-FCSN)	**44.4**

5   Conclusion

We have proposed a simple yet novel framework `Adaptive-H-FCSN` for adaptive highlight detection using the user history which has received less attention in the literature. Different from commonly applied ranking model, we introduced a convolutional model for highlight detection that is computationally efficient as it can process an entire video of any length at once and also does not require expensive shot detection computation. We proposed temporal-adaptive normalization (T-AIN) that has affine parameters which is adaptively computed using the user history information. The proposed T-AIN leads to high-performing and user-specific highlight detection. Our empirical results on a large-scale dataset indicate that the proposed framework outperforms alternative approaches. Lastly, we further demonstrate an application of our learned model in video summarization where the learning is currently limited to small datasets.

Acknowledgements. The work was supported by NSERC. We thank NVIDIA for donating some of the GPUs used in this work.

References

1. Agnihotri, L., Kender, J., Dimitrova, N., Zimmerman, J.: Framework for personalized multimedia summarization. In: ACM SIGMM International Workshop on Multimedia Information Retrieval (2005)
2. Babaguchi, N., Ohara, K., Ogura, T.: Learning personal preference from viewer's operations for browsing and its application to baseball video retrieval and summarization. IEEE Trans. Multimed. **9**, 1016–1025 (2007)
3. Brock, A., Donahue, J., Simonyan, K.: Large scale GAN training for high fidelity natural image synthesis. In: International Conference on Learning Representations (2018)
4. Cai, S., Zuo, W., Davis, L.S., Zhang, L.: Weakly-supervised video summarization using variational encoder-decoder and web prior. In: European Conference on Computer Vision (2018)
5. Chen, T., Lucic, M., Houlsby, N., Gelly, S.: On self modulation for generative adversarial networks. In: International Conference on Learning Representations (2018)
6. De Vries, H., Strub, F., Mary, J., Larochelle, H., Pietquin, O., Courville, A.C.: Modulating early visual processing by language. In: Advances in Neural Information Processing Systems (2017)
7. Gong, B., Chao, W.L., Grauman, K., Sha, F.: Diverse sequential subset selection for supervised video summarization. In: Advances in Neural Information Processing Systems (2014)
8. Gygli, M.: Ridiculously fast shot boundary detection with fully convolutional neural networks. In: International Conference on Content-Based Multimedia Indexing (2018)
9. Gygli, M., Grabner, H., Riemenschneider, H., Van Gool, L.: Creating summaries from user videos. In: Fleet, D., Pajdla, T., Schiele, B., Tuytelaars, T. (eds.) ECCV 2014. LNCS, vol. 8695, pp. 505–520. Springer, Cham (2014). https://doi.org/10.1007/978-3-319-10584-0_33
10. Gygli, M., Grabner, H., Van Gool, L.: Video summarization by learning submodular mixtures of objectives. In: IEEE Conference on Computer Vision and Pattern Recognition (2015)
11. Gygli, M., Song, Y., Cao, L.: Video2GIF: automatic generation of animated gifs from video. In: IEEE Conference on Computer Vision and Pattern Recognition (2016)
12. Huang, X., Belongie, S.: Arbitrary style transfer in real-time with adaptive instance normalization. In: IEEE International Conference on Computer Vision (2017)
13. Jaimes, A., Echigo, T., Teraguchi, M., Satoh, F.: Learning personalized video highlights from detailed MPEG-7 metadata. In: International Conference on Image Processing (2002)
14. Jiao, Y., Yang, X., Zhang, T., Huang, S., Xu, C.: Video highlight detection via deep ranking modeling. In: Paul, M., Hitoshi, C., Huang, Q. (eds.) PSIVT 2017. LNCS, vol. 10749, pp. 28–39. Springer, Cham (2018). https://doi.org/10.1007/978-3-319-75786-5_3

15. Karras, T., Laine, S., Aila, T.: A style-based generator architecture for generative adversarial networks. In: IEEE Conference on Computer Vision and Pattern Recognition (2019)

16. Khosla, A., Hamid, R., Lin, C.J., Sundaresan, N.: Large-scale video summarization using web-image priors. In: IEEE Conference on Computer Vision and Pattern Recognition (2013)

17. Kim, G., Xing, E.P.: Reconstructing storyline graphs for image recommendation from web community photos. In: IEEE Conference on Computer Vision and Pattern Recognition (2014)

18. Kingma, D.P., Ba, J.: Adam: a method for stochastic optimization. In: International Conference on Learning Representations (2015)

19. Lea, C., Flynn, M.D., Vidal, R., Reiter, A., Hager, G.D.: Temporal convolutional networks for action segmentation and detection. In: IEEE Conference on Computer Vision and Pattern Recognition (2017)

20. Lee, Y.J., Ghosh, J., Grauman, K.: Discovering important people and objects for egocentric video summarization. In: IEEE Conference on Computer Vision and Pattern Recognition (2012)

21. Liu, M.Y., et al.: Few-shot unsueprvised image-to-image translation. In: IEEE International Conference on Computer Vision (2019)

22. Liu, W., Mei, T., Zhang, Y., Che, C., Luo, J.: Multi-task deep visual-semantic embedding for video thumbnail selection. In: IEEE Conference on Computer Vision and Pattern Recognition (2015)

23. Lu, Z., Grauman, K.: Story-driven summarization for egocentric video. In: IEEE Conference on Computer Vision and Pattern Recognition (2013)

24. Mahasseni, B., Lam, M., Todorovic, S.: Unsupervised video summarization with adversarial LSTM networks. In: IEEE Conference on Computer Vision and Pattern Recognition (2017)

25. del Molino, A.G., Gygli, M.: PHD-GIFs: personalized highlight detection for automatic gif creation. In: ACM Multimedia (2018)

26. Ngo, C.W., Ma, Y.F., Zhang, H.J.: Automatic video summarization by graph modeling. In: IEEE International Conference on Computer Vision (2003)

27. Panda, R., Das, A., Wu, Z., Ernst, J., Roy-Chowdhury, A.K.: Weakly supervised summarization of web videos. In: IEEE International Conference on Computer Vision (2017)

28. Panda, R., Roy-Chowdhury, A.K.: Collaborative summarization of topic-related videos. In: IEEE Conference on Computer Vision and Pattern Recognition (2017)

29. Park, T., Liu, M.Y., Wang, T.C., Zhu, J.Y.: Semantic image synthesis with spatially-adaptive normalization. In: IEEE Conference on Computer Vision and Pattern Recognition (2019)

30. Potapov, D., Douze, M., Harchaoui, Z., Schmid, C.: Category-specific video summarization. In: Fleet, D., Pajdla, T., Schiele, B., Tuytelaars, T. (eds.) ECCV 2014. LNCS, vol. 8694, pp. 540–555. Springer, Cham (2014). https://doi.org/10.1007/978-3-319-10599-4_35

31. Rochan, M., Wang, Y.: Video summarization by learning from unpaired data. In: IEEE Conference on Computer Vision and Pattern Recognition (2019)

32. Rochan, M., Ye, L., Wang, Y.: Video summarization using fully convolutional sequence networks. In: Ferrari, V., Hebert, M., Sminchisescu, C., Weiss, Y. (eds.) ECCV 2018. LNCS, vol. 11216, pp. 358–374. Springer, Cham (2018). https://doi.org/10.1007/978-3-030-01258-8_22

33. Sharghi, A., Gong, B., Shah, M.: Query-focused extractive video summarization. In: Leibe, B., Matas, J., Sebe, N., Welling, M. (eds.) ECCV 2016. LNCS, vol. 9912, pp. 3–19. Springer, Cham (2016). https://doi.org/10.1007/978-3-319-46484-8_1
34. Soleymani, M.: The quest for visual interest. In: ACM Multimedia (2015)
35. Song, Y., Vallmitjana, J., Stent, A., Jaimes, A.: TVSum: summarizing web videos using titles. In: IEEE Conference on Computer Vision and Pattern Recognition (2015)
36. Sun, M., Farhadi, A., Seitz, S.: Ranking domain-specific highlights by analyzing edited videos. In: Fleet, D., Pajdla, T., Schiele, B., Tuytelaars, T. (eds.) ECCV 2014. LNCS, vol. 8689, pp. 787–802. Springer, Cham (2014). https://doi.org/10.1007/978-3-319-10590-1_51
37. Takahashi, Y., Nitta, N., Babaguchi, N.: User and device adaptation for sports video content. In: IEEE International Conference on Multimedia and Expo (2007)
38. Tran, D., Bourdev, L., Fergus, R., Torresani, L., Paluri, M.: Learning spatiotemporal features with 3D convolutional networks. In: IEEE International Conference on Computer Vision (2015)
39. Truong, B.T., Venkatesh, S.: Video abstraction: a systematic review and classification. ACM Trans. Multimed. Comput. Commun. Appl. 3, 3-es (2007)
40. Ulyanov, D., Vedaldi, A., Lempitsky, V.: Improved texture networks: maximizing quality and diversity in feed-forward stylization and texture synthesis. In: IEEE Conference on Computer Vision and Pattern Recognition (2017)
41. Vasudevan, A.B., Gygli, M., Volokitin, A., Van Gool, L.: Query-adaptive video summarization via quality-aware relevance estimation. In: ACM Multimedia (2017)
42. Wang, X., Girshick, R., Gupta, A., He, K.: Non-local neural networks. In: IEEE Conference on Computer Vision and Pattern Recognition (2018)
43. Wang, X., Yu, K., Dong, C., Change Loy, C.: Recovering realistic texture in image super-resolution by deep spatial feature transform. In: IEEE Conference on Computer Vision and Pattern Recognition (2018)
44. Wei, Z., et al.: Sequence-to-segment networks for segment detection. In: Advances in Neural Information Processing Systems (2018)
45. Xiong, B., Kalantidis, Y., Ghadiyaram, D., Grauman, K.: Less is more: learning highlight detection from video duration. In: IEEE Conference on Computer Vision and Pattern Recognition (2019)
46. Yao, T., Mei, T., Rui, Y.: Highlight detection with pairwise deep ranking for first-person video summarization. In: IEEE Conference on Computer Vision and Pattern Recognition (2016)
47. Yu, Y., Lee, S., Na, J., Kang, J., Kim, G.: A deep ranking model for spatio-temporal highlight detection from a 360o video. In: AAAI Conference on Artificial Intelligence (2018)
48. Zhang, H., Goodfellow, I., Metaxas, D., Odena, A.: Self-attention generative adversarial networks. In: International Conference on Machine Learning (2019)
49. Zhang, K., Chao, W.L., Sha, F., Grauman, K.: Summary transfer: examplar-based subset selection for video summarization. In: IEEE Conference on Computer Vision and Pattern Recognition (2016)
50. Zhang, K., Chao, W.-L., Sha, F., Grauman, K.: Video summarization with long short-term memory. In: Leibe, B., Matas, J., Sebe, N., Welling, M. (eds.) ECCV 2016. LNCS, vol. 9911, pp. 766–782. Springer, Cham (2016). https://doi.org/10.1007/978-3-319-46478-7_47

51. Zhang, K., Grauman, K., Sha, F.: Retrospective encoders for video summarization. In: Ferrari, V., Hebert, M., Sminchisescu, C., Weiss, Y. (eds.) ECCV 2018. LNCS, vol. 11212, pp. 391–408. Springer, Cham (2018). https://doi.org/10.1007/978-3-030-01237-3_24

52. Zhang, Y., Kampffmeyer, M., Liang, X., Tan, M., Xing, E.P.: Query-conditioned three-player adversarial network for video summarization. In: British Machine Vision Conference (2018)

53. Zhao, B., Li, X., Lu, X.: Hierarchical recurrent neural network for video summarization. In: ACM Multimedia (2017)

54. Zhou, K., Qiao, Y., Xiang, T.: Deep reinforcement learning for unsupervised video summarization with diversity-representativeness reward. In: AAAI Conference on Artificial Intelligence (2018)

Improving 3D Object Detection Through Progressive Population Based Augmentation

Shuyang Cheng[1(✉)], Zhaoqi Leng[1], Ekin Dogus Cubuk[2], Barret Zoph[2], Chunyan Bai[1], Jiquan Ngiam[2], Yang Song[1], Benjamin Caine[2], Vijay Vasudevan[2], Congcong Li[1], Quoc V. Le[2], Jonathon Shlens[2], and Dragomir Anguelov[1]

[1] Waymo LLC, Mountain View, USA
shuyangcheng@waymo.com
[2] Google LLC, Mountain View, USA

Abstract. Data augmentation has been widely adopted for object detection in 3D point clouds. However, all previous related efforts have focused on manually designing specific data augmentation methods for individual architectures. In this work, we present the first attempt to automate the design of data augmentation policies for 3D object detection. We introduce the Progressive Population Based Augmentation (PPBA) algorithm, which learns to optimize augmentation strategies by narrowing down the search space and adopting the best parameters discovered in previous iterations. On the KITTI 3D detection test set, PPBA improves the StarNet detector by substantial margins on the moderate difficulty category of cars, pedestrians, and cyclists, outperforming all current state-of-the-art single-stage detection models. Additional experiments on the Waymo Open Dataset indicate that PPBA continues to effectively improve the StarNet and PointPillars detectors on a 20x larger dataset compared to KITTI. The magnitude of the improvements may be comparable to advances in 3D perception architectures and the gains come without an incurred cost at inference time. In subsequent experiments, we find that PPBA may be up to 10x more data efficient than baseline 3D detection models without augmentation, highlighting that 3D detection models may achieve competitive accuracy with far fewer labeled examples.

Keywords: Progressive population based augmentation · Data augmentation · Point cloud · 3D object detection · Data efficiency

Z. Leng—Work done while at Google LLC.

Electronic supplementary material The online version of this chapter (https://doi.org/10.1007/978-3-030-58589-1_17) contains supplementary material, which is available to authorized users.

A. Vedaldi et al. (Eds.): ECCV 2020, LNCS 12366, pp. 279–294, 2020.
https://doi.org/10.1007/978-3-030-58589-1_17

1 Introduction

LiDAR is a prominent sensor for autonomous driving and robotics because it provides detailed 3D information critical for perceiving and tracking real-world objects [2,29]. The 3D localization of objects within LiDAR point clouds represents one of the most important tasks in visual perception, and much effort has focused on developing novel network architectures for point clouds [1,16,21,25,31–33,37]. Following the image classification literature, such modeling efforts have employed manually designed data augmentation schemes for boosting performance [1,16,22,25,31,33,36,37].

In recent years, much work in the 2D image literature has demonstrated that investing heavily into data augmentation may lead to gains comparable to those obtained by advances in model architectures [4,5,11,20,38]. Despite this, 3D detection models have yet to significantly leverage automated data augmentation methods (but see [18]). Naively porting ideas that are effective for images to point cloud data presents numerous challenges, as the the types of augmentations appropriate for point clouds differ tremendously. Transformations appropriate for point clouds are typically geometric-based and may contain a large number of parameters. Thus, the search space proposed in [4,38] may not be naively reused for an automated search in point cloud augmentation space. Finally, because the search space is far larger, employing a more efficient search method becomes a practical necessity. Several works have attempted to significantly accelerate the search for data augmentation strategies [5,11,20], however it is unclear if such methods transfer successfully to point clouds.

In this work, we demonstrate that automated data augmentation significantly improves the prediction accuracy of 3D object detection models. We introduce a new search space for point cloud augmentations in 3D object detection. In this search space, we find the performance distribution of augmentation policies is quite diverse. To effectively discover good augmentation policies, we present an evolutionary search algorithm termed *Progressive Population Based Augmentation* (PPBA). PPBA works by narrowing down the search space through successive iterations of evolutionary search, and by adopting the best parameters discovered in past iterations. We demonstrate that PPBA is effective at finding good data augmentation strategies across datasets and detection architectures. Additionally, we find that a model trained with PPBA may be up to 10x more data efficient, implying reduced human labeling demands for point clouds.

Our main contributions can be summarized as follows: (1) We propose an automated data augmentation technique for localization in 3D point clouds. (2) We demonstrate that the proposed search method effectively improves point cloud 3D detection models compared to random search with less computational cost. (3) We demonstrate up to a 10x increase in data efficiency when employing PPBA. (4) Beyond 3D detection, we also demonstrate that PPBA generalizes to 2D image classification.

2   Related Work

Data augmentation has been an essential technique for boosting the performance of 2D image classification and object detection models. Augmentation methods typically include manually designed image transformations, to which the labels remain invariant, or distortions of the information present in the images. For example, elastic distortions, scale transformations, translations, and rotations are beneficial on models trained on MNIST [3,24,26,30]. Crops, image mirroring and color shifting/whitening [14] are commonly adopted on natural image datasets like CIFAR-10 and ImageNet. Recently, cutout [6] and mixup [34] have emerged as data augmentation methods that lead to good improvements in natural image datasets. For object detection in 2D images, image mirroring and multi-scale training are popular distortions [10]. Dwibedi et al. add new objects on training images by cut-and-paste [7].

While the augmentation operations mentioned above are designed by domain experts, there are also automated approaches to designing data augmentation strategies for 2D images. Early attempts include Smart Augmentation, which uses a network to generate augmented data by merging two or more image samples [17]. Ratner et al. use GANs to output sequences of data augmentation operations [23]. AutoAugment uses reinforcement learning to optimize data augmentation strategies for classification [4] and object detection [38]. More recently, improved search methods are able to find data augmentation strategies more efficiently [5,11,20].

While all the mentioned work so far is on 2D image classification and object detection, automated data augmentation methods have not been explored for 3D object detection tasks to the best of our knowledge. Models trained on KITTI use a wide variety of manually designed distortions. Due to the small size of the KITTI training set, data augmentation has been shown to improve performance significantly (common augmentations include horizontal flips, global scale distortions, and rotations) [1,16,25,31,33]. Yan et al. add new objects in training point clouds by pasting points inside ground truth 3D bounding boxes [31]. Despite its effectiveness for KITTI models, data augmentation was not used on some of the larger point cloud datasets [22,36]. Very recently, an automated data augmentation approach was studied for point cloud classification [18].

Historically, 2D vision research has focused on architectural modifications to improve generalization. More recently, it was observed that improving data augmentation strategies can lead to comparable gains to a typical architectural advance [4,8,34,38]. In this work, we demonstrate that a similar type of improvement can also be obtained by an effective automated data augmentation strategy for 3D object detection over point clouds.

3   Methods

We formulate the problem of finding the right augmentation strategy as a special case of hyperparameter schedule learning. The proposed method consists of two

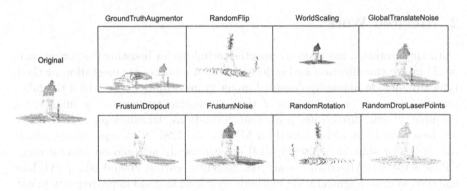

Fig. 1. Visualization of the augmentation operations in the proposed search space. An augmentation policy is defined by a list of distinct augmentation operations and the corresponding augmentation parameters. Details of these operations are in Table 7 and Table 8 in the Appendix.

components: a specialized data augmentation search space for point cloud inputs and a search algorithm for the optimization of data augmentation parameters. We describe these two components below.

3.1 Search Space for 3D Point Cloud Augmentation

In the proposed search space, an augmentation policy consists of N augmentation operations. Additionally, each operation is associated with a probability and some specialized parameters. For example, the ground-truth augmentation operation has parameters denoting the probability for sampling vehicles, pedestrians, cyclists, etc.; the global translation noise operation has parameters for the distortion magnitude of the translation operation on x, y and z coordinates. To reduce the size of the search space and increase the diversity of the training data, these different operations are always applied according to some learned probabilities in the same, pre-determined order to point clouds during training.

The basic augmentation operations in the proposed search space fall into two main categories: *global operations*, which are applied to all points in a frame (such as rotation along Z-axis, coordinate scaling, etc.), and *local operations*, which are applied to points locally (such as dropping out points within a frustum, pasting points within bounding boxes from other frames, etc.). Our list of augmentation operations (see Fig. 1) consists of GroundTruthAugmentor, RandomFlip, WorldScaling, GlobalTranslateNoise, FrustumDropout, FrustumNoise, RandomRotation and RandomDropLaserPoints. In total, there are 8 augmentation operations and 29 operation parameters in the proposed search space.

3.2 Learning Through Progressive Population Based Search

The proposed search process is maximizing a given metric Ω on a model θ by optimizing a schedule of augmentation operation parameters $\boldsymbol{\lambda} = (\lambda_t)_{t=1}^T$, where

t represents the number of iterative updates for the augmentation operation parameters during model training. For point cloud detection tasks, we use mean average precision (mAP) as the performance metric. The search process for the best augmentation schedule $\boldsymbol{\lambda}^*$ optimizes:

$$\boldsymbol{\lambda}^* = \arg \max_{\boldsymbol{\lambda} \in \Lambda^T} \Omega(\theta) \tag{1}$$

During training, the objective function L (which is used for optimization of the model θ given data and label pairs $(\mathcal{X}, \mathcal{Y})$) is usually different from the actual performance metric Ω, since the optimization procedure (i.e. stochastic gradient descent) requires a differentiable objective function. Therefore, at each iteration t the model θ is optimizing:

$$\boldsymbol{\theta}_t^* = \arg \min_{\theta \in \Theta} L(\mathcal{X}, \mathcal{Y}, \lambda^t) \tag{2}$$

During search, the training process of the model is split into $\mathcal{N}$ iterations. At every iteration, $\mathcal{M}$ models with different λ_t are trained in parallel and are afterwards evaluated with the metric Ω. Models trained in all previous iterations are placed in a population $\mathcal{P}$. In the initial iteration, all model parameters and augmentation parameters are randomly initialized. After the first iteration, model parameters are determined through an *exploit* phase - inheriting from a better performing parent model by exploiting the rest of the population $\mathcal{P}$. The *exploit* phase is followed by an *exploration* phase, in which a subset of the augmentation operations will be explored for optimization by mutating the corresponding augmentation parameters in the parent model, while the remaining augmentation parameters will be directly inherited from the parent model.

Similar to Population Based Training [12], the *exploit* phase will keep the good models and replace the inferior models at the end of every iteration. In contrast with Population Based Training, the proposed method focuses only on a subset of the search space at each iteration. During the *exploration* phase, a successor might focus on a different subset of the parameters than its predecessor. In that case, the remaining parameters (parameters that the predecessor does not focus on) are mutated based on the parameters of the corresponding operations with the best overall performance. In Fig. 2, we show an example of Progressive Population Based Augmentation. The complete PPBA algorithm is described in detail in Algorithm 1 in the Appendix.

3.3   Schedule Optimization with Historical Data

The search spaces for data augmentation are different between 2D images and 3D point clouds. For example, each operation in the AutoAugment [4] search space for 2D images has a single parameter. Furthermore, any value of this parameter within the predefined range leads to a reasonable image. For this reason, even sampling random augmentation policies for 2D images leads to some improvement in generalization [4,5]. On the other hand, the augmentation operations for 3D point clouds are much harder to optimize. Each operation

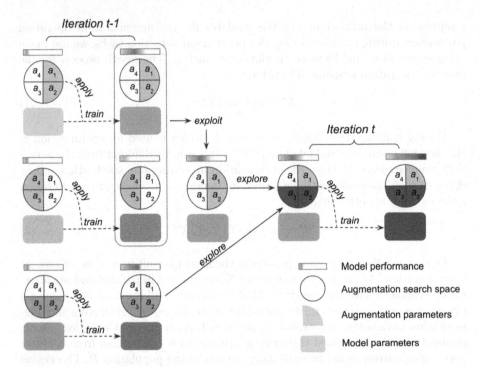

Fig. 2. An example of Progressive Population Based Augmentation. Four augmentation operations (a_1, a_2, a_3, a_4) are applied to the input data during training; their parameter set comprises the full search space. During progressive population based search, only two augmentation operations out of the four are explored for optimization at every iteration. For example, at the beginning of iteration $t-1$, augmentation parameters of (a_1, a_2) are selected for exploration for the blue model while augmentation parameters of (a_3, a_4) are selected for exploration for the purple model. At the end of training in iteration $t-1$, an inferior model (the purple model) is *exploited* by a model with higher performance (the blue model). Afterwards, a successor will inherit both model parameters and augmentation parameters from the winner model - the blue model. During the *exploration* phase, the selected augmentation operations for exploration by the successor model are randomly sampled and become (a_2, a_3). Since augmentation parameters of a_3 have not been explored by the predecessor (the blue model), corresponding augmentation parameters of the best model (the green model), in which a_3 has been selected for exploration, will be adopted for exploration by the successor model.

has several parameters, and a good range for these parameters is not known a priori. For example, there are five parameters – theta_width, phi_width, distance, keep_prob and drop_type – in the FrustumDropout operation. The analogous operation for 2D images is cutout [6], which has only one parameter. Therefore it is more challenging to discover optimal parameters for point cloud operations with limited resources.

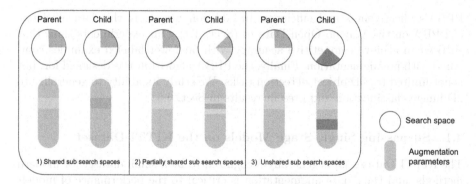

Fig. 3. Three types of scenarios for the subsets of the search space explored by the parent and the child models. 1) The subsets are the same. 2) The subsets are partially shared. 3) The subsets are unshared. In both 1) and 2), the overlapped augmentation parameters for exploration in the child model are mutated based on the corresponding parameters in the parent model (updating from light green to dark green). In both 2) and 3), the non-overlapped augmentation parameters for exploration in the child model are mutated based on the best augmentation parameters discovered in the past iterations (if available) or random sampling (updating from blue to yellow/red).

In order to learn the parameters for individual operations effectively, PPBA modifies only a small portion of the parameters in the search space at every iteration, and the historical information from the previous iterations are reused to optimize the augmentation schedule. By narrowing down the focus on certain subsets of the search space, it becomes easier to distinguish the inferior augmentation parameters. To mitigate the slowing down of search speed caused by the search space shrinkage at each training iteration, the best parameters of each operation discovered in the past iterations are adopted by the successors, when their focused subsets of the search space are different from their predecessors.

In Fig. 3, we show three types of scenarios for the subsets of the search space explored by a successor and its predecessor. The details of the exploration phase based on historical data are described in Algorithm 2 in the Appendix.

4 Experiments

In this section, we empirically investigate the performance of PPBA on predictive accuracy, computational efficiency and data efficiency. We focus on single-stage detection models due to their simplicity, speed advantages and widespread adoption [16,22,31].

We first benchmark PPBA on the KITTI object detection benchmark [9] and the Waymo Open Dataset [27] (Sects. 4.1 and 4.2). Our results show PPBA improves the baseline models and the improvement magnitude may be comparable to advances in 3D perception architectures. Next, we compare PPBA with random search and PBA [11] on the KITTI Dataset (Sect. 4.3) to demonstrate

PPBA's effectiveness and efficiency. In addition, we study the data efficiency of PPBA on the Waymo Open Dataset (Sect. 4.4). Our experiments show that PPBA can achieve competitive accuracy with far fewer labeled examples compared with no augmentation. Finally, the PPBA algorithm was designed for, but is not limited to, 3D object detection tasks. We study its ability to generalize to 2D image classification and present results in Sect. 4.5.

4.1 Surpassing Single-Stage Models on the KITTI Dataset

The KITTI Dataset [9] is generally recognized to be a small dataset for modern methods, and thus, data augmentation is critical to the performance of models trained on it [16,22,31]. We evaluate PPBA with StarNet [22] on the KITTI *test* split in Table 1. PPBA improves the detection performance of StarNet significantly, outperforming all current state-of-the-art single-stage point cloud detection models on the moderate difficulty category.

Table 1. Performance comparison of single-stage point cloud detection models on the KITTI *test* set using 3D evaluation. mAP is calculated with an IOU of 0.7, 0.5 and 0.5 for vehicles, cyclists and pedestrians, respectively

Method	Car			Pedestrian			Cyclist		
	Easy	Mod	Hard	Easy	Mod	Hard	Easy	Mod	Hard
ContFuse [19]	83.68	68.78	61.67	–	–	–	–	–	–
VoxelNet [37]	77.47	65.11	57.73	39.48	33.69	31.51	61.22	48.36	44.37
SECOND [31]	83.13	73.66	66.20	51.07	42.56	37.29	70.51	53.85	46.90
3D IoU Loss [35]	**84.43**	76.28	68.22	–	–	–	–	–	–
PointPillars [16]	79.05	74.99	68.30	52.08	43.53	41.49	75.78	59.07	52.92
StarNet [22]	81.63	73.99	67.07	48.58	41.25	39.66	73.14	58.29	52.58
StarNet [22] + PPBA	84.16	**77.65**	**71.21**	**52.65**	**44.08**	**41.54**	**79.42**	**61.99**	**55.34**

During the PPBA search, 16 trials are trained to optimize the mAP for car (30 iterations) and for pedestrian/cyclist (20 iterations), respectively. The same training and inference settings[1] as [22] are used, while all trials are trained on the *train* split (3,712 samples) and validated on the *validation* split (3,769 samples). We train the first iteration for 3,000 steps, and all subsequent iterations for 1,000 steps with batch size 64. The search is conducted in the search space described in Sect. 3.1.

Manually designed augmentation policies are typically kept constant during training. In contrast, stochasticity lies at the heart of the augmentation policies in PPBA, i.e. each operation is applied stochastically and its parameters evolve as the training progresses. We have found that simply using the final parameters discovered by PPBA gets worse results than PPBA.

---

[1] http://github.com/tensorflow/lingvo.

We use GroundTruthAugmentor to highlight the difference between the manually designed and learned augmentation policies. While training a StarNet vehicle detection model on KITTI with PPBA, the probability of applying the operation decreases from 100% to 16% and the probability of pasting vehicles reduces from 100% to 21%, while the probability of pasting pedestrians and cyclists increases from 0% to 28% and 8% respectively. This suggests that pasting the object of interest in every frame during training, as in manual designed policies, is not an optimal strategy and introducing a diverse set of objects from other classes is beneficial.

4.2   Automated Data Augmentation Benefits Large-Scale Data

The Waymo Open Dataset is a recently released, large-scale dataset for 3D object detection in point clouds [27]. The dataset contains roughly 20x more scenes than KITTI, and roughly 20x more human-annotated objects per scene. This dataset presents an opportunity to ask whether data augmentations – being critical to model performance on the KITTI dataset due to the small size of the dataset – continue to provide a benefit in a large-scale training setting more reflective of the self-driving conditions in the real world.

To address this question, we evaluate the proposed method on the Waymo Open Dataset. In particular, we evaluate PPBA with StarNet [22] and PointPillars [16] on the *test* split in Table 2 and Table 3 on both LEVEL_1 and LEVEL_2 difficulties at different ranges. Our results indicates that PPBA notably improves the predictive accuracy of 3D detection across architectures, difficulty levels and object classes. These results indicate that data augmentation remains an important method for boosting model performance even in large-scale dataset settings. Furthermore, the gains due to PPBA may be as large as changing the underlying architecture, without any increase in inference cost.

Table 2. Performance comparison on the Waymo Open Dataset *test* set for vehicle detection. Note that the results of PointPillars [16] on the Waymo Open Dataset are reproduced by [27]

Method	Difficulty	3D mAP (IoU = 0.7)				3D mAPH (IoU = 0.7)			
	Level	Overall	0–30	30–50	50m-Inf	Overall	0–30	30–50	50m-Inf
StarNet [22]	1	61.5	82.2	56.6	32.2	61.0	81.7	56.0	31.8
StarNet [22] + PPBA	1	**64.6**	**85.8**	**59.5**	**35.1**	**64.1**	**85.3**	**58.9**	**34.6**
StarNet [22]	2	54.9	81.3	49.5	23.0	54.5	80.8	49.0	22.7
StarNet [22] + PPBA	2	**56.2**	**82.8**	**54.0**	**26.8**	**55.8**	**82.3**	**53.5**	**26.4**
PointPillars [16]	1	63.3	82.3	59.2	35.7	62.8	81.9	58.5	35.0
PointPillars [16] + PPBA	1	**67.5**	**86.7**	**63.5**	**39.4**	**67.0**	**86.2**	**62.9**	**38.7**
PointPillars [16]	2	55.6	81.2	52.9	27.2	55.1	80.8	52.3	26.7
PointPillars [16] + PPBA	2	**59.6**	**85.6**	**57.6**	**30.0**	**59.1**	**85.1**	**57.0**	**29.5**

Table 3. Performance comparison on the Waymo Open Dataset *test* set for pedestrian detection. Note that the results of PointPillars [16] on the Waymo Open Dataset are reproduced by [27]

Method	Difficulty Level	3D mAP (IoU = 0.5)				3D mAPH (IoU = 0.5)			
		Overall	0–30	30–50	50m-Inf	Overall	0–30	30–50	50m-Inf
StarNet [22]	1	67.8	76.0	66.5	55.3	59.9	67.8	59.2	47.0
StarNet [22] + PPBA	1	**69.7**	**77.5**	**68.7**	**57.0**	**61.7**	**69.3**	**61.2**	**48.4**
StarNet [22]	2	61.1	73.1	61.2	44.5	54.0	65.2	54.5	37.8
StarNet [22] + PPBA	2	**63.0**	**74.8**	**63.2**	**46.5**	**55.8**	**66.8**	**56.2**	**39.4**
PointPillars [16]	1	62.1	71.3	60.1	47.0	50.2	59.0	48.3	35.8
PointPillars [16] + PPBA	1	**66.4**	**74.7**	**64.8**	**52.7**	**54.4**	**62.5**	**52.5**	**41.2**
PointPillars [16]	2	55.9	68.6	55.2	37.9	45.1	56.7	44.3	28.8
PointPillars [16] + PPBA	2	**60.1**	**72.2**	**59.7**	**42.8**	**49.2**	**60.4**	**48.2**	**33.4**

When performing the search with PPBA, 16 trials are trained to optimize the mAP for car and pedestrian, respectively. The list of augmentation operations described in Sect. 3.1, except for GroundTruthAugmentor and RandomFlip, are used during search. In our experiments, we have found RandomFlip has a negative impact on heading prediction for both car and pedestrian.

For both StarNet and PointPillars on the Waymo Open Dataset, the same training and inference settings[2] as [22] is used. All trials are trained on the full *train* set and validated on the 10% *validation* split (4,109 samples). For StarNet, we train the first iteration for 8,000 steps and the remaining iterations for 4,000 steps with batch size 128. The training steps for PointPillars are reduced by half in each iteration with batch size 64. We perform the search for 25 iterations on StarNet and for 20 iterations on PointPillars.

Even though StarNet and PointPillars are two distinct types of detection models, we have observed similar patterns in the evolution of their augmentation schedules. For StarNet and PointPillars, the probability of FrustumDropout is reduced from 100% to 23% and 56%, and the maximum rotation angle in RandomRotation is reduced from 0.785 to 0.54 and 0.42. These examples indicate that applying weaker data augmentation towards the end of training is beneficial.

4.3   Better Results with Less Computation

Above, we have verified the effectiveness of PPBA on improving 3D object detection on the KITTI Dataset and the Waymo Open Dataset. In this section, we analyze the computational cost of PPBA, and compare PPBA with random search and PBA [11] on the KITTI *test* split.

All searches are performed with StarNet [22] and the search space described in Sect. 3.1. For Random Search[3], 1,000 distinct augmentation policies are

---

[2] http://github.com/tensorflow/lingvo.

[3] Our initial experiment on random search shows the performance distribution of augmentation policies is spread on the KITTI *validation* split. In order to save computation resources, the random search here is performed on a fine-grained search space.

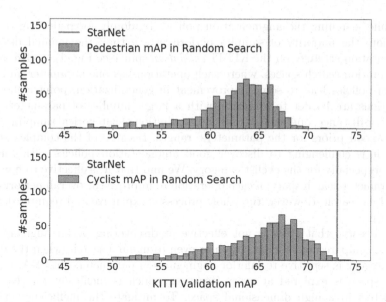

Fig. 4. 3D mAP of a population of 1,000 random augmentation policies for pedestrian and cyclist on the moderate difficulty on the KITTI *validation* split

randomly sampled and trained. PBA is run with 16 total trials while training the first iteration for 3,000 steps and the remaining iterations for 1,000 steps with batch size 64.

The baseline StarNet is trained for 8 hours with a TPU v3-32 Pod [13,15] on vehicle detection and pedestrian/cyclist detection models. Random search requires about $1,000 \times 8 = 8,000$ TPU hours for training. In comparison, both PBA and PPBA train with a much smaller cost of $8 \times 16 = 128$ TPU hours, with an additional real-time computation overhead of waiting for the evaluation result for $8 \times 16 = 128$ TPU hours. We observe that PPBA results in a more than 30x speedup compared to random search, while identifying better-performing augmentation strategies. Furthermore, PPBA outperforms PBA by a substantial margin with the same computational budget.

Table 4. Comparison of 3D mAP on StarNet on the KITTI *test* set across data augmentation methods

Method	TPU Hours	Car			Pedestrian			Cyclist		
		Easy	Mod	Hard	Easy	Mod	Hard	Easy	Mod	Hard
Manual design [22]	8	81.63	73.99	67.07	48.58	41.25	39.66	73.14	58.29	52.58
Random Search	8,000	81.89	74.94	67.39	**52.78**	**44.71**	41.12	73.71	59.92	54.09
PBA	256	83.16	75.02	69.72	41.28	34.48	32.24	76.8	59.43	52.77
PPBA	256	**84.16**	**77.65**	**71.21**	52.65	44.08	**41.54**	**79.42**	**61.99**	**55.34**

While searching the augmentation policies randomly for pedestrian/cyclist detection, the majority of samples perform worse than the manual designed augmentation strategy on the KITTI *validation* split (see Fig. 4). Unlike image augmentation search spaces, where each operation has one parameter and even random policies lead to some improvement in generalization, point cloud augmentations are harder to optimize, with a larger number of parameters (e.g. geometric distance, operation strength, distribution of categorical sampling, etc.) and no good priors for the parameters' ranges. Because of the complex search space, it is challenging to discover good augmentation policies with random search, especially for the cyclist category. We find that it is effective to fine tune the parameter search space of each operation to improve the overall performance of random search. However, the whole process is expensive and requires domain expertise.

We observe that PBA is not effective at discovering better augmentation policies, compared to random search or even to manual search, when the detection category is sensitive to inferior augmentation parameters. In PBA, the full search space is explored at every iteration, which is inefficient for searching parameters in a high dimensional space. To mitigate the inefficiency, PPBA progressively explores a subset of search space at every iteration, and the best parameters discovered in past iterations are adopted in the exploration phase. As in Table 4, PPBA shows much larger improvements on the car and cyclist categories, demonstrating the effectiveness of the proposed strategy.

4.4 Automated Data Augmentation Improves Data Efficiency

In this section, we conduct experiments to determine how PPBA performs when the dataset size grows. The experiments are performed with PointPillars on subsets of the Waymo Open Dataset with the following number of training examples: 10%, 30%, 50%, by randomly sampling run segments and single frames of sensor data, respectively. During training, the decay interval of the learning rate is linearly decreased accordingly to the percentile of data sampled (e.g., reduce the decay interval of learning rate by 50% when sampling 50% of the training examples), while the number of training epochs is set to be inversely proportional to the percentile of data sampled. As it is commonly known that smaller datasets need more regularization, we increase weight decay from 10^{-4} to 10^{-3}, when training on 10% of the examples.

Compared to downsampling from single frames of sensor data, performance degradation of PointPillars models is more severe when downsampling from run segments. This phenomenon is due to the relative lack of diversity in the run segments, which tend to contain the same set of distinct vehicles and pedestrians. As in Table 5, Fig. 5 and Fig. 6, we compare the overall 3D detection mAP on the Waymo Open Dataset *validation* set for all ground truth examples with ≥ 5 points and rated as LEVEL_1 difficulty for 3 sets of PointPillars models: with no augmentation, random augmentation policy and PPBA. While random augmentation policy can improve the PointPillars baselines and demonstrate the effectiveness of the proposed search space, PPBA pushes the limit even further.

Table 5. Compare 3D mAP of PointPillars with no augmentation, random augmentation and PPBA on the Waymo Open Dataset *validation* set as the dataset size grows

Method	Sample Unit	10%		30%		50%		100%	
		Car	Pedestrian	Car	Pedestrian	Car	Pedestrian	Car	Pedestrian
Baseline	run segment	42.5	46.1	49.5	56.4	52.5	59.1	57.2	62.3
Random	run segment	49.5	50.6	54.1	58.8	56.1	60.5	60.9	63.5
PPBA	run segment	**54.2**	**55.8**	**57.6**	**63.0**	**58.7**	**65.1**	**62.4**	**66.0**
Baseline	single frame	52.4	56.9	55.3	60.7	56.7	61.2	57.2	62.3
Random	single frame	58.3	59.8	59.4	61.9	59.7	62.1	60.9	63.5
PPBA	single frame	**59.8**	**64.2**	**60.7**	**65.5**	**61.2**	**66.2**	**62.4**	**66.0**

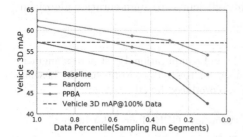 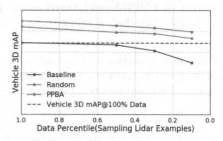

Fig. 5. Vehicle detection 3D mAP for PointPillars on Waymo Open Dataset *validation* set with no augmentation, random augmentation and PPBA as the dataset size changes

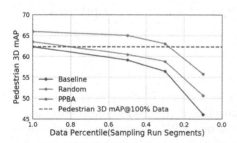 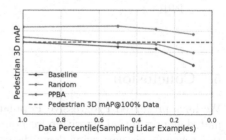

Fig. 6. Pedestrian detection 3D mAP for PointPillars on Waymo Open Dataset *validation* set with no augmentation, random augmentation and PPBA as the dataset size changes

PPBA is 10x more data efficient when sampling from single frames of sensor data, and 3.3x more data efficient when sampling from run segments. As we expected, the improvement from PPBA becomes larger when the dataset size is reduced.

4.5 Progressive Population Based Augmentation Generalizes on Image Classification

So far our experiments have demonstrated that PPBA consistently improves over alternatives for 3D object detection across datasets and architectures. However,

PPBA is a general algorithm, and in this section we validate its true versatility by applying it to a common 2D image classification problem.

To search for augmentation policies, we use the same reduced subset of the ImageNet training set with 120 classes and 6,000 samples as [4,20]. During the PPBA search, 16 trials are trained to optimize the Top-1 accuracy on the reduced validation set for 8 iterations while 4 operations are selected for exploration at every iteration. When replaying the learned augmentation schedule on the full training set, the ResNet-50 model is trained for 180 epochs with a batch size of 4096, a weight decay of 10^{-4} and a cosine decay learning rate schedule with learning rate of 1.6. The results on the ImageNet validation set, shown in Table 6, confirm that PPBA can be used as a highly efficient auto augmentation algorithm for tasks other than 3D object detection.

Table 6. Comparison of Top-1 accuracy (%) and computational cost across augmentation methods on the ImageNet validation set for ResNet-50. Note that the baseline results with Inception-style Pre-processing is reproduced by [4]

Method	Accuracy	GPU Hours	Hardware
Inception-style Pre-processing [28]	76.3	–	–
AutoAugment [4]	77.6	15000	GPU P100
Fast AutoAugment [20]	77.6	450	GPU V100
PPBA	77.5	16	GPU V100

5   Conclusion

We have presented Progressive Population Based Augmentation, a novel automated augmentation algorithm for point clouds. PPBA optimizes the augmentation schedule via narrowing down the search space and adopting the best parameters from past iterations. Compared with random search and PBA, PPBA can more effectively and more efficiently discover good augmentation policies in a rich search space for 3D object detection. Experimental results on the KITTI dataset and the Waymo Open Dataset demonstrate that the proposed method can significantly improve 3D object detection in terms of performance and data efficiency. While we have also validated the effectiveness of PPBA on a common task such as image classification, exploring the potential applications of the algorithm to more tasks and models remains an exciting direction of future work.

References

1. Chen, X., Ma, H., Wan, J., Li, B., Xia, T.: Multi-view 3D object detection network for autonomous driving. In: Proceedings of the IEEE Conference on Computer Vision and Pattern Recognition, pp. 1907–1915 (2017)

2. Cho, H., Seo, Y.W., Kumar, B.V., Rajkumar, R.R.: A multi-sensor fusion system for moving object detection and tracking in urban driving environments. In: 2014 IEEE International Conference on Robotics and Automation (ICRA), pp. 1836–1843. IEEE (2014)

3. Ciregan, D., Meier, U., Schmidhuber, J.: Multi-column deep neural networks for image classification. In: Proceedings of IEEE Conference on Computer Vision and Pattern Recognition, pp. 3642–3649. IEEE (2012)

4. Cubuk, E.D., Zoph, B., Mane, D., Vasudevan, V., Le, Q.V.: Autoaugment: learning augmentation policies from data. In: Proceedings of the IEEE Conference on Computer Vision and Pattern Recognition (2019)

5. Cubuk, E.D., Zoph, B., Shlens, J., Le, Q.V.: Randaugment: practical data augmentation with no separate search. arXiv preprint arXiv:1909.13719 (2019)

6. DeVries, T., Taylor, G.W.: Improved regularization of convolutional neural networks with cutout. arXiv preprint arXiv:1708.04552 (2017)

7. Dwibedi, D., Misra, I., Hebert, M.: Cut, paste and learn: surprisingly easy synthesis for instance detection. In: Proceedings of the IEEE International Conference on Computer Vision, pp. 1301–1310 (2017)

8. Fang, H.S., Sun, J., Wang, R., Gou, M., Li, Y.L., Lu, C.: Instaboost: boosting instance segmentation via probability map guided copy-pasting. In: The IEEE International Conference on Computer Vision (2019)

9. Geiger, A., Lenz, P., Stiller, C., Urtasun, R.: Are we ready for autonomous driving? The Kitti vision benchmark suite. In: Conference on Computer Vision and Pattern Recognition (CVPR) (2012)

10. Girshick, R., Radosavovic, I., Gkioxari, G., Dollár, P., He, K.: Detectron (2018)

11. Ho, D., Liang, E., Stoica, I., Abbeel, P., Chen, X.: Population based augmentation: Efficient learning of augmentation policy schedules. In: International Conference on Machine Learning, pp. 2731–2741 (2019)

12. Jaderberg, M., et al.: Population based training of neural networks. arXiv preprint arXiv:1711.09846 (2017)

13. Jouppi, N.P., et al.: In-datacenter performance analysis of a tensor processing unit. In: Proceedings of the 44th Annual International Symposium on Computer Architecture, pp. 1–12 (2017)

14. Krizhevsky, A., Sutskever, I., Hinton, G.E.: Imagenet classification with deep convolutional neural networks. In: Advances in Neural Information Processing Systems (2012)

15. Kumar, S., et al.: Scale MLPerf-0.6 models on Google TPU-v3 pods. arXiv preprint arXiv:1909.09756 (2019)

16. Lang, A.H., Vora, S., Caesar, H., Zhou, L., Yang, J., Beijbom, O.: Pointpillars: fast encoders for object detection from point clouds. In: Proceedings of the IEEE Conference on Computer Vision and Pattern Recognition, pp. 12697–12705 (2019)

17. Lemley, J., Bazrafkan, S., Corcoran, P.: Smart augmentation learning an optimal data augmentation strategy. IEEE Access 5, 5858–5869 (2017)

18. Li, R., Li, X., Heng, P.A., Fu, C.W.: Pointaugment: an auto-augmentation framework for point cloud classification. In: Proceedings of the IEEE/CVF Conference on Computer Vision and Pattern Recognition, pp. 6378–6387 (2020)

19. Liang, M., Yang, B., Wang, S., Urtasun, R.: Deep continuous fusion for multi-sensor 3D object detection. In: Proceedings of the European Conference on Computer Vision, pp. 641–656 (2018)

20. Lim, S., Kim, I., Kim, T., Kim, C., Kim, S.: Fast autoaugment. In: Advances in Neural Information Processing Systems (2019)

21. Luo, W., Yang, B., Urtasun, R.: Fast and furious: real time end-to-end 3D detection, tracking and motion forecasting with a single convolutional net. In: Proceedings of the IEEE Conference on Computer Vision and Pattern Recognition, pp. 3569–3577 (2018)
22. Ngiam, J., et al.: Starnet: targeted computation for object detection in point clouds. arXiv preprint arXiv:1908.11069 (2019)
23. Ratner, A.J., Ehrenberg, H., Hussain, Z., Dunnmon, J., Ré, C.: Learning to compose domain-specific transformations for data augmentation. In: Advances in Neural Information Processing Systems, pp. 3239–3249 (2017)
24. Sato, I., Nishimura, H., Yokoi, K.: Apac: augmented pattern classification with neural networks. arXiv preprint arXiv:1505.03229 (2015)
25. Shi, S., et al.: PV-RCNN: point-voxel feature set abstraction for 3D object detection. In: Proceedings of the IEEE/CVF Conference on Computer Vision and Pattern Recognition, pp. 10529–10538 (2020)
26. Simard, P.Y., Steinkraus, D., Platt, J.C., et al.: Best practices for convolutional neural networks applied to visual document analysis. In: Proceedings of International Conference on Document Analysis and Recognition (2003)
27. Sun, P., et al.: Scalability in perception for autonomous driving: Waymo open dataset. In: Proceedings of the IEEE/CVF Conference on Computer Vision and Pattern Recognition, pp. 2446–2454 (2020)
28. Szegedy, C., et al.: Going deeper with convolutions. In: Proceedings of the IEEE Conference on Computer Vision and Pattern Recognition (CVPR), pp. 1–9 (2015)
29. Thrun, S., et al.: Stanley: the robot that won the Darpa grand challenge. J. Field Robot. 23(9), 661–692 (2006)
30. Wan, L., Zeiler, M., Zhang, S., Le Cun, Y., Fergus, R.: Regularization of neural networks using dropconnect. In: International Conference on Machine Learning, pp. 1058–1066 (2013)
31. Yan, Y., Mao, Y., Li, B.: Second: sparsely embedded convolutional detection. Sensors 18(10), 3337 (2018)
32. Yang, B., Liang, M., Urtasun, R.: Hdnet: exploiting HD maps for 3D object detection. In: Proceedings of the 2nd Conference on Robot Learning, pp. 146–155 (2018)
33. Yang, B., Luo, W., Urtasun, R.: Pixor: real-time 3D object detection from point clouds. In: Proceedings of the IEEE Conference on Computer Vision and Pattern Recognition, pp. 7652–7660 (2018)
34. Zhang, H., Cisse, M., Dauphin, Y.N., Lopez-Paz, D.: mixup: beyond empirical risk minimization. In: International Conference on Learning Representations (2018)
35. Zhou, D., et al.: IoU loss for 2D/3D object detection. In: International Conference on 3D Vision (3DV). IEEE (2019)
36. Zhou, Y., et al.: End-to-end multi-view fusion for 3D object detection in lidar point clouds. In: Proceedings of the Conference on Robot Learning (2019)
37. Zhou, Y., Tuzel, O.: Voxelnet: end-to-end learning for point cloud based 3D object detection. In: Proceedings of the IEEE Conference on Computer Vision and Pattern Recognition, pp. 4490–4499 (2018)
38. Zoph, B., Cubuk, E.D., Ghiasi, G., Lin, T.Y., Shlens, J., Le, Q.V.: Learning data augmentation strategies for object detection. arXiv preprint arXiv:1906.11172 (2019)

DR-KFS: A Differentiable Visual Similarity Metric for 3D Shape Reconstruction

Jiongchao Jin[1,2], Akshay Gadi Patil[2], Zhang Xiong[1], and Hao Zhang[2(✉)]

[1] Beihang University, Beijing, China
{jinjiongchao,xiongz}@buaa.edu.cn
[2] Simon Fraser University, Burnaby, Canada
{akshay_gadi_patil,haoz}@sfu.ca

Abstract. We introduce a *differential visual similarity* metric to train deep neural networks for 3D reconstruction, aimed at improving reconstruction quality. The metric compares two 3D shapes by measuring distances between multi-view images *differentiably* rendered from the shapes. Importantly, the image-space distance is also differentiable and measures visual *similarity*, rather than pixel-wise distortion. Specifically, the similarity is defined by mean-squared errors over HardNet features computed from probabilistic keypoint maps of the compared images. Our differential visual shape similarity metric can be easily plugged into various 3D reconstruction networks, replacing their distortion-based losses, such as Chamfer or Earth Mover distances, so as to optimize the network weights to produce reconstructions with better structural fidelity and visual quality. We demonstrate this both objectively, using well-known shape metrics for retrieval and classification tasks that are independent from our new metric, and subjectively through a perceptual study.

Keywords: 3D Reconstruction · Visual similarity metric · Differentiablity

1 Introduction

Reconstructing 3D structures from 2D images is one of the most fundamental problems in computer vision. The problem is clearly ill-posed, hence most reconstruction algorithms are expected to incur errors against the ground truth. Commonly used error metrics are defined by *shape distortions*, such as Chamfer Distance (CD), Earth Mover Distance (EMD), Mean Square Error (MSE), and Intersection over Union (IoU). Most deep neural networks developed for 3D reconstruction [9–11,16,26,29,33] are trained by loss functions defined by one of these shape distortion metrics. To serve as a network loss to enable back propagation, the adopted metric needs to be *differentiable*. All of CD, EMD, and MSE

Electronic supplementary material The online version of this chapter (https://doi.org/10.1007/978-3-030-58589-1_18) contains supplementary material, which is available to authorized users.

A. Vedaldi et al. (Eds.): ECCV 2020, LNCS 12366, pp. 295–311, 2020.
https://doi.org/10.1007/978-3-030-58589-1_18

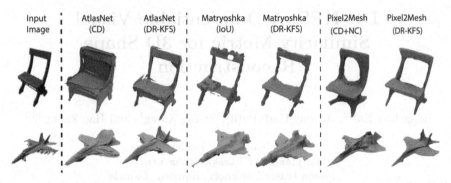

Fig. 1. Single-view 3D reconstruction by AtlasNet [10], Matryoshka Network [23], and Pixel2Mesh [29], trained with different losses. Left of each pair: trained by shape distortion in CD, IoU, and a combination of CD and normal consistency (NC). Right of each pair: using DR-KFS, our *differentiable visual similarity* metric, as train loss leads to results of higher visual fidelity in terms of shape structures and surface quality.

are differentiable. While the original IoU is not, there have been recent attempts to make it differentiable [16,35] (Fig. 1).

What is common about these well-adopted distortion metrics is that they are all designed to measure geometric distances between *aligned* 3D shapes, in the *object space*—they do not account for how the shapes are viewed by human observers, i.e., they are not "visual". Many recent works [7,9,19,22,23,27] have shown relative insensitivity of these object-space distortion metrics to *structural errors* such as missing parts and topological noise, and visual artifacts such as self intersections and poor surface quality. For example, thickening and elongating/shortening all four legs of a chair may result in a larger CD than removing one of the legs entirely, yet the latter alteration, a structural change, is more visually apparent; see Fig. 2. As a result, while some reconstructed 3D shapes do exhibit better visual quality, ratings based on distortion measures such as CD, EMD, MSE, or IoU may not reflect that superiority. This is not entirely surprising—a recent work by Blau and Michaeli [3] even suggests a trade-off between perceptual and distortion measures, albeit for image restoration.

Our key observation is that in contrast to shape distortion, measures of *visual similarity* are less sensitive to misalignment and slight shape distortion, but more sensitive to structural errors and visual quality. In computer graphics, the best known visual similarity metric is the light field descriptor (LFD) [6]. LFDs are computed for silhouette images of 3D shapes rendered from multiple viewpoints sampled around the shapes. Both a contour-based (Fourier descriptor) and a region-based (Zernike moment) image-space descriptor are employed, where rotational alignment is resolved via a discrete exhaustive search. However, due to the discrete rasterization during rendering and the use of truncated Fourier descriptors, LFD is non-differentiable. Some works in computer vision have also utilized multi-view, *projective* losses to train reconstruction networks, e.g., Perspective Transformer Net (PTN) [33], Neural Mesh Render [13], and

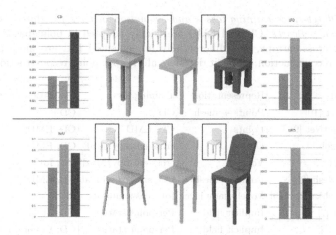

Fig. 2. A *trade-off* between shape distortion measures, e.g., CD and IoU, and a visual similarity metric such as LFD. Top: changing the shapes of the legs of a chair, e.g., by thickening and elongating (blue) or shortening (green), leads to larger shape distortions (reflected by larger CDs) compared to removing one leg (orange). The leg removal appears to make more of a visual impact due to a *structure* alteration and it is captured by a larger LFD (orange bars). Note that the bars show CDs/LFDs between the changed chair and the original, shown in pink in the boxes. Bottom: bending the legs (blue) or the back (green) leads to larger shape distortion, as reflected by *smaller* IoU, compared to leg removal. Again, the latter is captured by a larger LFD. (Color figure online)

SoftRas [16]. However, the projective- or image-space distances employed by these methods are still based on pixel-wise distortion, rather than similarity.

In this paper, we advocate the use of *differentiable visual shape similarity* metrics to train deep neural networks (DNNs) for 3D reconstruction. We introduce such a metric which compares two 3D shapes by measuring visual, image-space similarity between multi-view images rendered from the shapes, similar to LFD. However, one key difference is that the rendering process is differentiable, by employing a simplified *soft rasterization* [16]. In addition, we develop a differentiable similarity distance in the image space based on MSE defined over *probabilistic keypoint maps* of the compared images, rather than on RGB values. Furthermore, the MSE is defined over HardNet [20] features, rather than on the original keypoint maps. This choice is motivated in part by the finding in [3] that the perception-distortion trade-off appears less severe for distance between VGG features. Putting all these together, we arrive at a differentiable visual similarity metric, which we call DR-KFS to capture the use of Differentiable Rendering, Keypoint maps, and Feature-space image Similarity distances. DR-KFS better matches visual evaluation by humans, compared to object-space metrics.

To the best of our knowledge, existing 3D reconstruction networks all employ distortion-based reconstruction losses, either in the object or image spaces see Tables 1 and 2 for a summary. Our differential visual similarity metric can be

Table 1. *Object-space training losses and evaluation metrics for state-of-the-art 3D reconstruction networks.* CD and EMD are widely adopted in point-based networks, while MSE, cross-entropy, and differentially implemented IoU are more popular in voxel-based reconstruction. Implicit decoders all use per-point reconstruction loss.

Method	Representation	Training loss	Evaluation metrics
AtlasNet [10]	Points + mesh	CD	CD
PointOutNet [9]	Points	CD, EMD	CD, EMD
OGN [26]	Octree	cross-entropy/MSE	CD, EMD
HSP [11]	Voxel	cross-entropy/MSE	CD, IoU
Pixel2Mesh [29]	Points + mesh	CD, surface normal	CD, EMD
Matryoshka Net [23]	Voxel shape layer	IoU, cos-similarity	IoU
IM-Net [7]	Implicit field	Per-point status	CD, MSE, IoU, LFD
DISN [32]	Implicit field	Per-point status	CD, F-score, IoU

Table 2. *Multi-view training losses and evaluation metrics for state-of-the-art 3D reconstruction networks.* All the image-space errors are based on pixel-wise distortion such as MSE, IoU, or L1 Norm. All the evaluations are based on IoU.

Method	2D observation	Training loss	Evaluation
PTN [33]	Silhouettes	Pixel-wised MSE	IoU
NMR [13]	Silhouettes	Pixel-wised MSE	IoU
Soltani et al. [1]	Depth Maps and Silhouettes	L1 norm	IoU
SoftRas [16]	RGBA images	IoU	IoU

easily plugged into these networks, replacing the distortion losses so as to optimize the network weights to produce reconstruction results with better structural fidelity and visual quality. We demonstrate this both objectively, using well-known visual shape metrics for retrieval and classification tasks that are *independent* from DR-KFS, and subjectively through a user study.

Specifically, the 3D reconstruction networks tested for adaptation to DR-KFS training include OGN, AtlasNet [10], and Matryoshka Networks [23], which were picked as representative networks by Tatarchenko et al. [27] in their recent systematic study of single-view 3D reconstruction. We also consider Pixel2Mesh [29], NMR [13], SoftRas [16] and 3D-R2N2 [8]. The visual shape metrics we employ for evaluation include Shape Google [4], Multiview CNN or MVCNN [25], normal consistency or NC [19], F-score [23,27], as well as LFD. It is worth noting that the last three metrics have all been adopted by recent works on 3D reconstruction [7,19,23,27] to complement the distortion metrics.

2   Related Work

DNNs for 3D Reconstruction. One of the earlier DNNs for 3D reconstruction is 3D-R2N2 [8], which proposes the use of Recurrent Neural Networks (RNNs) to reconstruct a voxelized shape from a single-view image. Another notable work, PointOutNet [9], produces multiple reconstruction candidates for a single-view input image and demonstrates an improvement in the reconstruction quality over 3D-R2N2. OGN [26] generates volumetric 3D outputs in a compute-and-memory efficient manner by using an octree representation. AtlasNet [10] uses a collection of parametric surface elements to represent a 3D shape and to naturally infer a surface representation of the shape. Pixel2Mesh [29] produces 3D triangle meshes from a single image using Graph Convolution Networks (GCN) [14]. Matryoshka Network [23] introduces a novel and efficient 2D encoding scheme for 3D geometry, posing 3D reconstruction as a 2D prediction problem, while also speeding up the process.

Recent works on generative shape modeling using implicit functions have achieved state-of-the-art visual quality. IM-NET [7] and OccNet [19] represent 3D surfaces implicitly in the form of continuous binary decision functions. DeepSDF [12] learns a continuous Signed Distance Function (SDF) representation for a 3D shape. DISN [32] combines implicit SDF descriptors with the local input image features to generate 3D shape surfaces. Despite the good visual quality of the implicit-based generated surfaces, they often obtain lower scores on *shape distortion* metrics, such as CD.

Object-Space Metrics for 3D Reconstruction. Table 3 lists state-of-the-art 3D reconstruction DNNs trained using object-space losses, along with the shape metrics they employed. In principle, a network that is trained on a particular loss, e.g., CD, should be expected to perform better on that metric during testing, than methods that were not trained with the metric. Most methods that we are aware of train their networks using object-space distortion metrics. CD and EMD, and to a lesser degree, IoU, are the dominant measures applied for evaluating reconstruction quality. It is interesting to observe that some of the most recent methods, including IM-Net [7], OccNet [19], and Matryoshka Network [23], have opted to evaluate their methods using alternative measures such as LFD, Normal Consistency (NC), and F-score. Not surprisingly, each of these works pointed out the shortcomings of object-space distortion metrics in capturing visual similarity, which motivated their use of other alternatives. In terms of visual quality of the 3D reconstruction, the current state-of-the-art results are obtained by methods based on learning implicit fields [7,12,19,32].

Visual Shape Similarity Measures. For shape retrieval and classification, many visual shape similarity measures have been proposed. LFD [6] performs visual shape similarity by extracting local features from one-hundred orthogonal projections of each 3D model. ShapeGoogle [4] constructs compact and informative shape descriptor using a *bag-of-visual-features*. MVCNN [25] combines information from multiple views of a 3D shape into a single and compact shape descriptor with the help of CNNs. F-score [27] performs visual shape similarity explicitly by calculating the distance between object surfaces using a harmonic mean of precision and recall scores. And finally, normal consistency (NC) [19,29] measures how well any 3D reconstruction can capture higher order information by calculating a mean absolute dot product of face-normals of the given 3D models, represented as meshes. However, LFD, ShapeGoogle, F-score are not differentiable, and NC can only capture higher-order information.

Image-Space Similarity. Most common quantitative measures of image similarities include pixel-wise MSE and IoU. However, such measures are not designed to compare visual similarity between objects captured in the images, e.g., they are sensitive to misalignment. Hand-crafted image descriptors such as SIFT [17], SURF [2], and ORB [24] can be used to perform image retrieval to visually correspond closest image matches, but they are discrete and non-differentiable. On the other hand, the use of powerful image descriptors obtained from CNNs has been shown to make the image retrieval pipeline differentiable [18,21,25,34], allowing an end-to-end neural network approach for measuring image similarity. Generic image similarity has to account for rotation, translation, and occlusion of identical entities in the two images. Our aim is to develop a differentiable and visual image-similarity metric for rendered images of aligned 3D shapes.

Multi-view/Projective Approaches to 3D Reconstruction. Prior works rely on either a differentiable renderer or multi-view projections of 3D shapes and perform supervised learning on the obtained images rather than in the 3D object space, e.g., PTN [33]. Table 2 lists notable recent works along these lines and the losses and metrics they employed for training and evaluation. Specifically, NMR [13], which outperformed PTN, reconstructs a 3D mesh from a single image with silhouette-image supervision. Other works such as MarrNet [31] and Soltani et al. [1] use 2.5D sketch images and rendered silhouette images, respectively, for learning single-view 3D reconstruction. Most recently, SoftRas [16] reconstructs a 3D shape using shaded images, i.e., view-renderings of the input RGB image with a light source. Importantly, SoftRas converts the discrete raster operation to a probabilistic soft raster one, directly solving the indifferentiable raster operation of the traditional renderer present in [13,33]. However, in all these works, the image-space errors are still measured by distortion losses such as pixel-wise MSE or IoU, which do not capture shape similarity. In contrast, our metric, DR-KFS, is designed to capture visual similarity of the captured shapes. The renderer used in DR-KFS is a simplified version of SoftRas [16], which is more efficient than CNN-based methods for object rendering.

3   Differential Visual Shape Similarity Metric

Visual shape similarity metrics, such as LFD [6], rely on projections or renderings of multi-view images of 3D models from a set of view angles. We follow this theme of render-and-match-images to compare the visual similarity between two 3D shapes while make the framework *differentiable* so that it can be easily plugged into any 3D reconstruction DNN; see Fig. 4 for an overall pipeline.

Given two 3D models, one reconstructed using an existing approach such as AtlasNet [10], Matryoshka Network [23], Pixel2Mesh [29], or OGN [26], and the

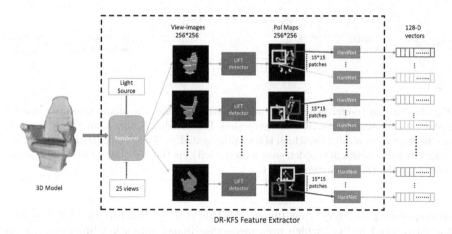

Fig. 3. Overall framework for obtaining DR-KFS descriptors using continuous patch features. The first step is the differentiable renderer, a *Soft Rasterizer*, which renders multi-view images. This renderer, coupled with the process of continuous patch feature extraction over the point-of-interest (PoI) maps, makes our framework differentiable.

other being the corresponding ground truth (GT) model, we first align them and render the models from twenty-five viewing angles, using the *Soft Rasterizer* [16] renderer, which is differentiable. For each view-image, a Point of Interest (PoI) map is obtained using a keypoint detection network. Visual similarity of the 3D models is determined using DR-KFS feature matching (see Fig. 4) obtained by extracting local features from PoI maps.

3.1 Differentiable Renderer

To measure 3D shape similarity using LFD, projected images from many angles are required. Due to hardware constraints in training our DNNs, we only use 25 viewpoints, sampled on a semi-sphere as an approximation to the ideal case. To achieve differentiability in the rendering process, we replace rasterization and z-buffering operations used in conventional rendering pipelines with soft rasterization and probabilistic map aggregation [16]. No color information is used in DR-KFS renderer as many reconstruction networks are incompatible with texture projections on the reconstructed model surfaces. A fixed position light source is aptly placed to provide information about shape surface quality when rendering the view-images. We make use of the rendered view-images to determine the similarity of given 3D models, as explained below.

Our visual similarity metric is based on matching rendered view-images by extracting discriminative local features over Point of Interest (PoI) maps, using Convolutional Neural Networks (CNNs).

PoI Map. A PoI map is a probabilistic map that gives a score for every pixel being a keypoint. It is generated by a CNN-based keypoint detection network

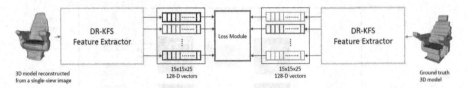

Fig. 4. An end-to-end pipeline of our visual, differentiable shape-similarity metric framework for single view 3D reconstruction. Feature descriptors from Fig. 3 are used to calculate the overall loss, which is backpropagated to the 3D reconstruction network (which reconstructs a 3D model, shown on the left) via the associated DR-KFS module, improving the reconstruction quality.

corresponding to every view-image. To this end, we borrow the LIFT detector introduced in [34], which uses piece-wise linear activation functions [28] in convolution layers to get a PoI map for an input image. It is formulated as:

$$PoI_{map} = f_{Net}(I) = \sum_n^N \delta_n \max_m^M(W_{mn} \circledast I + b_{mn}) \tag{1}$$

where $f_{Net}(I)$ is a non-linear function of the rendered view-image I, using a neural network Net, which is nothing but a CNN-based keypoint detector. N, M are hyperparameters controlling the complexity of the piece-wise linear activation function. δ is $+1$ if n is odd, and -1 otherwise. The parameters of the network Net to be learned are the convolution filter weights W_{mn} and biases b_{mn}, and $\circledast$ denotes the convolution operation. A detailed description of LIFT based keypoint detection can be found in [34].

Continuous Patch Features. After generating PoI maps for each of the rendered images, local patch features are extracted from them using a sliding window of size 32×32 with a stride of 16, resulting in 225 (15×15) patches per image I. These patches span the entire image and there exists an overlap between adjacent patches which provides continuity over the feature space.

Note that in existing image retrieval works, such as in [28,34], detected keypoints are projected back onto the original image and image features are extracted locally around these keypoints. Image retrieval is performed by matching these keypoint-based local features. This is a *discrete* process over the image space. As a result, this approach of matching images based on discrete keypoints, if employed, introduces non-differentiability into the latter parts. Unlike such methods, we avail continuous local patch features from the PoI maps by employing a sliding window technique spanning the entire map. Our framework is inspired by the recent work of [3] which shows that local patch features help alleviate the problem of "human perception *vs* image distortion" trade-off.

A sliding window on a PoI map, followed by the HardNet[20] feature extractor outputs a 128-D feature descriptor per patch (see Fig. 3). The HardNet [20] module we use is pretrained on UBC PhotoTour [30]. We coin these features as the DR-KFS feature descriptors. The continuous patch features, coupled with the early-stage render (*Soft Rasterizer*) allow our pipeline to be entirely differentiable, demonstrated further by our end-to-end learning framework in Fig. 4.

Localized Feature Matching. After extracting the DR-KFS feature descriptors for the two 3D models, we perform localized two-way feature matching. For every patch P_{recon} centered at (x, y) in the PoI maps corresponding to the reconstructed model, we consider a set of neighboring patches in the GT PoI maps, centered at $(x \pm \sigma, y \pm \sigma)$, and find a patch that yields the minimum MSE over the patch features. We then perform a weighted addition of all such minimal loss values to get a similarity score as given in Eq. 2.

$$Sim_{recon} = \sum w_i |P_{recon_i} - P_{gt_k}| \qquad (2)$$

where $w_i = average(P_{recon_i})$. P_{gt_k} is the kth patch in the PoI map corresponding to the GT model which has the least patch-feature MSE for P_{recon_i}. It should be noted that the search for the best match from the PoI patches of the ground truth model for a PoI patch of the reconstructed model is limited to PoI patches from the same viewpoint. The reconstructed features would not all converge to a single GT feature since our search is localized: search for a matching patch is done within a local neighborhood. However, the matching is many-to-one in general, which is by design. For example, due to scale discrepancies between the reconstructed and the GT models, e.g., a chair with long back vs. one with a short back, several patches of one model (e.g., on the long back) may be best matched with a single sampled patch on the other model (e.g., the short back).

We essentially repeat the above procedure for the PoI maps corresponding to the GT model and get a similarity score Sim_{GT}. Our final training loss is the sum of Sim_{recon} and Sim_{GT}, which is backpropagated through the DR-KFS feature descriptor module associated with the reconstructed 3D model (Fig. 4).

4 Results and Evaluation

We assess the impact of our differentiable visual similarity metric, DR-KFS, both qualitatively and quantitatively, on state-of-the-art single-view 3D reconstruction networks. A gallery of qualitative comparison on results generated when such networks are trained using DR-KFS is shown in Fig. 7.

4.1 Quantitative Evaluation

We consider existing single-view 3D reconstruction networks, including 3D-R2N2 [8], OGN [26], AtlasNet [10], Matryoshka Network [23], Pixel2Mesh [29], NMR [13] and SoftRas [16] to quantify both (object-space) shape distortion (CD and IoU) and the visual quality of the reconstructed models. The networks are trained either using their original losses or our metric DR-KFS. For visual quality measures, we choose representative tools including Shape Google, Normal Consistency, F-score, LFD, and MVCNN.

Table 3. Quantitative comparison for single-view 3D reconstruction networks when trained using their original distortion-based losses vs. DR-KFS, our differentiable visual similarity metric. Tested DNNs include AtlasNet25 [10] (Atlas25), Matryoshka Net [23] (MN), Pixel2Mesh [29] (P2M), Octree Generating Network [26] (OGN), 3D-R2N2 [8], and shape generation using NMR [13] and SoftRas [16]. Evaluation is done on both object-space (CD and IoU) and visual similarity metrics: SG = Shape Google, FS = F score, NC = Normal Consistency, LFD = Light Field Descriptor, MVCNN = multi-view CNN. Numbers reported represent average performance over four object categories.

Networks	Training strategies	Evaluation metrics							
		CD	IoU	SG	FS	NC	LFD	MVCNN	DR-KFS
Atlas25 [10]	Original	**6.53**	53.38	3.17	63.17	**0.792**	4338	0.532	0.288
	DR-KFS	7.11	**54.41**	**2.26**	**65.48**	0.790	**3796**	0.522	**0.213**
MN [23]	Original	**2.86**	65.33	4.32	**74.22**	0.825	3866	**0.637**	0.300
	DR-KFS	2.91	**66.08**	**2.08**	65.30	**0.829**	**3675**	0.660	**0.265**
P2M [29]	Original	**6.61**	54.08	6.08	58.15	**0.761**	4508	**0.648**	0.463
	DR-KFS	6.98	**54.32**	**3.84**	**60.23**	0.754	**4212**	0.662	**0.332**
OGN [26]	Original	6.13	**57.01**	9.94	**56.36**	0.695	4436	0.752	0.522
	DR-KFS	**6.08**	56.83	**8.73**	55.09	**0.702**	**4237**	**0.661**	**0.461**
3D-R2N2 [8]	Original	**7.42**	**53.98**	9.75	48.09	0.621	4692	0.841	0.539
	DR-KFS	7.56	52.74	**9.13**	**51.58**	**0.681**	**4416**	**0.713**	**0.441**
Shape	NMR[13]	11.95	48.90	8.13	51.72	0.522	7716	0.842	0.560
Generator	SoftRas[16]	9.75	54.53	7.11	57.08	0.685	5385	**0.631**	0.471
[13, 16]	DR-KFS	**8.06**	**55.24**	**5.48**	**62.02**	**0.634**	**4125**	0.741	**0.395**
IM-Net[7]	Original	7.02	66.01	1.38	69.43	0.740	3806	0.511	0.203

Note that the shape generator used in SoftRas [16] and NMR [13] are identical. We integrate DR-KFS training loss into this shape generator. Our method works well with mesh inputs. However, for voxel-based methods such as OGN/MN, we adopted a differentiable marching cubes layer (DMCL) proposed in [15] to convert voxels to meshes.

Dataset. All the tested networks are trained and evaluated on the large-scale 3D CAD model dataset, ShapeNet [5]. We use four shape categories to train and test all the networks: airplane (4,045 models), car (3,533), chair (6,778), and lamp (2,318), with an 80-20 train-test split. Input data is prepared accordingly as consumed by each one of the aforementioned networks.

Comparison Results. Table 3 shows the comparison results, where the numbers report *average* performance over all four object categories. In theory, DR-KFS can be employed for training IM-NET. Due to the limitation of our current of GPU memory, we only report the IM-NET numbers here for reference.

Our first observation is that when a network is trained using the new metric DR-KFS, it always performed better in DR-KFS, compared with the original network. Second, when trained using DR-KFS, some of the networks even outperformed their counterparts trained with the original (object-space) losses, when the results are evaluated using CD and IoU—this is the case 4 out of 10 or 40% of the time. In particular, Matryoshka was trained using IoU, but training with DR-KFS outperformed the original network when evaluated on IoU. Next, perhaps most importantly, we see that in majority of the cases (24 out of 30 or **80%** of the time), replacing the original object-space losses using DR-KFS, the networks improved their performance in terms of visual shape similarity metrics.

4.2 Qualitative Evaluation

To understand the visual quality of the reconstructed 3D models trained using DR-KFS as the network loss, we conduct two perceptual studies (PS) on Amazon Mechanical Turk (AMT), as described below.

PS-1. In this study, we present 50 questions, each containing a single-view image and a pair of reconstructed models, one using the original loss and one using DR-KFS. Each model is rendered along two views, with one of the views aligned to the input image. We use the reconstructed models from AtlasNet [10] and Matryoshka Net [23]. Models are selected at *random* from the *entire* reconstructed set. Turkers are presented with 50 questions in a random order, with randomized orderings of the reconstructed models as well as their view-images.

PS-2. We essentially repeat PS1, but instead of selecting the models from the entire reconstructed set, we select them *randomly* from the *top-20%* of the reconstructed set, filtered based on the scores obtained using three visual similarity measures: LFD, ShapeGoogle and MvCNN.

Both PS1 and PS2 are forced-choice responses involving 80 different participants per study. For PS-1, DR-KFS metric received 71% of the total votes (4000) and for PS-2, the percentage share shot up to 86.075%, indicating a positive trend when high quality visual samples are considered for perceptual evaluation.

4.3 Design Analysis

Reconstruction using "Visual Similarity Loss vs. Image Distortion Loss". In Sect. 2, we enlisted a sampler of representative works that address the task of single-view 3D reconstruction via image supervision. We argue that simply using a multi-view pipeline is insufficient in producing visually coherent results. To support this argument, we perform an experiment where we train the Shape Generator used in SoftRas [16] (state-of-the-art work for 3D reconstruction) using two different losses: (a) Image distortion loss (per-pixel MSE), and (b) Image-similarity loss (DR-KFS framework). We combine this analysis with the effect of visual-space representation described below.

Reconstruction from "Silhouette vs. Shaded Images". Silhouette images only capture the boundary, without any surface information of the 3D shape. It is straightforward to infer that the reconstructed 3D shapes using silhouette images are likely to have bad surface quality (see Fig. 5), whereas using shaded images yields better reconstruction. To understand the differences in the reconstruction quality, we employ the shape generator used in SoftRas [16] for 3D reconstruction from either silhouette images or shaded images, with and without integrating DR-KFS pipeline.

The quantitative results for the effect of the two loss functions (visual similarity *vs.* image distortion) are shown in Table 4. DR-KFS, which measures image similarity (not distortion), wins over per-pixel MSE loss, when trained on shaded images, measured using visual shape similarity metrics such as LFD and Shape Google. Using per-pixel MSE loss on shaded images, however, is much better than using DR-KFS on silhouette images, indicating that shaded images are friendlier for the task of 3D reconstruction. This is also underscored visually in Fig. 5. This supports our claim that for a method to be *truly visual*, it should not only make use of multi-view image pipeline, but should also employ a loss that captures visual similarity and not merely image distortion.

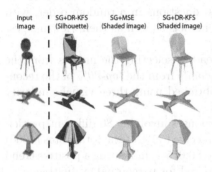

| Input Image | SG+DR-KFS (Silhouette) | SG+MSE (Shaded image) | SG+DR-KFS (Shaded image) |

Table 4. Training a simple Shape Generator (SG) adopted in SoftRas [16] with both DR-KFS and per-pixel MSE loss. DR-KFS, which performs image feature matching, essentially incorporates visual similarity, while per-pixel MSE is merely an image distortion loss on the image pixels. The reconstructed 3D models are shown in Fig. 5.

Fig. 5. 3D reconstruction results when the Shape Generator (SG) used in SoftRas [16] is trained using silhouettes vs. shaded images corresponding to an input RGB image. See Table 4 for the *average* similarity scores for the shapes shown above.

	LFD	Shape Google
SG+DR-KFS (Silhouette)	6989	8.98
SG+MSE (Shaded Image)	5156	7.33
SG+DR-KFS (Shaded Image)	**4125**	**5.48**

Features Capturing Perceptual Differences. Given a four-legged chair with a bar between adjacent pair of legs, humans can relate to it even after vertical shifts of the bars. However, if all the bars are removed, one cannot *uniquely* associate the new chair to the original one (see Fig. 6). Cognitively, the deletion operation is more visually apparent than the vertical positional shifts of the bars. It is desired that feature descriptors encompassing a visual similarity framework should be able to capture and reflect such perceptual differences. We investigate this by calculating the image-space distances on four different features of an image (rendered-view of a 3D shape): MSE on raw image pixels, MSE on LIFT image features [34], MSE on PoI maps, and DR-KFS approach. The input image and its corresponding PoI map are shown in Fig. 6. In our experiment, vertical shifts of the bars are obtained by moving them five pixels up/down. From Table 5, we observe that DR-KFS (differentiable approach) and LIFT features (non-differentiable) are more sensitive towards the deletion operation compared to others. Moreover, image matching using DR-KFS and LIFT based local image features seem to be tolerant to small part shifts (2nd and 4th row in Table 5), while per-pixel MSE on the image I and the POI maps are quite sensitive to such relatively less perceptual changes than the deletion operation.

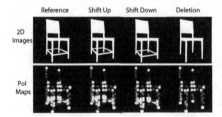

Table 5. Image similarity scores for the image-level operations shown in Fig. 6, using: MSE loss on image pixels (I), image features using LIFT descriptors [34], MSE loss on PoI map pixels, and DR-KFS framework. Italicized numbers (row-wise) indicate sensitivity to the respective image-level operation.

Fig. 6. Vertical positional shifts, and deletion of the leg bars. Binary images are shown on top and their corresponding PoI maps are on the bottom. Image similarity scores w.r.t the reference image for each operation using different image-level features are tabulated in Table 5.

Loss	Shift up	Shift down	Deletion
MSE (I)	0.2961	*0.3736*	0.2007
MSE(LIFT feats)	0.1174	0.1231	*0.1498*
MSE (PoI Maps)	0.0137	*0.0149*	0.0138
DR-KFS	0.0143	0.0160	*0.0256*

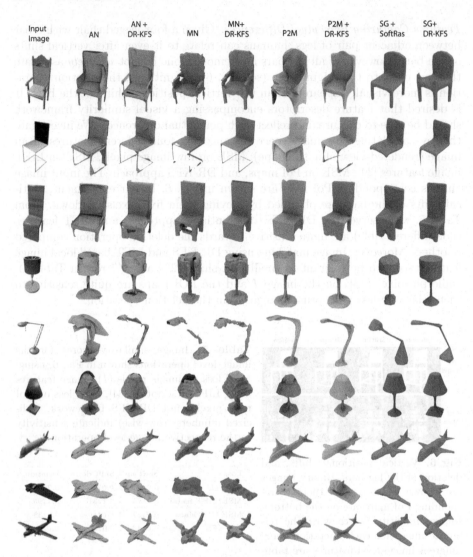

Fig. 7. A gallery of reconstructed 3D models obtained from AtlasNet (AN) [10], Matryoshka Net (MN) [23], Pixel2Mesh (P2M) [29] and Shape Generator (SG) [16], trained using the metrics as adopted in the respective works and by using our DR-KFS metric. Given an input image, replacing the original training loss with our DR-KFS loss results in an improvement in the visual quality of the reconstructed 3D models as shown above, and also supported by the numbers in Table 3.

5   Conclusion

We make a first step towards improving the visual quality of 3D reconstruction networks by making their training/reconstruction loss functions visual *and* driven by shape similarity rather than distortion. Overall, the new differentiable visual similarity metric we develop, DR-KFS, is shown to improve the reconstruction quality for all the tested networks (AtlasNet, Matryoshka, Pixel2Mesh, OGN, SoftRas, etc.), as judged by a variety of visual similarity measures (LFD, MVCNN, Shape Google, etc.). This is clearly a positive trend to motivate further investigation. However, the demonstrated advantage of DR-KFS is not yet "across-the-board". Our metric is still rather primitive as it does not utilize the most advanced and up-to-date tools that are available for multi-view rendering, feature extraction and learning, or image-space/perceptual assessment.

In general, the question of what the best 3D reconstruction is should depend on the application. For example, if the goal is to recognize or classify the shape, then visual similarity is more important. As well, functional understanding of an acquired shape hinges on accurate recovery of shape structures, which would support the use of visual similarity. On the other hand, if the reconstructed 3D shape is to reflect accurate physical measures, then object-space metrics should play a more prominent role.

One of the most obvious limitations of any visual similarity DR-KFS included, is that it is oblivious to errors that are *hidden* from the viewers due to occlusion. Such errors are not visible on the projected images, but they can be captured by object-space shape distances. One possibility to explore is to combine differentiable object- and image-space shape metrics as the training loss, e.g., as a weighted sum which would preserve the differentiability. This could be a good strategy to address the distortion-perception trade-off [3], but leaves the question of how to choose the weight.

References

1. Arsalan Soltani, A., Huang, H., Wu, J., Kulkarni, T.D., Tenenbaum, J.B.: Synthesizing 3D shapes via modeling multi-view depth maps and silhouettes with deep generative networks. In: Proceedings of the IEEE Conference on Computer Vision and Pattern Recognition, pp. 1511–1519 (2017)
2. Bay, H., Tuytelaars, T., Van Gool, L.: SURF: speeded up robust features. In: Leonardis, A., Bischof, H., Pinz, A. (eds.) ECCV 2006. LNCS, vol. 3951, pp. 404–417. Springer, Heidelberg (2006). https://doi.org/10.1007/11744023_32
3. Blau, Y., Michaeli, T.: The perception-distortion tradeoff. In: Proceedings of the IEEE Conference on Computer Vision and Pattern Recognition, pp. 6228–6237 (2018)
4. Bronstein, A.M., Bronstein, M.M., Guibas, L.J., Ovsjanikov, M.: Shape google: geometric words and expressions for invariant shape retrieval. Trans. Graph. **30**(1), 1:1–1:20 (2011)
5. Chang, A.X., et al.: Shapenet: an information-rich 3D model repository. arXiv preprint arXiv:1512.03012 (2015)

6. Chen, D.Y., Tian, X.P., Shen, Y.T., Ouhyoung, M.: On visual similarity based 3D model retrieval. In: Computer Graphics Forum, vol. 22, pp. 223–232. Wiley Online Library (2003)
7. Chen, Z., Zhang, H.: Learning implicit fields for generative shape modeling. In: CVPR (2019)
8. Choy, C.B., Xu, D., Gwak, J.Y., Chen, K., Savarese, S.: 3D-R2N2: a unified approach for single and multi-view 3D object reconstruction. In: Leibe, B., Matas, J., Sebe, N., Welling, M. (eds.) ECCV 2016. LNCS, vol. 9912, pp. 628–644. Springer, Cham (2016). https://doi.org/10.1007/978-3-319-46484-8_38
9. Fan, H., Su, H., Guibas, L.J.: A point set generation network for 3d object reconstruction from a single image. In: Proceedings of the IEEE Conference on Computer Vision and Pattern Recognition, pp. 605–613 (2017)
10. Groueix, T., Fisher, M., Kim, V.G., Russell, B., Aubry, M.: Atlasnet: a papier-mâché approach to learning 3D surface generation. In: Proceedings of IEEE Conference on Computer Vision and Pattern Recognition (CVPR) (2018)
11. Häne, C., Tulsiani, S., Malik, J.: Hierarchical surface prediction for 3D object reconstruction. In: 3DV (2017)
12. Joon Park, J., Florence, P., Straub, J., Newcombe, R., Lovegrove, S.: DeepSDF: learning continuous signed distance functions for shape representation. In: CVPR (2019)
13. Kato, H., Ushiku, Y., Harada, T.: Neural 3D mesh renderer. In: Proceedings of the IEEE Conference on Computer Vision and Pattern Recognition, pp. 3907–3916 (2018)
14. Kipf, T.N., Welling, M.: Semi-supervised classification with graph convolutional networks. arXiv preprint arXiv:1609.02907 (2016)
15. Liao, Y., Donne, S., Geiger, A.: Deep marching cubes: learning explicit surface representations. In: Proceedings of the IEEE Conference on Computer Vision and Pattern Recognition, pp. 2916–2925 (2018)
16. Liu, S., Li, T., Chen, W., Li, H.: Soft rasterizer: a differentiable renderer for image-based 3D reasoning. In: ICCV (2019)
17. Lowe, D.G., et al.: Object recognition from local scale-invariant features. In: ICCV, vol. 99, pp. 1150–1157 (1999)
18. Luo, Z., et al.: Contextdesc: local descriptor augmentation with cross-modality context. In: Proceedings of the IEEE Conference on Computer Vision and Pattern Recognition, pp. 2527–2536 (2019)
19. Mescheder, L., Oechsle, M., Niemeyer, M., Nowozin, S., Geiger, A.: Occupancy networks: learning 3D reconstruction in function space. In: CVPR (2019)
20. Mishchuk, A., Mishkin, D., Radenovic, F., Matas, J.: Working hard to know your neighbor's margins: local descriptor learning loss. In: Advances in Neural Information Processing Systems, pp. 4826–4837 (2017)
21. Mishkin, D., Radenovic, F., Matas, J.: Repeatability is not enough: learning affine regions via discriminability. In: Proceedings of the European Conference on Computer Vision (ECCV), pp. 284–300 (2018)
22. Pontes, J.K., Kong, C., Sridharan, S., Lucey, S., Eriksson, A., Fookes, C.: Image2Mesh: a learning framework for single image 3D reconstruction. In: Jawahar, C.V., Li, H., Mori, G., Schindler, K. (eds.) ACCV 2018. LNCS, vol. 11361, pp. 365–381. Springer, Cham (2019). https://doi.org/10.1007/978-3-030-20887-5_23
23. Richter, S.R., Roth, S.: Matryoshka networks: predicting 3D geometry via nested shape layers. In: CVPR (2018)
24. Rublee, E., Rabaud, V., Konolige, K., Bradski, G.R.: Orb: an efficient alternative to sift or surf. In: ICCV, vol. 11, p. 2. Citeseer (2011)

25. Su, H., Maji, S., Kalogerakis, E., Learned-Miller, E.G.: Multi-view convolutional neural networks for 3D shape recognition. In: Proceedings of ICCV (2015)
26. Tatarchenko, M., Dosovitskiy, A., Brox, T.: Octree generating networks: Efficient convolutional architectures for high-resolution 3D outputs. In: The IEEE International Conference on Computer Vision (ICCV), October 2017
27. Tatarchenko, M., Richter, S.R., Ranftl, R., Li, Z., Koltun, V., Brox, T.: What do single-view 3D reconstruction networks learn? In: CVPR (2019)
28. Verdie, Y., Yi, K., Fua, P., Lepetit, V.: Tilde: a temporally invariant learned detector. In: Proceedings of the IEEE Conference on Computer Vision and Pattern Recognition, pp. 5279–5288 (2015)
29. Wang, N., Zhang, Y., Li, Z., Fu, Y., Liu, W., Jiang, Y.G.: Pixel2mesh: generating 3D mesh models from single RGB images. In: Proceedings of the European Conference on Computer Vision (ECCV) (2018)
30. Winder, S.A., Brown, M.: Learning local image descriptors. In: 2007 IEEE Conference on Computer Vision and Pattern Recognition, pp. 1–8. IEEE (2007)
31. Wu, J., Wang, Y., Xue, T., Sun, X., Freeman, B., Tenenbaum, J.: Marrnet: 3D shape reconstruction via 2.5 D sketches. In: Advances in Neural Information Processing Systems, pp. 540–550 (2017)
32. Xu, Q., Wang, W., Ceylan, D., Mech, R., Neumann, U.: DISN: deep implicit surface network for high-quality single-view 3D reconstruction. CoRR abs/1905.10711 (2019). http://arxiv.org/abs/1905.10711
33. Yan, X., Yang, J., Yumer, E., Guo, Y., Lee, H.: Perspective transformer nets: learning single-view 3D object reconstruction without 3D supervision. In: Advances in Neural Information Processing Systems, pp. 1696–1704 (2016)
34. Yi, K.M., Trulls, E., Lepetit, V., Fua, P.: LIFT: learned invariant feature transform. In: Leibe, B., Matas, J., Sebe, N., Welling, M. (eds.) ECCV 2016. LNCS, vol. 9910, pp. 467–483. Springer, Cham (2016). https://doi.org/10.1007/978-3-319-46466-4_28
35. Yu, J., Jiang, Y., Wang, Z., Cao, Z., Huang, T.: Unitbox: an advanced object detection network. In: Proceedings of the 24th ACM International Conference on Multimedia, MM 2016, pp. 516–520 (2016)

SPAN: Spatial Pyramid Attention Network for Image Manipulation Localization

Xuefeng Hu[1]($\boxtimes$), Zhihan Zhang[1], Zhenye Jiang[1], Syomantak Chaudhuri[2], Zhenheng Yang[3], and Ram Nevatia[1]

[1] University of Southern California, Los Angeles, USA
{xuefengh,zhihanz,zhenyeji,nevatia}@usc.edu
[2] Indian Institute of Technology, Bombay, Mumbai, India
syomantak@iitb.ac.in
[3] Facebook AI, Menlo Park, USA
zhenheny@fb.com

Abstract. We present a novel framework, Spatial Pyramid Attention Network (SPAN) for detection and localization of multiple types of image manipulations. The proposed architecture efficiently and effectively models the relationship between image patches at multiple scales by constructing a pyramid of local self-attention blocks. The design includes a novel position projection to encode the spatial positions of the patches. SPAN is trained on a generic, synthetic dataset but can also be fine tuned for specific datasets; The proposed method shows significant gains in performance on standard datasets over previous state-of-the-art methods.

1 Introduction

Fast development of image manipulation techniques is allowing users to create modifications and compositions of images that look "authentic" at relatively low cost. Typical manipulation methods include splicing, copy-move, removal and various kinds of image enhancements to produce "fake" or "forged" images. We aim to localize the manipulated regions of different tampering types in images (Fig. 1).

There has been work on both detection and localization of manipulations in recent years. Among the methods that include localization, many deal with only one or a few types of manipulations such as splicing [8,16,18], copy-move [7,27,33,35,36], removal [44], and enhancement [5,6]. Some recent papers have proposed more general solutions that are not specific to manipulation types; these include RGB-Noise (RGB-N) Net [43] and Manipulation Tracing Network (ManTra-Net) [37]. The two methods differ in the granularity of their localization

Electronic supplementary material The online version of this chapter (https://doi.org/10.1007/978-3-030-58589-1_19) contains supplementary material, which is available to authorized users.

© Springer Nature Switzerland AG 2020
A. Vedaldi et al. (Eds.): ECCV 2020, LNCS 12366, pp. 312–328, 2020.
https://doi.org/10.1007/978-3-030-58589-1_19

| Manipulated Image | SPAN Prediction | GT Mask | Manipulated Image | SPAN Prediction | GT Mask |

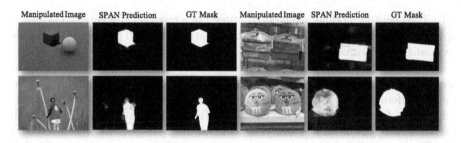

Fig. 1. Examples of images manipulated by splicing (left) and copy-move (right) techniques, SPAN predictions and Ground-truth Masks.

(bounding boxes in [43] *vs* masks in [37]). Our method is also designed for pixel-level masks predictions of multiple manipulation types.

A key assumption underlying forged region localization is that the color, intensity or noise distributions of manipulated regions are somehow different from those of the untampered ones. These distinctions are typically not transparent to a human observer but the expectation is that they can be learned by a machine. Ability to capture and model the relationship between tampered and authentic regions is crucial for this. RGB-N [43] proposes a Faster R-CNN [28] based method to detect tampered region; hence, its predictions are limited to the rectangular box. It also requires fine-tuning to adapt to a new dataset. ManTra-Net [37] combines the task of image manipulation type classification and localization into a unified framework and achieves comparable results to RGB-N but without fine-tuning on the target evaluation data. ManTra-Net makes pixel-level predictions. However, its localization module only models the "vertical" relationship of same points on different scales of the feature map but does not model the spatial relations between image patches. We propose to model both "vertical" relationship using multi-scale propagation and spatial relationship using a self-attention module in the Spatial Pyramid Attention Network (SPAN).

SPAN is composed of three blocks: a feature extractor, a spatial pyramid attention block and a decision module. We adopt the pre-trained feature extractor provided by [37] as our feature extractor. To build the relationship between different image patches, we construct a hierarchical structure of self-attention layers with the features as input.

The hierarchical structure is designed to first calculate local self-attention blocks, and the local information is propagated through a pyramid hierarchy. This enables an efficient and multi-scale modeling of neighbors. Inspired by [31], within each self-attention layer, we calculate a new representation for each pixel conditioned on its relationship to its neighborhood pixels. To better capture the spatial information of each neighbor in the self-attention block, we also introduce a positional projection which is more suitable for our localization task, instead of the original positional embedding methods used for machine translation task in

[31]. The decision module, which is composed of a few 2D convolutional layers, is applied on top of the output from pyramid spatial attention propagation module to predict the localization mask.

We conducted comprehensive experiments on five different popular benchmarks and compared with previous works on pixel-level manipulation localization of different tampering types. Our proposed SPAN model outperforms current state-of-the-art (SoTA) methods ManTra-Net [37] and RGB-N [43] with the same setting on almost every benchmark (e.g., 11.21% improvement over ManTra-Net on Columbia, without fine-tuning; 9.5% improvement over RGB-N on Coverage, with fine-tuning).

In summary we make three contributions in this work: (1) we designed a novel Spatial Pyramid Attention Network architecture that efficiently and explicitly compares patches through the local self-attention block on multiple scales; (2) we introduced positional projection in the self-attention mechanism, instead of classic positional embedding used in text processing; (3) We show that the ability to compare and model relationship between image patches at different scales results in higher accuracy compared to state-of-the-art methods.

2   Related Work

Manipulation Detection and Localization: Most of existing methods detect and localization a specific type of manipulation only, such as copy-move [7,27,33], removal [44], enhancement [5,6], and splicing [8,16,18,29]. Among these types, splice detection is the most studied topic. For this task, MFCN [29] has proposed a Multi-task Fully Convolutional Network [22] with enhancement on edge detection, jointly trained towards both edge detection and tampered region detection. MAG [18] proposed Generative-Adversarial based pipeline to train its model with augmented retouched samples, based on a new splicing dataset.

The types of manipulation, however, may not always be known in advance and a single image may contain multiple types of manipulations. Therefore, it is important to have systems that detect general types of manipulations. Some recent work [4,4,37,43] has addressed this problem, and shown success in building models that have robustness to detection of multiple manipulation techniques. J-LSTM [3] is an LSTM based patch comparison methods that finds tempered regions by detecting edges between the tempered patches and authentic patches. H-LSTM [4] improved this method by introducing a separate encoder-decoder structure to refine the predicted mask, and a Hilbert Curve route to process the image patches, to establish better context information. Both methods operate on fixed size image patches, which might cause failure when the tempered region size does not follow this assumption.

RGB-N [43] proposes a two-stream Faster R-CNN network [28], which combines the regular RGB-based Faster R-CNN with a parallel module that works on the noise information generated by the Steganalysis Rich Model (SRM) [11]. RGB-N is pre-trained on a synthetic tampering dataset generated base on MS-COCO [21]. Due to the R-CNN architecture, RGB-N is limited to localizing to a rectangular box whereas real objects are not necessarily rectangular. RGB-N also requires fine tuning on specific datasets to achieve high accuracy.

ManTra-Net [37] has proposed a system that jointly learns the Manipulation-Tracing feature for both image manipulation classification and forgery localization. ManTra-Net is composed of a VGG [30] based feature extractor and an LSTM [14] based detection module. The feature extractor is trained towards 385 types of image manipulation, then it is trained together with the detection module on a synthetic dataset for multi-type manipulation detection. While [3,4,43] require fine-tuning on different manipulation distributions to achieve state-of-the-art performance, ManTra-Net achieves comparable results without any training on the domain datasets.

Attention Mechanism. Attention mechanism, first introduced by [2], has brought tremendous benefit to the many fields, such as Machine Translation [2,31], Image Captioning [39], Image Question and Answering [40] and Object Detection [1,15,20]. Attention mechanism helps the neural network build input-aware connections to focus more on meaningful entities, such as words or regions, by replacing the classic learnable fixed weights with input dependent weights.

Self-Attention mechanism proposed in [31] calculates mutual attention between a group of input, has succeed in capturing long term word relationships within the same sentence, and has been dominating Machine Translation in the recent two years. Image transformer [26] has made an attempt to bring the self-attention mechanism to image generation, by adopting an encoder-decoder structure based on pixel-wise self-attention. The Non-local Neuron Network [32] uses self-attention mechanism to model non-local relationship between pixel features, and achieves an improvement in activity recognition, image classification and object detection. However, the non-local blocks are of $O(N^4)$ complexity in both times and space, for $N \times N$ size input, which implies that it could only be applied when the spatial resolution is reduced. For tasks addressed in [32], good results are obtained by applying the non-local layer to reduced resolution maps; however, for manipulation detection, it may be harmful to reduce spatial resolution as the small differences between pixels that are important in tracing forgery regions, may be lost during the pooling operations.

3   Method

We describe our proposed Spatial Pyramid Attention Network (SPAN) in this section. We first provide an overview of the framework, then the details of each module and lastly how the whole framework can be trained.

3.1   Overview

SPAN is composed of three blocks: feature extractor, pyramid spatial attention propagation and decision module. We adopt the pre-trained feature extractor as proposed in [37]. The feature extraction network is trained on synthesized data

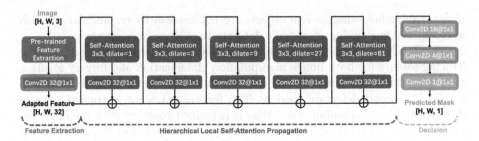

Fig. 2. Overview of the Spatial Pyramid Attention Network. There are three blocks in SPAN framework: feature extraction (blue), pyramid spatial attention propagation (green) and decision module (orange). (Color figure online)

generated based on images sampled from the *Dresden Image Database* [12], using classification supervision against 385-type image manipulation data. The feature extractor adopts the Wider & Deeper VGG Network [30,41] as the backbone architecture, along with Bayer [5] and SRM [11,42] Convolutional Layers to extract rich features from both visual artifacts and noise patterns.

On top of the embeddings extracted using the pre-trained feature extractor, we apply a convolution layer to adapt the feature to feed into the spatial attention module. The pyramid spatial attention propagation module is designed to establish the spatial relationship on multiple scales of the pixel representation. Five layers of the local self-attention blocks are recursively applied to extract information from neighboring pixels or patches. To preserve the details from multi-scale layers, we add the input back to the output after each self-attention block, through the residual link which is proposed in [13].

To generate the final mask output, 2D convolutional blocks are applied. A Sigmoid activation is applied to predict the soft tampering mask.

3.2 Local Self-attention Block

Consider an image feature tensor X of size $H \times W \times D$, where H and W are the spatial dimensions, and D is the feature dimension. For simplicity, we use $X_{i,j}$ to represent pixel at i-th row and j-th column, where each $X_{i,j} \in \mathbb{R}^D$. We calculate the Local Self-Attention value, $LSA(X_{i,j}|X,N,t)$, for each position $X_{i,j}$, by looking at it and its $(2N+1) \times (2N+1)$ neighborhood, dilated by distance t. For example, when $N = 1$, the 3×3 neighborhood looks like:

$$\begin{bmatrix} X_{i-t,j-t} & X_{i-t,j} & X_{i-t,j+t} \\ X_{i,j-t} & X_{i,j} & X_{i,j+t} \\ X_{i+t,j-t} & X_{i+t,j} & X_{i+t,j+t} \end{bmatrix}$$

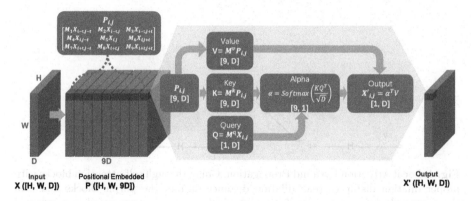

Fig. 3. Self-Attention Block: We perform self-attention mechanism over each target pixel $X_{i,j}$ and its local neighborhood $P_{i,j}$. With learnable projections M^q, M^k and M^v, we prepare Keys and Values from $P_{i,j}$ and Query from $X_{i,j}$ for the attention mechanism.

For simplicity, we rename the $(2N+1)^2$ neighbors linearly from top-left to bottom-right, as $\{Y_1, Y_2, ..., Y_{(2N+1)^2}\}$. Then, the LSA over $X_{i,j}$ is calculated as

$$LSA(X_{i,j}|X,N,t) = \frac{1}{C} \sum_{l=1}^{(2N+1)^2} \exp\left(\frac{\langle M^k Y_l, M^q X_{i,j}\rangle}{\sqrt{D}}\right) M^v Y_l \qquad (1)$$

where $\langle,\rangle$ represents the inner product, $C = \sum_{l=1}^{(2N+1)^2} \exp\left(\frac{\langle M^k Y_l, M^q X_{i,j}\rangle}{\sqrt{D}}\right)$ is the soft-max normalization factor, and $M^k, M^q, M^v \in \mathbb{R}^{D\times D}$ are learnable weights.

Equation 1 can be rewritten as

$$LSA(X_{i,j}|X,N,t)^\top = SoftMax\left(\frac{(M^q X_{i,j})^\top M^k P_{i,j}}{\sqrt{D}}\right)(M^v P_{i,j})^\top \qquad (2)$$

where $P_{i,j} = [Y_1, ..., Y_{(2N+1)^2}] \in \mathbb{R}^{D\times(2N+1)^2}$, and $M^k P_{i,j}, M^v P_{i,j}, M^q X_{i,j}$ correspond to the terminology of Keys, Values and Query in attention mechanism literature [2,31,32].

Relationships between neighborhood $\{Y_1, Y_2, ..., Y_{(2N+1)^2}\}$ and pixel $X_{i,j}$ can be explicitly modeled by the quadratic form $\langle M^k Y_l, M^q X_{i,j}\rangle$, and used to help build the feature for manipulation localization. Such information cannot be easily extracted by any convolution network with similar number of parameters, as convolution blocks only adds pixels together, rather than building their mutual relationship.

3.3 Positional Projection

As shown in the Transformer work [31], positional embedding is essential in self-attention calculation, to enable the model to tell the difference between inputs

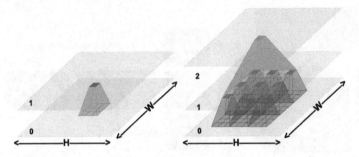

Fig. 4. Local Attention Pyramid Propagation: Going through self-attention blocks with proper dilation distances (e.g. dilation distance of 1 in the purple blocks and 3 in the orange blocks), information from each pixel location is propagated in a pyramid structure (e.g. each pixel on level-2 encodes information from 9 pixels on level-1 and 81 pixels on level-0).

with different temporal order. In our case, it is also important that the model know the relative spatial relationship between the neighbor pixel and query pixel. In machine translation models [31], positional embeddings are learnable weights that directly add to inputs at each input position. This is reasonable for language tasks, because the word embedding space is typically trained to support vector addition and subtraction corresponding to linguistic meaning. However, as our embedding space does not follow these requirements, we propose to use learnable matrix projections $\{M_l\}_{1 \leq l \leq (2N+1)^2}$ to represent the $(2N+1)^2$ possible relative spatial relationships. With positional projection, the LSA block now becomes:

$$LSA(X_{i,j}|X,N,t) = \frac{1}{C} \sum_{l=1}^{(2N+1)^2} \exp\left(\frac{\langle M^k M_l Y_l, M^q X_{i,j}\rangle}{\sqrt{D}}\right) M^v Y_l \qquad (3)$$

with soft-max normalization factor $C = \sum_{l=1}^{(2N+1)^2} \exp\left(\frac{\langle M^k M_l Y_l, M^q X_{i,j}\rangle}{\sqrt{D}}\right)$.

In Fig. 3 we illustrate a local self-attention block when $N = 1$. Given an image tensor X, we first prepare the neighborhood $P_{i,j}$ with 9 positional projected neighbors for each pixel $X_{i,j}$. Note that for edge and corner pixels, the neighborhood sizes are 6 and 4 respectively. The neighbors are then projected into Keys, Values and Query through matrix projections M^k, M^v and M^q. Finally, Keys, Values and Query are assembled into output $X'_{i,j}$ (Eq. 3).

3.4   Pyramid Propagation

The local self-attention blocks compare pixels and their neighbors with limited distances Nt. However, images might have large tampered regions, which require comparison between pixels far away from each other. Instead of simply enlarging N to have better pixel coverage, we iteratively apply our local self-attention blocks to propagate the information in a pyramid structure.

As shown in Fig. 4, the purple and orange stacks represent local self-attention blocks, each with $N = 1$, $t = 1$ and $N = 1$, $t = 3$. Through the self-attention block, each pixel on layer 1 represents 9 pixels from layer 0; each pixel on layer 2 represents 9 pixels from layer 0, and therefore 81 pixels from layer 1. With properly set dilation distances, pixels from the top of h-layer local self-attention structure can reach to $(2N + 1)^{2h}$ pixels from the bottom layers.

There are two benefits of the pyramidal design: 1) We can use a small value for N, which makes each self-attention block efficient to compute; 2) in upper layers, pixels not only represent themselves, but encode information from their local regions. Therefore, by comparing those pixels, the self-attention blocks compare information from two different patches.

Analysis of Block Size. As demonstrated in Sect. 3.4, a h-layer $M \times M$ self-attention block structure with dilation distances $\{1, M, ..., M^{h-1})\}$ helps the output pixels related to most pixels from the input, where $M = 2N + 1$ is the neighborhood size of each self-attention block. Each pixel in the final layer covers a $M^h \times M^h$ region on the input features. Therefore, to cover a $S \times S$ size image, $h = O(\log_M S)$ layers would be needed. As S^2 self-attention calculation is needed at each layer, and each self-attention block is of $O(M^2)$ complexity in both time and memory, the total complexity with respect to M is $O(S^2 M^2 \log_M S)$, which reaches to its minimum with $M = 2N + 1 = 3$, among possible integer values.

The 3×3 blocks not only offer the largest coverage under the same memory and time budget, but also offer comparisons over more scales. In our SPAN implementation, we adopt a 5-layer, 3×3 self-attention blocks structure, with dilation distances $1, 3, 9, 27, 81$, as demonstrated in Fig. 2. The proposed structure compares and models the relationship between neighbor regions at five different scales, of 3, 9, 27, 81 and 243.

3.5 Framework Training

To train SPAN, we freeze the feature extractor and train the following parts in an end-to-end fashion, supervised by binary ground-truth mask with 1 labels tampered pixels and 0 labels authentic pixels. We adopt Binary Cross-Entropy (BCE) as our loss function:

$$Loss(X^{pred}, B) = \frac{1}{HW} \sum_{i=1}^{H} \sum_{j=1}^{H} \left(-B_{i,j} \log X_{i,j}^{pred} - (1 - B_{i,j}) \log \left(1 - X_{i,j}^{pred} \right) \right)$$

where X^{pred} is the output predicted mask, and B is the binary ground-truth mask. $B_{i,j}$ and $X_{i,j}^{pred}$ represent the corresponding pixel value at location i, j.

4   Experiments

In this section, we describe experiments on five different datasets to explore the effectiveness of SPAN model. These datasets include (1) base training dataset as proposed in [37]; (2) NIST16 [25]; (3) Columbia [24]; (4) Coverage [33]; (5) CASIA [9]. We follow the evaluation protocols as in [18,37,43]. We compare with other state-of-the-art (SoTA) methods under the setup with and without finetuning for specific datasets. We compare only with methods that attempt to detect general manipulations and not tuned to a specific manipulation type, such as splicing [18,29].

4.1 Datasets

Following the practices in [37], we used the synthetic dataset proposed in [37] for training and validation. Four other popular benchmarks for manipulation detection are used for fine-tuning and evaluation.

- **Synthetic Dataset** [37] is composed of four subsets: the removal and enhancements datasets from [37], the splicing dataset from [34] and the copy-move dataset from [35]. All samples from the synthetic dataset are of size 224×224.
- **NIST16** [25] contains 584 samples with binary ground-truth masks. Samples from NIST16 are manipulated by one of the types from Splicing, Copy-Move and Removal, and are post-processed to hide visible traces. A 404:160 training-testing split is provided by [43].
- **Columbia** [24] is a Splicing based dataset, containing 180 images, with provided edge masks. We transform the edge masks into binary region masks, with positive tampered pixels and negative authentic pixels. All images in Columbia are used for testing only.
- **Coverage** [33] is a Copy-Move based dataset, containing only 100 samples, with provided binary ground-truth masks, with post-processing to remove the visible traces of manipulation. A 75:25 training-testing split is provided by [43].
- **CASIA** [9] is composed of CASIAv1 and CASIAv2 splits. CASIAv2 contains 5123 images, CASIAv1 contains 921 images; both with provided binary ground-truth masks. Samples from both subsets are manipulated by either Splicing or Copy-Move operations. Image enhancement techniques including filtering and blurring are applied to the samples for post-processing. According to [43], CASIAv2 is the train split, and CASIAv1 is the test split.

4.2 Implementation Details

As discussed in Sect. 3.4 and demonstrated in Fig. 2, we adopt the 5-layer 3×3 self-attention block structure, with dilation distance $1, 3, 9, 27, 81$. We use the residual link to preserve information from each level, and use the new proposed positional projection to capture the spatial relationships between neighbors. We report the performance using this setting, unless otherwise specified; ablation on different model choices is given in Sect. 4.3.

To train SPAN, we set batch size to 4 with 1000 batches per epoch. The batches are uniformly sampled from the synthetic dataset described above. The

Table 1. Pixel-level localization performance under the metrics of AUC and F_1 comparison on validation set (SynVal) of the pre-training data and four different benchmarks. We could not find the model that ManTra-Net reported in the paper. Hence, we also report the performance of ManTra-Net's default GitHub model where we share the same feature extractor.

	SynVal	Columbia	Coverage	CASIA	NIST16
	F_1	AUC	AUC	AUC	AUC
ManTra-Net [37]	48.39	82.4	81.9	**81.7**	79.5
ManTra-Net (GitHub)	–	77.95	76.87	79.96	78.05
SPAN	**81.45**	**93.61**	**92.22**	79.72	**83.95**

models are optimized by the Adam optimizer [17] with initial learning 10^{-4} without decay. The validation loss, precision, recall and F_1 are evaluated at each epoch. Learning rate is halved if the validation loss fails to decrease per 10 epochs, until it reaches 10^{-7}. The training is stopped early if the validation loss fails to decrease for 30 epochs. The best model is picked according to the validation loss.

Following the setting from [3,35,43], we also fine-tune our model on the training splits from the four standard benchmarks for some of our experiments, after it is trained on the synthetic data. For NIST16 and Coverage, we use the exact same split provided by RGB-N [43]; for CASIA, we use CASIAv2 as training and CASIAv1 as testing, as stated in RGB-N [43].

Stopping criteria for fine-tuning are set as following: For NIST16 and Converage, we use cross validation instead of using a fixed validation set, due to the small training split sizes. For CASIA, we randomly sampled 300 images from its training split to construct a validation split. The model with the best validation AUC scores are picked and used for evaluation on test set.

Note that the pre-trained feature extractor achieves its peak performance on original size images, as described in [37]. However, a fixed input size is required for the detection module, for fine-tuning and generalization. Therefore, during fine-tuning and inference stage, we extract the features from original sized image first, and then resize the feature tensor into fixed resolution of 224×224, for optimal performance. Total run time for a 256×384 image is around 0.11 s.

4.3   Evaluation and Comparison

We evaluate and compare SPAN with current SoTA methods [3,4,10,19,23,37, 43] on the four benchmarks under different setups: (1) pre-training only; (2) pre-training + fine-tuning. We also explored the effectiveness of each proposed module by conducting ablation studies on the four standard benchmarks. We use pixel-level Area Under the Receiver Operating Characteristic Curve (AUC) and F_1 score for comparison against the state-of-the-art general type forgery localization methods, according to [43] and [37].

Pre-training Only. We compare the performances of SPAN and ManTra-Net [37] under the same setup: both models are trained on synthetic dataset as in [37] and evaluated on validation split of the synthetic dataset (SynVal) and four different datasets. To fairly compare with [37], all images in these four datasets are used as test data. As presented in Table 1, SPAN outperforms ManTra-Net by a large margin on SynVal. With the same feature extraction backbone, our SPAN is more effective of modeling the spatial relationship among different patches and thus generates better localization accuracy. Comparing with ManTra-Net [37] on other four datasets, SPAN also shows better performance under the metric of AUC, demonstrating that our proposed model has good capability of generalizing to other datasets without extra adaptation.

SPAN achieves performance gains of over 10% in AUC compared to ManTra-Net on Columbia [24] and Coverage [33]. The performance boost on CASIA [9] is not as significant. There are two possible reasons: 1) Tampered regions in Columbia and Coverage have more varied scales than those in CASIA dataset and our SPAN also benefits from our more effective multi-scale modeling ; 2) CASIA samples have lower average resolution (256×384) compare to the other datasets (NIST: 2448×3264; Columbia: 666×1002; Coverage: 209×586). As we show in Table 5, the performance gap between SPAN and ManTra-Net decreases when test images are resized to lower resolution.

Table 2. Pixel-level localization performance under the metrics of AUC and F_1 comparison on validation set of the pre-training data and four different benchmarks. For NIST16, Coverage and CASIA, all models are fine-tuned on the corresponding training splits unless specifically stated. *We found there is an overlap of images between training and test data of NIST16.

	Supervision	Columbia		Coverage		CASIA		NIST16*	
		AUC	F_1	AUC	F_1	AUC	F_1	AUC	F_1
ELA [19]	Unsupervised	58.1	47.0	58.3	22.2	61.3	21.4	42.9	23.6
NOI1[23]	Unsupervised	54.6	57.4	58.7	26.9	61.2	26.3	48.7	28.5
CFA1 [10]	Unsupervised	72.0	46.7	48.5	19.0	52.2	20.7	50.1	17.4
J-LSTM [3]	Fine-tuned	–	–	61.4	–	–	–	76.4	–
H-LSTM [4]	Fine-tuned	–	–	71.2	–	–	–	79.4	–
RGB-N [43]	Fine-tuned	85.8	69.7	81.7	43.7	79.5	**40.8**	93.7*	**72.2***
SPAN	Pre-training	**93.6**	**81.5**	91.2	53.5	81.4	33.6	83.6	29.0
SPAN	Fine-tuned	–	–	**93.7**	**55.8**	**83.8**	38.2	**96.1***	58.2*

Pre-training + Fine-Tuning. We now compare SPAN with other SoTA methods [3,4,43] under the fine-tuning setup. We also report scores of traditional unsupervised signal analysis models (ELA [19], NOI1 [23] and CFA1 [10]) here as they are also evaluated over the testing splits rather than over the whole dataset. For fair comparison, we follow the same practices as in [43]: (1) directly

Table 3. Pixel-level AUC and F_1 comparison on multiple manipulation types evaluated NIST16 [25] dataset.

	Splicing		Removal		Copy-Move	
	AUC	F_1	AUC	F_1	AUC	F_1
ManTra-Net [37] (GitHub)	85.89	38.56	65.52	14.86	79.84	15.03
SPAN (pre-training)	90.27	42.66	77.15	15.73	82.82	13.81
SPAN (fine-tuned)	**99.15**	**82.94**	**90.95**	**49.95**	**90.94**	**40.54**

evaluate on Columbia dataset; (2) finetune our model on Coverage and CASIA training split and evaluate on test split; (3) finetune the model on training split of NIST16 provided by [43] and evaluate on test split. The results as shown in Table 2 demonstrate that SPAN without fine-tuning already outperformed RGB-N and other methods by a large margin on Columbia and Coverage, further proving that our spatial attention module has the strong capability of generalization. With fine-tuning, the performances on all four datasets further improve.

J-LSTM and H-LSTM also make predictions by comparing image patches. Two possible reasons our method achieves better performance are: 1) J-LSTM and H-LSTM look at patches at a single scale only, which might limit them from detecting tempered regions that are very large or very small; 2) J-LSTM treats patches independently and adopts an LSTM structure to process patches linearly, which might lose track of the spatial and context information of those patches; H-LSTM attempts to alleviate this problem by taking in patches along a specifically designed Hilbert-Curve route, but it could still be hard for a linear LSTM to explicitly model spatial information. SPAN considers patches with its context and neighbor pixels, and has the ability to model spatial relationship though the positional projection.

There is not a large performance gain on CASIA dataset compared to RGB-N [43], which is pre-trained on object detection dataset (MS-COCO dataset [21]). One observation on CASIA dataset is that compared to Columbia and Coverage, where there is a combination of both tampered objects and random tampered region, most tampering on CASIA images occur on objects. We put NIST16 numbers in the last column as we observed there are visually almost identical images in both training and testing splits follow the same protocol in [43].

Manipulation Type Analysis. In Table 3 we present SPAN's performance on different manipulation types when evaluated on NIST16 [25] dataset. For comparison, we also generate the per-class results by directly evaluating the model provided in ManTra-Net GitHub[1]. Without fine-tuning on NIST16 (comparing the first tow rows), our SPAN model performs consistently better than ManTra-Net on all three manipulation types, demonstrating that our proposed spatial attention model is effective agnostic to tampering types. SPAN results can be further improved with adaptation onto this specific dataset.

---

[1] https://github.com/ISICV/ManTraNet.

Table 4. Comparison of SPAN variants evaluated with F_1 metric on SynVal, Columbia, Coverage, CASIA and NIST16

F1	SynVal	Columbia	Coverage	CASIA	NIST16
SPAN (Res)	79.36	68.76	44.53	31.56	27.92
SPAN (Res+PE)	79.72	68.38	49.13	26.01	28.05
SPAN (LSTM+PP)	78.99	80.56	49.09	29.78	27.70
SPAN (Res+PP)	**81.76**	**81.45**	**53.47**	**33.63**	**28.99**

Table 5. Robustness analysis of SPAN over NIST16 and Columbia dataset

Pixel AUC	NIST16		Columbia	
	SPAN	ManTra-Net	SPAN	Mantra-Net
No manipulation	83.95	78.05	93.6	77.95
Resize (0.78x)	83.24	77.43	89.99	71.66
Resize (0.25x)	80.32	75.52	69.08	68.64
No manipulation	83.95	78.05	93.6	77.95
GaussianBlur (kernal size = 3)	83.1	77.46	78.97	67.72
GaussianBlur (kernal size = 15)	79.15	74.55	67.7	62.88
No manipulation	83.95	78.05	93.6	77.95
GaussianNoise (sigma = 3)	75.17	67.41	75.11	68.22
GaussianNoise (sigma = 15)	67.28	58.55	65.8	54.97
No manipulation	83.95	78.05	93.6	77.95
JPEGCompress (quality = 100)	83.59	77.91	93.32	75
JPEGCompress (quality = 50)	80.68	74.38	74.62	59.37

Ablation Studies. We explored how much each proposed component contributes to the final performance. We explore: (1) how to combine the outputs from different layers of multi-scale attention module (convolution LSTM (LSTM) [38] and Residual link (Res)); (2) how to model self-attention (position projection (PP) and position embedding (PE)). Besides the variants in the two modules, the number of self-attention hierarchy is set to 5 and $N = 1$. The comparison of different variants is presented in Table 4. Comparing the first two rows, Residual link performs better than LSTM on fusing the multi-scale attention feature. One possible explanation is Residual link is easier to optimize compared to the more complex LSTM; with a strong feature encoded from our attention module, SPAN with Residual link converges better. We also compare the performances of SPAN using two types of position modeling: position embedding as in [31] and our proposed position projection. We compare three variants Res+PP, Res+PE and Res only in Table 4. Under the setup of SPAN model, Res and Res+PE perform similarly across different datasets. Replacing PE with PP as the positional modeling achieves significant performance improvement on all datasets.

Coverage	CASIA	NIST16	Columbia

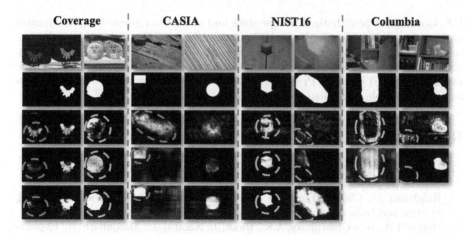

Fig. 5. Comparison of SPAN prediction results on Coverage, CASIA, Columbia NIST16 datasets, with ManTra-Net predictions. From top to bottom: Manipulated Image, Ground-truth mask, ManTra-Net prediction, SPAN prediction and fine-tuned SPAN prediction. There is no fine-tuning result on Columbia dataset as there is no training split in Columbia. (Color figure online)

Robustness. We conducted experiments to explore the robustness of SPAN to various manipulation types in Table 5. To produce modified samples, we apply standard OpenCV built-in functions `AREAResize`, `GaussianBlur`, `GaussianNoise`, `JPEGCompress` on NIST16 and Columbia. SPAN demonstrates more robust performance to compression but is more sensitive to resizing.

Qualitative Results. Figure 5 shows some SPAN and MantraNet [37] results. The examples show that SPAN produces better predictions compared to Man-Tra Net in three commonly observed cases: 1) interior of tempered regions (green circles); 2) the correct manipulated object (blue circles) and 3) noisy false negative predictions (orange circles).

5   Conclusion

We presented a Spatial Pyramid Attention Network (SPAN) that models the relationships between patches on multiple scales through a pyramid structure of local self-attention blocks, to detect and localize multiple image manipulation types with or without fine tuning. SPAN outperforms the state-of-the-art models [37,43]. The method is both accurate and robust in general type manipulation detection and localization, indicating that modeling patch relationship at different scales help capture the essential information in manipulation localization. However, SPAN may be less effective with lower image resolution.

Acknowledgement. This work is based on research sponsored by the Defense Advanced Research Projects Agency under agreement number FA8750-16-2-0204. The

U.S. Government is authorized to reproduce and distribute reprints for governmental purposes notwithstanding any copyright notation thereon. The views and conclusions contained herein are those of the authors and should not be interpreted as necessarily representing the official policies or endorsements, either expressed or implied, of the Defense Advanced Research Projects Agency or the U.S. Government. We thank Arka Sadhu for valuable discussions and suggestions.

References

1. Ba, J., Mnih, V., Kavukcuoglu, K.: Multiple object recognition with visual attention. arXiv preprint arXiv:1412.7755 (2014)
2. Bahdanau, D., Cho, K., Bengio, Y.: Neural machine translation by jointly learning to align and translate. arXiv preprint arXiv:1409.0473 (2014)
3. Bappy, J.H., Roy-Chowdhury, A.K., Bunk, J., Nataraj, L., Manjunath, B.: Exploiting spatial structure for localizing manipulated image regions. In: Proceedings of the IEEE International Conference on Computer Vision, pp. 4970–4979 (2017)
4. Bappy, J.H., Simons, C., Nataraj, L., Manjunath, B., Roy-Chowdhury, A.K.: Hybrid LSTM and encoder-decoder architecture for detection of image forgeries. IEEE Trans. Image Process. **28**(7), 3286–3300 (2019)
5. Bayar, B., Stamm, M.C.: A deep learning approach to universal image manipulation detection using a new convolutional layer. In: Proceedings of the 4th ACM Workshop on Information Hiding and Multimedia Security, pp. 5–10 (2016)
6. Bayar, B., Stamm, M.C.: Constrained convolutional neural networks: a new approach towards general purpose image manipulation detection. IEEE Trans. Inf. Forensics Secur. **13**(11), 2691–2706 (2018)
7. Cozzolino, D., Poggi, G., Verdoliva, L.: Efficient dense-field copy-move forgery detection. IEEE Trans. Inf. Forensics Secur. **10**(11), 2284–2297 (2015)
8. Cozzolino, D., Poggi, G., Verdoliva, L.: Splicebuster: a new blind image splicing detector. In: 2015 IEEE International Workshop on Information Forensics and Security (WIFS), pp. 1–6. IEEE (2015)
9. Dong, J., Wang, W., Tan, T.: Casia image tampering detection evaluation database. In: 2013 IEEE China Summit and International Conference on Signal and Information Processing, pp. 422–426. IEEE (2013)
10. Ferrara, P., Bianchi, T., De Rosa, A., Piva, A.: Image forgery localization via fine-grained analysis of CFA artifacts. IEEE Trans. Inf. Forensics Secur. **7**(5), 1566–1577 (2012)
11. Fridrich, J., Kodovsky, J.: Rich models for steganalysis of digital images. IEEE Trans. Inf. Forensics Secur. **7**(3), 868–882 (2012)
12. Gloe, T., Böhme, R.: The dresden image database for benchmarking digital image forensics. J. Digit. Forensic Pract. **3**(2–4), 150–159 (2010)
13. He, K., Zhang, X., Ren, S., Sun, J.: Deep residual learning for image recognition. In: Proceedings of the IEEE Conference on Computer Vision and Pattern Recognition, pp. 770–778 (2016)
14. Hochreiter, S., Schmidhuber, J.: Long short-term memory. Neural Comput. **9**(8), 1735–1780 (1997)
15. Hu, J., Shen, L., Sun, G.: Squeeze-and-excitation networks. In: Proceedings of the IEEE Conference on Computer Vision and Pattern Recognition, pp. 7132–7141 (2018)

16. Huh, M., Liu, A., Owens, A., Efros, A.A.: Fighting fake news: image splice detection via learned self-consistency. In: Proceedings of the European Conference on Computer Vision (ECCV), pp. 101–117 (2018)
17. Kingma, D.P., Ba, J.: Adam: a method for stochastic optimization. arXiv preprint arXiv:1412.6980 (2014)
18. Kniaz, V.V., Knyaz, V., Remondino, F.: The point where reality meets fantasy: mixed adversarial generators for image splice detection. In: Advances in Neural Information Processing Systems, pp. 215–226 (2019)
19. Krawetz, N., Solutions, H.F.: A picture's worth. Hacker Factor Solutions 6(2), 2 (2007)
20. Li, X., Wang, W., Hu, X., Yang, J.: Selective kernel networks. In: Proceedings of the IEEE Conference on Computer Vision and Pattern Recognition, pp. 510–519 (2019)
21. Lin, T.-Y., et al.: Microsoft COCO: common objects in context. In: Fleet, D., Pajdla, T., Schiele, B., Tuytelaars, T. (eds.) ECCV 2014. LNCS, vol. 8693, pp. 740–755. Springer, Cham (2014). https://doi.org/10.1007/978-3-319-10602-1_48
22. Long, J., Shelhamer, E., Darrell, T.: Fully convolutional networks for semantic segmentation. In: Proceedings of the IEEE Conference on Computer Vision and Pattern Recognition, pp. 3431–3440 (2015)
23. Mahdian, B., Saic, S.: Using noise inconsistencies for blind image forensics. Image Vis. Comput. 27(10), 1497–1503 (2009)
24. Ng, T.T., Hsu, J., Chang, S.F.: Columbia image splicing detection evaluation dataset. DVMM lab. Columbia Univ CalPhotos Digit Libr (2009)
25. NIST: NIST nimble 2016 datasets (2016). https://www.nist.gov/itl/iad/mig/
26. Parmar, N., et al.: Image transformer. arXiv preprint arXiv:1802.05751 (2018)
27. Rao, Y., Ni, J.: A deep learning approach to detection of splicing and copy-move forgeries in images. In: 2016 IEEE International Workshop on Information Forensics and Security (WIFS), pp. 1–6. IEEE (2016)
28. Ren, S., He, K., Girshick, R., Sun, J.: Faster R-CNN: towards real-time object detection with region proposal networks. In: Advances in Neural Information Processing Systems, pp. 91–99 (2015)
29. Salloum, R., Ren, Y., Kuo, C.C.J.: Image splicing localization using a multi-task fully convolutional network (MFCN). J. Vis. Commun. Image Represent. 51, 201–209 (2018)
30. Simonyan, K., Zisserman, A.: Very deep convolutional networks for large-scale image recognition. arXiv preprint arXiv:1409.1556 (2014)
31. Vaswani, A., et al.: Attention is all you need. In: Advances in Neural Information Processing Systems, pp. 5998–6008 (2017)
32. Wang, X., Girshick, R., Gupta, A., He, K.: Non-local neural networks. In: Proceedings of the IEEE Conference on Computer Vision and Pattern Recognition, pp. 7794–7803 (2018)
33. Wen, B., Zhu, Y., Subramanian, R., Ng, T.T., Shen, X., Winkler, S.: Coverage–a novel database for copy-move forgery detection. In: 2016 IEEE International Conference on Image Processing (ICIP), pp. 161–165. IEEE (2016)
34. Wu, Y., Abd-Almageed, W., Natarajan, P.: Deep matching and validation network: an end-to-end solution to constrained image splicing localization and detection. In: Proceedings of the 25th ACM International Conference on Multimedia, pp. 1480–1502 (2017)
35. Wu, Y., Abd-Almageed, W., Natarajan, P.: Busternet: detecting copy-move image forgery with source/target localization. In: Proceedings of the European Conference on Computer Vision (ECCV), pp. 168–184 (2018)

36. Wu, Y., Abd-Almageed, W., Natarajan, P.: Image copy-move forgery detection via an end-to-end deep neural network. In: 2018 IEEE Winter Conference on Applications of Computer Vision (WACV), pp. 1907–1915. IEEE (2018)

37. Wu, Y., AbdAlmageed, W., Natarajan, P.: Mantra-net: manipulation tracing network for detection and localization of image forgeries with anomalous features. In: Proceedings of the IEEE Conference on Computer Vision and Pattern Recognition, pp. 9543–9552 (2019)

38. Xingjian, S., Chen, Z., Wang, H., Yeung, D.Y., Wong, W.K., Woo, W.C.: Convolutional LSTM network: a machine learning approach for precipitation nowcasting. In: Advances in Neural Information Processing Systems, pp. 802–810 (2015)

39. Xu, K., et al.: Show, attend and tell: neural image caption generation with visual attention. In: International Conference on Machine Learning, pp. 2048–2057 (2015)

40. Yang, Z., He, X., Gao, J., Deng, L., Smola, A.: Stacked attention networks for image question answering. In: Proceedings of the IEEE Conference on Computer Vision and Pattern Recognition, pp. 21–29 (2016)

41. Zagoruyko, S., Komodakis, N.: Wide residual networks. arXiv preprint arXiv:1605.07146 (2016)

42. Zhou, P., Han, X., Morariu, V.I., Davis, L.S.: Two-stream neural networks for tampered face detection. In: 2017 IEEE Conference on Computer Vision and Pattern Recognition Workshops (CVPRW), pp. 1831–1839. IEEE (2017)

43. Zhou, P., Han, X., Morariu, V.I., Davis, L.S.: Learning rich features for image manipulation detection. In: Proceedings of the IEEE Conference on Computer Vision and Pattern Recognition, pp. 1053–1061 (2018)

44. Zhu, X., Qian, Y., Zhao, X., Sun, B., Sun, Y.: A deep learning approach to patch-based image inpainting forensics. Sig. Process. Image Commun. **67**, 90–99 (2018)

Adversarial Learning for Zero-Shot Domain Adaptation

Jinghua Wang(ID) and Jianmin Jiang[(✉)](ID)

Research Institute for Future Media Computing, College of Computer Science
and Software Engineering, and Guangdong Laboratory of Artificial Intelligence
and Digital Economy (SZ), Shenzhen University, Shenzhen, China
{wang.jh,jianmin.jiang}@szu.edu.cn

Abstract. Zero-shot domain adaptation (ZSDA) is a category of
domain adaptation problems where neither data sample nor label is available for parameter learning in the target domain. With the hypothesis
that the shift between a given pair of domains is shared across tasks,
we propose a new method for ZSDA by transferring domain shift from
an *irrelevant task* (*IrT*) to the *task of interest* (*ToI*). Specifically, we
first identify an *IrT*, where dual-domain samples are available, and capture the domain shift with a coupled generative adversarial networks
(CoGAN) in this task. Then, we train a CoGAN for the *ToI* and restrict
it to carry the same domain shift as the CoGAN for *IrT* does. In addition,
we introduce a pair of co-training classifiers to regularize the training
procedure of CoGAN in the *ToI*. The proposed method not only derives
machine learning models for the non-available target-domain data, but
also synthesizes the data themselves. We evaluate the proposed method
on benchmark datasets and achieve the state-of-the-art performances.

Keywords: Transfer learning · Domain adaptation · Zero-shot
learning · Coupled generative adversarial networks

1 Introduction

When a standard machine learning technique learns a model with the training
data and applies the model on the testing data, it implicitly assumes that the
testing data share the distribution with the training data [16,37,38]. However,
this assumption is often violated in applications, as the data in real-world are
often from different domains [32]. For example, the images captured by different cameras follow different distributions due to the variations of resolutions,
illuminations, and capturing views.

Domain adaptation techniques tackle the problem of domain shift by transferring knowledge from the label-rich source domain to the label-scarce target
domain [2,6,15]. They have a wide range of applications, such as person
re-identification [43], semantic segmentation [24], attribute analysis [44], and
medical image analysis [7]. Most domain adaptation techniques assume that
the data in target domain are available at the training time for model learning

© Springer Nature Switzerland AG 2020
A. Vedaldi et al. (Eds.): ECCV 2020, LNCS 12366, pp. 329–344, 2020.
https://doi.org/10.1007/978-3-030-58589-1_20

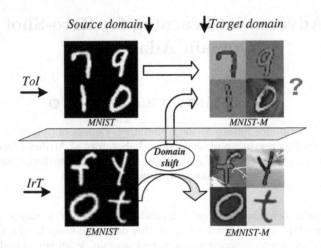

Fig. 1. An intuitive example of ZSDA (best viewed in color). The *ToI* is digit image analysis and the *IrT* is letter image analysis. The source domain consists of gray scale images and the target domain consists of color images. In order to learn the model for the unseen *MNIST-M* (*i.e.,* target-domain data in *ToI*), we first learn the domain shift based on the dual-domain samples in the *IrT*, then transfer it to the *ToI*. (Color figure online)

[6,23,29,42]. However, this is not always the case in the real-world. For example, we may want a artificial intelligence system to provide continuous service with a newly installed camera [19]. This involves a domain adaptation task, where the source domain consists of the images captured by the old camera and the target domain consists of the non-accessible images captured by the new camera. Such a task is referred to as domain generalization [8] or zero-shot domain adaptation (ZSDA) [28,36].

In this paper, we propose a new method to tackle the challenging ZSDA tasks, where only the source-domain data is available in the *Task of Interest (ToI)*. It is recognized that the existence of domain shift does not allow us to learn a model for the target domain based on the source-domain data alone. To solve this problem, we establish a hypothesis that the domain shift, which intrinsically characterizes the difference between the domains, is shared by different tasks. For successful ZSDA, we firstly learn the domain shift from an *irrelevant task (IrT)* where many data in both domains are available, then transfer this domain shift to the *ToI* and learn the model for the target domain.

We illustrate an example of ZSDA in Fig. 1, which learns a model for the color digit images (*i.e., MNIST-M* [6]), given the grayscale digit images (*i.e., MNIST* [25]), the grayscale letter images (*i.e., EMNIST* [3]), and the color letter images (*i.e., EMNIST-M*). In this example, the *ToI* and *IrT* are digit and letter image analysis, respectively. The source domain consists of grayscale images, and the target domain consists of color images. We consider these two tasks to have the same domain shift, which transforms grayscale images to color images.

With the available dual-domain data in the *IrT*, we can train a coupled generative adversarial networks (CoGAN) [21] (*i.e., CoGAN-IrT*) to model the joint distribution of images in these two domains. This *CoGAN-IrT* not only shows the sharing of two domains in high-level concepts, but also implicitly encodes the difference between them, which is typically referred to as domain shift. We consider one source-domain sample and one target-domain to be *paired samples* if they are realizations of the same thing and correspond to each other. Figure 1 shows eight grayscale images and their correspondences in the color domain. The RGB image and depth image of the same scene are also paired samples [39]. Based on the observation that it is the domain shift that introduces the difference between *paired samples*, we define the domain shift to be the distribution of representation difference between *paired samples*.

For successful ZSDA in the *ToI*, we train a *CoGAN-ToI* to capture the joint distribution of dual-domain samples and use it to synthesize the unseen samples in the target domain. Besides the available samples in the source domain, we introduce two supervisory signals for *CoGAN-ToI* training. Firstly, we transfer the domain shift from *IrT* to *ToI* and enforce the *CoGAN-ToI* to encode the same domain shift with *CoGAN-IrT*. In other words, we restrict that the representation difference between paired samples to follow the same distribution in two tasks. To improve the quality of the synthesized target-domain samples, we also take a pair of co-training classifiers to guide the training procedure of *CoGAN-ToI*. The predictions of these two classifiers are trained to be (i) consistent when receiving samples from both *IrT* and *ToI*, and (ii) different as much as possible when receiving samples which are not from these two tasks. In the training procedure, we guide the *CoGAN-ToI* to synthesize such target-domain samples that the classifiers produce consistent predictions when taking them as the input. With domain shift preservation and co-training classifiers consistency as the supervisory signals, our *CoGAN-ToI* can synthesize high quality data for the non-accessible target domain and learn well-performed models.

To summary, we propose a new method for ZSDA by learning across both domains and tasks and our contributions can be highlighted in two folds.

- Firstly, we propose a new strategy for domain adaptation through domain shift transferring across tasks. For the first time, we define the domain shift to be the distribution of representation difference between paired samples in two domains. We learn the domain shift from a *CoGAN-IrT* that captures the joint distribution of dual-domain samples in the *IrT* and design a method for shift transferring to the *ToI*, where only source domain is seen. In addition to domain shift preservation, we also take the consistency of two co-training classifiers as another supervisory signal for *CoGAN-IoT* training to better explore the non-accessible target domain.
- Secondly, our method has a broader range of applications than existing methods [28,36]. While our method is applicable when paired samples in the *IrT* are non-accessible, the work [28] is not. While our method can learn the domain shift from one *IrT* and transfer it to multiple different *ToIs*, the work [36] is only applicable to a given pair of *(IrT, ToI)*.

2   Related Work

While standard machine learning methods involves with a single domain [13,35], domain adaptation uses labeled data samples in one or more source domains to learn a model for the target domain. For transferable knowledge learning, researchers normally minimize the discrepancy between domains by learning domain-invariant features [6,22,33,40]. Ganin and Lempitsky [6] introduced gradient reversal layer to extract features that can confuses the domain classifier. Long et al. [22] introduced residual transfer network to bridge the source domain and the target domain by transferable features and adaptive classifiers. Taking maximum mean discrepancy (MMD) as the measurement between domains, Tzeng et al. [33] introduced an adaptation layer to learn representations which are not only domain invariant but also semantically meaningful. In order to solve the problem of class weight bias, Yan et al. [40] introduced weighted MMD and proposed a classification EM algorithm for unsupervised domain adaptation.

These methods achieve good performances in various computer vision tasks. However, none of them can solve the ZSDA problem as they rely on the target-domain data at the training time. The existing techniques for ZSDA can be summarized into three categories based on their strategies.

The first strategy learns domain-invariant features which not only work in the available source domains but also generalize well to the unseen target domain. Domain-invariant component analysis (DICA) [26] is a kernel-based method that learns a common feature space for different domains while preserving the posterior probability. For cross domain object recognition, multi-task autoencoder (MTAE) [8] extends the standard denoising autoencoder framework by reconstructing the analogs of a given image for all domains. Conditional invariant deep domain generalization (CIDDG) [20] introduces an invariant adversarial network to align the conditional distributions across domains and guarantee the domain-invariance property. With a structured low-rank constraint, deep domain generalization framework (DDG) [5] aligns multiple domain-specific networks to learn sharing knowledge across source domains.

The second strategy assumes that a domain is jointly determined by a sharing latent common factor and a domain specific factor [14,18,41]. This strategy identifies the common factor through decomposition and expects it to generalize well in the unseen target domain. Khosla et al. [14] model each dataset as a biased observation of the visual world and conduct the decomposition via max-margin learning. Li et al. [18] develop a low-rank parameterized CNN model to simultaneously exploit the relationship among domains and learn the domain agnostic classifier. Yang and Hospedales [41] parametrise the domains with continuous values and propose a solution to predict the subspace of the target domain via manifold-valued data regression. Researchers also correlate domains with semantic descriptors [15] or latent domain vectors [17].

The third strategy first learns the correlation between domains from an assistant task, then accomplishes ZSDA based on the available source-domain data and the domain correlation [28,36]. Normally, this strategy relies on an IrT where data from both source and target domain are sufficiently available.

In comparison with the first two strategies, this strategy can work well with a single source domain. Zero-shot deep domain adaptation (ZDDA) [28] aligns representations from source domain and target domain in the IrT and expect the alignment in the ToI. CoCoGAN [36] aligns the representation across tasks in the source domain and takes the alignment as the supervisory signal in the target domain.

3   Background

Generative Adversarial Networks (GAN) consists of two competing models, *i.e.*, the generator and the discriminator [9]. Taking a random vector $z \sim p_z$ as the input, the generator aims to synthesize images $g(z)$ which are resemble to the real image as much as possible. The discriminator tries to distinguish real images from the synthesized ones. It takes an image x as the input and outputs a scalar $f(x)$, which is expected to be large for real images and small for synthesized images. The following objective function formulates the adversarial training procedure of the generator and the discriminator:

$$\max_g \min_f V(f, g) \equiv E_{x \sim p_x}[-\log f(x)] + E_{z \sim p_z}[-\log(1 - f(g(z)))], \quad (1)$$

where E is the empirical estimate of expected value of the probability. In fact, Eq. (1) measures the Jensen-Shannon divergence between the distribution of real images and that of the synthesized images [9].

Coupled Generative Adversarial Networks (CoGAN) consists of a pair of GANs (*i.e.*, GAN$_1$ and GAN$_2$) which are closely related with each other. With each GAN corresponds to a domain, CoGAN captures the joint distribution of images from two different domains [21]. Let $x_i \sim p_{x_i}(i = 1, 2)$ be the images from the ith domain. In GAN$_i(i = 1, 2)$, we denote the generator as g_i and the discriminator as f_i. Based on a sharing random vector z, the generators synthesize image pairs $(g_1(z), g_2(z))$ which not only are indistinguishable from the real ones but also have correspondences. We can formulate the objective function of the CoGAN as follows:

$$\max_{g_1, g_2} \min_{f_1, f_2} V(f_1, f_2, g_1, g_2) \equiv E_{x_1 \sim p_{x_1}}[-\log f_1(x_1)] + E_{z \sim p_z}[-\log(1 - f_1(g_1(z)))]$$
$$+ E_{x_2 \sim p_{x_2}}[-\log f_2(x_2)] + E_{z \sim p_z}[-\log(1 - f_2(g_2(z)))], \quad (2)$$

subject to two constraints: (i) $\theta_{g_1^j} = \theta_{g_2^j}$, $1 \le j \le n_g$; and (ii) $\theta_{f_1^{n_1-k}} = \theta_{f_2^{n_2-k}}$, $0 \le k \le n_{fs} - 1$. The parameter $n_i(i = 1, 2)$ denotes the number of layers in the discriminator f_i. While the first constraint restricts the generators to have n_g sharing bottom layers, the second restricts the discriminators to have n_f sharing top layers. These two constraints force the generators and discriminators to process the high-level concepts in the same way, so that the CoGAN is able to discover the correlation between two domains. CoGAN is effective in dual-domain analysis, as it is capable to learn the joint distribution of data samples (*i.e.*, p_{x_1, x_2}) based on the samples drawn individually from the marginal distributions (*i.e.*, p_{x_1} and p_{x_2}).

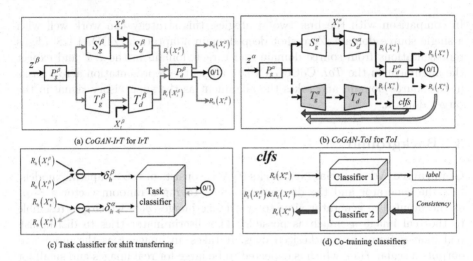

Fig. 2. The network structures of our method. The *CoGAN-IrT* in (a) models the joint distribution of (x_s^β, x_t^β) in the *IrT*. The *CoGAN-ToI* in (b) models the joint distribution of (x_s^α, x_t^α) in the *ToI*. In the discriminators of these two CoGANs, we use $R_l(\cdot)$ to denote the lower level representation produced by the non-sharing layers, and $R_h(\cdot)$ to denote higher level representations produced by the sharing layers, respectively. The task classifier in (c) discriminates $\delta_h^\beta = R_h(x_t^\beta) \ominus R_h(x_s^\beta)$ from $\delta_h^\alpha = R_h(x_t^\alpha) \ominus R_h(x_s^\alpha)$. We maximize the loss of this task classifier to align the domain shift. The co-training classifiers in (d) produce the labels for X_s^α and consistent predictions for X_s^β and X_t^β. To train the *CoGAN-ToI*, we use domain shift preservation to regularize the higher level features and co-training classifiers to regularize the lower level features. The back-propagation directions of these two signals are marked by orange and red, respectively.

4   Approach

4.1   Problem Definition

We define a domain $D = \{X, P(X)\}$ to be the data sample space X and its marginal probability distribution $P(X)$ [27]. Given the data samples X, a task $T = \{Y, P(Y|X)\}$ consists of a label space Y and the conditional probability distribution $P(Y|X)$. This work considers two tasks to be the same as long as they have sharing label space. In *ToI*, the label space is Y^α, the source domain is $D_s^\alpha = \{X_s^\alpha, P(X_s^\alpha)\}$, and the target domain is $D_t^\alpha = \{X_t^\alpha, P(X_t^\alpha)\}$. Then, the *ToI* is denoted by $T^\alpha = \{Y^\alpha, P_s(Y^\alpha|X_s^\alpha)\} \cup \{Y^\alpha, P_t(Y^\alpha|X_t^\alpha)\}$.

Given the labeled data samples in the source domain (*i.e.*, (x_s^α, y_s^α), $x_s^\alpha \in X_s^\alpha$ and $y_s^\alpha \in Y^\alpha$), our ZSDA task aims to derive the conditional probability distribution $P(Y^\alpha|X_t^\alpha)$ in the target domain. In general, the main challenge of this task is induced by non-accessibility of the target domain, as well as the domain shift, *i.e.*, $P(X_s^\alpha) \neq P(X_t^\alpha)$ and $P_s(Y^\alpha|X_s^\alpha) \neq P_t(Y^\alpha|X_t^\alpha)$.

4.2 Main Idea

In order to accomplish the ZSDA task, we identify an irrelevant task (IrT) that satisfies two constraints: (i) the IrT involves the same pair of domains with the ToI; and (ii) the dual-domain samples in the IrT are available. Under the hypothesis that the shift between a given pair of domains maintains across tasks, we propose to learn the domain shift from IrT and transfer it to ToI.

Let the label space of IrT be Y^β. We denote the IrT as $T^\beta = \{Y^\beta, P_s(Y^\beta|X_s^\beta)\} \cup \{Y^\beta, P_t(Y^\beta|X_t^\beta)\}$, where $D_s^\beta = \{X_s^\beta, P(X_s^\beta)\}$ is the source domain and $D_t^\beta = \{X_t^\beta, P(X_t^\beta)\}$ is the target domain, respectively. Note that, the source-domain samples (X_s^α and X_s^β) are in the same sample space. It is also true in the target domain. In the example of Fig. 1, while the source-domain data X_s^α=$MNIST$ and X_s^β=$EMNIST$ are grayscale images, the target-domain data X_t^α=$MNIST$-M and X_t^β=$EMNIST$-M are color images.

In this work, we define two corresponding samples from source domain and target domain as *paired samples*. In most cases, two *paired samples* are different views of the same object. For example, a grayscale image in $MNIST$ and its corresponding color image in $MNIST$-M are paired samples in Fig. 1. The depth image and RGB image of the same scene are also *paired samples*. While the similarity between *paired samples* is determined by the object itself, their difference is mainly introduced by the domain shift. Our work only assumes the existences of correspondence between dual-domain samples. Nevertheless, the correspondences between them in the IrT are not required.

For correlation analysis between *paired samples*, we train CoGANs to capture the joint distribution of source-domain and target-domain data. As both X_s^β and X_t^β are available, we can easily train $CoGAN$-IrT (Fig. 2 (a)) for the IrT using the standard method [21]. The main difficulty lies in the training of $CoGAN$-ToI (Fig. 2 (b)) for the ToI, as the target-domain data X_t^α is not available. To tackle this problem, we propose two kinds of supervisory information for $CoGAN$-ToI training, which are *domain shift preservation* and *co-training classifiers consistency*.

For easier transferring across tasks, we define domain shift to be the distribution of element-wise difference between *paired samples* in the representation space. We can learn the domain shift from the $CoGAN$-IrT which carries the correlation between two domains by varying the inputting noise z^β. After that, we train $CoGAN$-ToI and enforce the representation difference between *paired samples* of ToI to follow the distribution learned in $CoGAN$-IrT by maximizing the loss of a task classifier. Figure 2 (c) visualizes the task classifier which aims to identify the task label of the representation difference. In this way, the domain shift is transferred from the IrT to ToI.

To better explore the unseen target domain of ToI, we also build two co-training classifiers (Fig. 2 (d)) and use their consistency to guide the training procedure of $CoGAN$-ToI. By enforcing the weights of the classifiers to be different from each other as much as possible, we aim to analyze data samples from distinct views. The classifiers are trained to: (i) predict the labels of X_s^α; (ii) produce consistent predictions when receiving X_s^β and X_t^β; and (iii) produce different predictions when receiving samples not involved with ToI or IrT. Thus, we can use the consistency of these two classifiers to evaluate whether a sample is involved with the two tasks. We guide the training procedure of $CoGAN\text{-}ToI$ to synthesize X_t^α as such that the classifiers also produce consistent predictions when receiving their representations.

4.3   Training

In $CoGAN\text{-}IrT$ (Fig. 2 (a)), the sharing layers are P_g^β and P_d^β, the non-sharing layers are S_g^β, T_g^β, S_d^β and T_d^β. The components of $CoGAN\text{-}ToI$ are denoted in similar way in Fig. 2 (b). For simplicity, we use $R_l(x)$ to denote the lower level representation of sample x produced by the non-sharing layers and $R_h(x)$ to denote the higher level representation produced by sharing layers. Note, the representation extraction procedures $R_l(\cdot)$ and $R_h(\cdot)$ vary with task and domain.

Domain Shift
We train $CoGAN\text{-}IrT$ based on the dual-domain samples (X_s^β and X_t^β) and let it carry the correlation between two domains. The $CoGAN\text{-}IrT$ can synthesize a set of *paired samples* for the IrT. For two *paired samples* ($x_s^\beta \in X_s^\beta, x_t^\beta \in X_t^\beta$), we characterize their shift by the element-wise difference between representations in a sharing layer, *i.e.*, $\delta_h^\beta = R_h(x_t^\beta) \ominus R_h(x_s^\beta)$. We then define the domain shift to be the distribution of δ_h^β, *i.e.*, $p_{\delta_h^\beta}$. Specifically, we can obtain a set of $\{\delta_h^\beta\}$ by feeding $CoGAN\text{-}IrT$ with different values of inputting noise z^β.

Co-training Classifiers
Both of the two co-training classifiers (denoted as clf_1 and clf_2) take the representations (*i.e.*, $R_l(X_s^\beta)$, $R_l(X_t^\beta)$, and $R_l(X_s^\alpha)$) as the input. With $R_l(x)$ as the input, the classifier clf_i ($i = 1, 2$) produces a c-dimensional vector $v_i(x)$, where c denotes the number of categories in X_s^α. We minimize the following loss to train the classifiers:

$$L(clf_1, clf_2) = \lambda_w L_w(w_1, w_2) + \lambda_{acc} L_{cls}(X_s^\alpha) - \lambda_{con} L_{con}(X^\beta) + \lambda_{diff} L_{diff}(\tilde{X}), \quad (3)$$

where L_w measures the similarity between the two classifiers, L_{cls} denotes the loss to classify the labeled source-domain samples of ToI, L_{con} assesses the consistency of the output scores when receiving dual-domain samples of IrT (*i.e.*, X_s^β and X_t^β) as the input, and L_{diff} assess the consistency when receiving samples $\tilde{X}$ which are not related ToI or IrT.

As in standard co-training methods [31], we expect the two classifiers to have diverse parameters so that they can analyze the inputs from different views. In this work, we implement these two classifiers with the same neural network structure and assess their similarity by the cosine distance between the parameters:

$$L_w = w_1^T * w_2 / ||w_1|| * ||w_2||, \quad (4)$$

where w_i is the vectored parameters of clf_i.

With the labeled source-domain data in *ToI*, we can easily formulate a multi-class classification problem and use the soft-max loss to define the second term of Eq. (3) as follows:

$$L_{cls} = - \sum_{x_s^\alpha \in X_s^\alpha} \sum_{i=1}^{2} \sum_{j=1}^{c} v_i^j(x_s^\alpha) l^j(x_s^\alpha), \qquad (5)$$

where $v_i^j(x_s^\alpha)$ is the jth element of the prediction $v_i(x_s^\alpha)$, the binary value $l^j(x_s^\alpha)$ denotes whether x_s^α belongs to the jth class or not. This item regularizes the classifiers to produce semantically meaningful vectors.

Different from X_s^α in *ToI*, the labels for dual-domain data in *IrT* are not available. It is impossible to predict their true labels. To gain supervisory signals from these label-missing data, we restrict the two classifiers to produce consistent predictions. The consistency for a given sample is measured by the dot product of its two predictions. Thus, we define the third term in Eq. (3) as:

$$L_{con} = \sum_{x^\beta \in X_s^\beta \cup X_t^\beta} v_1(x^\beta) \cdot v_2(x^\beta). \qquad (6)$$

The last term L_{diff} regularizes the classifiers to produce different predictions when receiving samples that are not related with the two tasks. It is defined in the same way to L_{con} and the only difference lies in the input $\tilde{X}$. Here, the samples in $\tilde{X}$ have two sources: (i) the samples in public datasets, *e.g.*, imageNet [4]; (ii) the corrupted images by replacing a patch of x_s^β, x_t^β, and x_s^α with random noise.

In principal, we can use the consistency of these two classifiers to assess whether a sample is properly involved with *IrT* or *ToI* in these two domains. Thus, we can guide the training procedure of *CoGAN-ToI* in such a way that the synthesized X_t^α should satisfy $v_1(X_t^\alpha) = v_2(X_t^\alpha)$, and take this as a complementary supervisory signal of domain shift preservation.

CoGAN-ToI

At this stage, we train *CoGAN-ToI* to capture the joint distribution of *paired samples* in the *ToI*. By correlating the two domains, a well-trained *CoGAN-ToI* is able to synthesize the non-available target-domain data. We use three constraints to train *CoGAN-ToI*, including (i) one branch captures the distribution of X_s^α; (ii) the domain shift is shared by the two tasks, *i.e.*, $p_{\delta_h^\beta} = p_{\delta_h^\alpha}$, where $\delta_h^\alpha = R_h(x_t^\alpha) \ominus R_h(x_s^\alpha)$; and (iii) the co-training classifiers have consistent predictions for the synthesized sample x_t^α, *i.e.*, $v_1(x_t^\alpha) = v_2(x_t^\alpha)$.

This work trains the two branches of *CoGAN-ToI* separately, unlike the standard method in [21] that trains them simultaneously. To satisfy the first constraint, we consider the source-domain branch (consisting of P_g^α, S_g^α, S_d^α, and P_d^α) as an independent GAN and train it using the available X_s^α.

Though involving in different tasks, both X_t^β and X_t^α are images from the target domain. Thus, they are composed of the same set of low-level details. In order to mimic the processing method learned in the *IrT*, we initialize the non-sharing components of *CoGAN-ToI* in the target domain as $T_g^\beta \rightarrow T_g^\alpha$ and $T_d^\beta \rightarrow T_d^\alpha$.

After initialization, we use the second and third constraints to train the non-sharing components (T_g^α and T_d^α) for the target domain and fine-tune the sharing components (P_g^α and P_d^α). Specifically, we minimize the following loss function:

$$V(P_g^\alpha, T_g^\alpha, T_d^\alpha, P_d^\alpha) \equiv \lambda_{con}^\alpha \sum_{x_t^\alpha = g_t^\alpha(z^\alpha)} v_1(x_t^\alpha) \cdot v_2(x_t^\alpha) - L_{clf}(\delta_h^\alpha, \delta_h^\beta), \qquad (7)$$

where $g_t^\alpha = P_g^\alpha + T_g^\alpha$ is the generator. While the first term assesses how the two classifiers agree with each other, the second term assesses how δ_h^α is distinguishable from δ_h^β.

With $CoGAN\text{-}ToI$, we train a classifier for the synthesized target-domain data by three steps. Firstly, we train a classifier $\Phi_s(\cdot)$ for the labeled source-domain data. Then, we synthesize a set of paired samples (x_s^α, x_t^α) and use $\Phi_s(\cdot)$ to predict their labels. Finally, we train a classifier $\Phi_t(\cdot)$ for x_t^α with the constraint $\Phi_s(x_s^\alpha) = \Phi_t(x_t^\alpha)$ and evaluate our method using the average accuracy.

5   Experiments

5.1   Adaptation Across Synthetic Domains

We conduct experiments on four gray image datasets, including MNIST (D_M) [25], Fashion-MNIST (D_F) [12], NIST (D_N) [10], and EMNIST (D_E) [3]. Both MNIST [25] and Fashion-MNIST have 70000 images from 10 classes. NIST is imbalance and has more than 40k images from 52 classes. EMNIST has more than 145k images from 26 classes.

These four datasets are in the gray domain ($G\text{--}dom$). We create three more domains for evaluation, i.e., the colored domain ($C\text{--}dom$), the edge domain ($E\text{--}dom$), and the negative domain ($N\text{--}dom$). The $C\text{--}dom$ is created using the method in [6], i.e., combining an image with a random color patch in BSDS500 [1]. We apply canny detector to create $E\text{--}dom$ and the operation of $I_n = 255 - I$ to create $N\text{--}dom$.

Implementation Details
In order to learn transferable domain shift across tasks, the two CoGANs (i.e., $CoGAN\text{-}IrT$ and $CoGAN\text{-}ToI$) have the same network structure. The two branches inside these CoGANs also share the same structure, and both generators and discriminators have seven layers. We transform the output of the last convolutional layer of discriminator into a column vector before feeding it into a single sigmoid function. The last two layers in generators and the first two layers in discriminators are non-sharing layers for low-level feature processing.

The task classifier has four convolutional layers to identify the task label of its input. We vary the input noise z^β of $CoGAN\text{-}IrT$ to extract $R_h(x)$ and thus obtain a set of δ_h^β. The parameters of the task classifier are initialized with zero-centered normal distribution. We adopt the stochastic gradient descent (SGD) method for optimization. The batch size is set to be 128 and the learning rate is set to be 0.0002.

Table 1. The accuracy of different methods with $(source, target) = (G\text{–}dom, C\text{–}dom)$

ToI	D_M			D_F			D_N		D_E	
IrT	D_F	D_N	D_E	D_M	D_N	D_E	D_M	D_F	D_M	D_F
ZDDA	73.2	92.0	94.8	51.6	43.9	65.3	34.3	21.9	71.2	47.0
CoCoGAN	78.1	92.4	**95.6**	56.8	56.7	**66.8**	41.0	44.9	**75.0**	54.8
ZDDA$_{dc}$	69.3	79.6	80.7	50.6	42.4	62.0	29.1	20.2	49.8	46.5
CTCC	68.5	74.9	77.6	42.0	52.9	60.9	37.0	43.6	47.3	45.2
Ours	**81.2**	**93.3**	95.0	**57.4**	**58.7**	62.0	**44.6**	**45.5**	72.4	**58.9**

Table 2. The accuracy of different methods with $(source, target) = (G\text{–}dom, E\text{–}dom)$

ToI	D_M			D_F			D_N		D_E	
IrT	D_F	D_N	D_E	D_M	D_N	D_E	D_M	D_F	D_M	D_F
ZDDA	72.5	91.5	93.2	54.1	54.0	65.8	42.3	28.4	73.6	50.7
CoCoGAN	79.6	**94.9**	95.4	61.5	57.5	71.0	48.0	36.3	77.9	58.6
ZDDA$_{dc}$	66.5	83.3	84.7	49.3	50.4	58.0	42.2	31.6	65.0	41.2
CTCC	65.5	73.9	80.5	44.0	40.8	37.3	40.0	31.4	57.7	48.2
Ours	**81.4**	93.5	**96.3**	**63.2**	**58.7**	**72.4**	**49.9**	**38.6**	**78.2**	**61.1**

The co-training classifiers are implemented as convolutional neural networks with three fully connected layers, with 200, 50, and c, respectively. We use c to denote the number of categories in X_s^α. We set the hyper-parameter as $\lambda_w = 0.01$, $\lambda_{acc} = 1.0$, $\lambda_{con} = 0.5$, and $\lambda_{diff} = 0.5$.

The source-domain branch in $CoGAN\text{-}ToI$ is firstly trained independently using the available data X_s^α. Based on the two supervisory signals, we use back-propagation method to train T_g^α and T_d^α. Simultaneously, the P_g^α and P_d^α are fine-tuned. In our experiment, we train the two branches of $CoGAN\text{-}ToI$ in an iterative manner to obtain the best results.

Results

With the above four datasets, we conduct experiments on ten different pairs of (IrT, ToI). Note, D_N and D_E are the same task, as both of them consist of letter images. We test four pairs of source domain and target domain, including $(G\text{–}dom, C\text{–}dom)$, $(G\text{–}dom, E\text{–}dom)$, $(C\text{–}dom, G\text{–}dom)$, and $(N\text{–}dom, G\text{–}dom)$.

We take two existing methods as the benchmarks, including ZDDA [28] and CoCoGAN [36]. In addition, we adapt ZDDA by introducing a domain classifier in order to learn from non-corresponding samples and denote it as ZDDA$_{dc}$. We also conduct ablation study by creating the baseline $CTCC$, which only uses $\underline{C}$o-$\underline{T}$raining $\underline{C}$lassifiers $\underline{C}$onsistency to train $CoGAN\text{-}ToI$.

As seen in Table 1, 2, 3 and 4, our method achieves the best performance in average. Taking D_E classification as an example, our method outperforms

Table 3. The accuracy of different methods with $(source, target) = (C\text{-}dom, G\text{-}dom)$

ToI	D_M			D_F			D_N		D_E	
IrT	D_F	D_N	D_E	D_M	D_N	D_E	D_M	D_F	D_M	D_F
ZDDA	67.4	85.7	87.6	55.1	49.2	59.5	39.6	23.7	75.5	52.0
CoCoGAN	73.2	89.6	**94.7**	61.1	50.7	70.2	47.5	57.7	80.2	67.4
ZDDA$_{dc}$	61.5	76.7	79.9	51.2	46.1	53.4	31.3	20.4	61.2	42.2
CTCC	62.1	76.9	68.6	47.2	45.6	57.6	27.5	33.6	58.0	49.9
Ours	**73.7**	**91.0**	93.4	**62.4**	**53.5**	**71.5**	**50.6**	**58.1**	**83.5**	**70.9**

Table 4. The accuracy of different methods with $(source, target) = (G\text{-}dom, C\text{-}dom)$

ToI	D_M			D_F			D_N		D_E	
IrT	D_F	D_N	D_E	D_M	D_N	D_E	D_M	D_F	D_M	D_F
ZDDA	78.5	90.7	87.6	56.6	57.1	67.1	34.1	39.5	67.7	45.5
CoCoGAN	80.1	92.8	93.6	63.4	61.0	72.8	47.0	43.9	78.8	58.4
ZDDA$_{dc}$	68.4	79.8	82.5	48.1	46.2	64.6	28.6	34.4	61.8	36.2
CTCC	68.4	80.0	80.2	50.1	55.1	61.3	37.6	33.9	56.1	33.9
Ours	**82.6**	**94.6**	**95.8**	**67.0**	**68.2**	**77.9**	**51.1**	**44.2**	**79.7**	**62.2**

ZDDA [28] by a margin of 8.9%, and outperforms CoCoGAN [36] by a margin of 4.1% when IrT is D_F in Table 1. In average, our method performs 7.38% better than ZDDA and 0.69% better than CoCoGAN in Table 1.

In each of the Table 1, 2, 3 and 4, the proposed method improves CTCC more than 10% in average. This means that the domain shift transferring are useful in the training procedure of *CoGAN-ToI*. Among the three tasks, the digit image classification is the easiest one. Out of all settings, the most successful one transfers knowledge from the $G\text{-}dom$ to the $E\text{-}dom$ with D_E as the IrT and D_M as the ToI. In this case, our method achieves the accuracy of 96.3%. For D_M classification in $G\text{-}dom$, our method achieve the accuracy of 95.8% with D_E as IrT and $N\text{-}dom$ as the source domain, outperforming other techniques (including 89.5% in [11], and 94.2% in [30]) which rely on the availability of the target-domain data in the training stage. With *CoGAN-ToI*, we not only derive models for the unseen target domain, but also synthesize data themselves. Figure 3 visualizes the generated images in $C\text{-}dom$ and $E\text{-}dom$ with $G\text{-}dom$ as the source domain.

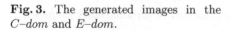

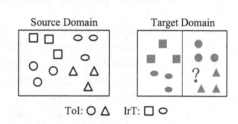

ToI: ○ △   IrT: □ ○

Fig. 3. The generated images in the C–dom and E–dom.

Fig. 4. An example. *ToI* represents a subset of the categories (*square* and *triangle*) and *IrT* represents the rest (*circle* and *ellipse*).

Table 5. The accuracy of different methods on Office-Home

Source	Ar			Cl			Pr			Rw		
Target	Cl	Pr	Rw	Ar	Pr	Rw	Ar	Cl	Rw	Ar	Cl	Pr
ZDDA$_{dc}$	53.2	61.4	68.8	67.4	57.0	68.4	60.9	40.6	62.4	68.1	43.4	50.3
CoCoGAN	62.2	69.5	74.5	66.7	74.0	66.4	57.6	53.4	71.7	69.2	51.3	65.8
CTCC	55.7	61.5	66.5	66.8	64.6	65.2	56.3	46.6	61.6	64.3	43.7	57.7
Ours	**62.7**	**71.9**	**76.3**	**72.6**	**75.1**	**73.9**	**70.3**	**60.8**	**74.8**	**72.2**	**61.4**	**72.2**

5.2 Adaptation in Public Dataset

We also evaluate our method on Office-Home [34], which has four different domains, *i.e.*, Art (**Ar**), Clipart (**Cl**), Product (**Pr**), and Real-world (**Rw**). It has more than 15k images from 65 categories.

As it is difficult to identify an analogous set for this dataset, we evaluate our method on adaptation across subsets. Give a pair of domains, we take a subset of the categories as the *ToI* and the rest as the *IrT*. An example is shown Fig. 4, where the *ToI* represents the classification of two categories (*square* and *triangle*), and *IrT* represents the classification of other two categories (*circle* and *ellipse*). Here, we set the parameters as $\lambda_w = 0.01$, $\lambda_{acc} = 1$, and $\lambda_{con} = \lambda_{diff} = 0.1$.

Let N_α denote the number of categories of *ToI*. We fix the value of N_α to be 10 and conduct experiments on all of the 12 possible different pairs of source domain and target domain. As seen in Table 5, our method achieves the best performances in all cases. This indicates that our method is applicable to a broad range of applications. Our method can beats both ZDDA and CoCoGAN by a margin larger than 10%, when source domain is *Rw* and target domain is *Cl*.

We use the parameter $\lambda_{con}^\alpha = 0.01$ to balance the two terms in Eq. (7). Generally, the CTCC mainly regularizes the training of T_s^α, which processes the low-level details. The detail-richer X_t^β means more knowledge are available for training, and the more transferable across tasks the T_s^α is. Thus, we set smaller

Table 6. The variation of accuracy against parameter λ_{con}^{α}

λ_{con}^{α}	ar→ cl	ar→ pr	ar→ rw	cl→ ar	cl→ pr	cl→ rw
0.001	59.3	68.5	73.3	65.7	68.3	69.3
0.005	61.6	70.3	75.7	70.6	74.6	71.1
0.01	62.7	71.9	76.3	72.6	75.1	73.9
0.02	62.1	71.0	74.7	72.1	76.1	72.8
0.1	53.0	64.8	66.1	60.4	60.4	63.5

Table 7. The variation of accuracy against number of supplementary samples

Num	ar→ cl	ar→ pr	ar→ rw	cl→ ar	cl→ pr	cl→ rw
0.8N	60.3	67.5	73.4	68.8	67.0	70.7
N	61.3	70.7	73.6	70.3	73.2	71.0
1.6N	62.5	71.5	76.0	71.5	74.3	73.5
2N	62.7	71.9	76.3	72.6	75.1	73.9
4N	62.7	71.9	76.3	72.7	75.1	73.9

value for λ_{con}^{α} when richer details are included in X_t^{β}. Table 6 lists the accuracy of our method on Office-Home with different parameter values of λ_{con}^{α}. As seen, our method performs well when $\lambda_{con}^{\alpha} \in [0.005, 0.01]$.

Let N_s be the number of samples in the $X = \{X_s^{\alpha}, X_s^{\beta}, X_t^{\beta}\}$. We use $2N_s$ supplementary samples to train the L_{diff} where (i) half are randomly cropped from the ImageNet and (ii) half are obtained by replacing patches of training samples with random noises. Table 7 lists the performance of our method with different number of supplementary samples. As seen, $2N_s$ supplementary samples are enough for model training.

6   Conclusion and Future Work

This paper proposes a new method for ZSDA based on the hypothesis that different tasks may share the domain shift for the given two domains. We learn the domain shift from one task and transfer it to the other by bridging two CoGANs with a task classifier. Our method takes the domain shift as the distribution of the representation difference between paired samples and transfers it across CoGANs. Our method is capable of not only learning the machine learning models for the unseen target domain, but also generate target-domain data samples. Experimental results on six datasets show the effectiveness of our method in transferring knowledge among images in different domains and tasks.

The proposed method learns the shift between domains and transfers it across tasks. This strategy makes our method to be applicable only when "large" shift exists across domains, such as (rgb, gray), (clipart, art) etc. Thus, our method cannot perform well on the datasets where the domain shift is "small", such as VLSC and Office-31. In the future, we will train a classifier to determine whether correspondence exists between a source-domain sample and a synthesized target-domain sample. Such a classifier can guide the training procedure of CoGAN, even when only samples from a single domain is available.

Acknowledgment. The authors wish to acknowledge the financial support from: (i) Natural Science Foundation China (NSFC) under the Grant no. 61620106008; (ii) Natural Science Foundation China (NSFC) under the Grant no. 61802266.

References

1. Arbelaez, P., Maire, M., Fowlkes, C., Malik, J.: Contour detection and hierarchical image segmentation. IEEE Trans. Pattern Anal. Mach. Intell. **33**(5), 898–916 (2011)
2. Chen, Y., Lin, Y., Yang, M., Huang, J.: Crdoco: pixel-level domain transfer with cross-domain consistency. In: CVPR, pp. 1791–1800 (2019)
3. Cohen, G., Afshar, S., Tapson, J., van Schaik, A.: EMNIST: an extension of MNIST to handwritten letters. arXiv (2017)
4. Deng, J., Dong, W., Socher, R., Li, L., Li, K., Li, F.: Imagenet: a large-scale hierarchical image database. In: CVPR, pp. 248–255 (2009)
5. Ding, Z., Fu, Y.: Deep domain generalization with structured low-rank constraint. IEEE Trans. Image Process. **27**(1), 304–313 (2018)
6. Ganin, Y., Lempitsky, V.: Unsupervised domain adaptation by backpropagation. In: ICML, vol. 37, pp. 1180–1189 (2015)
7. Ghassami, A., Kiyavash, N., Huang, B., Zhang, K.: Multi-domain causal structure learning in linear systems. In: NeurIPS, pp. 6269–6279 (2018)
8. Ghifary, M., Kleijn, W.B., Zhang, M., Balduzzi, D.: Domain generalization for object recognition with multi-task autoencoders. In: ICCV (2015)
9. Goodfellow, I.J., et al.: Generative adversarial nets. In: NIPS, pp. 2672–2680 (2014)
10. Grother, P., Hanaoka, K.: NIST special database 19 handprinted forms and characters database. In: National Institute of Standards and Technology (2016)
11. Haeusser, P., Frerix, T., Mordvintsev, A., Cremers, D.: Associative domain adaptation. In: ICCV, pp. 2784–2792 (2017)
12. Han, X., Kashif, R., Roland, V.: Fashion-MNIST: a novel image dataset for benchmarking machine learning algorithms. CoRR abs/1708.07747 (2017)
13. He, K., Zhang, X., Ren, S., Sun, J.: Deep residual learning for image recognition. In: CVPR, pp. 770–778 (2016)
14. Khosla, A., Zhou, T., Malisiewicz, T., Efros, A.A., Torralba, A.: Undoing the damage of dataset bias. In: ECCV (2012)
15. Kodirov, E., Xiang, T., Fu, Z., Gong, S.: Unsupervised domain adaptation for zero-shot learning. In: ICCV (2015)
16. Krizhevsky, A., Sutskever, I., Hinton, G.E.: Imagenet classification with deep convolutional neural networks. In: NIPS, pp. 1106–1114 (2012)
17. Kumagai, A., Iwata, T.: Zero-shot domain adaptation without domain semantic descriptors. CoRR abs/1807.02927 (2018)
18. Li, D., Yang, Y., Song, Y.Z., Hospedales, T.M.: Deeper, broader and artier domain generalization. In: ICCV (2017)
19. Li, D., Yang, Y., Song, Y.Z., Hospedales, T.M.: Learning to generalize: Meta-learning for domain generalization. In: AAAI (2018)

20. Li, Y., et al.: Deep domain generalization via conditional invariant adversarial networks. In: ECCV (2018)
21. Liu, M.Y., Tuzel, O.: Coupled generative adversarial networks. In: NIPS (2016)
22. Long, M., Zhu, H., Wang, J., Jordan, M.I.: Unsupervised domain adaptation with residual transfer networks. In: NIPS, pp. 136–144 (2016)
23. Lopez-Paz, D., Hernández-Lobato, J., Schölkopf, B.: Semi-supervised domain adaptation with non-parametric copulas. In: NIPS (2012)
24. Luo, Y., Zheng, L., Guan, T., Yu, J., Yang, Y.: Taking a closer look at domain shift: Category-level adversaries for semantics consistent domain adaptation. In: CVPR (2019)
25. Lécun, Y., Bottou, L., Bengio, Y., Haffner, P.: Gradient-based learning applied to document recognition. In: Proceedings of the IEEE (1998)
26. Muandet, K., Balduzzi, D., Schölkopf, B.: Domain generalization via invariant feature representation. In: ICML (2013)
27. Pan, S.J., Yang, Q.: A survey on transfer learning. IEEE TKDE 22(10), 1345–1359 (2010)
28. Peng, K.C., Wu, Z., Ernst, J.: Zero-shot deep domain adaptation. In: ECCV (2018)
29. Pinheiro, P.O.: Unsupervised domain adaptation with similarity learning. In: CVPR (2018)
30. Saito, K., Ushiku, Y., Harada, T.: Asymmetric tri-training for unsupervised domain adaptation. In: ICML, pp. 2988–2997 (2017)
31. Saito, K., Watanabe, K., Ushiku, Y., Harada, T.: Maximum classifier discrepancy for unsupervised domain adaptation. In: CVPR, pp. 3723–3732 (2018)
32. Torralba, A., Efros, A.A.: Unbiased look at dataset bias. In: CVPR (2011)
33. Tzeng, E., Hoffman, J., Zhang, N., Saenko, K., Darrell, T.: Deep domain confusion: maximizing for domain invariance. Computer Science (2014)
34. Venkateswara, H., Eusebio, J., Chakraborty, S., Panchanathan, S.: Deep hashing network for unsupervised domain adaptation. In: CVPR, pp. 5385–5394 (2017)
35. Wang, J., Jiang, J.: An unsupervised deep learning framework via integrated optimization of representation learning and GMM-based modeling. In: ACCV, vol. 11361, pp. 249–265 (2018)
36. Wang, J., Jiang, J.: Conditional coupled generative adversarial networks for zero-shot domain adaptation. In: ICCV (2019)
37. Wang, J., Jiang, J.: SA-net: a deep spectral analysis network for image clustering. Neurocomputing 383, 10–23 (2020)
38. Wang, J., Wang, G.: Hierarchical spatial sum-product networks for action recognition in still images. IEEE Trans. Circuits Syst. Video Techn. 28(1), 90–100 (2018)
39. Wang, J., Wang, Z., Tao, D., See, S., Wang, G.: Learning common and specific features for RGB-D semantic segmentation with deconvolutional networks. In: ECCV, pp. 664–679 (2016)
40. Yan, H., Ding, Y., Li, P., Wang, Q., Xu, Y., Zuo, W.: Mind the class weight bias: weighted maximum mean discrepancy for unsupervised domain adaptation. In: IEEE, pp. 945–954 (2017)
41. Yang, Y., Hospedales, T.: Zero-shot domain adaptation via kernel regression on the grassmannian (2015). https://doi.org/10.5244/C.29.DIFFCV.1
42. Yao, T., Pan, Y., Ngo, C.W., Li, H., Tao, M.: Semi-supervised domain adaptation with subspace learning for visual recognition. In: CVPR (2015)
43. Zhong, Z., Zheng, L., Luo, Z., Li, S., Yang, Y.: Invariance matters: Exemplar memory for domain adaptive person re-identification. In: CVPR (2019)
44. Zhu, P., Wang, H., Saligrama, V.: Learning classifiers for target domain with limited or no labels. In: ICML, pp. 7643–7653 (2019)

YOLO in the Dark - Domain Adaptation Method for Merging Multiple Models

Yukihiro Sasagawa[1]([✉])(iD) and Hajime Nagahara[2](iD)

[1] Socionext Inc., Kyoto, Japan
sasagawa.yukihiro@socionext.com
[2] Osaka University, Osaka, Japan
nagahara@ids.osaka-u.ac.jp

Abstract. Generating models to handle new visual tasks requires additional datasets, which take considerable effort to create. We propose a method of domain adaptation for merging multiple models with less effort than creating an additional dataset. This method merges pre-trained models in different domains using glue layers and a generative model, which feeds latent features to the glue layers to train them without an additional dataset. We also propose a generative model that is created by distilling knowledge from pre-trained models. This enables the dataset to be reused to create latent features for training the glue layers. We apply this method to object detection in a low-light situation. The YOLO-in-the-Dark model comprises two models, Learning-to-See-in-the-Dark model and YOLO. We present the proposed method and report the result of domain adaptation to detect objects from RAW short-exposure low-light images. The YOLO-in-the-Dark model uses fewer computing resources than the naive approach.

Keywords: Knowledge distillation · Domain adaptation · Object detection

1 Introduction

Performing visual tasks in a low-light situation is a difficult problem. Short-exposure images to not have enough features for visual processing, and the brightness enhancement of the image causes noise that affects visual tasks. In contrast, long-exposure images also contain noise that affects visual tasks owing to motion blur.

In previous work, image processing that handles extreme low-light photography has been developed using an additional dataset (the See-in-the-Dark dataset) [2]. This dataset contains RAW images captured under various exposure conditions. This approach is a straightforward way to create models to perform visual tasks in low-light conditions. However, creating a new dataset requires considerable effort. In particular, end-to-end training for models to perform visual tasks requires many images with annotation.

© Springer Nature Switzerland AG 2020
A. Vedaldi et al. (Eds.): ECCV 2020, LNCS 12366, pp. 345–359, 2020.
https://doi.org/10.1007/978-3-030-58589-1_21

Knowledge distillation is an excellent way to reuse models trained for other visual tasks [5]. We propose a new method to generate models for performing new visual tasks without the need for an additional dataset. Similar approaches known as unsupervised domain adaptation [1,15] have been proposed. Those methods are effective for changing the domain of a single model (e.g., a classifier). In contrast, our research focuses on merging other models trained on different domains.

We apply the proposed method to achieve object detection in a low-light situation. The well-known object detection model YOLO (You Only Look Once) [10,11] uses public datasets PASCAL VOC and COCO [4,7], and Fig. 1 shows its detection results for low-light images based on the See-in-the-Dark (SID) dataset. Figure 1(a) is the result of a long-exposure (10 s) image, which is sufficient to obtain good results. In contrast, Fig. 1(b) is the result of a short-exposure (100 ms) image that has been brightness enhanced so it the same as that of the long-exposure image. The results of the short-exposure image are degraded.

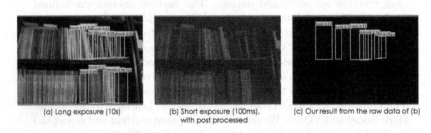

(a) Long exposure (10s)   (b) Short exposure (100ms), with post processed   (c) Our result from the raw data of (b)

Fig. 1. Object detection in low-light situations. (a) Detection result of a long-exposure (10 s) image. (b) Detection result of a short-exposure (100 ms) image. (c) Our detection result of the RAW data of (b).

We use the Learning-to-See-in-the-Dark (SID) model to handle low-light photography [2]. It improves object detection in the case of short-exposure images. Figure 1(c) shows the result of our method; it can detect objects from the RAW data of the image in Fig. 1(b). Usually, such object detection requires end-to-end training using a dataset containing RAW images with annotations. In contrast, our proposal can import the knowledge of pre-trained models and apply them to models for new visual tasks easily to improve performance. In the remainder of this paper, we present the results of object detection using the proposed method, and we analyze critical elements of the method and discuss directions for future research.

2   Related Work

The lack of datasets in low-light situations has been discussed by researchers. In this section, we review previous research on low-light datasets. Our proposal uses the technique of knowledge distillation, and various methods related to knowledge distillation have been extensively studied in the literature. Hence, we also provide a short review of proposed approaches to knowledge distillation.

Dark Image Dataset. After [2] was presented, [8] discussed datasets for low-light situations. The authors created the Exclusively Dark (ExDark) dataset for research on low-light visual tasks. They found that noise is a notable component in low-light images that affects training. Their results also indicate that denoising improves the edge features of objects, but increases artifacts in the image. The authors also compared features learned by Resnet-50 in bright and low-light images. They initially believed that training data normalization and the progression of data through the layers of a Convolutional Neural Network (CNN) toward high-level abstraction should normalize the data and cause the brightnesses of high- and low-light data to be disregarded because brightness is not a crucial feature for the classification of objects. However, the results of the evaluation indicate that the high- and low-light features are different in the t-SNE embedding space. This result indicates that a model for low-light situations should be trained by an appropriate dataset.

The ExDark low-light dataset is much smaller than the COCO dataset and is too small to create training data for visual tasks in general. However, we refer to this research to compare it with the approach of artificially making each image in the COCO data dataset darker.

Inverse Mapping. Inverse mapping is a method for image-to-image translation and its aim is to find a mapping between a source ($\mathcal{A}$) and target ($\mathcal{B}$). AEGAN [9] is a well-known approach to obtaining an inverse generator using an AutoEncoder based on generative adversarial nets. Like AutoEncoder, AEGAN contains an inverse generator (IG) and generator (G) and generates image x' from original image x in image domain $\mathcal{A}$ via latent space vector z'. The IG compresses a generated image x into a latent space vector z' and G reconstructs the z' into a new image x'. AEGAN minimizes the difference between generated image x and reconstructed image x'. This structure produces latent space vector z' to represent image-to-image translation.

Invertible AutoEncoder (InvAuto) [14] is another method for image-to-image translation. The translators $F_{\mathcal{AB}}$ ($\mathcal{A}$ to $\mathcal{B}$) and $F_{\mathcal{BA}}$ ($\mathcal{B}$ to $\mathcal{A}$) share InvAuto as part of the encoder (E) and decoder (D). This method is used to convert between the features corresponding to two different image domains ($\mathcal{A}$ and $\mathcal{B}$). Encoder E realizes an inversion of decoder D (and vice versa) and shares parameters with D. This introduces a strong correlation between the two translators.

We use a concept similar to that of AEGAN to create a generative model, as described in Sect. 3.2.

Hints for Knowledge Distillation. The use of hint information in teacher networks is a well-known approach in knowledge distillation. FitNet [12] is a popular method that uses hint information for model compression. This method chooses hidden layers in the teacher network as hint layers and chooses guided layers in the student network corresponding to the hidden layers. The parameters of the guided layers are optimized by a loss function (e.g., L2 loss) that measures the difference between the outputs of the hint layers and guided layers.

The optimization method for object detection proposed in [3] also uses hint information; this method chooses hint and guided layers and defines the L2 loss as an optimization target. It also refers to the prediction result (the classification and bounding boxes of objects) of a teacher network as a soft target.

We use guided layers during the training, as described in Sect. 3.3.

3 Proposed Model: YOLO in the Dark

3.1 Overview

Figure 2 shows an overview of our method, which merges two models trained in different domains ($\mathcal{A}$ and $\mathcal{B}$). The model for domain $\mathcal{A}$ predicts data Ya from data X. The other model for domain $\mathcal{B}$ predicts data Z from data Yb. Data Ya and Yb are assumed to be the same data type. For example, model $\mathcal{A}$ predicts an RGB image from a RAW image, and the model $\mathcal{B}$ predict an object class and location from the RGB image. After training both models $\mathcal{A}$ and $\mathcal{B}$, this method extracts model fragments with the boundaries of the latent features A and B. The new model is composed of fragments of models $\mathcal{A}$ and $\mathcal{B}$, which are combined through a glue layer.

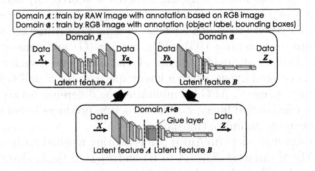

Fig. 2. Our method of domain adaptation, which merges two models trained in different domains $\mathcal{A}$ and $\mathcal{B}$.

This glue layer helps transform latent feature A to latent feature B in each model fragment. The SID model performs well on low-light images [2], so we use the SID model for model $\mathcal{A}$. We also use the object detection model YOLO [10,11] for model $\mathcal{B}$.

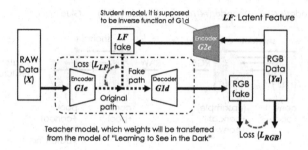

Fig. 3. Scheme for knowledge distillation. The red part $G2e$ is the student model. (Color figure online)

3.2   Generative Model for Domain Adaptation

Training the glue layer requires additional data for domain $\mathcal{A} + \mathcal{B}$; however, creating a dataset requires considerable effort. Our method defines a generative model for training the glue layer using knowledge distillation.

Figure 3 illustrates the scheme of knowledge distillation for the generative model. The generative model outputs the latent features of A from data Ya, as described in Fig. 2. The SID model is an encoder–decoder structure, so the generative model is the inverse function of the decoder. In Fig. 3, encoder–decoder $G1e$-$G1d$ is a teacher model and train student model $G2e$ by feeding RAW and RGB image pair from the SID dataset and SID model $G1e$-$G1d$.

This training uses the loss between RGB data and "fake" RGB data reconstructed via the $G2e$-$G1d$ combination, expressed as follows:

$$L_{RGB} = \|RGB_{data} - RGB_{fake}\|_1 \tag{1}$$

The training also uses another loss between the latent features LF from $G2e$ and latent features from $G1e$, as follows:

$$L_{LF} = \sum_i \|LF^i{}_{G2e} - LF^i{}_{G1e}\|_1, \tag{2}$$

where i is the index number of the layers starting from the outputs of each encoder ($G2e$ or $G1e$).

These two loss functions help define $G2e$ as the inverse function of $G1d$, and the total loss function is as follows:

$$L_{total} = L_{RGB} + L_{LF}. \tag{3}$$

Figure 4 shows the structure of the glue layer for the latent features of the SID encoder. Figure 4 (a) is the SID network structure. As explained in [2], the SID network is based on U-net [13], which is composed of an encoder and a decoder. The encoder extract features using convolution and pooling layers. The pooling layer behaves as a low-pass filter for spatial frequency so that the

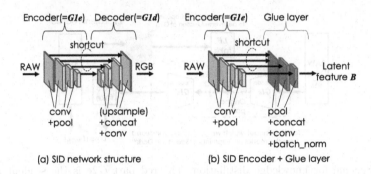

Fig. 4. Structure of the glue layer. (a) SID network structure. (b) Glue layer structure after the SID encoder.

features contain different frequency information as a result of each pooling layer. The SID Encoder has four levels of features corresponding to the pooling scales 1/1, 1/2, 1/4, and 1/8. Each layer in the encoder feeds the latent features to the corresponding layers in the decoder. The glue layer must be sufficiently expressive with respect to frequency information for the subsequent network (i.e., the object detection network).

Figure 5 shows the reconstructed RGB images using the latent features of the SID encoder. Figure 5 (a) illustrates images reconstructed using all features. The images have a Peak Signal-to-Noise Ratio (PSNR) of 31.81 and Structural Similarity (SSIM) of 0.752 with respect to the original images. Figures 5 (b), (c), and (d) are images reconstructed using fewer features, which removes high spatial frequency information. The quality of these images is worse than that of Fig. 5 (a). To detect objects, the detailed shape of the object must be identified, so we decided to use all latent features for the glue layer.

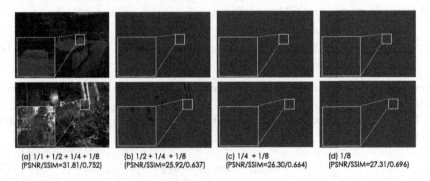

Fig. 5. Reconstruction using latent features, with the quality metrics (PSNR/SSIM) to original images. Reconstruct results using (a) 1/8-, 1/4-, 1/2-, and 1/1-scale features, (b) 1/8-, 1/4-, and 1/2-scale features, (c) 1/8- and 1/4-scale features, and (d) 1/8-scale features.

Figure 4 (b) shows the glue layer structure following the SID encoder. The glue layer is composed of pooling, concatenation, convolution, and batch normalization. The pooling and concatenate functions help gather latent features. The convolution and batch normalization functions help convert a new latent feature for domain $\mathcal{B}$.

The knowledge in the generative model is distilled, as described in Fig. 3, according to the structure of the glue layer. Figure 6 shows the RGB images reconstructed using the distilled knowledge. The RGB images from the $G2e$-$G1d$ combination in Fig. 6(b) and the RGB images generated by the SID model in Fig. 6(a) seem similar. The SID-model generated RGB images have a PSNR of 30.18 and SSIM of 0.917 with respect to the "fake" RGB images. These values indicate good image similarity, so we conclude that the behavior of $G2e$ is sufficiently similar to that of the inverse function of $G1d$.

We also fine-tune the $G2e$ to improve the transformation of latent feature A to latent feature B in each model fragment. We focused on the result of the classifier network in the YOLO model to optimize the generative model. According to [10], the YOLO model contains a classifier network at the beginning of the network itself. This classifier network learns the feature map for processing in the succeeding detection network so it should effectively optimize the generative model corresponding to the SID dataset. We use cosine similarity as the finetuning loss L_{G2e-FT} between the results of the classifier network via $G2e$ ($LF_{G2e-cls}$) and the original YOLO ($LF_{YOLO-cls}$) as follows:

$$L_{G2e-FT} = cos(\overrightarrow{LF_{G2e-cls}}, \overrightarrow{LF_{YOLO-cls}}), \tag{4}$$

where $\overrightarrow{LF_{G2e-cls}}$ and $\overrightarrow{LF_{YOLO-cls}}$ are vectors reshaped from the feature tensors of the classifier networks. In addition, $LF_{G2e-cls}$ and $LF_{YOLO-cls}$ are described in Fig. 7.

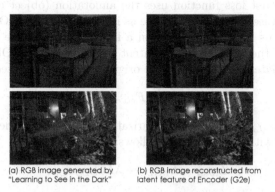

(a) RGB image generated by "Learning to See in the Dark"

(b) RGB image reconstructed from latent feature of Encoder (G2e)

Fig. 6. Reconstructed RGB images. (a) Image generated by the SID model, which is equivalent to $G1e$-$G1d$. (b) Image generated by $G2e$-$G1d$.

3.3   Training Environment

Figure 7 shows the training environment for the proposed YOLO-in-the-Dark model $\mathcal{A} + \mathcal{B}$. Figure 7(a) shows a complete view of the environment, where the dotted boundary shows the parts used for training the new model. The glue layer is the target of training, which uses the RGB data via the encoder $G2e$ generated by knowledge distillation. The training environment uses the original YOLO model, which uses the same RGB data as $G2e$. We use the COCO dataset [7] for training.

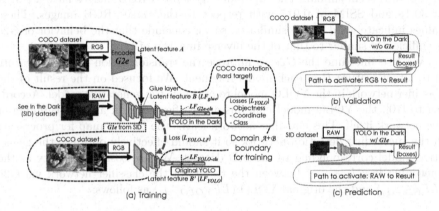

(a) Training   (c) Prediction

Fig. 7. Training environment of the YOLO-in-the-Dark model. (a) Complete view of the environment. (b) Validation behavior using the RGB data for inference. (c) Prediction behavior using the RAW data for inference.

During the training phase, the glue layer is optimized according to the loss functions. The first loss function uses the annotation (object coordinates and class) of the dataset, which is the same as in the original YOLO scheme [11].

The other loss function is based on a latent feature in the original YOLO model, which is the L2 loss between latent feature B in the YOLO-in-the-Dark model and the latent feature B' in the original YOLO, expressed as follows:

$$L_{YOLO-LF} = \|LF_{glue} - LF_{YOLO}\|_2. \tag{5}$$

Loss $L_{YOLO-LF}$ works as a regularization term so that the total loss uses the second loss with a coefficient as follows:

$$L_{total} = L_{YOLO} + \lambda L_{YOLO-LF}, \tag{6}$$

where L_{YOLO} is the same as the loss function used in the original YOLO scheme [11].

Figure 7(b) shows the dataflow during validation. The validation uses the same path as training, which uses the RGB data and evaluates enough samples from the dataset to confirm the glue layer is behaving correctly. Figure 7(c) shows the dataflow during prediction. The prediction uses the other path, using the RAW data via the encoder $G1e$ transferred from the SID model. This stage is for evaluating the proposed YOLO-in-the-Dark model, which will improve object detection in short-exposure RAW images.

4   Experiments

4.1   Object Detection in RAW Images

Figure 8 shows the object detection results for the SID dataset. Figure 8 (a) is the result obtained by the original YOLO model, which used a brightness enhanced RGB image. The brightness enhancement of the RGB image makes it easier for the original YOLO model to detect objects. The original YOLO model detects the objects in image a1 well. However, this model cannot detect the objects in image a2. This is because the brightness enhancement adds noise and affects the inference. Our proposed YOLO-in-the-Dark model detects the objects in the RAW images directly. The detection results are shown in images b1 and b2. Images c1 and c2 are the baseline detection results in which the original YOLO model uses the SID ground truth (long-exposure) image. In image b1, the proposed model performs as well as the original YOLO model (image a1). In addition, the proposed model can detect objects in image b2.

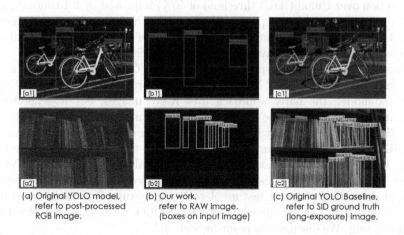

(a) Original YOLO model, refer to post-processed RGB image.

(b) Our work, refer to RAW image. (boxes on input image)

(c) Original YOLO Baseline, refer to SID ground truth (long-exposure) image.

Fig. 8. Detection results. (a) Results of the original YOLO model on a brightness enhanced RGB image. (b) Results of our proposed model on a RAW image. (c) Baseline detection results of the original YOLO model on a long-exposure image.

The SID dataset [2]contains both indoor and outdoor images, and the illuminance at the camera in the indoor scenes is between 0.03 lux and 0.3 lux. Hence, the results for image b2 indicate that the YOLO-in-the-Dark model can handle scenes illuminated by less than 1 lux.

As explained in Sect. 1, the YOLO-in-the-Dark model uses the encoder in SID at the front end to process RAW images. The new model creates a latent feature using the SID encoder, which processes the low-light image and outputs the results to the back end of the YOLO model via the glue layer. This improves object detection performance in low-light situations.

We also evaluated other images in the SID dataset. Figure 9 shows the prediction results (F-measure) using the SID dataset categorized by the size of the bounding boxes. We created annotation data for this evaluation by detecting objects from reference images using the original YOLO model. The reference images were captured under a long exposure so that the original YOLO could detect the objects easily. We used these detection results as the ground truth for this evaluation.

Figure 9 shows the results for the original YOLO model, which detected objects in brightness-enhanced RGB images. Figure 9 also shows the YOLO model trained by the dark COCO dataset, in which the brightness was scaled to emulate low-light situations. This trained YOLO performs worse than the original YOLO model. Previous work [8] has found that noise is a notable component in low-light images, and it hence affects training. The study [8] also found that denoising improves the edge features of the objects but increases the artifacts. SID [2] should hence be a good solution to this problem. In addition, Fig. 9 shows the results for the YOLO-in-the-Dark model. The F-measure is better for all sizes of bounding box than those of the first two models. The mAP (at an Intersection over Union (IOU) threshold of 50%) improved by 2.1 times ($0.26 \rightarrow 0.55$) compared with the original YOLO model. The latent features defined in Sect. 3.2 seem to be sufficient to detect any size of object, and the distillation of knowledge by the generative model has been successful.

Finally, Fig. 9 shows the results for the SID+YOLO model, which is a simple combination of SID and YOLO (the naive approach). Like the YOLO-in-the-Dark model, the naive approach uses RAW images, but it generates an RGB image using SID. The YOLO-in-the-Dark model reuses parameters of both SID and YOLO so that it ideally should achieve the same performance as the SID+YOLO model. Figure 9 instead indicates that the YOLO-in-the-Dark model has a performance that is close but not equal to that of the SID+YOLO model. We finetuned both the glue layer and the generative model to improve the transformation of latent feature A to latent feature B in each model fragment. Figure 9 also shows the results for the YOLO-in-the-Dark model without the finetuning. We discuss this point in Sect. 4.2.

We also evaluated the detection performance in low-light situations. The SID dataset does not cover various levels of illuminance, so we collected additional indoor images that contain objects (e.g., cars, fruit, and animals). Each image was captured using the same camera settings (f/5.6 and ISO-6400). Both the original YOLO model and the proposed YOLO-in-the-Dark model can detect objects at an illuminance of 0.055 lux with similar exposure times (case (a)). In contrast, the original YOLO model requires a longer exposure time (810 ms) when the illuminance is 0.013 lux (case (b)). The YOLO-in-the-Dark model

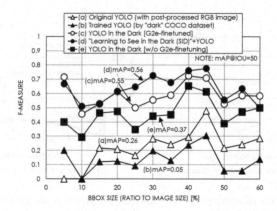

Fig. 9. Prediction results (F-measure) using the SID dataset for the original YOLO model, the YOLO model trained by dark COCO, the YOLO-in-the-Dark model, the SID+YOLO model, and the YOLO-in-the-Dark model without finetuning the generative model.

still detects objects even when the exposure time is shorter (810 → 333 ms). The minimum exposure time indicates the sensitivity in low-light situations. According to the result of case (b), the YOLO-in-the-Dark model can reduce exposure time by a factor of 0.4; this means that the sensitivity is improved by 2.4 times when compared with the sensitivity of the original YOLO model using brightness-enhanced RGB images.

4.2 Ablation Study

As explained in Sect. 3.2 and 3.3, the training environment of the YOLO-in-the-Dark model uses the COCO dataset with the generative model trained by the SID dataset. Our method supposes that the encoder $G2e$ outputs the latent feature A from the COCO dataset, which emulates the relationship between a pair of RAW and RGB images in the SID dataset. In this section, we describe a sequence of controlled experiments that evaluate the effect of different elements in the training environment.

Input Images. Figure 10(a) shows the histograms of the average image in both the COCO and SID datasets. Images in the SID dataset tend to have low pixel levels; this is because of the low-light situations. The distribution of pixel levels for the images in the COCO dataset differs from that of the SID dataset, and this difference might affect encoder $G2e$.

We evaluate the training with different preprocessing parameters (gamma correction) for the input images. Figure 10(b) shows the histograms of the average image in the COCO dataset with gamma correction. A large gamma leads to low pixel levels, as in the SID dataset. Figure 11 reports the results for different levels of gamma correction of 0.67, 1.25, and 1.5.

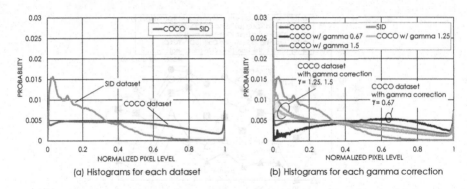

(a) Histograms for each dataset   (b) Histograms for each gamma correction

Fig. 10. Histograms of average dataset images. (a) Histograms for the average image of each dataset and (b) histograms after each image has been gamma corrected.

The histograms of the preprocessed COCO dataset show that gamma correction with a value of 1.25 or 1.5 causes the histogram to become close to the histogram of the SID dataset. Hence, gamma correction of the training images should improve the results of the model. When gamma is 0.67, Fig. 11 shows that the mAP slightly degrades (0.37 → 0.36). When gamma is 1.25, the mAP improves (0.37 → 0.40), and when 1.5, the mAP does not change.

These results show that gamma correction can mitigate the effects due to differences in the datasets. However, the model is sensitive to the value of gamma, and this leads to side effects. Our evaluation indicates that the gamma correction requires fine adjustment corresponding to the histogram of pixel levels in the dataset.

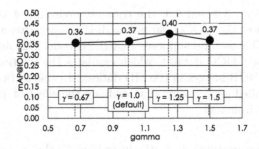

Fig. 11. Detection results (mAP) corresponding to the level of gamma correction in the training images.

Data Augmentation. Data augmentation is an effective way to improve the generalization of the model [6]. We also evaluate the effect of the data augmentation as well as gamma correction for input images.

Table 1 reports the case in which the data has been augmented by changing brightness and contrast. This approach randomly changes the brightness and contrast by 10% of the maximum pixel levels in each image. The mAP slightly degrades (0.37 → 0.35) compared with the default condition (no augmentation). According to previous reports, that found that the results are sensitive to the value of gamma, data augmentation may have a negative effect on this training. The proposed method reuses the pre-trained model, which had been generalized well. Hence, it needs generalization only for the glue layer. As a result, data augmentation does not seem effective for this method.

Table 1. Results for data augmentation conditions (mAP).

Condition	mAP@IOU=50
Default dataset processing (No augumentation)	0.37
Contrast/brightness augmentation	0.35

Finetuning the Generative Model. As described in Sect. 3.2, we finetune the generative model to improve the training of the glue layer. We use the fine-tuning loss function L_{G2e-FT} between the results of the classifier network via $G2e$ ($LF_{G2e-cls}$) and the original YOLO ($LF_{YOLO-cls}$), as described in Fig. 7. This finetuning optimizes $G2e$ with respect to both features $LF_{G2e-cls}$ and $LF_{YOLO-cls}$ so that the optimized $G2e$ can output a better latent feature A. This helps optimize the glue layer. Table 2 reports the L_{G2e-FT} and mAP, corresponding to the cases (c) and (e) in Fig. 9. The result shows that finetuning the generative model substantially improves the mAP with a lower cosine similarity.

Table 2. Results for $G2e$ finetuning L_{G2e-FT} and mAP under each condition.

Condition	L_{G2e-FT}	mAP@IOU=50
Default dataset processing (not finetuned)	0.21	0.37
$G2e$ finetuned	0.15	0.55

Computing Resources. The other contribution of this method is that it reduces the amount of computing resources required. Table 3 compares the resources used by the proposed method and the SID+YOLO model (MAC operations) to process an 832×832 RAW image created by resampling from the original RAW image ($4,240 \times 2,832$). Our method can omit the SID encoder by merging models via latent features. The SID decoder is computationally costly (192.63 GMACs), so the glue layer reduces the total amount of computational resources needed.

Table 3. Comparison of the computing resources used by the proposed and naive approaches (on a 832 × 832 RAW image).

Processing category	MAC operations [GMACs]	
	SID+YOLO	Our work
SID Encoder	45.45	45.45
SID Decoder	192.63	N/A
Glue Layer	N/A	36.68
YOLO	32.71	31.77
Total	270.79	113.90

5   Conclusion

We proposed a method of domain adaptation for merging multiple models that is less effort than creating an additional dataset. This method merges models pre-trained in different domains using glue layers and a generative model, which outputs latent features to train the glue layers without the need for an additional dataset. We also propose a generative model that is created by knowledge distillation from the pre-trained models. It also enables datasets to be reused to create latent features for training the glue layers.

The proposed YOLO-in-the-Dark model, which is a combination of the YOLO model and SID model, is able to detect objects in low-light situations. Our evaluation result indicates that the YOLO-in-the-Dark model can work in scenes illuminated by less than 1 lux. The proposed model is also 2.4 times more sensitive than the original YOLO model at 0.013 lux. Simply combining SID and YOLO (the naive approach) also improves object detection in low-light situations, and this naive approach demonstrates the ideal performance of the YOLO-in-the-Dark model. The performance of our model still needs to be improved so it is closer to this ideal. We also presented an evaluation of the YOLO-in-the-Dark model under various conditions. Preprocessing training input images and finetuning the generative model are effective in improving the optimization of the glue layer.

The other contribution of this method is the reduction in computing resources. In contrast to the naive approach, our method can omit the SID decoder by merging models via latent features. The SID decoder is computationally expensive (192.63 GMACs), so the glue layer can reduce the total amount of computational resources required.

In future work, we plan to apply this method to other tasks including multimodal tasks. We hope that this method can be applied to various models using public datasets. It will extend the functionalities of models more efficiently.

References

1. Chang, W.G., You, T., Seo, S., Kwak, S., Han, B.: Domain-specific batch normalization for unsupervised domain adaptation. In: CVPR 2019 (2019)
2. Chen, C., Chen, Q., Xu, J., Koltun, V.: Learning to see in the dark. In: CVPR 2018 (2018)
3. Chen, G., Choi, W., Yu, X., Han, T., Chandraker, M.: Learning efficient object detection models with knowledge distillation. In: Advances in Neural Information Processing Systems 30, pp. 742–751 (2017)
4. Everingham, M., Gool, L.V., Williams, C., Winn, J., Zisserman, A.: The PASCAL visual object classes challenge 2012 (VOC2012) results. In: VOC 2012 (2012)
5. Hinton, G., Vinyals, O., Dean, J.: Distilling the knowledge in a neural network (2015). arXiv:1503.02531
6. Krizhevsky, A., Sutskever, I., Hinton, G.: Imagenet classification with deep convolutional neural networks. In: Proceedings of the Advances Neural Information Processing Systems 25 (NIPS), pp. 1097–1105 (2012)
7. Lin, T.Y., et al.: Microsoft coco: common objects in context (2014). arXiv:1405.0312
8. Loh, Y.P., Chan, C.S.: Getting to know low-light images with the exclusively dark dataset. Comput. Vis. Image Underst. **178**, 30–42 (2019). https://doi.org/10.1016/j.cviu.2018.10.010
9. Luo, J., Xu, Y., Tang, C., Lv, J.: Learning inverse mapping by autoencoder based generative adversarial nets (2017). arXiv:1703.10094
10. Redmon, J., Divvala, S., Girshick, R., Farhadi, A.: You only look once: unified, real-time object detection. In: CVPR 2016 (2016)
11. Redmon, J., Farhadi, A.: Yolov3: an incremental improvement (2018). arXiv:1804.02767
12. Romero, A., Ballas, N., Kahou, S.E., Chassang, A., Gatta, C., Bengio, Y.: Fitnets: hints for thin deep nets (2014). arXiv:1412.6550
13. Ronneberger, O., Fischer, P., Brox, T.: U-net: convolutional networks for biomedical image segmentation. In: MICCAI (2015)
14. Teng, Y., Choromanska, A., Bojarski, M.: Invertible autoencoder for domain adaptation (2018). arXiv:1802.06869
15. Xie, S., Zheng, Z., Chen, L., Chen, C.: Learning semantic representations for unsupervised domain adaptation. In: ICML 2018 (2018)

Identity-Aware Multi-sentence Video Description

Jae Sung Park[1(✉)], Trevor Darrell[2], and Anna Rohrbach[2]

[1] Paul G. Allen School of Computer Science and Engineering,
University of Washington, Seattle, USA
jspark96@cs.washington.edu
[2] University of California, Berkeley, Berkeley, USA

Abstract. Standard video and movie description tasks abstract away
from person identities, thus failing to link identities across sentences.
We propose a multi-sentence Identity-Aware Video Description task,
which overcomes this limitation and requires to re-identify persons locally
within a set of consecutive clips. We introduce an auxiliary task of Fill-in
the Identity, that aims to predict persons' IDs consistently within a set of
clips, when the video descriptions are given. Our proposed approach to
this task leverages a Transformer architecture allowing for coherent joint
prediction of multiple IDs. One of the key components is a gender-aware
textual representation as well an additional gender prediction objective
in the main model. This auxiliary task allows us to propose a two-stage
approach to Identity-Aware Video Description. We first generate multi-
sentence video descriptions, and then apply our Fill-in the Identity model
to establish links between the predicted person entities. To be able to
tackle both tasks, we augment the Large Scale Movie Description Challenge (LSMDC) benchmark with new annotations suited for our problem statement. Experiments show that our proposed Fill-in the Identity
model is superior to several baselines and recent works, and allows us to
generate descriptions with locally re-identified people.

1 Introduction

Understanding and describing videos that contain multiple events often requires
establishing "who is who": who are the participants in these events and who is
doing what. Most of the prior work on automatic video description focuses on
individual short clips and ignores the aspect of participants' identity. In particular, prior works on movie description tend to replace all character identities with
a generic label SOMEONE [35,41]. While reasonable for individual short clips,
it becomes an issue for longer video sequences. As shown in Fig. 1, descriptions
that contain SOMEONE would not be satisfying to visually impaired users, as
they do not unambiguously convey who is engaged in which action in video.

Electronic supplementary material The online version of this chapter (https://
doi.org/10.1007/978-3-030-58589-1_22) contains supplementary material, which is
available to authorized users.

© Springer Nature Switzerland AG 2020
A. Vedaldi et al. (Eds.): ECCV 2020, LNCS 12366, pp. 360–378, 2020.
https://doi.org/10.1007/978-3-030-58589-1_22

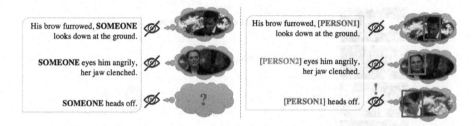

Fig. 1. Compare video description with SOMEONE labels vs. **Identity-Aware Video Description**: in the first case it may be difficult for a visually impaired person to follow what is going on in the video, while in the second case it becomes clear who is performing which action.

Several prior works attempt to perform person re-identification [5,36,39] in movies and TV shows, sometimes relying on associated subtitles or textual descriptions [27,31,40]. Most such works take the "linking tracks to names" problem statement, i.e. trying to name all the detected tracks with proper character names. Others like [29] aim to "fill in" the character proper names in the given ground-truth movie descriptions.

In this work, we propose a different problem statement, which does not require prior knowledge of movie characters and their appearance. Specifically, we group several consecutive movie clips into sets and aim to establish person identities *locally* within each set of clips. We then propose the following two tasks. First, given ground-truth descriptions of a set of clips, the goal is to fill in person identities in a coherent manner, i.e. to predict the same ID for the same person within a set (see Fig. 2). Second, given a set of clips, the goal is to generate video descriptions that contain corresponding local person IDs (see Fig. 1). We refer to these two tasks as **Fill-in the Identity** and **Identity-Aware Video Description**. The first (auxiliary) task is by itself of interest, as it requires to establish person identities in a multi-modal context of video and description.

We experiment with the Large Scale Movie Description Challenge (LSMDC) dataset and associated annotations [34,35], as well as collect more annotations to support our new problem statement. We transform the global character information into local IDs within each set of clips, which we use for both tasks.

Fill-in the Identity. Given textual descriptions of a sequence of events, we aim to fill in the person IDs in the blanks. In order to do that, two steps are necessary. First, we need to attend to a specific person in the video by relating visual observations to the textual descriptions. Second, we need to establish links within a set of blanks by relating corresponding visual appearances and textual context. We learn to perform both steps jointly, as the only supervision available to us is that of the identities, not the corresponding visual tracks. Our key idea is to consider an entire set of blanks jointly, and exploit the mutual relations between the attended characters. We thus propose a Transformer model [42] which jointly infers the identity labels for all blanks. Moreover, to support this

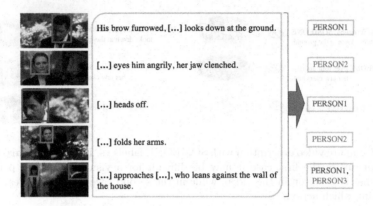

Fig. 2. Example of the **Fill-in the Identity** task.

process we make use of one additional cue available to us: the gender of the person in question. We train a text-based gender classifier, which we integrate in our model, along with an additional visual gender prediction objective, which aims to recognize gender based on the attended visual appearance.

Identity-Aware Video Description. Given a set of clips, we aim to generate descriptions with local IDs. Here, we take a two-stage approach, where we first obtain the descriptions with SOMEONE labels as a first step, and next apply our Fill-in the Identity method to give SOMEONEs their IDs. We believe there is a potential in exploring other models that would incorporate the knowledge of identities into generation process, and leave this to future work.

Our contributions are as follows. (1) We introduce a new task of **Identity-Aware Video Description**, which extends prior work in that it aims to obtain multi-sentence video descriptions with local person IDs. (2) We also introduce a task of **Fill-in the Identity**, which, we hope, will inform future directions of combining identity information with video description. (3) We propose a Transformer model for this task, which learns to attend to a person and use gender evidence along with other visual and textual cues to correctly fill in the person's ID. (4) We obtain state-of-art results in the **Fill-in the Identity** task, compared to several baselines and two recent methods. (5) We further leverage this model to address **Identity-Aware Video Description** via a two-stage pipeline, and show that it is robust enough to perform well on the generated descriptions.

2   Related Work

Video Description. Automatic video description has attracted a lot of interest in the last few years, especially since the arrival of deep learning techniques [2,9, 22,23,33,44,54,58]. Here are a few trends present in recent works [20,30,47,62]. Some works formulate video description as a reinforcement learning problem [19, 28,49]. Several methods address grounding semantic concepts during description generation [52,59,61]. A few recent works put an emphasis on the use of syntactic information [14,45]. New datasets have also been proposed [12,50], including a work that focuses on dense video captioning [17], where the goal is to temporally localize and caption individual events.

In this work, we generate multi-sentence descriptions for long video sequences, as in [32,37,55]. This is different from dense video captioning, where one does not need to obtain one coherent multi-sentence description. Recent works that tackle multi-sentence video description include [56], who generate fine-grained sport narratives, [53], who jointly localize events and decide when to generate the following sentence, and [25], who introduce a new inference method that relies on multiple discriminators to measure the quality of multi-sentence descriptions.

Person Re-identification. Person re-identification aims to recognize whether two images of a person are the same individual. This is a long standing problem in computer vision, with numerous deep learning based approaches introduced over the years [5,26,36,39,46]. We rely on [36] as our face track representation.

Connections to Prior Work. Next, we detail how our work compares to the most related prior works.

Identity-Aware Video Description: Closely related to ours is the work of [34]. They address video description of *individual* movie clips with grounded and co-referenced (re-identified) people. In their problem statement re-identification is performed w.r.t. *a single previous clip* during description generation. Unlike [34], we address *multi-sentence* video description, which requires consistently re-identifying people over *multiple clips at once* (on average 5 clips).

Fill-in the Identity: Our task of predicting *local* character IDs for a set of clips given ground-truth descriptions with blanks, is related to the work of [29]. However, they aim to fill in *global* IDs (proper names). In order to learn the global IDs, they use 80% of each movie for training. Our problem statement is different, as it requires no access to the movie characters' appearance during training: we maintain disjoint training, validation and test movies. A number of prior works attempt to link all the detected face tracks to global character IDs in TV shows and movies [3,11,15,21,27,31,38,40], which is different from our problem statement that tries to fill character IDs locally with textual guidance. We compare to two recent approaches to Fill-in the Identity in Sect. 5.2.

3   Connecting Identities to Video Descriptions

An integral part of understanding a story depicted in a video is to establish who are the key participants and what actions they perform over the course of time. Being able to correctly link the repeating appearances of a certain person could potentially help follow the story line of this person. We first address the task of **Fill-in the Identity**, where we aim to solve a related problem: fill in the persons' IDs based on the given video descriptions with blanks.

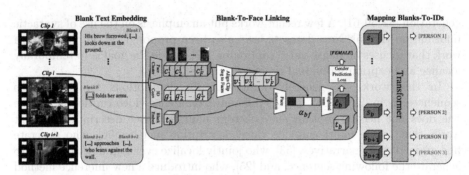

Fig. 3. Overview of our approach to **Fill-in the Identity** task. See Sect. 3.1.

Our approach is centered around two key ideas: joint prediction of IDs via a Transformer architecture [42], supported by gender information inferred from textual and visual cues (see Fig. 3). We then present our second task, **Identity-Aware Video Description**, which aims to generate multi-sentence video descriptions with local person IDs. We present a two-stage pipeline, where our baseline model gives us multi-sentence descriptions with SOMEONE labels, after which we leverage the **Fill-in the Identity** auxiliary task to link the predicted SOMEONE entities.

3.1   Fill-in the Identity

For a set of video clips V_i and their descriptions with blanks $D_i, i = 1, 2..., N$, we aim to fill in character identities that are locally consistent within the set. We first detect faces that appear in each clip and cluster them based on their visual appearance. Then, for every blank in a sentence, we attend over the face cluster centers using visual and textual context, to find the cluster best associated with the blank. We process all the blanks sequentially and pass their visual and textual representations to a Transformer model [42], which analyzes the entire set at once and predicts the most probable sequence of character IDs (Fig. 3).

Visual Representation. Now, we describe the details of getting local face descriptors and other global visual features for a clip V_i. We detect all the faces in every frame of the clip, using the face detector by [60]. Then, we extract 512-dim face feature vectors with the FaceNet model [36][1] trained on the VGGFace2 dataset [5]. The feature vectors are clustered using DBSCAN algorithm [10], which does not require specifying a number of clusters as a parameter. We take the mean of face features in each cluster, resulting in F face feature vectors $(c_1^i, ..., c_F^i)$.

In addition to the face features, we extract spatio-temporal features that describe the clip semantically. These features help the model locate where to look for the relevant face cluster for a given blank. We extract I3D [6] features and apply mean pooling across a fixed number of T segments following [48],

---

[1] https://github.com/davidsandberg/facenet.

giving us a sequence $(g_1^i, ..., g_T^i)$. We then associate each face cluster with the best temporally aligned segment as follows. For each face cluster c_f^i, we keep track of its frame indices and get a "center" index. Dividing this index by the total number of frames in a clip gives us a relative temporal position of the cluster r_f^i, $0 <= r_f^i < 1$. We get the corresponding segment index $t_f^i = \lfloor r_f^i * T \rfloor$ and obtain the global visual context $v_f^i = g_{t_f^i}^i$. We concatenate face cluster features c_f^i with the associated global visual context v_f^i as our final visual representation.

Filling in the Blanks. Suppose there are B blanks in the set of N sentences $D_i{}^2$. One way to fill in these blanks is to train a language model, such as BERT [8], by masking the blanks to directly predict the character IDs. As we aim to incorporate visual information in our model, we take the following approach.

First, each blank b in a sentence D_i has to receive a designated textual encoding. We use with a pretrained BERT model, which has been shown effective for numerous NLP tasks. Instead of relying on a generic pretrained model, we train it to predict the *gender* corresponding to each blank, which often can be inferred from text. For example, in Fig. 2, one can infer that the person in the first clip is male due to the phrase "His brow". We process all sentences in the set jointly. To get a representation for each blank, we access output embedding from the [CLS] token, a special sentence classification token in [8] whose representation captures the meaning of the entire sentence, over all sentences, and a hidden state of the last layer associated with the specific blank token. Note, that the same [CLS] token is used for all blanks in the set. The final representation t_b is a concatenation of the [CLS] and the blank token representation.

For each clip V_i, we obtain F face cluster representations $(c_1^i, ...c_F^i)$, which we combine with the corresponding clip level representations $(v_1^i, ...v_F^i)$. To find the best matching face cluster for the blank b, we predict the attention weights α_{bf} over all clusters in the clip based on the t_b, and compute a weighted sum over the face clusters, $\hat{c}_b$:

$$e_{bf} = W_{\alpha_2} \tanh(W_{\alpha_1}[c_f^i; v_f^i; t_b]),$$

$$\alpha_{bf} = \frac{\exp(e_{bf})}{\sum_{k=1}^{F} \exp(e_{bk})}, \hat{c}_b = \sum_{f=1}^{F} \alpha_{bf} c_f \qquad (1)$$

We concatenate the visual representation $\hat{c}_b$ with t_b as the final representation for the blank: $s_b = [\hat{c}_b; t_b]$.

Given a set of B blanks represented by $(s_1, ..., s_B)$ we now aim to link the corresponding identities to each other. Instead of making pairwise decisions w.r.t. matching and non-matching blanks, we want to predict an entire sequence of IDs jointly. We thus choose the Transformer [42] architecture to let the self-attention mechanisms model multiple pairwise relationships at once. Specifically, we train a Transformer with $(s_1, ..., s_B)$ as inputs and local person IDs $(l_1, ..., l_B)$ as outputs. As we fill in the blanks in a sequential manner, we prevent the future blanks from

---

^{2} Some sentences may have multiple blanks, others may have none.

impacting the previous blanks by introducing a causal masking in the encoder. We train the entire model end-to-end and learn the attention mechanism in Eq. 1 jointly with the ID prediction. Denoting θ as all the parameters in the model, the loss function we minimize is:

$$L_{character}(\theta) = -\sum_{b=1}^{B} p_\theta(l_b \mid s_1, ..., s_{b-1}, l_1, ..., l_{b-1}) \tag{2}$$

We explore the effect of an additional component, a *gender prediction loss* L_{gender}, that forces the attended visual representation to be gender-predictive. We add a single layer perceptron that takes the predicted feature $\hat{c}_b$ and aims to recognize the gender g_b for the blank b. The final loss function we minimize is as follows:

$$L_{gender}(\theta) = -\sum_{b=1}^{B} p_\theta(g_b \mid \hat{c}_b)$$
$$L(\theta) = L_{character} + \lambda_{gen} L_{gender} \tag{3}$$

where $\hat{c}_b$ is calculated in Eq. 1 and λ_{gen} is a hyperparameter.

We also notice that it is possible to boost the performance of our Transformer model by a simple training data augmentation. Note that there are various ways to split the training data into clip sequences with length N: one can consider a non-overlapping segmentation, i.e. $\{1, ...N\}, \{N + 1, ...2N\}, ...$ or additionally add all the overlapping sets $\{2, ...N + 1\}, \{3, ...N + 2\}, ...$ Since we predict the local IDs, every such set would result in a unique data point, meaning using all the possible sets can increase the amount of training data by a factor of N.

3.2 Identity-Aware Video Description

Here, given a set of N clips V_i, we aim to predict their descriptions D_i that would also contain the relevant local person IDs. First, we follow prior works [12,25] to build a multi-sentence description model with SOMEONE labels. It is an LSTM based decoder that takes as input a visual representation and a sentence generated for a previous clip. Here, our visual representation for V_i is I3D [6] and Resnet-152 [13] mean pooled temporal segments. Once we have obtained a multi-sentence video description with SOMEONE labels, we process the generated sentences with our Fill-in the Identity model. We demonstrate that this approach is effective, although the Fill-in the Identity model is only trained on ground-truth descriptions. Note, that this two-stage pipeline could be applied to any video description method.

4   Dataset

As our main test-bed, we choose the Large Scale Movie Description Challenge (LSMDC) [35], while leveraging and extending the character annotations

Table 1. Statistics for our tasks, based on the LSMDC dataset. See Section 4.

	Movies	Sentences	Sets	Blanks
Training	153	101,079	20,283	87,604
Validation	12	7,408	1,486	6,457
Public Test	17	10,053	2,018	8,431

from [34]. They have marked every sentence in the MPII Movie Description (MPII-MD) dataset where a person's proper name (e.g. *Jane*) is mentioned, and labeled the person specific pronouns *he, she* with the associated names (e.g. *he is John*). For each one out of 94 MPII-MD movies, we are given a list of all unique identities and their genders. We extend these annotations to 92 additional movies, covering the entire LSMDC dataset (except for the Blind Test set).

We use these annotations as follows. (1) We drop the pronouns and focus on the underlying IDs. (2) We split each movie into sets of consecutive 5 clips (the last set in a movie may contain less than 5 clips). (3) We relabel global IDs into local IDs *within each set*. E.g. if we encounter a sequence of IDs *Jane, John, Jane, Bill*, it will become *PERSON1, PERSON2, PERSON1, PERSON3*. This relabeling is applied for both tasks that we introduce in this work.

We provide dataset statistics, including number of movies, individual sentences, sets and blanks in Table 1[3]. Around 52% of all blanks correspond to PERSON1, 31% – to PERSON2, 12% – to PERSON3, 4% – to PERSON4, 1% or less – to PERSON5, PERSON6, etc. This shows that the clip sets tend to focus on a single person, but there is still a challenge in distinguishing PERSON1 from PERSON2, PERSON3, ... (up to PERSON11 in training movies).

In our experiments we use the LSMDC Validation set for development, and LSMDC Public Test set for final evaluation.

5   Experiments

5.1   Implementation Details

For each clip, we extract I3D [6] pre-trained on the Kinetics [16] dataset and Resnet-152 [13] features pre-trained on the ImageNet [7] dataset. We mean pool them temporally to get $T = 5$ segments [48]. We detect on average 79 faces per clip (and up to 200). In the DBSCAN algorithm used to cluster faces, we set $\epsilon = 0.2$, which is the maximum distance between two samples in a cluster. Clustering the faces within each clip results in about 2.2 clusters per clip, and clustering over a set results in 4.2 clusters per set. BERT [8] models use the BERT-base model architecture with default settings as in [51]. Transformer [42]

---

[3] Note, that the reported number of training clip sets reflects the default non-overlapping "segmentation", as done for validation and test movies. One is free to define the training clip sets as arbitrary sets of 5 consecutive clips.

has a feedforward dimension of 2048 and 6 self-attention layers. We train the Fill-in the Identity model for 40 epochs with learning rate 5e−5 with hyperparameter $\lambda_{gender} = 0.2$. We train the baseline video description model for 50 epochs with learning rate 5e−4. We fix batch size as 16 across all experiments, where each batch contains a set of clips and descriptions.

5.2   Fill-in the Identity

Evaluation Metrics. First, we discuss the metrics used to evaluate the **Fill-in the Identity** task. Given a sequence of blanks and corresponding ground-truth IDs, we consider all unique pairwise comparisons between the IDs. A pair is labeled as "Same ID" if the two IDs are the same, and "Different ID" otherwise. We obtain such labeling for the ground-truth and predicted IDs. Then, we can compute a ratio of the matching labels between the ground-truth and predicted pairs, e.g. if 6 out of 10 pair labels match, the prediction gets an accuracy 0.6.[4] The final accuracy is averaged across all sets. When define like this, we obtain an *instance-level* accuracy over ID pairs ("Inst-Acc"). Note, that it is important to correctly predict both "Same ID" and "Different ID" labels, which can be seen as a 2-class prediction problem. The instance-level accuracy does not distinguish between these two cases. Thus, we introduce a *class-level* accuracy, where we separately compute accuracy over the two subsets of ID pairs ("Same-Acc", "Diff-Acc") and report the harmonic mean between the two ("Class-Acc").

Baselines and Ablations. Table 2 summarizes our experiments on the LSMDC Validation set. We include two simple baselines: "The same ID" (all IDs are the same) and "All different IDs" (all IDs are distinct: 1, 2, ...). "GT Gender as ID" directly uses *ground truth* male/female gender as a character ID (Person 1/2), and serves as an upper-bound for gender prediction. We consider two vision-only baselines, where we cluster all the detected faces within a set of 5 clips, and pick a random cluster ("Random Face Cluster") or the most frequent cluster ("Most Frequent Face Cluster") within a clip for each blank. We also consider a language-only baseline "BERT Character LM", which uses a pretrained BERT model to directly fill in all the blanks. Then we include our Transformer model with our BERT Gender pretrained model. We show the effect of our training data augmentation and use augmentation in the following versions. Finally, we study the impact of adding each visual component ("+ Face." and "+ Video"), and introduce our vision-based gender loss (full model).

We make the following observations. (1) Instance accuracy for all the same/all distinct IDs provides the insight into how the pairs are distributed (40.7% of all pairs belong to "Same ID" class, 59.3.7% – to "Different ID" class). Neither is a good solution, getting Class-Acc 0. (2) Our Transformer model with BERT Gender representation improves over the vanilla BERT Character model (57.9

---

[4] This resembled pairwise precision/recall used in clustering [1]. However, these are not applicable in our scenario as they can not handle singleton clusters (with one element). Thus, we compute pairwise accuracy instead.

Table 2. Fill-in the Identity accuracy of several baselines, our full method and its ablations on the LSMDC Validation set. We report the predicted ID accuracy at class and instance level, as well as gender accuracy. See Section 5.2 for details.

Method	Same Acc	Diff Acc	Class Acc	Inst Acc	Gen Acc
The same ID	100.0	0.0	0.0	40.7	-
All different IDs	0.0	100.0	0.0	59.3	-
GT Gender as ID	100.0	43.0	60.1	65.5	-
V: Random Face Cluster	54.8	52.0	53.4	55.3	-
V: Most Frequent Face Cluster	48.3	68.8	56.7	61.6	-
L: BERT Character LM	57.9	65.1	57.9	66.4	-
L: Transf. + BERT Gen. LM	57.3	68.9	62.6	67.9	80.3
L: Transf. + BERT Gen. LM + Aug.	60.7	68.7	64.4	69.6	81.8
L+V: Transf. + BERT Gen. LM + Aug. + Face.	62.8	68.0	65.3	69.7	81.8
L+V: Transf. + BERT Gen. LM + Aug. + Face + Video.	64.8	66.6	65.7	69.6	81.8
L+V: Transf. + BERT Gen. LM + Aug. + Face + Video + Gen. Loss	63.5	68.4	65.9	69.8	83.0

vs. 62.6 in Class-Acc). (3) This is also higher than 60.1 of "GT Gender as ID", i.e. our model relies on other language cues besides gender. (4) Training with our data augmentation scheme further improves Class-Acc to 64.4. (5) Introducing face features boosts the Class-Acc to 65.3, and adding video features improves it to 65.7. (7) Finally, visual gender prediction loss leads to the overall Class-Acc of 65.9. Note, that the instance-level accuracy (Inst-Acc) does not always reflect the improvements as it may favor the majority class ("Different ID").

We also report gender accuracy (Gen Acc) for the variants of our model (last 4 rows in Table 2). For models without the visual gender loss, we report the accuracy based on their BERT language model trained for gender classification. We see that data augmentation on the language side helps improve gender accuracy (80.3 vs 81.8). Incorporating visual representation with the gender loss boosts the accuracy further (81.8 vs 83.0).

Human Performance. We also assess human performance in this task in two scenarios: with and without seeing the video. The former provides an overall upper-bound accuracy, while the latter gives an upper-bound for the text-only models. We randomly select 200 sets of Test clips and ask 3 different Amazon Mechanical Turk

Table 3. Fill-in the Identity human performance and our method evaluated on 200 random Test clips sets.

Method	Same Acc	Diff Acc	Class Acc	Inst Acc
Human w/o video	56.5	78.5	65.7	70.0
Human	83.9	90.0	86.8	87.0
Ours	64.6	70.7	67.5	70.3

(AMT) workers to assign the IDs to the blanks. For each set we compute a *median* accuracy across the 3 workers to remove the outliers and report the average accuracy over all sets. Table 3 reports the obtained human performance and the corresponding accuracy of our model on the same 200 sets. Human performance gets a significant boost when the workers can see the video, indicating that *video provides many valuable cues*, and not everything can be inferred from text. Our full method outperforms "Human w/o video" but falls behind the final "Human" performance.

Comparison to State-of-the-Art. Since the data for our new tasks has been made public, other research groups have reported results on it, including the works by Yu et al. [57] and Brown et al. [4][5]. Yu et al. propose an ensemble over two models: Text-Only and Text-Video. The Text-Only model builds a pairwise matrix for a set of blanks and learns to score each pair by optimizing a binary cross entropy loss. The Text-Video model considers two tasks: linking blanks to tracks and linking tracks to tracks. The two tasks are trained separately, with the help of additional supervision from an external dataset [29], using triplet margin loss. While Yu et al. use gender loss to pre-train their language model, we introduce gender loss on the visual side.

Brown et al. train a Siamese network on positive (same IDs) and negative (different IDs) pairs of blanks. The network relies on an attention mechanism over the face features to identify the relevant person given the blank encoding.

Table 4 reports the results on the LSMDC Test set. As we see, our approach significantly outperforms both methods. To gain further

Table 4. Fill-in the Identity accuracy of our method and two recent works on the LSMDC Test set.

Method	Same Acc	Diff Acc	Class Acc	Inst Acc
Yu et al. [57]	26.4	87.3	40.6	65.9
Brown et al. [4]	33.6	81.0	47.5	64.8
Ours text only	56.0	71.2	62.7	68.0
Ours	60.6	70.0	64.9	69.6

insights, we analyze some differences in behavior between these methods and our approach.

We take a closer look at the distribution of the predicted IDs (PERSON1, PERSON2, ...) by our method, and the two other approaches. Figure 4 provides a histogram over the reference data and the compared approaches. We can see that our predictions align well with the true data distribution, while the two other methods struggle to capture the data distribution. Notably, both methods favor more diverse IDs (2, 3, ...), failing to link many of the re-occurring character appearances. This may be in part due to the difference in the objective used in our approach vs. the others. While they implement a binary classifier that selects the best matching face track for each blank, we use Transformer to fill in the blanks jointly, allowing us to better capture both local and global context.

Figure 5 provides a qualitative example, comparing our approach, Yu et al. and Brown et al. As suggested by our analysis, these two methods often predict diverse IDs instead of recognizing the true correspondences.

---

[5] Note, that we have corrected some errors, affecting about 3% of the annotations. While Yu et al. [57] and Brown et al. [4] have trained their models on the old annotations, all reported results are obtained on the *corrected* test set.

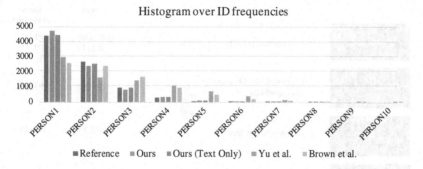

Fig. 4. Fill-in the Identity: histogram over the frequencies of predicted IDs for our method, its text-only version and two SOTA works. See Sect. 5.2.

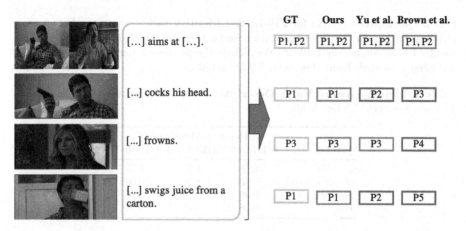

Fig. 5. Qualitative example for **Fill-in the Identity** task, comparison between our approach and two recent methods. Correct/incorrect predictions are labeled with green/red, respectively. P1, P2, ... are person IDs. See Sect. 5.2 for details. (Color figure online)

Ours vs. Ours Text Only. In Fig. 6, we compare our full model vs. its text-only version. After having seen the two characters in the first clip (P1 and P2), our full model recognizes that the man and woman appearing in a next set of clips are the same two characters, and successfully links them as P1 in the second and P2 in the third and fourth clip with the correct genders. In the last clip with two characters, the full model is also able to visually ground the same woman as "heading out" and assign the ID as P2. On the other hand, the text-only model cannot tell that the first two characters appear in the next set of clips without a visual signal, and incorrectly assigns the blanks as different character IDs, P3 and P4, after the second blank. The text only model also fails to link the character in third and last clip as the same ID due to the limited information available in textual descriptions alone. This shows that the task is hard for the

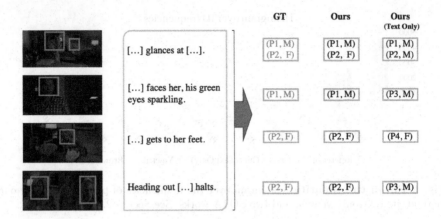

Fig. 6. Qualitative example for **Fill-in the Identity** task, comparison between our final approach with visual representation and a text-only ablation ("Transf. + BERT Gender LM + Augm." in Table 2). We include the predicted character ID (P1, P2, ...) and gender for each blank. See Sect. 5.2 for details.

Table 5. Identity-Aware Video Description scores for our method on the LSMDC Test set. See Sect. 5.3 for details.

Method	Per set, MAX score		
	METEOR	BLEU@4	CIDEr-D
Same ID	9.22	1.59	6.31
All different IDs	8.84	1.38	6.06
Ours Text-Only	10.29	1.74	6.88
Ours	10.38	1.75	6.95

text-only model, while *our final model learns to incorporate visual information successfully.*

5.3   Identity-Aware Video Description

Finally, we evaluate our two-stage approach for the **Identity-Aware Video Description** task. One issue with directly evaluating predicted descriptions with the character IDs is that the evaluation can strongly penalize predictions that do not closely follow the ground-truth. Consider an example with ground-truth *[P1] approaches [P2]. [P2] gets up.* vs. prediction *[P1] is approached by [P2]. [P1] stands up.* As we see, direct comparison would yield a low score for this prediction, due to different phrasing which leads to different local person IDs. If we instead consider an ID permutation *[P2] approaches [P1]. [P1] gets up.*,

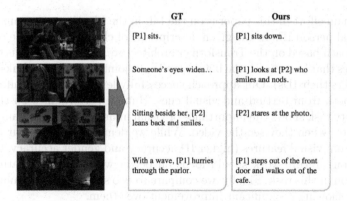

	GT	**Ours**
	[P1] sits.	[P1] sits down.
	Someone's eyes widen...	[P1] looks at [P2] who smiles and nods.
	Sitting beside her, [P2] leans back and smiles.	[P2] stares at the photo.
	With a wave, [P1] hurries through the parlor.	[P1] steps out of the front door and walks out of the cafe.

Fig. 7. Qualitative example for the **Identity-Aware Video Description** task. P1, P2, ... are person IDs. See Sect. 5.3 for details.

we can get a better match with the prediction. Thus, we consider all the possible ID permutations as references, evaluate our prediction w.r.t. all of them, and choose the reference that gives the highest BLEU@4 to compute the final scores. The caption with the best ID permutation is then used to evaluated at set level using the standard automatic scores (METEOR [18], BLEU@4 [24], CIDEr-D [43]).

Results. In Table 5, we compare results from our captioning model (Sect. 3.2) with different Fill-in the Identity approaches to fill in the IDs on the LSMDC Test set. Our model outperforms the baseline approaches, including Ours Text-Only model. This confirms that our Fill-in the Identity model successfully uses visual signal to perform well, on both ground truth and predicted sentences.

Figure 7 provides an example output of our two-stage approach. We show the ground truth descriptions with person IDs on the left, and our generated descriptions with the predicted IDs on the right. Our Fill-in the Identity model consistently links [P1] as the woman who "sits down", "looks at [P2]", and "steps out", while [P2] as the man who "smiles" and "stares" across the clips.

While we experiment with a fairly straightforward video description model, our two-stage approach can enable character IDs if applied to any model.

6   Conclusion

In this work we address the limitation of existing literature on automatic video and movie description, which typically ignores the aspect of person identities.

Our main effort in this paper is on the Fill-in the Identity task, namely filling in the local person IDs in the given descriptions of clip sequences. We propose a new approach based on the Transformer architecture, which first learns to attend to the faces that best match the blanks, and next jointly establishes links between them (infers their IDs). Our approach successfully leverages gender information, inferred both from textual and visual cues. Human performance on the Fill-in the Identity task shows the importance of visual information, as humans perform much better when they see the video. While we demonstrate that our approach benefits from visual features (higher ID accuracy and gender accuracy, better ID distribution), future work should focus on better ways of incorporating visual information in this task. Finally, we compare to two state-of-the-art multi-modal methods, showing a significant improvement over them.

We also show that our Fill-in the Identity model enables us to use a two-stage pipeline to tackle the Identity-Aware Video Description task. While this is a simple approach, it is promising to see that our model can handle automatically generated descriptions, and not only ground-truth descriptions. We hope that our new proposed tasks will lead to more research on bringing together person re-identification and video description and will encourage future works to design solutions that go beyond a two-stage pipeline.

Acknowledgements. The work of Trevor Darrell and Anna Rohrbach was in part supported by the DARPA XAI program, the Berkeley Artificial Intelligence Research (BAIR) Lab, and the Berkeley DeepDrive (BDD) Lab.

References

1. Banerjee, A., Krumpelman, C., Ghosh, J., Basu, S., Mooney, R.J.: Model-based overlapping clustering. In: Proceedings of the Eleventh ACM SIGKDD International Conference on Knowledge Discovery in Data Mining, pp. 532–537 (2005)
2. Baraldi, L., Grana, C., Cucchiara, R.: Hierarchical boundary-aware neural encoder for video captioning. In: Proceedings of the IEEE Conference on Computer Vision and Pattern Recognition (CVPR) (2017)
3. Bojanowski, P., Bach, F., Laptev, I., Ponce, J., Schmid, C., Sivic, J.: Finding actors and actions in movies. In: Proceedings of the IEEE International Conference on Computer Vision (ICCV) (2013)
4. Brown, A., Albanie, S., Liu, Y., Nagrani, A., Zisserman, A.: LSMDC V2 challenge presentation. In: 3rd Workshop on Closing the Loop Between Vision and Language (2019)
5. Cao, Q., Shen, L., Xie, W., Parkhi, O.M., Zisserman, A.: Vggface2: a dataset for recognising faces across pose and age. In: IEEE International Conference on Automatic Face & Gesture Recognition (FG 2018). IEEE (2018)
6. Carreira, J., Zisserman, A.: Quo vadis, action recognition? a new model and the kinetics dataset. In: Proceedings of the IEEE Conference on Computer Vision and Pattern Recognition, pp. 6299–6308 (2017)
7. Deng, J., Dong, W., Socher, R., Li, L.J., Li, K., Fei-Fei, L.: Imagenet: a large-scale hierarchical image database. In: Proceedings of the IEEE Conference on Computer Vision and Pattern Recognition (CVPR), pp. 248–255. IEEE (2009)

8. Devlin, J., Chang, M.W., Lee, K., Toutanova, K.: Bert: pre-training of deep bidirectional transformers for language understanding. arXiv preprint arXiv:1810.04805 (2018)
9. Donahue, J., et al.: Long-term recurrent convolutional networks for visual recognition and description. In: Proceedings of the IEEE Conference on Computer Vision and Pattern Recognition (CVPR) (2015)
10. Ester, M., Kriegel, H.P., Sander, J., Xu, X.: A density-based algorithm for discovering clusters a density-based algorithm for discovering clusters in large spatial databases with noise. In: Proceedings of the Second International Conference on Knowledge Discovery and Data Mining, pp. 226–231. ACM (1996)
11. Everingham, M., Sivic, J., Zisserman, A.: "hello! my name is... buffy" - automatic naming of characters in TV video. In: Proceedings of the British Machine Vision Conference (BMVC) (2006)
12. Gella, S., Lewis, M., Rohrbach, M.: A dataset for telling the stories of social media videos. In: Proceedings of the Conference on Empirical Methods in Natural Language Processing (EMNLP), pp. 968–974 (2018)
13. He, K., Zhang, X., Ren, S., Sun, J.: Deep residual learning for image recognition. In: Proceedings of the IEEE Conference on Computer Vision and Pattern Recognition (CVPR), pp. 770–778 (2016)
14. Hou, J., Wu, X., Zhao, W., Luo, J., Jia, Y.: Joint syntax representation learning and visual cue translation for video captioning. In: Proceedings of the IEEE International Conference on Computer Vision, pp. 8918–8927 (2019)
15. Jin, S., Su, H., Stauffer, C., Learned-Miller, E.: End-to-end face detection and cast grouping in movies using Erdos-Renyi clustering. In: Proceedings of the IEEE International Conference on Computer Vision, pp. 5276–5285 (2017)
16. Kay, W., et al.: The kinetics human action video dataset. arXiv preprint arXiv:1705.06950 (2017)
17. Krishna, R., Hata, K., Ren, F., Fei-Fei, L., Niebles, J.C.: Dense-captioning events in videos. In: Proceedings of the IEEE International Conference on Computer Vision (ICCV), pp. 706–715 (2017)
18. Lavie, M.D.A.: Meteor universal: language specific translation evaluation for any target language. In: Proceedings of the Annual Meeting of the Association for Computational Linguistics (ACL), p. 376 (2014)
19. Li, L., Gong, B.: End-to-end video captioning with multitask reinforcement learning. In: IEEE Winter Conference on Applications of Computer Vision (WACV) (2019)
20. Li, Y., Yao, T., Pan, Y., Chao, H., Mei, T.: Jointly localizing and describing events for dense video captioning. In: Proceedings of the IEEE Conference on Computer Vision and Pattern Recognition (CVPR), pp. 7492–7500 (2018)
21. Miech, A., Alayrac, J.B., Bojanowski, P., Laptev, I., Sivic, J.: Learning from video and text via large-scale discriminative clustering. In: Proceedings of the IEEE International Conference on Computer Vision, pp. 5257–5266 (2017)
22. Pan, P., Xu, Z., Yang, Y., Wu, F., Zhuang, Y.: Hierarchical recurrent neural encoder for video representation with application to captioning. In: Proceedings of the IEEE Conference on Computer Vision and Pattern Recognition (CVPR) (2016)
23. Pan, Y., Yao, T., Li, H., Mei, T.: Video captioning with transferred semantic attributes. In: Proceedings of the IEEE Conference on Computer Vision and Pattern Recognition (CVPR) (2017)
24. Papineni, K., Roukos, S., Ward, T., Zhu, W.J.: BLEU: a method for automatic evaluation of machine translation. In: Proceedings of the Annual Meeting of the Association for Computational Linguistics (ACL) (2002)

25. Park, J.S., Rohrbach, M., Darrell, T., Rohrbach, A.: Adversarial inference for multi-sentence video description. In: Proceedings of the IEEE Conference on Computer Vision and Pattern Recognition (CVPR) (2019)
26. Parkhi, O.M., Vedaldi, A., Zisserman, A.: Deep face recognition. In: Proceedings of the British Machine Vision Conference (BMVC) (2015)
27. Parkhi, O.M., Rahtu, E., Zisserman, A.: It's in the bag: stronger supervision for automated face labelling. In: Proceedings of the IEEE International Conference on Computer Vision Workshops (ICCV Workshops) (2015)
28. Pasunuru, R., Bansal, M.: Reinforced video captioning with entailment rewards. In: Proceedings of the Annual Meeting of the Association for Computational Linguistics (ACL) (2017)
29. Pini, S., Cornia, M., Bolelli, F., Baraldi, L., Cucchiara, R.: M-VAD names: a dataset for video captioning with naming. Multimedia Tools Appl. **78**(10), 14007–14027 (2019)
30. Rahman, T., Xu, B., Sigal, L.: Watch, listen and tell: multi-modal weakly supervised dense event captioning. In: Proceedings of the IEEE International Conference on Computer Vision, pp. 8908–8917 (2019)
31. Ramanathan, V., Joulin, A., Liang, P., Fei-Fei, L.: Linking people in videos with "their" names using coreference resolution. In: Proceedings of the European Conference on Computer Vision (ECCV) (2014)
32. Rohrbach, A., Rohrbach, M., Qiu, W., Friedrich, A., Pinkal, M., Schiele, B.: Coherent multi-sentence video description with variable level of detail. In: Proceedings of the German Conference on Pattern Recognition (GCPR) (2014)
33. Rohrbach, A., Rohrbach, M., Schiele, B.: The long-short story of movie description. In: Proceedings of the German Conference on Pattern Recognition (GCPR) (2015)
34. Rohrbach, A., Rohrbach, M., Tang, S., Oh, S.J., Schiele, B.: Generating descriptions with grounded and co-referenced people. In: Proceedings of the IEEE Conference on Computer Vision and Pattern Recognition (CVPR) (2017)
35. Rohrbach, A., et al.: Movie description. Int. J. Comput. Vis. (IJCV) **123**(1), 94–120 (2017)
36. Schroff, F., Kalenichenko, D., Philbin, J.: Facenet: a unified embedding for face recognition and clustering. In: Proceedings of the IEEE Conference on Computer Vision and Pattern Recognition (CVPR) (2015)
37. Shin, A., Ohnishi, K., Harada, T.: Beyond caption to narrative: video captioning with multiple sentences. In: Proceedings of the IEEE IEEE International Conference on Image Processing (ICIP) (2016)
38. Sivic, J., Everingham, M., Zisserman, A.: "who are you?"-learning person specific classifiers from video. In: Proceedings of the IEEE Conference on Computer Vision and Pattern Recognition (CVPR) (2009)
39. Taigman, Y., Yang, M., Ranzato, M., Wolf, L.: Deepface: closing the gap to human-level performance in face verification. In: Proceedings of the IEEE Conference on Computer Vision and Pattern Recognition (CVPR) (2014)
40. Tapaswi, M., Baeuml, M., Stiefelhagen, R.: "knock! knock! who is it?" probabilistic person identification in TV-series. In: Proceedings of the IEEE Conference on Computer Vision and Pattern Recognition (CVPR) (2012)
41. Torabi, A., Pal, C., Larochelle, H., Courville, A.: Using descriptive video services to create a large data source for video annotation research. arXiv:1503.01070 (2015)
42. Vaswani, A., et al.: Attention is all you need. In: Advances in Neural Information Processing Systems (NIPS) (2017)

43. Vedantam, R., Zitnick, C.L., Parikh, D.: Cider: consensus-based image description evaluation. In: Proceedings of the IEEE Conference on Computer Vision and Pattern Recognition (CVPR) (2015)
44. Venugopalan, S., Rohrbach, M., Donahue, J., Mooney, R., Darrell, T., Saenko, K.: Sequence to sequence - video to text. In: Proceedings of the IEEE International Conference on Computer Vision (ICCV) (2015)
45. Wang, B., Ma, L., Zhang, W., Jiang, W., Wang, J., Liu, W.: Controllable video captioning with POS sequence guidance based on gated fusion network. In: Proceedings of the IEEE International Conference on Computer Vision, pp. 2641–2650 (2019)
46. Wang, F., Xiang, X., Cheng, J., Yuille, A.L.: NormFace: L2 hypersphere embedding for face verification. In: Proceedings of the 25th ACM International Conference on Multimedia. ACM (2017)
47. Wang, J., Jiang, W., Ma, L., Liu, W., Xu, Y.: Bidirectional attentive fusion with context gating for dense video captioning. In: Proceedings of the IEEE Conference on Computer Vision and Pattern Recognition (CVPR), pp. 7190–7198 (2018)
48. Wang, L., et al.: Temporal segment networks: towards good practices for deep action recognition. In: Leibe, B., Matas, J., Sebe, N., Welling, M. (eds.) ECCV 2016. LNCS, vol. 9912, pp. 20–36. Springer, Cham (2016). https://doi.org/10.1007/978-3-319-46484-8_2
49. Wang, X., Chen, W., Wu, J., Wang, Y.F., Wang, W.Y.: Video captioning via hierarchical reinforcement learning. In: Proceedings of the IEEE Conference on Computer Vision and Pattern Recognition (CVPR), pp. 4213–4222 (2018)
50. Wang, X., Wu, J., Chen, J., Li, L., Wang, Y.F., Wang, W.Y.: Vatex: a large-scale, high-quality multilingual dataset for video-and-language research. In: Proceedings of the IEEE International Conference on Computer Vision (ICCV) (2019)
51. Wolf, T., et al.: Huggingface's transformers: state-of-the-art natural language processing. arXiv abs/1910.03771 (2019)
52. Wu, X., Li, G., Cao, Q., Ji, Q., Lin, L.: Interpretable video captioning via trajectory structured localization. In: Proceedings of the IEEE Conference on Computer Vision and Pattern Recognition (CVPR) (2018)
53. Xiong, Y., Dai, B., Lin, D.: Move forward and tell: a progressive generator of video descriptions. In: Proceedings of the European Conference on Computer Vision (ECCV) (2018)
54. Yao, L., et al.: Describing videos by exploiting temporal structure. In: Proceedings of the IEEE International Conference on Computer Vision (ICCV) (2015)
55. Yu, H., Wang, J., Huang, Z., Yang, Y., Xu, W.: Video paragraph captioning using hierarchical recurrent neural networks. In: Proceedings of the IEEE Conference on Computer Vision and Pattern Recognition (CVPR) (2016)
56. Yu, H., Cheng, S., Ni, B., Wang, M., Zhang, J., Yang, X.: Fine-grained video captioning for sports narrative. In: Proceedings of the IEEE Conference on Computer Vision and Pattern Recognition (CVPR), pp. 6006–6015 (2018)
57. Yu, Y., Chung, J., Kim, J., Yun, H., Kim, G.: LSMDC V2 challenge presentation. In: 3rd Workshop on Closing the Loop Between Vision and Language (2019)
58. Yu, Y., Ko, H., Choi, J., Kim, G.: End-to-end concept word detection for video captioning, retrieval, and question answering. In: Proceedings of the IEEE Conference on Computer Vision and Pattern Recognition (CVPR) (2017)
59. Zanfir, M., Marinoiu, E., Sminchisescu, C.: Spatio-temporal attention models for grounded video captioning. In: Proceedings of the Asian Conference on Computer Vision (ACCV) (2016)

60. Zhang, K., Zhang, Z., Li, Z., Qiao, Y.: Joint face detection and alignment using multitask cascaded convolutional networks. IEEE Signal Process. Lett. **23**(10), 1499–1503 (2016)
61. Zhou, L., Kalantidis, Y., Chen, X., Corso, J.J., Rohrbach, M.: Grounded video description. In: Proceedings of the IEEE Conference on Computer Vision and Pattern Recognition (CVPR), pp. 6578–6587 (2019)
62. Zhou, L., Zhou, Y., Corso, J.J., Socher, R., Xiong, C.: End-to-end dense video captioning with masked transformer. In: Proceedings of the IEEE Conference on Computer Vision and Pattern Recognition (CVPR), pp. 8739–8748 (2018)

VQA-LOL: Visual Question Answering Under the Lens of Logic

Tejas Gokhale(✉)[ID], Pratyay Banerjee[ID], Chitta Baral[ID], and Yezhou Yang[ID]

Arizona State University, Tempe, USA
{tgokhale,pbanerj6,chitta,yz.yang}@asu.edu

Abstract. Logical connectives and their implications on the meaning of a natural language sentence are a fundamental aspect of understanding. In this paper, we investigate whether visual question answering (VQA) systems trained to answer a question about an image, are able to answer the logical composition of multiple such questions. When put under this *Lens of Logic*, state-of-the-art VQA models have difficulty in correctly answering these logically composed questions. We construct an augmentation of the VQA dataset as a benchmark, with questions containing logical compositions and linguistic transformations (negation, disjunction, conjunction, and antonyms). We propose our Lens of Logic (LOL) model which uses question-attention and logic-attention to understand logical connectives in the question, and a novel Fréchet-Compatibility Loss, which ensures that the answers of the component questions and the composed question are consistent with the inferred logical operation. Our model shows substantial improvement in learning logical compositions while retaining performance on VQA. We suggest this work as a move towards robustness by embedding logical connectives in visual understanding.

Keywords: Visual question answering · Logical robustness

1 Introduction

Theories about logic in human understanding have a long history. In modern times, Piaget and Fodor [34] studied the representation of logical hypotheses in the human mind. George Boole [7] formalized conjunction, disjunction, and negation into an "algebra of thought" as a way to improve, systemize, and mathematize Aristotle's Logic [12]. Horn regarded negation to be a fundamental and defining characteristic of human communication [19], following the traditions of Sankara [35], Spinoza [42], and Hegel [18]. Recent studies [11] have suggested

---

T. Gokhale and P. Banerjee—Equal Contribution.

Electronic supplementary material The online version of this chapter (https://doi.org/10.1007/978-3-030-58589-1_23) contains supplementary material, which is available to authorized users.

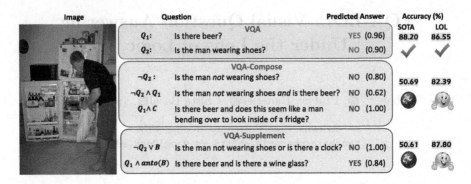

Fig. 1. State-of-the-art models answer questions from the VQA dataset (Q_1, Q_2) correctly, but struggle when asked a logical composition including negation, conjunction, disjunction, and antonyms. We develop a model that improves on this metric substantially, while retaining VQA performance.

that infants can formulate intuitive and stable logical structures to interpret dynamic scenes and to entertain and rationally modify hypotheses about the scenes. As such we argue that understanding logical structures in questions, is a fundamental requirement for any question-answering system.

If a question can be put at all, then it can be answered. [44]

In the above proposition, Wittgenstein linked the process of asking a question with the existence of an answer. While we do not comment on the existence of an answer, we suggest the following softer proposition -

If questions $Q_1 \ldots Q_n$ can be answered, then so should all composite questions created from $Q_1 \ldots Q_n$

Visual question answering (VQA) [3] is an intuitive, yet challenging task that lies at a crucial intersection of vision and language. Given an image and a question about it, the goal of a VQA system is to provide a free-form or open-ended answer. Consider the image in Fig. 1 which shows a person in front of an open fridge. When asked the questions Q_1 (*Is there beer?*) and Q_2 (*Is the man wearing shoes?*) independently, the state-of-the-art model LXMERT [43] answers both correctly. However when we insert a negation in Q_2 (*Is the man not wearing shoes?*) or for a conjunction of two questions $\neg Q_2 \wedge Q1$ (*Is the man not wearing shoes and is there beer?*), the system makes wrong predictions. Our motivation is to reliably answer such logically composed questions. In this paper, we analyze VQA systems under this *Lens of Logic (LOL)* and develop a model that can answer such questions reflecting human logical inference. We offer our work as the first investigation into the logical structure of questions in visual question-answering and provide a solution that *learns* to interpret logical connectives in questions.

The first question is: can models pre-trained on the VQA dataset answer logically composed questions? It turns out that these models are unable to do so, as illustrated in Fig. 1 and Table 2. An obvious next experiment is to *split the question* into its component questions, predict the answer to each, and combine the answers logically. However language parsers (either oracle or trained parsers) are not accurate at understanding negation, and as such this approach does not yield correct answers for logically composed questions. The question then arises: can the model answer such questions, if we explicitly train it with data that also contains logically composed questions? For this investigation, we construct two datasets, VQA-Compose and VQA-Supplement, by utilizing annotations from the VQA dataset, as well as object and caption annotations from COCO [25]. We use these datasets to train the state-of-the-art model LXMERT [43] and perform multiple experiments to test for robustness towards logically composed questions.

After this investigation, we develop our LOL model architecture that jointly learns to answer questions while understanding the type of question and which logical connective exists in the question, through our attention modules, as shown in Fig. 3. We further train our model with a novel Fréchet-Compatibility loss that ensures compatibility between the answers to the component questions and the answer of the logically composed question. One key finding is that our models are better than existing models trained on logical questions, with a small deviation from state-of-the-art on VQA test set. Our models also exhibit better *Compositional Generalization* i.e. models trained to answer questions with a single logical connective are able to answer those with multiple connectives.

Our contributions are summarized below:

1. We conduct a detailed analysis of the performance of the state-of-the-art VQA model with respect to logically composed questions,
2. We curate two large scale datasets VQA-Compose and VQA-Supplement that contain logically composed binary questions.
3. We propose *LOL* – our end-to-end model with dedicated attention modules that answer questions by understanding the logical connectives in questions.
4. We show a capability of answering logically composed questions, while retaining VQA performance.

2   Related Work

Logic in Human Expression: Is logical thinking a natural feature of human thought and expression? Evidence in psychological studies [10,11,16] suggests that infants are capable of logical reasoning, toddlers understand logical operations in natural language and are able to compositionally compute meanings even in complex sentences containing multiple logical operators. Children are also able to use these meanings to assign truth values to complex experimental tasks. Given this, question-answering systems also need to answer compositional questions, and be robust to the manifestation of logical operators in natural language.

Logic in Natural Language Understanding: The task of understanding compositionality in question-answering (QA) can also be interpreted as understanding logical connectives in text. While question compositionality is largely unstudied, approaches in natural language understanding seek to transform sentences into symbolic formats such as first-order logic (FOL) or relational tables [24,30,47]. While such methods benefit from interpretability, they suffer from practical limitations like intractability, reliance on background knowledge, and failure to process noise and uncertainty. [8,39,41] suggest that better generalization can be achieved by learning embeddings to reason about semantic relations, and to simulate FOL behavior [40]. Recursive neural networks have been shown to learn logical semantics on synthetic English-like sentences by using embeddings [9,32].

Detection of negation in text has been studied for information extraction and sentiment analysis [31]. [22] have shown that BERT-based models [13,26] are incapable of differentiating between sentences and their negations. Concurrent to our work, [4] show the efficacy of FOL-guided data augmentation for performance improvements on natural language QA tasks that require reasoning. Since our work deals with both vision and language modalities, it encounters a greater degree of ambiguity, thus calling for robust VQA systems that can deal with logical transformations.

Visual Question Answering (VQA) [3] is a large-scale, human-annotated dataset for open-ended question-answering on images. VQA-v2[17] reduces the language bias in the dataset by collecting complementary images for each question-image pair. This ensures that the number of questions in the VQA dataset with the answer "YES" is equal to those with the answer "NO". This dataset contains 204k images from MS-COCO [25], and 1.1M questions.

Cross-modal pre-trained models [27,43,48] have proved to be highly effective in vision-and-language tasks such as VQA, referring expression comprehension, and image retrieval. While neuro-symbolic approaches [29] have been proposed for VQA tasks which require reasoning on synthetic images, their performance on natural images is lacking. Recent work seeks to incorporate reasoning in VQA, such as visual commonsense reasoning [14,46], spatial reasoning [20,21], and by integrating knowledge for end-to-end reasoning [1].

We take a step back and extensively analyze the pivotal task of VQA with respect to various aspects of generalization. We consider a rigorous investigation of a task, dataset, and models to be equally important as proposing new challenges that are arguably harder. In this paper we analyse existing state-of-the-art VQA models with respect to their robustness to logical transformations of questions.

Table 1. Illustration of question composition in `VQA-Compose`, for the same example as in Fig. 1. QF: Question Formula, AF: Answer Formula

QF	Question	AF	Answer
Q_1	Is there beer?	A_1	Yes
Q_2	Is the man wearing shoes?	A_2	No
$\neg Q_1$	Is there no beer?	$\neg A_1$	No
$\neg Q_2$	Is the man not wearing shoes?	$\neg A_2$	Yes
$Q_1 \wedge Q_2$	Is there beer and is the man wearing shoes?	$A_1 \wedge A_2$	No
$Q_1 \vee Q_2$	Is there beer or is the man wearing shoes?	$A_1 \vee A_2$	Yes
$Q_1 \wedge \neg Q_2$	Is there beer and is the man not wearing shoes?	$A_1 \wedge \neg A_2$	Yes
$Q_1 \vee \neg Q_2$	Is there beer or is the man not wearing shoes?	$A_1 \vee \neg A_2$	Yes
$\neg Q_1 \wedge Q_2$	Is there no beer and is the man wearing shoes?	$\neg A_1 \wedge A_2$	No
$\neg Q_1 \vee Q_2$	Is there no beer or is the man wearing shoes?	$\neg A_1 \vee A_2$	No
$\neg Q_1 \wedge \neg Q_2$	Is there no beer and is the man not wearing shoes?	$\neg A_1 \wedge \neg A_2$	No
$\neg Q_1 \vee \neg Q_2$	Is there no beer or is the man not wearing shoes?	$\neg A_1 \vee \neg A_2$	Yes

(a)

Is the lady holding the baby?

Is the man holding the baby?

Is the baby holding the man?

(b)

Are they in a restaurant?

Are they all girls?

Are they in a restaurant and are they all boys?

Fig. 2. Some questions in `VQA-Supplement` created with adversarial antonyms.

3 The Lens of Logic

A lens magnifies objects under investigation, by allowing us to zoom and focus on desired contents or processes. Our lens of logical composition of questions, allows us to magnify, identify, and analyze the problems in VQA models.

Consider Fig. 2(a), where we transform the first question *"Is the lady holding the baby"* by first replacing *"lady"* with an adversarial antonym *"man"* and observe that the system provides a wrong answer with very high probability. Swapping *"man"* with *"baby"* results in a wrong answer as well. In Fig. 2(b) a conjunction of two questions containing antonyms (*girls* vs *boys*) yields a wrong answer. We identify that the ability to answer composite questions created by negation, conjunction and disjunction of questions is crucial for VQA.

We use "closed questions" as defined in [6] to construct logically composed questions. Under this definition, if a closed question has a negative ("NO") answer then its negation must have an affirmative ("YES") answer. Of the three types of questions in the VQA dataset (yes/no, numeric, other), "yes-no" questions satisfy this requirement. Although, visual questions in the VQA dataset can have multiple correct answers [5], 20.91% of the questions (around 160k) in the VQA dataset are closed questions, i.e. questions with a single unambiguous yes-or-no answer, unanimously annotated by multiple human workers. This allows us to treat these questions as propositions and create a truth table for answers to compose logical questions as shown in Table 1.

3.1 Composite Questions

Let $\mathcal{D}$ be the VQA dataset. For closed questions Q_1 and Q_2 about image $I \in \mathcal{D}$, we define the composite question Q^* composed using connective $\circ \in \{\vee, \wedge\}$, as:

$$Q^* = \widehat{Q_1} \circ \widehat{Q_2}, \quad where \ \widehat{Q_1} \in \{Q_1, \neg Q_1\}, \ \widehat{Q_2} \in \{Q_2, \neg Q_2\}. \qquad (1)$$

3.2 Dataset Creation Process

Using the above definition we create two new datasets by utilizing multiple questions about the same image (VQA-Compose) and external object and caption annotations about the image from COCO to create more questions (VQA-Supplement). The seed questions for creating these datasets are all closed binary questions from VQA-v2 [17]. These datasets serve as test-beds, and enable experiments that analyze performance of models when answering such questions.

VQA-Compose: Consider the first two rows in Table 1. Q_1 and Q_2 are two questions about the image in Fig. 1 taken from the VQA dataset. Additional questions are composed from Q_1 and Q_2 by using the formulas in Table 1. Thus for each pair of closed questions in the VQA dataset, we get 10 logically composed questions. Using the same train-val-test split as the VQA-v2 dataset [17], we get *1.25 million samples* for our VQA-Compose dataset. The dataset is balanced in terms of the number of questions with affirmative and negative answers.

VQA-Supplement: Images in VQA-v2 follow identical train-val-test splits as their source MS-COCO [25]. Therefore, we use the object annotations from COCO to create additional closed binary questions, such as *"Is there a bottle"* for the example in Fig. 1. We also create "adversarial" questions about objects, like *"Is there a wine-glass?"* by using an object that is not present in the image (wine-glass), but is *semantically close* to an object in the image (bottle). We use Glove vectors [33] to find the adversarial object with the closest embedding. Following a similar strategy, we also convert captions provided in COCO to closed binary questions, for example *"Does this seem like a man bending over to look inside the fridge"*. Since we know what objects are present in the image, and the captions describe a "true" scene, we are able to obtain the ground-truth answers for questions created from objects and captions. Similar methods for creation of question-answer pairs have previously been used in [28,37].

Thus for every question, we obtain several questions from objects and captions, and use these to compose additional questions by following a process similar to the one for VQA-Compose. For each closed question in the VQA dataset, we get 20 additional logically composed questions by utilizing questions created from objects and captions, yielding a total of *2.55 million samples* as VQA-Supplement.

3.3   Analytical Setup

In order to test the robustness of our models to logically composed questions, we devise five key experiments to analyse baseline models and our methods. These experiments help us gain insights into the nuances of the VQA dataset, and allow us to develop strategies for promoting robustness.

Effect of Data Augmentation: In this experiment, we compare the performance of models on VQA-Compose and VQA-Supplement with or without logically composed training data. This experiment allows us to test our hypotheses about the robustness of any VQA model to logically composed questions. We first use models trained on VQA data to answer questions in our new datasets and record performance. We then explicitly train the same models with our new datasets, and make a comparison of performance with the pre-trained baseline.

Learning Curve: We train our models with an increasing number of logically composed questions and compare performance. This serves as an analysis of the number of logical samples needed by the model to understand logic in questions.

Training only with Closed Questions: In this ablation study, we restrict the training data to only closed questions i.e. "Yes-No" VQA questions, VQA-Compose and VQA-Supplement, allowing our model to focus solely on closed questions.

Compositional Generalization: We address whether training on closed questions containing single logical operation $(\neg Q_1, Q_1 \vee Q_2)$ can generalize to multiple operations $(Q_1 \wedge \neg Q_2, \neg Q_1 \vee Q_2)$. For instance, rows 1 through 6 in Table 1 are *single operation questions*, while rows 7 through 12 are *multi-operation questions*. Our aim is to have models that exhibit such compositional generalization.

Inductive Generalization: We investigate if training on compositions of two questions $(\neg Q_1 \vee Q_2)$ can generalize to compositions of more than two questions $(Q_1 \wedge \neg Q_2 \wedge Q_3 \ldots)$. This studies whether our models develop an understanding of logical connectives, as opposed to simply learning patterns from large data.

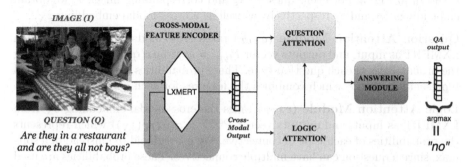

Fig. 3. LOL model architecture showing a cross-modal feature encoder followed by our Question-Attention (q_{ATT}) and Logic Attention (ℓ_{ATT}) modules. The concatenated output of is used by the Answering Module to predict the answer.

4   Method

In this section. we describe LXMERT [43] (a state-of-the-art VQA model), our Lens of Logic (LOL) model, attention modules which learn the question-type and logical connectives in the question, and the Fréchet-Compatibility (FC) Loss. This section refers to a composition of two questions, but applies to $n \geq 2$ questions.

4.1   Cross-Modal Feature Encoder

LXMERT (Learning Cross-Modality Encoder Representations from Transformers) [43] is one of the first cross-modal pre-trained frameworks for vision-and-language tasks, that combines a strong visual feature extractor [38] with a strong language model (BERT) [13]. LXMERT is pre-trained for key vision-and-language tasks, on a large corpus of ~9M image-sentence pairs, making it a powerful cross-modal encoder for vision+language tasks such as visual question answering, as compared to other models such as MCAN [45] and UpDn [2], and strong representative baseline for our experiments.

4.2   Our Model: Lens of Logic (LOL)

The design for our LOL model is driven by three key insights:

1. As logically composed questions are closed questions, understanding the type of question will guide the model to answer them correctly.
2. Predicted answers must be compatible with the predicted question type. For instance, a closed question can have an answer that is either "Yes" or "No".
3. The model must learn to identify the logical connectives in a question.

Given these insights, we develop the Question Attention module that encodes the type of question (*Yes-No, Number,* or *Other*), and the Logic Attention module that predicts the connectives (*AND, OR, NOT, no connective*) present in the question, and use these to learn representations. The overall model architecture is shown in Fig. 3. For every question Q and corresponding image I, we obtain embeddings z_Q and z_I respectively, as well as a cross-modal embedding z_X.

Question Attention Module (q_{ATT}) takes cross-modal embedding z_x from LXMERT as input, and outputs vector $P_{type} = softmax(\mathbf{q}_{ATT}(z_x))$, representing the probabilities of each question-type. These probabilities are used to get a final representation $\mathbf{z}^{\mathbf{type}}$ which combines the features for each question-type[1].

Logic Attention Module (ℓ_{ATT}) takes the cross-modal embedding z_X from LXMERT as input, and outputs vector $P^{conn} = \sigma(\ell_{ATT}(z_X))$ which represents the probabilities of each type of connective. We use sigmoid (σ) instead of a softmax, since a question can have multiple connectives. These probabilities are used to combine the features for each type of connective into a final representation $\mathbf{z}^{\mathbf{conn}}$ which encodes information about the connectives in the question.

---

[1] More training details in Supplementary Materials.

4.3   Loss Functions

We train our models jointly with the loss function given by:

$$\mathcal{L} = (1-\alpha_1-\alpha_2) \cdot \mathcal{L}_{ans} + \alpha_1 \cdot \mathcal{L}_{type} + \alpha_2 \cdot \mathcal{L}_{conn} + \beta \cdot \mathcal{L}_{FC}. \tag{2}$$

Answering Loss. ℓ_{ans} is conditioned on the type of question. We multiply the final prediction vector with the probability and the mask M_i for question-type i. M_i is a binary vector with 1 for every answer-index of type-i and 0 elsewhere:

$$\mathcal{L}_{ans} = \mathcal{L}_{BCE}\left(\sum_{i=1}^{3} \hat{y} \odot M_i \cdot P_i^{type}, y_{ans}\right). \tag{3}$$

Attention Losses: q_{ATT} is trained to minimize a Negative Log Likelihood (NLL) classification loss, ensuring a shrinkage of probabilities of the answer choices of the wrong type. ℓ_{ATT} is trained to minimize a multi-label classification loss, using Binary Cross-Entropy (BCE) given by:

$$\mathcal{L}_{type} = \mathcal{L}_{NLL}(\text{softmax}(z^{type}), y_{type}), \tag{4}$$
$$\mathcal{L}_{conn} = \mathcal{L}_{BCE}(\sigma(z^{conn}), y_{conn}), \tag{5}$$

where $y_{ans}, y_{type}, y_{conn}$ are labels for answer, question-type and connective.

Fréchet-Compatibility Loss: We introduce a new loss function that ensures compatibility between the answers predicted by the model for the component questions Q_1 and Q_2 and the composed question Q. Let A, A_1, A_2 be the respective answers predicted by the model for Q, Q_1, and Q_2. Q_i can have negation. Then Fréchet inequalities [7,15] provide us with bounds for the probabilities of the answers of the conjunction and disjunction of the two questions:

$$max(0, p(A_1) + p(A_2) - 1) \leq p(A_1 \wedge A_2) \leq min(p(A_1), p(A_2)). \tag{6}$$
$$max(p(A_1), p(A_2)) \leq p(A_1 \vee A_2) \leq min(1, p(A_1) + p(A_2)). \tag{7}$$

We define "Fréchet bounds" b_L and b_R to be the left and right bounds for the triplet A, A_1, A_2, and the "Fréchet Mean" m_A to be the average of the Fréchet bounds; $m_A = (b_L + b_R)/2$. Then, the Fréchet-Compatibility Loss given by:

$$\mathcal{L}_{FC} = (p(A) - \mathbb{1}(m_A > 0.5))^2, \tag{8}$$

ensures that the predicted answer and that determined by m_A match.

4.4   Implementation Details

The LXMERT feature encoder produces a vector z of length 768 which is used by our attention modules, each having sub-networks $\mathbf{f_i}, \mathbf{g_i}$ with 2 feed-forward layers. We first train our models without FC loss. Then we select the best models with a checkpoint of 10 epochs and finetune these further for 3 epochs with FC loss, since the FC loss is designed to work for a model whose predictions are not random. Thus our improvements in accuracy are attributable to the FC Loss and not more training epochs. We utilize the Adam optimizer [23] with a learning rate of 5e-5, batch size of 32 and train for 20 epochs. Our models are trained on 4 NVIDIA V100 GPUs, and take approximately 24 h for training 20 epochs (see footnote 1).

Table 2. Comparison of LXMERT and LOL trained on VQA data, combinations with Compose, Supplement, and our Frechet-Compatibility (FC) Loss (In all tables, best overall scores are bold, our best scores underlined)

Model	Trained on	Validation accuracy (%) ↑			
		VQA	YN	Comp	Supp
LXMERT	VQA	68.94	**86.65**	50.79	50.51
	VQA + Comp	67.85	85.32	85.03	80.85
	VQA + Comp + Supp	68.83	84.83	70.28	85.17
with FC Loss	VQA + Comp + Supp	67.84	84.92	75.31	85.25
LOL (qATT)	VQA	**69.08**	<u>85.32</u>	48.99	50.54
	VQA + Comp	67.51	84.82	84.85	79.62
	VQA + Comp + Supp	68.72	84.99	79.88	87.12
LOL (Full)	VQA + Comp	68.94	85.15	<u>85.13</u>	79.02
	VQA + Comp + Supp	68.86	84.87	81.07	87.54
with FC Loss	VQA + Comp + Supp	68.10	84.75	82.39	<u>**87.80**</u>
LXMERT	YN + Comp	–	84.13	84.44	79.39
	YN + Comp + Supp	–	84.09	82.63	88.15
LOL (ℓATT)	YN + Comp	–	85.22	<u>85.31</u>	79.87
	YN + Comp + Supp	–	85.26	84.37	<u>**89.00**</u>

5   Experiments

We first conduct analytical experiments to test for logical robustness and transfer learning capability. We use three datasets for our experiment: the VQA v2.0 [3] dataset, a combination of VQA and our VQA-Compose dataset, and a combination of VQA, VQA-Compose and VQA-Supplement. The size of the training dataset and the distribution of yes-no, number and other questions is kept the same as the original VQA dataset (~443k) for fair comparison. Since VQA-Supplement uses captions and objects from MS-COCO, we use is to analyze the ability of our models to generalize to a new source of data (MS-COCO) as well as questions containing adversarial objects. After training, our attention modules (q_{ATT} and ℓ_{ATT}) achieve an accuracy of 99.9% on average, showing almost perfect performance when it comes to learning the type of question and the logical connectives present in the question.

Table 3. Validation accuracies (%) for Compositional Generalization and Commutative Property. Note that 50% is random performance. (In all tables, best overall scores are bold, our best scores are underlined)

Model	YN	VQA-Compose		VQA-Supplement	
		Single	Multiple	Single	Multiple
LXMERT	85.07	83.95	61.99	86.65	60.00
LOL	85.12	84.60	66.03	87.42	66.05

Model	VQA-Compose		VQA-Supplement	
	$Q_1 \circ Q_2$	$Q_2 \circ Q_1$	$Q_1 \circ Q_2$	$Q_2 \circ Q_1$
LXMERT	82.34	80.44	85.57	81.78
LOL	84.91	83.64	85.62	83.41

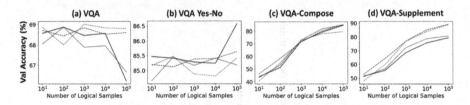

Fig. 4. Learning Curve comparison for models (Red: LXMERT, Blue: LOL) trained on our datasets (solid lines: VQA + Comp, dotted lines: VQA + Comp + Supp) (Color figure online)

5.1 Can't We Just Parse the Question into Components?

Since our questions are a composition of multiple questions, an obvious approach is to split the question into its components, and to discern the logical formula for composition. The answers to these component questions (predicted by VQA models) can be *re-combined* with the predicted logical formula to obtain the final answer. We use parsers to map components and logical operations to predefined slots in a logical function. The oracle parser uses the ground truth component questions and combines predicted answers using the true formula. However, at test time we do not have access to the true mapping and components. So we train a RoBERTa-Base [26] parser using B-I-O tagging [36] for a Named-Entity Recognition task with constituent questions as entities (see footnote 1).

The performance of the oracle parser serves as the upper bound as we have a perfect mapping, with the QA system being the only source of error. The trained parser has an exact-match accuracy of 85%, but only a 72% accuracy in determining the number of operands. The parser has an accuracy of 89% for questions with 3 or less operands, but only 78% for longer compositions. End-to-end (E2E) models do not need to parse questions and hence overcome these hurdles, but do require an understanding of logical operations. Table 4 shows that both oracle and trained parsers when used with LOL outperform parsers with LXMERT, by 6.82%) and 5.60% respectively. The LOL model without using any parsers is better than both LXMERT and LOL with the trained parser by 7.55% and 1.95% respectively.

5.2 Explicit Training with Logically Composed Questions

Can models trained on the VQA-v2 dataset answer logically composed questions? The first section of Table 2 shows that LXMERT, when trained only on questions from VQA-v2 has near random accuracy ($\sim$50%) on our logically composed datasets, thus exhibiting little robustness to such questions.

Can baseline model improve if trained explicitly with logically composed questions? We train the models with data containing a combination of samples from VQA-v2, VQA-Compose, and VQA-Supplement. The accuracy on VQA-Compose and VQA-Supplement improves, but there is a drop in performance on yes-no questions from VQA. Our models with our attention modules (q_{ATT} and ℓ_{ATT}) are able to retain performance on VQA-v2 while achieving improvements on all validation datasets.

5.3 Analysis

Training with Closed Questions only: We analyse the performance of models when trained only with closed questions from VQA, VQA + Comp and VQA + Comp + Supp and see that our model achieves the best accuracy on logically composed questions, as shown in Sects. 3 and 4 in Table 2. Since we train only closed questions, we do not use our question attention module for this experiment.

Effect of Logically Composed Questions: We increase the number of logical samples in the training data on a log scale from 10 to 100k. As can be seen from the learning curves in Fig. 4(a), models trained on VQA + Comp + Supp are able to retain performance on VQA validation data, while those trained only on VQA + Comp data deteriorate. Figure 4(b) shows that our models improve on VQA Yes-No performance after being trained on more logically composed samples, exhibiting transfer learning capabilities. In (c) both our models are comparable to the baseline, but our model shows improvements over the baseline when trained on VQA + Comp + Supp. In (d) for all levels of additional logical questions, our model trained on VQA + Comp + Supp is the best performing. From (c) and (d), we observe that a large number of logical questions are needed during training for the models to learn to answer them during inference. We also see that our model yields the best performance on VQA-Supplement.

Compositional Generalization: To test for compositional generalization, we train models on questions with a maximum of one connective (single) and test on those with multiple connectives. It can be seen from Table 3 that our models are better equipped than the baseline to generalize to multiple connectives and also to be able to generalize from VQA-Compose to Supplement.

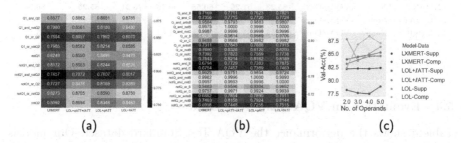

Fig. 5. Accuracy for each type of question in (a) VQA-Compose, (b) VQA-Supplement and for questions with number of operands greater than 2.

Inductive Generalization: We test our models on questions composed with more than two components. Parser-based models have this property by default. As shown by Fig. 5c our E2E models outperform the baseline LXMERT.

Commutative Property: Our models have identical answers when the question is composed either as $Q_1 \circ Q_2$ or $Q_2 \circ Q_1$, for logical operation $\circ$, as shown in Table 3. The parser-based models are agnostic to the order of components if the parsing is accurate, while our E2E models are robust to the order.

Accuracy per Category of Question Composition: In Fig. 5 we show a plot of accuracy versus question type for each model. Q, Q_1, Q_2 are questions from VQA, B, C are object-based and caption-based questions from COCO respectively. From the results, we interpret that questions such as $Q \wedge antonym(B), Q \wedge \neg B, Q \wedge \neg C$ are easy because the model is able to understand absence of objects, therefore can always answer these questions with a "NO". Similarly, $Q \vee B, Q \vee C$ are easily answered since presence of the object makes the answer always "YES". By simply understanding object presence many such questions can be answered. Figure 5 shows the model has the same accuracy for logically equivalent operations.

Table 4. Performance on 'test-standard' set of VQA-v2 and validation set of our datasets. LOL performance is close to SOTA on VQA-v2, but significantly better at logical robustness. *MCAN uses a fixed vocabulary that prohibits evaluation on VQA-Supplement which has questions created from COCO captions. #Test-dev scores, since MCAN does not report test-std single-model scores (In all tables, best overall scores are bold, our best scores underlined)

Model	Parser	Training data	Test-Std. Accuracy (%)↑				Val. Accuracy (%) ↑		
			Yes-No	Number	Other	Overall	Compose	Supplement	Overall
MCAN	None	VQA [45]	86.82#	53.26#	60.72#	70.90	52.42	*	*
LXMERT	None	VQA [43]	**88.20**	**54.20**	**63.10**	**72.50**	50.79	50.51	50.65
LOL (qATT)	None	VQA	87.33	54.03	62.40	72.03	48.99	50.54	49.77
LXMERT	Oracle	VQA	88.20	54.20	63.10	72.50	86.38	74.29	80.33
LXMERT	Trained	VQA	88.20	54.20	63.10	72.50	86.35	68.75	77.55
LOL (full)	Oracle	VQA+Ours	86.55	53.42	61.58	71.04	85.79	88.51	87.15
LOL (full)	Trained	VQA+Ours	86.55	53.42	61.58	71.04	82.13	84.17	83.15
LXMERT	None	VQA+Ours	85.23	51.25	60.58	69.78	75.31	85.25	80.28
LOL (qATT)	None	VQA+Ours	86.79	52.66	61.85	71.19	79.88	87.12	83.50
LOL (full)	None	VQA+Ours	86.55	53.42	61.58	71.04	**82.39**	**87.80**	85.10

5.4   Evaluation on VQA V2.0 Test Data

Table 4 shows the performance the VQA Test-Standard dataset. Our models maintain overall performance on the VQA test dataset, and at the same time substantially improve from random performance (∼50%) on logically composed questions to 82.39% on VQA-Compose and 87.80% on VQA-Supplement. This shows that logical connectives in questions can be learned while not degrading the overall performance on the original VQA test set (our models are within ∼1.5% of the state-of-the-art on all three types of questions on the VQA test-set).

6   Discussion

Consider the example, *"Is every boy who is holding an apple or a banana, not wearing a hat?"*, humans are able to answer it to be true if and only if each boy who is holding *at least one* of an apple or a banana is not wearing a hat [11]. Natural language contains such complex logical compositions, not to mention ambiguities and the influence of context. In this paper, we focus on the simplest – negation, conjunction, and disjunction. We have shown that existing VQA models are not robust to questions composed with these logical connectives, even when we train parsers to split the question into its components. When humans are faced with such questions, they may refrain from giving binary (Yes/No) answers. For instance, logically, the question *"Did you eat the pizza and did you like it?"* has a negative answer if either of the two component questions has a negative answer. However, humans might answer the same question with the answer *"Yes, but I did not like it"*. While human question-answering is indeed elaborate, explanatory, and clarifying, that is the scope of our future work; here we focus only on predicting a single binary answer.

We have shown how connectives in a question can be identified by enhancing LXMERT encoders with dedicated attention modules and loss functions. We would like to stress on the fact that we do not use knowledge of the connectives during inference, but instead train the network to be aware of it based on cross-modal features, instead of predicting purely based on language model embeddings which fail to capture these nuances. Our work is an attempt to modularize the understanding of logical components to train the model to utilize the outputs of the attention modules. We believe this work has potential implications on logic-guided data augmentation, logically robust question answering, and for conversational agents (with or without images). Similar strategies and learning mechanisms may be used in the future to operate "logically" in the image-space at the level of object classes, attributes, or semantic segments.

7   Conclusion

In this work, we investigate VQA in terms of logical robustness. The key hypothesis is that the ability to answer questions about an image, must be extendable to a logical composition of two such questions. We show that state-of-the-art models trained on VQA dataset lack this. Our solution involves the "Lens of Logic" model architecture that learns to answer questions with negation, conjunction, and disjunction. We provide VQA-Compose and VQA-Supplement, two datasets containing logically composed questions to serve as benchmarks. Our models show improvements in terms of answering these questions, while at the same time retaining performance on the original VQA test-set.

Acknowledgments. Support from NSF Robust Intelligence Program (1816039 and 1750082), DARPA (W911NF2020006) and ONR (N00014-20-1-2332) is gratefully acknowledged.

References

1. Aditya, S., Yang, Y., Baral, C.: Integrating knowledge and reasoning in image understanding. In: Proceedings of the 28th International Joint Conference on Artificial Intelligence, IJCAI 2019, pp. 6252–6259. AAAI Press (2019). http://dl.acm.org/citation.cfm?id=3367722.3367926
2. Anderson, P., et al.: Bottom-up and top-down attention for image captioning and visual question answering. In: CVPR (2018)
3. Antol, S., et al.: VQA: visual question answering. In: Proceedings of the IEEE International Conference on Computer Vision, pp. 2425–2433 (2015)
4. Asai, A., Hajishirzi, H.: Logic-guided data augmentation and regularization for consistent question answering. In: Proceedings of the 58th Annual Meeting of the Association for Computational Linguistics, pp. 5642–5650. Association for Computational Linguistics (2020). https://www.aclweb.org/anthology/2020.acl-main.499
5. Bhattacharya, N., Li, Q., Gurari, D.: Why does a visual question have different answers? In: Proceedings of the IEEE International Conference on Computer Vision, pp. 4271–4280 (2019)
6. Bobrow, D.G.: Natural language input for a computer problem solving system (1964)
7. Boole, G.: An investigation of the laws of thought: on which are founded themathematical theories of logic and probabilities. Dover Publications (1854)
8. Bordes, A., Usunier, N., Garcia-Duran, A., Weston, J., Yakhnenko, O.: Translating embeddings for modeling multi-relational data. In: Advances in Neural Information Processing Systems, pp. 2787–2795 (2013)
9. Bowman, S.R., Potts, C., Manning, C.D.: Recursive neural networks can learn logical semantics. arXiv preprint arXiv:1406.1827 (2014)
10. Carey, S.: Conceptual Change in Childhood. MIT Press, Cambridge (1985)
11. Cesana-Arlotti, N., Martín, A., Téglás, E., Vorobyova, L., Cetnarski, R., Bonatti, L.L.: Precursors of logical reasoning in preverbal human infants. Science **359**(6381), 1263–1266 (2018).https://doi.org/10.1126/science.aao3539. https://science.sciencemag.org/content/359/6381/1263
12. Corcoran, J.: Completeness of an ancient logic. J. Symb. Logic **37**(4), 696–702 (1972)
13. Devlin, J., Chang, M.W., Lee, K., Toutanova, K.: Bert: pre-training of deep bidirectional transformers for language understanding. arXiv preprint arXiv:1810.04805 (2018)
14. Fang, Z., Gokhale, T., Banerjee, P., Baral, C., Yang, Y.: Video2commonsense: generating commonsense descriptions to enrich video captioning. arXiv preprint arXiv:2003.05162 (2020)
15. Fréchet, M.: Généralisation du théoreme des probabilités totales. Fundamenta Mathematicae **1**(25), 379–387 (1935)
16. Gopnik, A., Meltzoff, A.N., Kuhl, P.K.: The Scientist in the Crib: Minds, Brains, and How Children Learn. William Morrow & Co (1999)
17. Goyal, Y., Khot, T., Summers-Stay, D., Batra, D., Parikh, D.: Making the V in VQA matter: elevating the role of image understanding in visual question answering. In: Proceedings of the IEEE Conference on Computer Vision and Pattern Recognition, pp. 6904–6913 (2017)
18. Hegel, G.W.F.: Hegel's science of logic (1929)

19. Horn, L.R., Kato, Y.: Negation and Polarity: Syntactic and Semantic Perspectives. OUP, Oxford (2000)
20. Hudson, D.A., Manning, C.D.: GQA: a new dataset for compositional question answering over real-world images. arXiv preprint arXiv:1902.09506 (2019)
21. Johnson, J., Hariharan, B., van der Maaten, L., Fei-Fei, L., Lawrence Zitnick, C., Girshick, R.: Clevr: a diagnostic dataset for compositional language and elementary visual reasoning. In: Proceedings of the IEEE Conference on Computer Vision and Pattern Recognition, pp. 2901–2910 (2017)
22. Kassner, N., Schütze, H.: Negated lama: birds cannot fly. arXiv preprint arXiv:1911.03343 (2019)
23. Kingma, D.P., Ba, J.: Adam: a method for stochastic optimization. arXiv preprint arXiv:1412.6980 (2014)
24. Lewis, M., Steedman, M.: Combined distributional and logical semantics. Trans. Assoc. Comput. Linguist. 1, 179–192 (2013). https://doi.org/10.1162/tacl_a_00219. https://www.aclweb.org/anthology/Q13-1015
25. Lin, T.-Y., et al.: Microsoft COCO: common objects in context. In: Fleet, D., Pajdla, T., Schiele, B., Tuytelaars, T. (eds.) ECCV 2014. LNCS, vol. 8693, pp. 740–755. Springer, Cham (2014). https://doi.org/10.1007/978-3-319-10602-1_48
26. Liu, Y., et al.: Roberta: a robustly optimized bert pretraining approach. arXiv preprint arXiv:1907.11692 (2019)
27. Lu, J., Batra, D., Parikh, D., Lee, S.: Vilbert: pretraining task-agnostic visiolinguistic representations for vision-and-language tasks. In: Advances in Neural Information Processing Systems, pp. 13–23 (2019)
28. Malinowski, M., Fritz, M.: A multi-world approach to question answering about real-world scenes based on uncertain input. In: Advances in Neural Information Processing Systems, pp. 1682–1690 (2014)
29. Mao, J., Gan, C., Kohli, P., Tenenbaum, J.B., Wu, J.: The neuro-symbolic concept learner: interpreting scenes, words, and sentences from natural supervision. In: International Conference on Learning Representations (2019). https://openreview.net/forum?id=rJgMlhRctm
30. Mintz, M., Bills, S., Snow, R., Jurafsky, D.: Distant supervision for relation extraction without labeled data. In: Proceedings of the Joint Conference of the 47th Annual Meeting of the ACL and the 4th International Joint Conference on Natural Language Processing of the AFNLP, vol. 2, pp. 1003–1011. Association for Computational Linguistics (2009)
31. Morante, R., Sporleder, C.: Modality and negation: an introduction to the special issue. Comput. Linguist. 38(2), 223–260 (2012)
32. Neelakantan, A., Roth, B., McCallum, A.: Compositional vector space models for knowledge base completion. arXiv preprint arXiv:1504.06662 (2015)
33. Pennington, J., Socher, R., Manning, C.: Glove: global vectors for word representation. In: Proceedings of the 2014 Conference on Empirical Methods in Natural Language Processing (EMNLP), pp. 1532–1543 (2014)
34. Piattelli-Palmarini, M.: Language and learning: the debate between jean piaget and noam chomsky (1980)
35. Raju, P.: The principle of four-cornered negation in Indian philosophy. Rev. Metaphys. 694–713 (1954)
36. Ramshaw, L., Marcus, M.: Text chunking using transformation-based learning. In: Third Workshop on Very Large Corpora (1995). https://www.aclweb.org/anthology/W95-0107

37. Ren, M., Kiros, R., Zemel, R.: Exploring models and data for image question answering. In: Advances in Neural Information Processing Systems, pp. 2953–2961 (2015)
38. Ren, S., He, K., Girshick, R., Sun, J.: Faster R-CNN: towards real-time object detection with region proposal networks. In: Advances in Neural Information Processing Systems, pp. 91–99 (2015)
39. Riedel, S., Yao, L., McCallum, A., Marlin, B.M.: Relation extraction with matrix factorization and universal schemas. In: Proceedings of the 2013 Conference of the North American Chapter of the Association for Computational Linguistics: Human Language Technologies, Atlanta, Georgia, pp. 74–84. Association for Computational Linguistics, June 2013. https://www.aclweb.org/anthology/N13-1008
40. Rocktäschel, T., Bošnjak, M., Singh, S., Riedel, S.: Low-dimensional embeddings of logic. In: Proceedings of the ACL 2014 Workshop on Semantic Parsing, pp. 45–49 (2014)
41. Socher, R., Chen, D., Manning, C.D., Ng, A.: Reasoning with neural tensor networks for knowledge base completion. In: Advances in Neural Information Processing Systems, pp. 926–934 (2013)
42. Spinoza, B.D.: Ethics, translated by andrew boyle, introduction by ts gregory (1934)
43. Tan, H., Bansal, M.: Lxmert: learning cross-modality encoder representations from transformers. arXiv preprint arXiv:1908.07490 (2019)
44. Wittgenstein, L.: Tractatus Logico-Philosophicus. Routledge (1921)
45. Yu, Z., Yu, J., Cui, Y., Tao, D., Tian, Q.: Deep modular co-attention networks for visual question answering. In: The IEEE Conference on Computer Vision and Pattern Recognition (CVPR), June 2019
46. Zellers, R., Bisk, Y., Farhadi, A., Choi, Y.: From recognition to cognition: visual commonsense reasoning. In: Proceedings of the IEEE Conference on Computer Vision and Pattern Recognition, pp. 6720–6731 (2019)
47. Zettlemoyer, L.S., Collins, M.: Learning to map sentences to logical form: structured classification with probabilistic categorial grammars. arXiv preprint arXiv:1207.1420 (2012)
48. Zhou, L., Palangi, H., Zhang, L., Hu, H., Corso, J.J., Gao, J.: Unified vision-language pre-training for image captioning and VQA. In: AAAI, pp. 13041–13049 (2020)

Piggyback GAN: Efficient Lifelong Learning for Image Conditioned Generation

Mengyao Zhai[1]($\boxtimes$), Lei Chen[1], Jiawei He[1], Megha Nawhal[1], Frederick Tung[2], and Greg Mori[1]

[1] Simon Fraser University, Burnaby, Canada
{mzhai,chenleic,jha203,mnawhal}@sfu.ca
mori@cs.sfu.ca
[2] Borealis AI, Vancouver, Canada
frederick.tung@borealisai.com

Abstract. Humans accumulate knowledge in a lifelong fashion. Modern deep neural networks, on the other hand, are susceptible to catastrophic forgetting: when adapted to perform new tasks, they often fail to preserve their performance on previously learned tasks. Given a sequence of tasks, a naive approach addressing catastrophic forgetting is to train a separate standalone model for each task, which scales the total number of parameters drastically without efficiently utilizing previous models. In contrast, we propose a parameter efficient framework, Piggyback GAN, which learns the current task by building a set of convolutional and deconvolutional filters that are factorized into filters of the models trained on previous tasks. For the current task, our model achieves high generation quality on par with a standalone model at a lower number of parameters. For previous tasks, our model can also preserve generation quality since the filters for previous tasks are not altered. We validate Piggyback GAN on various image-conditioned generation tasks across different domains, and provide qualitative and quantitative results to show that the proposed approach can address catastrophic forgetting effectively and efficiently.

Keywords: Lifelong learning · Generative adversarial networks

1 Introduction

Humans are lifelong learners: in order to function effectively day-to-day, we acquire and accumulate knowledge throughout our lives. The accumulated knowledge makes us efficient and versatile when we encounter new tasks. In contrast to human learning, modern neural network based learning algorithms usually fail to remember knowledge acquired from previous tasks when adapting to a new task (see Fig. 1(b)). In other words, it is difficult to generalize once a

© Springer Nature Switzerland AG 2020
A. Vedaldi et al. (Eds.): ECCV 2020, LNCS 12366, pp. 397–413, 2020.
https://doi.org/10.1007/978-3-030-58589-1_24

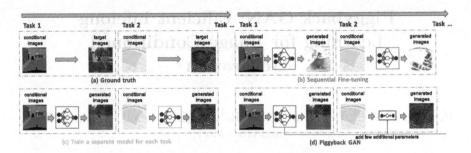

Fig. 1. Lifelong learning of image-conditioned generation. The goal of lifelong learning is to build a model capable of adapting to tasks that are encountered sequentially. Traditional fine-tuning methods are susceptible to catastrophic forgetting: when we add new tasks, the network forgets how to perform previous tasks (Fig. 1(b)). Storing a separate model for each task addresses catastrophic forgetting in an inefficient way as each set of parameters is only useful for a single task (Fig. 1(c)). Our Piggyback GAN achieves image-conditioned generation with high image quality on par with separate models at a lower parameter cost by efficiently utilizing stored parameters (Fig. 1(d)).

model is trained on a task. This is the well-known phenomenon of *catastrophic forgetting* [25]. Recent efforts [2,6,20,32] have demonstrated how discriminative models can continually learn a sequence of tasks. Despite the success of these efforts for discriminative models, lifelong learning for generative models remains a challenging and under-explored area.

Lifelong learning methods for discriminative models cannot be directly applied to generative models due to their intrinsic differences. *First,* it is well known that in classification tasks, the intermediate convolutional layers in deep neural networks are capable of providing generic features. These features can easily be reused by other models with varying and different classification goals. For generative models, the possibility of such reuse for new generative tasks has not been previously explored, to the best of our knowledge. *Second,* different from discriminative models, the output space of generative models is usually continuous, making it more challenging for the model to maintain the generation quality along the training of a sequence of tasks. *Third,* there could be conflicts between tasks under the generative setting. For discriminative models, it rarely happens that one image has different labels (appears in different tasks). However, for generative models, it is quite common that we want to translate the same image to different domains for different tasks [8,11,14,44].

One of the naive approaches to address catastrophic forgetting for generative models is to train a model for each task separately (see Fig. 1(c)). Unfortunately, this approach is not scalable in general: as new tasks are added, the storage requirements grow drastically. More importantly, setting the trained model aside without exploiting the benefit it can potentially provide facing a new task would be an inefficient use of resources. Recent works [37,42] have shown promising results on lifelong learning for generative models, but it is also revealed that image quality degradation and artifacts transfer from old to new tasks are inevitable. Therefore, it is valuable to have a continual learning framework designed that (1) is more parameter efficient, (2) preserves the generation quality of both current and previously learned tasks, and (3) can enable various conditional generation tasks across different domains.

In this paper, we introduce a generic continual learning framework *Piggyback GAN* that can perform various conditional image generation tasks across different domains (see Fig. 1(d)). The proposed approach addresses the catastrophic forgetting problem in generative continual learning models. Specifically, *Piggyback GAN* maintains a *filter bank*, the filters in which come from convolution and deconvolution layers of models trained on previous tasks. Facing a new task, Piggyback GAN learns to perform this task by reusing the filters in the filter bank by building a set of *piggyback filters* which are factorized into filters from the filter bank. Piggyback GAN also maintains a small portion of unconstrained filters for each task to ensure high-quality generation. Once the new task is learned, the unconstrained filters are appended to the filter bank to facilitate the training of subsequent tasks. Since the filters for the old task have not been altered, Piggyback GAN is capable of keeping the exact knowledge of previous tasks without forgetting any details.

To summarize, our contributions are as follows. We propose a continual learning framework for conditional image generation that is (1) *efficient*, learning to perform a new task by "piggybacking" on models trained on previous tasks and reusing the filters from previous models, (2) *quality-preserving*, maintaining the generation quality of current task and ensuring no quality degradation for previous tasks while saving more parameters, and (3) *generic*, enabling various conditional generation tasks across different domains. To the best of our knowledge, we are the first to make these contributions for generative continual learning models. We validate the effectiveness of our approach under two settings: (1) paired image-conditioned generation, and (2) unpaired image-conditioned generation. Extensive qualitative and quantitative comparisons with state-of-the-art models are carried out to illustrate the capability and efficiency of our proposed framework to learn new generation tasks without the catastrophic forgetting of previous tasks.

2   Related Work

Our work intersects two previously disjoint lines of research: lifelong learning of generative models, and parameter-efficient network learning.

Lifelong Learning of Generative Models. For discriminative tasks, e.g. classification, many works have been proposed recently for solving the problem of catastrophic forgetting. Knowledge distillation based approaches [4,20,32] work by minimizing the discrepancy between the output of the old and new network. Regularization-based approaches [2,5,6,19,22] regularize the network parameters when learning new tasks. Task-based approaches [3,28,31] adopt task-specific modules to learn each task.

For generative tasks, lifelong learning is an underexplored area and relatively less work studies the problem of catastrophic forgetting. Continual generative modeling was first introduced by Seff et al. [30], which incorporated the idea of *Elastic Weight Consolidation (EWC)* [19] into the lifelong learning for GANs. The idea of *memory replay*, in which images generated from a model trained on previous tasks are combined with the training images for the current task to allow for training of multiple tasks, is well explored by Wu et al. [37] for label-conditioned image generation. However, this approach is limited to label-conditioned image generation and is not applicable for image-conditioned image generation since no ground-truth conditional inputs of previous tasks are provided. *EWC* [19] has been adapted from classification tasks to generative tasks of label-conditioned image generation [30,37], but they present limited capability in both remembering previous tasks and generating realistic images. *Lifelong GAN* [42] is a generic knowledge-distillation based approach for conditioned image generation in lifelong learning. However, the image quality of generated images of previous tasks keeps decreasing when new tasks are learned. Moreover, all approaches mentioned above fail in the scenario when there are conflicts in the input-output space across tasks. To the best of our knowledge, Piggyback GAN is the first method that can preserve the generation quality of previous tasks, and enables various conditional generation tasks across different domains.

Parameter-Efficient Network Learning. In recent years, there has been increasing interest in better aligning the resource consumption of neural networks with the computational constraints of the platforms on which they will be deployed [7,36,38]. One important avenue of active research is parameter efficiency. Trained neural networks are typically highly over-parameterized [10]. There are many effective strategies for improving the parameter efficiency of trained networks: for example, pruning algorithms learn to remove redundant or less important parameters [12,40], and quantization algorithms learn to represent parameters using fewer bits while maintaining task performance [15,17,43]. Parameter efficiency can also be encouraged during training via architecture design [23,29], sparsity inducing priors [33], resource re-allocation [26], or knowledge distillation from a teacher network [34,41]. Our work is orthogonal to these methods, which may be applied on top of Piggyback GAN.

The focus of this work is parameter-efficient lifelong learning: we aim to leverage weight reuse and adaptation to improve parameter efficiency when extending a trained network to new tasks, while avoiding the catastrophic forgetting of previous tasks. Several approaches have been explored for parameter-efficient lifelong learning in a discriminative setting. For example, progressive networks [28] train a new network for each task and transfer knowledge from previous tasks by learning lateral connections, a linear combination of layer activations from previous tasks is computed and added to the inputs to the corresponding layer of the new task. While our approach computes a linear combination of filters to construct new filters and uses them the same as normal filters. Rebuffi et al. [27] aims to train a universal vector that are shared among all domains providing generic filters for all domains, and task-specific adapter modules are added to adjust the trained network to new tasks. While we are working on solving a sequence of tasks, the filters trained for the initial task are hardly generic enough for later tasks. Inspired by binary neural networks, Mallya et al. [24] introduced the piggyback masking method (which inspired the naming of Piggyback GAN). This method learns a binary mask over the network parameters for each new task. Tasks share the same backbone parameters but differ in the parameters that are enabled. Multi-task attention networks [21] learn soft attention modules for each new task. Task-specific features are obtained by scaling features with an input-dependent soft attention mask. To the best of our knowledge, Piggyback GAN is the first method for parameter-efficient lifelong learning in a generative setting (Fig. 2).

3   Method

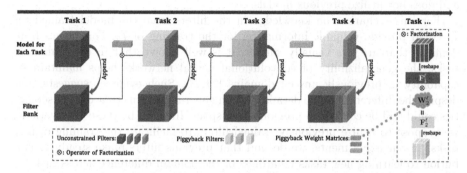

Fig. 2. Overview of Piggyback GAN. Piggyback GAN maintains a *filter bank* for composing convolutional filters to be used for the next task. Given a new task, most filters in the model are trained as *piggyback filters* that are factorized into filters in the filter bank. The remaining small portion of filters are learned without constraints and appended to the filter bank to benefit the training of subsequent tasks.

In this paper, we study the problem of lifelong learning for generative models. The overall goal is to learn the model for a given sequence of generation tasks $\{T_i\}_{i=1}^{N}$ under the assumption that the learning process has access to only one task at a time and training data for each task is accessible only once.

Assume that we have a learnt model M trained to perform the first task T_1. To proceed with the new task T_2, one naive approach is to adapt the learnt model M to the new task by fine-tuning its parameters. This approach suffers from the problem of catastrophic forgetting of task T_1. Another naive approach is to learn a separate standalone model M_i for each new task T_i and store all learnt models $\{M_i\}_{i=1}^{N}$, which drastically scales up total number of parameters without utilizing previous models.

The key idea of this paper is to build a parameter efficient lifelong generative model by "piggybacking" on existing models and reusing the filters from models trained on previous tasks while maintaining the generation quality to be similar to a single standalone model for all tasks. We achieve this by maintaining a filter bank, and compose the filters for new tasks mostly by factorizing it into filters in the filter bank, with a small portion of filters learnt without constraints. Once the new task is learnt, the filter bank is expanded by adding those unconstrained filters.

3.1   Piggyback Filter Learning

For lifelong learning of generative models, given an initial task T_1 and an upcoming new task T_2, there could be cases where T_1 and T_2 share similar inputs but different outputs, e.g. T_1 is *Photo → Monet Paintings* and T_2 is *Photo → Ukiyo-e Paintings*. These conflicts in input-output space do not occur when each task is trained separately. However, as already mentioned, learning a separate model for each task and storing all learnt models is memory consuming and does not utilize previous models.

For this scenario, the knowledge in the filters from the model trained for T_1 could provide valuable information for the training of T_2. For images from domains that are less visually similar, it is well known that in discriminative tasks, the generaliziblity of convolutional neural networks helps maintain its capability in previously unseen domains [24,27]. Such evidence suggests that despite the difference in fine details, certain general patterns that convey semantics could still be covered in previous filter space. Inspired by these observations, we propose to utilize previously trained filters to facilitate the training of new tasks. In our experiments, we do find that previous filters make steady contribution to learning new tasks regardless of the domain difference across tasks.

Consider a generator network G_1 learnt for initial task T_1 consisting of L layers of filters $\{F_1^\ell\}_{\ell=1}^L$ where ℓ denotes the index of layers. For the ℓth layer, let the kernel size be $s_w^\ell \times s_h^\ell$, number of input channels be c_{in}^ℓ and number of output channels be c_{out}^ℓ. Then F_1^ℓ is reshaped from 4D into 2D of size $s_w^\ell \times s_h^\ell \times c_{in}^\ell$ by c_{out}^ℓ, which we denote by $\mathcal{R}(F_1^\ell)$. For the upcoming task T_2, we learn the filters F_2^ℓ through factorization operation over filters of the corresponding layers in task T_1, namely

$$F_2^\ell = \mathcal{R}^{-1}(\mathcal{R}(F_1^\ell) \otimes W_2^\ell), \tag{1}$$

where $\otimes$ denotes the standard matrix multiplication operation and $\mathcal{R}^{-1}$ denotes the inverse of reshape operation from 2D back to 4D. The derived filters F_2^ℓ is denoted as *piggyback filters* and the corresponding learnable weights $W_2^\ell \in \mathbb{R}^{c_{out}^\ell \times c_{out}^\ell}$ is denoted as *piggyback weight matrix*. The resulting $F_2^\ell \in \mathbb{R}^{s_w^\ell \times s_h^\ell \times c_{in}^\ell \times c_{out}^\ell}$ is of the same size as F_1^ℓ. A bias term could be added to the factorized filters to adjust the final output.

Therefore, by constructing factorized filters, the number of parameters required to be learnt for this layer is reduced from $s_w^\ell \times s_h^\ell \times c_{in}^\ell \times c_{out}^\ell$ to only $c_{out}^\ell \times c_{out}^\ell$.

3.2   Unconstrained Filter Learning

Parameterizing a new task completely by making use of filters in the initial tasks, though saving substantial storage, cannot capture the key differences between tasks. The filters derived from previous tasks may only characterize certain attributes and may not generalize well to the new task, resulting in poor quality of generation details for the new model. Learning a small number of unconstrained filters gives the model greater flexibility in adapting to the new task. Moreover, learning these unconstrained filters could learn to generate patterns that do not exist in previous tasks, thereby, increasing the power of the model and helping the training of later tasks.

Specifically, for each layer in the generator, we allocate a small portion of unconstrained filters which are learned freely without the constraint to be constructed from previous filters. With a total of c_{out}^{ℓ} filters in F_2^{ℓ}, consider the proportion of unconstrained filters to be $\lambda\,(0 < \lambda \leq 1)^1$, then we have λc_{out}^{ℓ} unconstrained filters and $(1 - \lambda)c_{out}^{\ell}$ piggyback filters, and the size of *piggyback weight matrix* W_2^{ℓ} has changed to $c_{out}^{\ell} \times (1-\lambda)c_{out}^{\ell}$. Let the unconstrained filters at the lth layer of the nth task T_n be $F_n^{u,\ell}$, and piggyback filters at the lth layer of the nth task T_n be $F_n^{p,\ell}$.

For task T_1, all filters are unconstrained filters, namely

$$F_1^{\ell} = F_1^{u,\ell}. \tag{2}$$

For task T_2, we re-define the filters in F_l^2 to be the concatenations of unconstrained filters $F_2^{u,\ell}$ and piggyback filters $F_2^{p,\ell}$, namely F_2^{ℓ} is formulated as

$$\begin{aligned} F_2^{\ell} &= [F_2^{u,\ell}, F_2^{p,\ell}] \\ &= [F_2^{u,\ell}, \mathcal{R}^{-1}(\mathcal{R}(F_1^{u,\ell}) \otimes W_2^{\ell})], \end{aligned} \tag{3}$$

where $F_2^{u,\ell} \in \mathbb{R}^{s_w^{\ell} \times s_h^{\ell} \times c_{in}^{\ell} \times \lambda c_{out}^{\ell}}$, $F_2^{p,\ell} \in \mathbb{R}^{s_w^{\ell} \times s_h^{\ell} \times c_{in}^{\ell} \times (1-\lambda)c_{out}^{\ell}}$ and the resulting $F_2^{\ell} \in \mathbb{R}^{s_w^{\ell} \times s_h^{\ell} \times c_{in}^{\ell} \times c_{out}^{\ell}}$ is of the same size as F_1^{ℓ}.

3.3   Expanding Filter Bank

The introduction of unconstrained filters learnt from task T_2 brings extra storage requirement and it would be a waste if we set it aside when learning the filters for the upcoming new tasks. Therefore, instead of setting it aside, we construct the piggyback filters for task T_3 by making use of the unconstrained filters from both T_1 and T_2.

We refer to the full set of unconstrained filters as the *filter bank* which will be expanded every time new unconstrained filters are learnt for a new task. Expanding the *filter bank* could encode more diverse patterns that were not captured or did not exist in previous tasks. By doing so, the model increases

---

^{1} λc_{out}^{ℓ} could be rounded to the nearest integer. Or λ could be chosen to make λc_{out}^{ℓ} an integer.

the modality of the learned filters and provides more useful information for the subsequent tasks.

When learning the task T_3, the filter bank comprises filters corresponding to the task T_1 and T_2 and is given by $F_1^{u,\ell}$ and $F_2^{u,\ell}$, resulting in the size of the filter bank $s_w^\ell \times s_h^\ell \times c_{in}^\ell \times (1+\lambda)c_{out}^\ell$. Similarly, for task T_3 consider the proportion of unconstrained filters to be same as T_2, i.e. λ, the size of *piggyback weight matrix* W_3^ℓ would be $(1+\lambda)c_{out}^\ell \times (1-\lambda)c_{out}^\ell$. We construct F_3^ℓ as the concatenation of unconstrained filters $F_3^{u,\ell}$ and piggyback filters $F_3^{p,\ell}$ that are constructed from all filters in the *filter bank*, namely F_3^ℓ is formulated as

$$
\begin{aligned}
F_3^\ell &= [F_3^{u,\ell}, F_3^{p,\ell}] \\
&= [F_3^{u,\ell}, \mathcal{R}^{-1}(\mathcal{R}([F_1^{u,\ell}, F_2^{u,\ell}]) \otimes W_3^\ell)],
\end{aligned}
\tag{4}
$$

where $F_3^{u,\ell} \in \mathbb{R}^{s_w^\ell \times s_h^\ell \times c_{in}^\ell \times \lambda c_{out}^\ell}$, $F_3^{p,\ell} \in \mathbb{R}^{s_w^\ell \times s_h^\ell \times c_{in}^\ell \times (1-\lambda)c_{out}^\ell}$ and the resulting $F_3^\ell \in \mathbb{R}^{s_w^\ell \times s_h^\ell \times c_{in}^\ell \times c_{out}^\ell}$ is of the same size as F_1^ℓ and F_2^ℓ.

After task T_3 is learnt, the *filter bank* is expanded as $[F_1^{u,\ell}, F_2^{u,\ell}, F_3^{u,\ell}]$, whose size is $s_w^\ell \times s_h^\ell \times c_{in}^\ell \times (1+2\lambda)c_{out}^\ell$.

To summarize, when learning task T_n, F_n^ℓ could be written as

$$
\begin{aligned}
F_n^\ell &= [F_n^{u,\ell}, F_n^{p,\ell}] \\
&= [F_n^{u,\ell}, \mathcal{R}^{-1}(\mathcal{R}([F_1^{u,\ell}, F_2^{u,\ell}, ..., F_{n-1}^{u,\ell}]) \otimes W_n^\ell)].
\end{aligned}
\tag{5}
$$

After task T_n is learnt, the *filter bank* is expanded as $[F_1^{u,\ell}, F_2^{u,\ell}, ..., F_n^{u,\ell}]$, whose size is $s_w^\ell \times s_h^\ell \times c_{in}^\ell \times (1+(n-1)\lambda)c_{out}^\ell$, and the size of *piggyback weight matrix* W_{n+1}^ℓ for task T_{n+1} is $(1+(n-1)\lambda)c_{out}^\ell \times (1-\lambda)c_{out}^\ell$. It should be noted that the weights of filters in the *filter bank* remain fixed along the whole learning process.

3.4   Learning Piggyback GAN

We explore two conditional generation scenarios in this paper: (1) *paired image-conditioned generation*, in which training data contains M pairs of samples $\{(a_i, b_i)\}_{i=1}^M$, where $\{a_i\}_{i=1}^M$ represent conditional images, $\{b_i\}_{i=1}^M$ represent target images, and correspondence between a_i and b_i exists, namely for any conditional image a_i the corresponding target image b_i is also provided; (2) *unpaired image-conditioned generation*, in which training data contains images from two domains A and B, namely images $\{a_i\}_{i=1}^{M_a} \in A$ and images $\{b_i\}_{i=1}^{M_b} \in B$, and correspondence between a_i and b_i does not exist, namely for any conditional image a_i the corresponding target image b_i is not provided.

For both conditional generation scenarios, given a sequence of tasks and a state-of-the-art GAN model, we construct the convolutional and deconvolutional filters in the generator as described in Sect. 3.1, Sect. 3.2 and Sect. 3.3. The derived filters is updated by the standard gradient descent algorithm and the overall Piggyback GAN model is trained in the same way as any other existing generative models by adopting the desired learning objective for each task.

4   Experiments

We evaluate Piggyback GAN under two settings: (1) paired image-conditioned generation, and (2) unpaired image-conditioned generation. We first conduct an ablation study on the piggyback filters and unconstrained filters. We also demonstrate the generalization ability of our model by having same T_2 following different T_1. Finally, we compare our model with state-of-the-art approach [24] proposed for discriminative models (i.e., classification models), which also shares the idea of "piggybacking" on previously trained models by reusing the filters. Different from our approach, [24] learns a binary mask applying on the filters of a base network for each new task.

Training Details. All the sequential generation models are trained on images of size 256×256. We use the Tensorflow [1] framework with Adam optimizer [18]. We set the parameters $\lambda = \frac{1}{4}$ for all experiments. For paired image-conditioned generation, we use UNet architecture [14] and the last layer is set to be task-specific (a task-specific layer contains only unconstrained filters) for our approach and all baselines. For unpaired image-conditioned generation, we adopt the architecture [16,44] which have shown impressive results for neural style transfer, and last two layers are set to be task-specific for our approach and all baselines. For both conditional generative scenarios, bias terms are used to adjust the output after factorization.

Baseline Models. We compare Piggyback GAN to the following baseline models: (a) *Full*: The model is trained on single task, which could be treated as the "upper bound" for all approaches. (b) $\frac{1}{4}$*Full* and $\frac{1}{2}$*Full*: The model is trained on single task, and $\frac{1}{4}$ or $\frac{1}{2}$ number of filters used in baseline *Full* is used. (c) *Pure Factorization (PF)*: Model is trained with piggyback filters that are purely constructed from previously trained filters. (d) *Sequential Fine-tuning (SFT)*: The model is fine-tuned in a sequential manner, with parameters initialized from the model trained/fine-tuned on the previous task. (e) [24]: a state-of-the-art approach for discriminative model (i.e. classification model) which reuses the filters by learning and applying a binary mask on the filters of a base network for each new task.

Quantitative Metrics. In this work, we use two metrics *Acc* and *Fréchet Inception Distance (FID)* [13] to validate the quality of the generated data. *Acc* is the accuracy of the classifier network trained on real images and evaluated on generated images (higher Acc indicates better generation results). *FID* is an extensively used metric to compare the statistics of generated samples to samples from a real dataset. We use this metric to quantitatively evaluate the quality of generated images (lower FID indicates higher generation quality).

4.1   Paired Image-Conditioned Generation

We first demonstrate the effectiveness of Piggyback GAN on 4 tasks of paired image-conditioned generation, which are all image-to-image translations, on challenging domains and datasets with large variation across different modalities [9,14,35,39]. The first task is *semantic labels → street photos*, the second task is *maps → aerial photos*, the third task is *segmentations → facades*, and the fourth task is *edges → handbag photos*.

Ablation Study on Choice of λ. First we conduct an ablation study on the choice of different values of λ. For each upcoming task, we explore a set of 5 values of λ: 0, $\frac{1}{8}$, $\frac{1}{4}$, $\frac{1}{2}$ and 1. $\lambda = 0$ corresponds to the special case where all filters are piggyback filters, and $\lambda = 1$ corresponds to the special case where all filters are unconstrained filters. Figure 3 illustrates the FID for different λ. For example, $\lambda = \frac{1}{4}$ balances well the performance and number of parameters. While parameter size scales almost linearly with λ, the improvement in performance (lower FID is better) slows down gradually.

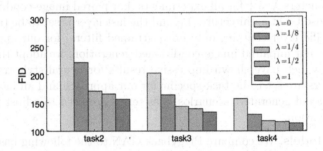

Fig. 3. FID for different λ. Ablation study of choice of different values of λ. For all upcoming new tasks, the performance improves as λ increases. However, while parameter size scales almost linearly with λ, the improvement of performance slows down gradually.

Ablation Study on Model Components. We also conduct an ablation study on the piggyback filters and unconstrained filters by comparing Piggyback GAN with the baseline models *Full*, $\frac{1}{4}$*Full* and *Pure Factorization (PF)* on Task 2 (*maps → aerial photos*) given the same model trained on Task 1 (*semantic labels → street photos*).

The quantitative evaluations of all approaches are summarized in Table 1. The baseline $\frac{1}{4}$*Full* produces the worst classification accuracy. Since the details like building blocks are hardly visible, the classifier sometimes mistakes category *facades* for category *maps*. The baseline *PF* generates images that contain more details as compared to the images generated using $\frac{1}{4}$*Full*. This suggests that *piggyback filters* can provide valuable information of the patterns in the training data. High *FID* scores for baselines $\frac{1}{4}$*Full* and *PF* indicate that generation qualities for both approaches are poor, and it is observed that the generated images are blurry, resulting in lots of missing details, and also contain lots of artifacts for both approaches. Our Piggyback GAN is parameter efficient and produces images having similar quality as the *Full* model.

Table 1. Model components. Ablation study of model components on paired image-conditioned generation tasks. Different models are trained and evaluated on task 2, FID and classification score on task 2 is reported.

	Full	$\frac{1}{2}$Full	$\frac{1}{4}$Full	PF	Piggyback GAN
Acc	97.90	88.10	84.87	99.82	97.99
FID	156.23	189.08	285.71	303.90	171.04

Generalization Ability. To demonstrate the generalization ability of our model, we learn the same T_2: *maps → aerial photos* with three different initial tasks T_1: *semantic labels → street photos*, *segmentations → facades*, and *edges → handbag photos* to evaluate whether Piggyback GAN could generalize well. The quantitative evaluation of all approaches are summarized in Table 2. The experimental results indicate that Piggyback GAN performs stable. The consistent Acc and FID scores indicate that the generated images on task T_2 from all three initial tasks T_1 have similar high image quality.

Table 2. Generalizability. Given different models trained for different initial tasks T_1, same new task T_2 is learnt. The consistent Acc and FID scores indicate that our model generalize well and perform consistently well given different set of filter banks achieved from different tasks.

T_1	semantic labels ↓ street photos	segmentations ↓ facades	edges ↓ handbags
Acc	97.99	97.08	98.09
FID	171.04	174.67	166.20

Comparison with SOTA Method and Baselines. We compare Piggyback GAN with the state-of-the-art approach [24], *Sequential Fine-tuning (SFT)* and baseline *Full*. Baseline *Full* can be considered as the "upper bound" approach, which provides the best performance that Piggyback GAN and [24] could achieve. For both Piggyback GAN and sequential fine-tuning, the model of *Task2* is initialized from the same model trained on *Task1*. The same model also serves as the backbone network for approach [24].

Generated images of each task for all approaches are shown in Fig. 4 and the quantitative evaluations of all approaches are summarized in Table 3. It is clear that the sequentially fine-tuned model completely forgets the previous task and can only generate incoherent edges2handbags-like (*edges → handbag photos*)-like patterns. The classification accuracy suggests that state-of-the-art approach [24] is able to adapt to the new task. However, it fails to capture the fine details of the new task, thus resulting in large artifacts in the generated images, especially for task 2 (*maps → aerial photos*) and task 4 (*edges → handbag photos*). In contrast, Piggyback GAN learns the current generative task while remembering the previous task and can better preserve the quality of generated images while shrinking the number of parameters by reusing the filters from previous tasks.

Table 3. Quantitative evaluation among different approaches for continual learning of paired image-conditioned generation tasks. Different models are trained and evaluated on tasks 1–4 based on the same model trained on task 1. The average FID score over 4 tasks is reported.

	Piggyback GAN	[24]	SFT	Full
Acc	89.80	88.09	24.92	91.10
FID	137.87	178.16	259.76	130.71

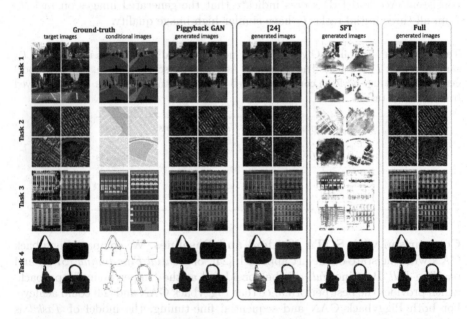

Fig. 4. Comparison among different approaches for continual learning of paired image-conditioned generation tasks. Piggyback GAN can learn the current task without forgetting the previous ones, and preserve the generation quality while shrinking the number of parameters by "piggybacking" on an existing model.

For paired image generation, we also compare our approach with *Lifelong GAN* [42] and *Progressive Network (PN)* [28]. The FID of *PN* over all tasks is 141.29. *Lifelong GAN* suffers from quality degradation: the FID on cityscapes increases from 122.52 to 160.28 then to 221.07 as tasks are added, while Piggyback GAN ensures no quality degradation since the filters for previous tasks are not altered. These results show the effectiveness of combining the two types of filters of Piggyback GAN.

4.2   Unpaired Image-Conditioned Generation

We also apply Piggyback GAN to another challenging scenario: unpaired image-conditioned generation domain $A \to$ domain B. The model is trained on unpaired data, where correspondence between domain A and domain B does not exist, namely for each image in domain A there is no corresponding ground-truth image in domain B. We apply our model on 2 sequences of tasks: tasks of image-to-image translation and tasks of style transfer.

Tasks of Image-to-Image Translation. We convert the same 4 tasks from the paired scenario to the unpaired scenario, and compared our approach with *Full*, the state-of-the-art approach [24] and *Sequential Fine-tuning (SFT)*. The results are shown in Table 4. We observed that the Sequential Fine-tuning (SFT) cannot remember previous tasks and suffers catastrophic forgetting. Our approach produces images with high quality on par with Full model, while [24] is capable of learning each new task, however the generation quality is poor.

Tasks of Style Transfer. Two tasks in a given sequence may share the same input domain but have different output domains, e.g. T_1 is *Photo $\to$ Monet Paintings* and T_2 is *Photo $\to$ Ukiyo-e Paintings*. While this is not a problem when training each task separately, it does cause problems for lifelong learning. We explore a sequence of tasks of unpaired image-conditioned generation as described above. The first task is unpaired image-to-image translation of *photos $\to$ Monet paintings*, the second task is unpaired image-to-image translation of *photos $\to$ Ukiyo-e paintings*.

We compare Piggyback GAN against the state-of-the-art approach [24], *Sequential Fine-tuning (SFT)* and baseline *Full* which serves as the "upper bound" approach. For both Piggyback GAN and sequential fine-tuning, the model of Task 2 is initialized from the same model trained on Task 1, which also serves as the backbone network for approach [24] to allow for fair comparison.

The quantitative and qualitative evaluations for comparison among different approaches are shown in Table 4 and Fig. 5, respectively. The baseline *SFT* completely forgets the previous *Monet* style and can only produce images in the *Ukiyo-e* style. Both the state-of-the-art approach [24] and Piggyback GAN are able to adapt to the new style. The classification accuracy and FID scores indicate that Piggyback GAN can better preserve the quality of generated images, and produce images with styles most closely resembling *Ukiyo-e* paintings.

Table 4. Unpaired image conditioned generation tasks. Different models are trained on all tasks based on the same model trained on task 1. The average score over all tasks is reported.

Tasks	4 Image-to-image Translation Tasks			
Methods	Piggyback GAN	[24]	SFT	Full
Acc	90.30	88.29	24.85	91.04
FID	113.89	149.51	262.37	109.46

Tasks	2 Style Transfer Tasks			
Methods	Piggyback GAN	[24]	SFT	Full
Acc	77.97	70.58	50.00	78.97
FID	106.95	118.26	135.05	101.09

Fig. 5. Comparison among different approaches for continual learning of unpaired style transfer tasks. Piggyback GAN can preserve the generation quality the current task most without forgetting the previous ones.

5   Conclusion

We proposed Piggyback GAN for lifelong learning of generative networks, which can handle various conditional generation tasks across different domains. Compared to the naive approach of training a separate standalone model for each task, our approach is parameter efficient since it learns to perform new tasks by making use of the model trained on previous tasks and combining with a small portion of unconstrained filters. At the same time, our model is able to maintain image generation quality comparable to the single standalone model for each task. Since the filters learned for previous tasks are preserved, our model is capable of preserving the exact generation quality of previous tasks. We validated our approach on various image-conditioned generation tasks across different domains, and the qualitative and quantitative results show our model addresses catastrophic forgetting effectively and efficiently.

References

1. Abadi, M., et al.: Tensorflow: a system for large-scale machine learning. In: Symposium on Operating Systems Design and Implementation (OSDI) (2016)
2. Aljundi, R., Babiloni, F., Elhoseiny, M., Rohrbach, M., Tuytelaars, T.: Memory aware synapses: learning what (not) to forget. In: European Conference on Computer Vision (ECCV) (2018)
3. Aljundi, R., Chakravarty, P., Tuytelaars, T.: Expert gate: lifelong learning with a network of experts. In: IEEE Conference on Computer Vision and Pattern Recognition (CVPR) (2017)
4. Castro, F.M., Marín-Jiménez, M.J., Guil, N., Schmid, C., Alahari, K.: End-to-end incremental learning. In: European Conference on Computer Vision (ECCV) (2018)
5. Chaudhry, A., Dokania, P.K., Ajanthan, T., Torr, P.H.: Riemannian walk for incremental learning: understanding forgetting and intransigence. In: European Conference on Computer Vision (ECCV) (2018)
6. Chaudhry, A., Ranzato, M., Rohrbach, M., Elhoseiny, M.: Efficient lifelong learning with a-gem. In: International Conference on Learning Representations (ICLR) (2019)
7. Chen, C., Tung, F., Vedula, N., Mori, G.: Constraint-aware deep neural network compression. In: European Conference on Computer Vision (ECCV) (2018)
8. Choi, Y., Choi, M., Kim, M., Ha, J.W., Kim, S., Choo, J.: Stargan: unified generative adversarial networks for multi-domain image-to-image translation. In: IEEE Conference on Computer Vision and Pattern Recognition (CVPR) (2018)
9. Cordts, M., et al.: The cityscapes dataset for semantic urban scene understanding. In: IEEE Conference on Computer Vision and Pattern Recognition (CVPR) (2016)
10. Denil, M., Shakibi, B., Dinh, L., Ranzato, M., de Freitas, N.: Predicting parameters in deep learning. In: Advances in Neural Information Processing Systems (NeurIPS) (2013)
11. Gatys, L.A., Ecker, A.S., Bethge, M.: Image style transfer using convolutional neural networks. In: IEEE Conference on Computer Vision and Pattern Recognition (CVPR) (2016)
12. He, Y., Liu, P., Wang, Z., Hu, Z., Yang, Y.: Filter pruning via geometric median for deep convolutional neural networks acceleration. In: IEEE Conference on Computer Vision and Pattern Recognition (CVPR) (2019)
13. Heusel, M., Ramsauer, H., Unterthiner, T., Nessler, B., Hochreiter, S.: Gans trained by a two time-scale update rule converge to a local nash equilibrium. In: Advances in Neural Information Processing Systems (NeurIPS) (2017)
14. Isola, P., Zhu, J.Y., Zhou, T., Efros, A.A.: Image-to-image translation with conditional adversarial networks. In: IEEE Conference on Computer Vision and Pattern Recognition (CVPR) (2017)
15. Jacob, B., et al.: Quantization and training of neural networks for efficient integer-arithmetic-only inference. In: IEEE Conference on Computer Vision and Pattern Recognition (CVPR) (2018)
16. Johnson, J., Alahi, A., Fei-Fei, L.: Perceptual losses for real-time style transfer and super-resolution. In: Leibe, B., Matas, J., Sebe, N., Welling, M. (eds.) ECCV 2016. LNCS, vol. 9906, pp. 694–711. Springer, Cham (2016). https://doi.org/10.1007/978-3-319-46475-6_43
17. Jung, S., et al.: Learning to quantize deep networks by optimizing quantization intervals with task loss. In: IEEE Conference on Computer Vision and Pattern Recognition (CVPR) (2019)

18. Kingma, D.P., Ba, J.: Adam: a method for stochastic optimization. In: International Conference on Learning Representations (ICLR) (2015)
19. Kirkpatrick, J., et al.: Overcoming catastrophic forgetting in neural networks. In: Proceedings of the National Academy of Sciences of the United States of America (2017)
20. Li, Z., Hoiem, D.: Learning without forgetting. IEEE Trans. Pattern Anal. Mach. Intell. **40**(12), 2935–2947 (2017)
21. Liu, S., Johns, E., Davison, A.J.: End-to-end multi-task learning with attention. In: IEEE Conference on Computer Vision and Pattern Recognition (CVPR) (2019)
22. Lopez-Paz, D., Ranzato, M.: Gradient episodic memory for continual learning. In: Advances in Neural Information Processing Systems (NeurIPS) (2017)
23. Ma, N., Zhang, X., Zheng, H.T., Sun, J.: Shufflenet v2: practical guidelines for efficient CNN architecture design. In: European Conference on Computer Vision (ECCV) (2018)
24. Mallya, A., Davis, D., Lazebnik, S.: Piggyback: adapting a single network to multiple tasks by learning to mask weights. In: European Conference on Computer Vision (ECCV) (2018)
25. McCloskey, M., Cohen, N.J.: Catastrophic interference in connectionist networks: the sequential learning problem. In: Bower, G.H., (eds.) Psychology of Learning and Motivation. Academic Press, Cambridge (1989)
26. Qiao, S., Lin, Z., Zhang, J., Yuille, A.: Neural rejuvenation: improving deep network training by enhancing computational resource utilization. In: IEEE Conference on Computer Vision and Pattern Recognition (CVPR) (2019)
27. Rebuffi, S.A., Bilen, H., Vedaldi, A.: Efficient parameterization of multi-domain deep neural networks. In: IEEE Conference on Computer Vision and Pattern Recognition (CVPR) (2018)
28. Rusu, A.A., et al.: Progressive neural networks. arXiv:1606.04671 (2016)
29. Sandler, M., Howard, A., Zhu, M., Zhmoginov, A., Chen, L.C.: Mobilenetv 2: inverted residuals and linear bottlenecks. In: IEEE Conference on Computer Vision and Pattern Recognition (CVPR) (2018)
30. Seff, A., Beatson, A., Suo, D., Liu, H.: Continual learning in generative adversarial nets. In: Advances in Neural Information Processing Systems (NeurIPS) (2017)
31. Serrà, J., Suris, D., Miron, M., Karatzoglou, A.: Overcoming catastrophic forgetting with hard attention to the task. In: International Conference on Machine Learning (ICML) (2018)
32. Shmelkov, K., Schmid, C., Alahari, K.: Incremental learning of object detectors without catastrophic forgetting. In: IEEE International Conference on Computer Vision (ICCV) (2017)
33. Tartaglione, E., Lepsøy, S., Fiandrotti, A., Francini, G.: Learning sparse neural networks via sensitivity-driven regularization. In: Advances in Neural Information Processing Systems (NeurIPS) (2018)
34. Tung, F., Mori, G.: Similarity-preserving knowledge distillation. In: IEEE International Conference on Computer Vision (ICCV) (2019)
35. Tyleček, R., Šára, R.: Spatial pattern templates for recognition of objects with regular structure. In: Weickert, J., Hein, M., Schiele, B. (eds.) GCPR 2013. LNCS, vol. 8142, pp. 364–374. Springer, Heidelberg (2013). https://doi.org/10.1007/978-3-642-40602-7_39
36. Wu, B., et al.: FBNet: hardware-aware efficient convnet design via differentiable neural architecture search. In: IEEE Conference on Computer Vision and Pattern Recognition (CVPR) (2019)

37. Wu, C., Herranz, L., Liu, X., Wang, Y., van de Weijer, J., Raducanu, B.: Memory replay gans: learning to generate images from new categories without forgetting. In: Advances in Neural Information Processing Systems (NeurIPS) (2018)
38. Yang, T.J., et al.: NetAdapt: platform-aware neural network adaptation for mobile applications. In: European Conference on Computer Vision (ECCV) (2018)
39. Yu, A., Grauman, K.: Fine-grained visual comparisons with local learning. In: IEEE Conference on Computer Vision and Pattern Recognition (CVPR) (2014)
40. Yu, R., et al.: NISP: pruning networks using neuron importance score propagation. In: IEEE Conference on Computer Vision and Pattern Recognition (CVPR) (2018)
41. Zagoruyko, S., Komodakis, N.: Paying more attention to attention: improving the performance of convolutional neural networks via attention transfer. In: International Conference on Learning Representations (ICLR) (2017)
42. Zhai, M., Chen, L., Tung, F., He, J., Nawhal, M., Mori, G.: Lifelong gan: continual learning for conditional image generation. In: IEEE International Conference on Computer Vision (ICCV) (2019)
43. Zhang, D., Yang, J., Ye, D., Hua, G.: LQ-Nets: learned quantization for highly accurate and compact deep neural networks. In: European Conference on Computer Vision (ECCV) (2018)
44. Zhu, J.Y., Park, T., Isola, P., Efros, A.A.: Unpaired image-to-image translation using cycle-consistent adversarial networks. In: IEEE International Conference on Computer Vision (ICCV) (2017)

TRRNet: Tiered Relation Reasoning for Compositional Visual Question Answering

Xiaofeng Yang[1], Guosheng Lin[1(✉)], Fengmao Lv[2], and Fayao Liu[3]

[1] Nanyang Technological University, Singapore, Singapore
xiaofeng001@e.ntu.edu.sg, gslin@ntu.edu.sg
[2] Center of Statistical Research, Southwestern University of Finance and Economics, Chengdu, China
[3] Institute for Infocomm Research A*STAR, Singapore, Singapore

Abstract. Compositional visual question answering requires reasoning over both semantic and geometry object relations. We propose a novel tiered reasoning method that dynamically selects object level candidates based on language representations and generates robust pairwise relations within the selected candidate objects. The proposed tiered relation reasoning method can be compatible with the majority of the existing visual reasoning frameworks, leading to significant performance improvement with very little extra computational cost. Moreover, we propose a policy network that decides the appropriate reasoning steps based on question complexity and current reasoning status. In experiments, our model achieves state-of-the-art performance on two VQA datasets.

Keywords: Visual question answering · Visual reasoning

1 Introduction

Visual Question Answering [3,10,15,17] is the task of answering a natural language question based on the content of an image. To precisely answer visual questions, the VQA models should be able to understand the language, the image and build a cross-modal mapping between the lingual and visual contents.

Current state-of-the-art VQA methods can be divided into two categories. The works from the first category [1,19,24] mainly focus on learning a multimodal joint representation of language and vision. The visual features, which are usually extracted from pre-trained networks, are combined with the language features through multiple self-attention and co-attention [7,21] modules. Although these methods are proven to work well on VQA [3] tasks, they usually don't generalize well on the compositional reasoning tasks [15] due to their lack of relation reasoning abilities. For example, given a question of *"What is the color of the food on the plate to the right of the girl?"*, the VQA model needs to first understand the mapping between language representation and visual features,

© Springer Nature Switzerland AG 2020
A. Vedaldi et al. (Eds.): ECCV 2020, LNCS 12366, pp. 414–430, 2020.
https://doi.org/10.1007/978-3-030-58589-1_25

and then reason over the relations between each pair of objects to decide the final answer. Hence, methods based on simple self-attention and co-attention are insufficient in performing relation reasoning. On the other hand, the methods from the second category mainly focus on designing neural modules [2,23] that can perform more diverse reasoning tasks. These methods perform extremely well on simulated datasets like CLEVR [17]. However, the design of reasoning modules is tricky and heavily relies on human efforts. Hence, they are not widely adopted in real-world datasets [3,15] that contain far more object classes and possible reasoning actions.

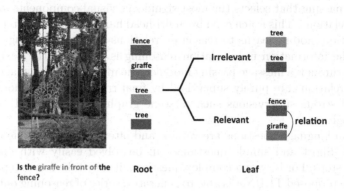

Fig. 1. The idea of tiered relation reasoning. In roots, there are many possible objects for answering questions. Based on the attention map of root, the objects can be classified as relevant or irrelevant objects. In leaf, pairwise relations are only generated between relevant objects.

The idea of reasoning over relations has been drawn in some of the previous works. In [26] the relations are generated for each pair of regions and then combined with language representations. In [4] dense pairwise relations are generated for multi-modal embeddings. Other works [34] also propose to generate relations based on geometry information e.g. generate relations for geometry close objects only. The dense object relations are usually noisy and computation consuming. Imagine the case that there are one hundred visual objects. There will be ten thousand possible pairwise relations. In fact, one sentence will not cover more than six objects. It brings in much difficulty to find relevant ones from all ten thousand relations. The geometry sparse relations are built on a strong assumption that relations are only valid for neighbor objects, which is not always true.

In this work, we propose a tiered relation reasoning method that dynamically selects object level candidates based on language representation and generates tidy object relations within the selected candidates objects only. The basic idea of tiered relation reasoning is illustrated in Fig. 1. We denote our structure as a tiered structure, because the reasoning is from coarse objects to fine candidate objects. In root, we have many objects grouped, and in leaf, the objects are split into relevant and irrelevant objects. The irrelevant objects could be eliminated from further processing. The tiered selection not only makes the network computational efficient but also improves relation reasoning performance.

Our proposed TRRNet (Tiered Relation Reasoning Network) consists of a series of TRR units. For each TRR unit, there are four basic components: root attention, root to leaf attention passing, leaf attention and a final message passing module to interact with the next TRR unit. The root attention is an object level attention attending to different visual features based on language representation. The functionality of this component could be achieved by many modern approaches [1,7,19]. After root attention, the network carries out root to leaf attention passing that selects the most significant visual components and builds pairwise relations. This is achieved by multi-head hard attentions. After that, the leaf attention module learns to reason on relations based on language features. Finally, the information from relation reasoning is passed to the next stage of reasoning through a message passing module. The mappings of language-objects, language-relations are purely supervised by final training loss, without seeking additional strong supervisions such as scene graph annotations [31] and functional programs.

Natural language questions are various and often require multiple reasoning steps [27]. Short and simple questions can be solved easily with one or two reasoning steps. For long and complex questions, it may take more steps to solve them. The proposed TRRNet is able to generate a series of reasoning outputs. On top of that, we design a policy network that decides the appropriate reasoning steps based on questions and current reasoning outputs. The policy network not only boosts final accuracy, but also improves overall processing speed.

In summary, our contributions are:

- We propose a novel tiered attention network for relation reasoning. The TRR network consists of a series of TRR units. Each TRR unit can be decomposed into four basic components: a root attention to model object level importance, a root to leaf attention passing module to select candidate objects based on root attention and generate pairwise relations, a leaf attention to model relation level importance and finally a message passing module for information communication between reasoning units.
- We propose a policy network that chooses the best reasoning steps based on natural language question and reasoning outputs.
- We achieve state-of-the-art performance on GQA dataset and competitive results on CLEVR datasets and VQAv2 dataset without functional program supervision.

2   Related Works

In this section, we categorize VQA tasks that do not require compositional reasoning skills as visual question answering [3] and VQA tasks that require multiple reasoning skills as visual reasoning [15,17].

2.1 Visual Question Answering

The improvement in visual question answering has mainly been done on two parts, namely better features and better attentions. Early VQA methods use pre-trained VGG or ResNet to extract visual features. The bottom-up and top-down network [1] proposed to extract visual features from an object detector [9]. The bottom-up features significantly improve VQA performance.

Better attentions also contribute a lot to improve VQA performance. Works in [20,32,37] further extended traditional attention methods for better visual groundings. Co-attention [21,24,29,36] performs attention based on a fused representation on image features and language features respectively. BAN [19] network further improved computation efficiency of co-attention through bilinear attention. Motivated by NLP works [28], self-attention is also widely adopted to improve VQA performance [7,35]. In these works, self-attention is used to generate intra-modality attention maps and a summation of the original features.

It is worth noticing that conventional VQA methods lack the ability of relation reasoning. Co-attention mechanism models correspondence between words and image regions. Relations between objects are largely missing. Self-attention models intra-modality importance, while it could not cover both semantic and spatial relations. Compared with transformer based methods which encode relations implicitly, our model encodes explicit relations directly, making it easier for the models to perform reasoning.

2.2 Visual Reasoning in VQA

Early visual reasoning works [26] generate dense relations for each pair of pixels. The dense relations are then combined with language features for language guided reasoning. State of the art reasoning methods are dominated by neural module networks [2]. In neural module networks, language representation is parsed into a series of logical steps. For each logical step, a different neural module for performance corresponding actions is designed. In [12,18,23], the questions are parsed into functional programs and specific executing engines are designed for the functional programs. The neural module approach [33] could achieve almost perfect performance on simulated datasets [17]. However, for real-world datasets [3,15], it's almost impossible to design specific modules for each type of reasoning due to the questions' high complexity.

In this work, we propose our TRRNet for relation reasoning. The model is purely trained with answer supervision. Strong functional program supervision is not used for training.

3 Our Approach

3.1 Overview

The task of VQA is described as follows: Given an image I and a question Q grounded on I, the purpose of VQA is to select the best answer from a set of all possible answers. The VQA task is usually defined as a classification problem.

Figure 2 shows a detailed illustration of our method. Following the standard VQA practice, the image is first processed with an object detector to extract n region features $V \in \mathbb{R}^{n \times d_v}$ and n bounding boxes $B \in \mathbb{R}^{n \times d_b}$, where the ith regional feature and box feature is denoted as $v_i \in \mathbb{R}^{d_v}$ and $b_i \in \mathbb{R}^{d_b}$. For language processing, we adopt Bert [6] word embeddings and input the embedding vector to a GRU [5] to better adapt the embeddings to VQA task. This step gets us a set of language embedding features $E \in \mathbb{R}^{m \times d_e}$ where m represents the number of words in one sentence. The proposed approach TRRNet consists of a series of TRR units. One TRR unit can be decomposed into four basic components, namely root attention (object level attention), root to leaf attention passing, leaf attention (relation level attention) and message passing module to pass message to the next TRR unit. Root attention is an object level attention, which learns attention weights for objects and generates a merged feature. Root to leaf attention passing module processes the object level attention map and produce pairwise relations. Based on the pairwise relations and language, leaf attention attends to different relations. Message passing module summarizes relation level attention maps and combines them with object level features, which will then be propagated to the next stage.

3.2 Root Attention

Given a set of image features $V \in \mathbb{R}^{n \times d_v}$ and bounding box features $B \in \mathbb{R}^{n \times d_b}$ extracted from Faster R-CNN and a set of question features from GRU $E \in \mathbb{R}^{m \times d_e}$, the *root attention* has two roles.

The first role is to generate an attention map for object level visual features based on language representations:

$$\alpha^{object} = softmax(Net(V, B, E)), \tag{1}$$

where "Net" denotes any possible networks that can generate object level attentions.

The second role of root attention is to generate a fused visual feature:

$$O^{root} = \alpha^{object} V^T, \tag{2}$$

where O^{root} denotes the output on root stage.

There are multiple choices for root attentions. Actually, most modern VQA methods that contain object level attention can be used as our root attention, such as Bottom-Up [1], DCN [24], BAN [19], Inter-intra [7]. In this work, we try both a simple attention method as used in Bottom-Up Attention [1] and a more complicated and advanced method as proposed in BAN [19].

This setting guarantees the flexibility of our work. Besides using one network as root attention, it's even possible to use several networks as the root attention and average the final attention maps to form a more robust object level features.

3.3 Root to Leaf Attention Passing

The *root to leaf attention passing* further processes the object level attention map from the root attention to produce pairwise relations. Note that the object level attention map might have multi-heads. We use multi-head hard attention to select relevant object candidates.

The idea of hard attention was first proposed in [30], where only the most related areas are selected for image captioning. When the image features are selected in a "hard" way, the network becomes undifferentiable. In [30] REIN-FORCE is used to find network parameters through sampling. In [22], hard attention is used to select grid features to answer visual questions. In this paper, we can format the hard attention problem in an easier and fully differentiable way.

For multi-head hard attention passing, we only select "hard" objects that are related to questions. Specifically, based on the attention maps from root attention, we select only the top k related objects based on the attention weights. After the visual features V_{Hard} are selected, we build pairwise relations by concatenating the features one-by-one and process them through a MLP to map them to size d_r. In experiment, we use $d_r = 256$, a value smaller than original visual feature dimension, because the relations usually contain less information than the objects and using smaller feature size can help to reduce overfitting. This step is represented as *Relation* in Eq. 4. Formally, for each attention head in root attention α^{object}, we perform:

$$V_{Hard} = Topk(\alpha, V, K), \tag{3}$$

$$R = Relation(V_{Hard}, B), \tag{4}$$

where K is a hyper-parameter for controlling how many objects for generating relations, B is the box features and R is the generated pairwise relations.

The refined object candidates also motivate us to generate triple-wise relations. For triple-wise relations, a set of relations $R_{triple} \in \mathbb{R}^{K^3 \times d_r}$ is generated for candidate objects. Ideally, the triple-wise relations can help the network to better understand long sentences with multiple objects. We show detailed ablation studies in Sect. 4.2.

3.4 Leaf Attention

The *leaf attention* performs reasoning over object relations. Same as root attention, the leaf attention also produces both an attention map and a feature merged by relation features and language features. Given the relation features $R \in \mathbb{R}^{K^2 \times d_r}$ and the final question embedding $e \in \mathbb{R}^{d_e}$, we first obtain a hidden feature by fusing language and visual features and then the fused features are further propagated through non-linear functions:

$$h = f(g(e) \cdot k(R)), \tag{5}$$

where g, k, and f are fully connected layers with activation function ReLU and "." operation represents element-wise multiplication. The h value is now regarded as a mapping correspondence. Based on the correspondence, an attention map can be calculated from h for each relation:

$$\alpha^{relation} = softmax(h). \tag{6}$$

Finally, a merged relation feature is generated by matrix multiplication:

$$O^{leaf} = \alpha^{relation} R^T. \tag{7}$$

If R is a multi-head feature, the final leaf feature O^{leaf} is a concatenation of all heads.

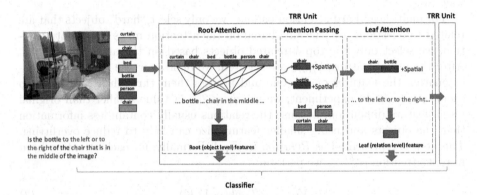

Fig. 2. A detailed illustration of a one layer TRR network with hard attention. Given an image and a natural question, the image is first processed with Faster R-CNN for feature extraction. In the root attention module, the object features are merged with language and an attention map of objects is generated. The root attention also produces a merged object level feature. The attention passing module processes the root attention map to select the most significant objects based on the attention map. Then pairwise relations are constructed and combined with spatial features (bounding boxes). The leaf attention performs attention on relations and generates relation level features. Root features and leaf features are finally combined with language features for predicting the final answer.

3.5   Message Passing Module for Units Interaction

To enable multi-stage reasoning, we propose a *message passing module*. In this module, we fuse the attended relation features O^{leaf} and object level features V through concatenation and a fully connected layer. The new visual feature is generated by:

$$V_{new} = f([O^{leaf}, V]), \tag{8}$$

where f is a linear function with activation.

3.6 Multi-stage Reasoning and Policy Network

Compositional natural language questions require multi-stage reasoning. Easy questions can be solved within two reasoning steps, while long and complex questions may require more reasoning steps. The proposed TRRNet is able to generate a series of reasoning outputs. For the tth TRR unit, the network takes the bounding box feature B, the tth visual features V_t and the question embedding E as input and output the root feature O_t^{root}, the leaf feature O_t^{leaf}, and the aggregated visual feature V_{t+1}:

$$O_t^{root}, O_t^{leaf}, V_{t+1} = TRR_t(B, V_t, E). \tag{9}$$

At the tth TRR unit, the policy network decides whether to proceed to the next stage of reasoning. The design of policy network follows two assumptions. First, complicated questions should be reasoned for more steps and second if the attended features from the previous steps are already stable, the network should stop.

We format the reasoning process as a sequential decision making process. At time step t, the "State" s_t is the question embedding E, question length l, current time step t and root feature O_{t-1}^{root} and O_t^{root}. The reason why we choose the root level features is because leaf features are calculated based on root, so root features itself can represent the reasoning status. The "Action" a_t is a binary action $[1, 0]$ where "1" denotes to proceed and "0" denotes not to proceed to the next reasoning step. The "Reward" r_t is the number of correct VQA predictions of current batch. In order to minimize the total number of reasoning steps, we also introduce a small scalar penalty term $p_t = 0.1$ for each time the policy network chooses to proceed. The "Policy" is a function $\pi(a_t|s_t, \theta)$. The policy function decides the next action based on State s_t. The decision process is regarded as a Markov Decision Process and can be trained with reinforcement learning settings.

To be more specific about the policy network structure, we first calculate the L2 distance of root features from the previous two time steps: $d = L2(O_{t-1}^{root}, O_t^{root})$. The distance and question embedding are then processed by two separate mlps for final prediction:

$$d = L2(O_{t-1}^{root}, O_t^{root}), \tag{10}$$

$$z = MLP[MLP(d), MLP(E, t, l)], \tag{11}$$

$$\pi(a_t|s_t, \theta) = softmax(z). \tag{12}$$

The policy network is trained with policy gradient method REINFORCE. During training, we alternatively train the main network and the policy network. We first train the main network and then fix the parameters of main network and train the policy network. After that, we go back to train the main network again. We set the maximum number of reasoning steps to be $N = 3$. Since at time step $t = 1$, the O_0^{root} value is $None$, the policy network is only used after $t = 2$. The loss of policy network is defined as: $L = -E_{s \sim \pi}[r - p]$.

3.7 The Readout Layer

In readout layer, the final stage root features O^{root} and leaf features O^{leaf} are further combined and processed with a linear function to represent the final visual features:

$$O^{all} = f([O_t^{root}, O_t^{leaf}]). \tag{13}$$

Question embedding features are processed through a fully connected layer with an activation function ReLU to represent the final language features:

$$E^{final} = g(E). \tag{14}$$

Finally, the question feature and visual features are combined via an element-wise product for classification:

$$Answer = softmax(h(O^{all} \cdot E^{final})), \tag{15}$$

where h is a fully connected layer with an activation function.

4   Experiments

4.1   Experimental Setup

We use GQA [15] dataset for our experiments. The GQA dataset contains 22M compositional questions and 140 K images. Compared to VQA [3], GQA dataset contains questions that require multiple reasoning skills. In our experiment, we use pre-trained Faster-RCNN in [1] to extract region features of size 36×2048. We use pre-trained Bert word embedding and GRU to extract language features of size 20×1024, where 20 is the length of questions. There are two training and testing splits in GQA, one split "all-split" contains all images and the other split "balanced-split" contains balanced-split with re-sampled answer distribution. For ablation studies, all models are trained on the "balanced-split" and tested on the "testdev" set. The final model is trained on the "all-split" and finetuned on the "balanced-split".

4.2   Ablation Study

We perform extensive ablation studies on the GQA "balanced-set". The overall results are shown in Table 1.

TRRNet vs. Basic Attention Models. We first investigate the improvement of our proposed TRRNet over basic attention models. For each basic attention model, we perform two experiments. First, we train the network alone on "balanced-split" and second we use the model as the root attention component in a one layer TRRNet with hyper-meter $K = 6$. All the models are trained with the same hyper-parameter settings. Evaluation is done on the "testdev" dataset. The experiments are carried out on two attention methods, a simple attention model used in Bottom-Up attention [1], a more advanced model used in BAN [19]. Results show that when using weaker attention models, our TRRNet significantly outperforms the baseline by 6%. Even when the baseline model is a strong and complicated attention model, our one layer TRRNet could improve performance by 0.6%. For simplicity, all remaining ablation experiments use Bottom-Up attention as root attention.

Component Analysis. We then study the influence of root to leaf attention passing. The experiments are done in three folds: first, we build a simple baseline relation module where the relations are generated for each pair of objects. Then we compare the baseline model and root to leaf attention passing with different K choices. Finally, we demonstrate the result of triple-wise relation attention. By simply adding a relation module to Bottom-Up attention, the accuracy is improved by 5%. Surprisingly, by choosing top 6 most significant objects, hard attention with more than 36 times less computation could achieve 0.5% better performance than the relation baseline. This observation further proves the fundamental idea of our networks. Dense relations are usually noisy and using only important objects could generate even better results. By increasing the K value to 12, the model could achieve 0.2% improvement. We continue to generate triple-wise relations for hard attentions. The attention features of triple-wise relations are combined with pairwise features and object level features to generate the final results. The triple-wise relations also help to improve accuracy by 0.2% compared with pairwise relations. Due to the complexity of triple-wise relations, it's not used in our final models.

The Number of TRR Units. Also we investigate the effect of reasoning steps i.e. the number of TRR units. We start from a 1 layer TRR network and increase the length to 3 layers. We observe that a 2 layer TRRNet significantly improves model performance by almost 1% compared with 1 layer reasoning model. While adding more TRR units to above 3 layers does not help to further improve the performance. With our proposed policy network, the accuracy could be increased further by 0.3% compared with a simple 2 layers TRRNet.

Length of GRU Embeddings. Finally, we study the length of GRU embeddings. The feature dimension is increased from 512 to 2048. The network achieves the highest accuracy at feature dimension 1024. Longer or shorter feature dimensions all reduce overall performance.

4.3   Experimental Results on GQA

Training Details. In final experiments, we use dual root attentions, one Bottom-Up attention and one BAN attention to better capture object importance. Both of the root attentions are appended with their own leaf attentions. For both leaf attentions, we adopt attention passing with $K = 6$. The final root

Table 1. Ablation experiments of TRRNet on the GQA balanced dataset. K stands for the number of objects chosen for hard attention.

Method	Accuracy (%)
Bottom-up [1]	50.20
TRR (Bottom-up)	**56.13**
BAN [19]	55.82
TRR (BAN)	**56.43**
Bottom-up + Relation	55.60
TRR hard attention (K = 6)	56.13
TRR hard attention (K = 12)	56.28
TRR triple attention (K = 6)	56.29
TRR 1 layer	56.13
TRR 2 layers	57.00
TRR 3 layers	56.88
TRR Policy	**57.32**
Embedding 512	55.85
Embedding 1024	**56.13**
Embedding 2048	55.90

Table 2. Comparisons with state-of-the-art methods of GQA on the blind test2019 set. Our model achieves a state-of-the-art performance of 60.74% for a single model without using functional programs and new image features.

Method	Binary	Open	Consistency	Plausibility	Validity	Accuracy
Bottom-up [1]	66.64	34.83	78.71	84.57	96.18	49.74
MAC [14]	71.23	38.91	81.59	84.48	96.16	54.06
BAN [19]	76.00	40.41	91.70	85.58	96.16	57.10
GRN [11]	74.93	41.24	87.41	84.68	96.14	57.04
LCGN [13]	73.77	42.33	84.68	84.81	96.48	57.07
TRRNet (ours)	77.83	45.65	90.95	85.15	96.40	**60.74**

features and leaf features are concatenated before processed by the classifier. For training, we first use the "all-split" to train our model for 4 epochs with learning rate of 1×10^{-4} and further fine-tune the trained model on the "balanced-split" with a learning rate of 5×10^{-5}. A step by step result on the "testdev" split is shown in Table 3.

Comparison with State-of-the-Art of GQA. As illustrated in Table 2, our single model achieves state-of-the-art performance on the GQA "test2019" split without new image features and functional program. We notice that [16] proposes to generate symbolic representation and new visual features with scene graph annotations. To further improve performance, we also train Faster-CNN

Table 3. Step-by-step results of the TRRNet Dual Root Attention on GQA "testdev" split.

Method	Accuracy(%)
Dual Root trained on "balanced-split"	56.10
TRR Dual Root trained on "balanced-split"	57.86
TRR Dual Root trained on "all-split"	57.89
TRR Dual Root fine-tuned on " balanced-split" (ours)	60.32

models for feature extraction using the scene graph annotation provided. With new visual features, our model could be further improved by 3% to **63.20%**. Moreover, we use model ensemble and the tiny evaluation trick mentioned in [8] to further improve model performance. An ensemble version of model achieves an accuracy of **74.03%** , ranked the 2nd place on GQA 2019 leaderboard.

4.4 Experimental Results on VQAv2 and CLEVR

We also evaluate our proposed TRRNet on VQAv2 dataset and simulated visual reasoning dataset CLEVR.

VQAv2 Dataset. For VQAv2 dataset, we use root attention similar to the network structure mentioned in [35] to encode visual and language features and generate root level attentions. Due to the nature of VQAv2 dataset that it does not contain questions that require compositional reasoning skills, we remove the policy network and simply adopt a two layer TRRNet. Testing results are shown in Table 4. Our model achieves better results compared with models trained with the same training data and same visual language features. It does extremely well on yes or no questions.

CLEVR Dataset. For CLEVR dataset, we use grid features generated from pretrained Resnet, instead of region proposal features from Faster-RCNN. We use [35] as root attention. Since the visual feature is grid feature, we increase the K value to 36. Testing results are shown on Table 5. Functional programs work as a very strong supervision for visual reasoning. For a fair comparison, our model is only compared with those methods using natural languages only.

4.5 Visualization

In this section, we show visualizations of the attention maps in our network. Similar to ablation studies, we use Bottom-Up attention as the root attention. Both root attention and leaf attention visualizations are shown in Fig. 3. The images are chosen from the "testdev" set. Interestingly, the majority of the test images are reasoned for two stages. For root attentions, we plot objects with attention scores more than 0.1. For leaf attentions, we only show the relation with the highest attention score.

Table 4. Single model performance on *test-dev* and *test-standard* splits of VQAv2 dataset.

Method	Test-dev				Test-std
	All	Y/N	Num	Other	All
Bottom-up [1]	65.32	81.82	44.21	56.05	65.67
DCN [24]	66.87	83.51	46.61	57.26	66.97
MFH [10]	68.76	84.27	49.56	59.89	–
BAN [19]	70.04	85.42	54.04	60.52	70.35
DFAF [7]	70.22	86.09	53.32	60.49	70.34
MCAN [35]	70.63	85.82	53.26	60.72	70.90
TRRNet (ours)	**70.80**	**87.27**	51.89	61.02	**71.20**

Table 5. Single model performance on CLEVR dataset.

Method	All	Count	Exist	Compare numbers	Query attribute	Compare attribute
Human [17]	92.6	86.7	96.6	86.5	95.0	96.0
CNN+LSTM+SA [17]	73.2	59.7	77.9	75.1	80.9	70.8
RN [26]	95.5	90.1	97.8	93.6	97.9	97.1
FiLM [25]	97.7	94.3	99.1	96.8	99.1	99.1
MAC [14]	**98.9**	97.1	99.5	99.1	99.5	99.5
TRRNet (ours)	**98.8**	96.8	99.5	98.9	99.6	99.3

The first example shows an easy question that contains only one relation. In stage1, the root attention could already identify the most important area for answering questions. The second example shows a difficult question that contains more than one relations. Although the name "cake" is not directly displayed in the question, the root attention could successfully focus on the object. The third example is asking for abstraction. There is no clue from the words to guide where the attention should look and no relations can be found from the sentence. Visualizations show that the model learns to locate meaningful areas: the sky. The final example shows an example of three stage reasoning.

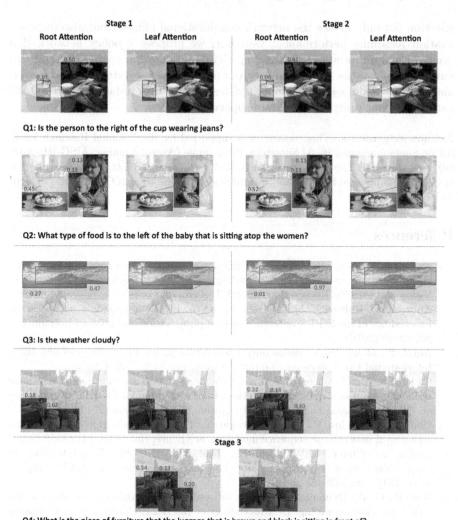

Fig. 3. Visualization of attention maps. For root attention, we plot all object regions with attention score larger than 0.1. For leaf attention, we only display the most confident relation attended. The first two examples show the model's ability to identifying the right objects. Example 3 shows the network's power of learning common sense: the correlation of weather and sky. Example 4 demonstrates an example of three stage reasoning.

5   Conclusion

In this work, we propose a tiered relation reasoning method that dynamically selects object level candidates based on language representation and generates tidy object relations within the selected candidates objects only. The tiered selection not only makes the network computational efficient but also improves relation reasoning performance. Moreover, we propose a policy network that decides the appropriate reasoning steps based on question complexity and current reasoning status. In experiments, our model achieves state-of-the-art performance on GQA dataset and competitive results on CLEVR datasets and VQAv2 dataset without functional program supervision.

Acknowledgements. This research was supported by the National Research Foundation Singapore under its AI Singapore Programme (Award Number: AISG-RP-2018-003) and the MOE Tier-1 research grants: RG28/18 (S) and RG22/19 (S). F. Lv's participation is supported by National Natural Science Foundation of China (No. 11829101 and 11931014).

References

1. Anderson, P., et al.: Bottom-up and top-down attention for image captioning and visual question answering. In: Proceedings of the IEEE Conference on Computer Vision and Pattern Recognition, pp. 6077–6086 (2018)
2. Andreas, J., Rohrbach, M., Darrell, T., Klein, D.: Neural module networks. In: Proceedings of the IEEE Conference on Computer Vision and Pattern Recognition, pp. 39–48 (2016)
3. Antol, S., et al.: VQA: visual question answering. In: Proceedings of the IEEE International Conference on Computer Vision, pp. 2425–2433 (2015)
4. Cadene, R., Ben-Younes, H., Cord, M., Thome, N.: MUREL: multimodal relational reasoning for visual question answering. In: Proceedings of the IEEE Conference on Computer Vision and Pattern Recognition, pp. 1989–1998 (2019)
5. Chung, J., Gulcehre, C., Cho, K., Bengio, Y.: Empirical evaluation of gated recurrent neural networks on sequence modeling. arXiv preprint arXiv:1412.3555 (2014)
6. Devlin, J., Chang, M.W., Lee, K., Toutanova, K.: BERT: pre-training of deep bidirectional transformers for language understanding. arXiv preprint arXiv:1810.04805 (2018)
7. Gao, P., et al.: Dynamic fusion with intra-and inter-modality attention flow for visual question answering. In: Proceedings of the IEEE Conference on Computer Vision and Pattern Recognition, pp. 6639–6648 (2019)
8. Geng, S., Zhang, J., Zhang, H., Elgammal, A., Metaxas, D.N.: 2nd place solution to the GQA challenge 2019. arXiv preprint arXiv:1907.06794 (2019)
9. Girshick, R.: Fast R-CNN. In: Proceedings of the IEEE International Conference on Computer Vision, pp. 1440–1448 (2015)
10. Goyal, Y., Khot, T., Summers-Stay, D., Batra, D., Parikh, D.: Making the V in VQA matter: elevating the role of image understanding in visual question answering. In: Proceedings of the IEEE Conference on Computer Vision and Pattern Recognition, pp. 6904–6913 (2017)
11. Guo, D., Xu, C., Tao, D.: Graph reasoning networks for visual question answering. arXiv preprint arXiv:1907.09815 (2019)
12. Hu, R., Andreas, J., Rohrbach, M., Darrell, T., Saenko, K.: Learning to reason: end-to-end module networks for visual question answering. In: The IEEE International Conference on Computer Vision (ICCV), October 2017
13. Hu, R., Rohrbach, A., Darrell, T., Saenko, K.: Language-conditioned graph networks for relational reasoning. arXiv preprint arXiv:1905.04405 (2019)

14. Hudson, D.A., Manning, C.D.: Compositional attention networks for machine reasoning. arXiv preprint arXiv:1803.03067 (2018)
15. Hudson, D.A., Manning, C.D.: GQA: a new dataset for real-world visual reasoning and compositional question answering. In: Proceedings of the IEEE Conference on Computer Vision and Pattern Recognition, pp. 6700–6709 (2019)
16. Hudson, D.A., Manning, C.D.: Learning by abstraction: the neural state machine. arXiv preprint arXiv:1907.03950 (2019)
17. Johnson, J., Hariharan, B., van der Maaten, L., Fei-Fei, L., Lawrence Zitnick, C., Girshick, R.: CLEVR: a diagnostic dataset for compositional language and elementary visual reasoning. In: Proceedings of the IEEE Conference on Computer Vision and Pattern Recognition, pp. 2901–2910 (2017)
18. Johnson, J., et al.: Inferring and executing programs for visual reasoning. In: Proceedings of the IEEE International Conference on Computer Vision, pp. 2989–2998 (2017)
19. Kim, J.H., Jun, J., Zhang, B.T.: Bilinear attention networks. In: Advances in Neural Information Processing Systems, pp. 1564–1574 (2018)
20. Kim, J.H., et al.: Multimodal residual learning for visual QA. In: Advances in Neural Information Processing Systems, pp. 361–369 (2016)
21. Lu, J., Yang, J., Batra, D., Parikh, D.: Hierarchical question-image co-attention for visual question answering. In: Advances In Neural Information Processing Systems, pp. 289–297 (2016)
22. Malinowski, M., Doersch, C., Santoro, A., Battaglia, P.: Learning visual question answering by bootstrapping hard attention. In: Proceedings of the European Conference on Computer Vision (ECCV), pp. 3–20 (2018)
23. Mascharka, D., Tran, P., Soklaski, R., Majumdar, A.: Transparency by design: Closing the gap between performance and interpretability in visual reasoning. In: Proceedings of the IEEE Conference on Computer Vision and Pattern Recognition, pp. 4942–4950 (2018)
24. Nguyen, D.K., Okatani, T.: Improved fusion of visual and language representations by dense symmetric co-attention for visual question answering. In: Proceedings of the IEEE Conference on Computer Vision and Pattern Recognition, pp. 6087–6096 (2018)
25. Perez, E., Strub, F., De Vries, H., Dumoulin, V., Courville, A.: FiLM: visual reasoning with a general conditioning layer. In: Thirty-Second AAAI Conference on Artificial Intelligence (2018)
26. Santoro, A., et al.: A simple neural network module for relational reasoning. In: Advances in Neural Information Processing Systems, pp. 4967–4976 (2017)
27. Sukhbaatar, S., Weston, J., Fergus, R., et al.: End-to-end memory networks. In: Advances in Neural Information Processing Systems, pp. 2440–2448 (2015)
28. Vaswani, A., et al.: Attention is all you need. In: Advances in Neural Information Processing Systems, pp. 5998–6008 (2017)
29. Xiong, C., Zhong, V., Socher, R.: Dynamic coattention networks for question answering. arXiv preprint arXiv:1611.01604 (2016)
30. Xu, K., et al.: Show, attend and tell: neural image caption generation with visual attention. In: International Conference on Machine Learning, pp. 2048–2057 (2015)
31. Yang, X., Tang, K., Zhang, H., Cai, J.: Auto-encoding scene graphs for image captioning. In: Proceedings of the IEEE/CVF Conference on Computer Vision and Pattern Recognition (CVPR), June 2019
32. Yang, Z., He, X., Gao, J., Deng, L., Smola, A.: Stacked attention networks for image question answering. In: Proceedings of the IEEE Conference on Computer Vision and Pattern Recognition, pp. 21–29 (2016)

33. Yi, K., Wu, J., Gan, C., Torralba, A., Kohli, P., Tenenbaum, J.: Neural-symbolic VQA: disentangling reasoning from vision and language understanding. In: Advances in Neural Information Processing Systems, pp. 1031–1042 (2018)
34. Yu, L., et al.: MattNet: modular attention network for referring expression comprehension. In: Proceedings of the IEEE Conference on Computer Vision and Pattern Recognition, pp. 1307–1315 (2018)
35. Yu, Z., Yu, J., Cui, Y., Tao, D., Tian, Q.: Deep modular co-attention networks for visual question answering. In: Proceedings of the IEEE Conference on Computer Vision and Pattern Recognition, pp. 6281–6290 (2019)
36. Yu, Z., Yu, J., Xiang, C., Fan, J., Tao, D.: Beyond bilinear: generalized multimodal factorized high-order pooling for visual question answering. IEEE Trans. Neural Netw. Learn. Syst. **29**(12), 5947–5959 (2018)
37. Zhu, C., Zhao, Y., Huang, S., Tu, K., Ma, Y.: Structured attentions for visual question answering. In: Proceedings of the IEEE International Conference on Computer Vision, pp. 1291–1300 (2017)

Mining Inter-Video Proposal Relations
for Video Object Detection

Mingfei Han[1,2], Yali Wang[1], Xiaojun Chang[2], and Yu Qiao[1(✉)]

[1] Guangdong-Hong Kong-Macao Joint Laboratory of Human-Machine
Intelligence-Synergy Systems, Shenzhen Institutes of Advanced Technology,
Chinese Academy of Sciences, Shenzhen 518055, China
{yl.wang,yu.qiao}@siat.ac.cn
[2] Faculty of Information Technology, Monash University, Melbourne, Australia
hmf282@gmail.com, cxj273@gmail.com

Abstract. Recent studies have shown that, context aggregating information from proposals in different frames can clearly enhance the performance of video object detection. However, these approaches mainly exploit the intra-proposal relation within single video, while ignoring the intra-proposal relation among different videos, which can provide important discriminative cues for recognizing confusing objects. To address the limitation, we propose a novel Inter-Video Proposal Relation module. Based on a concise multi-level triplet selection scheme, this module can learn effective object representations via modeling relations of hard proposals among different videos. Moreover, we design a Hierarchical Video Relation Network (HVR-Net), by integrating intra-video and inter-video proposal relations in a hierarchical fashion. This design can progressively exploit both intra and inter contexts to boost video object detection. We examine our method on the large-scale video object detection benchmark, i.e., ImageNet VID, where HVR-Net achieves the SOTA results. Codes and models are available at https://github.com/youthHan/HVRNet.

Keywords: Video object detection · Inter-Video Proposal Relation · Multi-level triplet selection · Hierachical Video Relation Network

1 Introduction

Video object detection has emerged as a new challenge in computer vision [1,5,27,28,35,41]. The traditional image object detectors [12,22,24,25] often fail

---

M. Han and Yali Wang—Equal contribution.

Electronic supplementary material The online version of this chapter (https://doi.org/10.1007/978-3-030-58589-1_26) contains supplementary material, which is available to authorized users.

© Springer Nature Switzerland AG 2020
A. Vedaldi et al. (Eds.): ECCV 2020, LNCS 12366, pp. 431–446, 2020.
https://doi.org/10.1007/978-3-030-58589-1_26

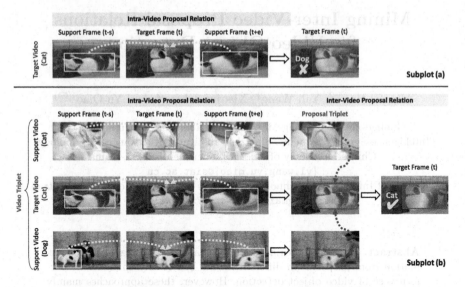

Fig. 1. Motivation. Subplot (a): the intra-video proposal relation often captures what *Cat* looks like and how it moves in this video. But it has little clue about object variations among different videos. As a result, the detector mistakenly recognizes *Cat* as *Dog* at the target frame *t*, even though it leverages spatio-temporal contexts from other support frames $t - s$ and $t + e$. Subplot (b): to tackle such problem, we design a novel inter-video proposal relation module, which can adaptively mine hard object proposals (i.e., proposal triplet) from highly-confused videos (i.e., video triplet), and effectively learn and correct their relation to reduce object confusion among videos.

in this task, due to the fact that objects in videos often contain motion blur, sudden occlusion, rare pose, etc. Recent studies [5,28] have shown that, modeling the relation of object proposal from different frames can effective aggregate spatio-temporal context and yield better representation for detection task. These approaches, however, only utilize the proposal-relations within the same video, and still face the difficulty to distinguish the confusing objects which have similar appearance and/or motion characteristics from different videos.

As shown in Fig. 1(a), the detector mistakenly recognizes *Cat* as *Dog* at the target frame *t*, even though it leverages spatio-temporal contexts from other support frames $t - s$ and $t + e$ to enhance the proposal representation of current frame. The main reason is that, such intra-video relation only describes what this *Cat* looks like and how it moves in this video. It has little clue about object relations or variations among different videos, e.g., in Fig. 1(b), *Cat* in the target video looks similar to *Dog* in the support video, while it looks different from *Cat* in the support video. In this case, the detector would get confused to distinguish *Cat* from *Dog*, if it only focuses on each individual video but without understanding object relations among different videos.

To address such difficulty, we design a novel Inter-Video Proposal Relation method, which can effectively leverage inter-video proposal relation to learn discriminative representations for video object detection. Specifically, we first introduce a multi-level triplet selection scheme to select hard training proposals among confused videos. Since these proposal triplets are the key factors to avoid confusion, we can exploit the relation on each of them to construct better object features. Moreover, we propose a powerful Hierarchical Video Relation Network (HVR-Net), by integrating intra-video and inter-video proposal relation modules in a unified framework. In this case, it can progressively utilize both intra-video and inter-video contextual dependencies to boost video object detection. We investigate extensive experiments on the large-scale video object detection benchmark, i.e., ImageNet VID. Our HVR-Net shows its superiority with **83.8 mAP** by ResNet101 and **85.4 mAP** by ResNeXt101 32x4d.

2 Related Works

Object Detection in Still Images. Object detection in still images [3,8,9,14, 22,24,26] has recently achieved remarkable successes, with the fast development of deep neural networks [13,18,29,30,36] and large-scale well-annotated datasets [21,27]. The existing approaches can be mainly categorized into two frameworks, i.e., two-stage frameworks (such as RCNN [9], Fast-RCNN [8], Faster-RCNN [26]) and one-stage frameworks (such as YOLO [24], SSD [22], RetinaNet [20]). Two-stage detectors often achieve a better detection accuracy, while one-stage detectors often maintain computation efficiency. Recently, anchor-free detectors also show the impressive performance [6,19,38,39], inspired by key point detection. However, these image-based detectors often fail in video object detection, since they often ignore the challenging spatio-temporal characteristics in videos such as motion blur, object occlusion, etc.

Object Detection in Videos. To improve object detection in still images, the previous works of video object detection often leverage temporal information, by box-level association and feature aggregation. Box-level association mainly designs the post-processing steps on image object detector, which can effectively produce tubelet of objects in videos. Such approaches [7,11,16] have been widely-used to boost performance. On the other hand, feature aggregation mainly takes nearby frames as video context to enhance feature representation of current frame [1,5,28,35,41]. In particular, recent studies have shown that, learning proposal relations among different frames can alleviate difficulty in detecting objects in videos [5,28], via long-term dependency modeling [31,33]. However, these approaches mainly focus on intra-video relations of object proposals, while neglecting object relations among different videos. As a result, they often fail to detect objects which contain highly-confused appearance or motion characteristics in videos. Alternatively, we design a novel HVR-Net, which can progressively integrate intra-video and inter-video proposal relations, based on a concise multi-level triplet selection scheme. This allows to effectively alleviate object confusion among videos to boost detection performance.

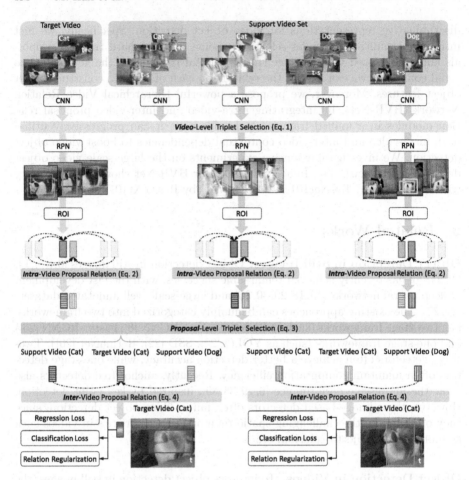

Fig. 2. Our HVR-Net framework. It can effectively boost video object detection, by integrating intra-video and inter-video proposal relation progressively within a multi-level triplet selection scheme. More explanations can be found in Sect. 3.

3   Our HVR-Net

Overview. In this section, we systematically introduce our Hierachical Video Relation Network (HVR-Net), which can boost video object detection by leveraging both intra-video and inter-video contexts within a multi-level triplet selection scheme. The whole framework is shown in Fig. 2. **First**, we design a video-level triplet selection module. For a target video, it can flexibly select two confused videos from a set of support videos, i.e., the most dissimilar video in the same category, the most similar video in the different categories, according to their CNN features. As a result, we obtain a triplet of confused videos in each training batch, which can guide our HVR-Net to model object confusion among videos. **Second**, we introduce an intra-video proposal relation module. For each video in the triplet, we feed its sampled frames (e.g., $t - s$, t and $t + e$) into RPN and ROI layers of Faster RCNN. This produces feature vectors of object proposals for each frame. Subsequently, we aggregate proposals from support frames (e.g., $t - s$, $t + e$) to enhance proposals in the target frame t. As a result, each proposal feature in the target frame t integrates long-term dependencies in the corresponding video, which can address intra-video problem such as motion blur, occlusion, etc. **Third**, we develop a proposal-level triplet selection module. Note that, the intra-video-enhanced proposals mainly contain object semantics in each individual video, while ignoring object variations among videos. To model such variations, we select hard proposal triplets from the video triplet, according to the intra-video-enhanced features. **Finally**, we design an inter-video proposal relation module. For each proposal triplet, it can aggregate proposals from support videos to enhance proposals in the target video. In this case, each proposal feature further leverages inter-video dependencies to tackle object confusion among videos.

3.1 Video-Level Triplet Selection

To effectively alleviate inter-video confusions, we start finding a triplet of hard videos for training. Specifically, we randomly sample K object categories from training set, and randomly sample N videos per category. Hence, there are $K \times N$ videos in a batch. Then, we randomly select one video as *target video*, and use other $(K \times N - 1)$ videos as a set of *support videos*. For each video, we randomly sample one frame as *target frame* t, and sample other $T - 1$ frames as *support frames*, e.g., frame $t - s$ and frame $t + e$ in Fig. 2.

For each video, we feed its T frames individually into the CNN backbone of Faster RCNN for feature extraction. As a result, the feature tensor of this video is with size of $H \times W \times C \times T$, where $H \times W$ and C are respectively the spatial size and the number of feature channels. Then, we perform global average pooling along spatial and temporal dimensions of this tensor, which produces a C-dimension video representation. According to cosine similarity between video representations, we find the video triplet

$$\mathcal{V}^{triplet} = \{\mathcal{V}^{target}, \mathcal{V}^+, \mathcal{V}^-\}, \tag{1}$$

where $\mathcal{V}^+$ is the most dissimilar support video in the class which $\mathcal{V}^{target}$ belongs to, and $\mathcal{V}^-$ is the most similar support video in the other classes.

3.2 Intra-video Proposal Relation

After finding $\mathcal{V}^{triplet}$, we generate object proposals for each video in this triplet. Specifically, we feed the sampled T frames of each video into RPN and ROI layers of Faster RCNN, which produces M proposal features per frame.

Recent studies have shown that, spatio-temporal proposal aggregation among different frames [5,28] can boost video object detection. Hence, we next introduce intra-video proposal relation module, which builds up proposal dependencies between target frame and support frames in each video. Specifically, we adapt a concise non-local-style relation module for $\mathcal{V}^v$ in the video triplet $(v \in \{target, +, -\})$,

$$\boldsymbol{\alpha}_{t,m}^v = \mathbf{x}_{t,m}^v + \sum_{i \in \Omega} \sum_j g(\mathbf{x}_{t,m}^v, \mathbf{x}_{i,j}^v) \times \mathbf{x}_{i,j}^v, \tag{2}$$

where $\mathbf{x}_{t,m}^v$ is the m-th proposal feature in the target frame t, $\mathbf{x}_{i,j}^v$ is the j-th proposal feature in the support frame i, and i belongs to the set of support frames Ω (e.g., $\Omega = \{t - s, t + e\}$). As shown in Eq. (2), we first compare similarity between $\mathbf{x}_{t,m}^v$ and $\mathbf{x}_{i,j}^v$, by a kernel function $g(\cdot, \cdot)$, e.g., Embedded Gaussian in [33]. Then, we aggregate $\mathbf{x}_{t,m}^v$ by weighted sum over all the proposal features of support frames. As a result, $\boldsymbol{\alpha}_{t,m}^v$ is an enhance version of $\mathbf{x}_{t,m}^v$, which contains video-level object semantics to tackle motion blur, object occlusion, etc.

3.3 Proposal-Level Triplet Selection

After intra-video relation module, $\boldsymbol{\alpha}_{t,m}^v$ in Eq. (2) can integrate spatio-temporal object contexts from video $\mathcal{V}^v$ itself. However, it contains little clue to describe object relation among confused videos. To discover such inter-video relation, we propose to further select hard proposal triplets, from intra-video-enhanced proposals in the video triplet $\mathcal{V}^{triplet}$. Specifically, we compare cosine similarity between these proposals, according to their features in Eq. (2). For a proposal $\mathcal{P}_{t,m}^{target}$ in the target video, we obtain its corresponding proposal triplet,

$$\mathcal{P}^{triplet} = \{\mathcal{P}_{t,m}^{target}, \mathcal{P}^+, \mathcal{P}^-\}, \tag{3}$$

where $\mathcal{P}^+$ is the most dissimilar proposal in the same category, $\mathcal{P}^-$ is the most similar proposal in the other categories.

3.4 Inter-Video Proposal Relation

After finding all the proposal triplets, we model relation on each of them in order to describe object variation among videos. To achieve this goal, we use a concise non-local-style relation module for each proposal triplet,

$$\boldsymbol{\beta}_{t,m}^{target} = \boldsymbol{\alpha}_{t,m}^{target} + f(\boldsymbol{\alpha}_{t,m}^{target}, \boldsymbol{\alpha}^+) \times \boldsymbol{\alpha}^+ + f(\boldsymbol{\alpha}_{t,m}^{target}, \boldsymbol{\alpha}^-) \times \boldsymbol{\alpha}^-, \tag{4}$$

where $f(\cdot, \cdot)$ is a kernel function (e.g., Embedded Gaussian) for similarity comparison, $\boldsymbol{\alpha}^+$ is the intra-video-enhanced feature of hard positive proposal $\mathcal{P}^+$, and $\boldsymbol{\alpha}^-$ is the intra-video-enhanced feature of hard negative proposal $\mathcal{P}^-$. By Eq. (4), we further aggregate the proposal $\mathcal{P}_{t,m}^{target}$ in the target video, with inter-video object relationships. Finally, to effectively reduce object confusions when performing detection, we introduce the follow loss for a target video,

$$\mathcal{L} = \mathcal{L}_{detection} + \gamma \mathcal{L}_{relation}, \tag{5}$$

where $\mathcal{L}_{detection} = \mathcal{L}_{regression} + \mathcal{L}_{classification}$ is the tradition detection loss (i.e., bbox regression and object classification) on the final proposal features $\boldsymbol{\beta}_{t,m}^{target}$

in the target frame. γ is a weight coefficient. $\mathcal{L}_{relation}$ is a concise triplet-style metric loss to regularize Eq. (4),

$$\mathcal{L}_{relation} = \max(d(\alpha_{t,m}^{target}, \alpha^-) - d(\alpha_{t,m}^{target}, \alpha^+) + \lambda, 0). \qquad (6)$$

Via this loss, we emphasize a discriminative relation constraint when computing $\beta_{t,m}^{target}$ in Eq. (4), i.e., $d(\alpha_{t,m}^{target}, \alpha^+) > d(\alpha_{t,m}^{target}, \alpha^-) + \lambda$, where d is Euclidean distance. In this case, $\beta_{t,m}^{target}$ becomes more discriminative to alleviate inter-video object confusion, by increasing relation between $\mathcal{P}_{t,m}^{target}$ and $\mathcal{P}^+$ as well as decreasing relation between $\mathcal{P}_{t,m}^{target}$ and $\mathcal{P}^-$.

4   Experiments

We mainly evaluate our HVR-Net on the large-scale ImageNet VID dataset [18]. It consists of 3862 training videos (1,122,397 frames) and 555 validation videos (176,126 frames), with bbox annotations across 30 object categories. Moreover, we train our model on intersection of ImageNet VID and DET dataset [5,28], and report mean Average Precision (mAP) on validation set of VID.

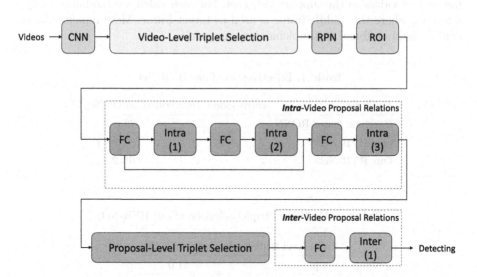

Fig. 3. HVR-Net Architecture. We flexibly adapt the widely-used Faster RCNN architecture as our HVR-Net. More implementation details can be found in Sect. 4.1.

4.1   Implementation Details

Architecture. We flexibly adapt Faster RCNN to our HVR-Net with the following details. The architecture is shown in Fig. 3. We use ResNet-101 [13] as backbone for ablation studies, and also report the results on ResNeXt-101-32x4d [36] for SOTA comparison. We extract the feature of each sampled frame after the *conv4* stage, in order to select video triplet in a training batch. Region Proposal Network (RPN) is used to generate proposals from each frame of the selected video triplet, by using the feature maps after the *conv4* stage. We introduce three intra modules in Fig. 3. Before each of them, we add 1024-dim fully-connected (FC) layer. Additionally, we use a skip connection between intra (1) and intra (3), to increase learning flexibility. In this case, the intra (3) module can use both initial and transformed proposals of support frames to enhance proposals in the target frame.

We introduce one inter module which is added upon a 1024-dim FC layer. Additionally, for both intra and inter modules, the kernel function is set as Embedded Gaussian in [33], where each embedding transformation in this kernel is a 1024-dim FC layer.

Training Details. We implement our HVR-Net on Pytorch, by 8 GPUs of 1080Ti. In each training batch, we randomly sample $K = 3$ object categories from training set, and randomly sample $N = 3$ videos per category. Hence, there are 9 videos in a batch. Then, we randomly select one video as target video, and use other 8 videos as the support video set. For each video, we randomly sample 3 frames, where the middle frame is used as target frame. More implementation details could be found in supplementary materials.

Table 1. Effectiveness of our HVR-Net.

Methods	Intra-video	Inter-video	mAP (%)
Baseline: faster-RCNN	–	–	73.2
Our HVR-Net	✓	–	$80.6_{\uparrow 7.4}$
Our HVR-Net	✓	✓	$\mathbf{83.2}_{\uparrow 10.0}$

Table 2. Multi-level triplet selection of our HVR-Net.

Multi-level triplet selection	mAP (%)
Simple	81.0
Our	**83.2**

4.2 Ablation Studies

Effectiveness of HVR-Net. We first compare our HVR-Net with the baseline architecture, i.e., Faster RCNN. As shown in Table 1, our HVR-Net significantly outperforms Faster RCNN, indicating its superiority in video object detection. More importantly, HVR-Net with both intra-video and inter-video is better than

Table 3. Supervision in Our HVR-Net.

Detection loss	Relation regularization	mAP (%)
✓	–	80.0
✓	✓	**83.2**

Table 4. Number of Intra and Inter Modules in Our HVR-Net.

No. of intra	No. of inter	mAP (%)
2	1	81.8
3	1	**83.2**
3	2	82.1

that with intra-video only (83.2 vs. 80.6). It demonstrates that, learning proposal interactions inside each single video is not sufficient to describe category differences among videos. When adding inter-video proposal relation module, our HVR-Net can flexibly select hard proposals from confused videos, and effectively build up relations among these proposals to distinguish object confusions (Table 5).

Multi-level Triplet Selection. Our HVR-Net is built upon a multi-level triplet selection scheme, including video-level and proposal-level proposal selection. To demonstrate the effectiveness, we replace these two selection modules with a simple approach, i.e., selecting random videos and using all proposals in each video. In Table 2, when using the straightforward selection, the performance of HVR-Net is getting worse. The main reason is that, blindly selected videos and proposals do not guide our HVR-Net to focus on object confusion in videos. Alternatively, when we add our video and proposal triplet selection, HVR-Net can effectively leverage hard proposals of confused videos to learn and correct inter-video object relations, in order to boost video object detection.

Supervision in HVR-Net. As mentioned in Sect. 3.4, we introduce a relation regularization in Eq. (6), in order to emphasize the correct relation constraint on inter-video relation module in Eq. (4). We investigate it in Table 3. As expected, this regularization can boost HVR-Net by a large margin, by enhancing similarity between proposals in the same category, and reducing similarity between proposals in the different categories.

Table 5. Number of Testing Frames in Our HVR-Net.

Testing frames	5	11	17	21	31
mAP (%)	80.5	81.6	82.0	82.9	**83.2**

Table 6. Comparison with the state-of-the-art methods on ImageNet VID (mAP).

Methods	Backbone	Post-processing	Base detector	mAP(%)
D& T [7]	ResNet101	–	R-FCN	75.8
MANet [32]	ResNet101	–	R-FCN	78.1
LWDN [15]	ResNet101	–	R-FCN	76.3
RDN [5]	ResNet101	–	Faster-RCNN	81.8
LongRange [28]	ResNet101	–	FPN	81.0
Deng [4]	ResNet101	–	R-FCN	79.3
PSLA [10]	ResNet101+DCN	–	R-FCN	80.0
THP [40]	ResNet101+DCN	–	R-FCN	78.6
STSN [1]	ResNet101+DCN	–	R-FCN	78.9
Ours	ResNet101	–	Faster-RCNN	**83.2**
TCNN [17]	DeepID+Craft [23, 37]	Tublet Linking	RCNN	73.8
STMN [35]	ResNet101	Seq-NMS	R-FCN	80.5
FGFA [41]	Align. Incep.-ResNet	Seq-NMS	R-FCN	80.1
D& T($\tau = 10$) [7]	ResNet101	Viterbi	R-FCN	78.6
D& T($\tau = 1$) [7]	ResNet101	Viterbi	R-FCN	79.8
MANet [32]	ResNet101	Seq-NMS	R-FCN	80.3
ST-Lattice [2]	ResNet101	Tublet-rescore	R-FCN	79.6
SELSA [34]	ResNet101	Seq-NMS	Faster-RCNN	82.5
Deng [4]	ResNet101	Seq-NMS	R-FCN	80.8
PSLA [10]	ResNet101+DCN	Seq-NMS	R-FCN	81.4
STSN+[1]	ResNet101+DCN	Seq-NMS	R-FCN	80.4
Ours	ResNet101	Seq-NMS	Faster-RCNN	**83.8**
D& T [7]	ResNeXt101	Viterbi	Faster-RCNN	81.6
D& T [7]	Inception-v4	Viterbi	R-FCN	82.1
LongRange [28]	ResNeXt101-32×8d	–	FPN	83.1
RDN [5]	ResNeXt101-64×4d	–	Faster-RCNN	83.2
RDN [5]	ResNeXt101-64×4d	Seq-NMS	Faster-RCNN	84.5
SELSA [34]	ResNeXt101-32×4d	–	Faster-RCNN	84.3
SELSA [34]	ResNeXt101-32×4d	Seq-NMS	Faster-RCNN	83.7
Ours	ResNeXt101-32×4d	–	Faster-RCNN	**84.8**
Ours	ResNeXt101-32×4d	Seq-NMS	Faster-RCNN	**85.5**

Table 7. Comparison with state-of-the-art methods in mAP.

Methods	Fast (mAP)	Medium (mAP)	Slow (mAP)
FGFA [41]	57.6	75.8	83.5
MANet [32]	56.7	76.8	86.9
Deng [4]	61.1	78.7	86.2
LongRange [28]	64.2	79.5	86.7
Ours	**66.6**	**82.3**	**88.7**

Number of Intra and Inter Relation Modules. We investigate the performance of our HVR-Net, with different number of intra-video and inter-video proposal modules. When changing the number of intra modules (or inter modules), we fix the number of inter modules (or intra modules). The results are shown in Table 4. As expected, when increasing the number of both modules, the performance of HVR-Net is getting better and tends to become flat. Hence, in our experiment, we set the number of intra modules as three, and set the number of inter module as one.

Number of Testing Frames. We investigate the performance of HVR-Net, w.r.t., the number of sampled frames in a testing video. As expected, when increasing the number of testing frames, the performance of HVR-Net is getting better and tends to become stable. Hence, we choose the number of testing frames as 31 in our experiment. Besides, we test HVR-Net by unloading inter-video proposal relation module in the testing phase, which achieves the comparable mAP.

4.3   SOTA Comparison

We compare our HVR-Net with a number of recent state-of-the-art approaches on ImageNet VID validation set. As shown in Table 6 and Table 7, HVR-Net achieves the best performance among various settings and object categories.

In Table 6, we first make comparison without any video-level post-processing techniques. Under the same backbone, We significantly outperform the well-known approaches such as FGFA [41] and MANet [32], which uses expensive optical flow as guidance of feature aggregation. More importantly, our HVR-Net outperform the recent approaches [5, 28] that mainly leverage proposal relations among different frames for spatio-temporal context aggregation. This further confirms the effectiveness of learning inter-video proposal relation. Second, we equip HVR-Net with the widely-used post-processing approach Seq-NMS. Once again, we outperform other state-of-the-art approaches under the same backbone. It shows that, our HVR-Net is compatible and complementary with post-processing of video object detection, which can further boost performance.

Additionally, we follow FGFA [41] to evaluate detection performance on the categories of slow, medium, and fast objects, where these three categories are divided by their average IoU scores between objects across nearby frames, i.e., Slow (score > 0.9), Medium (score $\in$ [0.7, 0.9]), Fast (Others). As shown in Table 7, our HVR-Net boost the detection performance on all these three categories, showing the importance of inter-video proposal relation for confusion reduction.

4.4   Visualization

Detection Visualization. We show the detection result of HVR-Net in Fig. 4. Specifically, we compare two settings, i.e., baseline with only intra-video proposal

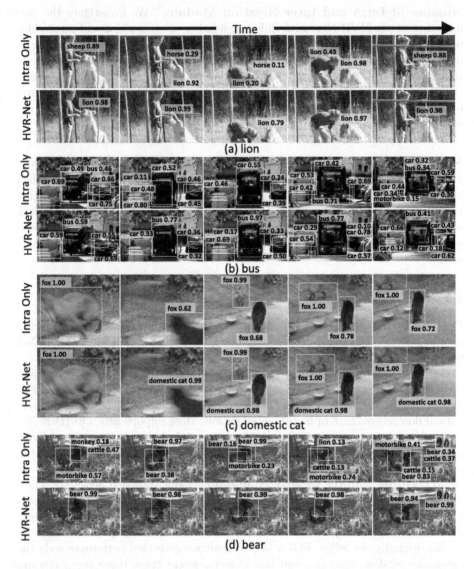

Fig. 4. Detection visualization. For each video, the first row shows the baseline with only intra-video proposal relation module. The second row shows HVR-Net with both intra-video and inter-video proposal relation modules. Clearly, our inter-video can effectively guide HVR-Net to tackle object confusion in videos. For example, a female lion in Subplot (a) looks quite similar to a horse, due to its color and its motion in this video. As a result, the baseline mistakenly recognizes it as a horse, when only using intra-video relation aggregation. By introducing inter-video proposal relation, our HVR-Net successfully distinguish such object confusion in videos. Other subplots also exhibit the similar result, i.e., it is necessary and important to learn inter-video proposal relations to boost video object detection.

relation module, and HVR-Net with both intra-video and inter-video proposal relation modules. As expected, when only using intra-video relation aggregation, baseline fails to recognize the object in the video, e.g., a female lion in Subplot (a) is mistakenly recognized as a horse with confidence larger than 0.9. The main reason is that, intra-video relation mainly focuses on what the object looks like and how it moves in this video. For the video in Subplot (a), the appearance and motion of this lion are quite similar to a horse, leading to high confusion. Alternatively, when we introducing inter-video proposal relation module, HVR-Net successfully distinguish such object confusion in videos. Hence, it is necessary and important to learn inter-video proposal relations for video object detection.

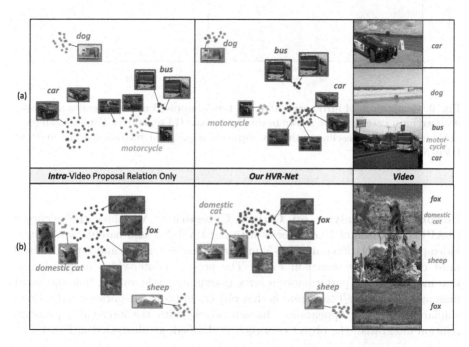

Fig. 5. Proposal feature visualization of video triplet by t-SNE. With intra-video relation only, proposals of confusing objects mistakenly stay together as a cluster (i.e. domestic cats and foxes in (b), cars and motobikes in (a)). Our HVR-Net can learn the discriminative cues and clarify those proposals of confusing objects. For each video triplet, three target frames and their proposals are shown.

Video and Proposal Feature Visualization in HVR-Net. We visualize the proposal features of target frames in video triplets with t-SNE in Fig. 5. As expected, with inter-video proposal relation integrated, the proposal features of confusing objects can be clarified, while baseline, with intra-video proposal relations only, mistakenly clusters the proposals not belong to same category, e.g., in Fig. 5(b), proposals of domestic cat mistakenly stay with proposals of fox together as a cluster, while our HVR-Net can learn a compact cluster (e.g., proposals of fox) and assign proposals of domestic cat correctly. The reason is that the object confusion is clarified with inter-video proposal relation integrated, leading to enlarged difference of confused proposals in feature embedding.

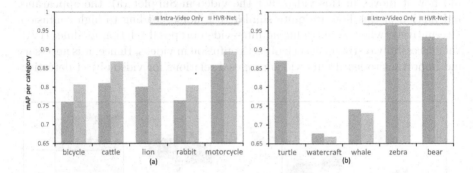

Fig. 6. Comparison of mAP per Category. Top-5 improved most categories and top-5 declined most categories are shown in subplot (a) and (b) separately. For each category, mAP is shown for baseline with only intra-video proposal relation module and our HVR-Net.

Performance Analysis on Object Categories. We show the accuracy (mAP) comparison of 10 categories with our HVR-Net and baseline with intra-video proposal relation only. Top-5 improved most categories and top-5 declined most categories are shown in Fig. 6. The proposed inter-video proposal relation module boosts performance a large margin in cattle, rabbit, lion and other mammal categories. The reason is that objects in those categories usually share similar motion and appearance characteristics. With the inter-video proposal relation integrated, the object confusion is clarified, as illustrated in Fig. 4.

5 Conclusion

In this work, we propose to learn inter-video object relations for video object detection. Based on a flexible multi-level triplet selection scheme, we develop a Hierachical Video Relation Network (HVR-Net), which can effectively leverage intra-video and inter-video relation in a unified manner, in order to progressively tackle object confusions in videos. We perform extensive experiments on the large-scale video object detection benchmark, i.e., ImageNet VID. The results show that our HVR-Net is effective and important for video object detection.

Acknowledgement. This work is partially supported by Science and Technology Service Network Initiative of Chinese Academy of Sciences (KFJ-STS-QYZX-092),

Guangdong Special Support Program (2016TX03X276), Shenzhen Basic Research Program (CXB201104220032A), National Natural Science Foundation of China (61876176, U1713208), the Joint Lab of CAS-HK. This work is also partially supported by Australian Research Council Discovery Early Career Award (DE190100626).

References

1. Bertasius, G., Torresani, L., Shi, J.: Object detection in video with spatiotemporal sampling networks. In: ECCV, pp. 331–346 (2018)
2. Chen, K., et al.: Optimizing video object detection via a scale-time lattice. In: CVPR, pp. 7814–7823 (2018)
3. Dai, J., Li, Y., He, K., Sun, J.: R-FCN: object detection via region-based fully convolutional networks. In: Advances in Neural Information Processing Systems, pp. 379–387 (2016)
4. Deng, H., et al.: Object guided external memory network for video object detection. In: ICCV, pp. 6678–6687 (2019)
5. Deng, J., Pan, Y., Yao, T., Zhou, W., Li, H., Mei, T.: Relation distillation networks for video object detection. In: ICCV (2019)
6. Duan, K., Bai, S., Xie, L., Qi, H., Huang, Q., Tian, Q.: CenterNet: keypoint triplets for object detection. In: ICCV, pp. 6569–6578 (2019)
7. Feichtenhofer, C., Pinz, A., Zisserman, A.: Detect to track and track to detect. In: ICCV, pp. 3038–3046 (2017)
8. Girshick, R.: Fast R-CNN. In: ICCV, pp. 1440–1448 (2015)
9. Girshick, R., Donahue, J., Darrell, T., Malik, J.: Rich feature hierarchies for accurate object detection and semantic segmentation. In: CVPR, pp. 580–587 (2014)
10. Guo, C., et al.: Progressive sparse local attention for video object detection. In: ICCV, pp. 3909–3918 (2019)
11. Han, W., et al.: SEQ-NMS for video object detection. arXiv preprint arXiv:1602.08465 (2016)
12. He, K., Gkioxari, G., Dollár, P., Girshick, R.: Mask R-CNN. In: CVPR, pp. 2961–2969 (2017)
13. He, K., Zhang, X., Ren, S., Sun, J.: Deep residual learning for image recognition. In: CVPR, pp. 770–778 (2016)
14. Hu, H., Gu, J., Zhang, Z., Dai, J., Wei, Y.: Relation networks for object detection. In: CVPR, pp. 3588–3597 (2018)
15. Jiang, Z., Gao, P., Guo, C., Zhang, Q., Xiang, S., Pan, C.: Video object detection with locally-weighted deformable neighbors. In: AAAI, vol. 33, pp. 8529–8536 (2019)
16. Kang, K., et al.: Object detection in videos with tubelet proposal networks. In: CVPR, pp. 727–735 (2017)
17. Kang, K., et al.: T-CNN: tubelets with convolutional neural networks for object detection from videos. TCSVT **28**(10), 2896–2907 (2017)
18. Krizhevsky, A., Sutskever, I., Hinton, G.E.: ImageNet classification with deep convolutional neural networks. In: Advances in Neural Information Processing Systems, pp. 1097–1105 (2012)
19. Law, H., Deng, J.: CornerNet: detecting objects as paired keypoints. In: ECCV, pp. 734–750 (2018)
20. Lin, T.Y., Goyal, P., Girshick, R., He, K., Dollár, P.: Focal loss for dense object detection. In: ICCV, pp. 2980–2988 (2017)

21. Lin, T.-Y., et al.: Microsoft COCO: common objects in context. In: Fleet, D., Pajdla, T., Schiele, B., Tuytelaars, T. (eds.) ECCV 2014. LNCS, vol. 8693, pp. 740–755. Springer, Cham (2014). https://doi.org/10.1007/978-3-319-10602-1_48

22. Liu, W., et al.: SSD: single shot multibox detector. In: Leibe, B., Matas, J., Sebe, N., Welling, M. (eds.) ECCV 2016. LNCS, vol. 9905, pp. 21–37. Springer, Cham (2016). https://doi.org/10.1007/978-3-319-46448-0_2

23. Ouyang, W., et al.: DeepID-net: multi-stage and deformable deep convolutional neural networks for object detection. arXiv preprint arXiv:1409.3505 (2014)

24. Redmon, J., Divvala, S., Girshick, R., Farhadi, A.: You only look once: unified, real-time object detection. In: CVPR, pp. 779–788 (2016)

25. Ren, S., He, K., Girshick, R., Sun, J.: Faster R-CNN: towards real-time object detection with region proposal networks. In: NIPS, pp. 91–99 (2015)

26. Ren, S., He, K., Girshick, R., Sun, J.: Faster R-CNN: towards real-time object detection with region proposal networks. In: IEEE-TPAMI (2017)

27. Russakovsky, O., et al.: Imagenet large scale visual recognition challenge. IJCV 115(3), 211–252 (2015)

28. Shvets, M., Liu, W., Berg, A.C.: Leveraging long-range temporal relationships between proposals for video object detection. In: ICCV (2019)

29. Simonyan, K., Zisserman, A.: Very deep convolutional networks for large-scale image recognition. arXiv preprint arXiv:1409.1556 (2014)

30. Szegedy, C., et al.: Going deeper with convolutions. In: CVPR, pp. 1–9 (2015)

31. Vaswani, A., et al.: Attention is all you need. In: NIPS, pp. 5998–6008 (2017)

32. Wang, S., Zhou, Y., Yan, J., Deng, Z.: Fully motion-aware network for video object detection. In: ECCV, pp. 542–557 (2018)

33. Wang, X., Girshick, R., Gupta, A., He, K.: Non-local neural networks. In: CVPR, pp. 7794–7803 (2018)

34. Wu, H., Chen, Y., Wang, N., Zhang, Z.: Sequence level semantics aggregation for video object detection. In: ICCV (2019)

35. Xiao, F., Jae Lee, Y.: Video object detection with an aligned spatial-temporal memory. In: ECCV, pp. 485–501 (2018)

36. Xie, S., Girshick, R., Dollár, P., Tu, Z., He, K.: Aggregated residual transformations for deep neural networks. In: CVPR, pp. 1492–1500 (2017)

37. Yang, B., Yan, J., Lei, Z., Li, S.Z.: Craft objects from images. In: CVPR, pp. 6043–6051 (2016)

38. Zhou, X., Zhuo, J., Krahenbuhl, P.: Bottom-up object detection by grouping extreme and center points. In: CVPR, pp. 850–859 (2019)

39. Zhu, C., He, Y., Savvides, M.: Feature selective anchor-free module for single-shot object detection. In: CVPR, pp. 840–849 (2019)

40. Zhu, X., Dai, J., Yuan, L., Wei, Y.: Towards high performance video object detection. In: CVPR, pp. 7210–7218 (2018)

41. Zhu, X., Wang, Y., Dai, J., Yuan, L., Wei, Y.: Flow-guided feature aggregation for video object detection. In: ICCV, pp. 408–417 (2017)

TVR: A Large-Scale Dataset
for Video-Subtitle Moment Retrieval

Jie Lei$^{(\boxtimes)}$, Licheng Yu, Tamara L. Berg, and Mohit Bansal

University of North Carolina at Chapel Hill, Chapel Hill, USA
{jielei,licheng,tlberg,mbansal}@cs.unc.edu

Abstract. We introduce TV show Retrieval (**TVR**), a new multimodal retrieval dataset. TVR requires systems to understand both videos and their associated subtitle (dialogue) texts, making it more realistic. The dataset contains 109K queries collected on 21.8K videos from 6 TV shows of diverse genres, where each query is associated with a tight temporal window. The queries are also labeled with query types that indicate whether each of them is more related to video or subtitle or both, allowing for in-depth analysis of the dataset and the methods that built on top of it. Strict qualification and post-annotation verification tests are applied to ensure the quality of the collected data. Additionally, we present several baselines and a novel Cross-modal Moment Localization (**XML**) network for multimodal moment retrieval tasks. The proposed XML model uses a late fusion design with a novel Convolutional Start-End detector (**ConvSE**), surpassing baselines by a large margin and with better efficiency, providing a strong starting point for future work. (TVR dataset and code are publicly available: https://tvr.cs.unc.edu/. We also introduce TVC for multimodal captioning at https://tvr.cs.unc.edu/tvc.html).

1 Introduction

Enormous numbers of multimodal videos (with audio and/or text) are being uploaded to the web every day. To enable users to search through these videos and find relevant moments, an efficient and accurate method for retrieval of video data is crucial. Recent works [8,13] introduced the task of Single Video Moment Retrieval (SVMR), whose goal is to retrieve a moment from a single video via a natural language query. Escorcia *et al.* [7] extended SVMR to Video Corpus Moment Retrieval (VCMR), where a system is required to retrieve the most relevant moments from a large video corpus instead of from a single video. However, these works rely on a single modality (visual) as the context source for retrieval, as existing moment retrieval datasets [8,13,19,25] are based on videos. In practice, videos are often associated with other modalities such as audio or text, e.g., subtitles for movie/TV-shows or audience discourse accompanying live

Electronic supplementary material The online version of this chapter (https://doi.org/10.1007/978-3-030-58589-1_27) contains supplementary material, which is available to authorized users.

© Springer Nature Switzerland AG 2020
A. Vedaldi et al. (Eds.): ECCV 2020, LNCS 12366, pp. 447–463, 2020.
https://doi.org/10.1007/978-3-030-58589-1_27

stream videos. These associated modalities could be equally important sources for retrieving user-relevant moments. Figure 1 shows a query example in the VCMR task, in which both videos and subtitles are vital to the retrieval process.

Query: **Rachel** explains to her dad on the phone why she can't marry her fiancé.
Query Type: video + subtitle

Fig. 1. A TVR example in the VCMR task. Ground truth moment is shown in *green box*. Colors in the query indicate whether the words are related to video (*blue*) or subtitle (*magenta*) or both (*black*). To better retrieve relevant moments from the video corpus, a system needs to comprehend both videos and subtitles (Color figure online)

Hence, to study multimodal moment retrieval with both video and text contexts, we propose a new dataset - TV show Retrieval (**TVR**). Inspired by recent works [18,20,30] that built multimodal datasets based on Movie/Cartoon/TV shows, we select TV shows as our data resource as they typically involve rich social interactions between actors, involving both activities and dialogues. During data collection, we present annotators with videos and associated subtitles to encourage them to write multimodal queries. A tight temporal timestamp is labeled for each video-query pair. We do not use predefined fixed segments (as in [13]) but choose to freely annotate the timestamps for more accurate localization. Moreover, query types are collected for each query to indicate whether it is more related to the video, the subtitle, or both, allowing deeper analyses of systems. To ensure data quality, we set up strict qualification and post-annotation quality verification tests. In total, we have collected 108,965 high-quality queries on 21,793 videos from 6 TV shows, producing the largest dataset of this kind. Compared to existing datasets [8,13,19,25], we show TVR has greater linguistic diversity (Fig. 3) and involves more actions and people in its queries (Table 2).

With the TVR dataset, we extend the moment retrieval task to a more realistic multimodal setup where both video and subtitle text need to be considered (i.e., 'Video-Subtitle Moment Retrieval'). In this paper, we focus on the corpus-level task VCMR , as SVMR can be viewed as a simplified version of VCMR in which the ground-truth video is given beforehand. Prior works [7–9,13,14,32] explore the moment retrieval task as a ranking problem over a predefined set of moment proposals. These proposals are usually generated using handcrafted heuristics [13,14] or sliding windows [7–9,32] and are usually not temporally precise, leading to suboptimal performance. Furthermore, these methods may not be easily scaled to long videos: the number of proposals often increase quadratically with video length, making computational costs infeasible. Recent methods [10,21] adapt start-end span predictors [3,28] from the reading comprehension task to moment retrieval, by early fusion of video and language (query) features, then applying neural networks on the fused features to predict start-end probabilities. It has been shown [10] that using span predictors outperforms several proposal-based methods. Additionally, start-end predictors allow a hassle-free extension to long videos, with only linearly increased computational cost. While [10] has shown promising results in SVMR, it is not scalable to VCMR as it uses early fusion. Consider retrieving N queries in a corpus of M videos. This requires running several layers of LSTM [15] on $M \cdot N$ early fused representations to generate the probabilities, which is computationally expensive for large values of M and N.

To address these challenges, we propose Cross-modal Moment Localization (**XML**), a late fusion approach for VCMR. In XML, videos (or subtitles) and queries are encoded independently, thus only $M+N$ neural network operations are needed. Furthermore, videos can be pre-encoded and stored. At test time, one only needs to encode new user queries, which greatly reduces user waiting time. Late fusion then integrates video and query representations with highly optimized matrix multiplication to generate 1D query-clip similarity scores over the temporal dimension of the videos. To produce moment predictions from these similarity scores, a naive approach is to rank the aforementioned sliding window proposals with confidence scores computed as the average of the similarity scores inside each proposal region. Alternatively, one can use TAG [37] to progressively group top-scored clips. However, these methods rely on handcrafted rules and are not trainable. Inspired by image edge detectors [29], we propose Convolutional Start-End detector (**ConvSE**) that learns to detect start (up) and end (down) edges in the similarity signals with two trainable 1D convolution filters. Using the same backbone net, we show ConvSE has better performance than both approaches. With late fusion and ConvSE, we further show XML outperforms previous methods [7,10,13], and does this with better computational efficiency.

To summarize, our contributions are 2-fold: (*i*) We introduce **TVR** dataset, a large-scale multimodal moment retrieval dataset with 109K high-quality queries of great linguistic diversity.[1] (*ii*) We propose **XML**, an efficient approach that uses a late fusion design for the VCMR task. The core of XML is our novel **ConvSE** module which learns to detect start-end edges in 1D similarity signals. Comprehensive experiments and analyses show XML surpasses all presented baselines by a large margin and runs with better efficiency.

---

[1] We also collected a new multimodal captioning dataset with 262K captions, named as TV show Caption (**TVC**) https://tvr.cs.unc.edu/tvc.html.

2 Related Work

The goal of natural language-based moment retrieval is to retrieve relevant moments from a single video [8,13] or from a large video corpus [7]. In the following, we present a brief overview of the community efforts on these tasks and make distinctions between existing works and ours.

Datasets. Several datasets have been proposed for the task, e.g., DiDeMo [13], ActivityNet Captions [19], CharadesSTA [8], and TACoS [25], where queries can be localized solely from video. TVR differs from them by requiring additional text (subtitle) information in localizing the queries. Two types of data annotation have been explored in previous works: (*i*) uniformly chunking videos into segments and letting an annotator pick one (or more) and write an unambiguous description [13]. For example, moments in DiDeMo [13] are created from fixed 5-s segments. However, such coarse temporal annotations are not well aligned with natural moments. In TVR, temporal windows are freely selected to more accurately capture important moments. (*ii*) converting a paragraph written for a whole video into separate query sentences [8,19,25]. While it is natural for people to use temporal connectives (e.g., 'first', 'then') and anaphora (e.g., pronouns) [27] in a paragraph, these words make individual sentences less suitable as retrieval queries. In comparison, the TVR annotation process encourages annotators to write queries individually without requiring the context of a paragraph. Besides, TVR also has a larger size and greater linguistic diversity, see Sect. 3.2.

Methods. Existing works [7–9,13,14,32] pose moment retrieval as ranking a predefined set of moment proposals. These proposals are typically generated with handcrafted rules [13,14] or sliding windows [7–9,32]. Typically, such proposals are not temporally precise and are not scalable to long videos due to high computational cost. [8,9,32] alleviate the first with a regression branch that offsets the proposals. However, they are still restricted by the coarseness of the initial proposals. Inspired by span predictors in reading comprehension [3,28] and action localization [22], we use start-end predictors to predict start-end probabilities from early fused query-video representations. Though these methods can be more flexibly applied to long videos and have shown promising performance on single video moment retrieval, the time cost of early fusion becomes unbearable when dealing with the corpus level moment retrieval problem: they require early fusing every possible query-video pair [7]. Proposal based approaches MCN [13] and CAL [7] use a late fusion design, in which the video representations can be pre-computed and stored, making the retrieval more efficient. The final moment predictions are then made by ranking the Squared Euclidean Distances between the proposals w.r.t. a given query. However, as they rely on predefined proposals, MCN and CAL still suffer from the aforementioned drawbacks, leading to less precise predictions and higher costs (especially for long videos). Recent works [4,35,36] consider word-level early fusion with the videos, which can be even more expensive. In contrast, XML uses a late fusion design with a novel Convolutional Start-End (ConvSE) detector, which produces more accurate moment predictions while reducing the computational cost.

3   Dataset

Our TVR dataset is built on 21,793 videos from 6 long-running TV shows across 3 genres (*sitcom, medical, crime*), provided by TVQA [20]. Videos are paired with subtitles and are on average 76.2 s in length. In the following, we describe how we collected TVR and provide a detailed analysis of the data.

3.1   Data Collection

We used Amazon Mechanical Turk (AMT) for TVR data collection. Each AMT worker was asked to write a query using information from the video and/or subtitle, then mark the start and end timestamps to define a moment that matches the written query. This query-moment pair is required to be a unique match within the given video, i.e., the query should be a referring expression [13,17] that uniquely localizes the moment. We additionally ask workers to select a query type from three types: *video-only* - queries relevant to the visual content only, *sub-only* - queries relevant to the subtitles only, and *video+sub* - queries that involve both. In our pilot study, we found workers preferred to write *sub-only* queries. A similar phenomenon was observed in TVQA [20], where people can achieve 72.88% QA accuracy by reading the subtitles only. Therefore, to ensure that we collect a balance of queries requiring one or both modalities, we split the data annotation into two rounds - *visual* round and *textual* round. For the visual round, we encourage workers to write queries related to the visual content, including both *video-only* and *video+sub* queries. For the textual round, we encourage *sub-only* and *video+sub* queries. We ensure data quality with the following strategies:[2]

Qualification Test. We designed a set of 12 multiple-choice questions as our qualification test and only let workers who correctly answer at least 9 questions participate in our annotation task, ensuring that workers understand our task requirements well. In total, 1,055 workers participated in the test, with a pass rate of 67%. Adding this qualification test greatly improved data quality.

Automatic Check. During collection, we used an automatic tool checking that all required annotations (query, timestamps, etc) have been performed and each query contains at least 8 words and is not copied from the subtitle.

Manual Check. Additional manual check of the collected data was done in house throughout the collection process. Those disqualified queries were re-annotated and workers with disqualified queries were removed from our worker list.

Post-annotation Verification. To verify the quality of the collected data, we performed a post-annotation verification experiment. We set up another AMT task where workers were required to rate the quality of the collected query-moment pairs based on *relevance, is the query-moment pair a unique-match*, etc.

---

[2] We present a pipeline figure of our data collection procedure in the supplementary.

The rating was done in a *likert-scale* manner with 5 options: *strongly agree, agree, neutral, disagree* and *strongly disagree*. Results show that 92% of the pairs have a rating of at least *neutral*. We further analyzed the group of queries that were rated as *strongly disagree*, and found that 80% of them were still of acceptable quality: e.g., slightly mismatched timestamps (≤ 1 s). This verification was conducted on 3,600 query-moment pairs. Details are presented in the supplementary file.

Table 1. Comparison of TVR with existing moment retrieval datasets. Q stands for query. Q *context* indicate which modality the queries are related. *Free st-ed* indicates whether the timestamps are freely annotated. *Individual Q* means the queries are collected as individual sentences, rather than sentences in paragraphs

Dataset	Domain	#Q/#videos	Vocab. size	Avg. Q len.	Avg. len. (s) Moment/video	Q context Video	Text	Free st-ed	Q type anno.	Individual Q
TACoS [25]	Cooking	16.2K/0.1K	2K	10.5	5.9/287	✓	–	✓	–	–
DiDeMo [13]	Flickr	41.2K/10.6K	7.6K	8.0	6.5/29.3	✓	–	–	–	✓
ActivityNet Captions [19]	Activity	72K/15K	12.5K	14.8	36.2/117.6	✓	–	✓	–	–
CharadesSTA [8]	Activity	16.1K/6.7K	1.3K	7.2	8.1/30.6	✓	–	✓	–	–
TVR	TV show	109K/21.8K	57.1K	13.4	9.1/76.2	✓	✓	✓	✓	✓

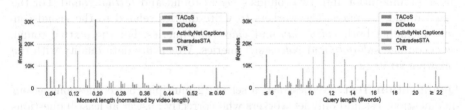

Fig. 2. Distributions of moment (*left*) and query (*right*) lengths. Compared to existing moment retrieval datasets [8,13,19,25], TVR has relatively shorter moments (normalized) and longer queries. Best viewed digitally with zoom

3.2 Data Analysis and Comparison

Table 1 shows an overview of TVR and its comparisons with existing moment retrieval datasets [8,13,19,25]. TVR contains 109K human annotated query-moment pairs on 21.8K videos, making it the largest of its kind. Moments have an average length of 9.1 s, and are annotated with tight start and end timestamps, enabling training and evaluating on more precise localization. Compared to existing datasets, TVR has relatively shorter (video-length normalized) moments and longer queries (Fig. 2). It also has greater linguistic diversity (Fig. 3): it has more unique 4-grams and almost every query is unique, making the textual understanding of TVR more challenging. As TVR is collected on TV

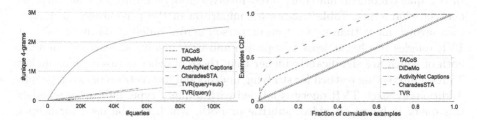

Fig. 3. *Left*: #unique 4-gram as a function of #queries. *Right*: CDF of queries ordered by frequency, to obtain this plot, we sampled 10K queries from each dataset, we consider two queries to be the same if they exact match, after tokenization and lemmatization, following [34]. Compared to existing moment retrieval datasets [8,13,19,25], TVR has greater diversity, i.e., it has more unique 4-grams and almost every TVR query is unique. Best viewed digitally with zoom

Table 2. Percentage of queries that have multiple actions or involve multiple people. Statistics is based on 100 manually labeled queries from each dataset. We also show query examples, with unique person mentions underlined and actions in **bold**. Compared to existing datasets, TVR queries typically have more people and actions and require both *video* and *sub* (subtitle) context

Dataset	#actions ≥2 (%)	#people ≥2 (%)	Query examples (*query type*)
TACoS [25]	20	0	She **rinses** the peeled carrots off in the sink. (*video*)
			The person **removes** roots and outer leaves and **rewashes** the leek. (*video*)
CharadesSTA [8]	6	12	A person is **eating** food slowly. (*video*)
			A person is **opening** the door to a bedroom. (*video*)
ActivityNet	44	44	He then **grabs** a metal mask and **positions** himself correctly on the floor. (*video*)
Caption [19]			The same man **comes** back and **lifts** the weight over his head again. (*video*)
DiDeMo [13]	6	10	A dog **shakes** its body. (*video*)
			A lady in a cowboy hat **claps** and **jumps** excitedly. (*video*)
TVR	67	66	Bert **leans** down and **gives** Amy a hug who is **standing** next to Penny. (*video*)
			Taub **argues** with the patient that fighting in Hockey **undermines** the sport. (*sub*)
			Chandler **points** at Joey while **describing** a woman who wants to **date** him. (*video+sub*)

shows, query-moment matching often involves understanding rich interactions between characters. Table 2 shows a comparison of the percentages of queries that involve more than one action or person across different datasets. 66% of TVR queries involve at least two people and 67% involve at least two actions, both of which are significantly higher than those of other datasets. This makes TVR an interesting testbed for studying multimodal interactions between people. Additionally, each TVR query is labeled with a query type, indicating whether this query is based on video, subtitle or both, which can be used for deeper analyses of the systems.

4   Cross-Modal Moment Localization (XML)

In VCMR, the goal is to retrieve a moment from a large video corpus $V = \{v_i\}_{i=1}^{n}$ given a query q_j. Each video v_i is represented as a list of consecutive short clips, i.e., $v_i = [c_{i,1}, c_{i,2}, ..., c_{i,l}]$. In TVR, each short clip is also associated with temporally aligned subtitle sentences. The retrieved moment is denoted as $v_i[t_{st}{:}t_{ed}] = [c_{i,t_{st}}, c_{i,t_{st}+1}, ..., c_{i,t_{ed}}]$. To address VCMR, we propose a hierarchical Cross-modal Moment Localization (XML) network. XML performs video retrieval (VR) in its shallower layers and more fine-grained moment retrieval in its deeper layers. It uses a late fusion design with a novel Convolutional Start-End (ConvSE) detector, making the moment predictions efficient and accurate.

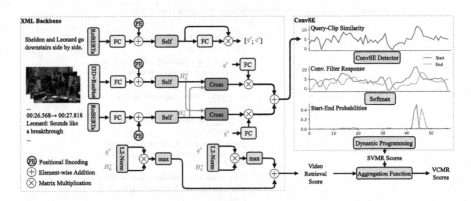

Fig. 4. Cross-modal Moment Localization (XML) model overview. *Self = Self Encoder, Cross = Cross Encoder.* We describe *XML Backbone* in Sect. 4.1, *ConvSE* module in Sect. 4.2 and show XML's training and inference procedure in Sect. 4.3

4.1   XML Backbone Network

Input Representations. To represent videos, we consider both appearance and motion features. For appearance, we extract 2048D ResNet-152 [12] features at 3FPS and max-pool the features every 1.5 s to get a clip-level feature. For motion, we extract 1024D I3D [2] features every 1.5 s. The ResNet-152 model is pre-trained on ImageNet [5] for image recognition, and the I3D model is pre-trained on Kinetics-600 [16] for action recognition. The final video representation is the concatenation of the two features after L2-normalization, denoted as $E^v \in \mathbb{R}^{l \times 3072}$, where l is video length (#clips). We extract contextualized text features using a 12-layer RoBERTa [23]. Specifically, we first fine-tune RoBERTa using the queries and subtitle sentences in TVR train-split with MLM objective [6], then fix the parameters to extract contextualized token embeddings from its second-to-last layer [21]. For queries, we directly use the extracted token embeddings, denoted as $E^q \in \mathbb{R}^{l_q \times 768}$, where l_q is query length (#words). For subtitles, we first extract token-level embeddings, then max-pool them every 1.5 s to get a 768D clip-level feature vector. We use a 768D zero vector if encountering no subtitle. The final subtitle embedding is denoted as $E^s \in \mathbb{R}^{l \times 768}$. The extracted features are projected into a low-dimensional space via a linear layer with ReLU [11]. We then add learned positional encoding [6] to the projected features. Without ambiguity, we reuse the symbols by denoting the processed features as $E^v \in \mathbb{R}^{l \times d}$, $E^s \in \mathbb{R}^{l \times d}$, $E^q \in \mathbb{R}^{l_q \times d}$, where d is hidden size.

Query Encoding. As TVR queries can be related to either video or subtitle, we adopt a modular design to dynamically decompose the query into two modularized vectors. Specifically, the query feature is encoded using a *Self-Encoder*, consisting of a self-attention [31] layer and a linear layer, with a residual [12] connection followed by layer normalization [1]. We denote the encoded query as $H^q \in \mathbb{R}^{l_q \times d}$. Then, we apply two trainable modular weight vectors $\mathbf{w}_m \in \mathbb{R}^d$, $m \in \{v, s\}$ to compute the attention scores of each query word w.r.t. the video (v) or subtitle (s). The scores are used to aggregate the information of $H^q = \{\mathbf{h_r^q}\}_{r=1}^{l_q}$ to generate modularized query vectors $\mathbf{q}^m \in \mathbb{R}^d$ [33]:

$$a_r^m = \frac{\exp(\mathbf{w}_m^T \mathbf{h}_r^q)}{\sum_{k=1}^{l_q} \exp(\mathbf{w}_m^T \mathbf{h}_k^q)}, \quad \mathbf{q}^m = \sum_{r=1}^{l_q} a_r^m \mathbf{h_r^q}, \text{ where } m \in \{v, s\}. \quad (1)$$

Context Encoding. Given the video and subtitle features E^v, E^s, we use two Self-Encoders to compute their single-modal contextualized features $H_0^v \in \mathbb{R}^{l \times d}$ and $H_0^s \in \mathbb{R}^{l \times d}$. Then, we encode their cross-modal representations via *Cross-Encoder*. which takes as input the self-modality and cross-modality features, and encodes the two via cross-attention [31] followed by a linear layer, a residual connection, a layer normalization, and another Self-Encoder. We denote the final video and subtitle representations as $H_1^v \in \mathbb{R}^{l \times d}$ and $H_1^s \in \mathbb{R}^{l \times d}$, respectively.

4.2 Convolutional Start-End Detector

Given H_1^v, H_1^s and $\mathbf{q}^v, \mathbf{q}^s$, we compute query-clip similarity scores $S_{\text{query-clip}} \in \mathbb{R}^l$:

$$S_{\text{query-clip}} = \frac{1}{2}(H_1^v \mathbf{q}^v + H_1^s \mathbf{q}^s). \tag{2}$$

To produce moment predictions from $S_{\text{query-clip}}$, one could rank sliding window proposals with confidence scores computed as the average of scores in each proposal region, or use TAG [37] to progressively group top-scored regions. However, both methods require handcrafted rules and are not trainable. Inspired by edge detectors in image processing [29], we propose Convolutional Start-End detector (**ConvSE**) with two 1D convolution filters to learn to detect start (up) and end (down) edges in the score curves. Clips inside a semantically close span will have higher similarity to the query than those outside, naturally forming detectable edges around the span. Figure 4 (right) and Fig. 7 show examples of the learned ConvSE filters applied to the similarity curves. Specifically, we use two trainable filters (no bias) to generate the start (st) and end (ed) scores:

$$S_{\text{st}} = \text{Conv1D}_{\text{st}}(S_{\text{query-clip}}), \quad S_{\text{ed}} = \text{Conv1D}_{\text{ed}}(S_{\text{query-clip}}). \tag{3}$$

The scores are normalized with softmax to output the probabilities $P_{\text{st}}, P_{\text{ed}} \in \mathbb{R}^l$. In Sect. 5.3, we show ConvSE outperforms the baselines and is also interpretable.

4.3 Training and Inference

Video Retrieval. Given the modularized queries $\mathbf{q}^v, \mathbf{q}^s$ and the encoded contexts H_0^v, H_0^s, we compute the video-level retrieval (VR) score as:

$$s^{\text{vr}} = \frac{1}{2} \sum_{m \in \{v,s\}} \max\left(\frac{H_0^m}{\|H_0^m\|} \frac{\mathbf{q}^m}{\|\mathbf{q}^m\|} \right). \tag{4}$$

This essentially computes the cosine similarity between each clip and query and picks the maximum. The final VR score is the average of the scores from the two modalities. During training, we sample two negative pairs (q_i, v_j) and (q_z, v_i) for each positive pair of (q_i, v_i) to calculate a combined hinge loss as [33]:

$$L^{\text{vr}} = \frac{1}{n} \sum_i [\max(0, \Delta + s^{\text{vr}}(v_j|q_i) - s^{\text{vr}}(v_i|q_i))$$
$$+ \max(0, \Delta + s^{\text{vr}}(v_i|q_z) - s^{\text{vr}}(v_i|q_i))]. \tag{5}$$

Single Video Moment Retrieval. Given the start, end probabilities $P_{\text{st}}, P_{\text{ed}}$, we define single video moment retrieval loss as:

$$L^{\text{svmr}} = -\frac{1}{n} \sum_i [\log(P_{i,\text{st}}(t_{\text{st}}^i)) + \log(P_{i,\text{ed}}(t_{\text{ed}}^i))], \tag{6}$$

where t_{st}^i and t_{ed}^i are the ground-truth indices. At inference, predictions can be generated from the probabilities in linear time using dynamic programming [28]. The confidence score of a predicted moment $[t_{st}^{'}, t_{ed}^{'}]$ is computed as:

$$s^{svmr}(t_{st}^{'}, t_{ed}^{'}) = P_{st}(t_{st}^{'})P_{ed}(t_{ed}^{'}), \; t_{st}^{'} \leq t_{ed}^{'}. \tag{7}$$

To use length prior, we add an additional constraint $L_{min} \leq t_{ed}^{'} - t_{st}^{'} + 1 \leq L_{max}$. For TVR, we set $L_{min} = 2$ and $L_{max} = 16$ for clip length 1.5 s.

Video Corpus Moment Retrieval. Our final training loss combines both: $L^{vcmr} = L^{vr} + \lambda L^{svmr}$, where the hyperparameter λ is set as 0.01. At inference, we compute the VCMR score with the following aggregation function:

$$s^{vcmr}(v_j, t_{st}, t_{ed}|q_i) = s^{svmr}(t_{st}, t_{ed}|v_j, q_i)\exp(\alpha s^{vr}(v_j|q_i)), \tag{8}$$

where $s^{vcmr}(v_j, t_{st}, t_{ed}|q_i)$ is the retrieval score of moment $v_j[t_{st}{:}t_{ed}]$ w.r.t. the query q_i. The exponential term and the hyperparameter α are used to balance the importance of the two scores. A higher α encourages more moments from top retrieved videos. Empirically, we find $\alpha = 20$ works well. At inference, for each query, we first retrieve the top 100 videos based on s^{vr}, then rank all the moments in the 100 videos by s^{vcmr} to give the final predictions.

5 Experiments

5.1 Data, Metrics and Implementation Details

Data. TVR contains 109K queries from 21.8K videos. We split TVR into 80% *train*, 10% *val*, 5% *test-public* and 5% *test-private* splits such that videos and their associated queries appear in only one split. *test-public* will be used for a public leaderboard, *test-private* is reserved for future challenges.

Metrics. Following [7,8], we use average recall at K (R@K) over all queries as our metric. A prediction is correct if: (*i*) predicted video matches the ground truth; (*ii*) predicted span has high overlap with the ground truth where temporal intersection over union (IoU) is used to measure overlap.

Implementation Details. All baseline comparisons are configured to use the same hidden size as XML. We train the baselines following the original papers. We use the same features for all the models. To support retrieval using subtitle for the baselines, we add a separate subtitle stream and average the final predictions from both streams. Non-maximum suppression is not used as we do not observe consistent performance gain on the *val* set.

Table 3. Baseline comparison on TVR *test-public* set, VCMR task. Model references: *MCN* [13], *CAL* [7], *MEE* [24], *ExCL* [10]

Model	w/ video	w/ sub.	IoU = 0.5				IoU = 0.7				Runtime ↓
			R@1	R@5	R@10	R@100	R@1	R@5	R@10	R@100	(seconds)
Chance	–	–	0.00	0.02	0.04	0.33	0.00	0.00	0.00	0.07	
Proposal based methods											
MCN	✓	✓	0.02	0.15	0.24	2.20	0.00	0.07	0.09	1.03	–
CAL	✓	✓	0.09	0.31	0.57	3.42	0.04	0.15	0.26	1.89	–
Retrieval + Re-ranking											
MEE+MCN	✓	✓	0.92	3.69	5.58	17.91	0.42	1.89	2.98	10.84	66.8
MEE+CAL	✓	✓	0.97	3.75	5.80	18.66	0.39	1.69	2.98	11.52	161.5
MEE+ExCL	✓	✓	0.92	2.53	3.60	6.01	0.33	1.19	1.73	2.87	1307.2
XML	✓	✓	**7.25**	**16.24**	**21.65**	**44.44**	**3.25**	**8.71**	**12.49**	**29.51**	**25.5**

5.2   Baselines Comparison

In this section, we compare XML with baselines on TVR *test-public* set (5,445 queries and 1,089 videos). We report the runtime for top-performing methods, averaged across 3 runs on an RTX 2080Ti GPU. The time spent on data loading, pre-processing, backend model (i.e., ResNet-152, I3D, RoBERTa) feature extraction, etc, is ignored since they should be similar for all methods. We mainly focus on the VCMR task here. In the supplementary file, we include the following experiments: (1) model performance on single video moment retrieval and video retrieval tasks; (2) computation and storage cost comparison in a 1M videos corpus; (3) Temporal Endpoint Feature (TEF) [13] model results; (4) feature and model ablation studies; (5) VCMR results on DiDeMo [13] dataset, etc.

Proposal Based Methods. MCN [13] and CAL [7] pose the moment retrieval task as a ranking problem in which all moment proposal candidates are ranked based on their squared Euclidean Distance with the queries. For VCMR, they require directly ranking all the proposals (95K in the following experiments) in the video corpus for each query, which can be costly and difficulty. In contrast, XML uses a hierarchical design that performs video retrieval in its shallow layers and moment retrieval on the retrieved videos in its deeper layers. In Table 3, XML is showing to have significantly higher performance than MCN and CAL.

Retrieval+Re-ranking Methods. We also compare to methods under the retrieval+re-ranking setting [7] where we first retrieve a set of candidate videos using a given method and then re-rank the moment predictions in the candidate videos using another method. Specifically, we first use MEE [24] to retrieve 100 videos for each query as candidates. Then, we use MCN and CAL to rank all of the proposals in the candidate videos. ExCL [10] is an early fusion method designed for SVMR, with a start-end predictor. We adapt it to VCMR by combining MEE video-level scores with ExCL moment-level scores, using Eq. 8. The results are shown in Table 3. Compared to their purely proposal based counterparts (i.e., MCN and CAL), both MEE+MCN and MEE+CAL achieve significant performance gain, showing the benefit of reducing the number of proposals needed to rank (by reducing the number of videos). However, they are still far below XML as they use very coarse-grained, predefined proposals. In Sect. 5.3, we show our start-end detector performs consistently better than predefined proposals [7,37] under our XML framework. Compared to MEE+ExCL, XML achieves 9.85× performance gain (3.25 *vs.* 0.33, R@1 IoU = 0.7) and 51.3× speedup (25.5s *vs.* 1307.2s). In the supplementary file, we show that this speedup can be even more significant (287×) when retrieving on a larger scale video corpus (1M videos) with pre-encoded video representations. This huge speedup shows the effectiveness of XML's late fusion design over ExCL's early fusion design.

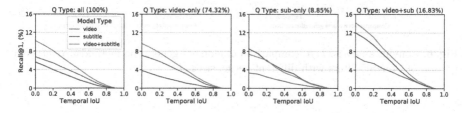

Fig. 5. Performance breakdown of XML models that use only *video*, *subtitle*, or both as inputs, by different query types (with percentage of queries shown in brackets). The performance is evaluated on TVR *val* set for VCMR

5.3   Model Analysis

Video vs. Subtitle. In Fig. 5, we compare to XML variants that use only video or subtitle. We observe that the full *video+subtitle* model has better overall performance than single modality models (*video* and *subtitle*), demonstrating that both modalities are useful. We also see that a model trained on one modality does not perform well on the queries tagged by another modality, e.g., the *video* model performs much worse on *sub-only* queries compared to the *subtitle* model.

ConvSE: Comparison and Analysis. To produce moment predictions from the query-clip similarity signals, we proposed ConvSE that learns to detect start (up) and end (down) edges in the 1D similarity signals. To show its effectiveness, we compare ConvSE with two baselines under our XML backbone network: (1) sliding window, where we rank proposals generated by multi-scale sliding windows, with proposal confidence scores calculated as the average of scores inside each proposal region. On average, it produces 87 proposals per video. The proposals used here are the same as the ones used for MCN and CAL in our previous experiments; (2) TAG [37] that progressively groups top-scored clips with the classical watershed algorithm [26]. Since these two methods do not produce start-end probabilities, we cannot train the model with the objective in Eq. 6. Thus, we directly optimize the query-clip similarity scores in Eq. 2 with Binary Cross Entropy loss: we assign a label of 1 if the clip falls into the ground-truth region,

0 otherwise. While both sliding window and TAG approaches rely on hand-crafted rules, ConvSE **learns from data**. We show in Fig. 6 (left), under the same XML backbone network, ConvSE has consistent better performance across all IoU thresholds on both VCMR and SVMR tasks.

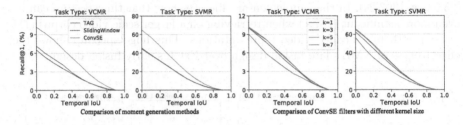

Fig. 6. ConvSE analysis. *Left*: comparison of moment generation methods. *Right*: comparison of ConvSE filters with different kernel sizes (k)

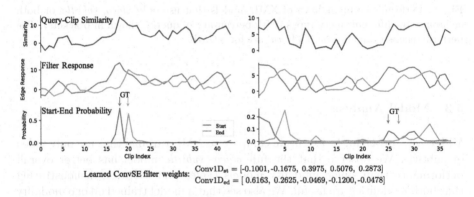

Fig. 7. Examples of learned ConvSE filters applying on query-clip similarity scores. Ground truth span is indicated by the two arrows labeled by *GT*. Note the two filters output stronger responses on the up (*Start*) and down (*End*) edges

In Fig. 6 (right), we vary the kernel size (k) of ConvSE filters. While the performance is reasonable when $k = 3, 5$ or 7, we observe a significant performance drop at $k = 1$. In this case, the filters essentially degrade to scaling factors on the scores. This comparison demonstrates that neighboring information is important. Figure 7 shows examples of using the **learned convolution filters**: the filters output stronger responses to the up (*Start*) and down (*End*) edges of the score curves and thus detect them. Interestingly, the learned weights $\text{Conv1D}_{\text{st}}$ and $\text{Conv1D}_{\text{ed}}$ in Fig. 7 are similar to the edge detectors in image processing [29].

Qualitative Analysis. Figure 8 shows XML example predictions on the TVR *val* set. In the top row, we also show the query word attention scores for video

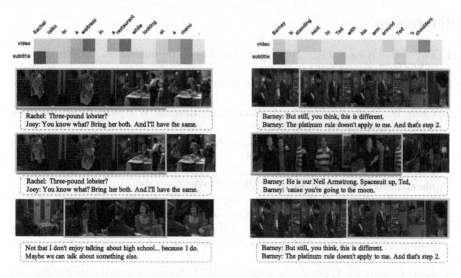

Fig. 8. XML prediction examples for VCMR, on TVR *val* set. We show top-3 retrieved moments for each query. *Top row* shows modular attention scores for query words. *Left column* shows a correct prediction, *right column* shows a failure. Text inside *dashed boxes* is the subtitles associated with the predicted moments. *Orange box* shows the predictions, *green bar* shows the ground truth

and subtitle, respectively. Figure 8 (left) shows a correct prediction. The top-2 moments are from the same video and are both correct. The third moment is retrieved from a different video. While incorrect, it is still relevant as it also happens in a 'restaurant'. Figure 8 (right) shows a failure. It is worth noting that the false moments are very close to the correct prediction with minor differences ('on the shoulder' *vs.* 'around the shoulder'). Besides, it is also interesting to see which words are important for video or subtitle. For example, the words 'waitress', 'restaurant', 'menu' and 'shoulder' get the most weight for video; while the words 'Rachel', 'menu', 'Barney', 'Ted' have higher attention scores for subtitle.

6   Conclusion

In this work, we present TVR, a large-scale dataset designed for multimodal moment retrieval tasks. Detailed analyses show TVR is of high quality and is more challenging than previous datasets. We also propose Cross-modal Moment Localization (XML), an efficient model suitable for the VCMR task.

Acknowledgements. We thank the reviewers for their helpful feedback. This research is supported by NSF Award #1562098, DARPA MCS Grant #N66001-19-2-4031, DARPA KAIROS Grant #FA8750-19-2-1004, ARO-YIP Award #W911NF-18-1-0336, and Google Focused Research Award.

References

1. Ba, J.L., Kiros, J.R., Hinton, G.E.: Layer normalization. arXiv preprint arXiv:1607.06450 (2016)
2. Carreira, J., Zisserman, A.: Quo vadis, action recognition? A new model and the kinetics dataset. In: CVPR (2017)
3. Chen, D., Fisch, A., Weston, J., Bordes, A.: Reading Wikipedia to answer open-domain questions. In: ACL (2017)
4. Chen, J., Chen, X., Ma, L., Jie, Z., Chua, T.S.: Temporally grounding natural sentence in video. In: EMNLP (2018)
5. Deng, J., Dong, W., Socher, R., Li, L.J., Li, K., Fei-Fei, L.: ImageNet: a large-scale hierarchical image database. In: CVPR (2009)
6. Devlin, J., Chang, M.W., Lee, K., Toutanova, K.: BERT: pre-training of deep bidirectional transformers for language understanding. In: NAACL (2019)
7. Escorcia, V., Soldan, M., Sivic, J., Ghanem, B., Russell, B.: Temporal localization of moments in video collections with natural language. arXiv preprint arXiv:1907.12763 (2019)
8. Gao, J., Sun, C., Yang, Z., Nevatia, R.: TALL: temporal activity localization via language query. In: ICCV (2017)
9. Ge, R., Gao, J., Chen, K., Nevatia, R.: MAC: mining activity concepts for language-based temporal localization. In: WACV (2019)
10. Ghosh, S., Agarwal, A., Parekh, Z., Hauptmann, A.: ExCL: extractive clip localization using natural language descriptions. In: NAACL (2019)
11. Glorot, X., Bordes, A., Bengio, Y.: Deep sparse rectifier neural networks. In: AISTATS (2011)
12. He, K., Zhang, X., Ren, S., Sun, J.: Deep residual learning for image recognition. In: CVPR (2016)
13. Hendricks, L.A., Wang, O., Shechtman, E., Sivic, J., Darrell, T., Russell, B.: Localizing moments in video with natural language. In: ICCV (2017)
14. Hendricks, L.A., Wang, O., Shechtman, E., Sivic, J., Darrell, T., Russell, B.: Localizing moments in video with temporal language. In: EMNLP (2018)
15. Hochreiter, S., Schmidhuber, J.: Long short-term memory. Neural Comput. 9, 1735–1780 (1997)
16. Kay, W., et al.: The kinetics human action video dataset. arXiv preprint arXiv:1705.06950 (2017)
17. Kazemzadeh, S., Ordonez, V., Matten, M., Berg, T.: ReferitGame: referring to objects in photographs of natural scenes. In: EMNLP (2014)
18. Kim, K.M., Heo, M.O., Choi, S.H., Zhang, B.T.: DeepStory: video story QA by deep embedded memory networks. In: IJCAI (2017)
19. Krishna, R., Hata, K., Ren, F., Fei-Fei, L., Niebles, J.C.: Dense-captioning events in videos. In: ICCV (2017)
20. Lei, J., Yu, L., Bansal, M., Berg, T.L.: TVQA: localized, compositional video question answering. In: EMNLP (2018)
21. Lei, J., Yu, L., Berg, T.L., Bansal, M.: TVQA+: spatio-temporal grounding for video question answering. In: ACL (2020)
22. Lin, T., Zhao, X., Su, H., Wang, C., Yang, M.: BSN: boundary sensitive network for temporal action proposal generation. In: ECCV (2018)
23. Liu, Y., et al.: RoBERTa: a robustly optimized BERT pretraining approach. arXiv preprint arXiv:1907.11692 (2019)

24. Miech, A., Laptev, I., Sivic, J.: Learning a text-video embedding from incomplete and heterogeneous data. arXiv preprint arXiv:1804.02516 (2018)
25. Regneri, M., Rohrbach, M., Wetzel, D., Thater, S., Schiele, B., Pinkal, M.: Grounding action descriptions in videos. TACL **1**, 25–36 (2013)
26. Roerdink, J.B., Meijster, A.: The watershed transform: definitions, algorithms and parallelization strategies. Fundam. Inf. **41**, 187–228 (2000)
27. Rohrbach, A., Rohrbach, M., Qiu, W., Friedrich, A., Pinkal, M., Schiele, B.: Coherent multi-sentence video description with variable level of detail. In: Jiang, X., Hornegger, J., Koch, R. (eds.) GCPR 2014. LNCS, vol. 8753, pp. 184–195. Springer, Cham (2014). https://doi.org/10.1007/978-3-319-11752-2_15
28. Seo, M., Kembhavi, A., Farhadi, A., Hajishirzi, H.: Bidirectional attention flow for machine comprehension. In: ICLR (2017)
29. Szeliski, R.: Computer Vision: Algorithms and Applications. Springer, Heidelberg (2010). https://doi.org/10.1007/978-1-84882-935-0
30. Tapaswi, M., Zhu, Y., Stiefelhagen, R., Torralba, A., Urtasun, R., Fidler, S.: MovieQA: understanding stories in movies through question-answering. In: CVPR (2016)
31. Vaswani, A., et al.: Attention is all you need. In: NeurIPS (2017)
32. Xu, H., He, K., Plummer, B.A., Sigal, L., Sclaroff, S., Saenko, K.: Multilevel language and vision integration for text-to-clip retrieval. In: AAAI (2019)
33. Yu, L., et al.: MattNet: modular attention network for referring expression comprehension. In: CVPR (2018)
34. Zellers, R., Bisk, Y., Farhadi, A., Choi, Y.: From recognition to cognition: visual commonsense reasoning. In: CVPR (2019)
35. Zhang, D., Dai, X., Wang, X., fang Wang, Y., Davis, L.S.: MAN: moment alignment network for natural language moment retrieval via iterative graph adjustment. In: CVPR (2018)
36. Zhang, Z., Lin, Z., Zhao, Z., Xiao, Z.: Cross-modal interaction networks for query-based moment retrieval in videos. In: SIGIR (2019)
37. Zhao, Y., Xiong, Y., Wang, L., Wu, Z., Tang, X., Lin, D.: Temporal action detection with structured segment networks. In: ICCV (2017)

Minimum Class Confusion for Versatile Domain Adaptation

Ying Jin, Ximei Wang, Mingsheng Long$^{(\boxtimes)}$, and Jianmin Wang

School of Software, BNRist, Research Center for Big Data,
Tsinghua University, Beijing, China
{jiny18,wxm17}@mails.tsinghua.edu.cn, {mingsheng,jimwang}@tsinghua.edu.cn

Abstract. There are a variety of Domain Adaptation (DA) scenarios subject to label sets and domain configurations, including closed-set and partial-set DA, as well as multi-source and multi-target DA. It is notable that existing DA methods are generally designed only for a specific scenario, and may underperform for scenarios they are not tailored to. To this end, this paper studies **Versatile Domain Adaptation (VDA)**, where one method can handle several different DA scenarios without any modification. Towards this goal, a more general inductive bias other than the *domain alignment* should be explored. We delve into a missing piece of existing methods: *class confusion*, the tendency that a classifier confuses the predictions between the correct and ambiguous classes for target examples, which is common in different DA scenarios. We uncover that reducing such pairwise class confusion leads to significant transfer gains. With this insight, we propose a general loss function: **Minimum Class Confusion (MCC)**. It can be characterized as (1) a *non-adversarial* DA method without explicitly deploying domain alignment, enjoying faster convergence speed; (2) a *versatile* approach that can handle four existing scenarios: Closed-Set, Partial-Set, Multi-Source, and Multi-Target DA, outperforming the state-of-the-art methods in these scenarios, especially on one of the largest and hardest datasets to date (7.3% on DomainNet). Its versatility is further justified by two scenarios proposed in this paper: Multi-Source Partial DA and Multi-Target Partial DA. In addition, it can also be used as a general regularizer that is orthogonal and complementary to a variety of existing DA methods, accelerating convergence and pushing these readily competitive methods to stronger ones. Code is available at https://github.com/thuml/Versatile-Domain-Adaptation.

Keywords: Versatile domain adaptation · Minimum class confusion

1 Introduction

The scarcity of labeled data hinders deep neural networks (DNNs) from use in real applications. This challenge gives rise to Domain Adaptation (DA) [27,33],

Electronic supplementary material The online version of this chapter (https://doi.org/10.1007/978-3-030-58589-1_28) contains supplementary material, which is available to authorized users.

an important technology that aims to transfer knowledge from a labeled source domain to an unlabeled target domain in the presence of dataset shift. A rich line of DNN-based methods [8,21–24,29,41–43,46,48,52] have been proposed for Unsupervised DA (UDA), a closed-set scenario with one source domain and one target domain sharing the same label set. Recently, several highly practical scenarios were proposed, such as Partial DA (PDA) [2,50] with the source label set subsuming the target one, Multi-Source DA (MSDA) [47,53] with multiple source domains, and Multi-Target DA (MTDA) [31] with multiple target domains. As existing UDA methods cannot be applied directly to these challenging scenarios, plenty of methods [2,3,31,47,50] have been designed for each specific scenario, which work quite well in each tailored scenario.

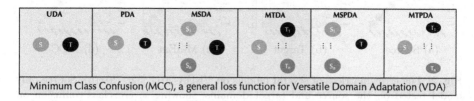

Fig. 1. Versatile Domain Adaptation (VDA) subsumes typical domain adaptation scenarios: (1) Unsupervised Domain Adaptation (UDA); (2) Partial Domain Adaptation (PDA); (3) Multi-Source Domain Adaptation (MSDA); (4) Multi-Target Domain Adaptation (MTDA); (5) Multi-Source Partial Domain Adaptation (MSPDA); (6) Multi-Target Partial Domain Adaptation (MTPDA). Note that scenarios (5)–(6) are newly proposed in this paper. Our **Minimum Class Confusion (MCC)** is a *versatile* method towards all these DA scenarios.

In practical applications, however, it is difficult to confirm the label sets and domain configurations in the data acquisition process. Therefore, we may be stuck in choosing a proper method tailored to the suitable DA scenario. The most ideal solution to escape from this dilemma is a *versatile* DA method that can handle various scenarios without any modification. Unfortunately, existing DA methods are generally designed only for a specific scenario and may underperform for scenarios they are not tailored to. For instance, PADA [3], a classic PDA method, excels at selecting out outlier classes but suffers from the internal domain shift in MSDA and MTDA, while DADA [31], an outstanding method tailored to MTDA, cannot be directly applied to PDA or MSDA. Hence, existing DA methods are not versatile enough to handle practical scenarios of complex variations.

In this paper, we define **Versatile Domain Adaptation (VDA)** as a line of *versatile* approaches able to tackle a variety of scenarios without any modification. Towards VDA, a more general inductive bias other than the domain alignment should be explored. In this paper, we delved into the error matrices of the target domain and found that the classifier trained on the source domain may confuse to distinguish the correct class from a similar class, such as cars and `trucks`. As shown in Fig. 2(b), the probability that a source-only model misclassifies `cars` as `trucks` on the target domain is over 25%. Further, we analyzed the error matrices in other DA scenarios and reached the same conclusion. These findings give us a fresh perspective to enable VDA: **class confusion**, the tendency that a classifier confuses the predictions between the correct and ambiguous classes for target examples. We uncover that less class confusion leads to more transfer gains for all the domain adaptation scenarios in Fig. 1.

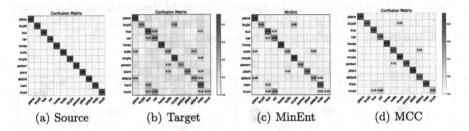

(a) Source   (b) Target   (c) MinEnt   (d) MCC

Fig. 2. The error matrices of several models on VisDA-2017 [32]. (a)–(b): Source-only model tested on the source and target domains, showing severe class confusion on target domain examples. (c)–(d): Models trained with entropy minimization (MinEnt) [10] and Minimum Class Confusion (MCC) on target domain examples, respectively. The proposed MCC loss substantially diminishes the class confusion.

Still, we need to address a new challenge that the ground-truth class confusion cannot be calculated if the labels in the target domain are inaccessible. Fortunately, the confusion between different classes can be naturally reflected by an example-weighted inner-product between the classifier predictions and their transposes. And we can define class confusion from this perspective, enabling it to be computed from well-calibrated classifier predictions. To this end, we propose a novel loss function: **Minimum Class Confusion (MCC)**. It can be characterized as a novel and versatile DA approach without explicitly deploying domain alignment [8,21], enjoying fast convergence speed. In addition, it can also be used as a general regularizer that is orthogonal and complementary to existing DA methods, further accelerating and improving those readily competitive methods. Our contributions are summarized as follows:

- We propose a practical setting, **Versatile Domain Adaptation (VDA)**, where one method can tackle many DA scenarios without modification.
- We uncover that the class confusion is a common missing piece of existing DA methods and that less class confusion leads to more transfer gains.
- We propose a novel loss function: **Minimum Class Confusion (MCC)**, which is versatile to handle four existing DA scenarios, including closed-set, partial-set, multi-source, and multi-target, as well as two proposed scenarios: multi-source partial DA and multi-target partial DA.
- We conduct extensive experiments on four standard DA datasets, and show that MCC outperforms the state-of-the-art methods in different DA scenarios,

especially on one of the largest and hardest datasets (**7.3%** on DomainNet), enjoying a faster convergence speed than mainstream DA methods.

2   Related Work

Unsupervised Domain Adaptation (UDA). Most of the existing domain adaptation researches focused on UDA, in which numerous UDA methods were proposed based on either *Moment Matching* or *Adversarial Training*.

Moment Matching methods aim at minimizing the distribution discrepancy across domains. Deep Coral [39] aligns second-order statistics between distributions. DDC [43] and DAN [21] utilize Maximum Mean Discrepancy [11], JAN [24] defines Joint Maximum Mean Discrepancy, SWD [18] introduces Sliced Wasserstein Distance and CAN [17] leverages Contrastive Domain Discrepancy.

Adversarial Training methods were inspired by the Generative Adversarial Networks (GANs) [9], aiming at learning domain invariant features in an adversarial manner. DANN [8] introduces a domain discriminator to distinguish source and target features, while the feature extractor strives to fool it. ADDA [42], MADA [29] and MCD [35] extend this architecture to multiple feature extractors and classifiers. Motivated by Conditional GANs [26], CDAN [22] aligns domain features in a class-conditional adversarial game. CyCADA [15] adapts features in both pixel and feature levels. TADA [46] introduces the first transferable attention mechanism. SymNet [51] uses a symmetric classifier, and DTA [19] learns discriminative features with a new adversarial dropout.

There are other approaches to domain adaptation. For instance, SE [7] is based on the teacher-student [40] architecture. TransNorm [45] tackles DA with a new transferable backbone. TAT [20] proposes transferable adversarial training to guarantee the adaptability. BSP [5] balances between the transferability and discriminability. AFN [48] enlarges feature norm to enhance feature transferability. Some methods [16,38,54,55] also utilize the less-reliable self-training or pseudo labeling, *e.g.* TPN [28] is based on pseudo class-prototypes.

Partial Domain Adaptation (PDA). In PDA, the target label set is a subset of the source label set. Representative methods include SAN [2], IWAN [50], PADA [3] and ETN [4], introducing different weighting mechanisms to select out outlier source classes in the process of domain feature alignment.

Multi-source Domain Adaptation (MSDA). In MSDA, there are multiple source domains of different distributions. MDAN [53] provides theoretical insights for MSDA, while Deep Cocktail Network [47] (DCTN) and M^3SDA [30] extend adversarial training and moment-matching to MSDA, respectively.

Multi-target Domain Adaptation (MTDA). In MTDA, we need to transfer a learning model to multiple unlabeled target domains. DADA [31] is the first approach to MTDA through disentangling domain-invariant representations.

3 Approach

In this paper, we study **Versatile Domain Adaptation (VDA)** where one method can tackle many scenarios without any modification. We justify the versatility of one method by four existing scenarios: **(1)** Unsupervised Domain Adaptation (**UDA**) [8], the standard scenario with a labeled source domain $\mathcal{S} = \{(\mathbf{x}_s^i, \mathbf{y}_s^i)\}_{i=1}^{n_s}$ and an unlabeled target domain $\mathcal{T} = \{\mathbf{x}_t^i\}_{i=1}^{n_t}$, where $\mathbf{x}^i$ is an example and $\mathbf{y}^i$ is the associated label; **(2)** Partial Domain Adaptation (**PDA**) [3], which extends UDA by relaxing the source domain label set to subsume the target domain label set; **(3)** Multi-Source Domain Adaptation (**MSDA**) [30], which extends UDA by expanding to S labeled source domains $\{\mathcal{S}_1, \mathcal{S}_2, ..., \mathcal{S}_S\}$; **(4)** Multi-Target Domain Adaptation (**MTDA**) [31], which extends UDA by expanding to T unlabeled target domains $\{\mathcal{T}_1, \mathcal{T}_2, ..., \mathcal{T}_T\}$. We further propose two scenarios to confirm the versatility: **(5)/(6)** Multi-Source/Multi-Target Partial Domain Adaptation (**MSPDA/MTPDA**), which extend PDA to multi-source/multi-target scenarios. Tailored to a specific scenario, existing methods fail to readily handle these scenarios. We propose **Minimum Class Confusion (MCC)** as a generic loss function for VDA. Hereafter, we denote by $\mathbf{y}_{i\cdot}$, $\mathbf{y}_{\cdot j}$ and $\mathbf{Y}_{ij}$ the i-row, the j-th column and the ij-th entry of matrix $\mathbf{Y}$, respectively (Fig. 3).

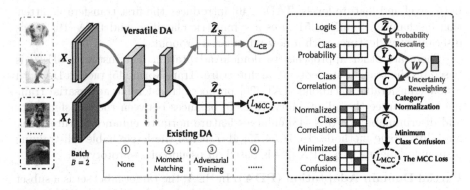

Fig. 3. The schematic of the **Minimum Class Confusion (MCC)** loss function. Given the shared feature extractor F, MCC is defined on the class predictions $\widehat{\mathbf{Y}}_t$ given by the source classifier G on the target data. MCC is versatile to address various domain adaptation scenarios standalone, or to be integrated with existing methods (moment matching, adversarial training, etc). (*Best viewed in color.*)

3.1 Minimum Class Confusion

To enable versatile domain adaptation, we need to find out a proper criterion to measure the pairwise class confusion on the target domain. Unlike previous methods such as CORAL [39] that focus on features, we explore the class predictions.

Denote the classifier output on the target domain as $\widehat{\mathbf{Y}}_t = G(F(\mathbf{X}_t)) \in \mathbb{R}^{B \times |\mathcal{C}|}$, where B is the batch size of the target data, $|\mathcal{C}|$ is the number of source classes, F is the feature extractor and G is the classifier. In our method, we focus on the classifier predictions $\widehat{\mathbf{Y}}$ and omit the domain subscript t for clarity.

Probability Rescaling. According to [12], DNNs tend to make overconfident predictions, hindering them from directly reasoning about the class confusion. Therefore, we adopt temperature rescaling [12,14], a simple yet effective technique, to alleviate the negative effect of overconfident predictions. Using temperature scaling, the probability $\widehat{Y}_{ij}$ that the i-th instance belongs to the j-th class can be recalibrated as

$$\widehat{Y}_{ij} = \frac{\exp\left(Z_{ij}/T\right)}{\sum_{j'=1}^{|\mathcal{C}|} \exp\left(Z_{ij'}/T\right)}, \tag{1}$$

where Z_{ij} is the logit output of the classifier layer and T is the temperature hyper-parameter for probability rescaling. Obviously, Eq. (1) boils down to the vanilla softmax function when $T = 1$.

Class Correlation. As $\widehat{Y}_{ij}$ reveals the relationship between the i-th instance and the j-th class, we define the class correlation between two classes j and j' as

$$\mathbf{C}_{jj'} = \widehat{\mathbf{y}}_{\cdot j}^{\mathsf{T}} \widehat{\mathbf{y}}_{\cdot j'}. \tag{2}$$

It is a coarse estimation of the class confusion. Lets delve into the definition of the class correlation in Eq. (2). Note that $\widehat{\mathbf{y}}_{\cdot j}$ denotes the probabilities that the B examples in each batch come from the j-th class. The class correlation measures the possibility that the classifier simultaneously classifies the B examples into the j-th and the j'-th classes. It is noteworthy that such pairwise class correlation is relatively safe: for false predictions with high confidence, the corresponding class correlation is still low. In other words, highly confident false predictions will negligibly impact the class correlation.

Uncertainty Reweighting. We note that examples are not equally important for quantifying class confusion. When the prediction is closer to a uniform distribution, showing no obvious peak (obviously larger probabilities for some classes), we consider the classifier as ignorant of this example. On the contrary, when the prediction shows several peaks, it indicates that the classifier is reluctant between several ambiguous classes (such as car and truck). Obviously, these examples that make the classifier ambiguous across classes will be more suitable for embodying class confusion. As defined in Eq. (2), these examples can be naturally highlighted with higher probabilities on the several peaks. Further, we introduce a *weighting* mechanism based on uncertainty such that we can quantify class confusion more accurately. Here, those examples with higher certainty in class predictions given by the classifier are more reliable and should contribute more to the pairwise class confusion. We use the *entropy* function $H(p) \triangleq -\mathbb{E}_p \log p$ in information theory as an uncertainty measure of distribution p. The entropy (uncertainty) $H(\widehat{\mathbf{y}}_{i\cdot})$ of predicting the i-th example by the classifier is defined as

$$H(\widehat{\mathbf{y}}_{i\cdot}) = -\sum_{j=1}^{|\mathcal{C}|} \widehat{Y}_{ij} \log \widehat{Y}_{ij}. \tag{3}$$

While the entropy is a measure of uncertainty, what we want is a probability distribution that places larger probabilities on examples with larger certainty of class predictions. A *de facto* transformation to probability is the softmax function

$$W_{ii} = \frac{B\left(1 + \exp(-H(\widehat{\mathbf{y}}_{i\cdot}))\right)}{\sum\limits_{i'=1}^{B}\left(1 + \exp(-H(\widehat{\mathbf{y}}_{i'\cdot}))\right)}, \tag{4}$$

where W_{ii} is the probability of quantifying the importance of the i-th example for modeling the class confusion, and $\mathbf{W}$ is the corresponding diagonal matrix. Note that we take the opposite value of the entropy to reflect the *certainty*. Laplace Smoothing [37] (*i.e.* adding a constant 1 to each addend of the softmax function) is used to form a *heavier-tailed* weight distribution, which is suitable for highlighting more certain examples as well as avoiding overly penalizing the others. For better scaling, the probability over the examples in each batch of size B is rescaled to sum up to B such that the average weight for each example is 1. With this weighting mechanism, the preliminary definition of *class confusion* is

$$\mathbf{C}_{jj'} = \widehat{\mathbf{y}}_{\cdot j}^{\mathsf{T}} \mathbf{W} \widehat{\mathbf{y}}_{\cdot j'}. \tag{5}$$

Category Normalization. The batch-based definition of the class confusion in Eq. (5) is native for the mini-batch SGD optimization. However, when the number of classes is large, it will run into a severe *class imbalance* in each batch. To tackle this problem, we adopt a category normalization technique widely used in Random Walk [25]:

$$\widetilde{\mathbf{C}}_{jj'} = \frac{\mathbf{C}_{jj'}}{\sum_{j''=1}^{|\mathcal{C}|} \mathbf{C}_{jj''}}. \tag{6}$$

Taking the idea of Random Walk, the normalized class confusion in Eq. (6) has a neat interpretation: It is probable to walk from one class to another (resulting in the wrong classification) if the two classes have a high class confusion.

Minimum Class Confusion. Given the aforementioned derivations, we can formally define the loss function to enable Versatile Domain Adaptation (VDA). Recall that $\widetilde{\mathbf{C}}_{jj'}$ well measures the confusion between each class pair j and j'. We only need to minimize the cross-class confusion, *i.e.* $j \neq j'$. Namely, the ideal situation is that no examples are ambiguously classified into two classes at the same time. In this sense, the Minimum Class Confusion (MCC) loss is defined as

$$L_{\mathrm{MCC}}(\widehat{\mathbf{Y}}_t) = \frac{1}{|\mathcal{C}|} \sum_{j=1}^{|\mathcal{C}|} \sum_{j' \neq j}^{|\mathcal{C}|} \left| \widetilde{\mathbf{C}}_{jj'} \right|. \tag{7}$$

Since the class confusion in Eq. (6) has been normalized, minimizing the *between-class* confusion in Eq. (7) implies that the *within-class* confusion is maximized. Note that Eq. (7) is a general loss that is pluggable to existing approaches.

We want to emphasize that the inductive bias of *class confusion* in this work is more general than that of *domain alignment* in previous work [8,21,22,25,35]. As discussed in Sect. 2, many previous methods explicitly align features from the source and target domains, at the risk of deteriorating the feature discriminability and impeding the transferability [5]. Further, the inductive bias of *class confusion* is general and applicable to a variety of domain adaptation scenarios, while that of *domain alignment* will suffer when the domains cannot be aligned naturally (*e.g.* the partial-set DA scenarios [2,3,50]).

3.2 Versatile Approach to Domain Adaptation

The main motivation of this work is to design a versatile approach to a variety of DA scenarios. As the class confusion is a common inductive bias of many DA scenarios, combining the cross-entropy loss on the source labeled data and the MCC loss on the target unlabeled data will enable these DA scenarios.

Denote by $\widehat{\mathbf{y}}_s = G(F(\mathbf{x}_s))$ the class prediction for a source example $\mathbf{x}_s$, and by $\widehat{\mathbf{Y}}_t = G(F(\mathbf{X}_t))$ the class predictions for a batch of B target examples $\mathbf{X}_t$. The versatile approach (also termed by **MCC** for clarity) proposed in this paper for a variety of domain adaptation scenarios is formulated as

$$\min_{F,G} \mathbb{E}_{(\mathbf{x}_s,\mathbf{y}_s)\in\mathcal{S}} L_{\text{CE}}(\widehat{\mathbf{y}}_s, \mathbf{y}_s) + \mu \mathbb{E}_{\mathbf{X}_t\subset\mathcal{T}} L_{\text{MCC}}(\widehat{\mathbf{Y}}_t), \tag{8}$$

where L_{CE} is the cross-entropy loss and μ is a hyper-parameter for the MCC loss. With this joint loss, feature extractor F and classifier G of the deep DA model can be trained end-to-end by back-propagation. Note that, Eq. (8) is a *versatile* approach to many DA scenarios without any modifications to the loss.

- **Unsupervised Domain Adaptation (UDA).** Since Eq. (8) is formulated natively for this vanilla domain adaptation scenario, MCC can be directly applied to this scenario without any modification.
- **Partial Domain Adaptation (PDA).** Without explicit domain alignment, we need not to worry about the *misalignment* between source outlier classes and target classes, which is the technical bottleneck of PDA [3]. Meanwhile, compared to the confusion between the target classes, the confusion between the source outlier classes on the target domain is negligible in the MCC loss. Therefore, we can directly apply Eq. (8) to PDA.
- **Multi-source Domain Adaptation (MSDA).** Prior methods of MSDA consider multiple source domains as different domains, capturing the internal source domain shifts, and a simple merge of source domains proves fragile. However, since MCC is based on class confusion instead of domain alignment, we can safely merge S source domains as $\mathcal{S} \leftarrow \mathcal{S}_1 \cup \cdots \cup \mathcal{S}_S$.
- **Multi-target Domain Adaptation (MTDA).** Though a simple merge of target domains is risky for existing methods, for MCC applied to MTDA, we can safely merge T target domains as $\mathcal{T} \leftarrow \mathcal{T}_1 \cup \cdots \cup \mathcal{T}_T$.
- **Multi-source/Multi-target Partial Domain Adaptation (MSPDA/ MTPDA).** As MCC can directly tackle PDA and MSDA/MTDA, it can handle these derived scenarios by simply merging multiple sources or targets.

3.3   Regularizer to Existing DA Methods

Since the inductive bias of *class confusion* is orthogonal to the widely-used *domain alignment*, our method is well complementary to the previous methods. The MCC loss Eq. (7) can serve as a regularizer pluggable to existing methods.

We take as an example the standard domain alignment framework [8, 22] based on domain-adversarial training. Integrating the MCC loss as a regularizer yields

$$\min_{F,G} \max_{D} \ \mathbb{E}_{(\mathbf{x}_s, \mathbf{y}_s) \in \mathcal{S}} L_{\mathrm{CE}} \left(\widehat{\mathbf{y}}_s, \mathbf{y}_s \right) + \mu \, \mathbb{E}_{\mathbf{x}_t \subset \mathcal{T}} L_{\mathrm{MCC}} (\widehat{\mathbf{Y}}_t) - \lambda \, \mathbb{E}_{\mathbf{x} \in \mathcal{S} \cup \mathcal{T}} L_{\mathrm{CE}} (D(\widehat{\mathbf{f}}), \mathbf{d}), \quad (9)$$

where the third term is the domain-adversarial loss for the domain discriminator D striving to distinguish the source from the target, and $\mathbf{d}$ is the domain label, $\widehat{\mathbf{f}} = F(\mathbf{x})$ is the feature representation learned to confuse the domain discriminator. The overall framework is a *minimax* game between two players F and D, in which λ and μ are trade-off hyper-parameters between different loss functions. Generally, the MCC loss is also readily pluggable to other representative domain adaptation frameworks, *e.g.* moment matching [21] and large norm [48].

4   Experiments

We evaluate MCC as a standalone approach with many methods for six domain adaptation scenarios (UDA, MSDA, MTDA, PDA, MSPDA and MTPDA). We also evaluate MCC as a regularizer to existing domain adaptation methods.

4.1   Setup

We use four standard datasets: **(1) Office-31** [34]: a classic domain adaptation dataset with 31 categories and 3 domains: *Amazon* (**A**), *Webcam* (**W**) and *DSLR* (**D**); **(2) Office-Home** [44]: a more difficult dataset (larger domain shift) with 65 categories and 4 domains: *Art* (**A**), *Clip Art* (**C**), *Product* (**P**) and *Real World* (**R**); **(3) VisDA-2017** [32]: a dataset with 12 categories and over 280,000 images; **(4) DomainNet** [30]: the largest and hardest domain adaptation dataset, with approximately 0.6 million images from 345 categories and 6 domains: *Clipart* (**c**), *Infograph* (**i**), *Painting* (**p**), *Quickdraw* (**q**), *Real* (**r**) and *Sketch* (**s**).

Our methods are implemented based on **PyTorch**. ResNet [13] pre-trained on ImageNet [6] is used as the network backbone. We use Deep Embedded Validation (DEV) [49] to select hyper-parameter T and provide parameter sensitivity analysis. A balance between the cross-entropy and MCC, *i.e.* $\mu = 1.0$ works well for all experiments. We run each experiment for 5 times and report the average results.

4.2 Results and Discussion

Multi-target Domain Adaptation (MTDA). We evaluate the MTDA tasks following the protocol of DADA [31], which provides six tasks on DomainNet, the most difficult dataset to date. We adopt the strategy that directly merges multiple target domains. As shown in Table 1, many competitive methods are not effective in this challenging scenario. However, our simple method outperforms the current state-of-the-art method DADA [31] by a big margin (**7.3%**). Note that the source-only accuracy is rather low on this dataset, validating that our method, with well-designed mechanisms, is sufficiently robust to wrong predictions.

Multi-source Domain Adaptation (MSDA). When running our method for MSDA, we similarly merge multiple source domains in MCC and compare it to existing DA algorithms that are specifically designed for MSDA on DomainNet. As shown in Table 1, based on the inductive bias of minimizing the class confusion, MCC significantly outperforms M^3SDA [30], the state-of-the-art method by a big margin (**5.0%**). Note that these specific methods are of very complex architecture and loss designs and may be hard to use in practical applications.

Table 1. Accuracy (%) on DomainNet for MTDA and MSDA (ResNet-101).

	(a) MTDA							(b) MSDA						
Method	c:	i:	p:	q:	r:	s:	Avg	Method	:c	:i	:p	:q	:r	:s	Avg
ResNet [13]	25.6	16.8	25.8	9.2	20.6	22.3	20.1	ResNet [13]	47.6	13.0	38.1	13.3	51.9	33.7	32.9
SE [7]		21.3	8.5	14.5	13.8	16.0	19.7	15.6	MCD [35]	54.3	22.1	45.7	7.6	58.4	43.5	38.5
MCD [35]	25.1	19.1	27.0	10.4	20.2	22.5	20.7	DCTN [47]	48.6	23.5	48.8	7.2	53.5	47.3	38.2
DADA [31]	26.1	20.0	26.5	12.9	20.7	22.8	21.5	M^3SDA [30]	58.6	26.0	52.3	6.3	62.7	49.5	42.6
MCC	**33.6**	**30.0**	**32.4**	**13.5**	**28.0**	**35.3**	**28.8**	**MCC**	**65.5**	**26.0**	**56.6**	**16.5**	**68.0**	**52.7**	**47.6**

Partial Domain Adaptation (PDA). Due to the existence of source outlier classes, PDA is known as a challenging scenario because of the misalignment between the source and target classes. For a fair comparison, we follow the protocol of PADA [3] and AFN [48], where the first 25 categories (in alphabetic order) of the Office-Home dataset are taken as the target domain. As shown in Table 2, on this dataset, MCC outperforms AFN [48], the *ICCV'19 honorable-mention* entry and the state-of-the-art method for PDA, by a big margin (**3.3%**).

Unsupervised Domain Adaptation (UDA). We evaluate MCC for the most common UDA scenario on several datasets. (1) *VisDA-2017.* As reported in Table 3, MCC surpasses state-of-the-art UDA algorithms and yields the highest accuracy to date among methods of no complex architecture and loss designs. (2) *Office-31.* As shown in Table 4, MCC performs the best. (3) *Two Moon* [20]. We train a shallow MLP from scratch and plot the decision boundaries of MCC and Minimum Entropy (MinEnt) [10]. MCC yields much better boundaries in Fig. 4.

Table 2. Accuracy (%) on Office-Home for PDA (ResNet-50).

Method (S:T)	A:C	A:P	A:R	C:A	C:P	C:R	P:A	P:C	P:R	R:A	R:C	R:P	Avg
ResNet [13]	38.6	60.8	75.2	39.9	48.1	52.9	49.7	30.9	70.8	65.4	41.8	70.4	53.7
DAN [21]	44.4	61.8	74.5	41.8	45.2	54.1	46.9	38.1	68.4	64.4	51.5	74.3	56.3
JAN [24]	45.9	61.2	68.9	50.4	59.7	61.0	45.8	43.4	70.3	63.9	52.4	76.8	58.3
PADA [3]	51.2	67.0	78.7	52.2	53.8	59.0	52.6	43.2	78.8	73.7	56.6	77.1	62.0
AFN [48]	58.9	76.3	81.4	70.4	**73.0**	77.8	72.4	55.3	80.4	75.8	60.4	79.9	71.8
MCC	**63.1**	**80.8**	**86.0**	**70.8**	72.1	**80.1**	**75.0**	**60.8**	**85.9**	**78.6**	**65.2**	**82.8**	**75.1**

Table 3. Accuracy (%) on VisDA-2017 for UDA (ResNet-101).

Method	Plane	Bcybl	Bus	Car	Horse	Knife	Mcyle	Persn	Plant	Sktb	Train	Truck	Mean
ResNet [13]	55.1	53.3	61.9	59.1	80.6	17.9	79.7	31.2	81.0	26.5	73.5	8.5	52.4
MinEnt [10]	80.3	75.5	75.8	48.3	77.9	27.3	69.7	40.2	46.5	46.6	79.3	16.0	57.0
DANN [8]	81.9	77.7	82.8	44.3	81.2	29.5	65.1	28.6	51.9	54.6	82.8	7.8	57.4
DAN [21]	87.1	63.0	76.5	42.0	90.3	42.9	85.9	53.1	49.7	36.3	85.8	20.7	61.1
MCD [35]	87.0	60.9	83.7	64.0	88.9	79.6	84.7	76.9	88.6	40.3	83.0	25.8	71.9
CDAN [22]	85.2	66.9	83.0	50.8	84.2	74.9	88.1	74.5	83.4	**76.0**	81.9	**38.0**	73.9
AFN [48]	**93.6**	61.3	**84.1**	70.6	**94.1**	79.0	**91.8**	**79.6**	89.9	55.6	**89.0**	24.4	76.1
MCC	88.1	**80.3**	80.5	**71.5**	90.1	**93.2**	85.0	71.6	89.4	73.8	85.0	36.9	**78.8**

Multi-source/Multi-target Partial Domain Adaptation (MSPDA / MTPDA).

Table 5 shows that MCC is versatile to handle these hard scenarios.

4.3 Empirical Analyses

General Regularizer. MCC can be used as a general regularizer for existing DA methods. We compare its performance with entropy minimization (MinEnt) [10] and Batch Spectral Penalization (BSP) [5]. As shown in Tables 6 and 7, MCC yields larger improvements than MinEnt and BSP to a variety of DA methods.

Ablation Study. It is interesting to investigate the contribution of each part of the MCC loss: Class Correlation (**CC**), Probability Rescaling (**PR**), and Uncertainty Reweighting (**UR**). Results in Table 8 justify that each part has its

Table 4. Accuracy (%) on Office-31 for UDA (ResNet-50).

Method	A→W	D→W	W→D	A→D	D→A	W→A	Avg
ResNet [13]	68.4 ± 0.2	96.7 ± 0.1	99.3 ± 0.1	68.9 ± 0.2	62.5 ± 0.3	60.7 ± 0.3	76.1
DAN [21]	80.5 ± 0.4	97.1 ± 0.2	99.6 ± 0.1	78.6 ± 0.2	63.6 ± 0.3	62.8 ± 0.2	80.4
RTN [23]	84.5 ± 0.2	96.8 ± 0.1	99.4 ± 0.1	77.5 ± 0.3	66.2 ± 0.2	64.8 ± 0.3	81.6
DANN [8]	82.0 ± 0.4	96.9 ± 0.2	99.1 ± 0.1	79.7 ± 0.4	68.2 ± 0.4	67.4 ± 0.5	82.2
JAN [24]	85.4 ± 0.3	97.4 ± 0.2	99.8 ± 0.2	84.7 ± 0.3	68.6 ± 0.3	70.0 ± 0.4	84.3
GTA [36]	89.5 ± 0.5	97.9 ± 0.3	99.8 ± 0.4	87.7 ± 0.5	72.8 ± 0.3	71.4 ± 0.4	86.5
CDAN [22]	94.1 ± 0.1	**98.6 ± 0.1**	100.0 ± 0.0	92.9 ± 0.2	71.0 ± 0.3	69.3 ± 0.3	87.7
AFN [48]	88.8 ± 0.5	98.4 ± 0.3	99.8 ± 0.1	87.7 ± 0.6	69.8 ± 0.4	69.7 ± 0.4	85.7
MDD [52]	94.5 ± 0.3	98.4 ± 0.3	**100.0 ± 0.0**	93.5 ± 0.2	**74.6 ± 0.3**	72.2 ± 0.1	88.9
MCC	**95.5 ± 0.2**	**98.6 ± 0.1**	**100.0 ± 0.0**	**94.4 ± 0.3**	72.9 ± 0.2	**74.9 ± 0.3**	**89.4**

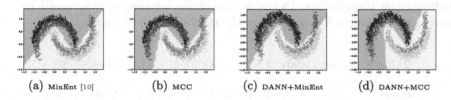

| (a) MinEnt [10] | (b) MCC | (c) DANN+MinEnt | (d) DANN+MCC |

Fig. 4. Decision boundaries on the Two Moon dataset. Blue points indicate target data, and different classes of the source data are depicted in purple and yellow. (Color figure online)

Table 5. Accuracy (%) on Office-Home for MSPDA and MTPDA.

(a) MSPDA

Method	:A	:C	:P	:R	Avg
DANN [8]	58.3	43.6	60.7	71.2	58.5
PADA [3]	62.8	51.8	71.7	79.2	66.4
M³SDA [30]	67.4	55.3	72.2	80.4	68.8
AFN [48]	77.1	61.2	79.3	82.5	75.0
MCC	**79.6**	**67.5**	**80.6**	**85.1**	**78.2**

(b) MTPDA

Method	A:	C:	P:	R:	Avg
DANN [8]	44.6	44.8	39.1	44.1	43.1
PADA [3]	59.9	53.7	51.1	61.4	56.5
DADA [31]	65.1	63.0	60.4	63.0	62.9
AFN [48]	68.7	65.6	63.4	67.5	66.3
MCC	**73.1**	**72.1**	**69.4**	**68.3**	**70.7**

Table 6. Accuracy (%) on VisDA-2017 as *regularizer* for UDA (ResNet-101).

Method	Plane	Bcybl	Bus	Car	Horse	Knife	Mcyle	Persn	Plant	Sktb	Train	Truck	Mean
DANN [8]	81.9	77.7	82.8	44.3	81.2	29.5	65.1	28.6	51.9	54.6	82.8	7.8	57.4
DANN + MinEnt [10]	87.4	55.0	75.3	**63.8**	87.4	43.6	**89.3**	72.5	82.9	**78.6**	85.6	27.4	70.7
DANN + BSP [5]	**92.2**	72.5	**83.8**	47.5	87.0	54.0	86.8	72.4	80.6	66.9	84.5	37.1	72.1
DANN + MCC	90.4	**79.8**	72.3	55.1	**90.5**	**86.8**	86.6	**80.0**	**94.2**	76.9	**90.0**	**49.6**	**79.4**
CDAN [22]	85.2	66.9	**83.0**	50.8	84.2	74.9	88.1	74.5	83.4	76.0	81.9	38	73.9
CDAN + MinEnt [10]	90.5	65.8	79.1	62.2	89.8	28.7	**92.8**	75.4	86.8	65.3	85.2	35.3	71.4
CDAN + BSP [5]	92.4	61.0	81.0	57.5	89.0	**80.6**	90.1	77.0	84.2	77.9	82.1	38.4	75.9
CDAN + MCC	**94.5**	**80.8**	78.4	**65.3**	**90.6**	79.4	87.5	**82.2**	**94.7**	**81.0**	**86.0**	**44.6**	**80.4**

Table 7. Accuracy (%) on Office-31 as *regularizer* for UDA (ResNet-50).

Method	A→W	D→W	W→D	A→D	D→A	W→A	Avg
DANN [8]	82.0±0.4	96.9±0.2	99.1±0.1	79.7±0.4	68.2±0.4	67.4±0.5	82.2
DANN + MinEnt [10]	91.7±0.3	98.3±0.1	100.0±0.0	87.9±0.3	68.8±0.3	68.1±0.3	85.8
DANN + BSP [5]	93.0±0.2	98.0±0.2	100.0±0.0	90.0±0.4	71.9±0.3	73.0±0.3	87.7
DANN + MCC	**95.6±0.3**	**98.6±0.1**	99.3±0.0	**93.8±0.4**	**74.0±0.3**	**75.0±0.4**	**89.4**
CDAN [22]	94.1±0.1	**98.6±0.1**	100.0±0.0	92.9±0.2	71.0±0.3	69.3±0.3	87.7
CDAN + MinEnt [10]	91.7±0.2	98.5±0.1	100.0±0.0	90.4±0.3	72.3±0.2	69.5±0.2	87.1
CDAN + BSP [5]	93.3±0.2	98.2±0.2	100.0±0.0	93.0±0.2	**73.6±0.3**	72.6±0.3	88.5
CDAN + MCC	**94.7±0.2**	**98.6±0.1**	100.0±0.0	**95.0±0.1**	73.0±0.2	**73.6±0.3**	**89.2**
AFN [48]	88.8±0.5	98.4±0.3	99.8±0.1	87.7±0.6	69.8±0.4	69.7±0.4	85.7
AFN + MinEnt [10]	90.3±0.4	**98.7±0.2**	100.0±0.0	92.1±0.5	73.4±0.3	71.2±0.3	87.6
AFN + BSP [5]	89.7±0.4	98.0±0.2	99.8±0.1	91.0±0.4	71.4±0.3	71.4±0.2	86.9
AFN + MCC	**95.4±0.3**	98.6±0.2	100.0±0.0	**96.0±0.2**	**74.6±0.3**	**75.2±0.2**	**90.0**

indispensable contribution. To enable ease of use, we seamlessly integrate these parts into a coherent loss and reduce the number of hyper-parameters.

Table 8. Ablation study of MCC on Office-31 for UDA (ResNet-50).

Method	A→W	D→W	W→D	A→D	D→A	W→A	Avg
MCC (CC Only)	92.2	96.9	**100.0**	88.6	73.2	64.5	85.9
MCC (CC + PR)	93.1	98.5	**100.0**	91.6	70.9	69.0	87.2
MCC (CC + PR + UR)	93.7	**98.6**	**100.0**	93.2	72.1	73.7	88.4
MCC (all)	**95.5**	**98.6**	**100.0**	**94.4**	**72.9**	**74.9**	**89.4**

Further, we analyze how the specially designed Uncertainty Reweighting (UR) mechanism works. Figure 5 shows three typical examples as well as their weights and the confusion values before and after reweighting. The classifier prediction on the first image shows no obvious peak, while the one on the third image shows two obvious peaks on classes calculator and phone. The third image is more suitable for embodying class confusion. Naturally, its confusion value is higher than the first one, and our reweighting mechanism further highlights the suitable one. On the other hand, as the reweighting mechanism is defined with entropy, we recognize that it will improperly assign high weights to examples with highly confident predictions, including the wrong ones. As shown in the second image, its ground truth label is a lamp, but it is classified as a bike. In our method, the confusion value of such an example is so low that the influence of higher weight can be neglected. Therefore, our reweighting mechanism is effective and reliable.

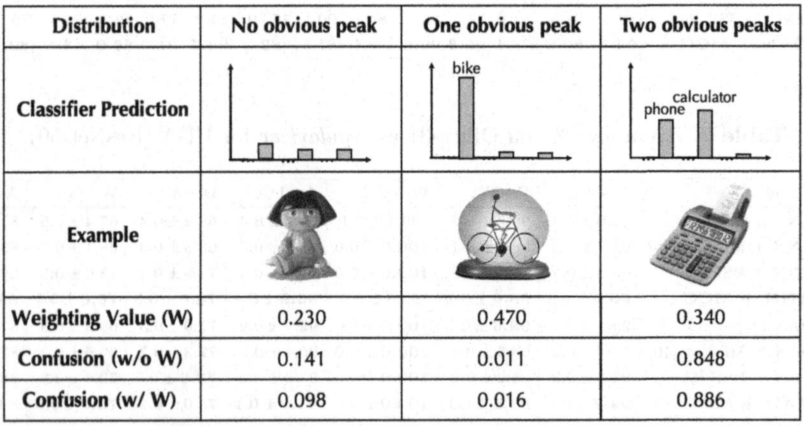

Distribution	No obvious peak	One obvious peak	Two obvious peaks
Classifier Prediction		bike	calculator phone
Example			
Weighting Value (W)	0.230	0.470	0.340
Confusion (w/o W)	0.141	0.011	0.848
Confusion (w/ W)	0.098	0.016	0.886

Fig. 5. Three typical samples and the corresponding weights and confusion values.

Theoretical Insight. Ben-David *et al.* [1] derived the expected error $\mathcal{E}_T(h)$ of a hypothesis h on the target domain $\mathcal{E}_T(h) \le \mathcal{E}_S(h) + \frac{1}{2}d_{\mathcal{H}\Delta\mathcal{H}}(\mathcal{S}, \mathcal{T}) + \epsilon_{\text{ideal}}$ by: **(a)** expected error of h on the source domain, $\mathcal{E}_S(h)$; **(b)** the A-distance $d_{\mathcal{H}\Delta\mathcal{H}}(\mathcal{S}, \mathcal{T})$, a measure of domain discrepancy; and **(c)** the error ϵ_{ideal} of the ideal joint hypothesis h^* on both source and target domains. As shown in Fig. 6, MCC has the lowest A-distance [1], which is close to the oracle one (*i.e.* supervised learning on both domains). In Fig. 7, the ϵ_{ideal} value of MCC is also lower than that of mainstream DA methods. Both imply better generalization.

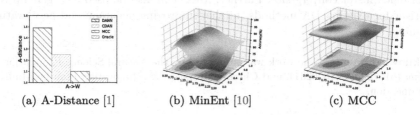

(a) A-Distance [1]   (b) MinEnt [10]   (c) MCC

Fig. 6. (a): A-Distance of the last fc-layer features of task $A \to W$ on Office-31 (UDA); (b)–(c): Hyper-parameter sensitivity of task $A \to W$ on Office-31 (UDA).

Parameter Sensitivity. Temperature factor T and MCC coefficient μ are the two hyper-parameters of MCC and MinEnt [10] when applying them standalone or with existing methods. We traverse hyper-parameters around their optimal values $[T^*, \mu^*]$, as shown in Fig. 6, MCC is much less sensitive to its hyper-parameters.

Convergence Speed. We show the training curves throughout iterations in Fig. 7. Impressively, MCC takes only 1000 iterations to reach an accuracy of 95%, while at this point the accuracies of CDAN and DANN are below 85%. When used as a regularizer for existing domain adaptation methods, MCC largely accelerates convergence. In general, MCC is 3× faster than the existing methods.

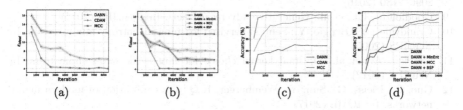

(a)   (b)   (c)   (d)

Fig. 7. The ϵ_{ideal} error values (%) and training curves throughout iterations.

5   Conclusion

This paper studies a more practical paradigm, Versatile Domain Adaptation (VDA), where one method tackles many scenarios. We uncover that less class confusion implies more transferability, which is the key insight to enable VDA. Based on this, we propose a new loss function: Minimum Class Confusion (MCC). MCC can be applied as a versatile domain adaptation approach to a variety of DA scenarios. Extensive results justify that our method, without any modification, outperforms state-of-the-art scenario-specific domain adaptation methods with much faster convergence. Further, MCC can also be used as a general regularizer for existing DA methods, further improving accuracy and accelerating training.

Acknowledgments. The work was supported by the Natural Science Foundation of China (61772299, 71690231), and China University S&T Innovation Plan by Ministry of Education.

References

1. Ben-David, S., Blitzer, J., Crammer, K., Kulesza, A., Pereira, F., Vaughan, J.W.: A theory of learning from different domains. Mach. Learn. **79**(1–2), 151–175 (2010)
2. Cao, Z., Long, M., Wang, J., Jordan, M.I.: Partial transfer learning with selective adversarial networks. In: CVPR (2018)
3. Cao, Z., Ma, L., Long, M., Wang, J.: Partial adversarial domain adaptation. In: ECCV (2018)
4. Cao, Z., You, K., Long, M., Wang, J., Yang, Q.: Learning to transfer examples for partial domain adaptation. In: CVPR (2019)
5. Chen, X., Wang, S., Long, M., Wang, J.: Transferability vs. discriminability: batch spectral penalization for adversarial domain adaptation. In: ICML (2019)
6. Deng, J., Dong, W., Socher, R., Li, L.J., Li, K., Fei-Fei, L.: ImageNet: a large-scale hierarchical image database. In: CVPR (2009)
7. French, G., Mackiewicz, M., Fisher, M.: Self-ensembling for visual domain adaptation. In: ICLR (2018)
8. Ganin, Y., et al.: Domain-adversarial training of neural networks. JMLR **17**(1), 2096–2130 (2016)
9. Goodfellow, I.J., et al.: Generative adversarial nets. In: NeurIPS (2014)
10. Grandvalet, Y., Bengio, Y.: Semi-supervised learning by entropy minimization. In: NeurIPS (2005)
11. Gretton, A., et al.: Optimal kernel choice for large-scale two-sample tests. In: NeurIPS (2012)
12. Guo, C., Pleiss, G., Sun, Y., Weinberger, K.Q.: On calibration of modern neural networks. In: ICML (2017)
13. He, K., Zhang, X., Ren, S., Sun, J.: Deep residual learning for image recognition. In: CVPR (2016)
14. Hinton, G., Vinyals, O., Dean, J.: Distilling the knowledge in a neural network. arXiv preprint arXiv:1503.02531 (2015)
15. Hoffman, J., : CyCADA: cycle-consistent adversarial domain adaptation. In: ICML (2018)

16. Inoue, N., Furuta, R., Yamasaki, T., Aizawa, K.: Cross-domain weakly-supervised object detection through progressive domain adaptation. In: CVPR (2018)
17. Kang, G., Jiang, L., Yang, Y., Hauptmann, A.: Contrastive adaptation network for unsupervised domain adaptation. In: CVPR (2019)
18. Lee, C., Batra, T., Baig, M.H., Ulbricht, D.: Sliced wasserstein discrepancy for unsupervised domain adaptation. In: CVPR (2019)
19. Lee, S., Kim, D., Kim, N., Jeong, S.G.: Drop to adapt: learning discriminative features for unsupervised domain adaptation. In: ICCV (2019)
20. Liu, H., Long, M., Wang, J., Jordan, M.: Transferable adversarial training: a general approach to adapting deep classifiers. In: ICML (2019)
21. Long, M., Cao, Y., Wang, J., Jordan, M.I.J.: Learning transferable features with deep adaptation networks. In: ICML (2015)
22. Long, M., Cao, Z., Wang, J., Jordan, M.I.: Conditional adversarial domain adaptation. In: NeurIPS (2018)
23. Long, M., Zhu, H., Wang, J., Jordan, M.I.: Unsupervised domain adaptation with residual transfer networks. In: NeurIPS (2016)
24. Long, M., Zhu, H., Wang, J., Jordan, M.I.: Deep transfer learning with joint adaptation networks. In: ICML (2017)
25. von Luxburg, U.: A tutorial on spectral clustering. Stat. Comput. **17**(4), 395–416 (2007)
26. Mirza, M., Osindero, S.: Conditional generative adversarial nets. arXiv preprint arXiv:1411.1784 (2014)
27. Pan, S.J., Yang, Q.: A survey on transfer learning. TKDE **22**(10), 1345–1359 (2010)
28. Pan, Y., Yao, T., Li, Y., Wang, Y., Ngo, C.W., Mei, T.: Transferrable prototypical networks for unsupervised domain adaptation. In: CVPR (2019)
29. Pei, Z., Cao, Z., Long, M., Wang, J.: Multi-adversarial domain adaptation. In: AAAI (2018)
30. Peng, X., Bai, Q., Xia, X., Huang, Z., Saenko, K., Wang, B.: Moment matching for multi-source domain adaptation. In: ICCV (2019)
31. Peng, X., Huang, Z., Sun, X., Saenko, K.: Domain agnostic learning with disentangled representations. In: ICML (2019)
32. Peng, X., Usman, B., Kaushik, N., Hoffman, J., Wang, D., Saenko, K.: VisDA: the visual domain adaptation challenge. arXiv preprint arXiv:1710.06924 (2017)
33. Quionero-Candela, J., Sugiyama, M., Schwaighofer, A., Lawrence, N.D.: Dataset Shift in Machine Learning. The MIT Press, Cambridge (2009)
34. Saenko, K., Kulis, B., Fritz, M., Darrell, T.: Adapting visual category models to new domains. In: Daniilidis, K., Maragos, P., Paragios, N. (eds.) ECCV 2010. LNCS, vol. 6314, pp. 213–226. Springer, Heidelberg (2010). https://doi.org/10.1007/978-3-642-15561-1_16
35. Saito, K., Watanabe, K., Ushiku, Y., Harada, T.: Maximum classifier discrepancy for unsupervised domain adaptation. In: CVPR (2018)
36. Sankaranarayanan, S., Balaji, Y., Castillo, C.D., Chellappa, R.: Generate to adapt: aligning domains using generative adversarial networks. In: CVPR (2018)
37. Schutze, H., Manning, C.D., Raghavan, P.: Introduction to information retrieval. In: Proceedings of the International Communication of Association for Computing Machinery Conference (2008)
38. Shu, R., Bui, H.H., Narui, H., Ermon, S.: A DIRT-T approach to unsupervised domain adaptation. In: ICLR (2018)
39. Sun, B., Saenko, K.: Deep CORAL: correlation alignment for deep domain adaptation. In: Hua, G., Jégou, H. (eds.) ECCV 2016. LNCS, vol. 9915, pp. 443–450. Springer, Cham (2016). https://doi.org/10.1007/978-3-319-49409-8_35

40. Tarvainen, A., Valpola, H.: Weight-averaged consistency targets improve semi-supervised deep learning results. In: NeurIPS (2017)
41. Tzeng, E., Hoffman, J., Darrell, T., Saenko, K.: Simultaneous deep transfer across domains and tasks. In: ICCV (2015)
42. Tzeng, E., Hoffman, J., Saenko, K., Darrell, T.: Adversarial discriminative domain adaptation. In: CVPR (2017)
43. Tzeng, E., Hoffman, J., Zhang, N., Saenko, K., Darrell, T.: Deep domain confusion: maximizing for domain invariance. CoRR abs/1412.3474 (2014)
44. Venkateswara, H., Eusebio, J., Chakraborty, S., Panchanathan, S.: Deep hashing network for unsupervised domain adaptation. In: CVPR (2017)
45. Wang, X., Jin, Y., Long, M., Wang, J., Jordan, M.I.: Transferable normalization: towards improving transferability of deep neural networks. In: Advances in Neural Information Processing Systems, pp. 1953–1963 (2019)
46. Wang, X., Li, L., Ye, W., Long, M., Wang, J.: Transferable attention for domain adaptation. In: AAAI (2019)
47. Xu, R., Chen, Z., Zuo, W., Yan, J., Lin, L.: Deep cocktail network: multi-source unsupervised domain adaptation with category shift. In: CVPR (2018)
48. Xu, R., Li, G., Yang, J., Lin, L.: Larger norm more transferable: an adaptive feature norm approach. In: ICCV (2019)
49. You, K., Wang, X., Long, M., Jordan, M.: Towards accurate model selection in deep unsupervised domain adaptation. In: ICML (2019)
50. Zhang, J., Ding, Z., Li, W., Ogunbona, P.: Importance weighted adversarial nets for partial domain adaptation. In: CVPR (2018)
51. Zhang, Y., Tang, H., Jia, K., Tan, M.: Domain-symmetric networks for adversarial domain adaptation. In: CVPR (2019)
52. Zhang, Y., Liu, T., Long, M., Jordan, M.I.: Bridging theory and algorithm for domain adaptation. In: ICML (2019)
53. Zhao, H., Zhang, S., Wu, G., Moura, J.M., Costeira, J.P., Gordon, G.J.: Adversarial multiple source domain adaptation. In: NeurIPS (2018)
54. Zou, Y., Yu, Z., Kumar, B., Wang, J.: Unsupervised domain adaptation for semantic segmentation via class-balanced self-training. In: ECCV (2018)
55. Zou, Y., Yu, Z., Liu, X., Kumar, B.V., Wang, J.: Confidence regularized self-training. In: ICCV (2019)

Large Batch Optimization for Object Detection: Training COCO in 12 minutes

Tong Wang[1,2(✉)], Yousong Zhu[1,3], Chaoyang Zhao[1], Wei Zeng[4,5], Yaowei Wang[5], Jinqiao Wang[1,2,6], and Ming Tang[1]

[1] National Laboratory of Pattern Recognition, Institute of Automation, Chinese Academy of Sciences, Beijing, China
{tong.wang,yousong.zhu,chaoyang.zhao,jqwang,tangm}@nlpr.ia.ac.cn
[2] School of Artificial Intelligence, University of Chinese Academy of Sciences, Beijing, China
[3] ObjectEye Inc., Beijing, China
[4] Peking University, Beijing, China
weizeng@pku.edu.cn
[5] Peng Cheng Laboratory, Shenzhen, China
wangyw@pcl.ac.cn
[6] NEXWISE Co., Ltd., Guangzhou, China

Abstract. Most of existing object detectors usually adopt a small training batch size (*e.g.* 16), which severely hinders the whole community from exploring large-scale datasets due to the extremely long training procedure. In this paper, we propose a versatile large batch optimization framework for object detection, named *LargeDet*, which successfully scales the batch size to larger than 1K for the first time. Specifically, we present a novel Periodical Moments Decay LAMB (PMD-LAMB) algorithm to effectively reduce the negative effects of the lagging historical gradients. Additionally, the Synchronized Batch Normalization (SyncBN) is utilized to help fast convergence. With LargeDet, we can not only prominently shorten the training period, but also significantly improve the detection accuracy of sparsely annotated large-scale datasets. For instance, we can finish the training of ResNet50 FPN detector on COCO within 12 min. Moreover, we achieve 12.2% mAP@0.5 absolute improvement for ResNet50 FPN on Open Images by training with batch size 640.

Keywords: Object detection · Large batch optimization · Periodical moments decay

1 Introduction

With the advent of Neural Networks, researchers have made great progress in various domains, including computer vision [11], speech [1], natural language processing [21] and so on. Among the many factors contributing to their success, the available of massive data is definitely one of the most important one. It is

© Springer Nature Switzerland AG 2020
A. Vedaldi et al. (Eds.): ECCV 2020, LNCS 12366, pp. 481–496, 2020.
https://doi.org/10.1007/978-3-030-58589-1_29

widely accepted that large-scale dataset helps the convolutional neural networks (CNN) generalize well. Consequently, there is a tendency that the size of dataset is growing larger and larger.

Object detection, as a fundamental computer vision task, also benefits from large-scale datasets. From PASCAL VOC [5] to MS COCO [14], the volume of the dataset becomes larger in terms of the number of images and the diversity of categories. Nowadays, datasets like Open Images [12] and Objects365 [20] come in an unprecedented scale. However, previous classical object detectors such as Faster R-CNN [19], Mask R-CNN [7] and RetinaNet [13] *etc.* usually adopt a relatively small mini-batch (16, for example) to train. Such small batch size restrains researchers from making full use of rapidly increased computational power. And the unbearable long training procedure of large-scale datasets severely impedes its development both in research and industry.

Nevertheless, it is not trivial to directly increase the batch size as large batch size training always suffers from performance degradation [8]. A rule of thumb for training neural network is the Linear Scaling Rule (LSR) [10], which suggests that when the batch size becomes K times, the learning rate should also be multiplied by K. However, since the LSR requests the learning rate to grow proportional to the batch size, it has divergence issue when the batch size increases to a certain value, *e.g.* 256. To tackle this issue, [23] proposed Layer-wise Adaptive Rate Scaling (LARS) algorithm to adjust each layer's learning rate based on the norm of its weights and the norm of its gradients. Another similar algorithm is LAMB which is first proposed in [24] for the fast training of BERT [4]. Both LARS and LAMB have the same design philosophy and yield good performance in the large batch training for image classification.

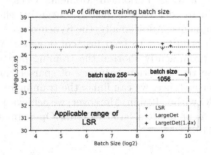

Fig. 1. The performance of ResNet50 FPN detector on COCO dataset when using different training batch sizes. LSR means the Linear Scaling Rule (LSR). LargeDet is the proposed framework. 1.4x indicates the training iterations is extended to 1.4 times. The horizontal black dotted line is the performance of baseline with batch size 16

Although researchers have achieved promising results on large batch training for image classification, there are fewer works that concentrate on large batch training for object detection. Since the emergence of those large-scale datasets like Open Images and Objects365, the need to explore large batch training for object detection has become more urgent than ever before. MegDet [17] successfully trains a detector with batch size 256 by using LSR and some other measures like Warm-up and Cross-GPU Batch Normalization, but the exploration to larger batch size still remains blank. *What happens if the batch size exceeds 256? Is there any chance to train a detector with an even larger batch size?*

In this paper, we systematically analyze the large batch optimization and propose a large batch training framework called LargeDet for object detection. To be specific, we first theoretically analyze the problems of LSR and give its applicable range of training batch size. Then, motivated by LAMB [24], we propose a Periodical Moments Decay LAMB algorithm (PMD-LAMB), which can further improve the performance under larger batch size setting. Additionally, we combine the PMD-LAMB with SyncBN to improve the stability and generalization of the network especially when the batch size is extremely large. As showed in Fig. 1, with the aid of LargeDet, we successfully scale the batch size from 256 to 1056 with ResNet50 FPN detector on COCO dataset without much accuracy loss, which is the first work to train an object detector using a batch size larger than 1K. The experimental results demonstrate that the proposed LargeDet not only speeds up the training cycle, but also achieves higher accuracy on large-scale datasets, such as Open Images.

In summary, our contributions are three-fold:

1. We propose PMD-LAMB algorithm to effectively eliminate the optimization difficulty of large batch size training, which significantly expands the applicable range of LSR.
2. We present a simple yet effective large batch training framework, named LargeDet, with which we can successfully scale the batch size up to 1056 without much accuracy drop.
3. LargeDet helps us to shrink the training time of ResNet50 FPN on COCO to 12 min. Besides, it also helps us achieve 12.2% and 10.8% absolute improvement with ResNet50 and ResNet101 FPN detector respectively on Open Images V5 with much less training time.

2   Related Work

2.1   CNN-Based Detectors

For object detection, current deep learning based approaches can be roughly divided into two categories, single-stage and two-stage approaches. Classical two-stage algorithms, including Faster R-CNN [19], R-FCN [2], Mask R-CNN [7], Deformable Convolutional Networks [3], *etc.* all generate numerous but less accurate proposals in the first stage, and then refine them in the second stage. Single-stage approaches such as YOLO [18], SSD [15], RetinaNet [13], *etc.* work in a more concise manner. They make predictions on the whole feature map directly without proposal generating process, thus enjoying higher speed. Nowadays, some single-stage detectors, like ExtremeNet [25] and FCOS [22] can also achieve competitive results as two-stage detectors.

2.2 Large Batch Optimization

In the past few years, large batch size training for image classification has attracted a lot of researchers' attention. Linear Scaling Rule (LSR) has been considered as a rule of thumb for training CNNs, which guarantees roughly equivalent accuracy when using different batch sizes to train. It is mathematically explained in [6,17] and widely applied in daily practice. By utilizing LSR and Warm-up strategy, [6] is able to scale the batch size to 8192 for ResNet50. [23] proposes Layer-wise Adaptive Rate Scaling (LARS) algorithm to successfully scale the batch size for ResNet50 to 32768. The successor of LARS algorithm is LAMB, which is first proposed in [24]. Both of LARS and LAMB all leverage the norm of weights and the norm of gradients to adjust each layer's learning rate. The only difference lies in that LARS originates from the most commonly used SGD algorithm, while LAMB is a variant of ADAM [9].

Different from the flourish in the research of large batch training for image classification, there are few works which focus on the large batch training for object detection. [17] analyzes the effect of Batch Normalization in object detection and provides a solution for large batch training. Specifically, they implement Cross-GPU Batch Normalization to enable collecting sufficient statistics from more samples, making it possible to train a detector with a large batch size. By combining LSR, Cross-GPU Batch Normalization and Warm-up strategy, they are able to train the Faster R-CNN detector with batch size 256. So far, the object detection with larger batch size is still the terra incognita which deserves to be further researched.

3 Method

In this section, we first analyze the problems of Linear Scaling Rule (LSR) for detectors with large batch size. Then we introduce the proposed PMD-LAMB algorithm. Last, we present the object detection framework LargeDet designed for large batch training and propose three practical guidelines.

3.1 Problems of Linear Scaling Rule

MegDet [17] is the first successful attempt to train a detector using a large batch size of 256. It mainly adopts the Linear Scaling Rule (LSR) [6,17] to obtain roughly equivalent accuracy when different batch sizes are adopted. In addition, the Warm-up strategy and Cross-GPU Batch Normalization [17] are further used to guarantee the early convergence of the model and make the training more stable. However, the empirical limits of LSR guideline are not well explored in [17] since the learning rate can not be scaled infinitely. Consequently, it is a problem whether LSR is still suitable for a larger batch size.

Figure 2 shows the experimental results when LSR is applied to different batch sizes. When the training batch size is less than or equal to 256, LSR can maintain almost the same level of accuracy as the 16 mini-batch baseline (with a

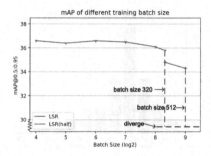

Fig. 2. The performance of ResNet50 FPN detector on COCO dataset when we use LSR. Half means we lower the learning rate to a half. The black dotted line represents the network diverges and the accuracy is 0

learning rate of 0.02). Nevertheless, if the training batch size is larger than 256, increasing the learning rate proportionally will lead to performance degradation or even network divergence. The LSR no longer holds. When we compromise to use a lower learning rate, *e.g.* 0.32 for batch size 512, the network will not diverge, but the final accuracy drops dramatically (34.3 *vs.* 36.7). This is mainly because a small learning rate leads to insufficient training, and the model can not converge to an optimal value. Therefore, how to tackle the optimization difficulty is the main issue with the larger batch size training.

3.2 Periodical Moments Decay LAMB

As the discussion in Sect. 3.1, the main issue which hinders us from using a larger batch size is that the network diverges with a large learning rate. Previous optimization algorithms usually set a global learning rate for the network which is shared by all layers. In fact, the neural network is composed of a great many layers with various feature channel numbers, feature map sizes and weight distributions. This diversity endows each layer diverse tolerance to the learning rate. To make it more intuitive, we calculate the ratio between each layer's weight norm and gradient norm of a ResNet50 FPN detector at iteration 10000. We observe that there are significant differences in the ratios among different layers from Fig. 3. When the learning rate is too large, the layers which have less tolerance to the big learning rate collapse at first, and then the whole network diverges. To tackle this issue, [24] proposed LAMB to adaptively adjust each layer's learning rate based on the ratio of its weights' norm and gradients' norm. This algorithm endows each layer a proper learning rate, thus making it possible to train a network with a larger batch size. For LAMB, each update of the network parameters is determined by the exponential moving averages of the gradient (m_t) and the squared gradient (v_t). m_t and v_t are the estimate of the first and second order moments of the gradients, respectively, as in Eq. (1) and (2).

$$m_t = \beta_1 m_{t-1} + (1 - \beta_1)g_t \tag{1}$$

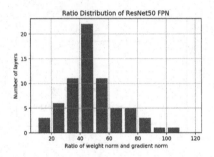

Fig. 3. The ratio distribution of ResNet50 FPN detector at iteration 10000

$$v_t = \beta_2 v_{t-1} + (1 - \beta_2)g_t^2 \qquad (2)$$

For large batch optimization, what we concern most is the fast convergence of the algorithm. As the batch size grows larger, the training iterations reduce proportionally. How to achieve expected accuracy in limited iterations is the core. For LAMB, the exponential moving average incorporates the knowledge of previously observed data, making the training process more stable. Nevertheless, overindulging in previous gradients may be harmful for fast convergence. With the iterative training, the network becomes more and more intelligent. Thus, the generated gradients become more precise to reflect the correct optimizing direction. While the heavily dependence on previous knowledge limits the network's update towards the more accurate direction. Historical gradients will dominate the training process and the network is prone to get trapped in saddle points. To help the network jump out of the saddle points, we argue to appropriately reduce the strong dependence of current update step on previously acquired information.

Algorithm 1. Periodical Moments Decay LAMB

Require: network parameters $x_1 \in \mathcal{R}^d$, classification loss function L_{cls}, box regression
   loss function L_{reg}, weight decay λ, learning rate η, hyper-parameters $0 < \beta_1, \beta_2 < 1$, scaling factor α, $\epsilon > 0$, cycle length T
1: Set $m_0 = 0, v_0 = 0$
2: **for** $t = 1$ to N **do**
3:     Draw b samples S_t from $\mathcal{P}$.
4:     Compute $g_t = \frac{1}{|S_t|} \sum_{s_t \in S_t} (\nabla L_{cls}(x_t, s_t) + \nabla L_{reg}(x_t, s_t))$.
5:     $\xi = \cos(\frac{\pi}{2} \cdot \frac{t\%T}{T})$
6:     $m_t = \xi \beta_1 m_{t-1} + (1 - \beta_1)g_t$
7:     $v_t = \xi \beta_2 v_{t-1} + (1 - \beta_2)g_t^2$
8:     Compute update step $r_t = \frac{m_t}{\sqrt{v_t} + \epsilon}$
9:     $x_{t+1} = x_t - \eta \frac{\alpha \|x_t\|}{\|r_t + \lambda x_t\|}(r_t + \lambda x_t)$
10: **end for**

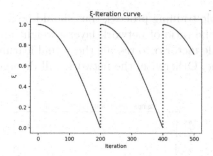

Fig. 4. An example of ξ-Iteration curve (T = 200)

Therefore, we propose a novel periodical moments decay LAMB to reduce the network's heavy dependence on previous acquired knowledge. To be specific, coefficient ξ is introduced to control the dependence on the accumulated moments. As formulated in Eq. (4) and (5), ξ is multiplied to the accumulated first- and second-order moments. When ξ equals to one, previously acquired knowledge is fully utilized to participate in current network parameter update. A smaller ξ indicates that the accumulated moments influence less to current update step. We perform such moments decay in a periodical way. The whole training process is divided into several consecutive time windows with cycle length T iterations. In each cycle, the coefficient ξ decays from 1 to 0, as illustrated in Fig. 4. Equation (3) formulates the computation of ξ, where t indicates the current iteration. T is an introduced hyper-parameter of the cycle length. The pseudo-code of PMD-LAMB is showed in Algorithm 1. To explore the effect of PMD-LAMB, we train two ResNet50 FPN detectors with LAMB and PMD-LAMB, respectively. They are both trained on COCO dataset with batch size 1056. The loss curves are visualized in Fig. 5. We can see that these two loss curves tangle with each other at the initial training stage. While the loss curve of PMD-LAMB has lower loss values when the training process levels off. This phenomenon indicates that PMD-LAMB helps the network converge faster and optimize to a better point with the same number of iterations, thus yielding higher accuracy.

$$\xi = \cos(\frac{\pi}{2} \cdot \frac{t\%T}{T}) \tag{3}$$

$$m_t = \xi\beta_1 m_{t-1} + (1 - \beta_1)g_t \tag{4}$$

$$v_t = \xi\beta_2 v_{t-1} + (1 - \beta_2)g_t^2 \tag{5}$$

There are two vital hyper-parameters in PMD-LAMB. Hyper-parameter T controls the cycle length. If T is set to a small value, the relaxation of current update step on previous moments is performed at a high frequency. Despite our intuition is to reduce the dependence on previous gradients, directly abandoning the historical knowledge is inadvisable. A too small cycle length T leads to fluctuated and unstable training process, thus should be avoided. Another is the scale factor α. Since the ratio of network layer's weight and gradient is usually large, we need the scale factor α to rescale the actual learning rate for each layer to a proper magnitude. Otherwise, the network will diverge.

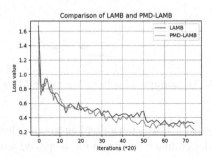

Fig. 5. The loss curves of ResNet50 FPN trained with LAMB and PMD-LAMB, respectively

3.3   LargeDet Framework and Guidelines

With the core of PMD-LAMB, we propose LargeDet for large batch training by combining LSR and SyncBN. In object detection, it is a common practice to adopt frozen BN which inherits the batch statistics from ImageNet pre-trained models and doesn't update the parameters during training. Frozen BN forces the network's convolutional parameters to adapt to the batch statistics inherited from ImageNet. Such adaption usually requires a long training procedure, thus is harmful for fast convergence. Out of this reason, we use SyncBN to allow the network to collect its own data statistics during training. With the aid of LargeDet, we successfully scale the batch size from 256 to 1056. We summary our findings into the following three guidelines.

G1) LSR works well at a batch size smaller than 256, but is not suitable for a larger batch size. An intrinsic problem of LSR lies in that the network will diverge under a large learning rate. As LSR suggests, a large learning rate is essential to keep accuracy when a large batch size is applied. While the divergence of network is inevitable when the learning rate is higher than certain threshold. By carefully tunning the warm-up hyper-parameters, one can reduce the risk of divergence and make it work at a larger batch size. However, as the learning rate continues to grow, the network diverges no matter how to change the configurations of warm-up.

G2) SyncBN is beneficial for large batch optimization. As analyzed before, frozen BN requires a long training procedure to force the network's convolutional parameters to adapt to the batch statistics inherited from ImageNet. For a larger batch size, the training iterations is reduced proportionally, which gives the network less time to perform this adaption. Insufficient adaption leads to inferior performance. SyncBN enables the network to collect accurate batch statistics from the target dataset. Thus it is beneficial to achieve satisfactory performance in less iterations.

G3) PMD-LAMB guarantees the convergence of detectors under large learning rate and paves the way for large batch optimization. PMD-LAMB dynamically adjusts each layer's learning rate based on the ratio of the weights' norm and the gradients' norm. Such layer-wise learning rate successfully addresses the divergence issue of LSR when large batch size is adopted. In addition to this, PMD-LAMB reduces the strong dependence on previous gradients by cyclically decaying the weight of accumulated first and second moments to reduce the negative effect of historical gradients. PMD-LAMB helps the network generalize well and achieves better performance compared to LAMB. In general, these two features make it the key component of our framework.

In summary, LargeDet addresses the divergence issue of LSR and makes it feasible to train detectors with a larger batch size. Consequently, it endows us the ability to fully utilize the increasing computational power to reduce the training time and provides a practical way to tackle with large-scale datasets.

4   Experiments

In this section, we conduct comprehensive experiments on MS COCO [14] to validate the effectiveness of our proposed framework. With the aid of LargeDet, we are able to shrink the training time of ResNet50 FPN detector on COCO dataset to 12 min and finish the training of Mask R-CNN with ResNet50 backbone in 17 min with 160 NVIDIA Tesla V100 GPUs. Besides, we perform experiments on large-scale dataset Open Images [12] as well. The results show that LargeDet is well suited for such large-scale datasets and it outperforms traditional small batch training both in training speed and accuracy.

All our experiments are conducted on the NVIDIA DGX2 server. In terms of software, we use the NVIDIA-optimized maskrcnn-benchmark[1] [16] to train the detectors.

4.1   Experiments on COCO

We first perform experiments on the challenging MS COCO2017 benchmark. We use the ~118K training images to train the detector and the final performance is reported on the 5000 validation images. For evaluation, we use standard COCO metric mAP@0.5:0.95, which averages mAP over IoUs from 0.5 to 0.95. We use the representative ResNet50 pre-trained on the ImageNet as the backbone and adopt the Feature Pyramid Network (FPN) as the detection framework.

For batch size 16, we use a base learning rate 0.02, with the total training iteration of 90000. We multiply the learning rate by factor 0.1 at iteration 60000,

---

[1] https://github.com/mlperf/training_results_v0.6/tree/master/NVIDIA/
benchmarks/maskrcnn/implementations/pytorch.

and again at 80000. For larger batch size, the global learning rate is set according to the LSR and the training iterations is reduced proportionally to keep the total training epochs constant. For SGD, the hyper-parameters, like weight decay and momentum, follow the default settings in [16]. For LAMB or PMD-LAMB algorithm, we adopt the weight decay 0.0005, β_1 0.9, β_2 0.999 by default.

Large Batch Size Training with LSR. We first explore the applicable range of LSR. Specifically, we conduct experiments using different batch sizes with and without SyncBN. When SyncBN is applied, we set the BN size to 32 as suggested in [17]. The results are summarized in Table 1.

Table 1. The results of LSR with different batch sizes. Half means we lower the learning rate to a half

Batch size	w.o. SyncBN	w. SyncBN
16	36.7	36.6
32	36.6	36.4
64	36.6	36.6
128	36.5	36.5
256	Failed	36.1
256 (half)	35.6	35.0
512	Failed	Failed
512 (half)	Failed	34.3

For experiments without BN, we observe that when the batch size is smaller than 128, the detector can achieve almost the same accuracy as small batch size (16) training. When the batch size becomes larger, the network fails to converge. SyncBN helps it converge at batch size 256, but still fails at batch size 512. Even if we can use a lower learning rate to make the network converge, it yields severe performance drops. Although we follow the settings as [17], the results we obtained is slightly different. We conjecture that the difference may be caused by different implementation details of SyncBN. It is worth noticing that when the batch size is larger than 64, the warm-up hyper-parameters need to be carefully tuned to avoid divergence whether there is SyncBN or not. Above experiments demonstrate that LSR can not handle a larger batch size. Thus it is of great significance to explore larger training batch size.

Larger Batch Size Training with LargeDet. As mentioned above, when the batch size is not that large (below 256), the LSR still holds with the help of SyncBN. However, it fails when the batch size becomes larger. To tackle this issue, we conduct experiments using LargeDet with different batch sizes. Since the performance with large batch size is our concern, we start our experiments from batch size 128.

Table 2. Comparisons of training with different batch sizes and methods

Batch size	GPUs	Method	mAP
16	8	SGD	36.7
128	64	LAMB	36.0
		PMD-LAMB	36.7
256	64	LAMB	36.2
		PMD-LAMB	36.7
512	64	LAMB	35.5
		PMD-LAMB	36.5
		PMD-LAMB (1.4x)	36.9
640	64	LAMB	35.6
		PMD-LAMB	36.2
		PMD-LAMB (1.4x)	36.7
1056	96	LAMB	34.8
		PMD-LAMB	35.3
		PMD-LAMB (1.4x)	36.1

We summary the main results in Table 2. The result of batch size 16 is the baseline. From the experimental results, we have the following observations. First, our proposed PMD-LAMB algorithm effectively solves the divergence issue when large batch size is applied and successfully expands the applicable range of LSR. As showed in Table 2, we successfully scale the batch size up to 1056. Besides, our proposed PMD-LAMB algorithm outperforms LAMB on different batch sizes. For batch size 128 and 256, the detector can recover the baseline performance by using PMD-LAMB without extending the training iterations. For larger batch size, PMD-LAMB can still provide consistent performance gain, 1% improvement at batch size 512 compared to LAMB, which shows the proposed PMD-LAMB is an effective method to further improve the performance.

Second, by slightly extending the training time, the performance can be further improved. For a relatively large batch size, it is hard to recover the performance in the same training epochs as the small batch size training baseline. For batch size 512 and 640, it still suffers little accuracy drop. And for batch size 1056, the accuracy drop is more, 1.4% lower than the baseline. We argue that this is mainly due to the inadequate statistics of BN with less iterations. By slightly increasing the training iterations, 1.4 times in our experiments, the performance can be further improved. For batch size 512, 1.4x training iterations yield 36.9 mAP, even higher than the small batch size training accuracy. In summary, our proposed framework is an effective approach to train an object detector when using a large batch size.

Table 3. The effect of hyper-parameters in PMD-LAMB

α	0.012					0.008	0.01	0.012	0.015	0.018
T	10	200	500	1000	∞	200				
mAP	35.6	36.7	36.7	36.6	36.2	36.4	36.5	36.7	36.2	35.2

The Effect of Hyper-parameters. In PMD-LAMB, there are two hyper-parameters introduced. One is the scale factor α, the other is the cycle length T. We perform ablation experiments to explore how they affect the performance. In the ablation study, we use LargeDet to train the detector with batch size 256. The results are summarized in Table 3. We can see that our method works well in a quite wide range for T. But a very small value will lead to obvious accuracy loss. For scale factor α, it should be set to a proper value. A small α leads to insufficient training and results in inferior performance. While a large α causes a fluctuated training process, which deteriorates the final performance as well. From the results in Table 3, we can know that setting α to 0.012, T to 200 is a good choice.

4.2 Training COCO in 12 min

With LargeDet, we are able to challenge the limit of fast training of COCO. We choose Faster R-CNN with FPN and Mask R-CNN detectors and explore how fast can we train a relatively strong detector. We adopt the ResNet50 as the backbone. For Faster R-CNN, when the detection mAP@0.5:0.95 reaches 36.6, we stop the training. For Mask R-CNN, the threshold is 37.6 detection mAP@0.5:0.95 and 33.9 segmentation mAP@0.5:0.95. We conduct experiments on 10 NVIDIA DGX-2 servers, 160 GPUs in total.

Table 4. The results of fast training experiments on COCO. Faster means Faster R-CNN with FPN. Mask represents the Mask R-CNN detector. Steps indicates when we lower the learning rate by factor 0.1

	Steps	Stop iter	Stop mAP	Time
Faster	2400,3200	3360	box:36.65	11 min 53 s
Mask	3000,4000	4120	box:37.61	16 min 55 s
			seg:34.98	

The total batch size is set to 320 for both Faster R-CNN and Mask R-CNN detectors. We set the global learning rate to 0.4 according to the LSR. Table 4 shows the results of fast training experiments. We are able to finish the training of Faster R-CNN in 12 min, and Mask R-CNN in 17 min, which will substantially accelerate the experimental cycle and facilitate the exploration of more effective strategies.

4.3 Experiments on Open Images

Open Images V5 is the largest existing dataset with object location annotations which contains a total of 16M bounding boxes for 600 object classes on 1.9M images. It is split into three subsets, train, validation and test subset. We use the train subset which contains 1.7M images to train and report the accuracy on validation subset. For evaluation, we use the official evaluation code for Open Images which evaluates only at IoU threshold 0.5.

In this section, we conduct experiments on Open Images V5 with batch size 16, 320 and 640, and compare the training time and accuracy. We adopt the FPN detector with ResNet50 and ResNet101 backbone. For experiment with batch size 16, the learning rate is set to 0.02 and other settings follows the default configurations for COCO. For large batch size 320 and 640, the learning rate is set to 0.4 and 0.8, respectively. We train the detector for 12 epochs, and multiply the learning rate by 0.1 and 0.01 at epoch 8 and 11, respectively. To make a fair comparison, we do not modify other settings such as anchor ratio and anchor number, *etc.* And we adopt multi-scale training for all experiments.

Table 5. Experiments on Open Images V5 with different batch sizes and backbones

Backbone	BatchSize	GPUs	mAP@0.5	Time
Res50	16	8	42.9	87 h
	320	32	53.1	23 h
	640	64	*55.1*	12 h
Res101	16	8	45.5	108 h
	320	32	54.0	27 h
	640	64	*56.3*	14 h

The results are summarized in Table 5. We achieve a nearly linear acceleration by utilizing more GPUs. With LargeDet, we reduce the training time of ResNet50 from 87 h to 12 h, approximately 8× acceleration. The same acceleration can be observed for ResNet101. Except for reducing the training time, we also obtain a considerable performance improvement when training with a larger batch size. For ResNet50 backbone, by using a batch size of 640, we achieve 12.2% mAP@0.5 absolute improvement, compared to the baseline trained with batch size 16. For ResNet101, we can achieve 10.8% improvement as well. We also observe that the improvements brought by LargeDet increase as the batch size grows.

We conjecture the performance gain may come from the following two aspects. On the one hand, large batch size increases the category richness in a mini-batch. Open Images V5 has object location annotations for 600 classes. According to official statistics, there are 8.4 boxes objects per image on average. If the batch size is 16, there are 135 objects in a mini-batch on average. Even if all these boxes are from different categories, the number of the total categories in a mini-batch is merely 135, which is far lower than the total category 600.

Positive Labels: Cat, Dog   Positive Labels: Bicycle, Vehicle, Bicycle helmet

Missing Labels: Person, Human face   Missing Labels: Person, Human face

Fig. 6. Example annotations in Open Images V5. Green boxes indicate the annotated ground truth. Red boxes are some missing instances which are not annotated in Open Images. Positive labels indicate some object classes are present in current image. Missing labels are those classes which are present in the image but no corresponding instances are annotated (Color figure online)

By increasing the batch size, the category richness can be greatly improved and the detector can achieve better performance.

On the other hand, large batch optimization is an effective strategy to fight against the label noise which is widespread in large-scale datasets. For instance, missing annotations frequently occur in Open Images, as illustrated in Fig. 6. During training, such missing instances are prone to be mistaken for negative samples, which will cause incorrect training signals and confuse the detector. By increasing batch size, these wrong signals can be suppressed to a great extent by numerous correctly annotated instances.

5   Conclusions

We present a large batch optimization framework for object detection called LargeDet, with which we successfully scale the batch size to 1056 with a ResNet50 backbone. Besides, LargeDet supports us to finish the training of ResNet50 FPN detector on COCO in 12 min. Moreover, large batch training provides a novel means to tackle the widespread label noise in large scale datasets. By training ResNet50 FPN with a batch size of 640, we achieve a remarkable 12.2% absolute improvement in terms of mAP@0.5 on Open Images V5. And the improvement for ResNet101 FPN is 10.8%. We hope that our research can serve for the future exploration of the large-scale datasets.

Acknowledgement. This work was supported by the Research and Development Projects in the Key Areas of Guangdong Province (No. 2019B010153001) and National Natural Science Foundation of China under Grants No. 61772527, No. 61976210, No. 61806200, No. 61702510, No. 61876086 and No. 61633002.

References

1. Deep neural networks for acoustic modeling in speech recognition: the shared views of four research groups. IEEE Signal Process. Mag. **29**(6), 82–97 (2012). https://doi.org/10.1109/MSP.2012.2205597

2. Dai, J., Li, Y., He, K., Sun, J.: R-FCN: object detection via region-based fully convolutional networks. In: Advances in Neural Information Processing Systems 29: Annual Conference on Neural Information Processing Systems 2016, 5–10 December 2016, Barcelona, Spain, pp. 379–387 (2016). http://papers.nips.cc/paper/6465-r-fcn-object-detection-via-region-based-fully-convolutional-networks

3. Dai, J., et al.: Deformable convolutional networks. In: IEEE International Conference on Computer Vision, ICCV 2017, Venice, Italy, 22–29 October 2017, pp. 764–773 (2017). https://doi.org/10.1109/ICCV.2017.89

4. Devlin, J., Chang, M., Lee, K., Toutanova, K.: BERT: pre-training of deep bidirectional transformers for language understanding. In: Proceedings of the 2019 Conference of the North American Chapter of the Association for Computational Linguistics: Human Language Technologies, NAACL-HLT 2019, Minneapolis, MN, USA, 2–7 June 2019, Volume 1 (Long and Short Papers), pp. 4171–4186 (2019). https://www.aclweb.org/anthology/N19-1423/

5. Everingham, M., Gool, L.V., Williams, C.K.I., Winn, J.M., Zisserman, A.: The pascal visual object classes (VOC) challenge. Int. J. Comput. Vis. **88**(2), 303–338 (2010). https://doi.org/10.1007/s11263-009-0275-4

6. Goyal, P., et al.: Accurate, large minibatch SGD: training ImageNet in 1 hour. CoRR abs/1706.02677 (2017). http://arxiv.org/abs/1706.02677

7. He, K., Gkioxari, G., Dollár, P., Girshick, R.B.: Mask R-CNN. In: IEEE International Conference on Computer Vision, ICCV 2017, Venice, Italy, 22–29 October 2017, pp. 2980–2988 (2017). https://doi.org/10.1109/ICCV.2017.322

8. Keskar, N.S., Mudigere, D., Nocedal, J., Smelyanskiy, M., Tang, P.T.P.: On large-batch training for deep learning: generalization gap and sharp minima. In: 5th International Conference on Learning Representations, ICLR 2017, Toulon, France, 24–26 April 2017, Conference Track Proceedings (2017). https://openreview.net/forum?id=H1oyRlYgg

9. Kingma, D.P., Ba, J.: Adam: a method for stochastic optimization. In: 3rd International Conference on Learning Representations, ICLR 2015, San Diego, CA, USA, 7–9 May 2015, Conference Track Proceedings (2015). http://arxiv.org/abs/1412.6980

10. Krizhevsky, A.: One weird trick for parallelizing convolutional neural networks. CoRR abs/1404.5997 (2014). http://arxiv.org/abs/1404.5997

11. Krizhevsky, A., Sutskever, I., Hinton, G.E.: ImageNet classification with deep convolutional neural networks. In: Advances in Neural Information Processing Systems 25: 26th Annual Conference on Neural Information Processing Systems 2012. Proceedings of a Meeting Held 3–6 December 2012, Lake Tahoe, Nevada, United States, pp. 1106–1114 (2012). http://papers.nips.cc/paper/4824-imagenet-classification-with-deep-convolutional-neural-networks

12. Kuznetsova, A., et al.: The open images dataset V4: unified image classification, object detection, and visual relationship detection at scale. CoRR abs/1811.00982 (2018). http://arxiv.org/abs/1811.00982

13. Lin, T., Goyal, P., Girshick, R.B., He, K., Dollár, P.: Focal loss for dense object detection. In: IEEE International Conference on Computer Vision, ICCV 2017, Venice, Italy, 22–29 October 2017, pp. 2999–3007 (2017). https://doi.org/10.1109/ICCV.2017.324

14. Lin, T.-Y., et al.: Microsoft COCO: common objects in context. In: Fleet, D., Pajdla, T., Schiele, B., Tuytelaars, T. (eds.) ECCV 2014. LNCS, vol. 8693, pp. 740–755. Springer, Cham (2014). https://doi.org/10.1007/978-3-319-10602-1_48

15. Liu, W., et al.: SSD: single shot multibox detector. In: Leibe, B., Matas, J., Sebe, N., Welling, M. (eds.) ECCV 2016. LNCS, vol. 9905, pp. 21–37. Springer, Cham (2016). https://doi.org/10.1007/978-3-319-46448-0_2

16. Massa, F., Girshick, R.: maskRCNN-benchmark: fast, modular reference implementation of instance segmentation and object detection algorithms in PyTorch (2018). https://github.com/facebookresearch/maskrcnn-benchmark

17. Peng, C., et al.: MegDet: a large mini-batch object detector. In: 2018 IEEE Conference on Computer Vision and Pattern Recognition, CVPR 2018, Salt Lake City, UT, USA, 18–22 June 2018, pp. 6181–6189 (2018). https://doi.org/10.1109/CVPR.2018.00647. http://openaccess.thecvf.com/content_cvpr_2018/html/Peng_MegDet_A_Large_CVPR_2018_paper.html

18. Redmon, J., Divvala, S.K., Girshick, R.B., Farhadi, A.: You only look once: unified, real-time object detection. In: 2016 IEEE Conference on Computer Vision and Pattern Recognition, CVPR 2016, Las Vegas, NV, USA, 27–30 June 2016, pp. 779–788 (2016). https://doi.org/10.1109/CVPR.2016.91, https://doi.org/10.1109/CVPR.2016.91

19. Ren, S., He, K., Girshick, R.B., Sun, J.: Faster R-CNN: towards real-time object detection with region proposal networks. IEEE Trans. Pattern Anal. Mach. Intell. **39**(6), 1137–1149 (2017). https://doi.org/10.1109/TPAMI.2016.2577031

20. Shao, S., Li, Z., Zhang, T., Peng, C., Yu, G., Zhang, X., Li, J., Sun, J.: Objects365: a large-scale, high-quality dataset for object detection. In: The IEEE International Conference on Computer Vision (ICCV), October 2019

21. Sutskever, I., Vinyals, O., Le, Q.V.: Sequence to sequence learning with neural networks. In: Advances in Neural Information Processing Systems 27: Annual Conference on Neural Information Processing Systems 2014, 8–13 December 2014, Montreal, Quebec, Canada, pp. 3104–3112 (2014). http://papers.nips.cc/paper/5346-sequence-to-sequence-learning-with-neural-networks

22. Tian, Z., Shen, C., Chen, H., He, T.: FCOS: fully convolutional one-stage object detection. In: The IEEE International Conference on Computer Vision (ICCV), October 2019

23. You, Y., Gitman, I., Ginsburg, B.: Scaling SGD batch size to 32k for ImageNet training. CoRR abs/1708.03888 (2017). http://arxiv.org/abs/1708.03888

24. You, Y., Li, J., Hseu, J., Song, X., Demmel, J., Hsieh, C.: Reducing BERT pretraining time from 3 days to 76 minutes. CoRR abs/1904.00962 (2019). http://arxiv.org/abs/1904.00962

25. Zhou, X., Zhuo, J., Krahenbuhl, P.: Bottom-up object detection by grouping extreme and center points. In: The IEEE Conference on Computer Vision and Pattern Recognition (CVPR), June 2019

Towards Practical and Efficient High-Resolution HDR Deghosting with CNN

K. Ram Prabhakar[✉], Susmit Agrawal, Durgesh Kumar Singh,
Balraj Ashwath, and R. Venkatesh Babu

Video Analytics Lab, Indian Institute of Science, Bangalore, India
kramprabhakar@gmail.com

Abstract. Generating High Dynamic Range (HDR) image in the presence of camera and object motion is a tedious task. If uncorrected, these motions will manifest as ghosting artifacts in the fused HDR image. On one end of the spectrum, there exist methods that generate high-quality results that are computationally demanding and too slow. On the other end, there are few faster methods that produce unsatisfactory results. With ever increasing sensor/display resolution, currently we are very much in need of faster methods that produce high-quality images. In this paper, we present a deep neural network based approach to generate high-quality ghost-free HDR for high-resolution images. Our proposed method is fast and fuses a sequence of three high-resolution images (16-megapixel resolution) in about 10 s. Through experiments and ablations, on different publicly available datasets, we show that the proposed method achieves state-of-the-art performance in terms of accuracy and speed.

1   Introduction

Natural scenes have a wide range of illumination that exceeds the dynamic range of standard digital camera sensors. The resulting image has undesirable saturated regions (too bright or too dark) while capturing an HDR scene. Widely followed software solution is to merge multiple low dynamic range (LDR) images into a single HDR image. Each LDR image in the input stack captures part of the brightness spectrum by varying exposure time (or aperture, ISO). Then, the HDR image is generated by combining the best regions of each LDR image. The generated HDR has a wider dynamic range than each LDR input image.

HDR fusion is a simple and straightforward process for static scenes: scenes without any camera or object motion [27]. However, in the presence of camera or object motion, such naive static fusion techniques result in artifacts. While the camera motion can be corrected using homography-based alignment procedures, the harder challenge is to address the object motion. If uncorrected, the

Electronic supplementary material The online version of this chapter (https:// doi.org/10.1007/978-3-030-58589-1_30) contains supplementary material, which is available to authorized users.

© Springer Nature Switzerland AG 2020
A. Vedaldi et al. (Eds.): ECCV 2020, LNCS 12366, pp. 497–513, 2020.
https://doi.org/10.1007/978-3-030-58589-1_30

Fig. 1. The input exposure sequence with motion is shown on the leftmost column. The result by our proposed method is shown in the second column. The zoomed regions of different methods are highlighted in (a) to (g). (a) Wu18 [40], (b) Kalantari17 [15], (c) SCHDR [25], (d) AHDR [42], (e) guide image generated by N_g, (f) ground truth and (g) proposed method. Best viewed in color monitor.

moving objects from all the input images appear mildly in the final result, resulting in a ghost-like perception, hence known as *ghosting artifact*. Several methods have been proposed in the literature to generate results without ghosting artifacts, a.k.a., HDR deghosting methods. The initially proposed rejection-based methods [4,5,8,9,11,23,24,26,28,41,45] are fast and easy to implement. Despite their better performance for mostly static scenes, they suffer from having only LDR content for moving objects. Alignment-based methods register input images to a selected reference image using rigid [36,39] or non-rigid [6,12,16,46] registration techniques. While the rigid registration techniques find complex object motion difficult to handle, the non-rigid techniques (such as optical flow) are inaccurate for deformable motions and occluded pixels. On the other hand, patch-based optimization methods [13,34] synthesize a static sequence from the dynamic input sequence. Despite their high-quality results, they suffer from huge computational complexity, making them not suitable for portable devices with limited computational resources.

Recently proposed deep learning-based HDR deghosting methods [15,40,42–44] generate visually pleasing results for the majority of the contents. However, they still suffer from artifacts in heavily saturated regions (see Fig. 1). Another limitation of the existing CNN-based methods is the need for more computational resources to process high-resolution images (more than 5 megapixels). For example, Yan *et al.*'s [43] approach can process a maximum of 2 megapixels (MP) on a GPU with 11 GB memory (RTX 2080Ti); their method requires GPU with more memory to process high-resolution images. Similarly, [40] can only process images up to 5 MP on a GPU with 11 GB memory.

To address the issues mentioned above, we propose a robust CNN based HDR deghosting method that can generate artifact-free results and can process up to 16 MP images within 11 s on a GPU with 11 GB memory. Unlike previous CNN based methods [15,40,42,43], in our approach, we avoid processing full resolution images with CNN. Alternatively, we process low-resolution images and upscale the result to the original full resolution. However, a simple bilinear or bicubic upsampling may introduce blur artifacts. Hence, we make use of bilateral guided upsampling [2, 7] to generate full resolution artifact-free output. The main contributions of our work is summarized as follows,

- We present a CNN based HDR deghosting algorithm that is robust to significant object motion and saturation.
- We propose an efficient HDR deghosting approach to process high-resolution images with Bilateral Guided Upsampler (BGU). We also demonstrate the use of BGU for a fusion task where the guide image is not readily available.
- We provide motion segmentation masks for UCSD dataset [15] to benefit HDR deghosting research community.
- We perform an extensive quantitative and qualitative evaluation on publicly available datasets. We also perform various ablation experiments on different choices in our model.

2   Related Works

In general, we can classify most of the deghosting algorithms into four categories: alignment-based, rejection-based, patch-based, and learning-based methods[1].

Alignment-Based Methods: The first class of algorithms register all input images to a chosen reference image using rigid or non-rigid registration techniques. Rigid registration methods handle global camera motion by matching feature descriptors such as SIFT [36], SURF [8] and Median Threshold Bitmap [39]. In [36], Tomaszewska and Mantuik perform RANSAC after SIFT to refine the matches. Non-rigid registration methods like Bogoni et al. [1], Kang et al. [16], Gallo et al. [6] and Zimmer et al. [46] make use of optical flow to align images. The non-rigid registration based methods have the advantage of producing results with moving HDR content. However, as some of the features used for rigid and non-rigid registration are not robust against huge brightness changes, they are more prone to fail for saturated regions or occluded regions.

Rejection-Based Methods: The second class of algorithms assume the input images to be fairly static and detects the pixels affected by motion [4, 18, 29], after which they combine only static images in those regions. To identify the moving objects in the registered images, many techniques have been proposed that make use of illumination constancy criteria, linear relationship between images [17, 35, 41], prediction and thresholding [9, 23, 29], thresholding background probability map [18], etc. Gallo et al. [5] perform patch wise comparison in logarithmic domain to identify moving regions. Raman and Chaudhuri [28] perform a comparison in super-pixels to improve motion segmentation accuracy along edges. Although this class of algorithms is fast, a major drawback is that they generate low dynamic range content in the moving object regions.

---

[1] Elaborate literature review can be found in [32, 33, 37].

Patch-Based Optimization Methods: The third class of algorithms handles both camera and object motion jointly by performing patch-based registration between the varying exposure images [13,21,34]. They pick one of the input images as the structural reference and find dense correspondences between the reference image and all other input source images in the stack. Using these correspondences, they synthesize modified source images that are structurally similar to the original reference image but resemble the corresponding source images in terms of the brightness. The synthesized images are fused using static fusion techniques like Debevec and Malik [3]. Despite the impressive performance, the patch-based methods [13,34] suffer from high computation time.

Deep Learning Methods: The fourth class of algorithms uses deep learning methods to fuse input images into an HDR image. In [15], a simple CNN network is trained to correct artifacts introduced by the optical flow method. Though [15] generates visually pleasing results for some images, it fails to correct warping errors in saturated regions, especially on moving regions. Wu et al. [40] treat HDR deghosting as an image translation problem. In their method, authors have shown that it is possible to fuse images without performing optical flow. In [43], Yan et al. use a multiscale CNN to extract features at different scales and fuse them. More recently, Yan et al. proposed an attention-guided method in [42]. In this approach, the authors identify regions with motion and exclude them during the fusion process. Each of the above-discussed methods addresses an important issue. However, existing CNN based methods require heavy computational resources to process high resolution images.

3   Proposed Method

Motivation for Using BGU: Existing CNN based HDR deghosting methods require a substantial amount of computational resources to process high-resolution images. The bottleneck is to process the images in their full resolution with CNNs. Hence these methods are not easily scalable to ever increasing image resolution. To circumvent this issue, we propose a method that performs all operations in low resolution and upscales the result to the required full resolution. For efficient and artifact-free upscaling, we utilize deep Bilateral Guided Upsampling (BGU) used by Gharbi et al. [7]. For upscaling, the BGU algorithm requires a guide image for structural reference (see Fig. 2). Such a guide image should satisfy the following two criteria: (1) must be in full high-resolution as the expected final HDR result, (2) should have similar structural details as the expected final HDR result. In other words, the guide image should be a high-resolution fused image without any ghosting artifacts. However, predicting the high-resolution guide image through a CNN will have the same challenges as discussed before.

   To address this challenge, we propose to generate a guide image through weight-map based strategy. The core idea is to predict weight maps at low resolution for downsampled inputs and upscale them using bicubic interpolation. The upscaled weight maps can be used to combine original high-resolution input images and generate guide image. As the upscaling is performed on the weight

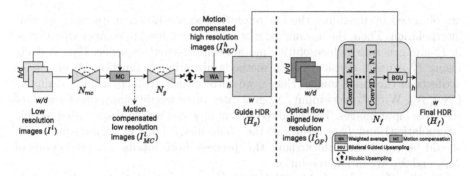

Fig. 2. Our method comprises of two components: guide and final HDR prediction. The input images in full-resolution (I^h) is downsampled by a factor of d to generate I^l. The optical flow aligned low-resolution images, I^l_{OF}, are fused by N_f model to generate final HDR result (H_f) in full-resolution using Bilateral Guided Upsampler (BGU). For efficient artifact-free upsampling, BGU requires a guide image (H_g) for structual guidance. H_g is generated with N_{mc} and N_g with I^l as input. For more details, see Sect. 3.

maps instead of the image domain, the predicted guide image will be void of any blurring artifacts. In this step, CNN processes only the low-resolution images, but the generated guide image is in full high-resolution. Finally, the generated guide image is used by BGU to produce the deghosted HDR result.

Pipeline: The objective of our HDR deghosting method is to fuse input varying exposure LDR input images ($\{I_1, \cdots, I_N\}$) into an HDR image (H_f) without any ghosting artifacts. Our approach is a reference-based HDR deghosting method, meaning, the generated HDR image will have identical structural details as that of the chosen reference image. Similar to [15,40,42,43], our method takes three different exposure images (I_-, I_0, I_+) as input with the middle image (I_0) as the reference, where (I_-, I_0, I_+) denotes the underexposed, normally exposed, and overexposed images.

We denote the full high-resolution input as $I^h = \{I^h_-, I^h_0, \text{and } I^h_+\}$ and the corresponding downsampled (by a factor d) images as $I^l = \{I^l_-, I^l_0, \text{and } I^l_+\}$ (see Fig. 2)[2]. As I^l may have camera and object motion, we align them using optical flow method by Liu et al. [20]. We align the non-reference images (I^l_- and I^l_+) to reference image (I^l_0) after exposure correction. The generated optical flow aligned sequence (I^l_{OF}) is fed to fusion (N_f) network, to generate a fused feature map at low resolution. Then, the fused feature map is passed to BGU to generate H_f in high-resolution. However, BGU requires a guide image for structural reference.

The guide image is generated in two steps: motion compensation followed by weight map based fusion. In the first step, I^l is passed to a CNN network (N_{mc}) to segment moving regions between reference and other non-reference images. The predicted segmentation maps are used to generate motion compensated sequence, I^l_{MC} in low-resolution. The corresponding high-resolution motion maps

---

[2] Throughout the rest of the paper, the full high-resolution images are denoted by superscript h, i.e., $\square^h$ and the downsampled low-resolution images by $\square^l$.

are obtained by upscaling the low-resolution segmentation maps using bicubic interpolation. Then, the upsampled motion maps are used to compensate motion in I^h and generate high-resolution motion compensated sequence, I^h_{MC}. Naively fusing I^h_{MC} using classical fusion methods like triangle function [3] may have artifacts in the guide image. Hence, we correct such artifacts using a trainable CNN, N_g. With I^l_{MC} as input, N_g generates three weight maps, one for each of the three input images. The weighted sum of upscaled weight maps (using bicubic interpolation) and I^h_{MC} results in the guide image (H_g) at high-resolution. It should be noted that, throughout this process, both inputs and predictions of N_{mc} and N_g are in low-resolution.

Finally, H_f is generated by BGU with H_g as structural guidance. We provide elaborate details of each step in further sections.

Guide Image Generation: The input images (I^l) to N_{mc} are assumed to be aligned for camera motion. If not, the low resolution (I^l) images can be aligned with simple homography based methods. The aligned images are passed to a CNN model, N_{mc}, to segment moving regions between reference image (I^l_0) and non-reference images (I^l_-, I^l_+) individually. To do so, we concatenate reference and a non-reference image, I^l_- (or I^l_+) together to form a six channel data and feed as input to N_{mc}. We use the U-net architecture [30] for N_{mc} model[3]. The model architecture consists of four encoders followed by four decoders with skip connections from encoders. The output of N_{mc} is a single channel binary segmentation map, M^l_- (M^l_+ for I^l_+ input) with same resolution as I^l. Each pixel in M^l_- has value between 0 and 1. Value '0' indicates the absence of motion at that pixel in either reference or the non-reference image, and '1' indicating the presence of motion in any one of them. Using the predicted motion segmentation maps, we perform motion compensation in the input sequence (MC box shown in Fig. 2). The motion compensation is achieved by replacing the moving regions in non-reference image with regions from reference image at the same location.

$$I^l_{mc,-} = M^l_- \times \mathcal{E}(I^l_0, t_0, t_-) + (1 - M^l_-) \times I^l_- \tag{1}$$

$$I^l_{mc,+} = M^l_+ \times \mathcal{E}(I^l_0, t_0, t_+) + (1 - M^l_+) \times I^l_+ \tag{2}$$

In the above equation, $\mathcal{E}()$ denotes exposure normalization function. For example, $\mathcal{E}()$ modifies the exposure of image A with exposure time t_A to exposure of image B with exposure time t_B by:

$$\mathcal{E}(A, t_A, t_B) = \left(A^\gamma \times \frac{t_B}{t_A} \right)^{\frac{1}{\gamma}} \tag{3}$$

where, $\gamma = 2.2$. At the end of this process, we generate motion compensated sequence $I^l_{MC} = \{I^l_{mc,-}, I^l_0, I^l_{mc,+}\}$. To generate motion compensated sequence in high-resolution, the predicted segmentation maps, M^l_- and M^l_+, are upscaled using bicubic interpolation to generate M^h_- and M^h_+. Then, we repeat the steps in Eqs. 1 and 2, but with high resolution motion maps and high resolution input images to create $I^h_{MC} = \{I^h_{mc,-}, I^h_0, I^h_{mc,+}\}$.

---

[3] More detailed model architecture is provided in supplementary material.

The motion compensated low resolution sequence I^l_{MC} is fed as input to N_g network to generate fusion weight maps. The input to N_g is obtained by concatenating $\{I^l_{mc,-}, I^l_0, I^l_{mc,+}\}$ images in channel dimension,

$$(w^l_-, w^l_0, w^l_+) = N_g(I^l_{mc,-}, I^l_0, I^l_{mc,+}) \tag{4}$$

The output of N_g consists of three weight maps: w^l_-, w^l_0, and w^l_+. Similar to N_{mc} model, we use U-net architecture for N_g network as well. Before using the weight maps for fusion, they are upscaled using bicubic interpolation to the required high resolution: w^h_-, w^h_0, and w^h_+. Then, the upscaled and normalized weight maps are used to combine motion corrected high-resolution images I^h_{MC} as,

$$H_g = w^h_- \times LH(I^h_{mc,-}) + w^h_0 \times LH(I^h_0) + w^h_+ \times LH(I^h_{mc,+}) \tag{5}$$

where, $LH()$ denotes the LDR to HDR conversion process. For a sample LDR image A, the corresponding HDR version is obtained by,

$$LH(A) = \frac{A^\gamma}{t_A} \tag{6}$$

N_f **Network:** The input images (I^l) are aligned using optical flow to compensate for camera and object motion. Instead of applying the optical flow on high resolution images, we apply it on downsampled low-resolution images, I^l. We use optical flow model proposed by Liu et al. [20] to align the non-reference images to the reference image. The optical flow aligned images, $I^l_{OF} = (I^l_{OF,-}, I^l_0, I^l_{OF,+})$, are passed as input data to N_f model to predict final HDR (H_f) result,

$$H_f = N_f(I^l_{OF}) \tag{7}$$

As shown in Fig. 3, three input images are passed to a small encoder with three convolution layers to extract individual image features. The extracted features of reference and a non-reference image $(I^l_{OF,-}$ or $I^l_{OF,+})$ is concatenated together and further processed by another convolutional layer. By doing so, the network has the ability to compare both feature maps and choose the properly exposed details from both. This operation is repeated for all three convolutional levels of the encoder to produce final feature maps for non-reference images. Then, the output of last layer in reference image encoder is concatenated with other two final feature maps of non-reference images. The concatenated feature maps are further processed by three blocks of Extended Stacked and Dilated Convolution (XSDC) blocks.

A higher receptive field is necessary for the network to gather information over a vast image region to faithfully reconstruct occluded and saturated regions. We achieve such a high receptive field by using XSDC blocks. XSDC is an simple extention to SDC block proposed by Schuster et al. [31]. The original SDC block has four dilated convolutions with dilation rate from 1 to 4, stacked parallely. The output of stacked convolutional layers is concatenated to produce output feature maps. We make small extention to SDC by increasing the number of consecutive convolutional layers from one (in SDC) to three and process the concatenated features with a 1×1 convolutional layer to aggregate the feature maps. In our implementation, we use three stacks of three convolutional layers (see Fig. 3) with dilation rate of 1, 2 and 4, and kernel size 3×3 for stacked convolutional layers. Through ablation experiments, we observed that our changes perform better than original SDC (see Table 4) by having better ability to transfer details in saturated regions. In addition, we also have dense connections among each of the three XSDC blocks. The output low-resolution feature map of final XSDC block is passed to the BGU, which takes high-resolution H_g generated from N_g model as the guide image to predict final fused HDR image, H_f in high-resolution.

Loss Functions: The total loss to train the model consists of three sub-losses at output of N_{mc}, N_g and N_f. The predicted segmentation maps (M_-^h and M_+^h) by N_{mc} is compared with ground truth segmentation maps ($\widetilde{M}_-^h$ and $\widetilde{M}_+^h$) using Binary Cross Entropy (BCE). The motion compensation loss ($\mathcal{L}_{mc}$) is computed by taking the sum of two individual BCE losses,

$$\mathcal{L}_{mc} = BCE(M_-^h, \widetilde{M}_-^h) + BCE(M_+^h, \widetilde{M}_+^h) \tag{8}$$

The ℓ_2 loss between guide image (H_g) and ground truth HDR ($\widetilde{H}$), final image (H_f) and $\widetilde{H}$ is used to train N_g and N_f.

$$\mathcal{L}_g = \ell_2(T(H_g), T(\widetilde{H})) \tag{9}$$

$$\mathcal{L}_f = \ell_2(T(H_f), T(\widetilde{H})) \tag{10}$$

where, $T(\square)$ denotes the tonemapping operation applied to HDR images with commonly used μ-law function,

$$T(H) = \frac{log(1 + \mu H)}{log(1 + \mu)} \tag{11}$$

where $\mu = 5000$. The total loss is taken as the weighted sum of three losses: $\mathcal{L}_{mc}$, $\mathcal{L}_g$ and $\mathcal{L}_f$.

$$\mathcal{L}_{total} = \alpha_1 \times \mathcal{L}_{mc} + \alpha_2 \times \mathcal{L}_g + \alpha_3 \times \mathcal{L}_f \tag{12}$$

where α_1, α_2 and α_3 denote the weight values assigned to individual loss functions. We found that assigning equal weight values from beginning takes longer time to stabilize the training and converge to optimum solution. Hence, we assign low value of $1e^{-4}$ for α_2 and α_3 for initial 25 epochs, thus making N_{mc} network learn better segmentation maps that would improve N_g. From 26th epoch till 50, we set α_2 as 1 and for epochs after 75, all three weights are assigned to one.

4   Experiments

4.1   Implementation

Datasets: We trained our model on the dataset provided by Kalantari and Ramamoorthi [15]. The dataset consists of 74 training and 15 testing images of

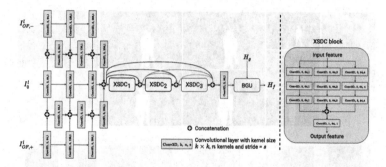

Fig. 3. Illustration of N_f architecture. The input to N_f is optical flow corrected low-resolution sequence $\{I^l_{OF}\}$ and H_g. The output is the final HDR image H_f in full-resolution.

1500 × 1000 resolution, with ground truth for evaluation. Additionally, we also tested our model on datasets provided by Prabhakar *et al.* [25], Sen *et al.* [34] and Tursun *et al.* [38].

Groundtruth for N_{mc}: For training N_{mc} model, we have generated ground truth by annotating the moving objects in each input image. For a sequence with three images: (I^h_-, I^h_0, I^h_+), we generated three segmentation maps, (S^h_-, S^h_0, S^h_+), by manually annotating only the moving regions (see Fig. 4). The ground truth is obtained by taking the union of reference and non-reference image segmentation mask. For example, $\widetilde{M}^h_-$ is generated by performing a union of S^h_- and S^h_0. Similarly, $\widetilde{M}^h_+$ is generated by taking a union of S^h_+ and S^h_0. In total, we have annotated 89 images of UCSD dataset including train and test set.

Data Pre-processing: To train our model, we extracted 50K patches of size 256 × 256 from the 74 training images and applied eight augmentations (rotation, flip, and noise) to increase the training sample size.

Network Details: For both N_{mc} and N_g, we use U-net model with four convolutional layers for encoder. The features are downsampled using the max pool after each block before passing to the next convolution layer. The decoder part consists of the same number of layers as encoders with the upsampling layer to increase the feature resolution. We use Leaky ReLu in all the layers except the final layer of N_{mc} and N_g, where we use sigmoid activation to predict weight maps in [0–1] range. Adam optimizer [19] is used to train our model with a learning rate of $2e^{-4}$ for 200 epochs.

4.2   Quantitative Evaluation

We compare our proposed method against seven state-of-the-art methods: 1. Hu13 [13], 2. Sen12 [34], 3. Kalantari17 [15], 4. Wu18 [40], 5. MSDN-HDR[4] [43],

---

[4] HDR-VDP-2 metric and scores for [25] dataset is not reported for Yan20 [44] and MSDN [43] as the codes are not publicly available. The numbers reported are taken from their paper.

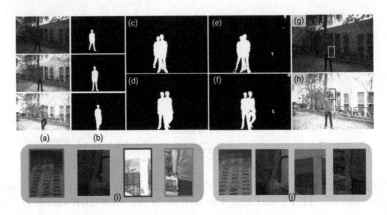

Fig. 4. (a) Input sequence, (b) ground truth motion masks (S_-^h, S_0^h, S_+^h), (c) $\widetilde{M}_-^h$, obtained by combining S_-^h and S_0^h, (d) $\widetilde{M}_+^h$, obtained by combining S_+^h and S_0^h, (e) predicted motion map, M_-^h, for inputs I_-^l and I_0^l, (f) predicted motion map M_+^h for inputs I_+^l and I_0^l, (g) motion corrected image $I_{mc,-}^h$, (h) motion corrected image $I_{mc,+}^h$, (i) zoomed regions in motion corrected images highlighting artifacts, and (j) corresponding regions at guide image (H_g) after refining with N_g.

6. AHDR [42], and 7. SCHDR [25]. In Table 1, we report the quantitative evaluation on two datasets: [15] and [25] using three full-reference image quality assessment metrics: PSNR computed in linear HDR domain (P_L), PSNR computed in tonemapped domain (P_T) after applying tonemapping (Eq. 11), and HDR-VDP-2 [22]. HDR-VDP-2 is a full-reference metric specifically designed for quantifying degradations in HDR images.

From results in Table 1, we observe that our proposed method with BGU (denoted as *Ours (w/ BGU)*) for downsampling factor of 8 performs better than almost all methods in both datasets. Additionally, we present results for another variation of our model without using BGU (*Ours (w/o BGU)*). By setting $d = 1$, means passing the input without downsampling, in full resolution to the network. Since the output of the XSDC block is already in full resolution, using BGU is redundant. Thus, the network is modified to predict three-channel RGB output at the end of the last XSDC module, excluding BGU. From the results reported in Table 1, we see that the network trained with $d = 1$ shows better performance than all methods, in all three metrics for both [15,25] datasets.

4.3 Ablation Experiments

Guide-Image Ablations: In Table 2, we present results for different simple baseline approaches to generate guide image. We report the PSNR-L (P_L) and PSNR-T (P_T) on test set of UCSD dataset.

[1] B1+[3]: A simple baseline to generate guide image would be to detect moving pixels by taking difference between images. We detect the moving pixels by

Table 1. Quantitative comparison of our proposed method against eight state-of-art HDR deghosting algorithms on [15] and [25] datasets. For [15] dataset, we report the scores averaged over all 15 test sequences, and for [25], over 116 test sequences with PSNR-L (P_L) and PSNR-T (P_T) metrics. The best performing method is highlighted in **bold** and the second best in blue color. P_T (in dB) is computed after tonemapping with Eq. 11.

Methods\Datasets	Metrics	Sen12 [34]	Hu13 [13]	Kalantari17 [15]	Wu18 [40]	Yan20 [44]	MSDN [43]	AHDR [42]	SCHDR [25]	Ours (w/o BGU)	Ours (w/ BGU)
[15]	P_L	38.57	30.83	41.07	40.91	–	41.01	41.01	39.68	**41.68**	41.33
	P_T	40.94	32.18	42.74	41.65	42.41	42.22	42.10	40.47	**43.08**	42.82
	HDR-VDP-2	61.98	60.12	66.64	**67.96**	–	–	66.67	66.80	67.21	66.94
[25]	P_L	29.57	28.87	32.08	30.72	–	–	31.83	31.44	**32.52**	32.11
	P_T	32.09	30.82	35.34	31.31	–	–	33.72	30.57	**35.84**	35.46
	HDR-VDP-2	62.43	60.47	64.47	64.03	–	–	64.32	62.20	**64.76**	64.57

thresholding (with 0.1) the image difference after brightness normalization. The threshold maps are used as M_-^h and M_+^h in Eq. 1 and 2 to generate motion compensated sequence. Then, the motion corrected HDR sequence is fused by using triangle weighing strategy [3].

[2] N_{mc}+[3]: Motion correction by N_{mc} followed by fusion with [3].

[3] N_g: Direct input without motion correction is passed to N_g model.

[4] B1+N_g: Motion compensation by simple difference and fusion with N_g model.

[5] N_{mc}+N_g: Motion compensation by N_{mc} and fusion with N_g model. This denotes the accuracy of H_g images.

From Table 2, we observe that while the individual performance of N_{mc} and N_g is low, using them together can result in significant boost in PSNR. Similarly, using N_{mc} for motion correction achieves a P_T score approximately 1.5 dB higher than B1. This is due to the fact that simple difference can produce false segmentation in heavily saturated regions. Whereas, N_{mc} network has learnt reliable features that can produce accurate motion segmentation maps.

Choice of Inputs: In Table 3, we present the accuracy obtained with different type of inputs to the fusion model (N_f). We report scores for the $d = 1$ setting.

[1] N_f: We pass direct input images without optical flow correction to N_f model.

[2] N_{mc}+N_f: We pass motion corrected sequence (I_{MC}^l) as input to N_f.

[3] [20]+N_f: optical flow aligned input is passed to N_f.

The low score for passing input images without optical flow correction is because of the fact that N_f has to perform both motion correction as well as fusion. For saturated regions in I_0, N_f has to choose details from under or over exposed image. However, in moving regions, it has to choose details from reference image. Hence, the network gets confused in heavily saturated regions, as to whether it is a saturated region or motion affected region. Most often, it chooses to identify

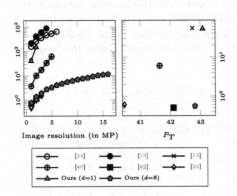

(a)   (b)

Fig. 5. Our proposed method can fuse images up to 16MP on a GPU with 11 GB memory for PSNR higher than state-of-the-art HDR deghosting algorithms. The y-axis for both the plots is in the logarithmic time scale in seconds. See Sect. 4.5 for details.

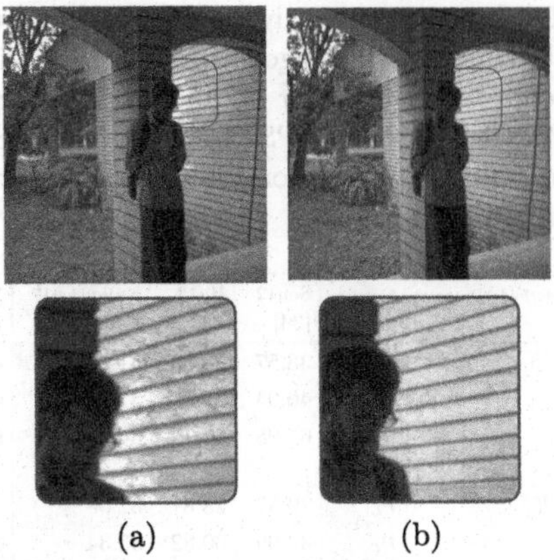

Fig. 6. A qualitative comparison between (a) Wu18 [40], (b) Proposed method on a test sequence from [25] dataset.

Table 2. Ablation experiments for different choice of guide image generation. See Sect. 4.3 for more details.

Methods	P_L	P_T
B1 + [3]	34.40	33.50
N_{mc} + [3]	37.55	37.98
N_g	36.38	37.24
B1 + N_g	38.32	39.06
N_{mc} + N_g	**40.73**	**41.56**

Table 3. Ablation on different type inputs to N_f model. See Sect. 4.3 for more details.

Methods	P_L	P_T
N_f	40.88	41.16
N_{mc} + N_f	41.02	41.54
[20] + N_f	**41.33**	**42.82**

Table 4. Base architecture ablation. See Sect. 4.3 for more details.

Methods	P_L	P_T
Vanilla	39.60	41.08
U-Net [30]	41.13	41.76
ResNet18 [10]	41.24	42.45
DenseNet [14]	41.03	41.78
SDC [31]	40.76	41.94
Ours (w/o dense)	41.21	42.34
Ours (w/ dense)	**41.33**	**42.82**

those regions as motion affected, hence losing HDR content in those pixels. By passing I_{MC} images to N_f, the fusion network has to focus only on merging them. However, as motion corrected sequence has low HDR content for moving objects and saturated regions, the final result has lower PSNR. Finally, passing optical flow corrected sequence results in higher PSNR as it can transfer details from neighbouring pixels for motion affected and saturated regions.

Architecture Ablation: In Table 4, we show results for different choices of network architecture. The vanilla network denotes using ten convolutional layers instead of XSDC blocks.

Fig. 7. Qualitative comparison between proposed method and state-of-the-art methods in a test set from UCSD dataset [15]. The input sequence is shown on the left column. (a) SCHDR [25], (b) Kalantari17 [15], (c) Sen12 [34], (d) H_g, (e) H_f and (f) ground truth.

4.4 Qualitative Evaluation

In Figs. 1 and 6, we show qualitative comparison against Wu18 [40] and AHDR [42] methods. Both Wu18 and AHDR fail to reconstruct details in saturated regions of the reference image that is occluded in other input images. As a result, the occluded regions appear as distinct structure in the result (zoomed regions in Fig. 1). Sen12 [34] suffers from color bleeding and over-smoothing artifacts introduced in heavily saturated regions of reference image (see Fig. 7). In Fig. 1 and 7, we show the results by Kalantari17 [15] and proposed method. Kalantari17 method introduces structural distortions for moving objects in saturated regions and hallucinates details in heavily saturated regions. Comparatively, our proposed method generates visually pleasing results without any artifacts[5]. All the results shown in the paper are generated with $d = 8$.

4.5 Running Time

In Fig. 5 left graph, we compare the running times of seven methods for image resolutions from 1 to 16 MP. The reported numbers are obtained from a machine with NVIDIA RTX 2080 Ti GPU, 32 GB RAM and an i7-8700 CPU. For each resolution, we report the average of 15 runs with three varying exposure images. The non-deep methods Sen12 [34] and Hu13 [13] take up to 468 s and 893 s respectively to fuse a 4 MP sequence. Among the deep learning-based methods, Kalantari17 [15] fuses a 1 MP sequence in 42 s. For any other higher resolutions, it throws an out-of-memory error. Similarly, Yan19 [42] can support a maximum

---

[5] For more results, please check supplementary material.

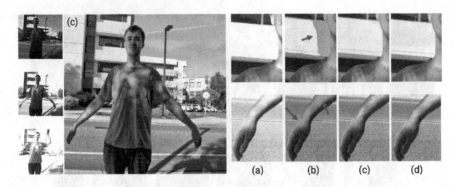

Fig. 8. H_g vs H_f: (a) Patches from I_0 image, (b) H_g, (c) H_f and (d) ground truth. The regions affected by motion and saturation is not corrected in the guide image (highlighted by the red arrows). In contrast, N_f network can generate plausible details in those regions as well. (Color figure online)

of 2 MP resolution in 0.83 s. Wu18 [40] method takes up to 10 s to process 1 MP images and can only support up to 5 MP. Comparatively, our proposed method with $d = 8$ can fuse 16 MP images in 11 s on a 11 GB GPU card. In Fig. 5 right graph, we report the average PSNR-T score for 15 test images of size 1.5 MP from dataset provided by Kalantari17 [15]. It should be noted that, for a fair comparison, we have used SIFT followed by the RANSAC method to align images for both Wu18 and our method. However, as the input images are downsampled by a factor of 8 before image alignment, it does not have significant overhead on our approach.

5   Discussion

Guide vs Final HDR: As shown in Figs. 1, 7 and 8, H_g image lacks details in moving regions affected by saturation. This effect is because H_g generation process is a weight map-based method; it does not have the freedom to move details from neighboring pixels. In contrast, as input to N_f are optical flow corrected images, the H_f has proper contents in those regions as well.

Limitations: Similar to [15,40,42], our method is also trained for a fixed number of three images. Hence, extending our method (including [15,41,42]) requires re-training with the new number of images. To overcome this issue, one can use the max-mean fusion strategy instead of feature concatenation [25].

6   Conclusion

In the current era of high-resolution imaging with smartphones and DSLRs, it is difficult to find the best performing HDR deghosting method that can support higher image resolutions with limited computational resources. We address this issue by performing all heavy processing in low resolution (including optical flow computation) and upscale the low-resolution output to full resolution using a guide image. The guide image is generated using a simple weight map based fusion of original full-resolution inputs. We have shown that our approach improves the state-of-the-art in generating high-quality artifact-free HDR images. Our approach can fuse 16-megapixel images in about 10 s on a GPU with 11 GB memory and achieves state-of-the-art results in publicly available datasets. We also have demonstrated the use of BGU for tasks where the guide image is not readily available. We hope that our paper will motivate the research community to explore BGU for other fusion tasks.

Acknowledgements. This work was supported by a project grant from MeitY (No.4(16)/2019-ITEA), Govt. of India and WIRIN.

References

1. Bogoni, L.: Extending dynamic range of monochrome and color images through fusion. In: Proceedings of the 15th International Conference on Pattern Recognition (2000)
2. Chen, J., Adams, A., Wadhwa, N., Hasinoff, S.W.: Bilateral guided upsampling. ACM Trans. Graph. (TOG) **35**(6), 203 (2016)
3. Debevec, P.E., Malik, J.: Recovering high dynamic range radiance maps from photographs. In: ACM SIGGRAPH 2008 Classes, p. 31. ACM (2008)
4. Eden, A., Uyttendaele, M., Szeliski, R.: Seamless image stitching of scenes with large motions and exposure differences. In: Conference on Computer Vision and Pattern Recognition. IEEE (2006)
5. Gallo, O., Gelfandz, N., Chen, W.C., Tico, M., Pulli, K.: Artifact-free high dynamic range imaging. In: ICCP, pp. 1–7. IEEE (2009)
6. Gallo, O., Troccoli, A., Hu, J., Pulli, K., Kautz, J.: Locally non-rigid registration for mobile HDR photography. In: Proceedings of the IEEE Conference on Computer Vision and Pattern Recognition Workshops, pp. 49–56 (2015)
7. Gharbi, M., Chen, J., Barron, J.T., Hasinoff, S.W., Durand, F.: Deep bilateral learning for real-time image enhancement. ACM Trans. Graph. (TOG) **36**(4), 1–12 (2017)
8. Granados, M., Kim, K.I., Tompkin, J., Theobalt, C.: Automatic noise modeling for ghost-free HDR reconstruction. ACM Trans. Graph. (TOG) **32**(6), 201 (2013)
9. Grosch, T.: Fast and robust high dynamic range image generation with camera and object movement. Vis. Model. Vis. RWTH Aachen 277–284 (2006)
10. He, K., Zhang, X., Ren, S., Sun, J.: Deep residual learning for image recognition. In: Proceedings of the IEEE Conference on Computer Vision and Pattern Recognition (2016)
11. Heo, Y.S., Lee, K.M., Lee, S.U., Moon, Y., Cha, J.: Ghost-free high dynamic range imaging. In: Kimmel, R., Klette, R., Sugimoto, A. (eds.) ACCV 2010. LNCS, vol. 6495, pp. 486–500. Springer, Heidelberg (2011). https://doi.org/10.1007/978-3-642-19282-1_39
12. Hu, J., Gallo, O., Pulli, K.: Exposure stacks of live scenes with hand-held cameras. In: Fitzgibbon, A., Lazebnik, S., Perona, P., Sato, Y., Schmid, C. (eds.) ECCV 2012. LNCS, vol. 7572, pp. 499–512. Springer, Heidelberg (2012). https://doi.org/10.1007/978-3-642-33718-5_36

13. Hu, J., Gallo, O., Pulli, K., Sun, X.: HDR deghosting: How to deal with saturation? In: IEEE Conference on Computer Vision and Pattern Recognition (2013)
14. Huang, G., Liu, Z., Van Der Maaten, L., Weinberger, K.Q.: Densely connected convolutional networks. In: Proceedings of the IEEE Conference on Computer Vision and Pattern Recognition, pp. 4700–4708 (2017)
15. Kalantari, N.K., Ramamoorthi, R.: Deep high dynamic range imaging of dynamic scenes. ACM Trans. Graph. (Proc. SIGGRAPH 2017) **36**(4), 144-1 (2017)
16. Kang, S.B., Uyttendaele, M., Winder, S., Szeliski, R.: High dynamic range video. ACM Trans. Graph. (TOG) **22**, 319–325 (2003)
17. Kao, W.C., Hsu, C.C., Chen, L.Y., Kao, C.C., Chen, S.H.: Integrating image fusion and motion stabilization for capturing still images in high dynamic range scenes. IEEE Trans. Consum. Electron. **52**(3), 735–741 (2006)
18. Khan, E.A., Akyüz, A., Reinhard, E.: Ghost removal in high dynamic range images. In: IEEE International Conference on Image Processing (2006)
19. Kingma, D.P., Ba, J.: Adam: a method for stochastic optimization. arXiv preprint arXiv:1412.6980 (2014)
20. Liu, C., et al.: Beyond pixels: exploring new representations and applications for motion analysis. Ph.D. thesis, Massachusetts Institute of Technology (2009)
21. Ma, K., Li, H., Yong, H., Wang, Z., Meng, D., Zhang, L.: Robust multi-exposure image fusion: a structural patch decomposition approach. IEEE Trans. Image Process. **26**(5), 2519–2532 (2017)
22. Mantiuk, R., Kim, K.J., Rempel, A.G., Heidrich, W.: HDR-VDP-2: a calibrated visual metric for visibility and quality predictions in all luminance conditions. ACM Trans. Graph. (TOG) **30**(4), 40 (2011)
23. Min, T.H., Park, R.H., Chang, S.: Histogram based ghost removal in high dynamic range images. In: 2009 IEEE International Conference on Multimedia and Expo, pp. 530–533. IEEE (2009)
24. Pece, F., Kautz, J.: Bitmap movement detection: HDR for dynamic scenes. In: 2010 Conference on Visual Media Production, pp. 1–8. IEEE (2010)
25. Prabhakar, K.R., Arora, R., Swaminathan, A., Singh, K.P., Babu, R.V.: A fast, scalable, and reliable deghosting method for extreme exposure fusion. In: 2019 IEEE International Conference on Computational Photography (ICCP), pp. 1–8. IEEE (2019)
26. Prabhakar, K.R., Babu, R.V.: Ghosting-free multi-exposure image fusion in gradient domain. In: IEEE International Conference on Acoustics, Speech and Signal Processing (2016)
27. Prabhakar, K.R., Srikar, V.S., Babu, R.V.: DeepFuse: a deep unsupervised approach for exposure fusion with extreme exposure image pairs. In: IEEE International Conference on Computer Vision (ICCV), pp. 4724–4732. IEEE (2017)
28. Raman, S., Chaudhuri, S.: Reconstruction of high contrast images for dynamic scenes. Vis. Comput. **27**(12), 1099–1114 (2011)
29. Reinhard, E., Heidrich, W., Debevec, P., Pattanaik, S., Ward, G., Myszkowski, K.: High Dynamic Range Imaging: Acquisition, Display, and Image-Based Lighting. Morgan Kaufmann, Burlington (2010)
30. Ronneberger, O., Fischer, P., Brox, T.: U-net: convolutional networks for biomedical image segmentation. In: Navab, N., Hornegger, J., Wells, W.M., Frangi, A.F. (eds.) MICCAI 2015. LNCS, vol. 9351, pp. 234–241. Springer, Cham (2015). https://doi.org/10.1007/978-3-319-24574-4_28

31. Schuster, R., Wasenmuller, O., Unger, C., Stricker, D.: SDC-stacked dilated convolution: a unified descriptor network for dense matching tasks. In: Proceedings of the IEEE Conference on Computer Vision and Pattern Recognition, pp. 2556–2565 (2019)
32. Sen, P.: Overview of state-of-the-art algorithms for stack-based high-dynamic range (HDR) imaging. Electron. Imaging **2018**(5), 1–8 (2018)
33. Sen, P., Aguerrebere, C.: Practical high dynamic range imaging of everyday scenes: photographing the world as we see it with our own eyes. IEEE Signal Process. Mag. **33**(5), 36–44 (2016)
34. Sen, P., Kalantari, N.K., Yaesoubi, M., Darabi, S., Goldman, D.B., Shechtman, E.: Robust patch-based HDR reconstruction of dynamic scenes. ACM Trans. Graph. **31**(6), 203 (2012)
35. Sidibé, D., Puech, W., Strauss, O.: Ghost detection and removal in high dynamic range images. In: 17th European Signal Processing Conference (2009)
36. Tomaszewska, A., Mantiuk, R.: Image registration for multiexposure high dynamic range image acquisition. In: Proceedings of the International Conference on Computer Graphics, Visualization and Computer Vision (2007)
37. Tursun, O.T., Akyüz, A.O., Erdem, A., Erdem, E.: The state of the art in HDR deghosting: a survey and evaluation. In: Computer Graphics Forum, vol. 34, pp. 683–707. Wiley Online Library (2015)
38. Tursun, O.T., Akyüz, A.O., Erdem, A., Erdem, E.: An objective deghosting quality metric for HDR images. In: Computer Graphics Forum, vol. 35, pp. 139–152. Wiley Online Library (2016)
39. Ward, G.: Fast, robust image registration for compositing high dynamic range photographs from hand-held exposures. J. Graph. Tools **8**(2), 17–30 (2003)
40. Wu, S., Xu, J., Tai, Y.W., Tang, C.K.: Deep high dynamic range imaging with large foreground motions. In: European Conference on Computer Vision, pp. 120–135 (2018)
41. Wu, S., Xie, S., Rahardja, S., Li, Z.: A robust and fast anti-ghosting algorithm for high dynamic range imaging. In: 2010 IEEE International Conference on Image Processing, pp. 397–400. IEEE (2010)
42. Yan, Q., et al.: Attention-guided network for ghost-free high dynamic range imaging. In: Proceedings of the IEEE Conference on Computer Vision and Pattern Recognition, pp. 1751–1760 (2019)
43. Yan, Q., et al.: Multi-scale dense networks for deep high dynamic range imaging. In: IEEE Winter Conference on Applications of Computer Vision (WACV), pp. 41–50. IEEE (2019)
44. Yan, Q., et al.: Deep HDR imaging via a non-local network. IEEE Trans. Image Process. **29**, 4308–4322 (2020)
45. Zhang, W., Cham, W.K.: Reference-guided exposure fusion in dynamic scenes. J. Vis. Commun. Image Represent. **23**(3), 467–475 (2012)
46. Zimmer, H., Bruhn, A., Weickert, J.: Freehand HDR imaging of moving scenes with simultaneous resolution enhancement. In: Computer Graphics Forum, vol. 30, pp. 405–414. Wiley Online Library (2011)

Monocular Differentiable Rendering for Self-supervised 3D Object Detection

Deniz Beker[1]([✉]), Hiroharu Kato[1], Mihai Adrian Morariu[1], Takahiro Ando[1], Toru Matsuoka[1], Wadim Kehl[2], and Adrien Gaidon[3]

[1] Preferred Networks, Inc., Chiyoda City, Japan
dbeker@preferred.jp
[2] Toyota Research Institute - Advanced Development, Chuo City, Japan
[3] Toyota Research Institute, Los Altos, USA

Abstract. 3D object detection from monocular images is an ill-posed problem due to the projective entanglement of depth and scale. To overcome this ambiguity, we present a novel self-supervised method for textured 3D shape reconstruction and pose estimation of rigid objects with the help of strong shape priors and 2D instance masks. Our method predicts the 3D location and meshes of each object in an image using differentiable rendering and a self-supervised objective derived from a pretrained monocular depth estimation network. We use the KITTI 3D object detection dataset to evaluate the accuracy of the method. Experiments demonstrate that we can effectively use noisy monocular depth and differentiable rendering as an alternative to expensive 3D ground-truth labels or LiDAR information.

1 Introduction

Autonomous driving relies heavily on 3D object perception for safe navigation. Most existing systems leverage active sensors (e.g., LiDAR, radar) for location estimation, yet they are either prohibitively costly for large-scale deployment or too sparse in spatial coverage. For these reasons, research in monocular 3D object detection has seen rising popularity in recent years.

Ongoing advancements have led to steadily improving detection accuracy [3,23,27]. Despite foregoing active sensing, they still require supervision in the form of 9D cuboid labels which encode 3D location, rotation and metric object dimensions. Most often, such labels are obtained with the help of annotation pipelines and 3D LiDAR point clouds, demanding the usage of costly human labour and expensive sensors. Alternatively, one can leverage self-supervised autolabeling techniques that employ strong shape priors and optimize 3D alignment in stereo point clouds [5,6,31] or mixed RGB/LiDAR setups [36].

In this paper, we explore a novel solution towards tackling 3D location and shape estimation from 2D instance mask detections, requiring only monocular

Electronic supplementary material The online version of this chapter (https://doi.org/10.1007/978-3-030-58589-1_31) contains supplementary material, which is available to authorized users.

A. Vedaldi et al. (Eds.): ECCV 2020, LNCS 12366, pp. 514–529, 2020.
https://doi.org/10.1007/978-3-030-58589-1_31

Render-and-Compare Loss

Fig. 1. We propose a self-supervised optimization pipeline for monocular 3D object estimation via analysis-by-synthesis over object pose, metric dimensions, shape and texture. Left: Starting from a 2D instance mask detection, we feed the latent variables h into our decoder network and use differentiable rendering to obtain 2D projections. We use various render-and-compare losses over multiple quantities for comparative analysis and back-propagate the error. Right: The fitting process over multiple iterations. Starting from a random initialization, we can recover the actual object properties quite well.

input and geometric priors for self-supervision. Recently, self-supervised depth estimation networks [9,26] have become interesting alternatives to LiDAR sensing. They are generally trained using only video sequences and are able to estimate depth with absolute scale when combined with weak supervision (e.g., velocity of the ego-camera [9]). However, they tend to be rather noisy, overall less accurate than LiDAR, and therefore not well-suited for precise 3D object localization and metric dimension estimation. We therefore incorporate additional strong priors over learned textured object shapes for regularization and run comparative scene analysis via differentiable rendering [2,11,19,20]. This allows for optimizing 3D variables against image evidence via self-supervision and back-propagation. While initial work focused mostly on toy examples, recent papers have successfully explored applications of analysis-by-synthesis in the wild [15,16,36,38] and we follow in their path.

We present an example optimization of our pipeline in Fig. 1. Starting from a 2D detector that produces 2D boxes with associated instance masks, we optimize over the 3D location, rotation, metric dimensions as well as the object's shape and texture. In our paper, we handle metric dimension as 3D scaling. For the actual alignment computation, we leverage complimentary cues in the form of instance masks, RGB values as well as monocular depth, and regularize with our textured shape space. While our initial estimation is quite rough, our method is able to converge to a good solution while traversing the possible space of shapes, textures, metric sizes and poses. This process is run sequentially until all 2D detections in the scene have been parsed and transformed into 3D estimates.

To summarize our main contributions: Firstly, we remove the necessity for 3D sensing or ground-truth information, making our 3D object estimation pipeline a truly monocular approach. Secondly, we propose the idea of regularizing noisy monocular depth maps at the instance level by strong geometric object priors. Lastly, we evaluate our pipeline on the KITTI 3D dataset [7] and show that we can achieve comparable accuracy to SoTA methods that rely on 3D supervision.

2   Related Work

Due to the huge body of related 3D detection work, we will focus our survey mostly on monocular methods. Mono3D [3] is an early work which introduces a region proposal method tailored to autonomous driving. This method computes the 3D bounding box by exploiting different cues such as semantic segmentation, instance segmentation, shape, context features and absolute locations in 2D/3D space. Importantly, they incorporate priors to limit the cuboid search space by using pre-defined object sizes and orientations as well as assuming a known ground plane. Deep3DBox [25] estimates cuboids from 2D bounding boxes, assuming that the former should be tightly fit to the latter when projected onto the image plane. This strong assumption can underperform in cases where the actual 2D bounding boxes are not tightly enclosing the objects due to occlusion or truncation. The advent of unsupervised/self-supervised monocular depth estimation [8,22,26] saw their inclusion in recent detection work. In Multi-Fusion [34] 3D region proposals are generated by backprojecting 2D proposals to the 3D space using monocular depth. Similarly, ROI-10D [23] employs such depth maps to robustify their bounding box lifting from the 2D to the 3D space.

Methods based on the idea of Pseudo-LiDAR [21,32] leverage reprojected monocular depth from off-the-shelf models and run detection networks that have been originally designed for LiDAR input with impressive accuracy improvements. All the mentioned approaches benefit greatly from the employed monocular depth modules but their analysis shows that their overall performance is heavily dominated by the accuracy of depth estimation. For this reason, regularization from additional priors is desirable.

In terms of analysis-by-synthesis, much novel work has been presented in respect to differentiable rendering. The works [2,11,17,20] propose different ways to produce gradients for the rasterization of triangle meshes. Building on such ideas, both [16] as well as [36] present render-and-compare optimization frameworks with learned shape spaces for automotive scenarios. The former uses 2D instance masks for estimating the shape and up-to-scale pose whereas the latter leverages LiDAR observations for full 3D estimation.

There exists other slightly-related work that leverages learned shape spaces for automotive object retrieval that we discuss for completeness. In [5] a detector initializes object instances which are further optimized for pose and shape priors over stereo depth. The authors later extended it [6] for temporal priors to also recover pose trajectories. In a similar vein, [31] runs full 3D object pose and shape recovery over stereo depth. The authors of [29] explore probabilistic 3D

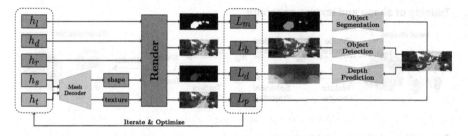

Fig. 2. Overall pipeline of our method where h_l, h_d, h_r are the variables for location, dimension and rotation, and h_s, h_t are the latent variables for shape and texture. We denote with L_m, L_b, L_d, L_p the losses for the silhouette, bounding box, depth map and pixel colors respectively.

object completion via a shape space and LiDAR as weak supervision. Their work assumes correct localization and focuses solely on the reconstruction quality.

3   Method

As mentioned, most existing monocular 3D detection methods rely on supervised learning, requiring 3D cuboid information. Instead, we propose to estimate all required 3D object properties via self-supervision and differentiable rendering, avoiding any form of 3D annotation. We depict our pipeline in Fig. 2 and will provide an overview before going into more detailed descriptions in the next subsections.

We initially run off-the-shelf object detectors on an image to produce masked 2D detections. For each detection, we place an initial estimate into 3D space by instantiating all object properties. Concretely, we capture metric location $h_{loc} \in \mathbb{R}^3$, metric dimension $h_{dim} \in \mathbb{R}^3$, rotation along the yaw axis $h_{rot} \in \mathbb{R}$, shape $h_{sh} \in \mathbb{R}^{D_{sh}}$ and texture $h_{tx} \in \mathbb{R}^{D_{tx}}$. Following the parameterization introduced in ROI-10D [23], we formulate the location h_{loc} as the difference between the center of 2D bounding box and the projected 3D centroid position $(h_u, h_v) \in \mathbb{R}^2$, together with the depth of the centroid h_{dep}. We initialize $h_u = h_v = 0$, whereas h_{dep} are set h_{dim} to the mean depth and dimensions. We also initialize h_{rot}, h_{sh}, h_{tx} to random numbers sampled from a Gaussian with mean $\mu = 0$ and standard deviation $\sigma = 0.1$.

After this initialization, we run an iterative pipeline of 1) rendering, 2) projective loss computation over multiple different cues and 3) backpropagation to our latent object variables. To produce a rendering, we feed h_{sh} and h_{tx} through generative models to produce object shape and texture. We then rescale and rotate using h_{dim} and h_{rot}, place the object in the 3D space using h_{loc} and finally render an RGB/D image pair with the differentiable renderer implementation from [11]. Note that we work at metric scale and therefore require known camera intrinsics during optimization.

Training of shape and texture generator

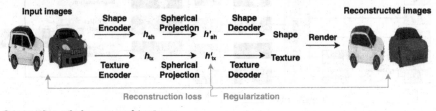

Generation of shapes and textures

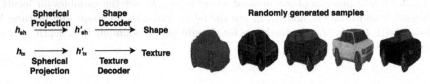

Fig. 3. Illustration of our shape and texture generator. During training, images of cars are encoded to shape and texture vectors that live on hyperspheres. These vectors can then be decoded to 3D shapes and texture maps, which in turn are rendered and used to compute reconstruction losses. After training, the decoder part can be used independently to generate novel textured shapes.

2D Object Detection and Segmentation. To identify all instances in the scene, we use an off-the-shelf Mask R-CNN [10] model with the X-152 backbone from detectron2 [33], trained on COCO [18] and thresholded at 0.1.

To compute our rendering losses, we need a clear foreground/background separation. In order to produce a background mask, we compute a union over all detection masks to first obtain a foreground mask, and then invert it. We will evaluate the majority of our experiments with this approach. Alternatively, one can also leverage semantic segmentation for the separation and we present an ablative comparison in the experimental section.

Rendering. We use the publicly available differentiable renderer from Kato et al. [11] and although they introduced their method for color and silhouettes only, we extend it for rendering depth maps in a differentiable way. We exploit the simple fact that a depth map is differentiable with respect to the z-coordinates of a surface without any approximations because the depth value at a pixel is computed by a weighted sum of z-coordinates of the surfaces at the pixel.

Shape and Texture Generation. Differentiable rasterization allows us to establish losses between renderings and an input image (*render-and-compare*). This approach was initially adopted by 3D-RCNN [16] in which shapes were projected to 2D silhouettes. Unfortunately, silhouettes retain only a fraction of the original information and have an especially strong impact on rotation estimation. To leverage image information as much as possible, we therefore additionally reconstruct textures and render a joint representation of textured shapes.

To encode a variety of object shapes and appearances, we employ a generative model that is able to produce a mesh and a texture map for given inputs. Concretely, we use the single-image 3D object reconstruction method presented in [13] and provide an illustration in Fig. 3. This method adopts an encoder-decoder architecture that transforms an image into a low-dimensional latent space which in turn can be used to generate a 3D object. Even though we train the entire encoder-decoder architecture beforehand, we discard the encoder and only use the decoder later on.

We would like our generator to produce plausible shapes and textures for any decoding input h_{sh} and h_{tx}, ensuring that our latent spaces are smoothly traversable without collapsing. Therefore, unlike [13], we project the hidden vectors onto a unit hypersphere, similar to [36]. To distribute h'_{sh} and h'_{tx} uniformly on the hypersphere, we employ a regularization technique where we take random samples on the sphere and pull the closest latent vectors towards those random samples. Let r_{sh} and r_{tx} be random samples uniformly distributed on the hyperspheres of D_{sh} and D_{tx} dimensions. We add the following term to the loss function:

$$L_h = \frac{1}{N_b} \left(\sum_{i=1}^{N_b} \min_j |r_{sh}^i - h_{sh}^j|_1 + \sum_{i=1}^{N_b} \min_j |r_{tx}^i - h_{tx}^j|_1 \right). \tag{1}$$

N_b is the size of minibatch, and $h_{sh}'^i, h_{tx}'^i, r_{sh}^i, r_{sh}^i$ represent i-th sample in the minibatch. The encoder-decoder architecture is trained using synthetic views of 3D car models taken from the ShapeNet dataset [1]. We found a latent dimensionality of 8 for the shape D_{sh} and texture D_{tx} spaces to work quite well.

Monocular Depth Estimation. To weaken the projective entanglement between object dimensions, locations and 2D appearance, we employ depth maps extracted by a pre-trained, state-of-the-art monocular depth estimator that is trained with self-supervision (PackNet [9]). The scale of the estimated depth is calibrated from weak supervision in the form of ego-camera velocity.

3.1   Loss Functions

For self-supervision we propose four loss functions between renderings and image evidence: L_p which penalizes pixel color deviations, L_b to maximize 2D bounding box overlap, L_m for silhouette alignment, and L_d for depth map differences. We notate an input image of $H \times W$ pixels as $I^{RGB} \in \mathbb{R}^{H \times W \times 3}$, the foreground map and depth map estimated from the image as $I^F \in \mathbb{R}^{H \times W}$ and $I^D \in \mathbb{R}^{H \times W}$, respectively. In addition, assuming that the number of detected objects is N_o, we notate the cropped and resized region of the i-th object from these maps as $I_i^{RGB} \in \mathbb{R}^{H_c \times W_c \times 3}$, $I_i^F \in \mathbb{R}^{H_c \times W_c}$, and $I_i^D \in \mathbb{R}^{H_c \times W_c}$. For the reconstructed counterparts, we use the same notations with hat.

Additionally, we use a regularization term L_{dim} to penalize unrealistic dimensions. We define it as the L1 distance between the estimated dimension and the mean dimension of objects in the dataset. The overall loss function is defined as the weighted sum of all the loss functions and the regularization term:

$$L = \lambda_p L_p + \lambda_b L_b + \lambda_m L_m + \lambda_d L_d + \lambda_{dim} L_{dim}. \tag{2}$$

Pixel Map Difference. The pixel map difference loss is defined as L1 distance between rendered image and input image by using equation

$$L_p = \frac{1}{H_c W_c} \sum_{i=1}^{N_b} \sum_{j=1}^{H_c} \sum_{k=1}^{W_c} |I_{ijk}^{RGB} - \hat{I}_{ijk}^{RGB}|. \tag{3}$$

Bounding Box Difference. We compute the bounding boxes of the objects from the rendered image. Let IoU_i^B be the intersection over union between input and rendered 2D bounding boxes of i-th object. We define the bounding box loss as

$$L_b = \sum_{i=1}^{N_b} \mathrm{maximum}(1 - IoU_i^B - t_b, 0). \tag{4}$$

As the predicted bounding boxes may be noisy, we allow the IoU to have a small error margin of up to t_b. This loss is differentiable with respect to the vertices of the reconstructed shapes as the bounding box is computed by differentiable projection of vertices to image coordinates. Fitting estimated 3D bounding boxes to detected 2D bounding boxes is a well-known approach [25], and some works [24] use IoU as a metric. Different from these works, we compute projected bounding boxes using 3D shapes instead of 3D bounding boxes.

Silhouette Difference. Concretely, assuming that $I_{ijk}^F = 1$ if the pixel jk of the i-th image is foreground and $I_{ijk}^F = 0$ if background, the IoU of i-th image is

$$IoU_i^F = \frac{\sum_{j=1}^{H_c} \sum_{k=1}^{W_c} I_{ijk}^F \hat{I}_{ijk}^F}{\sum_{j=1}^{H_c} \sum_{k=1}^{W_c} I_{ijk}^F + \hat{I}_{ijk}^F - I_{ijk}^F \hat{I}_{ijk}^F}, \tag{5}$$

$$L_m = \sum_{i=1}^{N_o} L_{m_i} = \sum_{i=1}^{N_o} (1 - IoU_i^F). \tag{6}$$

Depth Map Difference. The depth map loss is defined as the L1 difference between the rendered depth and model-predicted depth by using the formula

$$L_d = \frac{1}{H_c W_c} \sum_{i=1}^{N_b} \sum_{j=1}^{H_c} \sum_{k=1}^{W_c} I_{ijk}^F \hat{I}_{ijk}^F |I_{ijk}^D - \hat{I}_{ijk}^D|. \tag{7}$$

Note that during the optimization, all the pre-trained networks are used with fixed weights and are not optimized. We disable back-propagation for $\hat{I}^F_{ijk}$ and back-propagate gradients only through $\hat{I}^D_{ijk}$.

3.2 Escaping Rotational Local Minima

Estimating rotation with render-and-compare approaches can easily lead to local minima [12,30] due to non-linear objectives and visual ambiguities. To deal with this issue, we explore several rotations at every step of the optimization. We sample 4 different variations (h_{rot}, $h_{rot} \pm 15°$, $-h_{rot}$) as well as a random angle sampled from a uniform distribution. If an angle r_0 gives a lower error with the perceptual metric [37] than the one used at current iteration, we set h_{rot} to r_0. Also, the size of the input and the reconstructed images are adjusted to ensure that the area of the object bounding boxes is the same, and the background region is masked out using the estimated segmentation mask.

3.3 Detection Confidence Score

MonoDIS [27] proposes to learn confidences of 3D detections by weighting confidences of corresponding 2D detections. As their method requires ground-truth 3D bounding boxes for confidence training, it is not applicable in a straightforward manner to our problem. Instead, we propose to weight 2D confidences using our reconstruction errors after optimization. To this end, we compute the silhouette reconstruction loss L_{m_i} of the i-th image, and the ratio of protrusion b_i between rendered bounding box and detection bounding box. Our confidence score c_{3D} is then defined by using the formula

$$c_{3D} = c_{2D} e^{-\alpha_m L_{m_i}} e^{-\alpha_b b_i}, \tag{8}$$

with two hyperparameters α_m, α_b and the original 2D detection confidence c_{2D}.

4   Experiments

We evaluate the proposed method on KITTI 3D [7], one of the most popular benchmarks for 3D object detection in autonomous driving. KITTI 3D is composed of synchronized RGB images/LiDAR frames, for which annotations of 2D bounding boxes, 3D object location, 3D object dimension and 1D object rotation angle along the y-axis (yaw) are provided. Unlike other approaches, our method does not require 3D ground-truth or LiDAR observations in any part of the pipeline. For a fair comparison to similar work, we use the same training and validation splits proposed in [4] and focus on the *car* category.

Evaluation Metrics. KITTI 3D uses different AP metrics to measure 2D and 3D detection accuracy at different cut-off thresholds (usually 0.5 and 0.7). For each detection, the IoU overlap with a ground truth is computed either on the image plane (2D), in bird's eye view (BEV), or in volumetric 3D space. Following a recent suggestion by [27], we use the $AP_{|R_{40}}$ metric that is currently used in the official KITTI 3D benchmark leaderboard, instead of the older $AP_{|R_{11}}$ metric. For reference and a fair comparison to previously established work, we also evaluate against the $AP_{|R_{11}}$ metric and share the results in the supplement.

System Configuration. All experiments are performed on a Linux-based cluster with 8 V100 (32 GB) GPUs. Running the full pipeline on the KITTI 3D *validation* dataset takes approximately one day.

Hyperparameters Tuning. The hyperparameter values are set to $\lambda_b = \lambda_m = \lambda_p = 1$, $\lambda_d = 0.2$, $\lambda_{dim} = 0.1$, and $t_b = 0.1$. We use the Adam optimizer [14] with $\alpha = 0.03$, $\beta_1 = 0.5$ and $\beta_2 = 0.9$. For each detected object, we run the optimization pipeline for 150 iterations. The hyperparameters for confidence weighting are set to $\alpha_m = 0.1$ and $\alpha_b = 0.03$.

4.1   Comparison to SoTA

As a first experiment, we run our pipeline on the validation set and show our quantitative results in Table 1 where we compare our inferred 3D boxes against the ground truth labels. As can be seen, we position ourselves between the two leading monocular 3D detectors MonoDIS [27] and MoVI-3D [28] for the easy instances and fall slightly behind for the moderate and difficult cases. This shows that our self-supervised 3D estimation method can be competitive with fully supervised detectors trained on manually labeled 3D ground truth cuboids. For more difficult instances, the monocular depth maps become noisier at longer ranges, there is an increasing amount of occlusion, and there are smaller objects, all factors that challenge our losses for bounding box, silhouette, and pixel differences. The autolabeling solution from [36] leverages shape priors similar to us, yet has access to LiDAR observations that greatly benefit their scale and 3D location estimation. Nonetheless, especially for the more challenging 3D AP metric, we can retrieve much more accurate object estimates overall. One of our major differences over LiDAR equipped SDFLabel paper is the scale regression method. SDFLabel often fail the strict 3D AP metric due to small deviations in 3D scale regression, especially at range where very few LiDAR points are on the object. On the other hand, we leverage the dimension statistics of the cars as a strong additional prior over generic driving scene instances. While SDFLabel regress directly on metric scale, we regressed the deviation from the statistics.

We also present some qualitative results in Fig. 4. It can clearly be seen that our priors provide enough guidance to estimate the correct 3D locations and rotations of the objects even if the predicted depth map is noisy. The depth map provides weak supervision for the 2D loss to estimate the correct object scale and rough 3D location but is ultimately regularized by shape and texture. It is also evident that optimizing for the 2D bounding box IoU loss L_b results in 3D shape projections that tightly fit into the predicted 2D bounding boxes. Besides, the regularization loss L_{dim} on dimensions prevents the predicted shapes from being too different from the mean object shape.

4.2 Ablation Studies

Reconstruction Loss. Firstly, we analyze the contribution of each individual loss component on the final outcome. Table 2 shows that all losses have a significant impact. Among them, L_d is the most important, as the accuracy sharply drops by over 60% when it is removed from the loss function. This shows that using 2D loss functions alone is not as effective, and that even noisy knowledge of scene depth (either implicitly or explicitly) is essential for monocular object estimation tasks. On the other hand, optimizing only for L_d leads to an accuracy close to zero. This signifies that using depth alone is apparently not enough for accurate reconstruction. In such cases, the optimization is disregarding object boundaries and physical extents. For example, the detected and reconstructed silhouettes should overlap, but the loss function does not enforce this constraint. Additionally, the objects can warp into any possible size.

We experimented to integrate depth information by backprojecting depth maps into 3D point clouds similar to Pseudo-LiDAR [35]. We optimized the Hausdorff distance between predicted meshes and the point clouds, and also experimented with Procrustes variants to support rotation, translation and scale. However, both results were very similar to our current *depth-only* results.

Another strong impact can be seen from the silhouette loss L_m. This loss back-propagates 2D silhouette mismatches to penalize the reconstruction of the 3D shape. Along with the regularization loss of dimensions L_{dim}, it helps converging towards a realistic shape that matches the predicted mask.

On the other hand, the impact of bounding box IoU loss L_b and pixel color loss L_p have less impact on the final accuracy. L_b strongly enforces 2D projection

Table 1. Evaluation of different monocular 3D detection methods: We report $AP_{|R_{40}}$ on the KITTI 3D *validation* set. The values are calculated assuming an intersection-over-union (IoU) between the predicted and ground-truth bounding boxes (in bird's-eye view) of at least 0.7.

Method	Supervised	LiDAR	3D detection			Bird's eye view		
			Easy	Moderate	Hard	Easy	Moderate	Hard
MonoDIS [27]	✓		11.06	7.60	6.37	18.45	12.58	10.66
MoVI-3D [28]	✓		14.28	11.13	9.68	22.36	17.87	15.73
SDFLabel [36]		✓	1.23	0.54	n/a	15.7	10.52	n/a
MonoDR (ours)			12.50	7.34	4.98	19.49	11.51	8.72

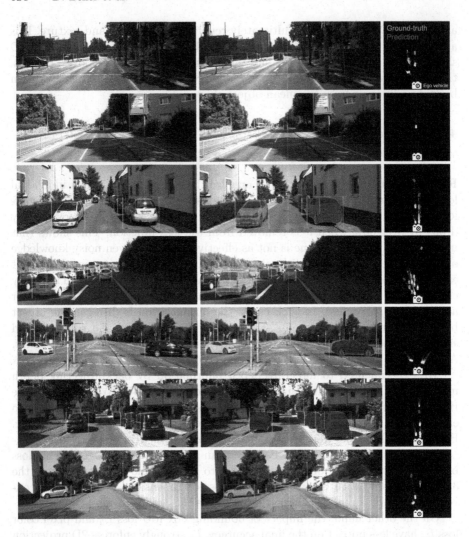

Fig. 4. Representative results obtained through our method: raw images with predicted 2D bounding boxes and instance masks are shown in the left column. Reconstructed images (with the projection of an estimated 3D shape onto the image plane) are shown in the middle column. Ground-truth (green)/predicted (red) 3D bounding boxes in the bird's-eye view (BEV) are shown in the right column. The white points in BEV represent projections of object point clouds generated from predicted depth maps. The ground truth information is only used for visualization purposes. (Color figure online)

of the reconstruction to be within the bounding box limits, which can be interpreted as an auxiliary support for the silhouette loss L_m. Similarly, the pixel loss impacts primarily the texturing of objects and helps mostly for minor textural misalignments but can hardly recover larger displacements or affect the shape.

Table 2. Evaluation of different reconstruction losses on KITTI 3D *val* set.

Reconstruction loss	3D Detection			Bird's eye view		
	Easy	Moderate	Hard	Easy	Moderate	Hard
Full (L)	12.50	7.34	4.98	19.49	11.51	8.72
w/o depth $(L - \lambda_d L_d)$	4.82	2.88	1.91	7.96	4.96	3.64
Depth only (L_d)	0.00	1.00	0.00	0.00	0.00	0.00

Table 3. Evaluation of different confidence weighting schemes on the *val* set.

Confidence weighting	3D detection			Bird's eye view		
	Easy	Moderate	Hard	Easy	Moderate	Hard
Full (c_{3D})	12.50	7.34	4.98	19.49	11.51	8.72
w/o silhouette recon. $(c_{2D}e^{-\alpha_b b_i})$	11.19	6.81	4.61	17.72	10.80	8.20
w/o box protrusion $(c_{2D}e^{-\alpha_m L m_i})$	12.43	7.29	4.89	19.29	11.34	8.64
w/o both (c_{2D})	10.74	6.48	4.40	17.19	10.36	7.95

Confidence Weighting. In Table 3, we illustrate the impact of our proposed confidence weighting schemes. In the last row, we present the accuracies calculated by using the scores obtained from 2D bounding box detection without applying any confidence weighting. In the first row, we present the result after applying the confidence weighting over all scores. Applying both weights improves the accuracy by 11%–16%. The second row shows that weighting the scores by using the silhouette reconstruction loss value significantly impacts the increase.

Table 4. Different segmentation networks, on KITTI 3D *validation* set.

Model	Box AP	Mask AP	2D detection			3D detection			Bird's eye view		
	COCO	COCO	Easy	Moderate	Hard	Easy	Moderate	Hard	Easy	Moderate	Hard
Mask R-CNN X152	50.2	44.0	82.24	74.82	59.34	12.50	7.34	4.98	19.49	11.51	8.72
Panoptic R101-FPN	42.4	–	78.59	72.17	59.66	9.26	5.84	4.35	15.42	9.97	7.44

Effect of the Segmentation Masks. In Table 4, we present the impact of the chosen 2D object detection and segmentation network on the final accuracy. We compared the results of Mask R-CNN X-152 [10] and Panoptic Segmentation R101-FPN, both taken from detectron2 [33] and pretrained on COCO [18]. Mask R-CNN has significantly better mAP than the panoptic segmentation model, especially for the easy and moderate instances. Even though panoptic segmentation provides better semantic masks, it provides worse 2D detection performance for occluded objects. This is due the general nature of panoptic segmentation, since it derives its bounding boxes tightly around the segmentations. As our method is strongly dependent on the tightness of 2D bounding boxes, panoptic segmentation has an overall negative impact on the accuracy.

(a) Example of failures caused by occlusion. The cars on the right side (top row) and the car on the left side (bottom row) are partially visible and the estimation of their 2D bounding boxes is not accurate, which leads to large errors in rotation estimation.

(b) Example of a failure caused by a non-tight 2D bounding box. The inaccurate 2D bounding box estimation leads to a large error in 3D shape estimation and pose translation.

Fig. 5. Two examples of cases where our method fails to estimate accurate object pose translations or rotations. Ground-truth (green)/predicted (red) 3D bounding boxes in the bird's-eye view (BEV) are shown in the right column. The white points in BEV represent projections of object point clouds generated from predicted depth maps. The ground truth information is only used for visualization purposes. (Color figure online)

4.3   Limitations and Failure Cases

The quality of the estimated bounding boxes and segmentation masks clearly impacts the rendering used for computing the loss. Inaccuracies in their estimation will be reflected in the quality of the estimated 3D shape and object pose. Figure 5a shows two example of failure cases due to stronger car occlusions. Because the 2D bounding boxes used in this work are not *amodal*, they only confine the visible parts of the cars. In this case, optimization is done by projecting the 3D shape onto a partial view of the car, which leads to a large error in the rotation estimation. This problem could be solved by fine-tuning the detector on a dataset with annotated amodal 2D bounding boxes.

The size of the estimated 3D shape is also constrained by the size of the estimated 2D bounding box. Figure 5b shows an example of a failure case where the 2D bounding box is larger than the object (non-tight). Because the optimization requires the projection of the 3D shape to fit the 2D bounding box, the shape is constrained to be larger than it should be. In this case, a large translation error can be seen due to a shift in the center of mass.

The computational cost is currently a bottleneck of our method. The current computation time is proportional to the number of detected objects and the mean computation time per image on the KITTI *validation* set is three minutes. The major time is spent on escaping the local minima of the rotation Sect. 3.2 as we had to render multiple times per image.

5 Conclusion

We presented a self-supervised approach, based on differentiable rendering, for 3D shape reconstruction and localization of rigid objects from monocular images. Although used in the context of autonomous driving, our method is generally applicable to many categories and scenarios. We showed that it is possible to use noisy monocular depth and differentiable rendering in conjunction with learned object priors as an alternative to expensive 3D ground-truth labels or LiDAR information.

Future work should investigate alternative approaches to estimating depth from monocular images and registering point clouds. Although significantly improved over the last years, accurate long-range depth estimation is still a challenging problem. This information is of vital importance in autonomous driving, where the system often needs to make quick decisions. Having accurate information about the surrounding environment, as early as possible, allows for better risk assessment, planning and control.

Accurate and fast point cloud registration, on the other hand, is necessary in our pipeline for achieving self-supervision. Many of the current techniques are either slow or require an initial guess for accurate registration. This makes them impractical for applications where real-time inference is required.

Acknowledgements. This work was supported by Toyota Research Institute Advanced Development, Inc. The authors would like to thank Richard Calland, Karim Hamzaoui, Rares Ambrus, Vitor Guizilini for their helpful comments and suggestions.

References

1. Chang, A.X., et al.: ShapeNet: an information-rich 3D model repository. In: CoRR (2015)
2. Chen, W., et al.: Learning to predict 3D objects with an interpolation-based differentiable renderer. In: NeurIPS (2019)
3. Chen, X., Kundu, K., Zhang, Z., Ma, H., Fidler, S., Urtasun, R.: Monocular 3D object detection for autonomous driving. In: CVPR (2016)
4. Chen, X., et al.: 3D object proposals for accurate object class detection. In: NIPS (2015)
5. Engelmann, F., Stückler, J., Leibe, B.: Joint object pose estimation and shape reconstruction in urban street scenes using 3D shape priors. In: Rosenhahn, B., Andres, B. (eds.) GCPR 2016. LNCS, vol. 9796, pp. 219–230. Springer, Cham (2016). https://doi.org/10.1007/978-3-319-45886-1_18
6. Engelmann, F., Stückler, J., Leibe, B.: SAMP: shape and motion priors for 4D vehicle reconstruction. In: WACV (2017)
7. Geiger, A., Lenz, P., Urtasun, R.: Are we ready for Autonomous Driving? The KITTI vision benchmark suite. In: CVPR (2012)

8. Godard, C., Mac Aodha, O., Brostow, G.J.: Unsupervised monocular depth estimation with left-right consistency. In: CVPR (2017)
9. Guizilini, V., Ambrus, R., Pillai, S., Raventos, A., Gaidon, A.: PackNet-SfM: 3D packing for self-supervised monocular depth estimation. In: CoRR (2019)
10. He, K., Gkioxari, G., Dollár, P., Girshick, R.: Mask R-CNN. In: ICCV (2017)
11. Ushiku, Y., Kato, H., Harada, T.: Neural 3D mesh renderer. In: CVPR (2018)
12. Insafutdinov, E., Dosovitskiy, A.: Unsupervised learning of shape and pose with differentiable point clouds. In: NIPS (2018)
13. Kato, H., Harada, T.: Learning view priors for single-view 3D reconstruction. In: CVPR (2019)
14. Kingma, D.P., Ba, J.: Adam: a method for stochastic optimization. In: ICLR (2015)
15. Kulkarni, N., Gupta, A., Tulsiani, S.: Canonical surface mapping via geometric cycle consistency. In: ICCV (2019)
16. Kundu, A., Li, Y., Rehg, J.M.: 3D-RCNN: instance-level 3D object reconstruction via render-and-compare. In: CVPR (2018)
17. Li, T.-M., Aittala, M., Durand, F., Lehtinen, J.: Differentiable Monte Carlo ray tracing through edge sampling. ACM Trans. Graph. (Proc. SIGGRAPH Asia) **37**(6), 222:1–222:11 (2018)
18. Lin, T.-Y., et al.: Microsoft COCO: common objects in context. In: Fleet, D., Pajdla, T., Schiele, B., Tuytelaars, T. (eds.) ECCV 2014. LNCS, vol. 8693, pp. 740–755. Springer, Cham (2014). https://doi.org/10.1007/978-3-319-10602-1_48
19. Liu, S., Li, T., Chen, W., Li, H.: Soft rasterizer: a differentiable renderer for image-based 3D reasoning. In: ICCV (2019)
20. Loper, M.M., Black, M.J.: OpenDR: an approximate differentiable renderer. In: Fleet, D., Pajdla, T., Schiele, B., Tuytelaars, T. (eds.) ECCV 2014. LNCS, vol. 8695, pp. 154–169. Springer, Cham (2014). https://doi.org/10.1007/978-3-319-10584-0_11
21. Ma, X., Wang, Z., Li, H., Ouyang, W., Zhang, P.: Accurate monocular 3D object detection via color-embedded 3D reconstruction for autonomous driving. In: ICCV (2019)
22. Mahjourian, R., Wicke, M., Angelova, A.: Unsupervised learning of depth and egomotion from monocular video using 3D geometric constraints. In: CVPR (2018)
23. Manhardt, F., Kehl, W., Gaidon, A.: ROI-10D: monocular lifting of 2D detection to 6D pose and metric shape. In: CVPR (2019)
24. Choi, H.M., Kang, H., Hyun, Y.: Multi-view reprojection architecture for orientation estimation. In: ICCVW (2019)
25. Mousavian, A., Anguelov, D., Flynn, J., Kosecka, J.: 3D bounding box estimation using deep learning and geometry. In: CVPR (2017)
26. Pillai, S., Ambruş, R., Gaidon, A.: SuperDepth: self-supervised, super-resolved monocular depth estimation. In: ICRA (2019)
27. Simonelli, A., Bulo, S.R., Porzi, L., López-Antequera, M., Kontschieder, P.: Disentangling monocular 3D object detection. In: ICCV (2019)
28. Simonelli, A., Bulò, S.R., Porzi, L., Ricci, E., Kontschieder,P.: Single-stage monocular 3D object detection with virtual cameras. In: CoRR (2019)
29. Stutz, D., Geiger, A.: Learning 3D shape completion under weak supervision. In: IJCV (2018)
30. Tulsiani, S., Efros, A.A., Malik, J.: Multi-view consistency as supervisory signal for learning shape and pose prediction. In: CVPR (2018)
31. Wang, R., Yang, N., Stueckler, J., Cremers, D.: DirectShape: photometric alignment of shape priors for visual vehicle pose and shape estimation. In: ICRA (2020)

32. Wang, Y., Chao, W.-L., Garg, D., Hariharan, B., Campbell, M., Weinberger, K.Q.: Pseudo-LiDAR from visual depth estimation: bridging the gap in 3D object detection for autonomous driving. In: CVPR (2019)
33. Wu, Y., Kirillov, A., Massa, F., Lo, W.-Y., Girshick, R.: Detectron2 (2019). https://github.com/facebookresearch/detectron2
34. Xu, B., Chen, Z.: Multi-level fusion based 3D object detection from monocular images. In: CVPR (2018)
35. You, Y., et al.: Pseudo-LiDAR++: accurate depth for 3D object detection in autonomous driving. In: ICLR (2020)
36. Zakharov, S., Kehl, W., Bhargava, A., Gaidon, A.: Autolabeling 3D objects with differentiable rendering of SDF shape priors. In: CVPR (2020)
37. Zhang, R., Isola, P., Efros, A.A., Shechtman, E., Wang, O.: The unreasonable effectiveness of deep features as a perceptual metric. In: CVPR (2018)
38. Zuffi, S., Kanazawa, A., Berger-Wolf, T., Black, M.J.: Three-D safari: learning to estimate zebra pose, shape, and texture from images "in the wild". In: ICCV (2019)

Shape Prior Deformation for Categorical 6D Object Pose and Size Estimation

Meng Tian[✉] [iD], Marcelo H. Ang Jr.[iD], and Gim Hee Lee[iD]

National University of Singapore, Kent Ridge, Singapore
tianmeng@u.nus.edu, {mpeangh,gimhee.lee}@nus.edu.sg

Abstract. We present a novel learning approach to recover the 6D poses and sizes of unseen object instances from an RGB-D image. To handle the intra-class shape variation, we propose a deep network to reconstruct the 3D object model by explicitly modeling the deformation from a pre-learned categorical shape prior. Additionally, our network infers the dense correspondences between the depth observation of the object instance and the reconstructed 3D model to jointly estimate the 6D object pose and size. We design an autoencoder that trains on a collection of object models and compute the mean latent embedding for each category to learn the categorical shape priors. Extensive experiments on both synthetic and real-world datasets demonstrate that our approach significantly outperforms the state of the art. Our code is available at https://github.com/mentian/object-deformnet.

Keywords: Category-level pose estimation · 3D object detection · Shape generation · Scene understanding

1 Introduction

Accurate 6D object pose estimation plays an important role in a variety of tasks, such as augmented reality, robotic manipulation, scene understanding, etc. In recent years, substantial progress has been made for instance-level 6D object pose estimation, where the exact 3D object models for pose estimation are given. Unfortunately, these methods [18,32,42] cannot be directly generalized to category-level 6D object pose estimation on new object instances with unknown 3D models. Consequently, the category, 6D pose and size of the objects have to be concurrently estimated. Although some other object instances from each category are provided as priors, the high variation of object shapes within the same category makes generalization to new object instances extremely challenging.

To the best of our knowledge, [24] is the first work to address the 6D object pose estimation problem at category-level. This approach defines 6D pose on semantically selected centers and trains a part-based random forest to recover the pose. However, building part representations upon 3D skeleton structures

Electronic supplementary material The online version of this chapter (https://doi.org/10.1007/978-3-030-58589-1_32) contains supplementary material, which is available to authorized users.

limits the generalization capability across unseen object instances. Another work [33] proposes the first data-driven solution and creates a benchmark dataset for this task. They introduce the Normalized Object Coordinate Space (NOCS) to represent different object instances within a category in a unified manner. A region-based network is trained to infer correspondences from object pixels to the points in NOCS. Class label and instance mask of each object are also obtained at the same time. These predictions are used together with the depth map to estimate the 6D pose and size of the object via point matching. However, the lack of explicit representation of shape variations limits their performance.

In this work, we propose to reconstruct the complete object models in the NOCS to capture the intra-class shape variation. More specifically, we first learn the categorical shape priors from the given object instances, and then train a network to estimate the deformation field of the shape prior (that is used to get the reconstructed object model) and the correspondences between object observation and the reconstructed model. The shape prior serves as prior knowledge of the category and encodes geometrical characteristics that are shared by objects of a given category. The deformation predicted by our network captures the instance-specific shape details, i.e. shape variation of that particular instance. We present a method which is applicable across different object categories and data representations to learn the shape priors. In particular, an autoencoder is trained on a collection of object models from various categories. For each category, we compute the mean latent embedding over all instances in the respective category. The categorical shape prior is constructed by passing the mean embedding through a decoder. Note that there is no restriction on the data representation (point cloud, mesh, or voxel) of shape priors or collected models as long as we choose a proper architecture for the encoder and decoder.

We use the Umeyama algorithm [29] to recover the 6D pose and metric size of the object from the correspondences estimated by our network that maps the point cloud obtained from the observed depth map to the points in NOCS. We evaluate our method on two standard benchmarks. Extensive experiments demonstrate the advantage of our network and prove the effectiveness of explicitly modeling the deformation. In summary, the main contributions of this work are:

- We propose a novel deep network for category-level 6D object pose and size estimation; our network explicitly models the deformation from the categorical shape prior to the object model.
- We present a learning-based method which utilizes the latent embeddings to construct the shape prior; our method is applicable across different categories and data representations.
- Our network achieves significantly higher mean average precisions on both synthetic and real-world benchmark datasets.

2   Related Work

Instance-Level Pose Estimation. Existing instance-level pose estimation approaches broadly fall into three categories. The first category casts votes in the pose space and further refines coarse pose with algorithms such as iterative closest point. LINEMOD [9] uses holistic template matching to find the nearest viewpoint. [26] generates a latent code for the input image and search for its nearest neighbor in the codebook. [11,27] aggregate the 6D votes cast by locally-sampled RGB-D patches. The second category directly maps input image to object pose. [10,37] extend 2D object detection network such that it can predict orientation as an add-on to the identity and 2D bounding box of the object. [15,32] regress 6D pose from RGB-D images in an end-to-end framework. The third category relies on establishing point correspondences. [2,12,17] regress the corresponding 3D object coordinate for each foreground pixel. [18,23,28] detect the keypoints of the object on image and then solve a Perspective-n-Point problem. [42] estimates a dense 2D-3D correspondence map between the input image and object model. Although our approach follows the approach from the third category, our task focuses on a more general setting where the object models are not available during inference.

Category-Level Object Detection. The task of 3D object detection aims to estimate 3D bounding boxes of objects in the scene. [25] runs sliding windows in 3D space and generates amodal proposals for objects. [6,14,21] first generate 2D object proposals and then lift the proposals to 3D space. [38,44] are single-stage detectors which directly detect objects from 3D data. Although the above mentioned methods address the problem at category-level, the considered objects are usually constrained to the ground surface, e.g.. instances of typical furniture classes in indoor scenes, cars, pedestrians, and cyclists in outdoor scenes. Consequently, the assumption that rotation is constrained to be only along the gravity direction has to be made. On the contrary, our approach can recover the full 6D pose of objects.

Category-Level Pose Estimation. There are only a few pioneering works focusing on estimating 6D pose of unseen objects. [3] leverages a generative representation of 3D objects and produces a multimodal distribution over poses with mixture density network. However, only rotation is considered in their work. [24] introduces a part-based random forest which employs simple depth comparison features, but our approach deals with RGB-D images. [33] proposes a canonical representation for all instances within an object category. Our approach also makes use of this representation. Instead of directly regressing the coordinates in NOCS, we account for intra-class shape variation by explicitly modeling the deformation from shape prior to object model. [4] trains a variational autoencoder to generate the complete object model. However, the reconstructed shape is not utilized for pose estimation. In our network, shape reconstruction and pose estimation are integrated together. [31] proposes the first category-level pose tracker, while our approach performs pose estimation without using temporal information.

Shape Deformation. 3D deformation is commonly applied for object reconstruction from a single image. [13,20,41] use free-form deformation in conjunction with voxel, mesh and point cloud representations, respectively. [34,36] starts from a coarse shape and predicts a series of deformations to progressively improve

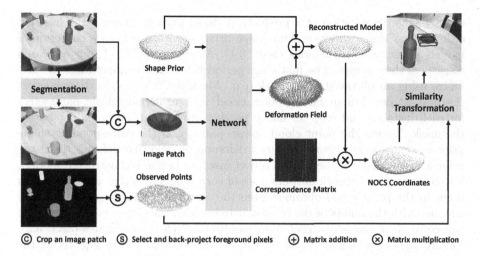

Fig. 1. Overview of our approach. We first obtain a foreground mask for each object instance. Next our network reconstructs the instance (*bowl* as an example) and establishes the correspondences between the observed points and the reconstructed model. Finally, the 6D pose is recovered by estimating a similarity transformation. Refer to Fig. 3 for the details of our network.

the geometry. Similar to us, [35] also supervise the deformation with global feature of the target. However, we circumvent the fixed topology assumption of mesh representation by using point clouds instead.

3   Our Method

Background. Given an RGB-D image as the input, our goal is to detect and localize all visible object instances in the 3D space. The object instances are not seen previously, but must come from known categories. Each object is represented by a class label and an amodal 3D bounding box parameterized by its 6D pose and size. The 6D pose is defined to be the rigid transformation (i.e. rotation and translation) that transforms the object from the reference to the camera coordinate frame. It is common to choose the coordinate frame of the given 3D object models as reference in instance-level 6D object pose estimation. Unfortunately, this is not viable for our category-level task since the instances of the 3D models are not available. To mitigate this problem, we leverage on the Normalized Object Coordinate Space (NOCS) – a shared canonical representation for all possible object instances within a category proposed in [33]. The categorical 6D object pose and size estimation problem is then reduced to finding the similarity transformation between the observed depth map of each object instance and its corresponding points in the NOCS (i.e. *NOCS coordinates*).

Overview. In contrast to [33] that directly outputs the NOCS coordinates from a Convolutional Neural Network (CNN), we propose an intermediate step to estimate the deformation of a pre-learned shape prior to improve the learning of intra-class shape variation. Our shape priors are learned from a collection of models spanning all categories (Sect. 3.1). As shown in Fig. 1, our approach consists of three stages. The first stage performs instance segmentation on color image using an off-the-shelf network (e.g.. Mask R-CNN [7]). Next we convert the masked depth map into a point cloud with camera intrinsic parameters for each instance and crop an image patch according to the bounding box of the mask. Taking the point cloud, image patch, and the corresponding shape prior as inputs, our network outputs a deformation field that deforms the shape prior into the shape of the desired object instance (a.k.a. reconstructed model). Furthermore, our network outputs a set of correspondences that associates each point in the point cloud obtained from the observed depth map of the object instance with the points of the reconstructed model. This set of correspondences is used to mask the reconstructed model into the NOCS coordinates (Sect. 3.2). Finally, the 6D pose and size of the object can be estimated by registering the NOCS coordinates and the point cloud obtained from the observed depth map (Sect. 3.3).

3.1 Categorical Shape Prior

Although object shape varies among different instances, an investigation over the 3D models reveals that objects of the same category (especially artificially generated objects) tend to have semantically and geometrically similar components. For example, cameras are usually made up of a nearly cuboid body and a cylindrical lens; and mugs are typically cylindrical with a handle. These categorical characteristics provide strong priors on the shape reconstruction of novel instances. We propose the learning of a mean shape to capture the high-level characteristics from all the available models for each respective category. To this end, we first train an autoencoder with all available object models and then compute the mean latent embedding of each object category with the encoder. These latent embeddings are passed into the decoder to get the mean shape priors for each object category. Unlike methods such as simple averaging [30] and principal component analysis (PCA) [3] that operate on voxel representations, our autoencoder framework can be easily altered to take any 3D representations.

Given a shape collection $\mathcal{M} = \{M_c^i \mid i = 1, 2, \cdots, N; c = 1, 2, \cdots, C\}$, where M_c^i is the 3D point cloud model of instance i from category c, we independently apply a similarity transformation to each model such that it is properly aligned in the NOCS. This step ensures that the learned shape prior has the same scale and orientation as the target shape to be reconstructed. The encoder Φ takes the point cloud and outputs a low-dimensional feature vector, i.e. the latent embedding $z_c^i \in \mathcal{R}^n$. The decoder Ψ takes this feature vector and outputs a point cloud that reconstructs the input:

$$\hat{M}_c^i = (\Psi \circ \Phi)(M_c^i) = \Psi(z_c^i). \tag{1}$$

Specifically, we adopt the PointNet-like encoder proposed in [40], and a three-layer fully-connected decoder as shown in Fig. 2a. The reconstruction error is

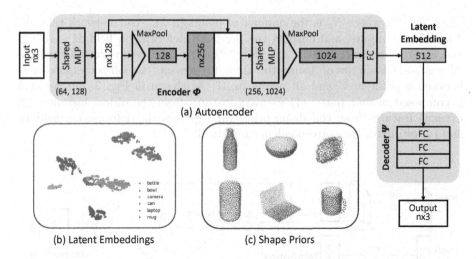

(a) Autoencoder

(b) Latent Embeddings   (c) Shape Priors

Fig. 2. (a) Architecture of the autoencoder. (b) The latent embeddings of all instances are mapped to $\mathcal{R}^2$ with T-SNE for visualization. These instances are from 6 categories - *bottle*, *bowl*, *camera*, *can*, *laptop* and *mug*. (c) Shape priors are reconstructed by passing mean latent embedding of each category through the decoder.

measured by the Chamfer distance:

$$d_{\mathrm{CD}}(M_c^i, \hat{M}_c^i) = \sum_{x \in M_c^i} \min_{y \in \hat{M}_c^i} \|x - y\|_2^2 + \sum_{y \in \hat{M}_c^i} \min_{x \in M_c^i} \|x - y\|_2^2. \qquad (2)$$

The autoencoder is trained on a shape collection by minimizing the reconstruction error. Once the training is converged, we obtain the latent embeddings $\{z_c^i\}$ of all instances in $\mathcal{M}$. Although not explicitly enforced during training, these latent embeddings form clusters in the latent space according to their categories. Figure 2b visualizes the clustering effect of the embeddings. We use T-SNE [16] to further embed these features in $\mathcal{R}^2$ for visualization. Similar clustering results are also observed on a different set of models [39]. Based on this observation, we compute the mean latent embedding for each category and then pass it through the decoder to construct the shape prior:

$$M_c = \Psi(z_c) = \Psi \left(\frac{1}{N_c} \sum_i z_c^i \right). \qquad (3)$$

The resulting categorical shape priors $\{M_c\}$ are shown in Fig. 2c.

3.2 Our Network Architecture

We denote the observation of an object instance as (V, I), where $V \in \mathcal{R}^{N_v \times 3}$ is the point cloud and $I \in \mathcal{R}^{H \times W \times 3}$ is the image patch. N_v denotes the number of foreground pixels with a valid depth value. The corresponding shape prior is

$M_c \in \mathcal{R}^{N_c \times 3}$, where N_c is the number of points in M_c. Our network takes V, I and M_c as inputs, and outputs the per-point deformation field $D \in \mathcal{R}^{N_c \times 3}$ and a correspondence matrix $A \in \mathcal{R}^{N_v \times N_c}$. The final reconstructed model is $M = M_c + D$. Each row of A sums to 1 since it represents the soft correspondences between a point in V and all points in M. As shown in Fig. 3, our network is composed of four parts: (1) extracts features from the object instance; (2) extracts features from the shape prior; (3) regresses the deformation field D; and (4) estimates the correspondence matrix A.

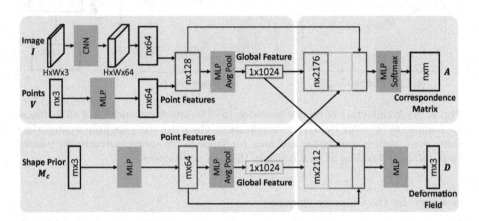

Fig. 3. Our Network Architecture. The upper-left and lower-left branches extract point and global features from the instance and the shape prior respectively. The upper-right branch estimates the correspondence matrix, and the lower-right branch predicts the deformation field. The exchange of global features is the key part of our network.

On the consideration that the depth and color are two different modalities, we follow the pixel-wise dense fusion approach proposed in [32] to effectively extract RGB-D features from the observation. For point cloud V, we use an embedding network similar to PointNet [22] to generate per-point geometric features by mapping each point in V to the d_g-dimensional feature space. The image patch I is processed with a fully convolutional network which follows an encoder-decoder architecture and maps I to $\mathcal{R}^{H \times W \times d_c}$. Next we associate the geometric feature of each point with its corresponding color feature and concatenate the feature pairs. Since each point in V has a corresponding pixel in I, not vice versa, redundant color features are discarded. The concatenated features are termed "instance point features" and fed to another shared multi-layer perceptron. An average pooling layer is used to generate the "instance global feature". The categorical shape prior M_c is a point cloud with purely geometric information. We apply a simpler embedding network to extract the "category point features" and the "category global feature".

The shape prior M_c provides the prior knowledge of the category, i.e. the coarse shape geometry and canonical pose. Although the observation (V, I) is partial, it provides instance-specific details of the target shape. A natural way to reconstruct the object in NOCS is to deform M_c under the guidance of (V, I). Consequently, we concatenate the category and instance global features, and enrich the category point features with the concatenated features. The obtained feature vectors are successively convolved with 1×1 kernels to generate the deformation field D. Similar intuition and feature concatenation strategy also apply to the estimation of A. We combine the instance point features and global feature to aggregate both local and global information for each point. Each point in V gets mapped to the points of the reconstructed model through concatenation with the category global feature. We obtain the NOCS coordinates, denoted as P, of the points in V by multiplying A and M, i.e.

$$P = A \times M = A(M_c + D) \in \mathcal{R}^{N_v \times 3}. \tag{4}$$

3.3 6D Pose Estimation

Our goal is to estimate the 6D pose and size of the object instance. Given depth observation V and its NOCS coordinates P, the optimal similarity transformation parameters (rotation, translation, and scaling) can be computed by solving the absolute orientation problem using Umeyama algorithm [29]. We also implement the RANSAC algorithm [5] for robust estimation.

3.4 Loss Functions

In this section, we define the loss functions used to train our network, and explain how we handle object symmetry during training.

Reconstruction Loss. We assume that ground-truth model M_{gt} is available during training. The deformation field D is supervised indirectly by minimizing the Chamfer distance (c.f. Eq. 2) between M and M_{gt}, i.e. $L_{cd} = d_{CD}(M, M_{gt}) = d_{CD}(M_c + D, M_{gt})$.

Correspondence Loss. It is impractical and unnecessary to pre-compute the ground-truth value for A. Alternatively, we supervise A indirectly via the NOCS coordinates P (which is a result of applying the correspondence matrix on the reconstructed model) since the ground-truth NOCS coordinates P_{gt} can be obtained easily from the object model and its 6D pose through image rendering. We use the smooth L_1 loss function:

$$L_{corr}(P, P_{gt}) = \frac{1}{N_v} \sum_{\mathbf{x} \in P} \sum_{i=1,2,3} \begin{cases} 5(x_i - y_i)^2, & \text{if } |x_i - y_i| \leq 0.1 \\ |x_i - y_i| - 0.05, & \text{otherwise} \end{cases}, \tag{5}$$

where $\mathbf{x} = (x_1, x_2, x_3) \in P$, and $\mathbf{y} = (y_1, y_2, y_3) \in P_{gt}$.

Object symmetry is an inevitable problem for pose estimation algorithms, especially for those that require supervised training. For symmetrical objects, there exists at least one rotation such that appearance of the object is preserved under this rotation. In other words, two observations of a symmetric object can be very similar but with different rotation labels. We follow the solution proposed by [19] to map ambiguous rotations to a canonical one. More specifically, the Map operator for an arbitrary rotation $R \in SO(3)$ is defined as:

$$\text{Map}(R) = R\hat{S}, \text{ with } \hat{S} = \underset{S \in \mathcal{S}(M_c^i)}{\arg\min} \|RS - I_3\|_F, \tag{6}$$

where the proper symmetry group $\mathcal{S}(M_c^i)$ is the set of rotations which preserve the appearance of a given object M_c^i. The experimental datasets assume continuous symmetry and the axis of symmetry is the y-axis of the NOCS. Hence, $\hat{S}$ takes the following form:

$$\hat{S} = \begin{bmatrix} \cos\hat{\theta} & 0 & -\sin\hat{\theta} \\ 0 & 1 & 0 \\ \sin\hat{\theta} & 0 & \cos\hat{\theta} \end{bmatrix}, \text{ with } \hat{\theta} = \arctan 2(R_{13} - R_{31}, R_{11} + R_{33}), \tag{7}$$

where R_{11}, R_{13}, R_{31}, and R_{33} are the elements of R. During training, we apply the Map operator to the rotation label: $R_{gt} \leftarrow R_{gt}\hat{S}$ to eliminate the rotation ambiguity of any symmetric object with the ground-truth pose (R_{gt}, T_{gt}). In practice, our network is supervised by ground-truth NOCS coordinates P_{gt}. Equivalently, we transform P_{gt} by $\hat{S}^T$: $P_{gt} \leftarrow \hat{S}^T P_{gt}$.

Regularization Losses. Row A_i of the matrix A represents the distribution over the correspondences between i-th point of V and the points in M. A_i can be understood as a relaxed one-hot vector, since each point of V usually can be well approximated by at most three points of M. We encourage A_i to be a peaked distribution by minimizing the average cross entropy: $L_{\text{entropy}} = \frac{1}{N_v} \sum_i \sum_j -A_{i,j} \log A_{i,j}$. We also regularize D to discourage large deformations: $L_{\text{def}} = \frac{1}{N_C} \sum_{\mathbf{d}_i \in D} \|\mathbf{d}_i\|_2$. Minimal deformation preserves the semantic consistency between shape prior and the reconstructed model. For example, we want that the point belongs to the handle of the "mug" prior remains in the handle after deformation. This consistency loss is beneficial for the prediction of the correspondence matrix A.

In summary, the overall objective is a weighted sum of all four losses:

$$L = \lambda_1 L_{\text{cd}} + \lambda_2 L_{\text{corr}} + \lambda_3 L_{\text{entropy}} + \lambda_4 L_{\text{def}}. \tag{8}$$

4   Experiments

4.1   Experimental Setup

Datasets. The CAMERA [33] dataset is generated by rendering and compositing synthetic objects into real scenes in a context-aware manner. In total, there are 300K composite images, where 25K are set aside for evaluation. The training set contains 1085 object instances selected from 6 different categories - *bottle, bowl, camera, can, laptop* and *mug*. The evaluation set contains 184 different instances. The REAL [33] dataset is complementary to the CAMERA. It captures 4300 real-world images of 7 scenes for training, and 2750 real-world images of 6 scenes for evaluation. Each set contains 18 real objects spanning the 6 categories. The two evaluation sets are referred to as CAMERA25 and REAL275.

Evaluation Metric. Following [33], we independently evaluate the performance of 3D object detection and 6D pose estimation. We report the average precision at different Intersection-Over-Union (IoU) thresholds for object detection. For 6D pose evaluation, the average precision is computed at $n°$ mcm. We ignore the rotational error around the axis of symmetry for symmetrical object categories (e.g. *bottle*, *bowl*, and *can*). Specially, we treat *mug* as symmetric object in the absence of the handle, and asymmetric object otherwise.

Baseline. Wang et al. [33] is currently the only publicly available code and datasets for the 6D object pose and size estimation task. Futhermore, it is also the state-of-the-art performance on the task. Hence, we choose [33] as our baseline for comparison.

4.2 Implementation Details

We collect all the instances in the CAMERA training dataset to train the autoencoder. Shape priors are learned from this collection and used in all experiments. Each prior consists of 1024 points. We use the Mask R-CNN implemented by matterport [1] for instance segmentation. For each detected instance, we resize the image patch to 192×192, and randomly sample 1024 points by repetition (if insufficient point count) or downsampling (if sufficient point count). To extract instance color features, we choose the PSPNet [43] with ResNet-18 [8] as backbone. We randomly select 5 point-pairs to generate a hypothesis for the RANSAC-based pose fitting. The maximum number of iteration is 128 and inlier threshold is set to 10% of the object diameter. For the hyperparameters of the total loss, we empirically find that $\lambda_1 = 5.0$, $\lambda_2 = 1.0$, $\lambda_3 = 1e-4$, and $\lambda_4 = 0.01$ are good choices.

4.3 Comparison to Baseline

We compare our approach to the Baseline [33] on CAMERA25 and REAL275. Quantitative results are summarized in Table 1.

CAMERA25. In the setting of estimating 6D object pose and size from an RGB-D image, we achieve a mAP of 83.1% for 3D IoU at 0.75, and a mAP of 54.3% for 6D pose at $5° 2$ cm. Our results are 14% and 22% higher than the Baseline [33], respectively. We naively remove the depth input and related sub-networks in our network (i.e. RGB image as the only input) to make fair comparisons with the Baseline [33], which takes an RGB image as its input. As shown in Table 1, our results without depth input are still significantly better than the Baseline [33] (i.e. +15.5% and +17.9%). On one hand, this experiment shows the advantage of explicit handling of the intra-class shape variation, and the effectiveness of our method which reconstructs the object via deformation. On the other hand, it also shows that adding depth to the network does help

Table 1. Comparisons on CAMERA25 and REAL275. We report the mAP w.r.t. different thresholds on 3D IoU, and rotation and translation errors.

Data	Method	mAP					
		$3D_{50}$	$3D_{75}$	$5° \, 2\,cm$	$5° \, 5\,cm$	$10° \, 2\,cm$	$10° \, 5\,cm$
CAMERA25	Baseline [33]	83.9	69.5	32.3	40.9	48.2	64.6
	Ours (RGB)	93.1	**84.6**	50.2	54.5	70.4	78.6
	Ours (RGB-D)	**93.2**	83.1	**54.3**	**59.0**	**73.3**	**81.5**
REAL275	Baseline [33]	**78.0**	30.1	7.2	10.0	13.8	25.2
	Ours (RGB)	75.2	46.5	15.7	18.8	33.7	47.4
	Ours (RGB-D)	77.3	**53.2**	**19.3**	**21.4**	**43.2**	**54.1**

Table 2. Evaluation of shape reconstruction with CD metric ($\times 10^{-3}$).

Data	Model	Bottle	Bowl	Camera	Can	Laptop	Mug	Average
CAMERA25	Reconstruction	1.81	1.63	4.02	0.97	1.98	1.42	1.97
	Shape Prior	3.41	2.20	9.01	2.21	3.27	2.10	3.70
REAL275	Reconstruction	3.44	1.21	8.89	1.56	2.91	1.02	3.17
	Shape Prior	4.99	1.16	9.85	2.38	7.14	0.97	4.41

to improve overall performance, although our improved performance does not rely on it solely. Given that depth image is required to uniquely determine the scale of the object, we recommend it in practical applications. The top row of Fig. 4 shows the average precision at different error thresholds for all 6 object categories. It provides independent analysis for 3D IoU, rotation, and translation error.

REAL275. The REAL training set only contains 3 object instances per category, we enlarge this training set such that the network can generalize well to unseen objects. Following the Baseline [33], we randomly select data from CAMERA and REAL training set according to a ratio of 3 : 1. In fair comparison to the Baseline [33], our approach improves the mAP by 23.1% for 3D IoU at 0.75 and 12.1% for 6D pose at $5° \, 2\,cm$. In strict comparison, we can still outperform by 16.4% and 8.5%, respectively. These results provide further evidence to support our approach. Figure 4 (bottom) shows more detailed analysis of the errors.

4.4   Evaluation of Shape Reconstruction

To evaluate the quality of the reconstruction, we compute the CD metric (c.f. Eq. 2) of the reconstructed model from our method with the ground truth model in the NOCS. We get a CD metric of 1.97 on CAMERA25 and 3.17 on REAL275. In comparison, the CD metrics are 3.70 and 4.41 on the respective dataset for the shape priors from our autoencoder. The better CD metrics of the reconstructed

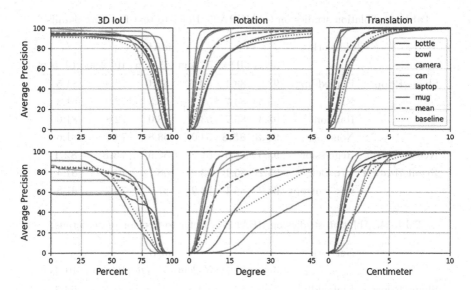

Fig. 4. Average precision vs. error thresholds on CAMERA25 (top row) and REAL275 (bottom row).

models compared to the shape priors show that the deformation estimation in our framework improves the quality of the 3D model reconstruction. Table 2 shows the CD metric of our reconstructed models and the shape priors for each category.

4.5 Ablation Studies

Different shape priors. We first evaluate how different shape priors influence the performance. All settings are kept the same in this experiment except for the choices of the priors. Results are summarized in Table 3 and 4. "Embedding" refers to the priors obtained from decoding the mean latent embeddings. We also try the instance whose latent embedding has the minimum L_2 distance to the mean latent embedding (denoted as "NN"). In addition, we explore random selection of one instance per category from the shape collection to compose our priors (denoted as "Random"). In general, our approach remains stable under different priors. Our network can adapt to different shape priors because the deformation is explicitly estimated. We achieve the best result for accurate pose (i.e. $5° \, 2$ cm) estimation when the learned categorical shape prior is used. Since our main target is to recover the 6D pose, we choose "Embedding" as our best model. To validate whether the priors are necessary, we use a point cloud uniformly sampled from a sphere of diameter one as our prior (denoted as "None"). The mAP decreases by 3.7% on real dataset when there are no priors, but the best result is achieved for object size estimation. Although shape priors are beneficial for estimating 6D pose, they sometimes bias shape reconstruction.

Table 3. Ablation studies on CAMERA25. Refer to text for more details.

	Network	mAP					
		$3D_{50}$	$3D_{75}$	5° 2cm	5° 5cm	10° 2cm	10° 5cm
Shape Priors	Embedding	93.2	83.1	**54.3**	**59.0**	73.3	81.5
	NN	93.3	85.7	52.7	57.3	72.9	81.3
	Random	93.3	85.7	53.4	58.0	72.8	81.0
	None	**93.3**	**85.8**	54.0	58.8	73.1	81.6
NOCS Coords	Regression	93.3	85.3	51.2	55.6	**73.8**	**82.1**
Regularization	w/o Def.	93.2	85.1	53.9	58.7	73.1	81.4
	w/o Entropy	93.2	85.1	53.2	57.9	73.2	81.8

Table 4. Ablation studies on REAL275. Refer to text for more details.

	Network	mAP					
		$3D_{50}$	$3D_{75}$	5° 2cm	5° 5cm	10° 2cm	10° 5cm
Shape Priors	Embedding	77.3	53.2	**19.3**	**21.4**	**43.2**	**54.1**
	NN	75.9	52.6	17.0	19.0	42.0	51.6
	Random	75.8	52.2	17.9	20.1	42.3	51.3
	None	77.2	**55.5**	15.6	19.8	38.4	53.6
NOCS Coords	Regression	**78.7**	54.9	13.7	14.9	42.5	51.4
Regularization	w/o Def.	77.1	50.2	13.4	15.4	37.3	49.8
	w/o Entropy	77.3	53.3	15.7	18.8	38.5	51.3

Directly regress the NOCS Coordinates? As indicated by Eq. 4, our approach decouples the NOCS coordinates P to shape reconstruction M and dense
correspondences A. However, both the network architecture and the training
will be much simpler when we follow [33] to regress P directly (denoted as
"Regression" in Table 3 and 4). For 6D pose estimation, the mAP of "Regression" at 5° 2 cm is notably lower than "Embedding" on CAMERA25 (−3.1%)
and REAL275 (−5.6%). This result further supports the benefit of handling
shape variation via reconstruction over naive regression of the NOCS coordinates. "Regression" achieves slightly better mAP for object size estimation since
it only finds the NOCS coordinates for the observed part, while "Embedding"
needs to complete the unknown part of the object.

Regularization losses. To validate the necessity of the two regularization
losses, we train the network without regularizing deformation or correspondence,
while keeping all the rest same as "Embedding". The mean average precisions of
both variants are still comparable to "Embedding" on synthetic dataset. However, mAP of 6D pose at 5° 2 cm drops noticeably (−5.9% and −3.6%) on the
more difficult real dataset.

4.6   Qualitative Results

In Fig. 5, we provide several qualitative results from both synthetic and real instances. The 6D pose and object size can be reliably recovered from noisy point correspondences using RANSAC-based pose fitting. Shape reconstruction can capture the variations between instances. The qualities of our predictions are generally better on synthetic data than real data, which is an indication that observation noise needs more attention in our future work. Out of the six categories, *camera* shows less accurate reconstruction due to its more complicated and varying geometry.

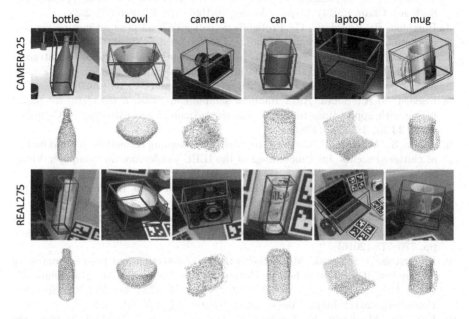

Fig. 5. Examples of qualitative results from CAMERA25 (top rows) and REAL275 (bottom rows). For each example, we visualize the results of pose estimation and the reconstructed model M. Our estimations are shown in red, while the ground truths are shown in green. (Color figure online)

5   Conclusions

We present a novel approach for category-level 6D object pose estimation. Our network explicitly models intra-class shape variation by the estimation of the deformation from a shape prior to the object model. Shape priors are learned form a collection of object models and constructed in the latent space. Experiments on synthetic and real datasets demonstrate the advantage of our proposed approach.

Acknowledgements. This research is supported in parts by the Singapore MOE Tier 1 grant R-252-000-A65-114, and the Agency for Science, Technology and Research (A*STAR) under its AME Programmatic Funding Scheme (Project #A18A2b0046).

References

1. Abdulla, W.: Mask r-cnn for object detection and instance segmentation on keras and tensorflow (2017). https://github.com/matterport/Mask_RCNN
2. Brachmann, E., Krull, A., Michel, F., Gumhold, S., Shotton, J., Rother, C.: Learning 6D object pose estimation using 3d object coordinates. In: Fleet, D., Pajdla, T., Schiele, B., Tuytelaars, T. (eds.) ECCV 2014. LNCS, vol. 8690, pp. 536–551. Springer, Cham (2014). https://doi.org/10.1007/978-3-319-10605-2_35
3. Burchfiel, B., Konidaris, G.: Probabilistic category-level pose estimation via segmentation and predicted-shape priors. arXiv preprint arXiv:1905.12079 (2019)
4. Chen, D., Li, J., Wang, Z., Xu, K.: Learning canonical shape space for category-level 6d object pose and size estimation. In: Proceedings of the IEEE/CVF Conference on Computer Vision and Pattern Recognition (CVPR) (2020)
5. Fischler, M.A., Bolles, R.C.: Random sample consensus: a paradigm for model fitting with applications to image analysis and automated cartography. Commun. ACM **24**(6), 381–395 (1981)
6. Gupta, S., Arbeláez, P., Girshick, R., Malik, J.: Aligning 3d models to rgb-d images of cluttered scenes. In: Proceedings of the IEEE Conference on Computer Vision and Pattern Recognition, pp. 4731–4740 (2015)
7. He, K., Gkioxari, G., Dollár, P., Girshick, R.: Mask r-cnn. In: Proceedings of the IEEE International Conference on Computer Vision, pp. 2961–2969 (2017)
8. He, K., Zhang, X., Ren, S., Sun, J.: Deep residual learning for image recognition. In: Proceedings of the IEEE Conference on Computer Vision and Pattern Recognition, pp. 770–778 (2016)
9. Hinterstoisser, S., et al.: Model based training, detection and pose estimation of texture-less 3D objects in heavily cluttered scenes. In: Lee, K.M., Matsushita, Y., Rehg, J.M., Hu, Z. (eds.) ACCV 2012. LNCS, vol. 7724, pp. 548–562. Springer, Heidelberg (2013). https://doi.org/10.1007/978-3-642-37331-2_42
10. Kehl, W., Manhardt, F., Tombari, F., Ilic, S., Navab, N.: Ssd-6d: Making rgb-based 3d detection and 6d pose estimation great again. In: Proceedings of the IEEE International Conference on Computer Vision, pp. 1521–1529 (2017)
11. Kehl, W., Milletari, F., Tombari, F., Ilic, S., Navab, N.: Deep learning of local rgb-d patches for 3d object detection and 6d pose estimation. In: Leibe, B., Matas, J., Sebe, N., Welling, M. (eds.) ECCV 2016. LNCS, vol. 9907, pp. 205–220. Springer, Cham (2016). https://doi.org/10.1007/978-3-319-46487-9_13
12. Krull, A., Brachmann, E., Michel, F., Ying Yang, M., Gumhold, S., Rother, C.: Learning analysis-by-synthesis for 6d pose estimation in rgb-d images. In: Proceedings of the IEEE International Conference on Computer Vision, pp. 954–962 (2015)
13. Kurenkov, A., et al.: Deformnet: Free-form deformation network for 3d shape reconstruction from a single image. In: 2018 IEEE Winter Conference on Applications of Computer Vision (WACV), pp. 858–866. IEEE (2018)
14. Lahoud, J., Ghanem, B.: 2d-driven 3d object detection in rgb-d images. In: Proceedings of the IEEE International Conference on Computer Vision, pp. 4622–4630 (2017)

15. Li, C., Bai, J., Hager, G.D.: A unified framework for multi-view multi-class object pose estimation. In: Proceedings of the European Conference on Computer Vision (ECCV), pp. 254–269 (2018)

16. Maaten, L.V.D.: Visualizing data using t-sne. J. Mach. Learn. Res. **9**, 2579–2605 (2008)

17. Michel, F., et al.: Global hypothesis generation for 6d object pose estimation. In: Proceedings of the IEEE Conference on Computer Vision and Pattern Recognition, pp. 462–471 (2017)

18. Peng, S., Liu, Y., Huang, Q., Zhou, X., Bao, H.: Pvnet: Pixel-wise voting network for 6dof pose estimation. In: Proceedings of the IEEE Conference on Computer Vision and Pattern Recognition, pp. 4561–4570 (2019)

19. Pitteri, G., Ramamonjisoa, M., Ilic, S., Lepetit, V.: On object symmetries and 6d pose estimation from images. In: 2019 International Conference on 3D Vision (3DV), pp. 614–622. IEEE (2019)

20. Pontes, J.K., Kong, C., Sridharan, S., Lucey, S., Eriksson, A., Fookes, C.: Image2mesh: a learning framework for single image 3d reconstruction. In: Jawahar, C.V., Li, H., Mori, G., Schindler, K. (eds.) ACCV 2018. LNCS, vol. 11361, pp. 365–381. Springer, Cham (2019). https://doi.org/10.1007/978-3-030-20887-5_23

21. Qi, C.R., Liu, W., Wu, C., Su, H., Guibas, L.J.: Frustum pointnets for 3d object detection from rgb-d data. In: Proceedings of the IEEE Conference on Computer Vision and Pattern Recognition, pp. 918–927 (2018)

22. Qi, C.R., Su, H., Mo, K., Guibas, L.J.: Pointnet: Deep learning on point sets for 3d classification and segmentation. In: Proceedings of the IEEE Conference on Computer Vision and Pattern Recognition, pp. 652–660 (2017)

23. Rad, M., Lepetit, V.: Bb8: a scalable, accurate, robust to partial occlusion method for predicting the 3d poses of challenging objects without using depth. In: Proceedings of the IEEE International Conference on Computer Vision, pp. 3828–3836 (2017)

24. Sahin, C., Kim, T.K.: Category-level 6d object pose recovery in depth images. In: Proceedings of the European Conference on Computer Vision (ECCV) Workshops (2018)

25. Song, S., Xiao, J.: Deep sliding shapes for amodal 3d object detection in rgb-d images. In: Proceedings of the IEEE Conference on Computer Vision and Pattern Recognition, pp. 808–816 (2016)

26. Sundermeyer, M., Marton, Z.C., Durner, M., Brucker, M., Triebel, R.: Implicit 3d orientation learning for 6d object detection from rgb images. In: Proceedings of the European Conference on Computer Vision (ECCV), pp. 699–715 (2018)

27. Tejani, A., Tang, D., Kouskouridas, R., Kim, T.-K.: Latent-class hough forests for 3d object detection and pose estimation. In: Fleet, D., Pajdla, T., Schiele, B., Tuytelaars, T. (eds.) ECCV 2014. LNCS, vol. 8694, pp. 462–477. Springer, Cham (2014). https://doi.org/10.1007/978-3-319-10599-4_30

28. Tekin, B., Sinha, S.N., Fua, P.: Real-time seamless single shot 6d object pose prediction. In: Proceedings of the IEEE Conference on Computer Vision and Pattern Recognition, pp. 292–301 (2018)

29. Umeyama, S.: Least-squares estimation of transformation parameters between two point patterns. IEEE Trans. Pattern Anal. Mach. Intell. **4**, 376–380 (1991)

30. Wallace, B., Hariharan, B.: Few-shot generalization for single-image 3d reconstruction via priors. In: Proceedings of the IEEE International Conference on Computer Vision, pp. 3818–3827 (2019)

31. Wang, C., et al.: 6-pack: Category-level 6d pose tracker with anchor-based keypoints. In: International Conference on Robotics and Automation (ICRA) (2020)

32. Wang, C., et al.: Densefusion: 6d object pose estimation by iterative dense fusion. In: Proceedings of the IEEE Conference on Computer Vision and Pattern Recognition, pp. 3343–3352 (2019)
33. Wang, H., Sridhar, S., Huang, J., Valentin, J., Song, S., Guibas, L.J.: Normalized object coordinate space for category-level 6d object pose and size estimation. In: Proceedings of the IEEE Conference on Computer Vision and Pattern Recognition, pp. 2642–2651 (2019)
34. Wang, N., Zhang, Y., Li, Z., Fu, Y., Liu, W., Jiang, Y.G.: Pixel2mesh: Generating 3d mesh models from single rgb images. In: Proceedings of the European Conference on Computer Vision (ECCV), pp. 52–67 (2018)
35. Wang, W., Ceylan, D., Mech, R., Neumann, U.: 3dn: 3d deformation network. In: Proceedings of the IEEE Conference on Computer Vision and Pattern Recognition, pp. 1038–1046 (2019)
36. Wen, C., Zhang, Y., Li, Z., Fu, Y.: Pixel2mesh++: Multi-view 3d mesh generation via deformation. In: Proceedings of the IEEE International Conference on Computer Vision, pp. 1042–1051 (2019)
37. Xiang, Y., Schmidt, T., Narayanan, V., Fox, D.: Posecnn: A convolutional neural network for 6d object pose estimation in cluttered scenes. Robotics: Science and Systems (RSS) (2018)
38. Yang, B., Luo, W., Urtasun, R.: Pixor: Real-time 3d object detection from point clouds. In: Proceedings of the IEEE conference on Computer Vision and Pattern Recognition, pp. 7652–7660 (2018)
39. Yang, Y., Feng, C., Shen, Y., Tian, D.: Foldingnet: Point cloud auto-encoder via deep grid deformation. In: Proceedings of the IEEE Conference on Computer Vision and Pattern Recognition, pp. 206–215 (2018)
40. Yuan, W., Khot, T., Held, D., Mertz, C., Hebert, M.: Pcn: Point completion network. In: 2018 International Conference on 3D Vision (3DV), pp. 728–737. IEEE (2018)
41. Yumer, M.E., Mitra, N.J.: Learning semantic deformation flows with 3d convolutional networks. In: Leibe, B., Matas, J., Sebe, N., Welling, M. (eds.) ECCV 2016. LNCS, vol. 9910, pp. 294–311. Springer, Cham (2016). https://doi.org/10.1007/978-3-319-46466-4_18
42. Zakharov, S., Shugurov, I., Ilic, S.: Dpod: 6d pose object detector and refiner. In: Proceedings of the IEEE International Conference on Computer Vision, pp. 1941–1950 (2019)
43. Zhao, H., Shi, J., Qi, X., Wang, X., Jia, J.: Pyramid scene parsing network. In: Proceedings of the IEEE Conference on Computer Vision and Pattern Recognition, pp. 2881–2890 (2017)
44. Zhou, Y., Tuzel, O.: Voxelnet: End-to-end learning for point cloud based 3d object detection. In: Proceedings of the IEEE Conference on Computer Vision and Pattern Recognition, pp. 4490–4499 (2018)

Dynamic and Static Context-Aware LSTM for Multi-agent Motion Prediction

Chaofan Tao[1,2], Qinhong Jiang[3], Lixin Duan[2], and Ping Luo[1(✉)]

[1] The University of Hong Kong, Pokfulam, Hong Kong, China
tcftrees@gmail.com, pluo@cs.hku.hk
[2] University of Electronic Science and Technology of China, Chengdu, China
lxduan@gmail.com
[3] SenseTime, Beijing, China
jiangqinhong@sensetime.com

Abstract. Multi-agent motion prediction is challenging because it aims to foresee the future trajectories of multiple agents (*e.g.* pedestrians) simultaneously in a complicated scene. Existing work addressed this challenge by either learning social spatial interactions represented by the positions of a group of pedestrians, while ignoring their temporal coherence (*i.e.* dependencies between different long trajectories), or by understanding the complicated scene layout (*e.g.* scene segmentation) to ensure safe navigation. However, unlike previous work that isolated the spatial interaction, temporal coherence, and scene layout, this paper designs a new mechanism, *i.e.*, Dynamic and Static Context-aware Motion Predictor (DSCMP), to integrates these rich information into the long-short-term-memory (LSTM). It has three appealing benefits. (1) DSCMP models the dynamic interactions between agents by learning both their spatial positions and temporal coherence, as well as understanding the contextual scene layout. (2) Different from previous LSTM models that predict motions by propagating hidden features frame by frame, limiting the capacity to learn correlations between long trajectories, we carefully design a differentiable queue mechanism in DSCMP, which is able to explicitly memorize and learn the correlations between long trajectories. (3) DSCMP captures the context of scene by inferring latent variable, which enables multimodal predictions with meaningful semantic scene layout. Extensive experiments show that DSCMP outperforms state-of-the-art methods by large margins, such as 9.05% and 7.62% relative improvements on the ETH-UCY and SDD datasets respectively.

Keywords: Motion prediction · Trajectory forecasting · Social model

Electronic supplementary material The online version of this chapter (https://doi.org/10.1007/978-3-030-58589-1_33) contains supplementary material, which is available to authorized users.

1   Introduction

Multi-agent motion prediction is an important task for many real-world applications such as self-driving vehicle, traffic surveillance, and autonomous mobile robot. However, it is challenging because it aims at foreseeing the future trajectories of multiple agents such as pedestrians simultaneously in a complicated scene. Existing work [1–3,5,15,21,29,35] that addressed this challenge can be generally partitioned into two categories. In the first category [1,2,5,29], previous work predicted the motions by learning social spatial interactions, which are represented by the positions of pedestrians. However, these approaches typically ignored the dependency between different long trajectories of pedestrians. In the second category [3,15,21,35], prior arts combined scene understanding to regularize the predicted trajectory, such as visual feature of the complicated scene layout.

Different from existing work that either model agents' interactions or the scene layout, we carefully designed novel mechanisms in LSTM to model dynamic interactions of pedestrians in both spatial and temporal dimensions, as well as modeling the semantic scene layout as latent probabilistic variable to constrain the predictions. These design principles enable our model to predict multiple trajectories for each agent that cohere in time and space with the other agents. We see that the proposed method outperforms its counterparts in many benchmarks as shown in Fig. 1(c).

We name the proposed method as Dynamic and Static Context-aware Motion Predictor (DSCMP), which has an encoder-decoder structure of LSTM that has carefully devised mechanisms to tackle multi-agent motion prediction. DSCMP has three appealing benefits that previous work did not have.

For the first benefit, unlike existing methods [1,5,21,29,35] that employed recurrent neural networks (RNNs) to learn motion by passing message frame by frame, DSCMP incorporates a queue mechanism in LSTM to explicitly propagate hidden features of multiple frames, enabling to capture long trajectories among pedestrians more explicitly than prior arts.

Specifically, the vanilla LSTM in previous approaches [4,22] attempts to learn a frame-by-frame predictor for each agent i, denoted as $m_{t+1}^i = p(m_t^i, h_{t-1}^i)$, where $p(\cdot)$ is a prediction function of LSTM, m_t^i represents the current motion state (*i.e.* the x, y positions) at the t-th frame and h_{t-1}^i is the hidden feature of the previous single frame. The frame-wise models hinders their capacity to capture the dependency between long trajectories of pedestrians.

The recent approaches of social-aware LSTM models [1,5,27] modified the above vanilla LSTM by using $m_{t+1}^i = p(m_t^i, \bigcup_i^{N(i)} h_{t-1}^i)$, where $\bigcup$ denotes combination of a set of hidden features of the spatial neighbors of the i-th pedestrian (denoted by $N(i)$) at the previous $(t-1)$-th frame.

However, the above methods are insufficient to consider the interactions across agents. For example, as shown in Fig. 1a, the agent 2 is heading towards agent 1 and agent 3. To avoid collision, the agent 2 tends to adjust his future trajectory by anticipating the intention of agent 1 and 3 based on their recent movement history, rather than their states at the previous one frame only.

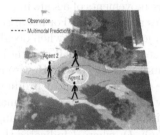
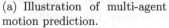
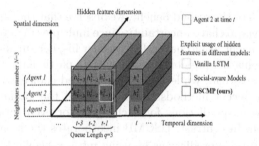

(a) Illustration of multi-agent motion prediction.

(b) Comparisons of different approaches.

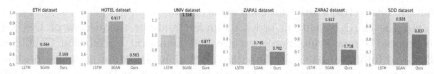

(c) Comparisons of the Relative Average Distance Error (ADE) on various datasets.

Fig. 1. (a) illustrates motion prediction in a real-world scenario, where both the dynamic motion context across agents and static context are involved. (b) compares various methods including LSTM, social-aware models (e.g. SGAN [5]) and DSCMP. The queue mechanism in DSCMP enriches the scope of dynamic context at each frame, enabling DSCMP to effectively learn long motions. (c) compares the Average Distance Error (ADE) of LSTM [6], SGAN [5] and DSCMP on ETH (two subsets: ETH, HOTEL) [17], UCY (three subsets: UNIV, ZARA1, ZARA2) [13], SDD [19] by using the performance of LSTM as baseline (i.e. unit 1; lower value is better). We see that DSCMP surpasses its counterparts by large margins. Best viewed in color. (Color figure online)

Different from the above existing approaches, a LSTM is carefully designed in DSCMP to learn both spatial dependencies and temporal coherence of moving pedestrians. The LSTM contains two modules, including an Individual Context Module (ICM) and a Social-aware Context Module (SCM). As shown in Fig. 1b, our model fully understands spatial and temporal contexts across agents by learning a predictor denoted as $m_{t+1}^i = p(m_t^i, \bigcup_i^{N(i)} Q_t^i)$, where Q_t^i denotes a set of features not only across agents at a certain frame, but also across multiple successive frames of different agents.

More specific, the ICM of DSCMP passes feature of the current motion state and the corresponding feature queue into LSTM cell. Multiple forget gates control the information flow of the frames in the queue. At each iteration, we update the queues by appending the features of the latest frame, and popping out the earliest features. Furthermore, the SCM of DSCMP refines the updated queues by using the queues of the neighboring agents. Since these queues preserve agent-specific motion cues in the past multiple frames, we are able to learn a long-range spatial-temporal interactions with the aggregation of queues.

For the second benefit, we observe that the future movements of agents in real scenarios have uncertainty, since multiple trajectories are plausible. For instance, an agent would naturally consider his/her surrounding scene layout when deciding his/her possible future paths. In particular, an agent could turn left or right in a crossroad, while he/she has limited choices around a street corner. However, the recent methods either neglected the guidance of scene layout to produce diverse predictions for each agent, or even totally ignored the scene information. In contrast, DSCMP incorporates the scene information into the learning of diverse predictions by using $m_{t+1}^i = p(m_t^i, \bigcup_i^{N(i)} Q_t^i, I)$, where I indicates the semantic scene feature after scene segmentation. In practice, this semantic scene feature is modeled as a latent variable of a probabilistic distribution to predict multiple future trajectories for each agent.

For the third benefit, in order to understand the uniqueness of DSCMP, we propose a new evaluation metric called Temporal Correlation Coefficient (TCC) to fully evaluate the temporal correlation of motion patterns, bridging the gap where the commonly used metric such as Average Distance Error (ADE) and Final Distance Error (FDE) are insufficient to evaluate temporal motion correlations. Extensive experiments on dataset ETH [17]-UCY [13], SDD [19] show that DSCMP surpasses state-of-the-art methods on all the above metrics by large margins, such as 9.05% and 7.62% relative improvements on metric ADE compared to the latest method STGAT [7] method.

To summarize the above benefits, this work has three main **contributions**. **(1)** We present a novel future motion predictor, named DSCMP, which is able to explicitly model both the spatial and temporal interactions between different agents, as well as producing multiple probabilistic predictions of future paths for each agent. **(2)** We carefully design the LSTM modules in DSCMP to achieve the above purposes, where all modules can be trained end-to-end including the Individual Context Module (ICM), the Social-aware Context Module (SCM), and a latent scene information module. **(3)** Extensive experiments on the ETH [17], UCY [13], and SDD [19] datasets demonstrate that DSCMP outperforms its counterparts by large margin in multiple evaluation metrics such as ADE and FDE, as well as a new metric, Temporal Correlation Coefficient (TCC), proposed by us to better examine the temporal correlation of motion prediction.

2   Related Work

Motion Prediction. Early approaches for motion prediction [20] like physics-based methods [17,18,30] and planning-based methods [9,12,25,31] are usually limited by hand-crafted kinematic equations and reward function respectively. With the development of recurrent neural networks, pattern-based methods [1, 5,8,10,11,11,16,27,29,33] have been studied recently. While most of the models consider the agents in world coordinates, some work [10,16] explore trajectory prediction with egocentric vision.

A pioneering pattern-based work that combine the LSTM and social interactions is introduced in [1]. The authors of [5] proposes an adversarial framework to sequentially generate predictions. A social pooling is employed to learn the spatial dependencies among agents. Spatio-temporal graphs [8,14,27] are adopted to model the relations on a complete graph, whereas these methods suffer from implicit modelling of dynamic edges or poor scalability with $O(N^2)$ complexity. STGAT [7] is the most relevant method to our work. It considers the temporal correlations of motion in multiple frames. Unlike STGAT deduces a single correlation representation from the whole observation process, we explicitly keep track of the temporal correlation for each iteration during observation. We also take scene context into consideration.

Contextual Understanding. Humans are capable of inferring and reasoning by understanding the context. Rich contextual information is proven to be valuable in sequential data modeling (e.g. video, language, speech). Attention mechanism [26,32] has shown great success in concentrating on the significant parts of visual or textual inputs at each time steps. Non-local operation [28] works as a generic block to capture the dependencies directly by computing the pair-wise relations in the long-range context. In the field of motion prediction, graph attention [7,11] assigns different importance to the neighboring pedestrians to involve the social-aware interactions. The authors of [21,35] encode visual features of scene context in the LSTM to predict physics-feasible trajectories.

Multimodal Predictions. Multi-modality is an important characteristic of motion prediction that implies the multiple plausible choices of future trajectories. To model this uncertainty, the model is required to generate diverse predictions. A common approach [5,21,35] is to fuse latent variables sampled from a predefined Gaussian distribution $\mathcal{N}(0,1)$ with hidden feature. However, predefined latent variables suffer from the absence of context reasoning. In this paper, the latent variable is learnable from the scene context, which enables our model to generate multimodal predictions with meaningful semantics.

3 Method

Overview. To be specific, during the observation from frame t_1 to t_{obs}, the motion states of all N agents $M_{t_1:t_{obs}} = \{M_t | t \in [t_1, \cdots, t_{obs}]\}$ in a scene and scene information I are given, where $M_t \in R^{N \times d}$. Symbol d is the dimension of input motion state. It refers to the x–y coordinates of agent's location in this paper, thereby $d = 2$. $M_t = \{M_t^i | i \in N\}$, where $M_t^i = (x_t^i, y_t^i) \in R^2$ denotes the location of agent i at frame t . Our goal is to predict the locations of all the agents $M_{t_{obs+1}:t_{obs+pred}}$ in the future frames. The workflow of our framework DSCMP is illustrated in Fig. 2. For each iteration during observation, we send the current motion states with proposed queues to the encoder, and then we update the queues via ICM and then refine them via SCM. The last hidden features in the encoder is concatenated with a scene-guided latent variable. Later, the fused feature is passed to a LSTM decoder to obtain predicted motion.

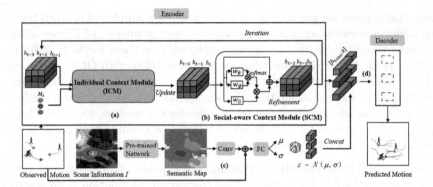

Fig. 2. The overview of our framework (DSCMP). Given a sequence of the observed motions, we construct agent-specific queue to store the LSTM features of previous frames within a queue length. The queue length is set to 3 as example. **(a)** For each iteration during observation, the current motion state and queues are encoded via ICM. **(b)** The queues are updated by appending the latest features and popping the earliest ones. In the SCM, the queues are adaptively refined by considering the pairwise relations of features in the neighbours' queues. **(c)** The semantic map of static context is incorporated with observed motion to generate a learnable latent variable z. **(d)** We concatenate the last hidden feature in the encoder and the latent variable to predict the motions via LSTM decoder.

3.1 The Function of Queues

With the setup of queues, the previous motion context and memory context for current frame t are temporarily stored. Specifically, we construct a hidden feature queue $Q_{h_t}^i = \left[h_{t-q}^i, \cdots, h_{t-1}^i \right] \in R^{1*q*h}$ and cell queue $Q_{c_t}^i = \left[c_{t-q}^i, \cdots, c_{t-1}^i \right] \in R^{1*q*h}$ for each agent i The size of LSTM feature is denoted as h. The queue length q describes a time bucket that the features are explicitly propagated. We initialize the hidden feature queues and cell queues with zero for each agent.

3.2 Individual Context Module

Based on the aforementioned queues, we firstly capture the temporal correlations of trajectories from individual level. In order to handle with multiple inputs in one iteration, we employ a tree-like LSTM cell [24]. The historical states in a time bucket $(t - q, \cdots, t - 1)$ are viewed as the children of the current state t. After an iteration, the queues are updated by appending features at frame t and popping features at frame $t - q$ We first average the hidden features in the queue to obtain a holistic representation $\tilde{h}_{t-1}^i = \sum_{l=1}^{q} h_{t-l}^i$ from the past frames. As the computational graph of ICM shown in Fig. 3a, the propagation is formulated as follows:

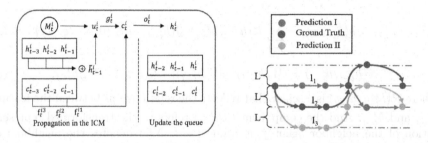

Fig. 3. (a): The computation graph of Individual Context Module (ICM). **(b):** An example to show Euclidean distance-based metric (e.g. ADE and FDE) cannot fully evaluate motion patterns. The prediction I and prediction II score the same ADE (5L/6) and FDE (L), whereas the prediction II captures the temporal correlation of motion pattern much better than the prediction I.

$$g_t^i = \sigma(W^g M_t^i + U^g \tilde{h}_{t-1}^i + b^g), \qquad f_t^{il} = \sigma(W^f M_t^i + U^f h_{t-l}^i + b^f),$$
$$o_t^i = \sigma(W^o M_t^i + U^o \tilde{h}_{t-1}^i + b^o), \qquad u_t^i = \tanh(W^u M_t^i + U^u \tilde{h}_{t-1}^i + b^u),$$
$$c_t^i = g_t^i \odot u_t^i + \sum_{l=1}^{q} f_t^{il} \odot c_{t-l}^i, \qquad\qquad h_t^i = o_t^i \odot \tanh(c_t^i), \tag{1}$$

where σ is sigmoid function and $\odot$ is element-wise multiplication. From the Eq. 1, we could observe that multiple previous frames pass message to the current cell. The contributions of these frames to the current state are controlled by multiple forget gates $f_t^{il}, l \in [1,q]$. In the case $q = 1$, ICM degenerates to a vanilla LSTM cell, which considers the previous single feature only at one iteration.

In practice, we assign each agent the queues of the fixed length. We point out that it is inappropriate in some cases. For example, the motion of some agents may be erratic that temporally incoherent with the past states. However, the adaptive forget gates could control the volume of information from the past frames. Hence, irrelevant motions could be filtered during the propagation.

3.3 Social-Aware Context Module

Social interactions works as an important part of dynamic context. Since the aggregation of neighbors' queues $\in R^{N(i)*q*h}$ store the surrounding historical information across agents, our model could learn spatio-temporal dependencies in a single operation. Here $N(i)$ denotes the number of neighbors of the agent i (include oneself). In the SCM, we compute the pair-wise relations of elements in the queues from neighbours. The refined queues can be viewed as a weighted sum from the neighbors' queues. Non-local block [28] is chosen for relation inference since it not only captures distant relations but also keeps the shape of input. The refinement of the hidden feature queue $Q_{h_{t+1}}^i = \left[h_{t-q+1}^i, \cdots, h_t^i \right]$ is computed as:

$$h_{t-l}^i = h_{t-l}^i \oplus \frac{1}{\mathcal{Z}_{t-l}^i} \sum_{j=1}^{N(i)} \mathcal{R}(h_{t-l}^i, h_{t-l}^j), \mathcal{G}(h_{t-l}^j), \ l \in \{0, \cdots, q-1\} \tag{2}$$

$$\mathcal{R}(h_{t-l}^i, h_{t-l}^j) = (W_\theta h_{t-l}^i)^\mathrm{T}(W_\phi h_{t-l}^j), \quad \mathcal{G}(h_{t-l}^j) = W_\mathcal{G} h_{t-l}^j, \tag{3}$$

where $\mathcal{R}(h_{t-l}^i, h_{t-l}^j)$ is a scalar that reflects the relationships between the feature h_{t-l}^i and h_{t-l}^j. And the component $\mathcal{G}(h_{t-l}^j)$ refers to the transformed representation of the neighbor agent j at frame $t-l$. $N(i)$ denotes the neighbors of agent i. $\mathcal{Z}_{t-l}^i = \sum_{j=1}^{N(i)} \mathcal{R}(h_{t-l}^i, h_{t-l}^j)$ is a normalization factor. The parameters of function $\mathcal{R}(\cdot, \cdot)$ and $\mathcal{G}(\cdot)$ are shared among agents. The cell queues stay invariant since we focus on motion interactions rather than memory at this step.

3.4   Semantic Guidance from Scene Context

Scene information is a valuable static context that provides the semantic of layout around the agents. In practice, we extract the semantic maps from resized 256×256 scene images I via pre-trained PSPNet [34,36] off-line. After that, we send the semantic maps to convolutional layers (Conv) and then combine them with the observed trajectories via a fully-connected (FC) layer. The latent variable z is obtained with reparameterization trick on the mean μ and variance σ as follows:

$$[\mu, \sigma] = \mathrm{FC}(\mathrm{Conv}(I) \oplus \sum_i^{N(i)} M_{t_1:t_{obs}}^i), \quad z \sim \mathcal{N}(\mu, \sigma), \tag{4}$$

where $\oplus$ denotes element-wise addition. The latent variable z enables multimodal predictions by going into the LSTM decoder with the last hidden features h_{obs} during observation. During the forecasting phase, the predicted motions $\hat{M}_{t_{obs}:t_{obs+pred}}^i$ are sequentially generated in the decoder.

3.5   Model Training

In order to encourage the coherence of temporal-neighboring features, we utilize a regularization loss L_c inspired by [23], which is defined as follows:

$$L_c = \begin{cases} 1 - \cos(h_{t_1}^i, h_{t_2}^i), & |t_1 - t_2| < q \\ \max(0, \ \cos(h_{t_1}^i, h_{t_2}^i) - \mathrm{margin}), & otherwise \end{cases} \tag{5}$$

where cos is cosine similarity and $margin$ is a hyperparameter. The pair-wise features $(h_{t_1}^i, h_{t_2}^i)$ are randomly sampled in a batch. L_c maximizes the similarity of features within a queue length (where frames are likely to strongly correlated with each other), while penalizing the similarity of features over a queue length (where frames are likely to belong to different motion patterns). The total loss function combines the regularization term L_c and the variety loss (the second term) followed by [5] as:

$$Loss = \lambda L_c + \min_m \left\| M_{t_{obs}:t_{obs+pred}}^i - \hat{M}_{t_{obs}:t_{obs+pred}}^{i(m)} \right\|_2, \tag{6}$$

where λ is a trade-off parameter. The variety loss computes the $L2$ distance between the best of m predictions and the ground truth, which encourages to cover the space of outputs that conform to the past trajectory.

4   Experiments

4.1   Datasets and Evaluation Metrics

We evaluate our method on three datasets (ETH [17], UCY [13], SDD [19]). ETH contains two subsets named ETH and HOTEL. UCY consists of three subsets, called UNIV, ZARA1 and ZARA2. Totally, there are 1,536 trajectories of pedestrian in the crowd collected in 5 scenes. We observe for 3.2 s (8 frames) and the predict the motions the next 4.8 s (12 frames) for every pedestrian simultaneously. For the data split and evaluation, we follow the leave-one-out method in [5]. SDD dataset has large volume with complex scenes. It contains 60 bird-eye-view videos with corresponding trajectories that involves multiple kinds of agents (pedestrian, bicyclist, .etc.). The observation duration is 3.2 s and the prediction duration ranged from 1 s to 4 s. We divide the dataset into 16,000 video clips and follow a 5-fold cross-validation setup.

Commonly used Euclidean-based metrics like ADE and FDE neglect the temporal correlation of motion patterns. An illustration example is shown in Fig. 3(b). In order to supplement this loophole, we introduce a new metric that requires no assumptions about the temporal distribution of trajectories, Temporal Correlation Coefficient (TCC). The TCC is defined as:

$$TCC = \frac{1}{2}(TCC_x + TCC_y). \tag{7}$$

$$TCC_x = \frac{1}{N}\sum_i^N \frac{Cov(\hat{x^i}, x^i)}{\sqrt{D(\hat{x^i})D(x^i)}}, \qquad TCC_y = \frac{1}{N}\sum_i^N \frac{Cov(\hat{y^i}, y^i)}{\sqrt{D(\hat{y^i})D(y^i)}}, \tag{8}$$

where the ground truth trajectory for the agent i is $M^i = (x^i, y^i)$. The corresponding predictions are denoted as $\hat{M^i} = (\hat{x^i}, \hat{y^i})$. From the equations above, we can observe that TCC ranges from $[-1, 1]$. A high TCC implies the predictions capture the time-varying motion patterns greatly, whereas a negative TCC denotes a weak capture of temporal correlation.

For the evaluation, the metric ADE (Average Distance Error) denotes the average L2 distance between the predictions and ground truth, and the metric FDE (Final Distance Error) reflects the L2 distance between the predictions and ground truth in the final frame. The TCC (Temporal Correlation Coefficient) is used to evaluation the temporal correlation of motion pattern in predictions.

4.2   Implementation Details

We preprocess the input motion state as the relative position. The size of hidden feature and the dimension of latent variable are set as 32 and 16 respectively.

Table 1. Quantitative comparisons on the ETH and UCY datasets. ADE/FDE are reported as in meters. All the models observe for 3.2 s and predict for the next 4.8 s.

Methods	ETH	HOTEL	UNIV	ZARA1	ZARA2	Avg.
Linear	0.91/1.97	0.42/0.81	0.70/1.33	0.58/1.23	0.64/1.21	0.65/1.31
LSTM	1.16/2.20	0.48/0.86	0.57/1.20	0.47/0.99	0.39/0.82	0.61/1.21
S-LSTM	0.84/1.85	0.45/0.86	0.55/1.14	0.35/0.76	0.36/0.77	0.51/1.08
SGAN	0.77/1.41	0.44/0.88	0.75/1.50	0.35/0.69	0.36/0.73	0.53/1.04
Sophie	0.70/1.43	0.76/1.67	0.54/1.24	0.30/0.63	0.38/0.78	0.54/1.15
MATF	1.01/1.75	0.43/0.80	0.44/0.91	0.26/0.45	0.26/0.57	0.48/0.90
STGAT	0.76/1.61	0.32/0.56	0.52/1.10	0.34/0.74	0.31/0.71	0.45/0.94
Ours$_{IC}$	0.70/1.35	0.30/0.52	0.52/1.09	0.37/0.76	0.30/0.63	0.44/0.87
Ours$_{IC-SC}$	0.70/1.36	0.27/0.48	0.51/1.08	0.35/0.74	0.28/0.61	0.42/0.85
Ours (full)	**0.66/1.21**	**0.27/0.46**	**0.50/1.07**	**0.33/0.68**	**0.28/0.60**	**0.41/0.80**
Rel. gain	+13.15%	+15.63%	+3.85%	+2.94%	+9.68%	+9.05%

The convolutional part for scene is three-layer with kernel size as 10, 10, 1. The subsequent FC layer is a 16×16 transformation with sigmoid activation. The queue length is set as 4, 2, 3 for the ETH dataset, ZARA datasets and otherwise datasets respectively. For the loss function, the λ and *margin* for the regularizer L_c are 0.1 and 0.5 respectively. The m in the variety loss is set as 20. The batch size is 64 and the learning rate is 0.001 with Adam optimizer.

4.3   Standard Evaluations

We choose two basic methods linear models and LSTM [6], and several representative state-of-the-art methods for comparison. S-LSTM [1] and SGAN [5] are the famous deterministic method and stochastic method that combine deep learning with spatial-only interaction respectively. The most recent approaches like Sophie [21], MATF [35] and STGAT [7] incorporate information from either static scenes or spatio-temporal dependencies. To verify the effectiveness of Individual Context Module and Social-aware Context Module, we adopt two variant of our methods, Ours$_{IC}$ and Ours$_{IC-SC}$. In accord with [5,7,21,35], the latent variables employed in the variants Ours$_{IC}$ and Ours$_{IC-SC}$ are sampled from $\mathcal{N}(0,1)$, and the results are reported by sampling 20 times to choose the best prediction. "Rel. gain" shows the relative ADE gain of our full model (marked in italic) compared with the latest method STGAT (marked in bold).

As presented in the Table 1 and 2, linear method and LSTM suffer from bad performance since they are too shallow to consider the surrounding context. Compared with the variant Ours$_{IC}$ with the state-of-the-methods, Ours$_{IC}$ has already shown advantages across different datasets. It indicates us that the explicit temporal dependencies extraction among multiple frames is valuable to enhance the performance. Ours$_{IC-SC}$ makes some improvement with social-aware features refinement. The performance gap between Ours$_{IC-SC}$ and our full model empirically shows that the guidance of static scene context is useful for multi-modal predictions.

Table 2. Quantitative comparisons on the SDD dataset. ADE/FDE are reported in pixel coordinates at 1/5 resolution followed [12]. All the models observe the agents 3.2 s and then make predictions in the next 1∼4 s.

Methods	1.0 s	2.0 s	3.0 s	4.0 s	Avg.
Linear	1.52/2.90	2.10/4.32	3.01/6.23	3.10/6.33	2.43/4.95
LSTM	1.38/2.05	2.04/3.48	3.02/5.91	3.07/5.96	2.38/4.35
S-LSTM	1.33/2.02	2.00/3.46	3.03/5.86	3.05/5.91	2.35/4.31
SGAN	1.37/2.26	2.50/4.95	2.82/5.54	2.85/5.78	2.39/4.63
STGAT	1.19/1.68	1.69/2.90	2.70/5.22	2.83/5.37	2.10/3.79
Ours$_{IC}$	1.10/1.66	1.70/2.90	2.55/5.02	2.65/5.13	2.00/3.68
Ours$_{IC\text{-}SC}$	1.09/1.65	1.68/2.87	2.52/4.96	2.61/5.08	1.98/3.64
Ours (full)	**1.08/1.63**	**1.64/2.83**	**2.48/4.91**	**2.57/5.02**	**1.94/3.60**
Rel. gain	+10.19%	+2.96%	+8.14%	+9.19%	+7.62%

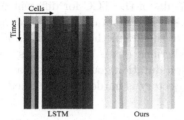
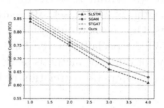

Fig. 4. (a): Memory cell visualization for LSTM and our method. Although the memory capacity decreases over time for both models (light blue → dark blue for LSTM, dark red → light red for ours), most of the cells in ours remain positive, which implies that they keep track of the evolving motion patterns from the historical context. **(b):** Comparison of TCC among state-of-the-art methods and ours. Our method enjoy high TCC, which indicates an effective capture of temporal correlation. (Color figure online)

5   Discussion

5.1   Memory Cell Visualization

As illustrated in Fig. 4(a), we compare the memory capacity of LSTM and our method via cell activation. Red denotes a positive cell state, and blue denotes a negative one. The most of cells in vanilla LSTM are negative. In contrast, our model keeps track of the context throughout the prediction. Although the memory capacity of our model recedes over time (from dark red to light red), it still stays active. These results inspire us that the frame-by-frame observation used by LSTM is prone to get short sight. Instead, explicit modeling on the dependencies in multiple frames improves the captures of long-term motion.

Table 3. Comparisons of non-linear trajectories on the ETH and UCY datasets. ADE/FDE are reported as in meters. "Ours" denotes the proposed full model.

Methods	ETH	HOTEL	UNIV	ZARA1	ZARA2	Avg.
Linear	1.01/2.21	0.50/0.96	0.80/1.56	0.62/1.33	0.99/1.95	0.78/1.60
LSTM	1.08/2.28	0.57/1.02	0.65/1.37	0.49/1.04	0.60/1.30	0.68/1.40
S-LSTM	0.92/2.01	0.48/0.92	0.61/1.29	0.38/0.79	0.50/1.06	0.58/1.21
SGAN	0.86/1.54	0.52/1.03	0.81/1.61	0.39/0.76	0.49/0.97	0.61/1.18
STGAT	0.86/1.67	0.39/0.71	0.60/1.33	0.39/0.85	0.50/1.14	0.55/1.14
Ours (q = 2)	0.73/1.37	0.34/0.63	0.61/1.29	**0.36/0.76**	**0.48/1.04**	0.50/1.02
Ours (q = 3)	0.71/1.35	**0.33/0.58**	**0.60/1.28**	0.38/0.81	0.49/1.09	**0.50/1.01**
Ours (q = 4)	**0.70/1.29**	0.34/0.60	0.60/1.31	0.39/0.83	0.49/1.09	0.50/1.02

5.2   The Capture of Motion Pattern

Figure 4(b) summarizes the quantitative results of the TCC for different methods in the SDD dataset. By learning the spatio-temporal context of dynamic agents, Our method outperforms the state-of-the-art methods (SLSTM, SGAN, STGAT), especially on the long-term predictions (4 s). TCC declines consistently for different methods as the prediction duration goes on. It is reasonable since the temporal correlation of longer trajectory is harder to be learned.

5.3   Exploration on the Queue Length

An important setup in our model is the proposed queue that understand the long-term motion. Hence, we study the effect of queue length on non-linear cases which are usually treated as hard cases. As shown in the Table 3, linear model and LSTM suffer from the unsatisfactory performance. Compared with the methods S-LSTM, SGAN and STGAT, our model enjoys relatively lower error. With the variation of queue length, our performance is robust and competitive against state-of-the-art methods. The model with long queue length ($q = 4$) is the most suitable for the ETH dataset, where most of the trajectories are highly non-linear. Short queue length ($q = 2$) works better in the ZARA1 and ZARA2 datasets, where the trajectories are faintly non-linear. Compared with linear trajectories, We speculate that the motion states of non-linear trajectory are temporally correlated within a relatively long range, where long queues capture long-term motions better than short queues.

5.4   Social Behaviors Understanding

In the scenario of real-world applications, it is imperative to handle the social interactions in the multi-agent system. Therefore, we verify whether our method well perceive the social behaviors in the crowd. As shown in the Fig. 5, we select three common scenarios involve social behaviors, "Walk in parallel", "Turing" and "Face-to-Face". From the comparison between the ground truth and various

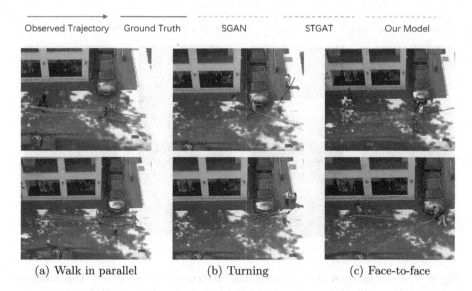

Observed Trajectory   Ground Truth   SGAN   STGAT   Our Model

(a) Walk in parallel   (b) Turning   (c) Face-to-face

Fig. 5. Comparison among our model, SGAN and STGAT in different scenarios. These visualizations show that 1): Our model is capable of generating convincing trajectories that are closer to the ground truth than other state-of-the-art methods. 2): The social interactions across agents are well captured by our model. Our predictions avoid collision during the whole forecasting period.

predictions, we could observe that the trajectories predicted by our method are close with ground truth. Moreover, our predictions are reliable that no collisions or large deviations appear during the whole forecasting period. It indicates that our model predict multiple trajectories for each agent that cohere in time and space with the other agents.

5.5   Analysis of Multimodal Predictions

In order to evaluate the quality of multimodal predictions, we visualize the diverse trajectories predicted by our model. The top row in the Fig. 6 reports the multimodal predictions for one agent of interest. Rather than using a predefined Gaussian noise, the learnable latent variable z benefits from the semantic cues of the static scene context. Our predictions (yellow lines) suggest plausible trajectories that are close to the ground truth (red line), instead of producing a wide spread of candidates randomly. For instance, in the scenario of "Crossroad" and "Intersection", it is possible for the target agent to turn left or right at the endpoint of observation. Our model provides predictions that in line with common sense. In the scenario of "Sideway" and "Corner", the target agent has limited choices for the future trajectories due to the constraint of scene layout. In these kinds of scenarios, all our predictions are moving towards reasonable directions. Hence, our model has good interpretability with the incorporation of scene information.

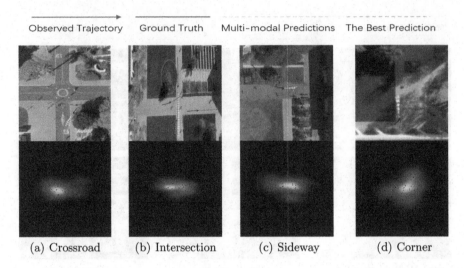

Observed Trajectory   Ground Truth   Multi-modal Predictions   The Best Prediction

(a) Crossroad   (b) Intersection   (c) Sideway   (d) Corner

Fig. 6. The visualization of the multimodal predictions in four scenes. **Top row:** we plot multiple possible future trajectories for one agent of interest. **Bottom row:** we visualize the distribution heatmap of destinations (location at the final frame) via kernel density estimation. The predicted destinations and ground truth are shown as black points and red point respectively. The distribution heatmap shows that our model not only provides semantically meaningful predictions, but also enjoys low uncertainty.

In the bottom row, we investigate the uncertainty of predictions with distribution heatmap. Here we estimate the distribution of the predicted destination (black point) via kernel density estimation, and then apply the true destination (red point) to this distribution. The brighter the location, the more possibility that the point belongs to the distribution. Our visualization shows that the true destination usually appear in the bright locations. It indicates our predictions enjoy low uncertainty.

6   Conclusions

In this paper, we proposed a novel method DSCMP to highlight the three core elements of contextual understanding, *i.e.* spatial interaction, temporal coherence and scene layout, for multi-agent motion prediction. We designed a differentiable queue mechanism embedded on LSTM to capture the spatial interactions across agents and temporal coherence in long-term motion. And a learnable latent variable was introduced to learn the semantics of scene layout. In order to understand the uniqueness of DSCMP, we also proposed a metric Temporal Correlation Coefficient (TCC) to evaluate the temporal correlation of predicted motion. Extensive experiments on three benchmark datasets demonstrate the effectiveness of our proposed method. For the future research on autonomous applications, this work sheds a light on the modelling of spatio-temporal dependencies in multiple frames and the semantic cues from scene layout.

Acknowledgments. This work is partially supported by the SenseTime Donation for Research, HKU Seed Fund for Basic Research, Startup Fund and General Research Fund No.27208720. This work is also partially supported by the Major Project for New Generation of AI under Grant No. 2018AAA0100400 and by the National Natural Science Foundation of China under Grant No. 61772118.

References

1. Alahi, A., Goel, K., Ramanathan, V., Robicquet, A., Fei-Fei, L., Savarese, S.: Social LSTM: human trajectory prediction in crowded spaces. In: Proceedings of the IEEE Conference on Computer Vision and Pattern Recognition, pp. 961–971 (2016)
2. Bisagno, N., Zhang, B., Conci, N.: Group LSTM: group trajectory prediction in crowded scenarios. In: Leal-Taixé, L., Roth, S. (eds.) ECCV 2018. LNCS, vol. 11131, pp. 213–225. Springer, Cham (2019). https://doi.org/10.1007/978-3-030-11015-4_18
3. Choi, C., Dariush, B.: Looking to relations for future trajectory forecast. In: Proceedings of the IEEE International Conference on Computer Vision, pp. 921–930 (2019)
4. Fragkiadaki, K., Levine, S., Felsen, P., Malik, J.: Recurrent network models for human dynamics. In: Proceedings of the IEEE International Conference on Computer Vision, pp. 4346–4354 (2015)
5. Gupta, A., Johnson, J., Fei-Fei, L., Savarese, S., Alahi, A.: Social GAN: socially acceptable trajectories with generative adversarial networks. In: Proceedings of the IEEE Conference on Computer Vision and Pattern Recognition, pp. 2255–2264 (2018)
6. Hochreiter, S., Schmidhuber, J.: Long short-term memory. Neural Comput. **9**(8), 1735–1780 (1997)
7. Huang, Y., Bi, H., Li, Z., Mao, T., Wang, Z.: STGAT: modeling spatial-temporal interactions for human trajectory prediction. In: Proceedings of the IEEE International Conference on Computer Vision, pp. 6272–6281 (2019)
8. Ivanovic, B., Pavone, M.: The trajectron: probabilistic multi-agent trajectory modeling with dynamic spatiotemporal graphs. In: Proceedings of the IEEE International Conference on Computer Vision, pp. 2375–2384 (2019)
9. Karasev, V., Ayvaci, A., Heisele, B., Soatto, S.: Intent-aware long-term prediction of pedestrian motion. In: 2016 IEEE International Conference on Robotics and Automation (ICRA), pp. 2543–2549. IEEE (2016)
10. Kim, B., Kang, C.M., Kim, J., Lee, S.H., Chung, C.C., Choi, J.W.: Probabilistic vehicle trajectory prediction over occupancy grid map via recurrent neural network. In: 2017 IEEE 20th International Conference on Intelligent Transportation Systems (ITSC), pp. 399–404. IEEE (2017)
11. Kosaraju, V., Sadeghian, A., Martín-Martín, R., Reid, I., Rezatofighi, H., Savarese, S.: Social-BIGAT: multimodal trajectory forecasting using bicycle-GAN and graph attention networks. In: Advances in Neural Information Processing Systems, pp. 137–146 (2019)
12. Lee, N., Choi, W., Vernaza, P., Choy, C.B., Torr, P.H., Chandraker, M.: Desire: distant future prediction in dynamic scenes with interacting agents. In: Proceedings of the IEEE Conference on Computer Vision and Pattern Recognition, pp. 336–345 (2017)

13. Lerner, A., Chrysanthou, Y., Lischinski, D.: Crowds by example. In: Computer Graphics Forum, vol. 26, pp. 655–664. Wiley Online Library (2007)
14. Ma, Y., Zhu, X., Zhang, S., Yang, R., Wang, W., Manocha, D.: Trafficpredict: trajectory prediction for heterogeneous traffic-agents. In: Proceedings of the AAAI Conference on Artificial Intelligence, vol. 33, pp. 6120–6127 (2019)
15. Manh, H., Alaghband, G.: Scene-LSTM: a model for human trajectory prediction. arXiv preprint arXiv:1808.04018 (2018)
16. Park, S.H., Kim, B., Kang, C.M., Chung, C.C., Choi, J.W.: Sequence-to-sequence prediction of vehicle trajectory via LSTM encoder-decoder architecture. In: 2018 IEEE Intelligent Vehicles Symposium (IV), pp. 1672–1678. IEEE (2018)
17. Pellegrini, S., Ess, A., Schindler, K., Van Gool, L.: You'll never walk alone: modeling social behavior for multi-target tracking. In: 2009 IEEE 12th International Conference on Computer Vision, pp. 261–268. IEEE (2009)
18. Petrich, D., Dang, T., Kasper, D., Breuel, G., Stiller, C.: Map-based long term motion prediction for vehicles in traffic environments. In: 16th International IEEE Conference on Intelligent Transportation Systems (ITSC 2013), pp. 2166–2172. IEEE (2013)
19. Robicquet, A., Sadeghian, A., Alahi, A., Savarese, S.: Learning social etiquette: human trajectory understanding in crowded scenes. In: Leibe, B., Matas, J., Sebe, N., Welling, M. (eds.) ECCV 2016. LNCS, vol. 9912, pp. 549–565. Springer, Cham (2016). https://doi.org/10.1007/978-3-319-46484-8_33
20. Rudenko, A., Palmieri, L., Herman, M., Kitani, K.M., Gavrila, D.M., Arras, K.O.: Human motion trajectory prediction: a survey. arXiv preprint arXiv:1905.06113 (2019)
21. Sadeghian, A., Kosaraju, V., Sadeghian, A., Hirose, N., Rezatofighi, H., Savarese, S.: Sophie: an attentive GAN for predicting paths compliant to social and physical constraints. In: Proceedings of the IEEE Conference on Computer Vision and Pattern Recognition, pp. 1349–1358 (2019)
22. Shah, R., Romijnders, R.: Applying deep learning to basketball trajectories. arXiv preprint arXiv:1608.03793 (2016)
23. Synnaeve, G., Dupoux, E.: A temporal coherence loss function for learning unsupervised acoustic embeddings. In: SLTU, pp. 95–100 (2016)
24. Tai, K.S., Socher, R., Manning, C.D.: Improved semantic representations from tree-structured long short-term memory networks. arXiv preprint arXiv:1503.00075 (2015)
25. Vasquez, D.: Novel planning-based algorithms for human motion prediction. In: 2016 IEEE International Conference on Robotics and Automation (ICRA), pp. 3317–3322. IEEE (2016)
26. Vaswani, A., et al.: Attention is all you need. In: Advances in Neural Information Processing Systems, pp. 5998–6008 (2017)
27. Vemula, A., Muelling, K., Oh, J.: Social attention: modeling attention in human crowds. In: 2018 IEEE international Conference on Robotics and Automation (ICRA), pp. 1–7. IEEE (2018)
28. Wang, X., Girshick, R., Gupta, A., He, K.: Non-local neural networks. In: Proceedings of the IEEE Conference on Computer Vision and Pattern Recognition, pp. 7794–7803 (2018)
29. Xu, Y., Piao, Z., Gao, S.: Encoding crowd interaction with deep neural network for pedestrian trajectory prediction. In: Proceedings of the IEEE Conference on Computer Vision and Pattern Recognition, pp. 5275–5284 (2018)
30. Yamaguchi, K., Berg, A.C., Ortiz, L.E., Berg, T.L.: Who are you with and where are you going? In: CVPR 2011, pp. 1345–1352. IEEE (2011)

31. Yi, S., Li, H., Wang, X.: Pedestrian behavior modeling from stationary crowds with applications to intelligent surveillance. IEEE Trans. Image Process. **25**(9), 4354–4368 (2016)
32. Zhang, H., Goodfellow, I., Metaxas, D., Odena, A.: Self-attention generative adversarial networks. arXiv preprint arXiv:1805.08318 (2018)
33. Zhang, P., Ouyang, W., Zhang, P., Xue, J., Zheng, N.: SR-LSTM: state refinement for LSTM towards pedestrian trajectory prediction. In: Proceedings of the IEEE Conference on Computer Vision and Pattern Recognition, pp. 12085–12094 (2019)
34. Zhao, H., Shi, J., Qi, X., Wang, X., Jia, J.: Pyramid scene parsing network. In: Proceedings of the IEEE Conference on Computer Vision and Pattern Recognition, pp. 2881–2890 (2017)
35. Zhao, T., et al.: Multi-agent tensor fusion for contextual trajectory prediction. In: Proceedings of the IEEE Conference on Computer Vision and Pattern Recognition, pp. 12126–12134 (2019)
36. Zhou, B., Zhao, H., Puig, X., Fidler, S., Barriuso, A., Torralba, A.: Scene parsing through ade20k dataset. In: Proceedings of the IEEE Conference on Computer Vision and Pattern Recognition (2017)

Image-Based Table Recognition: Data, Model, and Evaluation

Xu Zhong[✉][iD], Elaheh ShafieiBavani[iD], and Antonio Jimeno Yepes[iD]

IBM Research Australia, 60 City Road, Southgate, VIC 3006, Australia
peter.zhong@au1.ibm.com

Abstract. Important information that relates to a specific topic in a document is often organized in tabular format to assist readers with information retrieval and comparison, which may be difficult to provide in natural language. However, tabular data in unstructured digital documents, *e.g.* Portable Document Format (PDF) and images, are difficult to parse into structured machine-readable format, due to complexity and diversity in their structure and style. To facilitate image-based table recognition with deep learning, we develop and release the largest publicly available table recognition dataset PubTabNet (https://github.com/ibm-aur-nlp/PubTabNet.), containing 568k table images with corresponding structured HTML representation. PubTabNet is automatically generated by matching the XML and PDF representations of the scientific articles in PubMed Central^{TM} Open Access Subset (PMCOA). We also propose a novel attention-based encoder-dual-decoder (EDD) architecture that converts images of tables into HTML code. The model has a structure decoder which reconstructs the table structure and helps the cell decoder to recognize cell content. In addition, we propose a new Tree-Edit-Distance-based Similarity (TEDS) metric for table recognition, which more appropriately captures multi-hop cell misalignment and OCR errors than the pre-established metric. The experiments demonstrate that the EDD model can accurately recognize complex tables solely relying on the image representation, outperforming the state-of-the-art by 9.7% absolute TEDS score.

Keywords: Table recognition · Dual decoder · Dataset · Evaluation

1 Introduction

Information in tabular format is prevalent in all sorts of documents. Compared to natural language, tables provide a way to summarize large quantities of data in a more compact and structured format. Tables provide as well a format to assist readers with finding and comparing information. An example of the relevance

Electronic supplementary material The online version of this chapter (https://doi.org/10.1007/978-3-030-58589-1_34) contains supplementary material, which is available to authorized users.

of tabular information in the biomedical domain is in the curation of genetic databases in which just between 2% to 8% of the information was available in the narrative part of the article compared to the information available in tables or files in tabular format [17].

Tables in documents are typically formatted for human understanding, and humans are generally adept at parsing table structure, identifying table headers, and interpreting relations between table cells. However, it is challenging for a machine to understand tabular data in unstructured formats (*e.g.* PDF, images) due to the large variability in their layout and style. The key step of table understanding is to represent the unstructured tables in a machine-readable format, where the structure of the table and the content within each cell are encoded according to a pre-defined standard. This is often referred as *table recognition* [9].

This paper solves the following three problems in image-based table recognition, where the structured representations of tables are reconstructed solely from image input:

- **Data.** We provide a large-scale dataset PubTabNet, which consists of over 568k images and corresponding HTML representations of heterogeneous tables. PubTabNet is created by matching the PDF format and the XML format of the scientific articles contained in PMCOA[1].
- **Model.** We develop a novel attention-based encoder-dual-decoder (EDD) architecture (see Fig. 1) which consists of an encoder, a structure decoder, and a cell decoder. The EDD model is the first end-to-end table recognition model that supports joint training on table structure recognition and cell content recognition tasks. This model design allows the cell decoder to use information from the structure decoder to better focus on the local visual features of the cell being generated. This mechanism back-propagates cell content recognition loss to the structure decoder, which regularizes it to better locate table cells. EDD demonstrates superior performance on PubTabNet, compared to existing table recognition methods.
- **Evaluation.** By modeling tables as a tree structure, we propose a new tree-edit-distance-based evaluate metric for image-based table recognition. We demonstrate that our new metric is superior to the metric [16] commonly used in literature and competitions.

2   Related Work

Data. Analyzing tabular data in unstructured documents focuses mainly on three problems: i) *table detection*: localizing the bounding boxes of tables in documents, ii) *table structure recognition*: parsing only the structural (row and column layout) information of tables, and iii) *table recognition*: parsing both the structural information and content of table cells. Table 1 compares the datasets that have been developed to address one or more of these three problems. The PubTabNet dataset and the EDD model we develop in this paper aim at the

---

[1] https://www.ncbi.nlm.nih.gov/pmc/tools/openftlist/.

Table 1. Datasets for Table Detection (TD), Table Structure Recognition (TSR) and Table Recognition (TR).

Dataset	TD	TSR	TR	# Tables
Marmot [5]	✓	✗	✗	958
PubLayNet [39]	✓	✗	✗	113k
DeepFigures [33]	✓	✗	✗	1.4m
ICDAR2013 [9]	✓	✓	✓	156
ICDAR2019 [6]	✓	✓	✗	3.6k
UNLV [32]	✓	✓	✗	558
TableBank[a]	✓	✓	✗	417k (TD)
				145k (TSR)
SciTSR[b]	✗	✓	✓	15k
Table2Latex [3]	✗	✓	✓	450k
Synthetic data in [26]	✗	✓	✓	Unbounded
PubTabNet	✗	✓	✓	568k

[a] https://github.com/doc-analysis/TableBank
[b] https://github.com/Academic-Hammer/SciTSR

image-based table recognition problem. Comparing to other existing datasets for table recognition (*e.g.* SciTSR[2], Table2Latex [3], and TIES [26]), PubTabNet has three key advantages:

1. The tables are typeset by the publishers of over 6,000 journals, which offers considerably more diversity in table styles than other table datasets.
2. Cells are categorized into headers and body cells, which is important when retrieving information from tables.

Model. Traditional table detection and recognition methods rely on pre-defined rules [7,14,15,24,31,37] and statistical machine learning [1,4,18,20,34]. Recently, deep learning exhibit great performance in image-based table detection and structure recognition. Hao *et al.* used a set of primitive rules to propose candidate table regions and a convolutional neural network to determine whether the regions contain a table [10]. Fully-convolutional neural networks, followed by a conditional random field, have also been used for table detection [11,19,36]. In addition, deep neural networks for object detection, such as Faster-RCNN [28], Mask-RCNN [12], and YOLO [27] have been exploited for table detection and row/column segmentation [8,30,35,39]. Furthermore, graph neural networks are used for table detection and recognition by encoding document images as graphs [26,29].

There are several tools (see Table 2) that can convert tables in text-based PDF format into structured representations. However, there is limited work on image-based table recognition. Attention-based encoder-decoder was first proposed by Xu *et al.* for image captioning [38]. Deng *et al.* extended it by adding a

---

[2] https://github.com/Academic-Hammer/SciTSR

recurrent layer in the encoder for capturing long horizontal spatial dependencies to convert images of mathematical formulas into LATEX representation [2]. The same model was trained on the Table2Latex [3] dataset to convert table images into LATEX representation. As show in [3] and in our experimental results (see Table 2), the efficacy of this model on image-based table recognition is mediocre.

This paper considerably improves the performance of the attention-based encoder-decoder method on image-based table recognition with a novel EDD architecture. Our model differs from other existing EDD architectures [23,40], where the dual decoders are independent from each other. In our model, the cell decoder is triggered only when the structure decoder generates a new cell. In the meanwhile, the hidden state of the structure decoder is sent to the cell decoder to help it place its attention on the corresponding cell in the table image.

Evaluation. The evaluation metric proposed in [16] is commonly used in table recognition literature and competitions. This metric first flattens the ground truth and recognition result of a table are into a list of pairwise adjacency relations between non-empty cells. Then precision, recall, and F1-score can be computed by comparing the lists. This metric is simple but has two obvious problems: 1) as it only checks immediate adjacency relations between non-empty cells, it cannot detect errors caused by empty cells and misalignment of cells beyond immediate neighbors; 2) as it checks relations by exact match[3], it does not have a mechanism to measure fine-grained cell content recognition performance. In order to address these two problems, we propose a new evaluation metric: **T**ree-**E**dit-**D**istance-based **S**imilarity (TEDS). TEDS solves problem 1) by examining recognition results at the global tree-structure level, allowing it to identify all types of structural errors; and problem 2) by computing the string-edit-distance when the tree-edit operation is node substitution.

3   Automatic Generation of PubTabNet

PMCOA contains over one million scientific articles in both unstructured (PDF) and structured (XML) formats. A large table recognition dataset can be automatically generated if the corresponding location of the table nodes in the XML can be found in the PDF. In our previous work, we proposed an algorithm to match the XML and PDF representations of the articles in PMCOA [39]. We use this algorithm to extract the table regions from the PDF for the tables nodes in the XML. The table regions are converted to images with a 72 pixels per inch (PPI) resolution. We use this low PPI setting to relax the requirement of our model for high-resolution input images. For each table image, the corresponding table node (HTML) is extracted from the XML as the ground truth annotation.

It is observed that the algorithm generates erroneous bounding boxes for some tables, hence we use a heuristic to automatically verify the bounding boxes. For each annotation, the text within the bounding box is extracted from the PDF and compared with that in the annotation. The bounding box is considered to be

---

[3] Both cells are identical and the direction matches.

correct if the cosine similarity of the term frequency-inverse document frequency (Tf-idf) features of the two texts is greater than 90% and the length of the two texts differs less than 10%. In addition, to improve the learnability of the data, we remove rare tables which contains any cell that spans over 10 rows or 10 columns, or any character that occurs less than 50 times in all the tables. Tables of which the annotation contains `math` and `inline-formula` nodes are also removed, as we found they do not have a consistent XML representation.

After filtering the table samples, we curate the HTML code of the tables to remove unnecessary variations. First, we remove the nodes and attributes that are not reconstructable from the table image, such as hyperlinks and definition of acronyms. Second, table header cells are defined as `th` nodes in some tables, but as `td` nodes in others. We unify the definition of header cells as `td` nodes, which preserves the header identify of the cells as they are still descendants of the `thead` node. Third, all the attributes except 'rowspan' and 'colspan' in `td` nodes are stripped, since they control the appearance of the tables in web browsers, which do not match with the table image. These curations lead to consistent and clean HTML code and make the data more learnable.

Finally, the samples are randomly partitioned into 60%/20%/20% training/development/test sets. The training set contains 548,592 samples. As only a small proportion of tables contain spanning (multi-column or multi-row) cells, the evaluation on the raw development and test sets would be strongly biased towards tables without spanning cells. To better evaluate how a model performs on complex table structures, we create more balanced development and test sets by randomly drawing 5,000 tables with spanning cells and 5,000 tables without spanning cells from the corresponding raw set.

4   Encoder-Dual-Decoder (EDD) Model

Figure 1 shows the architecture of the EDD model, which consists of an encoder, an attention-based structure decoder, and an attention-based cell decoder. The use of two decoders is inspired by two intuitive considerations: i) table structure recognition and cell content recognition are two distinctively different tasks. It is not effective to solve both tasks at the same time using a single attention-based decoder. ii) information in the structure recognition task can be helpful for locating the cells that need to be recognized. The encoder is a convolutional neural network (CNN) that captures the visual features of input table images. The structure decoder and cell decoder are recurrent neural networks (RNN) with the attention mechanism proposed in [38]. The structure decoder only generates the HTML tags that define the structure of the table. When the structure decoder recognizes a new cell, the cell decoder is triggered and uses the hidden state of the structure decoder to compute the attention for recognizing the content of the new cell. This ensures a one-to-one match between the cells generated by the structure decoder and the sequences generated by the cell decoder. The outputs of the two decoders can be easily merged to get the final HTML representation of the table.

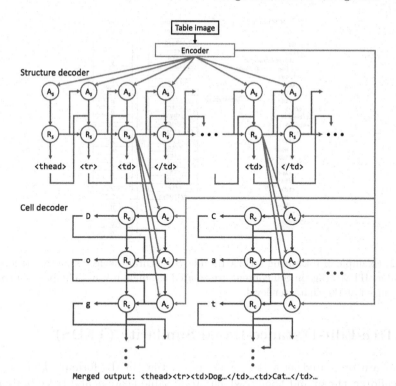

Fig. 1. EDD architecture. The encoder is a convolutional neural network which captures the visual features of the input table image. A_s and A_c are attention network for the structure decoder and cell decoder, respectively. R_s and R_c are recurrent units for the structure decoder and cell decoder, respectively. The structure decoder reconstructs table structure and helps the cell decoder to generate cell content. The output of the structure decoder and the cell decoder is merged to obtain the HTML representation of the input table image.

As the structure and the content of an input table image are recognized separately by two decoders, during training, the ground truth HTML representation of the table is tokenized into structural tokens, and cell tokens as shown in Fig. 2. Structural tokens include the HTML tags that control the structure of the table. For spanning cells, the opening tag is broken down into multiple tokens as '<td', 'rowspan' or 'colspan' attributes, and '>'. The content of cells is tokenized at the character level, where HTML tags are treated as single tokens.

Two loss functions can be computed from the EDD network: i) cross-entropy loss of generating the structural tokens (l_s); and ii) cross-entropy loss of generating the cell tokens (l_c). The overall loss (l) of the EDD network is calculated as,

$$l = \lambda l_s + (1 - \lambda)l_c, \tag{1}$$

where $\lambda \in [0, 1]$ is a hyper-parameter.

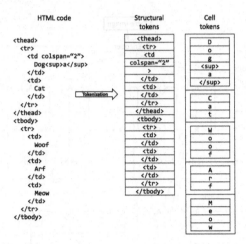

Fig. 2. Example of tokenizing a HTML table. Structural tokens define the structure of the table. HTML tags in cell content are treated as single tokens. The rest cell content is tokenized at the character level.

5   Tree-Edit-Distance-Based Similarity (TEDS)

Tables are presented as a tree structure in the HTML format. The root has two children thead and tbody, which group table headers and table body cells, respectively. The children of thead and tbody nodes are table rows (tr). The leaves of the tree are table cells (td). Each cell node has three attributes, *i.e.* 'colspan', 'rowspan', and 'content'. We measure the similarity between two tables using the tree-edit distance proposed by Pawlik and Augsten [25]. The cost of insertion and deletion operations is 1. When the edit is substituting a node n_o with n_s, the cost is 1 if either n_o or n_s is not td. When both n_o and n_s are td, the substitution cost is 1 if the column span or the row span of n_o and n_s is different. Otherwise, the substitution cost is the normalized Levenshtein similarity [22] ($\in [0, 1]$) between the content of n_o and n_s. Finally, TEDS between two trees is computed as

$$TEDS(T_a, T_b) = 1 - \frac{EditDist(T_a, T_b)}{max(|T_a|, |T_b|)},$$ (2)

where *EditDist* denotes tree-edit distance, and $|T|$ is the number of nodes in T. The table recognition performance of a method on a set of test samples is defined as the mean of the TEDS score between the recognition result and ground truth of each sample.

In order to justify that TEDS solves the two problems of the adjacency relation metric [16] described previously in Sect. 2, we add two types of perturbations to the validation set of PubTabNet and examine how TEDS and the adjacency relation metric respond to the perturbations.

1. To demonstrate the empty-cell and multi-hop misalignment issue, we shift some cells in the first row downwards[4], and pad the leftover space with empty cells. The shift distance of a cell is proportional to its column index. We tested 5 perturbation levels, i.e., 10%, 30%, 50%, 70%, or 90% of the cells in the first row are shifted. Figure S1 in supplemental material shows a perturbed example, where 90% of the cells in the first row are shifted.
2. To demonstrate the fine-grained cell content recognition issue, we randomly modify some characters into a different one. We tested 5 perturbation levels, i.e., the chance that a character gets modified is set to be 10%, 30%, 50%, 70%, or 90%. Figure S2 in supplemental material shows an example at the 10% perturbation level.

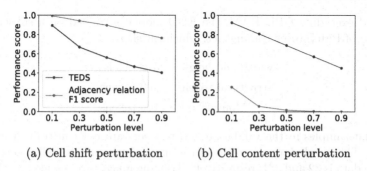

(a) Cell shift perturbation   (b) Cell content perturbation

Fig. 3. Comparison of the response of TEDS and the adjacency relation metric to cell shift perturbation and cell content perturbation. The adjacency relation metric is under-reacting to cell shift perturbation and over-reacting to cell content perturbation. Whereas TEDS demonstrates superiority at appropriately capturing the errors.

Figure 3 illustrates how TEDS and the adjacency relation F1-score respond to the two types of perturbations at different levels. The adjacency relation metric is under-reacting to the cell shift perturbation. At the 90% perturbation level, the table is substantially different from the original (see example in Fig. S1 in supplemental material). However, the adjacency relation F1-score is still nearly 80%. On the other hand, the perturbation causes a 60% drop on TEDS, demonstrating that TEDS is able to capture errors that the adjacency relation metric cannot.

When it comes to cell content perturbations, the adjacency relation metric is over-reacting. Even the 10% perturbation level (see example in Fig. S2 in supplemental material) leads to over 70% decrease in adjacency relation F1-score, which drops close to zero from the 50% perturbation level. In contrast, TEDS linearly decreases from 90% to 40% as the perturbation level increases from 10% to 90%, demonstrating the capability of capturing fine-grained cell content recognition errors.

---

[4] If the number of rows is greater than the number of columns, we shift the cells in the first column rightwards instead.

6 Experiments

The test performance of the proposed EDD model is compared with five off-the-shelf tools (Tabula[5], Traprange[6], Camelot[7], PDFPlumber[8], and Adobe Acrobat® Pro[9]) and the WYGIWYS model[10] [2]. We crop the test tables from the original PDF for Tabula, Traprange, Camelot, and PDFPlumber, as they only support text-based PDF as input. Adobe Acrobat® Pro is tested with both PDF tables and high-resolution table images (300 PPI). The outputs of the off-the-shelf tools are parsed into the same tree structure as the HTML tables to compute the TEDS score.

6.1 Implementation Details

To avoid exceeding GPU RAM, the EDD model is trained on a subset (399k samples) of PubTabNet training set, which satisfies

$$\text{width and height} \leq 512 \text{ pixels}$$
$$\text{structural tokens} \leq 300 \text{ tokens}$$
$$\text{longest cell} \leq 100 \text{ tokens.} \tag{3}$$

Note that samples in the validation and test sets are not constrained by these criteria. The vocabulary size of the structural tokens and the cell tokens of the training data is 32 and 281, respectively. Training images are rescaled to 448×448 pixels to facilitate batching and each channel is normalized by z-score.

We use the ResNet-18 [13] network as the encoder. The default ResNet-18 model downsamples the image resolution by 32. We modify the last CNN layer of ResNet-18 to study if a higher-resolution feature map improves table recognition performance. A total of five different settings are tested in this paper:

- EDD-S2: the default ResNet-18
- EDD-S1: stride of the last CNN layer set to 1
- EDD-S2S2: two independent last CNN layers for structure (stride = 2) and cell (stride = 2) decoder
- EDD-S2S1: two independent last CNN layers for structure (stride = 2) and cell (stride = 1) decoder
- EDD-S1S1: two independent last CNN layers for structure (stride = 1) and cell (stride = 1) decoder

---

[5] v1.0.4 (https://github.com/tabulapdf/tabula-java).

[6] v1.0 (https://github.com/thoqbk/traprange).

[7] v0.7.3 (https://github.com/camelot-dev/camelot).

[8] v0.6.0-alpha (https://github.com/jsvine/pdfplumber).

[9] v2019.012.20040.

[10] WYGIWYS is trained on the same samples as EDD by truncated back-propagation through time (200 steps). WYGIWYS and EDD use the same CNN in the encoder to rule out the possibility that the performance gain of EDD is due to difference in CNN.

We evaluate the performances of these five settings on the validation set (see Table S3 in supplemental material) and find that a higher-resolution feature map and independent CNN layers improve performance. As a result, the EDD-S1S1 setting provides the best validation performance, and is therefore chosen to compare with baselines on the test set.

The structure decoder and the cell decoder are single-layer long short-term memory (LSTM) networks, of which the hidden state size is 256 and 512, respectively. Both of the decoders weight the feature map from the encoder with soft-attention, which has a hidden layer of size 256. The embedding dimension of structural tokens and cell tokens is 16 and 80, respectively. At inference time, the output of both of the decoders are sampled with beam search (beam = 3).

The EDD model is trained with the Adam [21] optimizer with two stages. First, we pre-train the encoder and the structure decoder to generate the structural tokens only ($\lambda = 1$), where the batch size is 10, and the learning rate is 0.001 in the first 10 epochs and reduced by 10 for another 3 epochs. Then we train the whole EDD network to generate both structural and cell tokens ($\lambda = 0.5$), with a batch size 8 and a learning rate 0.001 for 10 epochs and 0.0001 for another 2 epochs. Total training time is about 16 days on two V100 GPUs.

6.2 Quantitative Analysis

Table 2 compares the test performance of the proposed EDD model and the baselines, where the average TEDS of simple[11] and complex (See footnote 11) test tables is also shown. By solely relying on table images, EDD substantially outperforms all the baselines on recognizing simple and complex tables, even the ones that directly use text extracted from PDF to fill table cells. Camelot is the best off-the-shelf tool in this comparison. Furthermore, the performance of Adobe Acrobat® Pro on image input is dramatically lower than that on PDF input, demonstrating the difficulty of recognizing tables solely on table images. When trained on the PubTabNet dataset, WYGIWYS also considerably outperform the off-the-shelf tools, but is outperformed by EDD by 9.7% absolute TEDS score. The advantage of EDD to WYGIWYS is more profound on complex tables (9.9% absolute TEDS) than simple tables (9.5% absolute TEDS). This proves the great advantage of jointly training two separate decoders to solve structure recognition and cell content recognition tasks.

6.3 Qualitative Analysis

To illustrate the differences in the behavior of the compared methods, Fig. 4 shows the rendering of the predicted HTML given an example input table. The table has 7 columns, 3 header rows, and 4 body rows. The table header has a complex structure, which consists of 4 multi-row (span = 3) cells, 2 multi-column (span = 3) cells, and three normal cells. Our EDD model is able to generate an extremely close match to the ground truth, making no error in

---

[11] Tables without multi-column or multi-row cells.

Table 2. Test performance of EDD and 7 baseline approaches. Our EDD model, by solely relying on table images, substantially outperforms all the baselines.

Input	Method	Average TEDS (%)		
		Simple[a]	Complex[b]	All
PDF	Tabula	78.0	57.8	67.9
	Traprange	60.8	49.9	55.4
	Camelot	80.0	66.0	73.0
	PDFPlumber	44.9	35.9	40.4
	Acrobat® Pro	68.9	61.8	65.3
Image	Acrobat® Pro	53.8	53.5	53.7
	WYGIWYS	81.7	75.5	78.6
	EDD	**91.2**	**85.4**	**88.3**

[a] Tables without multi-column or multi-row cells.
[b] Tables with multi-column or multi-row cells.

structure recognition and a single optical character recognition (OCR) error ('PF' recognized as 'PC'). The second header row is missing in the results of WYGIWYS, which also makes a few errors in the cell content. On the other hand, the off-the-shelf tools make substantially more errors in recognizing the complex structure of the table headers. This demonstrates the limited capability of these tools on recognizing complex tables.

Figures S4 (a)–(c) illustrate the attention of the structure decoder when processing an example input table. When a new row is recognized ('<tr>' and '</tr>'), the structure decoder focuses its attention around the cells in the row. When the opening tag ('<td>') of a new cell is generated, the structure decoder pays more attention around the cell. For the closing tag '</td>' tag, the attention of the structure decoder spreads across the image. Since '</td>' always follows the '<td>' or '>' token, the structure decoder relies on the language model rather than the encoded feature map to predict it. Figure S4 (d) shows the aggregated attention of the cell decoder when generating the content of each cell. Compared to the structure decoder, the cell decoder has more focused attention, which falls on the cell content that is being generated.

6.4 Error Analysis

We categorize the test set of PubTabNet into 15 equal-interval groups along four key properties of table size: width, height, number of structural tokens, and number of tokens in the longest cell. Figure 5 illustrates the number of tables in each group and the performance of the EDD model and the WYGIWYS model on each group. The EDD model outperforms the WYGIWYS model on all groups. The performance of both models decreases as table size increases. We train the models with tables that satisfy Eq. 3, where the thresholds are indicated with vertical dashed lines in Fig. 5. Except for width, we do not observe a steep

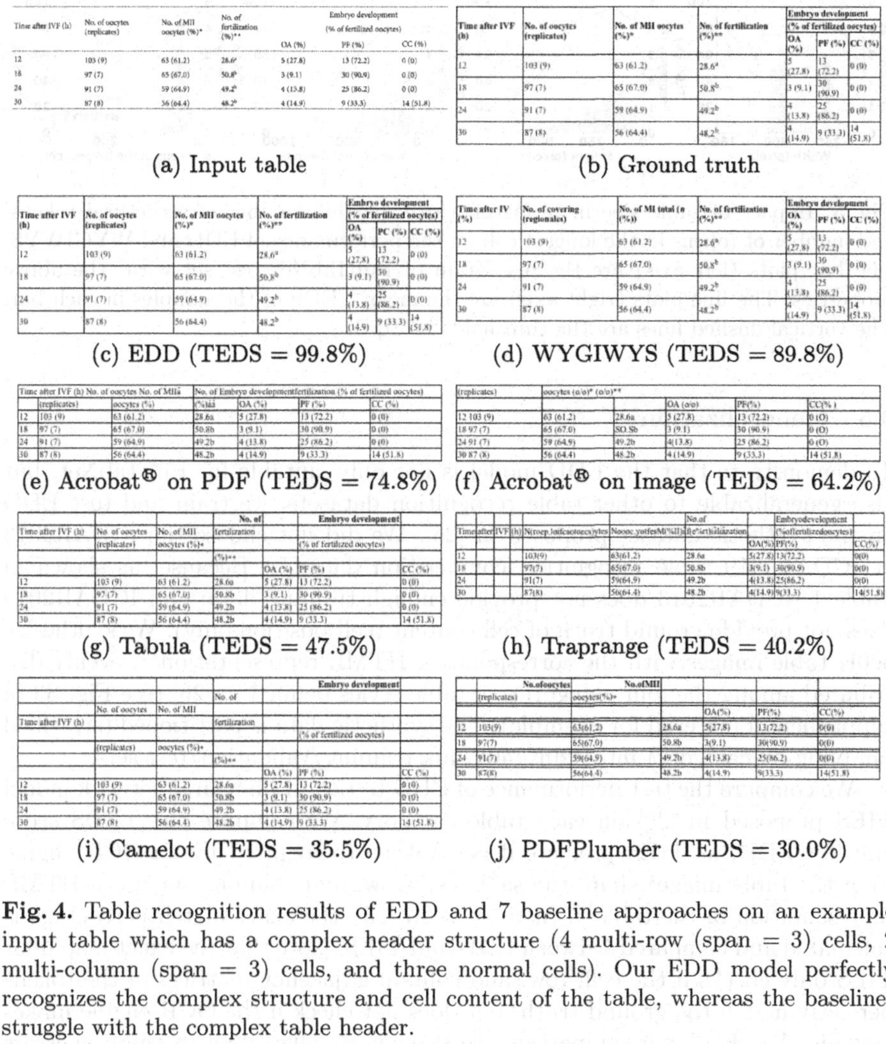

(a) Input table

(b) Ground truth

(c) EDD (TEDS = 99.8%)

(d) WYGIWYS (TEDS = 89.8%)

(e) Acrobat® on PDF (TEDS = 74.8%)

(f) Acrobat® on Image (TEDS = 64.2%)

(g) Tabula (TEDS = 47.5%)

(h) Traprange (TEDS = 40.2%)

(i) Camelot (TEDS = 35.5%)

(j) PDFPlumber (TEDS = 30.0%)

Fig. 4. Table recognition results of EDD and 7 baseline approaches on an example input table which has a complex header structure (4 multi-row (span = 3) cells, 2 multi-column (span = 3) cells, and three normal cells). Our EDD model perfectly recognizes the complex structure and cell content of the table, whereas the baselines struggle with the complex table header.

decrease in performance near the thresholds. We think the lower performance on larger tables is mainly due to rescaling images for batching, where larger tables are more strongly downsampled. The EDD model may better handle large tables by grouping table images into similar sizes as in [2] and using different rescaling sizes for each group.

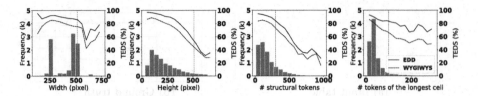

Fig. 5. Impact of table size in terms of width, height, number of structural tokens, and number of tokens in the longest cell on the performance of EDD and WYGIWYS. The bar plots (left axis) are the histogram of PubTabNet test set w.r.t. the above properties. The line plots (right axis) are the mean TEDS of the samples in each bar. The vertical dashed lines are the thresholds in Eq. 3.

6.5 Generalization

To demonstrate that the EDD model is not only suitable for PubTabNet, but also generalizable to other table recognition datasets, we train and test EDD on the synthetic dataset proposed in [26]. We did not choose the ICDAR2013 or ICDAR2019 table recognition competition datasets. Because, as shown in Table 1, ICDAR2013 does not provide enough training data; and ICDAR2019 does not provide ground truth of cell content (cell position only). We synthesize 500K table images with the corresponding HTML representation[12], evenly distributed among the four categories of table styles defined in [26] (see Fig. S3 in supplemental material for example). The synthetic data is partitioned (stratified sampling by category) into 420K/40k/40k training/validation/test sets.

We compare the test performance of EDD to the graph neural network model TIES proposed in [26] on each table category. We compute the TEDS score only for EDD, as TIES predicts if two tokens (recognized by an OCR engine from the table image) share the same cell, row, and column, but not a HTML representation of the table[13]. Instead, as in [26], the exact match percentage is calculated and compared between EDD and TIES. Note that the exact match for TIES only checks if the cell, row, and column adjacency matrices of the tokens perfectly match the ground truth, but does not check if the OCR engine makes any mistakes. For a fair comparison, we also ignore cell content recognition errors when checking the exact match for EDD, i.e., the recognized table is considered as an exact match as long as the structure perfectly matches the ground truth.

Table 3 shows the test performance of EDD and TIES, where EDD achieves an extremely high TEDS score (99.7+%) on all the categories of the synthetic dataset. This means EDD is able to nearly perfectly reconstructed both the structure and cell content from the table images. EDD outperforms TIES in terms of exact match on all table categories. In addition, unlike TIES, EDD does not show any significant downgrade in performance on category 3 or 4, in

---

[12] https://github.com/hassan-mahmood/TIES_DataGeneration.

[13] [26] does not describe how the adjacency relations can be converted to a unique HTML representation.

Table 3. Test performance of EDD and TIES on the dataset proposed in [26]. TEDS score is not computed for TIES, as it does not generate the HTML representation of input image.

Model	Average TEDS (%)				Exact match (%)			
	C1	C2	C3	C4	C1	C2	C3	C4
TIES	–	–	–	–	96.9	94.7	52.9	68.5
EDD	99.8	99.8	99.8	99.7	99.7	99.9	97.2	98.0

which the samples have a more complex structure. This demonstrates that EDD is much more robust and generalizable than TIES on more difficult examples.

7 Conclusion

This paper makes a comprehensive study of the image-based table recognition problem. A large-scale dataset PubTabNet is developed to train and evaluate deep learning models. By separating table structure recognition and cell content recognition tasks, we propose an attention-based EDD model. The structure decoder not only recognizes the structure of input tables, but also helps the cell decoder to place its attention on the right cell content. We also propose a new evaluation metric TEDS, which captures both the performance of table structure recognition and cell content recognition. Compare to the traditional adjacency relation metric, TEDS can more appropriately capture multi-hop cell misalignment and OCR errors. The proposed EDD model, when trained on PubTabNet, is effective on recognizing complex table structures and extracting cell content from image. PubTabNet has been made available and we believe that PubTabNet will accelerate future development in table recognition and provide support for pre-training table recognition models.

Our future works will focus on the following two directions. First, current PubTabNet dataset does not provide coordinates of table cells, which we plan to supplement in the next version. This will enable adding an additional branch to the EDD network to also predict cell location. We think this additional task will assist cell content recognition. In addition, when tables are available in text-based PDF format, the cell location can be used to extract cell content directly from PDF without using OCR, which might improve the overall recognition quality. Second, the EDD model takes table images as input, which implicitly assumes that the accurate location of tables in documents is given by users. We will investigate how the EDD model can be integrated with table detection neural networks to achieve end-to-end table detection and recognition.

References

1. Cesarini, F., Marinai, S., Sarti, L., Soda, G.: Trainable table location in document images. In: Object Recognition Supported by User Interaction for Service Robots, vol. 3, pp. 236–240. IEEE (2002)

2. Deng, Y., Kanervisto, A., Ling, J., Rush, A.M.: Image-to-markup generation with coarse-to-fine attention. In: Proceedings of the 34th International Conference on Machine Learning, vol. 70, pp. 980–989 (2017). JMLR.org

3. Deng, Y., Rosenberg, D., Mann, G.: Challenges in end-to-end neural scientific table recognition. In: 2019 International Conference on Document Analysis and Recognition (ICDAR), pp. 894–901. IEEE, September 2019. https://doi.org/10. 1109/ICDAR.2019.00166

4. Fan, M., Kim, D.S.: Table region detection on large-scale pdf files without labeled data. CoRR, abs/1506.08891 (2015)

5. Fang, J., Tao, X., Tang, Z., Qiu, R., Liu, Y.: Dataset, ground-truth and performance metrics for table detection evaluation. In: 2012 10th IAPR International Workshop on Document Analysis Systems, pp. 445–449. IEEE (2012)

6. Gao, L., et al.: ICDAR 2019 competition on table detection and recognition. In: 2019 International Conference on Document Analysis and Recognition (ICDAR), pp. 1510–1515. IEEE, September 2019. https://doi.org/10.1109/ICDAR.2019. 00166

7. Gatos, B., Danatsas, D., Pratikakis, I., Perantonis, S.J.: Automatic table detection in document images. In: Singh, S., Singh, M., Apte, C., Perner, P. (eds.) ICAPR 2005. LNCS, vol. 3686, pp. 609–618. Springer, Heidelberg (2005). https://doi.org/ 10.1007/11551188_67

8. Gilani, A., Qasim, S.R., Malik, I., Shafait, F.: Table detection using deep learning. In: 2017 14th IAPR International Conference on Document Analysis and Recognition (ICDAR). vol. 1, pp. 771–776. IEEE (2017)

9. Göbel, M., Hassan, T., Oro, E., Orsi, G.: ICDAR 2013 table competition. In: 2013 12th International Conference on Document Analysis and Recognition, pp. 1449–1453. IEEE (2013)

10. Hao, L., Gao, L., Yi, X., Tang, Z.: A table detection method for pdf documents based on convolutional neural networks. In: 2016 12th IAPR Workshop on Document Analysis Systems (DAS), pp. 287–292. IEEE (2016)

11. He, D., Cohen, S., Price, B., Kifer, D., Giles, C.L.: Multi-scale multi-task FCN for semantic page segmentation and table detection. In: 2017 14th IAPR International Conference on Document Analysis and Recognition (ICDAR), vol. 1, pp. 254–261. IEEE (2017)

12. He, K., Gkioxari, G., Dollár, P., Girshick, R.: Mask R-CNN. In: Proceedings of the IEEE International Conference on Computer Vision, pp. 2961–2969 (2017)

13. He, K., Zhang, X., Ren, S., Sun, J.: Deep residual learning for image recognition. In: Proceedings of the IEEE Conference on Computer Vision and Pattern Recognition, pp. 770–778 (2016)

14. Hirayama, Y.: A method for table structure analysis using DP matching. In: Proceedings of 3rd International Conference on Document Analysis and Recognition, vol. 2, pp. 583–586. IEEE (1995)

15. Hu, J., Kashi, R.S., Lopresti, D.P., Wilfong, G.: Medium-independent table detection. In: Document Recognition and Retrieval VII, vol. 3967, pp. 291–302. International Society for Optics and Photonics (1999)

16. Hurst, M.: A Constraint-based Approach to Table Structure Derivation (2003)

17. Jimeno Yepes, A., Verspoor, K.: Literature mining of genetic variants for curation: quantifying the importance of supplementary material. Database **2014** (2014)

18. Kasar, T., Barlas, P., Adam, S., Chatelain, C., Paquet, T.: Learning to detect tables in scanned document images using line information. In: 2013 12th International Conference on Document Analysis and Recognition, pp. 1185–1189. IEEE (2013)

19. Kavasidis, I., et al.: A saliency-based convolutional neural network for table and chart detection in digitized documents. In: International Conference on Image Analysis and Processing, pp. 292–302. Springer (2019)

20. Kieninger, T., Dengel, A.: The t-recs table recognition and analysis system. In: Lee, S.-W., Nakano, Y. (eds.) DAS 1998. LNCS, vol. 1655, pp. 255–270. Springer, Heidelberg (1999). https://doi.org/10.1007/3-540-48172-9_21

21. Kingma, D.P., Ba, J.: Adam: a method for stochastic optimization. In: Proceedings of International Conference on Learning Representations (ICLR) (2015)

22. Levenshtein, V.I.: Binary codes capable of correcting deletions, insertions, and reversals. Soviet Phys. Doklady **10**, 707–710 (1966)

23. Morais, R., Le, V., Tran, T., Saha, B., Mansour, M., Venkatesh, S.: Learning regularity in skeleton trajectories for anomaly detection in videos. In: Proceedings of the IEEE Conference on Computer Vision and Pattern Recognition, pp. 11996–12004 (2019)

24. Paliwal, S.S., Vishwanath, D., Rahul, R., Sharma, M., Vig, L.: TableNet: deep learning model for end-to-end table detection and tabular data extraction from scanned document images. In: 2019 International Conference on Document Analysis and Recognition (ICDAR), pp. 128–133. IEEE (2019)

25. Pawlik, M., Augsten, N.: Tree edit distance: robust and memory-efficient. Inf. Syst. **56**, 157–173 (2016)

26. Qasim, S.R., Mahmood, H., Shafait, F.: Rethinking Table Recognition Using Graph Neural Networks, pp. 142–147, September 2019. https://doi.org/10.1109/ICDAR. 2019.00166

27. Redmon, J., Divvala, S., Girshick, R., Farhadi, A.: You only look once: unified, real-time object detection. In: Proceedings of the IEEE Conference on Computer Vision and Pattern Recognition, pp. 779–788 (2016)

28. Ren, S., He, K., Girshick, R., Sun, J.: Faster R-CNN: towards real-time object detection with region proposal networks. In: Advances in Neural Information Processing Systems, pp. 91–99 (2015)

29. Riba, P., Dutta, A., Goldmann, L., Fornés, A., Ramos, O., Lladós, J.: Table detection in invoice documents by graph neural networks. In: 2019 International Conference on Document Analysis and Recognition (ICDAR), pp. 122–127. IEEE, September 2019. https://doi.org/10.1109/ICDAR.2019.00028

30. Schreiber, S., Agne, S., Wolf, I., Dengel, A., Ahmed, S.: Deepdesrt: deep learning for detection and structure recognition of tables in document images. In: 2017 14th IAPR International Conference on Document Analysis and Recognition (ICDAR), vol. 1, pp. 1162–1167. IEEE (2017)

31. Shafait, F., Smith, R.: Table detection in heterogeneous documents. In: Proceedings of the 9th IAPR International Workshop on Document Analysis Systems, pp. 65–72. ACM (2010)

32. Shahab, A., Shafait, F., Kieninger, T., Dengel, A.: An open approach towards the benchmarking of table structure recognition systems. In: Proceedings of the 9th IAPR International Workshop on Document Analysis Systems, pp. 113–120. ACM (2010)

33. Siegel, N., Lourie, N., Power, R., Ammar, W.: Extracting scientific figures with distantly supervised neural networks. In: Proceedings of the 18th ACM/IEEE on Joint Conference on Digital Libraries, pp. 223–232. ACM (2018)

34. e Silva, A.C.: Learning rich hidden Markov models in document analysis: table location. In: 2009 10th International Conference on Document Analysis and Recognition, pp. 843–847. IEEE (2009)

35. Staar, P.W., Dolfi, M., Auer, C., Bekas, C.: Corpus conversion service: a machine learning platform to ingest documents at scale. In: Proceedings of the 24th ACM SIGKDD International Conference on Knowledge Discovery & Data Mining, pp. 774–782. ACM (2018)

36. Tensmeyer, C., Morariu, V.I., Price, B., Cohen, S., Martinez, T.: Deep splitting and merging for table structure decomposition. In: 2019 International Conference on Document Analysis and Recognition (ICDAR), pp. 114–121. IEEE (2019)

37. Tupaj, S., Shi, Z., Chang, C.H., Alam, H.: Extracting Tabular Information from Text Files. EECS Department, Tufts University (1996)

38. Xu, K., et al.: Show, attend and tell: neural image caption generation with visual attention. In: International Conference on Machine Learning, pp. 2048–2057 (2015)

39. Zhong, X., Tang, J., Yepes, A.J.: Publaynet: largest dataset ever for document layout analysis. In: 2019 International Conference on Document Analysis and Recognition (ICDAR), pp. 1015–1022. IEEE, September 2019. https://doi.org/10.1109/ICDAR.2019.00166

40. Zhou, Y.F., Jiang, R.H., Wu, X., He, J.Y., Weng, S., Peng, Q.: Branchgan: unsupervised mutual image-to-image transfer with a single encoder and dual decoders. In: IEEE Transactions on Multimedia (2019)

Group Activity Prediction with Sequential Relational Anticipation Model

Junwen Chen[(✉)] [iD], Wentao Bao [iD], and Yu Kong [iD]

Golisano College of Computing and Information Sciences,
Rochester Institute of Technology, Rochester, USA
{jc1088,wb6219,yu.kong}@rit.edu

Abstract. In this paper, we propose a novel approach to predict group activities given the beginning frames with incomplete activity executions. Existing action (We define action as the behavior performed by a single person, and define activity as the behavior performed by a group of people) prediction approaches learn to enhance the representation power of the partial observation (We define *partial observation* as the beginning frames with incomplete activity execution, and *full observation* as the one with complete activity execution). However, for group activity prediction, the relation evolution of people's activity and their positions over time is an important cue for predicting group activity. To this end, we propose a sequential relational anticipation model (SRAM) that summarizes the relational dynamics in the partial observation and progressively anticipates the group representations with rich discriminative information. Our model explicitly anticipates both activity features and positions by two graph auto-encoders, aiming to learn a discriminative group representation for group activity prediction. Experimental results on two popularly used datasets demonstrate that our approach significantly outperforms the state-of-the-art activity prediction methods.

Keywords: Activity prediction · Group activity · Structured prediction · Relational model.

1 Introduction

Group activity prediction is to forecast an activity performed by a group of people before the activity ends. Different from group activity recognition, it only has access to the beginning frames of a video containing **incomplete** activity execution. It is useful in the scenarios where the intelligent systems have to make prompt decisions, such as surveillance and traffic accident avoidance where multiple people are present. Unfortunately, existing action prediction methods [21–23,51,52] are limited to actions performed by an individual. Even though some methods [23,24,45] attempt to predict actions performed by multiple people in standard databases such as UCF101 [39], they simply model the multiple people as a single entity and ignore their relations. This would undoubtedly result in a low prediction performance.

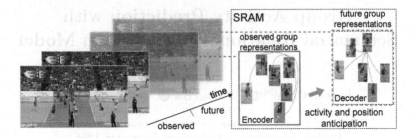

Fig. 1. Given the beginning frames, our method models the relational dynamics of a group, and predicts a group activity by anticipating the group activity representation and their positions occurring in the future unobserved frames.

As shown in [23], one of the major challenges in activity prediction is how to enhance the discriminative power of the features extracted from the partial observations. However, this is even more challenging to do so in group activity prediction as multiple people are present in the scene. Each person's individual action may vary and people's interactions frequently appear and change in a group activity. To this end, it is important to model the relations of multiple people in the observed frames and predict their future group representations. In addition, if only limited beginning frames are observed, it would be extremely difficult to directly anticipate the features of full observations at once. A temporally progressive anticipation model is desired for modeling activity evolution.

To address these challenges, we propose a novel sequential relational anticipation model (SRAM) for group activity prediction by anticipating group activities and positions in the future (see Fig. 1). SRAM is developed as an encoder-decoder framework, in which an *observation encoder* summarizes the relational dynamics in the beginning observed frames and a *sequential decoder* further anticipates the representations for group activities and positions occurring in the future. Specifically, the observation encoder naturally models the relational dynamics of people and complex interactions between people in the observed frames. To predict group activity, we propose a sequential decoder to anticipate the structured group representation in the future using several unrolling stages. Two graph auto-encoders are used in the sequential decoder to explicitly anticipate the activity and the position relations of people in the unobserved frames. We propose to make sequential prediction that *progressively* anticipates the future group representation by performing multiple unrolling stages guided by three novel loss functions. This allows us to better capture complex group activity evolution.

To our best knowledge, we are the first to investigate the challenging problem of group activity prediction. The benefit of our method is twofold. Firstly, it not only predicts people's group activities but also predicts individuals' positions in the future. Our experimental results show that predicting people's future positions significantly helps predict their group activities. Secondly, the proposed method progressively anticipates structured group representations, which has shown to be very powerful in prediction especially when limited frames are observed. This idea could be generalized to other prediction tasks, e.g. human motion prediction [31] and video prediction [47].

Our contributions can be summarized as follows:

- We propose a novel sequential decoder to anticipate the representations for multiple people's future positions and activity, aiming to learn a discriminative structural representation for group activity prediction.
- We progressively anticipate the structured group representations at several unrolling stages guided by novel loss functions. This improves the performance when only few frames are observed.
- Extensive experiments demonstrate that our method outperforms the existing state-of-the-arts by a large margin.

2 Related Work

Action Prediction aims to recognize the label of an action before the action is fully executed. Existing work [6,17,21,26,30,36,37,43,53] focuses on predicting actions performed by an individual. Ryoo et al. [34] used integral and dynamic bag-of-words to represent features variations over time. DeepSCN [23] and AAP-Net [24] make use of sequential context information by transferring knowledge in full videos to partial observations. Wang et al. [45] developed a teacher-student learning framework to distill knowledge from the action recognition task, in order to enhance action prediction. Gammulle et al. [11] presented a jointly learnt task for both action prediction and future motion representation inference.

Prediction on interactions between two people was studied in [51,52]. Yan et al. [51] developed a tri-coupled recurrent structure and an attention mechanism to address action prediction for two individuals' interactions. Yao et al. [52] predicted the motion of the interactions between two people, but did not predict their interaction labels. Different from them, we focus on the prediction of group activities involving multiple people. Our method elegantly captures complex relational dynamics between people for learning discriminative information.

Group Activity Recognition has been extensively studied in previous work [2,19,38,49]. Early work applies graphical models on the extracted hand-craft features [2,27,46] as group representations. Deep learning methods for multi-people activity recognition have shown excellent performance [4,8,12,18,38,41, 44]. HDTM [19] develops a two-stage LSTM model to firstly extract features of temporal individual motions and then aggregate neighborhood information. SSU [4] achieves the individual detection and group activity recognition in a unified framework. Recent work suggests that only part of people's motions contribute to the entire group activity [3,10,16,33,49], via suppressing the irrelevant actions. Previous work also shows that interactions between people are important in understanding group activity. For example, HRN [18] introduces a hierarchical spatial relational layer to learn the relational representations between two players. Other methods, including Stagnet [32], S-RNN [5], SBGAR [29] apply

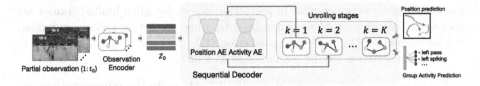

Fig. 2. Overall architecture. Our framework SRAM takes the beginning t_0 observed frames as input and predicts the group activity label. An observation encoder first summarizes the relational dynamics in partial observation as a latent variable Z_0. Then, a sequential decoder takes over Z_0 and progressively anticipates the group representation through K unrolling stages. The output of the last unrolling stage is expected to contain rich discriminative information for group activity prediction. Details can be seen in Fig. 3.

structural-RNN to obtain spatiotemporal features. ARG [48] explicitly models the interactions by employing graph convolution on a learnable graph.

The main difference between our work and group activity recognition methods is that we aim at predicting the group activity label given *incomplete* activity execution, while these methods are given complete activity executions. This prompts us to develop novel model architecture and loss functions in this work.

3   Our Approach

Problem Formulation. Our goal is to predict the activity label y of a group of people given a partial observation of a video containing incomplete activity execution. We define the *observation ratio* as the number of observed frames t_0 in a streaming video divided by the total number of frames T in the corresponding full video following [23,24], i.e. t_0/T. For instance, if a partial video contains 30 frames and the corresponding full video contains 100 frames, then the observation ratio of this activity is 30%.

During training, we have access to all full training videos containing complete group activity executions. These full videos are supposed to contain all the discriminative information for classification. During test, given a partial observation of a group activity execution, we encourage our model to anticipate the group representations that contain similar amount of discriminative information as the corresponding full observation. Thus, its prediction power can be enhanced.

Overall Architecture. The overall architecture is shown in Fig. 2. We formulate our group activity prediction model as an encoder-decoder framework that contains an *observation encoder* and a *sequential decoder*. Given a partial observation containing t_0 frames, the observation encoder summarizes the relational dynamics of the group from the partial observation and then the sequential decoder anticipates the group representation for activities and positions in the future unobserved frames.

Due to the large motion variations between a partial and a full observation, a novel sequential decoder is proposed in this work to progressively anticipate the structured group representation for the future unobserved frames by several unrolling stages. This is useful for enhancing the discriminative power of the anticipated representation, especially if limited frames are observed. Moreover, for group activity, relations between multiple people are discriminative information and they vary as time. To predict group activity, our sequential decoder uses two graph auto-encoders to concurrently perform relational anticipation on both people's activity features and their positions.

3.1   Relation Modeling for Group Activity

Given t_0 observed frames, we first extract features of all the observed t_0 frames, and then apply ROIAlign [15] to extract the feature vectors of multiple people based on their positions $\{B_1, B_2, \cdots, B_{t_0}\}(t \in \{1, \cdots, t_0\})$. Action features and positions of the i-th individual on the t-th frame are represented as $\mathbf{x}_t(i)$ and $\mathbf{b}_t(i)$ respectively. Afterwards, upon the individual dynamics, we follow [48] to explicitly model the pair-wise *position relations* and *action relations* of multiple people in the observed frames as two relation graphs $G_t^a \in \mathbb{R}^{N \times N}$ and $G_t^p \in \mathbb{R}^{N \times N}$, respectively. Both of the two graphs have N nodes representing N people in the t-th frame. Given the i-th and j-th individuals, the edge of the action similarity graph $G_t^a(i,j)$ is computed by the cosine similarity and normalized by Softmax function. The edge on the position relation graph $G_t^p(i,j)$ is computed by the normalized Euclidean distance (denoted by $d(\cdot, \cdot)$):

$$G_t^a(i,j) = \frac{\exp\left(\mathbf{x}_t(i)^T \cdot \mathbf{x}_t(j)\right)}{\sum_{j=1}^{N} \exp\left(\mathbf{x}_t(i)^T \cdot \mathbf{x}_t(j)\right)}, \; G_t^p(i,j) = \frac{1/d(\mathbf{b}_t(i), \mathbf{b}_t(j))}{\sum_{j=1}^{N} 1/d(\mathbf{b}_t(i), \mathbf{b}_t(j))}. \quad (1)$$

Once the graphs are built, we obtain the structured representations for the group activity in the observed frames. We will also perform anticipation on the two graphs representing the group activity in the unobserved frames, which will be discussed below.

3.2   Observation Encoder $\mathcal{E}$

The observation encoder $\mathcal{E}$ is proposed to summarize spatiotemporal information of the complex relational dynamics of multiple people in partial observations containing t_0 frames. $\mathcal{E}$ learns to map $G_{1:t_0}^a$, $G_{1:t_0}^p$, and $X_{1:t_0}$ to a latent variable Z_0, by the spatio-temporal graph convolution network ST-GCN [50]. Specifically, it first performs spatial graph convolution [20] on the two graphs G_t^p and G_t^a for the t-th frame

$$\sigma(G_t^p X_t W_p) + \sigma(G_t^a X_t W_a), \quad (2)$$

and then performing temporal convolution [28] on every three consecutive frames to learn the latent variable Z_0. Here, σ is ReLU activation, W_p and W_a are learnable weights, and X_t is the action features of N people. Latent variable Z_0 will be integrated in the sequential decoder, and guides its unrolling stages.

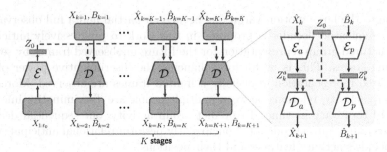

Fig. 3. Sequential decoder $\mathcal{D}$ is formulated as two auto-encoders $\mathcal{E}_a$-$\mathcal{D}_a$ and $\mathcal{E}_p$-$\mathcal{D}_p$ that progressively anticipate the group activity representation for future unobserved frames using multiple unrolling stages. At the k-th stage, $\mathcal{D}$ is fed with the summary of the partial observation encoded in the latent variable Z_0 as well as the action features $\hat{X}_k$ and the position features $\hat{B}_k$ from the previous stage. Then $\mathcal{D}$ anticipates the action features $\hat{X}_{k+1}$ and positions $\hat{B}_{k+1}$.

Different from ARG [48], our model captures the temporal patterns of people's relations, which is useful for group activity prediction.

3.3   Sequential Decoder $\mathcal{D}$

The performance of state-of-the-art action prediction methods [23,24] is still limited especially when few beginning frames are given. This is mainly because they use a direct mapping from partial observation to the corresponding full observation in one pass, which is not powerful enough to deal with large visual variations between partial and full observations. In this paper, we propose a sequential decoder that *progressively* anticipates the group representation that is expected to contain rich discriminative information as the fully observed activity using K unrolling stages (see Fig. 3). This allows us to create a more powerful model for group activity prediction.

Besides, different from individual action prediction methods [24,45], people's relations formulated as graphs using Eq. (1), are discriminative information for group activity. Moreover, the group activity varies overtime. It is necessary to predict group representations by anticipating relations in the unobserved stage. As described in Sect. 3.1, people's relations can be inferred from their action similarity and relative positions. For example, a partial observation of a volleyball activity is given, which contains run-up of ace spikers and waiting gestures of their opponents. Our model is supposed to predict it as "spiking" by the cue that the players are moving towards net with their actions. Therefore, we develop a sequential decoder as a mixture of two graph auto-encoders: an activity auto-encoder $\mathcal{E}_a$-$\mathcal{D}_a$ for predicting activity representations and a position auto-encoder $\mathcal{E}_p$-$\mathcal{D}_p$ for predicting positions of multiple people. The two auto-encoders are coupled by the shared latent variables Z_0 learned from partial observations.

Activity Auto-encoder $\mathcal{E}_a$-$\mathcal{D}_a$. Using K activity auto-encoders, the proposed sequential decoder progressively anticipates the activity representation by K unrolling stages. Each activity auto-encoder at the k-th stage ($k \in \{1, 2, \cdots, K\}$) is fed with the output $\hat{X}_k$ of the activity auto-encoder at the previous $(k-1)$-th stage. We use the spatiotemporal action features at the last observed frame t_0 as the input of the activity auto-encoder at stage $k = 1$. We encode the input $\hat{X}_k$ of current unrolling stage to a latent variable Z_k^a by

$$Z_k^a = \sigma(G_k^p \hat{X}_k U_{ep}) + \sigma(G_k^a \hat{X}_k U_{ea}), \tag{3}$$

and then decodes the activity representation $\hat{X}_{k+1}$:

$$\hat{X}_{k+1} = \sigma(G_k^a(Z_0 + Z_k^a))U_{da}) + \sigma(G_k^p(Z_0 + Z_k^a)U_{dp}), \tag{4}$$

where U_{ep}, U_{ea}, U_{dp}, U_{da} are learnable parameters. $\hat{X}_{k+1}$ is the anticipated group features at the k-th stage, and is served as the input for the activity auto-encoder at the $(k+1)$-th stage. The anticipation of $\hat{X}_{k+1}$ is conditioned on latent variables Z_k^a and Z_0, in order to both keep track of the short-term information of the previous unrolling stage and use the long-term spatiotemporal information in the partial observations. G_k^a and G_k^p are computed by the generated activity features and positions at the k-th stage using similar functions as Eq. (1), respectively (replacing time step t by the stage k).

The benefits of the progressive anticipation using K unrolling stages lie in two aspects. First, the temporal dependency of activity evolution is naturally built between successive stages. This allows us to naturally anticipate structured group activity representations for prediction purpose. Second, the prediction granularity can be controlled with the number of unrolling stages K. The case when $K = 1$ is equivalent to the existing one-pass solution used in [23,24].

Position Auto-encoder $\mathcal{E}_p$-$\mathcal{D}_p$. As described in Sect. 3.1, the interactions between two people also depend on their relative positions. Thus, it is necessary to explicitly anticipate the positions of these people in group activity prediction.

Similar to activity auto-encoder, the proposed sequential decoder also performs K unrolling stages for position prediction for a group of people using K position auto-encoders. Each position auto-encoder at stage k is fed with the output of its previous auto-encoder at stage $k - 1$, and outputs the positions of people. Experimental results in Sect. 4.4 show that the anticipated future positions of people help improve performance of group activity prediction.

The position auto-encoder first encodes the positions $\hat{B}_k$ of multiple people to a latent variable Z_k^p at stage k through graph convolution [20]:

$$Z_k^p = \sigma(G_k^p \hat{B}_k V_{ep}) + \sigma(G_k^a \hat{B}_k V_{ea}), \tag{5}$$

and then decodes the positions $\hat{B}_{k+1}$ for the next stage by

$$\hat{B}_{k+1} = \sigma(G_k^a(Z_0 + Z_k^p))V_{da}) + \sigma(G_k^p(Z_0 + Z_k^p)V_{dp}), \tag{6}$$

where V_{ep}, V_{ea}, V_{dp}, V_{da} are learnable parameters. G_k^p and G_k^a are the same graphs used in the activity auto-encoder. The anticipation of $\hat{B}_{k+1}$ is conditioned on latent variables Z_k^p and Z_0, in order to both keep track of the short-term position information of the previous unrolling stage and use the long-term spatiotemporal information in the partial observations.

Position prediction is also benefited by sequential prediction via several unrolling stages, since the prediction granularity can be controlled. Similar to the activity auto-encoder, the position auto-encoder at stage $k = 1$ also takes the positions B_{t_0} of people on the last observed frame as input. The activity auto-encoder and the position auto-encoder share the same graphs G_k^p and G_k^a and are both conditioned on the latent variable Z_0 (see Fig. 3).

3.4   Feature Aggregation for Prediction

SRAM returns both group activity and position representations at each of K unrolling stages. The K-th stage corresponds to the full observation status, which contains the most discriminative information of an activity. We disregard all the outputs given by the activity autoencoders from the 1-st to $(K - 1)$-th stages, and perform max-pooling on the output $\hat{X}_{K+1}$ given by the activity autoencoder at the K-th stage as the group activity representations. The resulting feature vector is used for group activity prediction. Similarly, we directly use the output $\hat{B}_{K+1}$ given by the K-th position autoencoder to perform position prediction.

3.5   Loss Functions and Model Learning

Adversarial Loss. Inspired by [13], we encourage SRAM to generate representations corresponding to ground-truth full observations. We use two discriminators for the features generated by the sequential decoder. Discriminator D_1 is an activity classifier implemented by a softmax layer. $\mathcal{L}_{cls}$ is computed on the output of D_1. Discriminator D_2 is an adversarial regularizer and tells the difference between the generated group features $\hat{X}_{1:K}$ and group features of full videos $F_{1:K}(X)$. Using the adversarial loss, SRAM is encouraged to generate features that are indistinguishable from the group features of the corresponding full videos:

$$\mathcal{L}_{GAN} = \mathbb{E}_{X_{(1:T)} \sim p_{data}(X_{(1:T)})} \log D_2\left(F_{1:K}(X)\right) \tag{7}$$
$$+ \mathbb{E}_{X_{(1:t_0)} \sim p_{data}(X_{(1:t_0)})} \log\left(1 - D_2(\hat{X}_{1:K})\right).$$

Note that the generated group representation $\hat{X}_{1:K}$ is computed by SRAM S from the partial observation $X_{1:t_0}$.

Sequential reconstruction loss is proposed to align the predicted activity representations to become close to the ground-truth activity representations at each unrolling stage. Since our method has K-stage sequential prediction, it is necessary to encourage the predicted group representations $\hat{X}_{1:K}$ on each of the K unrolling stages to become close to the ground-truth features at that timestamp. This is different from adversarial loss that only align the generated features to be close to ground-truth at full observation stage.

We train a separate ST-GCN $F(\cdot)$ as a recognition model to obtain the group activity representations of full videos X for training. The resulting frame-wise group representations $F(X)$ are used to encourage the activity features of the i-th person generated at the k-th unrolling stage to be similar to the ground-truth features using

$$\mathcal{L}_{\text{rec}} = \frac{1}{K \times N} \sum_{k=1}^{K} \sum_{i=1}^{N} \|\hat{\mathbf{x}}_k(i) - F_k(X, i)\|^2. \qquad (8)$$

Here, $F_k(X, i)$ is the features of the i-th person of the full video at the k-th stage. This loss function sequentially computes the difference between the predicted features $\hat{\mathbf{x}}_k(i)$ (the i-th row on $\hat{X}_{1:K}$) for the i-th person at unrolling stage k and the ground-truth features $F_k(X, i)$, mimicking how a partial observation is progressively approaching its corresponding full observation.

Position Regression Loss. We use the tracklets of individuals provided by [18] as the ground-truth of individuals positions. During training, we use the mean square error between the predicted positions and ground-truth positions at the K unrolling stages as loss function:

$$\mathcal{L}_{\text{reg}} = \frac{1}{K \times N} \sum_{k=1}^{K} \sum_{i=1}^{N} ||\hat{\mathbf{b}}_k(i) - \mathbf{b}_k(i)||^2, \qquad (9)$$

where the predicted position $\hat{\mathbf{b}}_k(i)$ is the i-th row of $\hat{B}_k$, i.e., the i-th person's position predicted by the sequential decoder at the k-th stage.

Model Learning. During training, the overall objective function is written as a sum of sequential reconstruction loss $\mathcal{L}_{\text{rec}}$, adversarial loss $\mathcal{L}_{\text{GAN}}$, classification loss $\mathcal{L}_{\text{cls}}$ implemented by softmax loss, and position regression loss $\mathcal{L}_{\text{reg}}$:

$$\min_{\mathcal{E}, \mathcal{D}} \max_{D_1, D_2} \mathcal{L}_{\text{rec}} + \mathcal{L}_{\text{GAN}} + \mathcal{L}_{\text{cls}} + \mathcal{L}_{\text{reg}}. \qquad (10)$$

Sequential relational anticipation model $(\mathcal{E}, \mathcal{D})$ and two discriminators (D_1, D_2) are alternatively trained until convergence.

3.6   Discussion

Group Activity Modeling and Anticipation. Our SRAM captures the interactions of multiple people in the observation encoder, and anticipates their future relations by a sequential decoder. This is different from existing action prediction methods [24,32] that can only predict the action of an individual. We believe such a novel method will pave the way for future research in other structured visual prediction.

Structured Sequential Prediction. Compared with group activity recognition methods [18,32,48], our method performs *sequential prediction* of group activity, in form of future positions and activity representations. Our activity prediction is also facilitated by explicitly predicting people's future positions.

Activity Evolution over Time. Our sequential decoder *progressively* predicts future representations through several unrolling stages, which boosts performance when only few frames are observed. It is guided by a sequential reconstruction loss, mimicking how a partial observation is sequentially approaching its full observation and an adversarial loss to make the generated full observation features to become indistinguishable from the real full observation features.

4   Experiments

4.1   Datasets

Volleyball Dataset [19] consists of 4830 video clips distributed in 8 group activities, such as *left spiking* and *right setting*. Each clip has 41 frames. [19] provides the players' tracklets and splits the dataset into training, validation and testing sets. Existing group activity recognition methods [18,19,32,48] use the middle 10 frames of each video. To generalize it to prediction task, we extend it to use the middle 20 frames as full observations, in order to model sequential dynamics. Note that the middle 20 frames contain complete group activity executions, because athletes generally move quickly to complete a group activity, such as direct spiking in a volleyball game.

Collective Activity Dataset (CAD) [7] contains 44 videos with 5 group activities, including *crossing*, *queueing*, *walking*, *talking* and *waiting*. The group activities in CAD are labeled as the majority of people's individual actions. We use the existing tracklet information and training/testing splits following [48]. The number of the frames in videos ranges from 100 to 2000. Following [4,32,48], we divide each video into 10-frame video clips. This expands training and testing data to 1746 and 765 clips, respectively. CAD mainly contains periodic activities such as *walking*, in which significant changes can be seen in 10 frames.

4.2   Implementation Details

Following [48], we extract a 1024-dimensional feature vector for each individual with tracklets provided by [19], using Inception-v3 [40] as backbone and ROIAlign [15]. We use three steps for training: First, Inception-v3 pretrained on ImageNet is fine-tuned on single frames by jointly predicting individual actions and group activities. Then, we freeze the backbone and finetune the recognition model $F(\cdot)$ given full videos in the training set. The recognition model contain two ST-GCN layers [50], both with 256-d hidden units. After that, we train the proposed model. The observation encoder has two layers ST-GCN with both 256-d hidden units. The activity auto-encoder's encoder has one graph convolution layer that encodes the input into 256-d latent feature space. The position auto-encoder has two layer graph convolution by encoding the 2-d positions into 64-d space and then 256-d latent space. During training, SRAM plus classifier D_1 and discriminator D_2 are alternatively updated.

The experiments are conducted with 10 different observation ratios ranging from 10% to 100% of full videos length. The number of unrolling stages K is set to 5. We use stochastic gradient descent for optimization. For Volleyball dataset, the three steps are trained for 30 epochs, 10 epochs and 20 epochs with learning rate 0.001, 0.001, 0.0001 respectively. For Collective Activity Dataset, the three steps are trained for 20 epochs, 50 epochs and 10 epochs with learning rate 0.0001, 0.0001, 0.0005, respectively.

4.3 Comparison with State-of-the-Art

We compare our method with the state-of-the-art prediction methods LRCN [42], DeepSCN [23], IBoW and DBoW [34], KD [45], AAPNet [24] and state-of-the-art group activity recognition methods, including HRN [18], HDTM [19] ARG [48], SSU [4]. Following these methods' original setting, LRCN and HDTM adopt the AlexNet [25] as the backbone. HRN, IBow, DBow, DeepSCN and original AAPNet use VGG-19. Our method follows ARG and SSU to use Inception-V3 method. HRN, HDTM, ARG, SSU and our method adopt the tracklets of players provided by [19]. To make a fair comparison, we extend state-of-the-art action prediction method AAPNet ("e-AAPNet" for simplification) to make use of tracking information and use Inception-V3 as backbone. We train all of the comparison methods using the parameters described in their original papers.

Results on Volleyball Dataset. Table 1 summarizes the prediction performance of the proposed method, existing action prediction methods and group activity recognition methods. Results demonstrate that our model outperforms the comparison methods. Existing action prediction methods, e.g., LRCN, IBoW, DBoW, DeepSCN, AAPNet, KD propose to improve the prediction performance by information transfer. However, they regard multiple people as a single entity and do not consider the interactions between multiple people. Thus, the extracted features do not contain informative cues of the interactions of people, resulting in a low prediction performance. The proposed method uses tracklets [19], while the existing predictors for individuals e.g. LRCN, IBoW, DBoW, DeepSCN, AAPNet, KD do not. To make a fair comparison, we extend AAPNet to use tracklet information. Experimental results show that our method can predict the dynamics of interactions and better enrich partial observations.

Group activity recognition methods such as HDTM, SSU, HRN, ARG do not have capability of gaining extra information from full activity executions. Thus, when the observation ratio is very low (10% or 20% observations), their performance is much lower than our method. Note that ARG applies random sampling strategy by sampling three frames from an entire video as input. In the comparison experiment, this strategy is applied in each of the partial observations as input. The proposed method consistently outperforms ARG, as our method captures the temporal dynamics of multiple people in the group, and sequentially generates features close to the corresponding full observations. It improves the representation power of the partial observations, and facilitates group activity prediction.

Table 1. Group activity prediction accuracy (%) on Volleyball dataset with observation ratios ranging from 10% to 100%. Group activity recognition results can be seen from the last column, in which 100% frames are observed.

Models	Tracklet	Backbones	10%	20%	30%	40%	50%	60%	70%	80%	90%	100%
LRCN [9]	No	AlexNet	48.17	51.61	54.67	57.44	59.76	61.23	63.75	64.32	64.77	65.37
HDTM [19]	Yes	AlexNet	52.43	59.09	66.04	76.37	80.48	81.82	84.07	84.47	84.60	84.06
IBoW [35]	No	VGG	58.03	60.72	64.84	65.26	67.51	70.80	73.45	74.24	74.29	75.63
DBoW [35]	No	VGG	58.03	55.56	56.16	58.93	59.90	61.97	63.79	63.06	63.88	64.78
DeepSCN [23]	No	VGG	59.46	62.23	65.52	70.38	72.55	77.37	79.75	80.35	80.31	80.78
HRN [18]	Yes	VGG	52.58	56.99	64.32	74.49	76.96	80.36	83.72	84.74	84.08	85.30
KD [45]	No	VGG	65.67	67.68	70.00	70.83	71.96	72.10	73.22	73.30	73.30	73.90
AAPNet [24]	No	VGG	59.53	65.37	68.29	72.25	75.24	77.79	79.91	80.25	80.18	80.78
e-AAPNet [24]	Yes	InceptionV3	62.98	70.31	77.64	83.55	84.91	85.86	87.54	87.23	87.92	89.01
SSU [4]	Yes	InceptionV3	63.20	70.65	79.66	84.07	87.13	87.65	88.30	88.18	88.41	89.01
ARG [48]	Yes	InceptionV3	64.82	69.41	76.07	79.43	82.70	83.99	85.04	85.19	85.86	85.94
Ours	Yes	InceptionV3	**77.86**	**82.57**	**84.97**	**87.06**	**88.63**	**88.93**	**89.08**	**88.93**	**88.48**	**91.97**

Results on Collective Activity Dataset. Comparison results are listed in Table 2. Our method outperforms existing methods ARG, DeepSCN, and AAPNet by a large margin. Given tracklets as input, our method is 6.54% higher than e-AAPNet at 50% observation ratio since the people's actions and relations are predicted in our model. Group

Table 2. Prediction accuracy (%) on Collective Activity Dataset.

Models	Tracklet	50%	100%
ARG [48]	Yes	88.10	88.37
DeepSCN [23]	No	81.31	82.22
AAPNet [24]	No	81.57	82.75
e-AAPNet [24]	Yes	86.01	86.67
Ours	Yes	**92.55**	**92.81**

activities such as *group walking* are cyclic, and thus the prediction performance of our method at 50% observation ratio is close to the one at 100% observation ratio.

4.4   Ablation Study

We perform detailed ablation studies on the Volleyball dataset to evaluate the contributions of the sequential prediction strategy, as well as the loss functions.

How Much Does the Prediction Loss Help? The impacts of loss functions are analyzed on Volleyball dataset in detail. The evaluation results can also validate the contributions of the proposed sequential prediction strategy. We compare the following the variants, including: (**1**) without the position regression loss $\mathcal{L}_{reg}$ defined in Eq. (9). In this variant, the position auto-encoder for predicting future positions is not used. During sequential prediction, we replace the individuals' positions in the future frames by the ones given by the last observed frame's. The positions are used for computing G^p for each unrolling stage. (**2**) without adversarial loss $\mathcal{L}_{GAN}$. (**3**) without sequential reconstruction loss $\mathcal{L}_{rec}$ for generated features of unrolling stages. (**4**) The proposed full network.

Compared with variant (**1**), the significant performance gains with all different observation ratios show that the prediction of people's positions is of high

Table 3. Ablation studies on volleyball dataset. We show the accuracy(%) given videos of observation ratio at 10%, 40%, 70%.

(a) Comparison of different loss functions, $\mathcal{L}_{GAN}$, $\mathcal{L}_{rec}$, $\mathcal{L}_{reg}$, and $\mathcal{L}_{cls}$.

Loss	10%	40%	70%	Average
(1)$\mathcal{L}_{GAN}+\mathcal{L}_{rec}+\mathcal{L}_{cls}$	75.09	85.59	87.06	84.85
(2) $\mathcal{L}_{rec}+\mathcal{L}_{reg}+\mathcal{L}_{cls}$	76.14	85.79	88.18	86.30
(3) $\mathcal{L}_{reg}+\mathcal{L}_{GAN}+\mathcal{L}_{cls}$	77.61	83.22	85.64	86.22
(4) Ours	**77.86**	**87.67**	**89.08**	**86.85**

(b) Comparison on the number of unrolling stages K.

K	10%	40%	70%	Average
1	70.38	80.02	86.14	81.47
2	72.36	86.59	89.07	85.22
5	77.86	**87.06**	89.08	**86.85**
10	**77.93**	86.69	**89.23**	86.79

importance for group activity prediction. Compared with variants (2) and (3), it shows that the adversarial loss $\mathcal{L}_{GAN}$ and the reconstruction loss $\mathcal{L}_{rec}$ in our method improve the performance by 0.55% and 0.63% on average, respectively. Therefore, the proposed sequential decoder guided by $\mathcal{L}_{GAN}$ and $\mathcal{L}_{rec}$ can generate more discriminative activity representations at each stage.

How Much Does the Sequential Prediction Help? Our sequential decoder predicts group activity representations at K unrolling stages. In this experiment, we evaluate the effect of the number of unrolling stages K on the prediction performance. We set K to 1, 2, 5, and 10, and compare the prediction performance. Table 3b indicates that the best overall prediction performance is achieved when $K = 5$. The prediction performance is slightly affected when $K = 10$, but the computational complexity of the prediction model is increased due to the extra unrolling stages. The average prediction performance drops to 81.47% if $K = 1$. The variant with $K = 1$ is the one that directly maps partial observation in one unrolling stage, similar to what [23,24] do. The result demonstrates the superiority of our progressive prediction in anticipating discriminative group representations given partial observations. If more stages are allowed ($K = 5$ or $K = 10$), the sequential decoder in our model can progressively generate discriminative features for group activity prediction even though it is given very limited frames. Therefore, its prediction performance is improved.

4.5 Position Prediction Evaluation

Visualization of Predicted Positions. As shown in Fig. 4, we visualize the movement of individuals learned by the position auto-encoder in SRAM. The position auto-encoder progressively predicts the positions of individuals at the unrolling stages. The visualization result shows our position auto-encoder can successfully predict the directions and step sizes of individuals in the future based on partial observations. Although Fig. 4 (bottom-right) shows the direction of the predicted movement is mostly accurate, the future position of a person is not accurate if the person moves very fast.

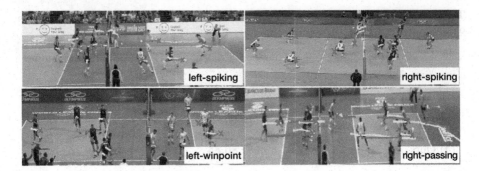

Fig. 4. Visualization of position predictions. The blue and the yellow lines denote the prediction positions and ground-truth positions [19], respectively. "×" indicates the starting point of the movement. Best viewed in color. (Color figure online)

Quantitative Evaluation. We quantitatively evaluate our position prediction results compared to two popular trajectory prediction methods SocialGAN [14] and SocialLSTM [1]. Following SocialGAN, Final Displacement Error (FDE) is used to compute the Euclidean distance between the predicted positions and ground-truth positions at the final timestamp and Average Displacement Error (ADE) is used to compute that at each unrolling stage.

As shown in Table 4, the results demonstrate that our method can accurately predict the future positions for a group of people, and our method outperforms the two trajectory prediction methods. This is mainly because we capture the relational action dynamics of multiple people while SocialGAN and SocialLSTM do not.

Table 4. Final Displacement Error (FDE) and Average Displacement Error (ADE) for position prediction.

Method	FDE	ADE
SocialGAN [14]	5.32	3.05
SocialLSTM [1]	6.44	4.44
Ours	**3.62**	**2.44**

5   Conclusion

We have proposed a novel sequential relational anticipation model (SRAM) to predict group activity given only the beginning frames of an activity execution. Our model captures complex relational dynamics of multiple people in the observed frames. It then anticipates the group representations including group activity features and position features. A novel sequential decoder is proposed to progressively anticipates the group representations through several unrolling stages. Extensive results on two datasets demonstrate that our method significantly outperforms the state-of-the-art methods. Results also validate that the progressive anticipation using multiple unrolling stages facilitates group activity prediction. Further experimental results show that the modeling and prediction of people's positions improves our performance on group activity prediction.

Acknowledgement. We thank Nvidia for the GPU donation. This research is supported in part by ONR Award N00014-18-1-2875.

References

1. Alahi, A., Goel, K., Ramanathan, V., Robicquet, A., Fei-Fei, L., Savarese, S.: Social LSTM: human trajectory prediction in crowded spaces. In: CVPR, pp. 961–971 (2016)
2. Amer, M.R., Lei, P., Todorovic, S.: HiRF: hierarchical random field for collective activity recognition in videos. In: Fleet, D., Pajdla, T., Schiele, B., Tuytelaars, T. (eds.) ECCV 2014. LNCS, vol. 8694, pp. 572–585. Springer, Cham (2014). https://doi.org/10.1007/978-3-319-10599-4_37
3. Azar, S.M., Atigh, M.G., Nickabadi, A., Alahi, A.: Convolutional relational machine for group activity recognition. In: CVPR, pp. 7892–7901 (2019)
4. Bagautdinov, T., Alahi, A., Fleuret, F., Fua, P., Savarese, S.: Social scene understanding: End-to-end multi-person action localization and collective activity recognition. In: CVPR, pp. 4315–4324 (2017)
5. Biswas, S., Gall, J.: Structural recurrent neural network (SRNN) for group activity analysis. In: WACV, pp. 1625–1632 (2018)
6. Cai, Y., Li, H., Hu, J.F., Zheng, W.S.: Action Knowledge Transfer for Action Prediction With Partial Videos, pp. 8118–8125 (2019)
7. Choi, W., Shahid, K., Savarese, S.: What are they doing? collective activity classification using spatio-temporal relationship among people. In: ICCV Workshops, pp. 1282–1289 (2009)
8. Deng, Z., Vahdat, A., Hu, H., Mori, G.: Structure inference machines: recurrent neural networks for analyzing relations in group activity recognition. In: CVPR, pp. 4772–4781 (2016)
9. Donahue, J., et al.: Long-term recurrent convolutional networks for visual recognition and description. In: CVPR, pp. 2625–2634 (2015)
10. Gammulle, H., Denman, S., Sridharan, S., Fookes, C.: Multi-level sequence GAN for group activity recognition. In: Jawahar, C.V., Li, H., Mori, G., Schindler, K. (eds.) ACCV 2018. LNCS, vol. 11361, pp. 331–346. Springer, Cham (2019). https://doi.org/10.1007/978-3-030-20887-5_21
11. Gammulle, H., Denman, S., Sridharan, S., Fookes, C.: Predicting the future: a jointly learnt model for action anticipation. In: ICCV, pp. 5562–5571 (2019)
12. Gavrilyuk, K., Sanford, R., Javan, M., Snoek, C.G.M.: Actor-transformers for group activity recognition. In: CVPR, June 2020
13. Goodfellow, I., et al.: Generative adversarial nets. In: NIPS, pp. 2672–2680 (2014)
14. Gupta, A., Johnson, J., Fei-Fei, L., Savarese, S., Alahi, A.: Social GAN: socially acceptable trajectories with generative adversarial networks. In: CVPR, pp. 2255–2264 (2018)
15. He, K., Gkioxari, G., Dollár, P., Girshick, R.: Mask R-CNN. In: ICCV, pp. 2961–2969 (2017)
16. Hu, G., Cui, B., He, Y., Yu, S.: Progressive relation learning for group activity recognition. In: CVPR (2020)
17. Hu, J.F., Zheng, W.S., Ma, L., Wang, G., Lai, J.H., Zhang, J.: Early action prediction by soft regression. In: TPAMI (2018)

18. Ibrahim, M.S., Mori, G.: Hierarchical relational networks for group activity recognition and retrieval. In: Ferrari, V., Hebert, M., Sminchisescu, C., Weiss, Y. (eds.) ECCV 2018. LNCS, vol. 11207, pp. 742–758. Springer, Cham (2018). https://doi.org/10.1007/978-3-030-01219-9_44

19. Ibrahim, M.S., Muralidharan, S., Deng, Z., Vahdat, A., Mori, G.: A hierarchical deep temporal model for group activity recognition. In: CVPR, pp. 1971–1980 (2016)

20. Kipf, T.N., Welling, M.: Semi-supervised classification with graph convolutional networks. arXiv preprint arXiv:1609.02907 (2016)

21. Kong, Y., Gao, S., Sun, B., Fu, Y.: Action prediction from videos via memorizing hard-to-predict samples. In: AAAI (2018)

22. Kong, Yu., Kit, D., Fu, Y.: A discriminative model with multiple temporal scales for action prediction. In: Fleet, D., Pajdla, T., Schiele, B., Tuytelaars, T. (eds.) ECCV 2014. LNCS, vol. 8693, pp. 596–611. Springer, Cham (2014). https://doi.org/10.1007/978-3-319-10602-1_39

23. Kong, Y., Tao, Z., Fu, Y.: Deep sequential context networks for action prediction. In: CVPR, pp. 1473–1481 (2017)

24. Kong, Y., Tao, Z., Fu, Y.: Adversarial action prediction networks. In: TPAMI (2020)

25. Krizhevsky, A., Sutskever, I., Hinton, G.E.: Imagenet classification with deep convolutional neural networks. In: NIPS, pp. 1097–1105 (2012)

26. Lan, T., Chen, T.-C., Savarese, S.: A hierarchical representation for future action prediction. In: Fleet, D., Pajdla, T., Schiele, B., Tuytelaars, T. (eds.) ECCV 2014. LNCS, vol. 8691, pp. 689–704. Springer, Cham (2014). https://doi.org/10.1007/978-3-319-10578-9_45

27. Lan, T., Sigal, L., Mori, G.: Social roles in hierarchical models for human activity recognition. In: CVPR, pp. 1354–1361 (2012)

28. Lea, C., Flynn, M.D., Vidal, R., Reiter, A., Hager, G.D.: Temporal convolutional networks for action segmentation and detection. In: CVPR, pp. 156–165 (2017)

29. Li, X., Choo Chuah, M.: SBGAR: semantics based group activity recognition. In: ICCV, pp. 2876–2885 (2017)

30. Ma, S., Sigal, L., Sclaroff, S.: Learning activity progression in LSTMs for activity detection and early detection. In: CVPR, pp. 1942–1950 (2016)

31. Martinez, J., Black, M.J., Romero, J.: On human motion prediction using recurrent neural networks. In: CVPR, pp. 2891–2900 (2017)

32. Qi, M., Qin, J., Li, A., Wang, Y., Luo, J., Van Gool, L.: stagNet: an attentive semantic RNN for group activity recognition. In: Ferrari, V., Hebert, M., Sminchisescu, C., Weiss, Y. (eds.) ECCV 2018. LNCS, vol. 11214, pp. 104–120. Springer, Cham (2018). https://doi.org/10.1007/978-3-030-01249-6_7

33. Ramanathan, V., Huang, J., Abu-El-Haija, S., Gorban, A., Murphy, K., Fei-Fei, L.: Detecting events and key actors in multi-person videos. In: CVPR, pp. 3043–3053 (2016)

34. Ryoo, M.S.: Human activity prediction: early recognition of ongoing activities from streaming videos. In: ICCV, pp. 1036–1043 (2011)

35. Ryoo, M., Aggarwal, J.: Stochastic representation and recognition of high-level group activities. IJCV 93(2), 183–200 (2011)

36. Sadegh Aliakbarian, M., Sadat Saleh, F., Salzmann, M., Fernando, B., Petersson, L., Andersson, L.: Encouraging LSTMs to anticipate actions very early. In: ICCV, pp. 280–289 (2017)

37. Shi, Y., Fernando, B., Hartley, R.: Action anticipation with RBF kernelized feature mapping RNN. In: Ferrari, V., Hebert, M., Sminchisescu, C., Weiss, Y. (eds.) ECCV 2018. LNCS, vol. 11214, pp. 305–322. Springer, Cham (2018). https://doi.org/10.1007/978-3-030-01249-6_19

38. Shu, T., Todorovic, S., Zhu, S.C.: CERN: confidence-energy recurrent network for group activity recognition. In: CVPR, pp. 5523–5531 (2017)

39. Soomro, K., Zamir, A.R., Shah, M.: Ucf101: a dataset of 101 human actions classes from videos in the wild. arXiv preprint arXiv:1212.0402 (2012)

40. Szegedy, C., Vanhoucke, V., Ioffe, S., Shlens, J., Wojna, Z.: Rethinking the inception architecture for computer vision. In: CVPR, pp. 2818–2826 (2016)

41. Tang, Y., Wang, Z., Li, P., Lu, J., Yang, M., Zhou, J.: Mining semantics-preserving attention for group activity recognition. In: Multimedia, pp. 1283–1291 (2018)

42. Tran, D., Bourdev, L., Fergus, R., Torresani, L., Paluri, M.: Learning spatiotemporal features with 3D convolutional networks. In: ICCV, pp. 4489–4497 (2015)

43. Vondrick, C., Pirsiavash, H., Torralba, A.: Anticipating visual representations from unlabeled video. In: CVPR, pp. 98–106 (2016)

44. Wang, M., Ni, B., Yang, X.: Recurrent modeling of interaction context for collective activity recognition. In: CVPR, pp. 3048–3056 (2017)

45. Wang, X., Hu, J.F., Lai, J.H., Zhang, J., Zheng, W.S.: Progressive teacher-student learning for early action prediction. In: CVPR, pp. 3556–3565 (2019)

46. Wang, Z., Shi, Q., Shen, C., Van Den Hengel, A.: Bilinear programming for human activity recognition with unknown mrf graphs. In: CVPR, pp. 1690–1697 (2013)

47. Wichers, N., Villegas, R., Erhan, D., Lee, H.: Hierarchical Long-term Video Prediction Without Supervision, pp. 6038–6046 (2018)

48. Wu, J., Wang, L., Wang, L., Guo, J., Wu, G.: Learning actor relation graphs for group activity recognition. In: CVPR, pp. 9964–9974 (2019)

49. Yan, R., Tang, J., Shu, X., Li, Z., Tian, Q.: Participation-contributed temporal dynamic model for group activity recognition. In: Multimedia, pp. 1292–1300 (2018)

50. Yan, S., Xiong, Y., Lin, D.: Spatial temporal graph convolutional networks for skeleton-based action recognition. In: AAAI (2018)

51. Yan, Y., Ni, B., Yang, X.: Predicting human interaction via relative attention model. In: IJCAI, pp. 3245–3251 (2017)

52. Yao, T., Wang, M., Ni, B., Wei, H., Yang, X.: Multiple granularity group interaction prediction. In: CVPR, pp. 2246–2254 (2018)

53. Zhao, H., Wildes, R.P.: Spatiotemporal feature residual propagation for action prediction. In: ICCV, pp. 7003–7012 (2019)

PiP: Planning-Informed Trajectory Prediction for Autonomous Driving

Haoran Song[1], Wenchao Ding[1], Yuxuan Chen[2], Shaojie Shen[1],
Michael Yu Wang[1], and Qifeng Chen[1(✉)]

[1] The Hong Kong University of Science and Technology, Hong Kong, China
{hsongad,wdingae,eeshaojie,mywang,cqf}@ust.hk
[2] University of Science and Technology of China, Hefei, China
cyuxuan@mail.ustc.edu.cn

Abstract. It is critical to predict the motion of surrounding vehicles for self-driving planning, especially in a socially compliant and flexible way. However, future prediction is challenging due to the interaction and uncertainty in driving behaviors. We propose planning-informed trajectory prediction (PiP) to tackle the prediction problem in the multi-agent setting. Our approach is differentiated from the traditional manner of prediction, which is only based on historical information and decoupled with planning. By informing the prediction process with the planning of the ego vehicle, our method achieves the state-of-the-art performance of multi-agent forecasting on highway datasets. Moreover, our approach enables a novel pipeline which couples the prediction and planning, by conditioning PiP on multiple candidate trajectories of the ego vehicle, which is highly beneficial for autonomous driving in interactive scenarios.

1 Introduction

Anticipating future trajectories of traffic participants is an essential capability of autonomous vehicles. Since traffic participants (agents) will affect the behavior of each other, especially in highly interactive driving scenarios, the prediction model is required to anticipate the *social interaction* among agents in the scene to achieve socially compliant and accurate prediction.

Despite the fact that the interaction among traffic agents is being investigated, far less attention is paid to how the uncontrollable (surrounding) agents interact with the controlled (ego) agent. Different future plans of the ego agent will largely affect the future behaviors of all surrounding agents, which leads to a significant difference in future predictions. Human drivers are accustomed to imagining *what* the situation will be *if* they are going to act in different ways. For example, they speculate whether the other vehicles will leave space if they insert aggressively or mildly, respectively. By considering the different future situations from multiple *"what-ifs"*, human drivers are adept at negotiating with

Electronic supplementary material The online version of this chapter (https://doi.org/10.1007/978-3-030-58589-1_36) contains supplementary material, which is available to authorized users.

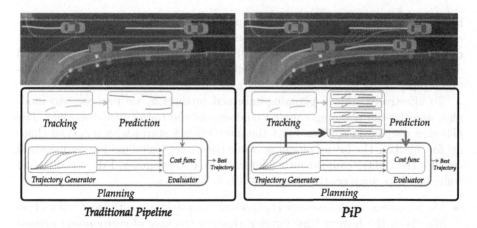

Fig. 1. Comparison between the traditional prediction approach (left) and PiP (right) under a lane merging scenario. Assume the ego vehicle (red) intends to merge to the left lane. It is required to predict the trajectories of surrounding vehicles (blue). To alleviate the uncertainty led by future interaction, PiP incorporates the future plans (dotted red curve) of ego vehicle in addition to the history tracks (grey curve). While the traditional prediction result is produced independently with the ego's future, PiP produces predictions one-to-one corresponding to the candidate future trajectories by enabling the novel planning-prediction-coupled pipeline. Therefore, PiP evaluates the planning safety more precisely and achieves more flexible driving behavior (solid red curve) compared with the traditional pipeline. (Color figure online)

other traffic participants while flexibly adapting their own driving behaviors. The key is that human drivers condition the prediction of surrounding vehicles on their own future intention. In this paper, we want to inform the interaction-aware prediction using the candidate plans of the controlled vehicle to mimic this thinking process.

To this end, we propose a novel planning-informed prediction framework (PiP). Note that PiP does not require the exact future trajectory, which is actually undetermined during prediction. PiP only conditions the prediction process on the candidate future trajectories proposed by the trajectory generator, like "insert aggressively" and "insert mildly" these kinds of "what-ifs". Accordingly, the best trajectory could be picked out after evaluating all the candidate plans by their corresponding predictions in the planning module.

There are two significant benefits of PiP. First, by incorporating the additional planning information, the interaction among agents can be better captured, which leads to a considerable improvement in prediction accuracy. Second, the planning-informed prediction will provide a highly valuable interface for the planning module during system integration. Explicitly, instead of evaluating multiple future plans under a fixed prediction result as most autonomous driving systems do, PiP conditions the prediction process on the ego vehicle's future plans, which uncovers how the other vehicles will interact with ego vehicle if the ego vehicle executes any specific planning trajectory. The PiP pipeline is especially suitable for planning in dense and highly interactive traffic (such as merging into a congested lane), which is hard to be handled using traditional decoupled prediction and planning pipeline. The comparison between the traditional pipeline for autonomous driving and PiP is illustrated in Fig. 1.

To effectively achieve planning-informed prediction, we propose two modules, namely, the planning coupled module and the target fusion module. The planning coupled module extracts the interaction features with a special channel for injecting the future planning, while the target fusion module encodes and decodes the tightly coupled future interaction among agents. PiP is end-to-end trainable. Our main contributions are listed as follows:

- The *planning coupled* module is proposed to model the multi-agent interaction from both the history time domain (history tracking of surrounding agents) and future time domain (future planning of controlled agent). By introducing the planning information into social context encoding, the uncertainty from the multi-modality of driving behavior is alleviated and thus leads to an improvement in prediction accuracy.
- The *target fusion* module is presented to capture the interdependence between target agents further. Since all the future states of targets are linked up together with the planning of the controlled agent, we apply a fully convolutional structure to model their future dependency at different spatial resolutions. The introduction of the target fusion module leads to further improvement for multi-agent forecasting.
- Our model outperforms state-of-the-art methods for multi-agent forecasting from tracking data. Moreover, the proposed planning-prediction-coupled pipeline extends the operational domain of planning by the integration with prediction, and some qualitative results are demonstrated.

2   Related Work

To accurately forecast the future trajectory of a specific vehicle, we need to discover the clues from its past observation and corresponding traffic configuration. In this paper, we focus on the data-driven trajectory prediction methods, which essentially learn the relationship between future trajectory and past motion states. Since vehicle behaviors are often inter-related, especially in dense traffic, it is crucial to consider interaction-aware trajectory prediction for autonomous driving, namely, in a multi-agent setting. In this section, we provide an overview of interaction-aware trajectory prediction methods and the common practice of integrating prediction with planning, which motivates our planning-informed prediction.

Interaction-aware trajectory prediction: Multi-agent learning and forecasting [9,16,18,28,31] is a challenging problem and Social LSTM [1] is one seminal work. In [1], the spatial interaction among pedestrians is learned using the proposed social pooling structure based on the hidden states generated by long short-term memory (LSTM) network, and [5] improves the social pooling strategy by applying convolutional layers. To better capture the multi-modal nature of future behaviors, some non-deterministic generative models are adopted based on generative adversarial networks (GANs) [10,11,25], and variational autoencoders (VAEs) [14,17]. Besides learning the interaction among agents, the agent-scene interaction is also modeled in [2,26,33]. The interaction-aware network structures are further extended to heterogeneous traffic [3,20] and applied to autonomous driving scenarios such as[5,6,17].

Trajectory prediction for control and planning: Targeting on the real-time driving, the popularly used vehicle motion planners [8,21,22,27,30] follow the workflow: first roll out multiple candidate ego trajectories; then score them using user-defined functions, in which the future trajectories of other vehicles predicted based on history tracks are considered; finally, pick out the best trajectory to execute. Note that the prediction result of other vehicles is fixed for different candidates from the trajectory generator of the ego vehicle. Namely, the traditional pipeline does not make "what-ifs", and think the reactions of other vehicles will be the same even given different ego actions. However, because the future planning of the ego vehicle in turn affects the behaviors of surrounding agents, the "predict-and-plan" workflow may be inadequate, especially in tightly coupled driving scenarios such as merging [13]. Differentiated from the traditional decoupled pipeline, PiP can be incorporated into a novel planning-prediction-coupled pipeline, which extends flexibility in dense traffic.

Planning-informed trajectory prediction: Incorporating planning information into prediction was attempted in some works on intelligent vehicles [29,32]. However, the frameworks were designed for specific scenarios, thereby constrained by specifically designed features [29] or prototype trajectories [32]. Rhinehart et al. proposed PRECOG [23] to condition prediction on the intentions of the ego vehicle. While even given the same intentions or goals of the ego vehicle, the specific time profile of how the ego vehicle reaches the goals significantly impacts the reaction of surrounding vehicles. It may pose restrictions to accurate prediction and accordingly motivates us to inform the prediction process by using the candidate plans from the planning module. Specifically, our proposed method is capable of providing accurate interaction-aware trajectory prediction for a large batch of different candidate planned trajectories efficiently, which facilitates planning in highly interactive environments [4,7].

3   Method

In PiP, the motion of each target vehicle is predicted by considering not only its own state and the other agents' states in the history time domain, but also the ego vehicle's planned trajectory. In this section, we first formulate the problem in Sect. 3.1, and describe the details of PiP in the following structure: the planning-coupled module which incorporates the ego vehicle's planned trajectory in the social tensors of neighboring vehicles' past motions (Sect. 3.2), the method of agent-centric target fusion (Sect. 3.3) and the maneuver-based decoding method for generating the probabilistic distribution of the location displacement between future frames (Sect. 3.4). Some implementation details are provided in Sect. 3.5.

3.1   Problem Formulation

Consider the driving scenario for an autonomous vehicle. The ego vehicle is commanded by the planning module, and the perception module senses the neighboring vehicles within a certain range. We formulate the trajectory prediction problem in the multi-agent setting as estimating the future states of a set of target vehicles around the ego vehicle v_{ego} conditioning on the tracking history of all surrounding vehicles and the planned future of the controllable ego vehicle. The objective is to learn the posterior distribution $P(\mathbf{Y}|\mathbf{X}, \mathcal{I})$ of multiple targets' future trajectories $\mathbf{Y} = \{Y_i|v_i \in V_{tar}\}$, where V_{tar} is the set of predicted targets selected within an ego-vehicle-centric area $\mathcal{A}_{tar}$. The conditional items contain the future planning of ego vehicle $\mathcal{I}$ and the past trajectories $\mathbf{X} = \{X_i|v_i \in V\}$, where V denotes the set of all vehicles involved around the ego vehicle, and $(v_{ego} \cup V_{tar}) \subseteq V$ as the ego vehicle is not required to be predicted. At any time t, the history trajectory and future trajectory of an agent i are denoted as $X_i = \left\{x_i^{t-T_{obs}+1}, x_i^{t-T_{obs}+2}..., x_i^t\right\}$ and $Y_i = \left\{y_i^{t+1}, y_i^{t+2}..., y_i^{t+T_{pred}}\right\}$, where the elements of $x_i, y_i \in \mathbb{R}^2$ represent waypoint coordinates in the past and future, respectively, while T_{obs} and T_{pred} refer to the number of frames for observation and prediction. Note that the planned trajectory $\mathcal{I} = Y_{ego} = \left\{y_{ego}^{t+1}, y_{ego}^{t+2}..., y_{ego}^{t+T_{pred}}\right\}$ is also used as a conditional item, since it's generated from ego vehicle's trajectory planner and thus can be accessible during prediction. Moreover, the introduction of $\mathcal{I}$ enables the planning-prediction-coupled pipeline as shown in Fig. 1.

3.2   Planning Coupled Module

In the planning coupled module, each predicted agent is processed in its own centric area $\mathcal{A}_{nbr}$, in which the ego vehicle v_{ego}, the target vehicle $v_{cent} \in V_{tar}$ and the other neighboring vehicles $V_{nbrs} \subseteq V$ located within $\mathcal{A}_{nbr}$ are included. There involve three encoding streams: the dynamic property of the target itself, the social interaction with the target's neighboring vehicles, and the spatial dependency with ego vehicle's future planning. Consequently, a target encoding $\mathcal{T}$ is generated by embedding these encodings together. In practice, we use relative trajectories in an agent-centric manner for capturing interdependencies between the centric agent and surrounding agents.

Trajectory Encoding: All trajectories contained in the planning coupled module could be classified into two types: observable and controllable. The history trajectories of traffic participants could be observed, and the planned trajectory to command the ego vehicle could be controlled. Before extracting the spatially interactive relationship between traffic agents, all trajectories are encoded independently

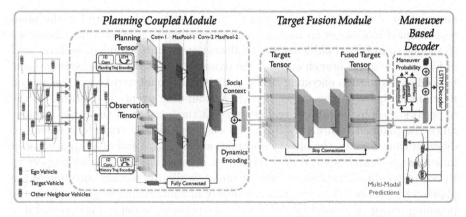

Fig. 2. The overview of *PiP* architecture: *PiP* consists of three key modules, including planning coupled module, target fusion module, and maneuver-based decoding module. Each predicted target is firstly encoded in the planning coupled module by aggregating all information within the target-centric area (blue square). A target tensor is then set up within the ego-vehicle-centric area (red square) by placing the target encodings into the spatial gird based on their locations. Afterward, the target tensors are passed through the following target fusion module to learn the interdependency between targets, and eventually, a fused target tensor is generated. Finally, the prediction of each target is decoded from the corresponding fused target encoding in the maneuver-based decoding module. The target vehicle marked with an ellipse is exemplified for planning coupled encoding and multi-modal trajectories decoding. (Color figure online)

to learn the temporal properties in their sequential locations. To better accomplish this work, each trajectory is preprocessed by converting its locations into relative coordinates with respect to the target vehicle and then fed into a temporal convolutional layer to obtain a motion embedding. After that, the Long Short-Term Memory (LSTM) networks are employed to encode the motion property for trajectories, and the hidden state $h(\cdot)$ therein is regarded as the motion encoding for the corresponding trajectory. Here, the LSTMs with different parameters are adopted for planned trajectory Y_{ego} and history trajectories including X_{ego}, X_{cent} and X_{nbr}, as they belong to the different time domains.

Planning and Observation Fusion: The use of LSTM encoder captures the temporal structure from the trajectory sequence, while it fails to handle the spatial interaction relationship with other agents in a scene. The social pooling strategy, proposed in [1], addresses this issue by pooling LSTM states of spatially proximal sequences in a target-centric grid named as "social tensor". The "convolutional social pooling" in [5] improves the strategy further by applying convolutional and max-pooling layers over the social tensor. Both of the methods are proposed for learning the spatial relationship among trajectories that takes place in the history period. In our proposed framework, we adopt the convolutional social pooling structure for modeling spatial interaction. In addition to interdependencies between target and neighbors in the past time, the spatial information of ego vehicle's planning in the future time is counted in the planning coupled module as an improvement. Accordingly, three encoding branches stemming from LSTM hidden states of all trajectories are included, as illustrated in Fig. 2. The lower branch encodes the dynamics property of the target vehicle by feeding its motion encoding $h(X_{cent})$ to a fully connected layer. The spatial relationship between the target and its surrounding agents is captured in the upper branches by building a grid centered at the location of the target vehicle. Since the planned future trajectory and observed history trajectory belong to different time domain, the history information of $h(X_{nbr})$ and $h(X_{ego})$ are placed into a target-centric spatial grid termed as observation tensor with respect to the corresponding locations at current time t, while the motion encoding of the planned trajectory $h(Y_{ego})$ is placed similarly in another spatial grid to form the planning tensor. It should be noted that the planning sequence is encoded in a reversed order because the planning of the near future is more reliable, and thus it should weight more in the encoding.

After that, both of the observation and planning tensors pass through convolutional layers and pooling layers in parallel and then are concatenated together before fed to the last max-pooling layer. Merging the information from the planning of ego vehicle and observation of surrounding vehicles, the resulting encoding $\mathcal{S}$ covers the social context for both the past and future time domain. Finally, the merged social encoding $\mathcal{S}$ concatenates with the target's dynamics encoding $\mathcal{D}$ to form a target encoding $\mathcal{T}$ that aggregates all the information accessible within the target-centric grid.

3.3   Target Fusion Module

In [1,5], the future states of each target is directly decoded from the agent-centric encoding result that aggregates history information. In this way, each trajectory is generated independently from the corresponding target encoding. However, the future states of targets are highly correlated, which indicates that the decoding process for a certain target also depends on the encoding of other targets. Therefore, we further fuse the encoding among different targets in the scene and decode the final trajectory from the fused encoding, which better captures the dependencies of future states of different targets in the same scene.

For jointly predicting the vehicles within the target area centered on the ego vehicle's location, each target vehicle $v_i \in V_{tar}$ represented by its encoding $\mathcal{T}_i$ is placed into an ego-vehicle-centric grid $\{\mathcal{T}_i | v_i \in V_{tar}\}$ based on their locations at the last time step of history trajectories. Inspired by some popular CNN architectures for segmentation [19,24] that produce correspondingly-sized output with hierarchical inference, we adopt the fully convolutional network (FCN) to learn the context of target tensor. The target tensor is fed into a symmetric FCN structure for capturing the spatial dependencies between target agents at different grid resolutions, where the skip-connected layers are combined by element-wise sum. The fused target tensor produced by this module retains its spatial structure the same as before fusion, from which the fused target encoding $\mathcal{T}_i^+$ of each target could be sliced out according to its grid location.

3.4 Maneuver Based Decoding

To address the inherent multi-modality nature of the driving behaviors, the maneuver based decoder built upon [5] is applied to predict the future trajectory for predefined maneuver classes $M = \{m_k | k = 1, 2, ..., 6\}$ together with the probability of each maneuver $P(m_k)$. The maneuvers are classified by lateral behaviors (including lane-keeping, left and right lane changes) and longitudinal behaviors (including normal driving and braking). Thereupon, the fused target encoding $\mathcal{T}_i^+$ of target vehicle $v_i \in V_{tar}$ is first fed into a pair of fully connected layers that followed by soft-max layers to get the lateral and longitudinal behavior probability respectively, and thus their multiplication produces the probability for each maneuver $P(m_k | \mathcal{I}, X)$. The trajectory under each maneuver class is generated by concatenating the fused target encoding with one-hot vectors of lateral behavior and longitudinal behavior together, followed by passing the resulted feature vector through an LSTM decoder. Instead of directly generating absolute future locations, our LSTM decoder operates in a residual learning manner that outputs displacement between predicted locations. The output vector contains the displacement $\delta y_i^{t+T} \in \mathbb{R}^2$ between neighboring predicted locations, the standard deviation vector $\sigma_i^{t+T} \in \mathbb{R}^2$ and correlation coefficient $\rho_i^{t+T} \in \mathbb{R}$ of predicted location $\hat{y}_i^{t+T}$ at the future time step $T \in \{1, 2, ..., T_{pred}\}$. The predicted location could be accordingly represented by a bivariate Gaussian distribution

$$\hat{y}_i^{t+T} \sim \mathcal{N}(\mu_i^{t+T}, \sigma_i^{t+T}, \rho_i^{t+T}), \tag{1}$$

where the mean vector is given by summing up all displacements along the future time steps T with the location at the last time step t of history trajectory

$$\mu_i^{t+T} = x_i^t + \sum_{\tau=1}^{T} \delta y_i^{t+\tau}. \tag{2}$$

For brevity, the Gaussian parameters for all future time steps of target v_i is written as Θ_i. Finally, the posterior probability of all target vehicles' future trajectories could be estimated from

$$P(\mathbf{Y}|\mathbf{X}, \mathcal{I}) = \prod_{v_i \in V_{tar}} \sum_{k=1}^{|M|} P_{\Theta_i}(Y_i | m_k, \mathbf{X}, \mathcal{I}) P(m_k | \mathbf{X}, \mathcal{I}). \tag{3}$$

3.5 Implementation Details

Our model is trained by minimizing the negative log likelihood of future trajectories under the true maneuver class m_{true} of all the target vehicles

$$- \sum_{v_i \in V_{tar}} \log \left(P_{\Theta_i}(Y_i | m_{true}, \mathbf{X}, \mathcal{I}) P(m_{true} | \mathbf{X}, \mathcal{I}) \right). \tag{4}$$

Each data instance contains a vehicle specified as the ego. The predicted targets are the vehicles located within the ego-vehicle-centric area $\mathcal{A}_{tar}$ with the size of 60.96×10.67 meters (200×35 feet), discretized as 25×5 spatial grid. The target-centric area $\mathcal{A}_{nbr}$ of each predicted vehicle is defined the same as $\mathcal{A}_{tar}$.

For the planning input $\mathcal{I}$ of the ego vehicle, its actual trajectory within the prediction horizon is directly used in training. While in evaluation and testing, $\mathcal{I}$ is fitted from its downsampled actual trajectory. It is handled in this way because we intend to restrict the prediction from accessing the complete information of planning trajectory, instead only a limited number of waypoints could be accessed. Furthermore, the ground-truth trajectories result from many planning cycles, while in practice, prediction can only be based on the current planning cycle. So the planning input is represented by a fitted quintic spline, which is a typically used representation for vehicle trajectory. This feature makes our planning-informed method easy to deploy in a real autonomous system. Although the fitted planning input cannot perfectly fit the actual future trajectory, it could be examined if our method can generalize well in practical use.

4   Experiments

In this section, we evaluate our method on two publicly available vehicle trajectory datasets, NGSI [12] and HighD [15]. Firstly, we compare the performance of our method against the existing state-of-the-art works quantitatively using the metrics of root mean squared error (RMSE) and negative log-likelihood (NLL). Next, as our method could anticipate different future configurations by performing different plans under the same historical situation, we evaluate PiP from more simulated future situations. Regarding the rationality and variety in generating feasible vehicle trajectories, we employ a model-based vehicle planner MPDM [4] to generate diverse vehicle trajectories with different lateral and longitudinal behaviors. In Sect. 4.4, a user study is conducted by comparing our generated results with the real situations to verify the rationalization of predicted outcomes, and more results are provided in Sect. 4.5 for qualitative analysis.

4.1   Datasets

We split all the trajectories contained in NGSIM and HighD separately, in which 70% are used for training with 20% and 10% for testing and evaluation. Each vehicle's trajectory is split into 8s segments composed of 3s of past and 5s of future positions 5 Hz. The 5s future of ego vehicle used as planning input is further downsampled 1 Hz in testing and evaluation. The objective is to predict all surrounding target vehicles' future trajectories over 5s prediction horizon.

NGSIM: NGSIM [12] is a real-world highway dataset which is commonly used in the trajectory prediction task. All vehicle trajectories over a 45-minute time span are captured 10 Hz, with each 15-minute segment under mild, moderate, and congested traffic conditions, respectively.

HighD: HighD [15] is a vehicle trajectories dataset released in 2018. The data is recorded from six different locations on Germany highways from the aerial perspective using a drone. It is composed of 60 recordings over areas of $400 \sim 420$ m span, with more than $110,000$ vehicles are contained.

4.2   Baseline Methods

We compare PiP with the following listed deterministic models and stochastic models. We also ablate the planning coupled module and target fusion module in PiP-noPlan and PiP-noFusion respectively, to study their effectiveness in improving prediction accuracy upon the baselines.

S-LSTM: Social LSTM [1] uses a fully connected layer for social pooling and produces a uni-modal distribution of future locations.

CS-LSTM: Convolutional Social LSTM [5] uses convolutional layers with social pooling and outputs a maneuver-based multi-modal prediction.

S-GAN: Social GAN [11] trains GAN based framework using the adversarial loss to generate diverse trajectories for multi-agent in a spatial-centric manner.

MATF: MATF-GAN [33] models spatial interaction of agents and scene context by convolutional fusion and uses GAN to produce stochastic predictions.

4.3   Quantitative Evaluation

Among all the above methods, S-GAN and MATF are stochastic models.[1] We report their RMSE by the best result among 3 samples (i.e., minRMSE). The others are all deterministic models that generate Gaussian distributions for all predicted locations along the trajectory, in which the means of Gaussian parameters are used as the predicted locations when calculating the RMSE for each time step t within the 5s prediction horizon: $RMSE(t) = \sqrt{\frac{1}{|V_{tar}|} \sum_{v_i \in V_{tar}} \|y_i - \hat{y}_i\|^2}$. For multi-modal distribution output by CS-LSTM, PiP and its variants, RMSE is evaluated using the predicted trajectory with the maximal maneuver probability $P(m_k)$. While RMSE is a concrete metric to measure prediction accuracy, it is limited to some extent since it tends to average all the prediction results and may fail to reflect the accuracy for distinct maneuvers. To overcome its limitation in evaluating multi-modal prediction, we adopt the same way from prior work [5] that additionally reports the negative log-likelihood (NLL) of the true trajectories under the prediction results represented by either uni-modal or multi-modal distributions.

The results of quantitative results are reported in Table 1. Our method significantly outperforms the deterministic models (S-LSTM and CS-LSTM) in both RMSE and NLL metrics on both datasets. Although sampling more trajectories and choosing the minimal error among all samples would undoubtedly lead to a lower RMSE for stochastic models (S-GAN and MATF), our deterministic model still achieves lower RMSE than stochastic models for sampling three times. The reason for not setting a larger sampling number for the stochastic models is that sampling too many times for prediction may not work well with planning and decision making since the probability of each sample is actually unknown.

---

[1] No NLL results of S-GAN and MATF, as they sample trajectories without generating probability. No RMSE result of MATF on the HighD dataset is reported in [33].

Table 1. Quantitative results on the NGSIM and HighD datasets are reported by RMSE and NLL metrics over 5s prediction horizon. The best results are marked by bold numbers. Note that for the stochastic methods (S-GAN and MATF), the minimal error from sampling three times reports their RMSE

Metric	Dataset	Time	S-LSTM [1]	CS-LSTM [5]	S-GAN [11]	MATF [33]	PiP-noPlan	PiP-noFusion	PiP
RMSE (m)	NGSIM	1s	0.60	0.58	0.57	0.66	**0.55**	**0.55**	**0.55**
		2s	1.28	1.26	1.32	1.34	1.20	1.19	**1.18**
		3s	2.09	2.07	2.22	2.08	2.00	1.95	**1.94**
		4s	3.10	3.09	3.26	2.97	3.01	2.90	**2.88**
		5s	4.37	4.37	4.40	4.13	4.27	4.07	**4.04**
	HighD	1s	0.19	0.19	0.30	-	0.18	**0.17**	**0.17**
		2s	0.57	0.57	0.78	-	0.53	0.53	**0.52**
		3s	1.18	1.16	1.46	-	1.09	**1.05**	**1.05**
		4s	2.00	1.96	2.34	-	1.86	**1.76**	**1.76**
		5s	3.02	2.96	3.41	-	2.81	**2.63**	**2.63**
Metric	Dataset	Time	S-LSTM	CS-LSTM	S-GAN	MATF	PiP-noPlan	PiP-noFusion	PiP
NLL (nats)	NGSIM	1s	2.38	1.91	-	-	**1.68**	1.71	1.72
		2s	3.86	3.44	-	-	**3.29**	**3.29**	3.30
		3s	4.69	4.31	-	-	4.20	**4.17**	**4.17**
		4s	5.33	4.94	-	-	4.87	4.81	**4.80**
		5s	5.89	5.48	-	-	5.42	5.33	**5.32**
	HighD	1s	0.42	0.37	-	-	0.20	0.20	**0.14**
		2s	2.58	2.43	-	-	2.28	2.28	**2.24**
		3s	3.93	3.65	-	-	3.53	3.53	**3.48**
		4s	4.87	4.51	-	-	4.39	4.37	**4.33**
		5s	5.57	5.17	-	-	5.05	5.01	**4.99**

The consistent improvements on NLL and RMSE metrics confirm that, by introducing the planning of ego vehicle into the prediction model and capturing the correlations between prediction targets, PiP is superior to all baselines in prediction accuracy. Additionally, the results of ablated models show that both the target fusion module and the planning coupled module lead to obvious improvement upon the CS-LSTM. By comparison, the inclusion of planning trajectory is more effective in improving the multi-agent forecasting accuracy.

4.4 User Study

To investigate if our prediction model generalizes to various future plans (different maneuver classes and aggressiveness) under different traffic configurations, we have also simulated diverse future scenarios by performing different planned trajectories for the ego vehicle. Accordingly, we conduct the user study that compares real and simulated traffic situations, as shown in the upper part of Fig. 3. Each pair of videos are derived from a segment of 8s traffic sequence recorded in the datasets. One video displays the complete recording of the real tracking data, while the other video shares the same 3s history sequence, and contains a different sequence in the last 5s which is composed by the predicted trajectories of targets (blue) under a different plan performed by ego vehicle (red). The other

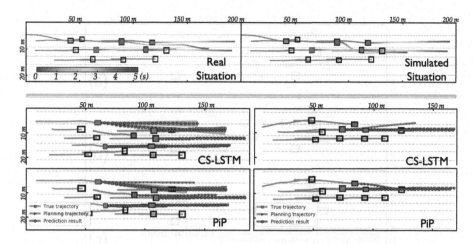

Fig. 3. Upper: user study example of comparing the real and simulated situations. Each comparison is visualized as video pair for users to choose the situation that violates their intuition. Lower: two example cases predicted by CS-LSTM and PiP. The ground truth (blue), planning (red) and predicted trajectories (green) are visualized by sets of locations with 0.2s time step. As both methods output maneuver-based multimodal distributions, only those trajectories with maneuver probability larger than 10% are shown for each target. The green circle denotes the mean value of distribution on each time step, and its radius is proportional to the maneuver probability of the corresponding trajectory. The green shadow area represents the variance of the distribution. (Color figure online)

agents (no color) outside the predictive range are hidden in the last 5s. Note that the same coloring scheme is used in the following experiments.

We display 20 pairs of videos with randomized order and ask participants to select the one in which the target vesicles' behavior looks unreasonable or against common sense. Totally 25 people participated in the user study, and our simulated results were selected as the unreasonable one with a rate of 52.2% (261/500), a bit higher than 50%. One reason is that the ego vehicle's planned trajectory in the simulated results is generated offline, but its real trajectory recorded in the datasets is resulted from replanning adaptively from time to time. Then it could be a clue for users to select the actual situation as the better one.

Nevertheless, our model still achieves a 47.8% rate of being selected as reasonable. It could also be noted in the upper part of Fig. 3, we generate an agile lane merging trajectory for the ego car, and the predicted outcome shows that the following vehicle reacts with deceleration while the leading vehicles maintain speed. Both of the forecastings make sense in real traffic, which indicates that our proposed method could be generalized to different plans.

4.5 Qualitative Analysis

In the following, we further investigate how the prediction is improved as well as explore how PiP enables the planning-prediction-coupled pipeline.

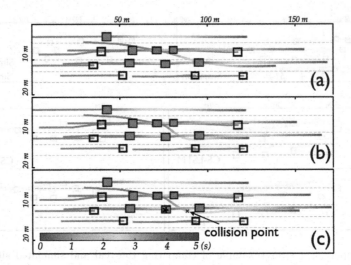

Fig. 4. Prediction results of performing diverse planned trajectories by ego vehicle: the history trajectories (grey) are from a traffic scene in NGSIM, and the future trajectories are visualized by gradient color varying over time. The target vehicle that collides with ego vehicle is marked with a star symbol, and the collision point is annotated by a cross symbol. (Color figure online)

Baseline Comparison: Since our method employs the same maneuver-based decoding as in CS-LSTM [5], the predictive distribution under the same traffic scenes is compared in the lower part of Fig. 3. In the left example, we notice that CS-LSTM outputs similar maneuver probability of keeping the lane and turning right for the left-rear target. At the same time, our method is more confident to target's actual maneuver of turning right. It is because that ego vehicle is planned to go straight under certain velocity, thereby leaving enough space for the target to merge to its right lane. By the same token, our method precisely predicts the right-rear target will keep lane but not turn left in the right example. At that moment, the ego vehicle intends to merge to the right lane gradually in a moderate manner, which blocks the way for the right-rear target to turn left in the near future. Both examples demonstrate that the planning-informed approach leads the prediction to be more accurate.

Active Planning: With PiP, it is feasible to explore how to plan in different traffic situations actively. In the following, we illustrate some challenging scenarios with history states acquired from datasets, and PiP produces diverse future states under different plans generated by the ego vehicle.

Figure 4 (a,b) shows prediction results when performing a moderate and aggressive lane changing in dense traffic. It could be noticed that the aggressive behavior in Fig. 4 (b) is risky as it is very close to the preceding vehicle after merging. Notably, when it merges aggressively a bit faster, as shown in Fig. 4 (c), a collision is forecasted between the controlled vehicle and the target with a star mark. The ability of forecasting collision further verifies the generalization

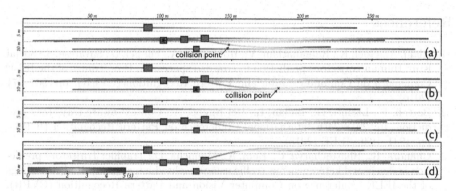

Fig. 5. Prediction results of performing diverse planned trajectories by ego vehicle: the history trajectories are from a highway scene in HighD. All the annotations are same with Fig. 4. The predicted future is shown with a collision in (a, b) and safe lane changing in (c,d). (Color figure online)

of our network as no collision occurred in the traffic recordings where the PiP model is trained. Figure 5 shows another example from the HighD dataset in which the vehicles go much faster than that in the NGSIM dataset. In this case, turning right is challenging. In Fig. 5 (a) the ego vehicle is planned to turn right and follow the right-front target. A prompt deceleration may cause the rear vehicle to fail to respond and results in a rear-end collision. PiP also anticipates in Fig. 5 (b) that a collision will occur if the ego vehicle plans to turn right and overtakes the right-font target. Nevertheless, it is still possible to find a proper way of merging to the right lane, as shown in Fig. 5 (c). Additionally, we also show a result of changing to the left lane in Fig. 5 (d), which is relatively easier as there exists larger space on the left for lane changing.

5   Conclusion

In this work, we present PiP for predicting future trajectories in a planning-informed approach. Leveraging on the fact that all traffic agents are tightly coupled throughout the time domain, the future prediction on surrounding agents is informed by incorporating history tracks with future planning of the controllable agent. PiP outperforms the state-of-the-art works for multi-agent forecasting on highway datasets. Furthermore, PiP enables a novel planning-prediction-coupled pipeline that produces future predictions one-to-one corresponding to candidate trajectories, and we demonstrate that it could act as a highly usable interface for planning in dense or fast-moving traffic. In the future, we plan to extend our approach to work under imperfect tracking or detection information, which is common in the perception module. Further, the future prediction and trajectory generation could be integrated into a motion planner that learns to generate optimal planning under interactive scenarios.

References

1. Alahi, A., Goel, K., Ramanathan, V., Robicquet, A., Fei-Fei, L., Savarese, S.: Social lstm: Human trajectory prediction in crowded spaces. In: Proceedings of the IEEE Conference on Computer Vision and Pattern Recognition (CVPR). pp. 961–971 (2016)
2. Bartoli, F., Lisanti, G., Ballan, L., Del Bimbo, A.: Context-aware trajectory prediction. In: 2018 24th International Conference on Pattern Recognition (ICPR). pp. 1941–1946. IEEE (2018)
3. Chandra, R., Bhattacharya, U., Bera, A., Manocha, D.: Traphic: Trajectory prediction in dense and heterogeneous traffic using weighted interactions. In: Proceedings of the IEEE Conference on Computer Vision and Pattern Recognition (CVPR). pp. 8483–8492 (2019)
4. Cunningham, A.G., Galceran, E., Eustice, R.M., Olson, E.: Mpdm: Multipolicy decision-making in dynamic, uncertain environments for autonomous driving. In: Proceedings IEEE International Conference on Robotics and Automation. pp. 1670–1677. IEEE (2015)
5. Deo, N., Trivedi, M.M.: Convolutional social pooling for vehicle trajectory prediction. In: Proceedings of the IEEE Conference on Computer Vision and Pattern Recognition Workshops (CVPR). pp. 1468–1476 (2018)
6. Ding, W., Chen, J., Shen, S.: Predicting vehicle behaviors over an extended horizon using behavior interaction network. In: Proceedings of IEEE International Conference on Robotics and Automation. pp. 8634–8640. IEEE (2019)
7. Ding, W., Zhang, L., Chen, J., Shen, S.: Safe trajectory generation for complex urban environments using spatio-temporal semantic corridor. IEEE Robot. Autom. Lett. 4(3), 2997–3004 (2019)
8. Fan, H., et al.: Baidu apollo em motion planner. arXiv preprint arXiv:1807.08048 (2018)
9. Felsen, P., Lucey, P., Ganguly, S.: Where will they go? predicting fine-grained adversarial multi-agent motion using conditional variational autoencoders. In: Proceedings of the European Conference on Computer Vision (ECCV). pp. 732–747 (2018)
10. Goodfellow, I., et al.: Generative adversarial nets. In: Advances in Neural Information Processing Systems. pp. 2672–2680 (2014)
11. Gupta, A., Johnson, J., Fei-Fei, L., Savarese, S., Alahi, A.: Social gan: Socially acceptable trajectories with generative adversarial networks. In: Proceedings of the IEEE Conference on Computer Vision and Pattern Recognition (CVPR). pp. 2255–2264 (2018)
12. Halkias, J., Colyar, J.: Next generation simulation fact sheet. Technical Report, Federal Highway Administration (FHWA), fHWA-HRT-06-135 (2006)
13. Hubmann, C., Schulz, J., Xu, G., Althoff, D., Stiller, C.: A belief state planner for interactive merge maneuvers in congested traffic. In: 2018 21st International Conference on Intelligent Transportation Systems (ITSC). pp. 1617–1624. IEEE (2018)
14. Kingma, D.P., Welling, M.: Auto-encoding variational bayes. arXiv preprint arXiv:1312.6114 (2013)
15. Krajewski, R., Bock, J., Kloeker, L., Eckstein, L.: The highd dataset: A drone dataset of naturalistic vehicle trajectories on german highways for validation of highly automated driving systems. In: 2018 21st International Conference on Intelligent Transportation Systems (ITSC). pp. 2118–2125. IEEE (2018)

16. Le, H.M., Yue, Y., Carr, P., Lucey, P.: Coordinated multi-agent imitation learning. In: Proceedings of the 34th International Conference on Machine Learning. vol. 70, pp. 1995–2003. JMLR. org (2017)
17. Lee, N., Choi, W., Vernaza, P., Choy, C.B., Torr, P.H., Chandraker, M.: Desire: Distant future prediction in dynamic scenes with interacting agents. In: Proceedings of the IEEE Conference on Computer Vision and Pattern Recognition (CVPR). pp. 336–345 (2017)
18. Lee, N., Kitani, K.M.: Predicting wide receiver trajectories in american football. In: 2016 IEEE Winter Conference on Applications of Computer Vision (WACV). pp. 1–9. IEEE (2016)
19. Long, J., Shelhamer, E., Darrell, T.: Fully convolutional networks for semantic segmentation. In: Proceedings of the IEEE Conference on Computer Vision and Pattern Recognition (CVPR). pp. 3431–3440 (2015)
20. Ma, Y., Zhu, X., Zhang, S., Yang, R., Wang, W., Manocha, D.: Trafficpredict: Trajectory prediction for heterogeneous traffic-agents. In: Proceedings of the AAAI Conference on Artificial Intelligence. vol. 33, pp. 6120–6127 (2019)
21. McNaughton, M., Urmson, C., Dolan, J.M., Lee, J.W.: Motion planning for autonomous driving with a conformal spatiotemporal lattice. In: Proceedings of IEEE International Conference on Robotics and Automation. pp. 4889–4895. IEEE (2011)
22. Pivtoraiko, M., Knepper, R.A., Kelly, A.: Differentially constrained mobile robot motion planning in state lattices. J. Field Robot. **26**(3), 308–333 (2009)
23. Rhinehart, N., McAllister, R., Kitani, K., Levine, S.: Precog: Prediction conditioned on goals in visual multi-agent settings. In: Proceedings of the IEEE International Conference on Computer Vision (ICCV). pp. 2821–2830 (2019)
24. Ronneberger, O., Fischer, P., Brox, T.: U-Net: convolutional networks for biomedical image segmentation. In: Navab, N., Hornegger, J., Wells, W.M., Frangi, A.F. (eds.) MICCAI 2015. LNCS, vol. 9351, pp. 234–241. Springer, Cham (2015). https://doi.org/10.1007/978-3-319-24574-4_28
25. Sadeghian, A., Kosaraju, V., Sadeghian, A., Hirose, N., Rezatofighi, H., Savarese, S.: Sophie: An attentive gan for predicting paths compliant to social and physical constraints. In: Proceedings of the IEEE Conference on Computer Vision and Pattern Recognition (CVPR). pp. 1349–1358 (2019)
26. Sadeghian, A., Legros, F., Voisin, M., Vesel, R., Alahi, A., Savarese, S.: Car-net: Clairvoyant attentive recurrent network. In: Proceedings of the European Conference on Computer Vision (ECCV). pp. 151–167 (2018)
27. Schwarting, W., Alonso-Mora, J., Rus, D.: Planning and decision-making for autonomous vehicles. Robotics, and Autonomous Systems, Annual Review of Control (2018)
28. Sun, C., Karlsson, P., Wu, J., Tenenbaum, J.B., Murphy, K.: Stochastic prediction of multi-agent interactions from partial observations. arXiv preprint arXiv:1902.09641 (2019)
29. Sun, L., Zhan, W., Tomizuka, M.: Probabilistic prediction of interactive driving behavior via hierarchical inverse reinforcement learning. In: 2018 21st International Conference on Intelligent Transportation Systems (ITSC). pp. 2111–2117. IEEE (2018)
30. Werling, M., Ziegler, J., Kammel, S., Thrun, S.: Optimal trajectory generation for dynamic street scenarios in a frenet frame. In: Proceedings of IEEE International Conference on Robotics and Automation. pp. 987–993. IEEE (2010)
31. Zhan, E., Zheng, S., Yue, Y., Sha, L., Lucey, P.: Generative multi-agent behavioral cloning. arXiv (2018)

32. Zhan, W., Sun, L., Hu, Y., Li, J., Tomizuka, M.: Towards a fatality-aware benchmark of probabilistic reaction prediction in highly interactive driving scenarios. In: 2018 21st International Conference on Intelligent Transportation Systems (ITSC). pp. 3274–3280. IEEE (2018)
33. Zhao, T., et al.: Multi-agent tensor fusion for contextual trajectory prediction. In: Proceedings of the IEEE Conference on Computer Vision and Pattern Recognition (CVPR). pp. 12126–12134 (2019)

PSConv: Squeezing Feature Pyramid into One Compact Poly-Scale Convolutional Layer

Duo Li[1,2], Anbang Yao[2(✉)], and Qifeng Chen[1(✉)]

[1] The Hong Kong University of Science and Technology, Kowloon, Hong Kong
duo.li@connect.ust.hk, cqf@ust.hk
[2] Intel Labs China, Beijing, China
anbang.yao@intel.com

Abstract. Despite their strong modeling capacities, Convolutional Neural Networks (CNNs) are often scale-sensitive. For enhancing the robustness of CNNs to scale variance, multi-scale feature fusion from different layers or filters attracts great attention among existing solutions, while the more granular kernel space is overlooked. We bridge this regret by exploiting multi-scale features in a finer granularity. The proposed convolution operation, named Poly-Scale Convolution (PSConv), mixes up a spectrum of dilation rates and tactfully allocates them in the individual convolutional kernels of each filter regarding a single convolutional layer. Specifically, dilation rates vary cyclically along the axes of input and output channels of the filters, aggregating features over a wide range of scales in a neat style. PSConv could be a drop-in replacement of the vanilla convolution in many prevailing CNN backbones, allowing better representation learning without introducing additional parameters and computational complexities. Comprehensive experiments on the ImageNet and MS COCO benchmarks validate the superior performance of PSConv. Code and models are available at https://github.com/d-li14/PSConv.

Keywords: Convolutional kernel · Multi-scale feature fusion · Dilated convolution · Categorization and detection

1 Introduction

With the booming development of CNNs, dramatic progress has been made in the field of computer vision. As an inherent feature extraction mechanism, CNNs naturally learn coarse-to-fine hierarchical image representations. To mimic human visual systems that could process instances and stuff concurrently, it is of

---

D. Li—Indicates intern at Intel Labs China.

Electronic supplementary material The online version of this chapter (https://doi.org/10.1007/978-3-030-58589-1_37) contains supplementary material, which is available to authorized users.

vital importance for CNNs to gather diverse information from objects of various sizes and understand meaningful contextual backgrounds. However, streamlined CNNs usually have fixed-sized receptive fields, lacking the ability to tackle this kind of issue. Such a deficiency restricts their performance on visual recognition tasks, especially scale-sensitive dense prediction problems. The advent of FCN [25] and Inception [35] demonstrates the privilege of multi-scale representation to perceive heterogeneous receptive fields with impressive performance improvement. Motivated by these pioneering works, follow-up approaches explore and upgrade multi-scale feature fusion with more intricate skip connections or parallel streams. However, we notice that most of the existing works capture these informative multi-scale features in a layer-wise or filter-wise style, laying emphasis on the architecture engineering of the entire network or their composed building blocks.

From a brand new perspective, we shift the focus of design from macro- to micro-architecture towards the target of easily exploiting multi-scale features without touching the overall network architecture. Expanding kernel sizes and extending the sampling window sizes via increasing dilation rates are two popular techniques to enlarge the receptive fields inside one convolution operation. Compared to large kernels that bring about more parameter storage and computational consumption, dilated convolution is an alternative to cope with objects in an array of scales without introducing extra computational complexities. In this paper, we present Poly-Scale Convolution (PSConv), a novel convolution operation, extracting multi-scale features from the more granular convolutional kernel space. PSConv respects two design principles: firstly, regarding one single convolutional filter, its constituent kernels use a group of dilation rates to extract features corresponding to different receptive fields; secondly, regarding all convolutional filters in one single layer, the group of dilation rates corresponding to each convolutional filter alternates along the axes of input and output channels in a cyclic fashion, extracting diverse scale information from the incoming features and mapping them into outgoing features in a wide range of scales. Through these atomic operations on individual convolutional kernels, we effectively dissolve the aforementioned deficiency of standard convolution and push the multi-scale feature fusion process to a much more granular level. This proposed approach tiles the *kernel lattice*[1] with hierarchically stacked pyramidal features defined in the previous methodologies [22]. Each specific feature scale in one pyramid layer can be grasped with a collection of convolutional kernels in a PSConv operation with the same corresponding dilation rate, thus the whole feature pyramid can be represented in a condensed fashion using one compact PSConv layer with a spectrum of dilation rates. *Poly-Scale* Convolution extends the conventional *mono-scale* convolution living on a homogeneous dilation space of kernel lattice, hence the name of this convolution form. In our PSConv, scale-aware features located in different channels collaborate as a unity to deal with

---

[1] *kernel lattice* refers to the two-dimensional flattened view of convolutional filters where the kernel space is reduced while the channel space is retained, thus each cell in the lattice represents an individual kernel (see Fig. 2 for intuitive illustration).

scale variance problems, which is critical for handling a single instance with a non-rigid shape or multiple instances with complex scale variations. For scale-variant stimuli, PSConv is capable of learning self-adaptive attention for different receptive fields following a dynamic routing mechanism, improving the representation ability without any additional parameters or memory cost.

Thanks to its plug-and-play characteristic, our PSConv can be readily used to replace the vanilla convolution of arbitrary state-of-the-art CNN architectures, e.g., ResNet [11], giving rise to PS-ResNet. We also build PS-ResNeXt featuring group convolutions to prove the universality of PSConv. These models are comprehensively evaluated on the ImageNet [7] dataset and show consistent gains over the baseline of plain CNN counterparts. More experiments on (semi-)dense prediction tasks, e.g., object detection and instance segmentation on the MS COCO dataset, further demonstrate the superiority of our proposed PSConv over the standard ones under the circumstances with severe scale variations. It should be noted that PSConv is also independent of other macro-architectural choices and thus orthogonal and complementary to existing multi-scale network designs at a coarser granularity, leaving extra room to combine them together for further performance enhancement.

Our core contributions are summarized as follows:

❑ We extend the scope of the conventional mono-scale convolution operation by developing our Poly-Scale Convolution, which effectively and efficiently aggregates multi-scale features via arranging a spectrum of dilation rates in a cyclic manner inside the kernel lattice.
❑ We investigate the multi-scale network design through the lens of kernel engineering instead of network engineering, which avoids the necessity of tuning network structure or layer configurations while achieves competitive performance, when adapted to existing CNN architectures.

2   Related Work

We briefly review previous relevant network and modular designs and clarify their similarities and differences compared to our proposed approach.

Multi-scale Network Design. Early works like AlexNet [18] and VGGNet [32] learn multi-scale features in a data-driven manner, which are naturally equipped with a hierarchical representation by the inherent design of CNNs. The shallow layers seek finer structures in the images like edges, corners, and texture, while deep layers abstract semantic information, such as outlines and categories. In order to break the limitation of fixed-sized receptive fields and enhance feature representation, many subsequent works based on explicit multi-scale feature fusion are presented. Within this scope, there exists a rich literature making innovations on **skip connection** and **parallel stream**.

The **skip connection** structure exploits features with multi-size receptive fields from network layers at different depths. The representative FCN [25] adds up feature maps from multiple intermediate layers with the skip connection.

Analogous techniques have also been applied to the field of edge detection, presented by HED [40]. In the prevalent encoder-decoder architecture, the decoder network could be a symmetric version of the encoder network, with skip connections over some mirrored layers [26] or concatenation of feature maps [31]. DLA [42] extends the peer-to-peer skip connections into a tree structure, aggregating features from different layers in an iterative and hierarchical style. Fish-Net [34] stacks an upsampling body and a downsampling head upon the backbone tail, refining features that compound multiple resolutions.

The **parallel stream** structure generates multi-branch features conditioned on a spectrum of receptive fields. Though too numerous to list in full, recent research efforts often attack conventional designs via either maintaining a feature pyramid virtually from bottom to top or repeatedly stacking split-transform-merge building blocks. The former pathway of design includes several exemplars like Multigrid [16] and HRNet [33], which operate on a stack of features with different resolutions in each layer. Similarly, Octave Convolution [5] decomposes the standard convolution into two resolutions to process features at different frequencies, removing spatial redundancy by separating scales. The latter pathway of design is more crowded with the following works. The Inception [13,35,36] family utilizes parallel pathways with various kernel sizes in one Inception block. BL-Net [2] is composed of branches with different computational complexities, where the features at the larger scale pass through fewer blocks to spare computational resources and the features from different branches at distinct scales are merged with a linear combination. Res2Net [8] and OSNet [45] construct a group of hierarchical residual-like connections or stacked Lite 3×3 layers along the channel axis in one single residual block. ELASTIC [38] and ScaleNet [20] learn a soft scaling policy to allocate weights for different resolutions in the paratactic branches. Despite distinct with respect to detailed designs, these works all extensively use down-sampling or up-sampling to resize the features to 2^n times and inevitably adjust the original architecture via the selection of new hyperparameters and layer configurations when plugged in. On the contrary, our proposed PSConv can be a straightforwardly drop-in replacement of the vanilla convolution, leading a trend towards more effective and efficient multi-scale feature representation. Conventionally, features with multi-size receptive fields are integrated via channel concatenation, weighted summation or attention models. In stark contrast, we suggest to explore multi-scale features in a finer granularity, encompassed in merely one single convolutional layer.

In addition to the aforementioned networks designed to enhance image classification, scale variance poses more challenges in (semi-)dense prediction tasks, *e.g.*, object detection and semantic segmentation. Faster R-CNN [9] uses predefined anchor boxes of different sizes to address this issue. DetNet [21], RFB-Net [24] and TridentNet [19] apply dilated convolutions to enlarge the receptive fields. DeepLab [4] and PSPNet [44] construct feature pyramid in a parallel fashion. FPN [22] is designed to fuse features at multiple resolutions through top-down and lateral connections and provides anchors specific to different scales.

Dynamic Convolution. All approaches above process multi-scale information without drilling down into the pure single convolutional layer. Complementarily, another line of research concentrates on injecting scale modules into the original network directly and handling various receptive fields in an automated fashion. STN [14] explicitly learns a parametric manipulation of the feature map conditioned on itself to improve the tolerance to spatial geometric transformations. ACU [15] and DCN [6,46] learn offsets at each sampling position of the convolutional kernel or the feature map to permit a flexible shape deformation during the convolution process. SAC [43] inserts an extra regression layer to densely infer the scale coefficient map and applies an adaptive dilation rate to the convolutional kernel at each spatial location of the feature map. POD [28] predicts a globally continuous scale and then converts the learned fractional scale to a channel-wise combination of integer scales for fast deployment. We respect the succinctness of these plugged-in modules and follow these approaches in their form. In this spirit, we formulate a novel convolution representation through cyclically alternating dilation rates along both input and output channel dimensions to address the scale variations. We also note that some of the aforementioned modules are designed specifically for (semi-)dense prediction problems, *e.g.*, SAC, DCN, and POD and others do not scale to large-scale classification benchmarks like ImageNet, *e.g.*, STN for MNIST and SVHN, ACU for CIFAR. In contrast, our proposed PSConv focuses on backbone engineering, empirically shows its effectiveness on ImageNet and generalizes well to other complicated tasks on MS COCO. Furthermore, while the offsets are learned efficiently in some methods (ACU, SAC, and DCN), the inference is time-consuming due to the dynamic grid sampling and the bilinear interpolation at each position. Aligning to the tenet of POD [28], it is unnecessary to permit too much freedom with floating-point offsets at each spatial location as DCN [6] and learning in such an aggressive manner places an extra burden on the inference procedure. We opt for a better accuracy-efficiency trade-off by constraining dilation rates in the integer domain and organizing them into repeated partitions. Last but not least, the recently proposed MixConv [37] may be the most related scale module compared to PSConv, which will be discussed at the end of the next section.

3   Method

Compared to previous multi-scale feature fusion solutions in a coarse granularity, we seek an alternative design with the finer granularity and stronger feature extraction ability, while maintaining a similar computational load.

3.1   Sketch of Convolution Operations

We initiate from elaborating the vanilla (dilated) convolution process to make the definition of our proposed PSConv self-contained. For a single convolutional layer, let the tensor $\mathcal{F} \in \mathbb{R}^{C_{in} \times H \times W}$ denotes its input feature map with the shape of $C_{in} \times H \times W$, where C_{in} is the number of channels, H and W are the height and width respectively. A set of C_{out} filters with the kernel size $K \times K$ are convolved with the input tensor individually to obtain the desired output feature map with C_{out} channels, where each filter has C_{in} kernels to match those channels in the input feature map. Denote the above filters as $\mathcal{G} \in \mathbb{R}^{C_{out} \times C_{in} \times K \times K}$, then the vanilla convolution operation can be represented as

$$\mathcal{H}_{c,x,y} = \sum_{k=1}^{C_{in}} \sum_{i=-\frac{K-1}{2}}^{\frac{K-1}{2}} \sum_{j=-\frac{K-1}{2}}^{\frac{K-1}{2}} \mathcal{G}_{c,k,i,j} \mathcal{F}_{k,x+i,y+j}, \tag{1}$$

where $\mathcal{H}_{c,x,y}$ is one element in the output feature map $\mathcal{H} \in \mathbb{R}^{C_{out} \times H \times W}$, $c = 1, 2, \cdots, C_{out}$ is the index of an output channel, $x = 1, 2, \cdots, H$ and $y = 1, 2, \cdots, W$ are indices of spatial positions in the feature map.

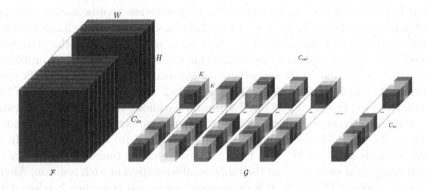

Fig. 1. Schematic illustration of our proposed PSConv operation. $\mathcal{F}$ represents the input feature map and $\mathcal{G}$ represents C_{out} convolutional filters in a set. Convolutional kernels with the same dilation rates in the set of filters $\mathcal{G}$ are rendered with the same color. Best viewed in color.

Dilated Convolution [41] enlarges sampling intervals in the spatial domain to cover objects of larger sizes. A dilated convolution with the dilation rate d can be represented as

$$\mathcal{H}_{c,x,y} = \sum_{k=1}^{C_{in}} \sum_{i=-\frac{K-1}{2}}^{\frac{K-1}{2}} \sum_{j=-\frac{K-1}{2}}^{\frac{K-1}{2}} \mathcal{G}_{c,k,i,j} \mathcal{F}_{k,x+id,y+jd}. \tag{2}$$

Noticing that a combination of dilation rates is conducive to extract both global and local information, we propose a new convolution form named Poly-Scale Convolution (PSConv) which scatters organized dilation rates over different kernels inside one convolutional filter. Furthermore, our PSConv integrates multi-scale features in a one-shot manner and brings characteristics of the dilated convolution into full play, thus without introducing additional computational cost. To gather multi-scale information from different input channels via a linear summation, dilation rates are varied at different kernels in one convolutional filter. To process an input channel with various receptive fields, dilation rates are also varied in different filters for a certain channel. It is written as

$$\mathcal{H}_{c,x,y} = \sum_{k=1}^{C_{in}} \sum_{i=-\frac{K-1}{2}}^{\frac{K-1}{2}} \sum_{j=-\frac{K-1}{2}}^{\frac{K-1}{2}} \mathcal{G}_{c,k,i,j} \mathcal{F}_{k,x+iD_{(c,k)},y+jD_{(c,k)}}, \tag{3}$$

where $D \in \mathbb{R}^{C_{out} \times C_{in}}$ is a matrix composed of channel-wise and filter-wise dilation rates in two orthogonal dimensions. An element $D_{(c,k)}$ is associated with a specific channel in one filter to support $\mathcal{G}_{c,k,\cdot,\cdot}$ as a unique convolutional kernel, thus the whole matrix D can be interpreted as a mathematical representation of the kernel lattice in its subspace of dilation rate.

3.2 Design Details

As stated above, our major work is to reformulate the dilation rate patterns in the subspace of kernel lattice. We ensure that each row and column of the matrix D have non-identical elements to achieve the desired properties of multi-scale feature fusion. On the contrary, if we avoid and retain identical elements in one row, then we would not collect multi-scale information to produce a new output channel in this operation, and it can be boiled down to multi-stream transformation before concatenation; if the similar event occurs in one column, the corresponding input channel would not have necessarily diverse receptive fields covered, and it reduces to the split-transform-summation design of multi-scale networks. These are both suboptimal according to our ablative experiments in Table 5. The illustration diagrams of these two simplified cases are provided in the supplementary materials.

Following the above analysis, the design philosophy of PSConv could be decomposed into two coupled ingredients. Firstly, we concentrate on a single filter. In order to constrain the number of different dilation rates in a reasonable range, we heuristically arrange them inside one filter with a cyclic layout, *i.e.*, dilation rates vary in a periodical manner along the axis of input channels. Specifically speaking, a total of C_{in} input channels are divided into P partitions. For each partition, $t = \lceil \frac{C_{in}}{P} \rceil$ channels are accommodated and a fixed pattern of dilation rates $\{d_1, d_2, \cdots, d_t\}$ is filled in to construct a row of the matrix D. Secondly, we broaden our horizons to all filters. In order to endow different filters with capacities to gather different kinds of scale combinations of input features, we adopt a shift-based strategy for dilation rates to flip the former filter to the latter one, *i.e.*, the pattern of dilation rates regarding a convolutional filter is shifted by one channel to build its adjacent filter. In the illustrative example of Fig. 2, $C_{in} = C_{out} = 16$ and the partition number P is set to 4, hence there leaves a blank of 4 dilation rates to be determined in the pattern $\{d_1, d_2, d_3, d_4\}$, where a specific colorization distinguishes one type of dilation rate from others. It is noted that viewed from the axis of output channels, dilation rates also present periodical variation. In other words, all types of dilation rates occur alternately along the vertical and horizontal axes in the trellis.

Furthermore, a comparison diagram is shown in Fig. 2, to achieve better intuitive comprehension about different convolution operations. The filters of PSConv are exhibited from the vertical view of $\mathcal{G}$ (with appropriate rotate transformation) in Fig. 1, where each tile in the grid represents a kernel of $K \times K$ shape and the grid corresponds to the dilation rate matrix D. The filters of (dilated) convolution and group convolution are likewise displayed. In the conventional filters, if a dilation rate is applied, it will dominate the whole kernel lattice, while our PSConv has clear distinctions compared with them. We claim that merely varying dilation rates in the axis of output channels equals to using split-transform-merge units spanning a spectrum of dilation rates in the different streams. Our method takes one step further to spread the scale information along both input and output channel axes, pushing the selection of scale-variant features into the entire kernel space. To the best of our knowledge, **it is the first attempt to mix up multi-scale information simultaneously in two orthogonal dimensions and leverage the complementary multi-scale benefits from such a fine granularity in the kernel space.**

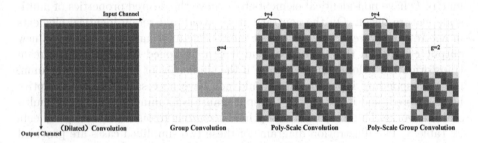

Fig. 2. Comparison between dilation space of kernel lattice in different convolution operations. Kernels of standard convolution (with or without dilation) are showcased in the leftmost, where each kernel is located at one cell in the lattice. Group convolution (group number $g = 4$) extensively utilized in the efficient network design is also included for reference. Poly-Scale convolution (cyclic interval $t = 4$) and Poly-Scale group convolution (group number $g = 2$ and cyclic interval $t = 4$) in the right show significant differences from the former two. Best viewed in color.

It is noteworthy that PSConv is a generalized form of dilated convolution: since the cyclic interval t decides how many types of dilation rates are contained in one partition, all kernels may share the same dilation rate once the partition number equals to that of input channels and then it degenerates into vanilla dilated convolution. The PSConv can also be applied to the group-wise convolution form by injecting the shared cyclic pattern into each group, as illustrated in the rightmost of Fig. 2. Owing to the interchangeability of channel indices, grouping channels with the same dilation rate together leads to an equivalent but efficient implementation, which is depicted in the supplementary materials.

The recently proposed MixConv [37] might be similar to PSConv at the first glimpse. However, they are distinct regarding both the design principle and the application focus. On the one hand, MixConv integrates multiple kernel sizes for different patterns of resolutions which inevitably increases the parameters and computational budget, while PSConv mixes up a spectrum of dilation rates with a unified kernel size to economically condense multi-scale features within one convolution operation. Thus, for these two convolution forms, the manipulations on the kernel lattice are shaped from orthogonal perspectives. On the other hand, MixConv is dedicatedly developed for depthwise convolution (DWConv), while PSConv is versatile to both standard and group convolution. Due to the inherent constraint of DWConv, each individual channel in a MixConv operation exploits feature representation of a certain scale. However, in our PSConv, multi-scale representations are scattered along both input and output channels in a periodical manner. Hence, an individual channel could gather multifarious feature resolutions from the view of either input or output channels. We attach a more in-depth discussion around their differences and an illustration of the DWConv-based variant of PSConv in the supplementary materials.

4   Experiments

We conduct extensive experiments from conceptual to dense prediction tasks on several large-scale visual recognition benchmarks. Experimental results empirically validate the effectiveness and efficiency of our proposed convolution form. All experiments are performed with the PyTorch [27] library.

4.1   ILSVRC 2012

ImageNet [7] is one of the most challenging datasets for image classification, which is served as the benchmark of the ILSVRC2012 competition. It includes 1,281,167 training images and 50,000 validation images, and each image is manually annotated as one of the 1,000 object categories.

We incorporate our PSConv layer into various state-of-the-art convolutional neural networks, including ResNet [11], ResNeXt [39] and SE-ResNet [12]. The training procedure is performed on the ImageNet training set by the SGD optimizer with the momentum of 0.9 and the weight decay of 1e-4. The mini-batch size is set to 256 and the optimization process lasts for a period of 120 epochs to achieve full convergence. The learning rate initiates from 0.1 and decays to zero following a half cosine function shaped schedule, the same as [2] and [5]. We adopt random scale and aspect ratio augmentation together with random horizontal flipping to process each training sample prior to feeding it into neural networks. We select the best-performing model along the training trajectory and report its performance on the ImageNet validation set. As is the common practice, we first resize the shorter side of validation images to 256 pixels and then crop the central region of 224 × 224 size for evaluation.

As shown in Table 1, network models equipped with PSConv layers demonstrate consistent improvement over counterpart baseline models mostly with over 1% gains of the top-1 error. We replace all the 3×3 standard convolutional layers in the middle of bottleneck blocks with our PSConv layers. In all of our main experiments, the cyclic interval is set to 4 and the dilation rate pattern is fixed as $\{d_1, d_2, d_3, d_4\} = \{1, 2, 1, 4\}$ which are determined by ablation studies, as detailed in the next subsection. It is observed that the PS-ResNet-50 model achieves 21.126% top-1 error, which is comparable to the vanilla ResNet-101 model with almost half of the trainable parameter storage and computational resource consumption. The PS-ResNeXt-50 (32 × 4d) model even achieves superior performance over the vanilla 101-layer ResNeXt model, which demonstrates the wide applicability of our PSConv in boosting both standard and group convolution. Furthermore, we integrate PSConv into the modern SE-ResNet models and obtain performance margins again, which showcases the compatibility of our proposed convolution operation to other advanced atomic operations such as the channel-attention modules. Notably, all the above gains are obtained without theoretically introducing any additional computational cost.

Table 1. Recognition error comparisons on the ImageNet validation set. The standard metrics of top-1/top-5 errors are measured using single center crop evaluation. The baseline results are re-implemented by ourselves.

Architecture	Conv Type	Top-1/Top-5 Err.(%)	Architecture	Conv Type	Top-1/Top-5 Err.(%)
ResNet-50	Standard	22.850/6.532	ResNet-101	Standard	21.102/5.696
	PSConv	**21.126/5.724**		PSConv	**19.954/5.052**
ResNeXt-50 (32 × 4d)	Standard	21.802/6.084	ResNeXt-101 (32 × 4d)	Standard	20.502/5.390
	PSConv	**20.378/5.296**		PSConv	**19.498/4.724**
SE-ResNet-50	Standard	22.192/6.040	SE-ResNet-101	Standard	20.732/5.406
	PSConv	**20.814/5.578**		PSConv	**19.786/4.924**

For horizontal comparison, we give a brief synopsis of some recent multi-scale networks in Table 2 for reference. Despite that discrepancies in model profiles and training strategies could lead to no apple-to-apple comparisons in most cases, our PS-ResNet-50 achieves competitive accuracy compared to other ResNet-50-based architectures under the similar level of parameters and computational complexities. Specifically, two variants of Dilated Residual Networks (DRN) increase the computation cost to a large extent due to the removed strides in the last two residual stages, but only achieves inferior or comparable performance.

Table 2. Performance comparison with state-of-the-art multi-scale network architectures on the ImageNet validation set.

Network	Params	GFLOPs	LR decay schedule	Top-1/Top-5 Err.(%)
ResNet-50 [11]	25.557M	4.089	Cosine (120 epoch)	22.850/6.532
PS-ResNet-50 (ours)	25.557M	4.089	Cosine (120 epoch)	**21.126/5.724**
DRN-A-50 [41]	25.557M	19.079	Stepwise (120 epoch)	22.9/6.6
DRN-D-54 [41]	35.809M	28.487	Stepwise (120 epoch)	21.2/5.9
FishNet-150 [34]	24.96M	6.45	Stepwise (100 epoch)	21.86/6.05
FishNet-150 [34]	24.96M	6.45	Cosine (200 epoch) w/ label smoothing	20.65/5.25
HRNet-W18-C [33]	21.3M	3.99	Stepwise (100 epoch)	23.2/6.6
OctResNet-50 [5] ($\alpha = 0.5$)	25.6M	2.4	cosine (110 epoch)	22.6/6.4
bL-ResNet-50 [2] ($\alpha = 2, \beta = 4$)	26.69M	2.85	cosine (110 epoch)	22.69/-
Res2Net-50 [8] ($26w \times 4s$)	25.70M	4.2	Stepwise (100 epoch)	22.01/6.15
ScaleNet-50 [20]	31.483M	3.818	Stepwise (100 epoch)	22.02/6.05

4.2 Ablation and Analysis

We first systematically probe the impact of partition numbers and dilation rate patterns in one cycle. We next assess the ability of PSConv to generalize to another classification benchmark beyond ImageNet, namely CIFAR-100.

Partition Number. On the one hand, provided that channels are divided into too many partitions, there leaves limited room for varied dilation rates within one partition and it frequently alternates around certain values. In the extreme case that the partition number equals to the number of channels, PSConv degenerates into the vanilla dilated convolution with a shared dilation rate. On the other hand, if there are too few partitions, each partition can accommodate a large number of heterogeneous dilation rates, which may have contradictory effects on extracting diverse features, hence we initially constrain the dilation rate in one basic pattern to toggle between 1 and 2 in this set of ablation experiments. Specifically, we set the dilation rate in one slot of a cycle to 2 and the other slots to 1. Under this constraint, features corresponding to large receptive fields will infrequently emerge with the growing cyclic interval, which may still impede the full utilization of multi-scale features.

Table 3. Performance comparison of PS-ResNet-50 with varied cyclic intervals on the ImageNet validation set. The best result is highlighted in **bold**, the same hereinafter.

Architecture	ResNet-50	PS-ResNet-50		
Cyclic Interval	$t = 1$ (baseline)	$t = 2$	$t = 4$	$t = 8$
Top-1/Top-5 Err.(%)	22.850/6.532	21.948/5.978	**21.476/5.720**	21.634/5.816

Table 4. Performance comparison of PS-ResNet-50 with various dilation patterns on the ImageNet validation set.

Dilation Pattern	$\{1,1,1,1\}$ (baseline)	$\{1,2,1,1\}$	$\{1,4,1,1\}$	$\{1,2,1,2\}$	$\{1,2,1,4\}$ (default)
Top-1/Top-5 Err.(%)	22.850/6.532	22.368/6.214	22.754/6.470	21.948/5.978	**21.126/5.724**

We use the ResNet-50 model on the ImageNet dataset for experiments and tune the partition numbers, giving rise to a spectrum of cyclic intervals. The corresponding results shown in Table 3 empirically support our speculation above. The PS-ResNet-50 ($t = 4$) achieves better performance when the cyclic interval increases from 2 to 4. The accuracy tends to decline when its cyclic interval gets further increment. Thus we set $t = 4$ as the default value in our main experiments. In each case, PS-ResNet-50 with a specific cyclic setting outperforms the vanilla ResNet-50 baseline result.

Pattern of Dilation Rates. Let the cyclic interval be 4. Noticing that the dilation rate pattern is an unordered set, we initially set any one of the dilation rate to a larger numeric value. For example, $\{d_1, d_2, d_3, d_4\}$ is set to $\{1, 2, 1, 1\}$, where the unique large dilation rate is placed in the second slot without loss of generality owing to its unordered nature. Next we assume that further increasing this large dilation rate (e.g., setting $\{d_1, d_2, d_3, d_4\} = \{1, 4, 1, 1\}$) would lead to intra-group separation of these two dilation rates and unsmoothed transition of the receptive fields. Then we tend to inject another large dilation rate into this pattern. Considering that the setting of $\{d_1, d_2, d_3, d_4\} = \{1, 2, 1, 2\}$ is equivalent to $t = 2$ in the above experiments, we change the pattern to $\{d_1, d_2, d_3, d_4\} = \{1, 2, 1, 4\}$ for the sake of perceiving larger receptive fields and interspacing the two different large dilation rates. This consequent PS-ResNet-50 achieves 21.126% top-1 error in the ImageNet evaluation, which is exactly the one reported in Table 1. For further exploration, we tentatively incorporate larger dilation rate to compose the combination of $\{d_1, d_2, d_3, d_4\} = \{1, 2, 4, 8\}$, but it shows much inferior performance (over 5% drop). We attribute this failure to the exclusively aggressive dilation rate arrangement, since inappropriately enlarging the receptive field can involve irrelevant pixels into spatial correlation (Table 4).

Apart from the static setting of dilation rates, we develop a learnable binary mask to distinguish the large dilation rate from the small one. This binary mask is decomposed via the Kronecker product, where the STE (Straight-Through Estimator) [29] technique is utilized to solve the discrete optimization problem. As a consequence, the dynamic version of PS-ResNet-50 with optional dilation rates of 1 and 2 reduces the top-1 error to 21.138%, that is close to the best-performing static PS-ResNet-50 ($t = 4$, $\{d_1, d_2, d_3, d_4\} = \{1, 2, 1, 4\}$) involving larger dilation rates in its PSConv pattern. Although extra parameters and computational complexity result in no fair comparison, it opens up a promising perspective deserving future research development.

Table 5. Performance comparison of PS-ResNet-50 on the ImageNet validation set, with the variation of dilation rates along different axes of kernel lattice.

Input Channel Axis	Output Channel Axis	Top-1/Top-5 Err.(%)
✓	✗	21.658/5.832
✗	✓	22.056/6.174
✓	✓	**21.126/5.724**

Table 6. Top-1 error comparisons on the CIFAR-100 test set. Our results were obtained by computing mean and standard deviation over 5 individual runs (denoted by mean ± std. in the table).

Architecture	Conv Type	Top-1 Error(%)	Architecture	Conv Type	Top-1 Error(%)
ResNeXt-29 (8 × 64d)	Standard (*official*)	17.77	ResNeXt-29 (16 × 64d)	Standard (*official*)	17.31
	Standard (*self impl.*)	18.074 ± 0.130		Standard (*self impl.*)	17.538 ± 0.094
	PSConv	**17.138 ± 0.286**		PSConv	**16.528 ± 0.353**

Following the searched optimal setting of $\{d_1, d_2, d_3, d_4\} = \{1, 2, 1, 4\}$, we remove the shift strategy among different filters, which means the variation of dilation rates only exists in the axis of input channels. In this setup, we observe a drop of around 1% regarding the top-1 validation accuracy. Symmetrically, we only vary the dilation rates in the axis of output channels with the same setting of $\{d_1, d_2, d_3, d_4\} = \{1, 2, 1, 4\}$, which indicates no cyclic operations inside each individual filter. As shown in Table 5, it also achieves inferior performance.

Beyond ImageNet. CIFAR-100 [17] is another widely-adopted benchmark for image classification, which consists of 50,000 training images and 10,000 test images. Each colorful image in the dataset is of 32×32 size and drawn from 100 classes, hence it is more challenging than CIFAR-10 with similar image qualities but a coarser taxonomy. We choose the high-performing ResNeXt [39] architecture as a strong baseline, and replace all the 3×3 convolutional layers in every bottleneck block with PSConv layers to build our PS-ResNeXt models for comparison. The data augmentation is the same as the preprocessing method in [11, 39], utilizing sequential zero padding, random cropping and standardization. The whole training regime strictly follows the original paper to isolate the contribution of our PSConv. For evaluation, we perform five independent runs of training the same architecture with different initialization seeds and report the mean top-1 error as well as the standard deviation.

We summarize the comparison results in Table 6. The performance of our reproduced ResNeXt-29 is slightly degraded, thus we list results from both the official release and our implementation, annotated as *official* and *self impl.* with the standard convolution respectively. It is evident that PS-ResNeXt-29 (8×64d) and PS-ResNeXt-29 (16×64d) outperform the original ResNeXts by around 1% accuracy gains. Even compared to the results from the original author, absolute gains of 0.632% and 0.782% are achieved using our PSConv neural networks. It is observed that using networks with various cardinalities on datasets with distinct characteristics (like thumbnails), PSConv could still yield satisfactory performance gains.

Speed Benchmark. For an input tensor with the size of $(N, C, H, W) = (200, 64, 56, 56)$, a standard 3×3 convolutional layer with 64 output channels takes 4.85ms to process on a single Titan X GPU, using CUDA v9.0 and cuDNN v7.0 as the backend. The dilated convolution with a dilation rate of 2 consumes 2.99 times of above and the inference time of our PSConv is $1.14\times$ of dilated convolution. There exist a similar trend in the comparison of their group convolution based counterparts. Thus, improved performance of inference speed can be achieved by optimizing vanilla dilated convolutions on GPU/CPU inference. The further optimized results for practical deployment are provided in the supplementary materials.

Scale Allocation. We dive into the PSConv kernels to analyze the law of scale-relevant feature distributions by dissecting the weight proportion with respect to different dilation rates, as is shown in the supplementary materials.

4.3 MS COCO 2017

To further demonstrate the generality of our proposed convolution, we apply the PSConv-based backbones to object detection and instance segmentation frameworks and finetune the PSConv-based detectors on the 2017 version of Microsoft COCO [23] benchmark. This large-scale dataset including 118,287 training images and 5,000 validation images is considered highly challenging owing to the huge number of objects within per image and large variation among these instances, which is suitable for inspecting the superiority of our PSConv models.

We use the popular MMDetection [3] toolbox to conduct experiments. ResNet-50/101 and ResNeXt-101 (32×4d) along with FPN [22] necks are selected as the backbone networks. For object detection and instance segmentation tasks, we adopt the main-stream Faster R-CNN [30] and Mask R-CNN [10] as the basic detectors respectively. We replace all the 3×3 convolutional layers in the pre-trained backbone network by PSConv layers, while the convolution layers in the FPN neck are kept as standard convolutions[2]. Then we finetune these detectors on the training set following the $1\times$ learning rate schedule, which indicates a total of 12 epochs with the learning rate divided by 10 at the epoch of 8th and 11st respectively. During this transfer learning process, we maintain the same data preparation pipeline and hyperparameter settings for our models as the baseline detectors. During evaluation, we test on the validation set and report the COCO-style Average Precision (AP) under IOU thresholds ranging from 0.5 to 0.95 with an increment of 0.05. We also keep track of scores for small, medium and large objects. These metrics comprehensively assess the qualities of detection and segmentation results from various views of different scales.

---

[2] Actually we have preliminary experiments by also replacing these layers with PSConv layers, but it achieves marginal benefit. For instance, AP^{bbox} of Faster R-CNN with ResNet-50 and FPN only increases from 38.4% to 38.6%.

Table 7. Bounding-box and mask Average Precision (AP) comparison on the COCO 2017 validation set for the instance segmentation track with different backbones.

Detector	Architecture	Conv Type	Box AP						Mask AP					
			AP	AP$_{50}$	AP$_{75}$	AP$_S$	AP$_M$	AP$_L$	AP	AP$_{50}$	AP$_{75}$	AP$_S$	AP$_M$	AP$_L$
Mask R-CNN	R50	Standard	37.3	59.0	40.2	21.9	40.9	48.1	34.2	55.9	36.2	15.8	36.9	50.1
		PSConv	39.4$_{(+2.1)}$	61.3	42.8	24.1	43.1	51.3	35.6$_{(+1.4)}$	57.9	37.9	17.2	38.4	52.4
	R101	Standard	39.4	60.9	43.3	23.0	43.7	51.4	35.9	57.7	38.4	16.8	39.1	53.6
		PSConv	41.6$_{(+2.2)}$	63.4	45.1	24.7	45.6	54.4	37.4$_{(+1.5)}$	60.0	39.8	17.8	40.4	55.1
	X101-32 × 4d	Standard	41.1	62.8	45.0	24.0	45.4	52.6	37.1	59.4	39.7	17.7	40.5	53.8
		PSConv	42.4$_{(+1.3)}$	64.4	46.1	25.4	46.5	55.7	38.0$_{(+0.9)}$	60.8	40.5	18.6	41.0	55.8
Cascade Mask R-CNN	R50	Standard	41.2	59.1	45.1	23.3	44.5	54.5	35.7	56.3	38.6	16.4	38.2	52.6
		PSConv	42.9$_{(+1.7)}$	61.7	46.9	24.2	46.5	57.2	36.9$_{(+1.2)}$	58.4	39.5	17.1	39.4	54.6
	R101	Standard	42.6	60.7	46.7	23.8	46.4	56.9	37.0	58.0	39.9	16.7	40.3	54.6
		PSConv	44.6$_{(+2.0)}$	63.2	48.6	25.9	48.7	59.6	38.4$_{(+1.4)}$	60.5	41.2	18.6	41.5	56.8
	X101-32 × 4d	Standard	44.4	62.6	48.6	25.4	48.1	58.7	38.2	59.6	41.2	18.3	41.4	55.6
		PSConv	45.3$_{(+0.9)}$	64.2	49.5	27.0	49.2	60.0	38.9$_{(+0.7)}$	61.1	41.8	19.4	41.9	56.6

The comparison results of Mask R-CNN are shown in Table 7 (similar comparisons of Faster R-CNN are provided in the supplementary materials), where the baseline results with standard backbone networks are extracted from the model zoo of MMDetection, and absolute gains of AP concerning our PSConv models are indicated in the parentheses. Since our ImageNet pre-trained backbones in Sect. 4.1 are trained using the cosine learning rate annealing, we would have an unfair accuracy advantage against the pre-trained backbones in MMDetection. In order to pursue fair comparison to its published baseline results, we first retrain backbones of ResNet and ResNeXt following the conventional stepwise learning rate annealing strategy [11] and then load these backbones to the detectors[3]. It is evident that PSConv brings consistent and considerable performance gains over the baseline results, across different tasks and various backbones. In addition, we introduce the Cascade (Mask) R-CNN [1] as a stronger baseline detector and reach the conclusion that our PSConv operation can benefit both basic detectors and more advanced cascade detectors.

Detectors with the ResNet-101 backbone consistently show larger margins among different tasks and frameworks compared to the other two backbones. Compared to ResNet-50, the 101-layer network almost quadruples the depth of the conv4_x stage, guaranteeing a higher capacity for performance amelioration. In addition, we come up with the hypothesis that its ResNeXt counterpart has already efficiently deployed the model capacity through adjusting the dimension of cardinality beyond network depth and width, leaving a bottleneck for further performance improvement in both classification and detection tasks. It is observed that the most significant improvement of Faster R-CNN and Mask R-CNN locates in the metric of AP_L among various object sizes, speaking to the theoretically enlarged receptive fields. Finally, representative visualization results of predicted bounding-boxes and masks are attached in the supplementary materials to raise the qualitative insight of our method.

5   Conclusion

In this paper, we have proposed a novel convolution operation named PSConv, which cyclically alternates dilation rates along the axes of input and output channels. PSConv permits to aggregate multi-scale features from a granular perspective and efficiently allocates weights to a collection of scale-specific features through dynamic execution. It is amenable to be plugged into arbitrary state-of-the-art CNN architectures in-place, demonstrating its superior performance on various vision tasks compared to the counterparts with regular convolutions.

References

1. Cai, Z., Vasconcelos, N.: Cascade R-CNN: delving into high quality object detection. In: CVPR (2018)

---

[3] If we adopt those unfair backbones pre-trained using cosine learning rate decay in Sect. 4.1, we can get even larger performance margins (*e.g.* 2.6% instead of 2.0% for Faster R-CNN with ResNet-50 and FPN).

2. Chen, C.F.R., Fan, Q., Mallinar, N., Sercu, T., Feris, R.: Big-little net: an efficient multi-scale feature representation for visual and speech recognition. In: ICLR (2019)
3. Chen, K., et al.: MMDetection: open MMLab detection toolbox and benchmark. arXiv e-prints arXiv:1906.07155 (Jun 2019)
4. Chen, L.C., Zhu, Y., Papandreou, G., Schroff, F., Adam, H.: Encoder-decoder with atrous separable convolution for semantic image segmentation. In: ECCV (2018)
5. Chen, Y., et al.: Drop an octave: reducing spatial redundancy in convolutional neural networks with octave convolution. In: ICCV (2019)
6. Dai, J., et al.: Deformable convolutional networks. In: ICCV (2017)
7. Deng, J., Dong, W., Socher, R., Li, L.J., Li, K., Fei-Fei, L.: ImageNet: a large-scale hierarchical image database. In: CVPR (2009)
8. Gao, S.H., Cheng, M.M., Zhao, K., Zhang, X.Y., Yang, M.H., Torr, P.: Res2net: a new multi-scale backbone architecture. IEEE TPAMI, 1 (2019)
9. Girshick, R.: Fast R-CNN. In: ICCV (2015)
10. He, K., Gkioxari, G., Dollar, P., Girshick, R.: Mask R-CNN. In: ICCV (2017)
11. He, K., Zhang, X., Ren, S., Sun, J.: Deep residual learning for image recognition. In: CVPR (2016)
12. Hu, J., Shen, L., Sun, G.: Squeeze-and-excitation networks. In: CVPR (2018)
13. Ioffe, S., Szegedy, C.: Batch normalization: accelerating deep network training by reducing internal covariate shift. In: ICML (2015)
14. Jaderberg, M., Simonyan, K., Zisserman, A., kavukcuoglu, k.: Spatial transformer networks. In: NIPS (2015)
15. Jeon, Y., Kim, J.: Active convolution: learning the shape of convolution for image classification. In: CVPR (2017)
16. Ke, T.W., Maire, M., Yu, S.X.: Multigrid neural architectures. In: CVPR (2017)
17. Krizhevsky, A., Hinton, G.: Learning multiple layers of features from tiny images. Master's thesis, Department of Computer Science, University of Toronto (2009)
18. Krizhevsky, A., Sutskever, I., Hinton, G.E.: Imagenet classification with deep convolutional neural networks. In: NIPS (2012)
19. Li, Y., Chen, Y., Wang, N., Zhang, Z.: Scale-aware trident networks for object detection. In: ICCV (2019)
20. Li, Y., Kuang, Z., Chen, Y., Zhang, W.: Data-driven neuron allocation for scale aggregation networks. In: CVPR (2019)
21. Li, Z., Peng, C., Yu, G., Zhang, X., Deng, Y., Sun, J.: Detnet: design backbone for object detection. In: ECCV (2018)
22. Lin, T.Y., Dollar, P., Girshick, R., He, K., Hariharan, B., Belongie, S.: Feature pyramid networks for object detection. In: CVPR (2017)
23. Lin, T.-Y., Maire, M., Belongie, S., Hays, J., Perona, P., Ramanan, D., Dollár, P., Zitnick, C.L.: Microsoft COCO: common objects in context. In: Fleet, D., Pajdla, T., Schiele, B., Tuytelaars, T. (eds.) ECCV 2014. LNCS, vol. 8693, pp. 740–755. Springer, Cham (2014). https://doi.org/10.1007/978-3-319-10602-1_48
24. Liu, S., Huang, D., Wang, Y.: Receptive field block net for accurate and fast object detection. In: ECCV (2018)
25. Long, J., Shelhamer, E., Darrell, T.: Fully convolutional networks for semantic segmentation. In: CVPR (2015)
26. Noh, H., Hong, S., Han, B.: Learning deconvolution network for semantic segmentation. In: ICCV (2015)
27. Paszke, A., et al.: Pytorch: an imperative style, high-performance deep learning library. In: NeurIPS (2019)

28. Peng, J., Sun, M., Zhang, Z., Yan, J., Tan, T.: POD: practical object detection with scale-sensitive network. In: ICCV (2019)
29. Rastegari, M., Ordonez, V., Redmon, J., Farhadi, A.: XNOR-Net: imagenet classification using binary convolutional neural networks. In: Leibe, B., Matas, J., Sebe, N., Welling, M. (eds.) ECCV 2016. LNCS, vol. 9908, pp. 525–542. Springer, Cham (2016). https://doi.org/10.1007/978-3-319-46493-0_32
30. Ren, S., He, K., Girshick, R., Sun, J.: Faster R-CNN: towards real-time object detection with region proposal networks. In: NIPS (2015)
31. Ronneberger, O., Fischer, P., Brox, T.: U-Net: convolutional networks for biomedical image segmentation. In: Navab, N., Hornegger, J., Wells, W.M., Frangi, A.F. (eds.) MICCAI 2015. LNCS, vol. 9351, pp. 234–241. Springer, Cham (2015). https://doi.org/10.1007/978-3-319-24574-4_28
32. Simonyan, K., Zisserman, A.: Very deep convolutional networks for large-scale image recognition. In: ICLR (2015)
33. Sun, K., et al.: High-resolution representations for labeling pixels and regions. arXiv e-prints arXiv:1904.04514, April 2019
34. Sun, S., Pang, J., Shi, J., Yi, S., Ouyang, W.: Fishnet: a versatile backbone for image, region, and pixel level prediction. In: NeurIPS (2018)
35. Szegedy, C., et al.: Going deeper with convolutions. In: CVPR (2015)
36. Szegedy, C., Vanhoucke, V., Ioffe, S., Shlens, J., Wojna, Z.: Rethinking the inception architecture for computer vision. In: CVPR (2016)
37. Tan, M., Le, Q.V.: Mixconv: Mixed depthwise convolutional kernels. In: BMVC (2019)
38. Wang, H., Kembhavi, A., Farhadi, A., Yuille, A.L., Rastegari, M.: ELASTIC: improving cnns with dynamic scaling policies. In: CVPR (2019)
39. Xie, S., Girshick, R., Dollar, P., Tu, Z., He, K.: Aggregated residual transformations for deep neural networks. In: CVPR (2017)
40. Xie, S., Tu, Z.: Holistically-nested edge detection. In: ICCV (2015)
41. Yu, F., Koltun, V., Funkhouser, T.: Dilated residual networks. In: CVPR (2017)
42. Yu, F., Wang, D., Shelhamer, E., Darrell, T.: Deep layer aggregation. In: CVPR (2018)
43. Zhang, R., Tang, S., Zhang, Y., Li, J., Yan, S.: Scale-adaptive convolutions for scene parsing. In: ICCV (2017)
44. Zhao, H., Shi, J., Qi, X., Wang, X., Jia, J.: Pyramid scene parsing network. In: CVPR (2017)
45. Zhou, K., Yang, Y., Cavallaro, A., Xiang, T.: Omni-scale feature learning for person re-identification. In: ICCV (2019)
46. Zhu, X., Hu, H., Lin, S., Dai, J.: Deformable convnets v2: more deformable, better results. In: CVPR (2019)

Hierarchical Context Embedding
for Region-Based Object Detection

Zhao-Min Chen[1,2], Xin Jin[2], Borui Zhao[2], Xiu-Shen Wei[2(✉)],
and Yanwen Guo[1(✉)]

[1] State Key Laboratory for Novel Software Technology, Nanjing University,
Nanjing, China
chenzhaomin123@gmail.com, ywguo@nju.edu.cn
[2] Megvii Research Nanjing, Megvii Technology, Beijing, China
{jinxin,zhaoborui}@megvii.com, weixs.gm@gmail.com

Abstract. State-of-the-art two-stage object detectors apply a classifier to a sparse set of object proposals, relying on region-wise features extracted by RoIPool or RoIAlign as inputs. The region-wise features, in spite of aligning well with the proposal locations, may still lack the crucial context information which is necessary for filtering out noisy background detections, as well as recognizing objects possessing no distinctive appearances. To address this issue, we present a simple but effective Hierarchical Context Embedding (HCE) framework, which can be applied as a plug-and-play component, to facilitate the classification ability of a series of region-based detectors by mining contextual cues. Specifically, to advance the recognition of context-dependent object categories, we propose an image-level categorical embedding module which leverages the holistic image-level context to learn object-level concepts. Then, novel RoI features are generated by exploiting hierarchically embedded context information beneath both whole images and interested regions, which are also complementary to conventional RoI features. Moreover, to make full use of our hierarchical contextual RoI features, we propose the early-and-late fusion strategies (*i.e.*, feature fusion and confidence fusion), which can be combined to boost the classification accuracy of region-based detectors. Comprehensive experiments demonstrate that our HCE framework is flexible and generalizable, leading to significant and consistent improvements upon various region-based detectors, including FPN, Cascade R-CNN and Mask R-CNN.

Keywords: Object detection · Context embedding · Region-based CNNs

1 Introduction

The region-based object detectors [4,13,14,23,33,39] popularized by R-CNN framework [14] are conceptually intuitive and flexible, and have achieved top

Electronic supplementary material The online version of this chapter (https://doi.org/10.1007/978-3-030-58589-1_38) contains supplementary material, which is available to authorized users.

© Springer Nature Switzerland AG 2020
A. Vedaldi et al. (Eds.): ECCV 2020, LNCS 12366, pp. 633–648, 2020.
https://doi.org/10.1007/978-3-030-58589-1_38

| Baseline | Ours | Baseline | Ours |

(a) Filtering out noisy detections.   (b) Recognizing indistinctive objects.

Fig. 1. Motivation and example results of our Hierarchical Context Embedding (HCE) framework. By incorporating discriminative context information, our framework can effectively filter out the noisy false positive background detections, and correctly classify objects (*e.g.*, "skateboard") which possess no distinctive appearances.

accuracies on challenging benchmarks like MS-COCO [25]. Region-based detectors first generate a sparse set of object proposals, and then refine the proposal locations and classify them as one of the foreground classes or as background using a detection head. One crucial module in such a proposal-driven pipeline is the RoIPool [13] or RoIAlign [18] operator, which is responsible for extracting RoI (Region of Interests) features aligned with the proposal locations for the detection head.

In this paper, we revisit the RoI features in region-based detectors from the perspective of context information embedding. Our key motivation relies on the fact that while each RoI in very deep CNNs may have a very large theoretical receptive field that often spans the entire input image [13]. However, the effective receptive field [28] may only occupy a fraction of the full theoretical receptive field, making the RoI features insufficient for characterizing objects that are highly dependent on context information, such as "bowl", "skateboard" etc. Here, the contextual information means any auxiliary information that can assist in suppressing the false positive detections in noisy backgrounds, or recognizing objects that have no distinctive appearances themselves. For example, as shown in Fig. 1(a), the semantic features of "motorcycle" are strong evidences for filtering out the activations of irrelevant object categories like "spoon", "bowl", and "sink". On the other hand, as shown by Fig. 1(b), the scene and even the human pose are useful clues for correctly classifying a proposal as "skateboard", rather than "tennis racket".

Recently, several works exploited the region-level context information to improve the localization ability of two-stage detectors. Chen *et al.* [6] demonstrated that rich contextual information from neighboring regions can better refine the proposal locations for two-stage detectors. Kantorov *et al.* [20] leveraged the surrounding context regions to improve weakly supervised object localization. However, to the best of our knowledge, currently there is no enabling framework which is systematically designed for embedding context information to improve the *classification ability* of region-based detectors.

In this paper, we present a novel Hierarchical Context Embedding (HCE) framework for region-based object detectors. Our framework consists of three modules. Firstly, we consider that the simplest way to break the contextual limit in object detection, is to partially cast the object-level feature learning as an image-level multi-label classification task. Building upon this realization, we design an image-level categorical embedding module, which in essence is a multi-label classifier upon the detection backbone, in parallel with the existing region-based detection head. It enables the backbone to exploit the whole image context to learn discriminative features for context-dependent object categories. Even as a standalone enhancement, our image-level categorical embedding module can lead to improvements over existing region-based detectors.

Upon the image-level categorical embedding module, at the instance-level, we design a simple but effective process to generate hierarchical contextual RoI features which can be directly utilized by the region-wise detection head. Because our contextual RoI features are enhanced by image-level categorical supervisions and exploit larger contexts, they are by nature complementary to conventional RoI features, which is trained by region-based detectors and only exploits limited context. Later, the early-and-late strategies, *i.e.*, feature fusion and confidence fusion, are designed to make full use of our contextual RoI features. By quantitative experiments, we demonstrate that they can be combined to further boost the classification accuracy of the detection head.

In general, our proposed HCE framework is easy to implement and is end-to-end trainable. We conduct extensive experiments on MS-COCO 2017 [25] to validate the effectiveness of our HCE framework. Without bells and whistles, our HCE framework delivers consistent accuracy improvements for almost all existing mainstream region-based detectors, including FPN [23], Mask R-CNN [18] and Cascade R-CNN [4]. We also conduct ablation studies to verify the effectiveness of each module involved in our HCE framework. Figure 1 gives the example images of the baseline method and our method, which demonstrates that our framework can effectively filter out the noisy background detections and correctly classify indistinctive objects by leveraging the context information it exploited.

2   Related Work

2.1   Region-Based Object Detection

Convolutional neural networks have lead to a paradigm shift of object detection in the past decades [26]. Among a large number of approaches, the two-stage R-CNN series [4,13,14,23,33] have become the leading detection framework. The pioneer work, *i.e.*, R-CNN [14], extracts region proposals from image with selective search [36], and applies a convolutional network to classify each region of interests independently. Fast R-CNN [13] improves R-CNN by sharing convolutional features among RoIs, which enables fast training and inference. Then, Faster R-CNN [33] advances the region proposal generation with a Region Proposal Network (RPN). RPN shares the feature extraction backbone with the detection head, which in essence is a Fast R-CNN [33]. Faster R-CNN is a famous two-stage detection framework, and is the foundation for many follow-up works [8, 23].

Over very recent years, several algorithms have been proposed to further improve the two-stage Faster R-CNN framework. For example, Feature Pyramid Networks (FPN) [23] constructed inherent CNN feature pyramids, which can largely improve the detection performance of small objects. Mask R-CNN [18] extended Faster RCNN by constructing the mask branch, and boosted the performance of both object detection and instance segmentation. Cascade R-CNN [4] utilized multi-stage training strategy to progressively improve the quality of region proposals, and demonstrated significant gains for high quality (measured by higher IoUs) object detection. Complementary to these works, in this paper, we focus on developing a Hierarchical Context Embedding (HCE) framework to improve the *classification ability* of all region-based detectors. Thanks to the simplicity and generalization ability of our HCE framework, it brings consistent and significant improvements over aforementioned leading region-based detectors, *e.g.*, FPN, Mask R-CNN and Cascade R-CNN.

2.2   Context Information for Object Detection

In object detection, both global context [12] and local context [30] are widely exploited for improving performance, especially when object appearances are insufficient due to small object size, occlusion, or poor image quality. Our work is inspired by some of previous works, but the key motivation or implementation significantly differ with these works. Next, we review several topics in object detection, which are closely related to our work.

Combined Localization and Classification. Before the era of deep learning, Harzallah *et al.* [16] proposed to combine two closely related tasks, *i.e.*, object localization and image classification. They demonstrated that classification can improve detection by a contextual combination and vice versa. Similar in spirit, we utilize the fully image-level context to learn object-level concepts. But differently, we utilize global context to learn CNN features rather than hand-crafted features adopted in [16]. Furthermore, we integrate hierarchical contextual clues beneath both whole images and interested regions to modern region-based CNN detectors, rather than the traditional sliding window detector used by [16].

Region Proposal Refinement. Recently, Chen *et al.* [6] explored the rich contextual information to refine the region proposals for object detection. The neighboring regions with useful contexts can benefit the localization quality of region proposals, which further lead to better detection performance. Instead of refining proposals, we focus on improving the *classification ability* of region-based detectors by embedding hierarchical contextual clues.

Weakly-Supervised Object Detection. In weakly supervised object detection, the bounding box annotations are not provided, and only image-level categorical labels are available. The common practice [1, 2, 7, 20] in this area is to first generate a set of noisy object proposals, and then learn from these noisy proposals with specially designed robust algorithms. Among them, Kantorov *et al.* [20] proposed a context-aware deep network which leverages the surrounding context regions to improve localization. Unlike the usage of region-level context information [20] for weakly supervised detection, we focus on the task of fully-supervised object detection, and particularly exploit global image-level context to advance the recognition of context-dependent object categories.

2.3   Context Information for Other Vision Tasks

Beyond object detection, context information has also been utilized to improve other vision tasks. For example, Wang *et al.* [37] leveraged attention mechanisms and LSTMs to discover semantic-aware regions and capture the long-range contextual dependencies for multi-label image recognition. He *et al.* [17] proposed an adaptive context module to generate multi-scale context representations for semantic segmentation. Qu *et al.* [29] embedded multi-context information (the appearance of the input image and semantic understanding) to obtain the shadow matte. Byeon *et al.* [3] leveraged the LSTM units to capture the entire available past context on video prediction. Li *et al.* [21] adopted the dilated convolution to acquire more contextual information for single image deraining.

3   Approach

3.1   Framework Overview

We begin by briefly describing our Hierarchical Context Embedding (HCE) framework (see Fig. 2) for region-based object detection. Firstly, an image-level categorical embedding module is employed to advance the feature learning of the objects that are highly dependent on larger context clues. Then, hierarchical contextual RoI features are generated by fusing both instance-level and global-level information derived from the image-level categorical embedding module. Finally, early-and-late fusion modules are designed to make full use of the contextual RoI features to improve the classification performance. Our HCE framework is flexible and generalizable, as it can be applied as a plug-and-play component for almost all mainstream region-based object detectors.

3.2   Image-Level Categorical Embedding

As aforementioned, conventional RoI-based training for region-based detectors may lack the context information, which is crucial for learning discriminative filters for context-dependent objects. To break this limitation, in parallel with the RoI-based branch, we exploit image-level categorical embedding upon the detection backbone, enabling the backbone to learn object-level concepts adaptively from *global-level* context. Our image-level categorical embedding module does

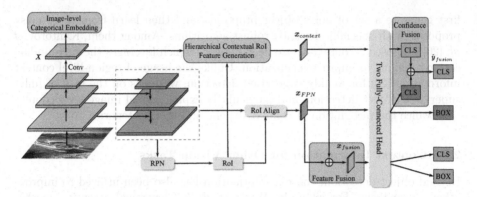

Fig. 2. Overview of our Hierarchical Context Embedding (HCE) framework. At the image-level, we design an *image-level categorical embedding* module upon the detection backbone, which enables the network to learn object-level concepts from global-level context. At the instance-level, we generate *hierarchical contextual RoI features* that are complementary to conventional RoI features, and design the early-and-late fusion strategies (*i.e., feature fusion* and *confidence fusion*) to make full use of the contextual RoI features for improving the classification accuracy of the detection head.

not require additional annotations, as the image-level labels can be conveniently obtained by collecting all instance-level categories in an image.

Essentially, our image-level categorical embedding module is based on a multi-label classifier. As shown in Fig. 2 and Fig. 3(a), we first apply a 3×3 convolution layer on the output of ResNet $conv_5$ to obtain the input feature map, and then employ both global max-pooling (GMP) and global average-pooling (GAP) for feature aggregation (as in [38]). Here, the additional 3×3 convolution layer aims to alleviate the possible slide effects over the original detection backbone.

We refer to the input feature map to our image-level embedding module as *context-embedded image feature*. This is because the input feature map conveys whole image context for learning all object categories that appear in the image, and in turn, each location of the feature map is supervised by all object categories. By contrast, conventional RoI-based trained by region-based detectors only exploits limited context for learning each object category.

Formally, let $\boldsymbol{X} \in \mathbb{R}^{d \times h \times w}$ denote the input feature map, where d is the channel dimensionality, h and w are the height and width, respectively. Then, the multi-label classifier is constructed by C binary classifiers for all categories:

$$\hat{\boldsymbol{y}}_{cls} = f_{cls}((f_{gmp}(\boldsymbol{X}) + f_{gap}(\boldsymbol{X}))) \in \mathbb{R}^C , \qquad (1)$$

where C denotes the number of categories, each element of $\hat{\boldsymbol{y}}_{cls}$ is a confidence score (logits), and f_{cls} is binary classifier modeled as one fully-connected layer. We assume that the ground truth label of an image is $\boldsymbol{y} \in \mathbb{R}^C$, where $y^i = \{0, 1\}$ denotes whether object of category i appears in the image or not. The multi-label loss can be formulated as follows

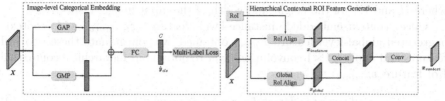

(a) Image-level categorical embedding.   (b) Hierarchical contextual RoI feature.

Fig. 3. The design of our image-level categorical embedding module and hierarchical contextual RoI feature generation module.

$$\mathcal{L}_{mll} = - \sum_{c=1}^{C} y^c \log(\sigma(\hat{y}_{cls}^c)) + (1 - y^c) \log(1 - \sigma(\hat{y}_{cls}^c)), \tag{2}$$

where $\sigma(\cdot)$ is the sigmoid function.

Because the global feature learning strategy is complementary to RoI-based training, our image-level categorical embedding module standalone can boost the performance of existing region-based detectors (demonstrated later by experiments, cf. Table 2). However, one limitation of image-level categorical embedding might be that the derived context-embedded image feature can not be directly used by the detection head.

3.3 Hierarchical Contextual RoI Feature Generation

To further benefit region-wise classification, we generate hierarchical contextual RoI features by combining the instance-level and global-level information from the context-embedded image features. The hierarchical contextual RoI feature generation process is shown in Fig. 3(b).

Context-Embedded Instance-Level Feature. We apply RoIAlign [18] with proposals generated by RPN on the context-embedded feature map X to obtain RoI features $x_{instance}$:

$$x_{instance} = f_{RoIAlign}(X; h', w') \in \mathbb{R}^{d \times 7 \times 7}, \tag{3}$$

where $f_{RoIAlign}(\cdot)$ is the RoIAlign operation and h' and w' are the height and width of the RoI, respectively. As $x_{instance}$ is extracted from the context-embedded image feature X, we term it as *context-embedded instance-level feature*.

Context-Aggregated Global-Level Feature. To leverage larger context, we exploit RoIAlign on the context-embedded image feature X to aggregate the global-level context. We refer to the derived RoI feature as *context-aggregated global-level feature* x_{global}:

$$x_{global} = f_{RoIAlign}(X; H, W) \in \mathbb{R}^{d \times 7 \times 7}, \tag{4}$$

where H and W are the height and width of the input image, respectively.

Once context-embedded instance-level feature $x_{instance}$ and context-aggregated global-level feature x_{global} obtained, we concatenate these two RoI features and apply a 1×1 convolution layer to obtain our hierarchical contextual RoI feature $x_{context}$:

$$x_{context} = f_{conv}([x_{instance} : x_{global}]) \in \mathbb{R}^{d \times 7 \times 7}, \tag{5}$$

where $f_{conv}(\cdot)$ denotes the 1×1 convolution operation, [:] refers to concatenation and the ReLU nonlinearity operations are performed following the convolution layer. As the resulting hierarchical contextual RoI feature $x_{context}$ absorbs rich context information from the context-embedded image feature X, it is by nature complementary to the conventional RoI feature extracted from the feature pyramid network (FPN) [23].

3.4   Early-and-Late Fusion and Inference

To make full use of our contextual RoI feature $x_{context}$, we design the early-and-late fusion strategies, *i.e.*, feature fusion and confidence fusion, which has been proven effective in many applications [11,15]. We show that early-and-late fusion is also well suited to improve region-wise detectors, as it can fully absorb hierarchically embedded information from different levels.

Feature Fusion. To incorporate our contextual RoI features $x_{context}$ into region-based detection pipeline, the simplest way is fusing them with the original RoI features extracted from the feature pyramid network (FPN) [23] with element addition. Formally, let x_{FPN} denote the original RoI feature extracted from FPN, and x_{fusion} denote the fused RoI feature, then we have:

$$x_{fusion} = x_{context} + x_{FPN} \in \mathbb{R}^{d \times 7 \times 7}. \tag{6}$$

As shown in Fig. 2, the fused feature map x_{fusion} is then fed into the $2fc$ detection head to produce refined bounding boxes and classification scores.

Confidence Fusion. We also consider a simple confidence fusion strategy which is complementary to feature fusion. We apply the $2fc$ head on our hierarchical contextual RoI feature $x_{context}$ to produce a classification confidence (logits), and then fuse it with that from the corresponding FPN RoI feature x_{FPN} by addition. Formally, let $\hat{y}_{fusion}$ denote the fused the confidence:

$$\hat{y}_{fusion} = f_{2fc}(x_{context}) + f_{2fc}(x_{FPN}) \in \mathbb{R}^{C}. \tag{7}$$

The fused confidence is transformed by a soft-max layer to produce a novel classification score.

For each proposal, the classification score $\hat{y}_{fusion}$, paired with the refined bounding box predicted the FPN RoI feature, forms another prediction in parallel with the prediction from the feature fusion branch. It is worth mentioning that the weights of the $2fc$ head applied on different RoI features are shared.

Inferences. Our early-and-late fusion strategy produces two different predictions for a single object proposal. To obtain the final result, as shown by the pipeline in Fig. 2, we firstly collect all the boxes and confidences from two prediction branches (*i.e.*, feature fusion and confidence fusion), and then perform NMS over all these boxes. Furthermore, as demonstrated later in experiments, while our two fusion strategies are complementary during training, using only one prediction branch during inference will not cause obvious performance drop but reduce computational cost. However, the performance by only using one fusion strategy for training is inferior to that by using two fusion strategies together.

Loss Function. The whole network is trained end-to-end, and the overall loss is computed as follows:

$$\mathcal{L} = \mathcal{L}_{feat} + \mathcal{L}_{conf} + \mathcal{L}_{mll} + \mathcal{L}_{rpn}, \tag{8}$$

where $\mathcal{L}_{feat}$ and $\mathcal{L}_{conf}$ are the losses for the feature fusion and confidence fusion branches, respectively. All loss terms are considered equally important, without extra hyper-parameters to characterize the trade-off between them, which reveals HCE is generalized and not tricky.

4   Experiments

We conduct extensive experiments on the MS-COCO 2017 dataset [25] to demonstrate the effectiveness and generalization ability of our hierarchical context embedding framework. MS-COCO 2017 is the most popular benchmark for general object detection, which contains 80 object categories, 118K images for training, 5K images for validation (`val`) and 20K for testing (`test-dev`). We report the standard COCO-style Average Precision (AP) with different IoU thresholds from 0.5 to 0.95 with an interval of 0.05 as metric. All models are trained on COCO training set and evaluate on the `val` set. For fair comparisons with the state-of-the-art, we also report the results on the `test-dev` set.

4.1   Implementation Details

We implement our method and re-implement all baseline methods based on MMDetection codebase [5]. The re-implementations of the baselines strictly follow the default settings of MMDetection. Images are resized such that the short edge has 800 pixels while the long edge has less than 1333 pixels. We use no data augmentation except horizontal flipping for training. The ResNet is exploited as backbone, which is pre-trained on ImageNet [10]. Models are trained in a batch size of 16 on 8 GPUs. We train all models with SGD optimizer for 12 epochs in the total, with the initial learning rate as 0.02 and decreased by a factor of 0.1 at 8th epoch and 11th epoch. Weight decay and momentum are set as 0.0001 and 0.9, respectively. We also adopt the linear warming up strategy to begin the training of our model.

Table 1. Compared with baselines (FPN [23], Mask R-CNN [18] and Cascade R-CNN [4]) on MS-COCO 2017 val. "HCE" denotes that the models are trained and inferred on both feature fusion and confidence fusion. Clearly, our HCE framework achieves consistent accuracy gains overall all baseline detectors on all evaluation metrics.

Backbone	Method	HCE	AP	AP^{50}	AP^{75}	AP^S	AP^M	AP^L
ResNet-50-FPN	FPN		36.3	58.3	39.1	21.6	40.2	46.9
		✓	**38.4**	**61.0**	**41.8**	**22.9**	**42.5**	**49.1**
	Mask R-CNN		37.3	59.1	40.3	22.2	41.1	48.3
		✓	**38.8**	**61.3**	**42.1**	**23.2**	**42.8**	**49.7**
	Cascade R-CNN		40.5	58.7	44.1	22.3	43.6	53.8
		✓	**41.7**	**60.5**	**45.0**	**23.4**	**44.9**	**55.2**
ResNet-101-FPN	FPN		38.3	60.1	41.7	22.8	42.8	49.8
		✓	**40.0**	**62.3**	**43.4**	**24.0**	**44.1**	**51.9**
	Mask R-CNN		39.4	60.9	43.0	23.3	43.7	51.5
		✓	**40.5**	**62.6**	**44.0**	**24.4**	**44.5**	**53.4**
	Cascade R-CNN		41.9	60.1	45.7	23.2	45.9	56.2
		✓	**43.0**	**61.6**	**46.9**	**24.6**	**46.6**	**57.4**

4.2 Comparisons with Baselines

To demonstrate the generality of our HCE framework, we consider three well-known region-based object detectors as our baseline systems, including Feature Pyramid Network (FPN) [23], Mask R-CNN [18] and Cascade R-CNN [4]. All detectors are instantiated with two different backbones, *i.e.*, ResNet-50 and ResNet-101 with FPN. Integrating our framework with Mask R-CNN and Cascade R-CNN is as straightforward as with FPN. For example, we apply our framework within each training stage of Cascade R-CNN.

Comparison results on MS-COCO 2017 val are shown in Table 1. Our HCE framework achieves consistent accuracy gains overall all baseline detectors on all evaluation metrics. Specifically, without the bells and whistles, our method improves 2.1% and 1.7% AP for FPN with ResNet-50 and ResNet-101 backbones, respectively. While for more advanced Mask R-CNN and Cascade R-CNN, our method also brings more than 1% AP improvement on both ResNet-50 and ResNet-101 backbones, *e.g.*, improving the AP for Mask R-CNN with ResNet-50-FPN from 37.3% to 38.8%.

Additionally, it can be observed that our improvements for Mask R-CNN and Cascade R-CNN baselines are not as significant as FPN. We conjecture that this is because Mask R-CNN and Cascade R-CNN themselves integrate mechanisms for better feature learning, which might overlap with the performance gains with our method. Specifically, Mask R-CNN benefits from extra accurate instance-level mask supervisions, while Cascade R-CNN enjoys IoU-specific multi-stage training to progressively refine object proposals and learn discriminative features

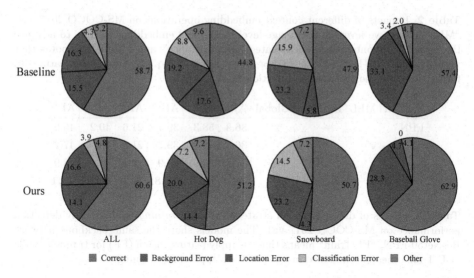

Fig. 4. Error Analyses: These illustrations show the percentage of different error types in the top N detections (N = # objects in that category).

for IoU-specific proposals. However, even in these cases, our method can also obtain +1% AP improvement over these competing baseline methods.

4.3 Error Analyses

In the following, we perform error analyses to further understand in what aspects our HCE framework improves the region-based object detectors. Following the settings of [31], we choose the top N predictions for each category during inference time. Each prediction is classified based on the type of error:

- Correct: correct class and IOU > 0.5
- Location Error: correct class and 0.1 < IOU < 0.5
- Background Error: IOU < 0.1 for any object
- Classification Error: class is wrong and IOU > 0.5
- Other: class is wrong and 0.1 < IOU < 0.5

We compare different error types between the FPN baseline and our method with ResNet-50 as backbone on MS-COCO 2017 val. Figure 4 shows the results of each error type averaged across all 80 categories, and each error type for "hot dog", "snowboard" and "baseball glove" which are highly dependent on context information. Obviously, our method can effectively improve the classification ability of region-based detector and reduce the background errors to a large extent, without compromising the localization performance or increasing other type of errors. Our improvements are particularly noticeable for context-dependent object categories. For example, the (normalized) correctly recognized instances of "hot dog" increase from 44.8% to 51.2%, while the background false positive detections reduce from 17.6% to 14.4%. These observations validate that our HCE framework can indeed improve the classification ability.

Table 2. Impacts of different context embedding operations on MS-COCO 2017 `val`. "MLL" means we leverage the image-level categorical embedding module to advance the learning of context-dependent categories. "Instance" and "Global" denotes that we utilize instance-level (cf. Eq. (3)) or global-level (cf. Eq. (4)) contextual features to further improve the region-wise detection head.

Method	MLL	Instance	Global	AP	AP^{50}	AP^{75}	AP^S	AP^M	AP^L
FPN				36.3	58.3	39.1	21.6	40.2	46.9
	✓			36.8	58.9	39.7	21.9	40.5	47.2
	✓	✓		37.8	59.9	40.9	22.2	41.4	48.9
	✓	✓	✓	**38.4**	**61.0**	**41.8**	**22.9**	**42.5**	**49.1**

Table 3. Effects of different fusion strategies during *training*, evaluated by detection performance on MS-COCO 2017 `val`. The models share the same backbone network ResNet50-FPN. "FF Train" means that we apply feature fusion (FF) for training, while "CF Train" means confidence fusion (CF) are applied for training.

Method	FF Train	CF Train	AP	AP^{50}	AP^{75}	AP^S	AP^M	AP^L
FPN			36.8	58.9	39.7	21.9	40.5	47.2
	✓		37.6	60.3	40.7	22.5	41.4	48.2
		✓	37.4	60.2	40.1	23.0	41.1	47.6
	✓	✓	**38.4**	**61.0**	**41.8**	**22.9**	**42.5**	**49.1**

4.4 Ablation Studies

In this section, we conduct three series of ablation experiments to analyze the proposed method, using ResNet-50 as backbone on MS-COCO 2017 `val`.

Context Embedding Operations. We first investigate the impacts of different context embedding operations in our HCE framework. Specifically, there are three context embedding operations involved in our framework. Firstly, the image-level categorical embedding module employs multi-label learning (denoted as "MLL") to embed global-level context to advance the learning of context-dependent categories. Then, for further improving region-based classification, both the context-embedded instance-level feature (denoted as "Instance") and the context-aggregated global-level feature (denoted as "Global") are combined to generate hierarchical contextual RoI feature.

Table 2 shows the performance improvements by progressively integrating more context embedding operations. Solely applying MLL on the detection backbone gives 0.5% AP improvement. This verifies that image-level categorical embedding advances the feature learning for context-dependent object categories. Then, the context-embedded instance-level feature which can be directly utilized by the detection head brings another 1.0% AP improvement. Finally, global-level context embedding for contextual RoI feature improves 0.6% AP.

Table 4. Effects of different fusion strategies in testing, which are evaluated by the inference time and detection performance on MS-COCO 2017 **val**. Note that all models are trained with both fusion strategies. "FF Test" denotes that we evaluate the feature fusion (FF) strategy during inference, while "CF Test" means the results are evaluated by confidence fusion (CF) strategy. Inference speed is evaluated on a single 1080ti GPU.

Method	FF Test	CF Test	Speed	AP	AP^{50}	AP^{75}	AP^S	AP^M	AP^L
FPN			0.087 s	36.3	58.3	39.1	21.6	40.2	46.9
	✓		0.090 s	38.2	60.8	41.5	22.6	42.2	49.0
		✓	0.094 s	38.3	60.8	41.6	22.8	42.3	49.0
	✓	✓	0.100 s	**38.4**	**61.0**	**41.8**	**22.9**	**42.5**	**49.1**

These results suggest that the context embedding operations in our framework are complementary with each other.

Fusion Strategies in Training. We consider the proposed two fusion strategies, feature fusion and confidence fusion, are complementary to each other. To verify this, we evaluate the performance by training the model with feature fusion and confidence fusion individually, as well as both of them. Table 3 shows the results of different fusion strategies. Specifically, "FF Train" means that we apply feature fusion (FF) for training, while "CF Train" means confidence fusion (CF) are applied for training. Utilizing feature fusion and confidence fusion individually for training can outperform the baseline (FPN with MLL) by 0.8% and 0.6% AP, respectively. Training with both fusion strategies achieves the best result, and is clearly better than using each individual fusion strategy separately.

Fusion Strategies in Testing. We also evaluate each fusion strategy independently during inference, with all HCE models trained with both fusion strategies. Table 4 shows the results of each fusion strategy and the combined fusion strategies. "FF Test" denotes that we evaluate the feature fusion (FF) strategy during inference, while "CF Test" means that the results are evaluated by confidence fusion (CF) strategy. We can see that once the model is trained with both fusion strategies, using only one fusion branch for inference will not cause obvious accuracy drop, but brings computational economy. For example, using the feature fusion branch for inference adds very minimal time cost (0.003s) to the baseline, but increases the AP from 36.3% to 38.2%. These results also prove the complementarity of the proposed two fusion strategies.

4.5 Comparisons with State-of-the-Art

We compare our proposed method with state-of-the-art on MS-COCO 2017 **test-dev**. For fair comparisons, we report the performance of all methods with single-model inference. Specifically, we apply our method on FPN, Mask R-CNN and Cascade R-CNN in 2× training scheme without bells and whistles. Table 5 shows all comparison results.

Table 5. Comparisons with the state-of-the-art single-model detectors on MSCOCO 2017 `test-dev`. "*" denotes using tricks (with bells and whistles) during inference.

Method	Backbone	AP	AP^{50}	AP^{75}	AP^S	AP^M	AP^L
YOLOv3 [32]	Darknet-53	33.0	57.9	34.4	18.3	35.4	41.9
SSD513 [27]	Res101	31.2	50.4	33.3	10.2	34.5	49.8
RetinaNet [24]	Res101-FPN	39.1	59.1	42.3	21.8	42.7	50.2
FCOS [35]	Res101-FPN	41.5	60.7	45.0	24.4	44.8	51.6
FPN [23]	Res101-FPN	36.2	59.1	39.0	18.2	39.0	48.2
Mask R-CNN [18]	Res101-FPN	38.2	60.3	41.7	20.1	41.1	50.2
Cascade R-CNN [4]	Res101-FPN	42.8	62.1	46.3	23.7	45.5	55.2
Deformable R-FCN* [9]	Aligned-Inception-ResNet	37.5	58.0	40.8	19.4	40.1	52.5
DCNv2* [40]	Res101-DeformableV2	46.0	**67.9**	**50.8**	**27.8**	49.1	**59.5**
IoU-Net [19]	Res101-FPN	40.6	59.0	–	–	–	–
TridentNet [22]	Res101	42.7	63.6	46.5	23.9	46.6	56.6
Cascade +Rank-NMS [34]	Res101-FPN	43.2	61.8	47.0	24.6	46.2	55.4
HCE FPN	Res101-FPN	41.0	63.5	44.7	23.4	44.2	52.2
HCE Mask R-CNN	Res101-FPN	41.6	63.9	45.4	23.7	44.7	53.1
HCE Cascade R-CNN	Res101-FPN	44.1	63.2	47.9	25.2	46.9	57.0
HCE Cascade R-CNN*	Res101-FPN	**46.5**	65.6	50.6	27.4	**49.9**	59.4

Our hierarchical context embedding framework, when integrated with FPN, Mask R-CNN and Cascade R-CNN object detectors, consistently outperforms state-of-the-art object detectors using the same backbone network. For fairly comparisons with Deformable R-FCN* and DCNv2* which adopt multi-scale 3x training scheme and multi-scale testing, we follow the same experimental setting to train our HCE Cascade R-CNN*. It gives an AP of 46.5%, which surpasses R-FCN* and DCNv2*. These results demonstrate the superior performance of the proposed context embedding framework.

5   Conclusions

In this paper, we investigated the limitation of context information on conventional region-based detectors, and proposed a novel and effective Hierarchical Context Embedding (HCE) framework to facilitate the classification ability of current region-based detectors. Comprehensive experiments demonstrated the consistent outperforming accuracy on almost all existing mainstream region-based detectors, include FPN, Mask R-CNN and Cascade R-CNN. In the future, we will concentrate in extending the usage scope of our HCE framework and adapting it to one-stage detection paradigm.

Acknowledgements. Z.-M. Chen's contribution was made when he was an intern in Megvii Research Nanjing. This research was supported by the National Key Research and Development Program of China under Grant 2017YFA0700800, the National Natural Science Foundation of China under Grants 61772257 and the Fundamental Research Funds for the Central Universities 020914380080.

References

1. Bilen, H., Pedersoli, M., Tuytelaars, T.: Weakly supervised object detection with convex clustering. In: CVPR, pp. 1081–1089 (2015)
2. Bilen, H., Vedaldi, A.: Weakly supervised deep detection networks. In: CVPR, pp. 2846–2854 (2016)
3. Byeon, W., Wang, Q., Srivastava, R.K., Koumoutsakos, P.: Contextvp: fully context-aware video prediction. In: ECCV, pp. 753–769 (2018)
4. Cai, Z., Vasconcelos, N.: Cascade R-CNN: delving into high quality object detection. In: CVPR, pp. 6154–6162 (2018)
5. Chen, K., et al.: MMDetection: open mmlab detection toolbox and benchmark. arXiv preprint arXiv:1906.07155 (2019)
6. Chen, Z., Huang, S., Tao, D.: Context refinement for object detection. In: ECCV, pp. 71–86 (2018)
7. Cinbis, R.G., Verbeek, J., Schmid, C.: Weakly supervised object localization with multi-fold multiple instance learning. TPAMI **39**(1), 189–203 (2016)
8. Dai, J., Li, Y., He, K., Sun, J.: R-FCN: object detection via region-based fully convolutional networks. In: NIPS, pp. 379–387 (2016)
9. Dai, J., et al.: Deformable convolutional networks. In: ICCV, pp. 764–773 (2017)
10. Deng, J., Dong, W., Socher, R., Li, L.J., Li, K., Fei-Fei, L.: ImageNet: a large-scale hierarchical image database. In: CVPR, pp. 248–255 (2009)
11. Ebersbach, M., Herms, R., Eibl, M.: Fusion methods for ICD10 code classification of death certificates in multilingual corpora. In: CLEF (2017)
12. Galleguillos, C., Belongie, S.: Context based object categorization: a critical survey. CVIU **114**(6), 712–722 (2010)
13. Girshick, R.: Fast R-CNN. In: ICCV, pp. 1440–1448 (2015)
14. Girshick, R., Donahue, J., Darrell, T., Malik, J.: Rich feature hierarchies for accurate object detection and semantic segmentation. In: CVPR, pp. 580–587 (2014)
15. Gunes, H., Piccardi, M.: Affect recognition from face and body: early fusion vs. late fusion. In: SMC, vol. 4, pp. 3437–3443. IEEE (2005)
16. Harzallah, H., Jurie, F., Schmid, C.: Combining efficient object localization and image classification. In: CVPR, pp. 237–244 (2009)
17. He, J., Deng, Z., Zhou, L., Wang, Y., Qiao, Y.: Adaptive pyramid context network for semantic segmentation. In: CVPR, pp. 7519–7528 (2019)
18. He, K., Gkioxari, G., Dollár, P., Girshick, R.: Mask R-CNN. In: ICCV, pp. 2961–2969 (2017)
19. Jiang, B., Luo, R., Mao, J., Xiao, T., Jiang, Y.: Acquisition of localization confidence for accurate object detection. In: ECCV, pp. 784–799 (2018)
20. Kantorov, V., Oquab, M., Cho, M., Laptev, I.: ContextLocNet: context-aware deep network models for weakly supervised localization. In: Leibe, B., Matas, J., Sebe, N., Welling, M. (eds.) ECCV 2016. LNCS, vol. 9909, pp. 350–365. Springer, Cham (2016). https://doi.org/10.1007/978-3-319-46454-1_22
21. Li, X., Wu, J., Lin, Z., Liu, H., Zha, H.: Recurrent squeeze-and-excitation context aggregation net for single image deraining. In: ECCV, pp. 254–269 (2018)
22. Li, Y., Chen, Y., Wang, N., Zhang, Z.: Scale-aware trident networks for object detection. In: ICCV, pp. 6054–6063 (2019)
23. Lin, T.Y., Dollár, P., Girshick, R., He, K., Hariharan, B., Belongie, S.: Feature pyramid networks for object detection. In: CVPR, pp. 2117–2125 (2017)
24. Lin, T.Y., Goyal, P., Girshick, R., He, K., Dollár, P.: Focal loss for dense object detection. In: ICCV, pp. 2980–2988 (2017)

25. Lin, T.-Y., et al.: Microsoft COCO: common objects in context. In: Fleet, D., Pajdla, T., Schiele, B., Tuytelaars, T. (eds.) ECCV 2014. LNCS, vol. 8693, pp. 740–755. Springer, Cham (2014). https://doi.org/10.1007/978-3-319-10602-1_48
26. Liu, L., et al.: Deep learning for generic object detection: a survey. IJCV **128**(2), 1–30 (2019). https://doi.org/10.1007/s11263-019-01247-4
27. Liu, W., et al.: SSD: single shot MultiBox detector. In: Leibe, B., Matas, J., Sebe, N., Welling, M. (eds.) ECCV 2016. LNCS, vol. 9905, pp. 21–37. Springer, Cham (2016). https://doi.org/10.1007/978-3-319-46448-0_2
28. Luo, W.: Understanding the effective receptive field in deep convolutional neural networks. In: NIPS (2016)
29. Qu, L., Tian, J., He, S., Tang, Y., Lau, R.W.: Deshadownet: a multi-context embedding deep network for shadow removal. In: CVPR, pp. 4067–4075 (2017)
30. Rabinovich, A., Vedaldi, A., Galleguillos, C., Wiewiora, E., Belongie, S.: Objects in context. In: ICCV, pp. 1–8 (2007)
31. Redmon, J., Divvala, S., Girshick, R., Farhadi, A.: You only look once: unified, real-time object detection. In: CVPR, pp. 779–788 (2016)
32. Redmon, J., Farhadi, A.: Yolov3: an incremental improvement. arXiv preprint arXiv:1804.02767 pp. 1–6 (2018)
33. Ren, S., He, K., Girshick, R., Sun, J.: Faster R-CNN: towards real-time object detection with region proposal networks. In: NIPS, pp. 91–99 (2015)
34. Tan, Z., Nie, X., Qian, Q., Li, N., Li, H.: Learning to rank proposals for object detection. In: ICCV, pp. 8273–8281 (2019)
35. Tian, Z., Shen, C., Chen, H., He, T.: FCOS: fully convolutional one-stage object detection. In: ICCV, pp. 9627–9636 (2019)
36. Uijlings, J., Van De Sande, K.E., Gevers, T., Smeulders, A.W.: Selective search for object recognition. IJCV **104**(2), 154–171 (2013). https://doi.org/10.1007/252Fs11263-013-0620-5
37. Wang, Z., Chen, T., Li, G., Xu, R., Lin, L.: Multi-label image recognition by recurrently discovering attentional regions. In: ICCV, pp. 464–472 (2017)
38. Woo, S., Park, J., Lee, J.Y., So Kweon, I.: CBAM: convolutional block attention module. In: ECCV, pp. 3–19 (2018)
39. Xu, C.D., Zhao, X.R., Jin, X., Wei, X.S.: Exploring categorical regularization for domain adaptive object detection. In: CVPR, pp. 11724–11733 (2020)
40. Zhu, X., Hu, H., Lin, S., Dai, J.: Deformable convnets v2: more deformable, better results. In: CVPR, pp. 9308–9316 (2019)

Attention-Driven Dynamic Graph Convolutional Network for Multi-label Image Recognition

Jin Ye[1], Junjun He[1,2], Xiaojiang Peng[1], Wenhao Wu[1], and Yu Qiao[1(✉)]

[1] ShenZhen Key Lab of Computer Vision and Pattern Recognition, Shenzhen Institutes of Advanced Technology, Chinese Academy of Sciences, Shenzhen, China
yu.qiao@siat.ac.cn
[2] School of Biomedical Engineering, the Institute of Medical Robotics, Shanghai Jiao Tong University, Shanghai, China

Abstract. Recent studies often exploit Graph Convolutional Network (GCN) to model label dependencies to improve recognition accuracy for multi-label image recognition. However, constructing a graph by counting the label co-occurrence possibilities of the training data may degrade model generalizability, especially when there exist occasional co-occurrence objects in test images. Our goal is to eliminate such bias and enhance the robustness of the learnt features. To this end, we propose an Attention-Driven Dynamic Graph Convolutional Network (ADD-GCN) to dynamically generate a specific graph for each image. ADD-GCN adopts a Dynamic Graph Convolutional Network (D-GCN) to model the relation of content-aware category representations that are generated by a Semantic Attention Module (SAM). Extensive experiments on public multi-label benchmarks demonstrate the effectiveness of our method, which achieves mAPs of 85.2%, 96.0%, and 95.5% on MS-COCO, VOC2007, and VOC2012, respectively, and outperforms current state-of-the-art methods with a clear margin.

Keywords: Multi-label image recognition · Semantic attention · Label dependency · Dynamic graph convolutional network

1 Introduction

Nature scenes usually contains multiple objects. In the computer vision community, multi-label image recognition is a fundamental computer vision task and plays a critical role in wide applications such as human attribute recognition [19], medical image recognition [9] and recommendation systems [15,33]. Unlike single-label classification, multi-label image recognition needs to assign

---

J. Ye, J. He, X. Peng—Equally-contributed first authors.

Electronic supplementary material The online version of this chapter (https://doi.org/10.1007/978-3-030-58589-1_39) contains supplementary material, which is available to authorized users.

© Springer Nature Switzerland AG 2020
A. Vedaldi et al. (Eds.): ECCV 2020, LNCS 12366, pp. 649–665, 2020.
https://doi.org/10.1007/978-3-030-58589-1_39

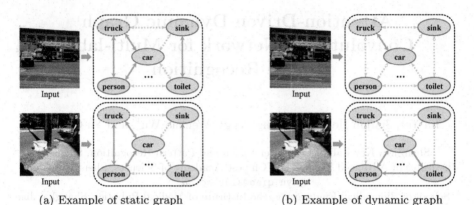

(a) Example of static graph   (b) Example of dynamic graph

Fig. 1. Static graph and dynamic graph. Solid line indicates higher relation and dashed line indicates lower relation of the categories. (a) illustrates all images share a static graph [2,3]. (b) shows our motivation that different image has its own graph that can describe the relations of co-occurred categories in the image.

multiple labels to a single image. Therefore it is reasonable to take account of the relationships of different labels to enhance recognition performance.

Recently, Graph Convolutional Network (GCN) [16] achieves great success in modeling relationship among vertices of a graph. Current state-of-the art methods [2,4] build a complete graph to model the label correlations between each two categories by utilizing prior frequency of label co-occurrence of the target dataset and achieved remarkable results. However, building such a global graph for the whole dataset could cause the frequency-bias problem in most common datasets. As highlighted in [25,26], most prominent vision datasets are afflicted with the co-occur frequency biases despite the best efforts of their creators. Let us consider a common category "car", which always appears with different kind of vehicles such as "truck", "motorbike", and "bus". This may inadvertently cause the frequency-bias in these datasets, which would guide the model to learn higher relations among them. Specifically, as shown in Fig. 1(a), each image share a static graph which is built by calculating the co-occurrence frequency of categories in target dataset. The static graph gives higher relation values between "car" and "truck" and lower ones between "car" and "toilet"in each image. This may result in several problems as follows: 1) failing to identify "car" in a different context such as in the absence of "truck", 2) hallucinating "truck" even in a scene containing only "car", and 3) ignoring "toilet" when "car" co-occurs with "toilet".

Given these issues, our goal is to build a dynamic graph that can capture the content-aware category relations for each image. Specifically, as shown in Fig. 1(b), we construct the image-specific dynamic graph in which "car" and "toilet" has strong connections for the image that "car" and "toilet" appear together and vice versa. To this end, we propose a novel Attention-Driven Dynamic Graph Convolutional Network (ADD-GCN) for multi-label image recognition which leverages content-aware category representations to construct dynamic graph representation. Unlike previous graph based methods [2,4], ADD-GCN models semantic relation for each input image by estimating an image-specific dynamic graph. Specifically, we first decompose the convolutional feature map into multiple content-aware category representations through the Semantic Attention Module (SAM). Then we feed these representations into a Dynamic GCN (D-GCN) module which performs feature propagation via two joint graphs: static graph and dynamic graph. Finally discriminative vectors are generated by D-GCN for multi-label classification. The static graph mainly captures coarse label dependencies over the training dataset and learns such semantic relations as shown in Fig. 1(a). The correlation matrix of dynamic graph is the output feature map of a light-weight network applied upon content-aware category representations for each image, and is used to capture fine dependencies of those content-aware category representations as illustrated in Fig. 1(b).

Our main contributions can be summarized as follows,

- The major contribution of this paper is that we introduce a novel dynamic graph constructed from content-aware category representations for multi-label image recognition. The dynamic graph is able to capture category relations for a specific image in an adaptive way, which further enhance its representative and discriminative ability.
- We elaborately design an end-to-end Attention-Driven Dynamic Graph Convolutional Network (ADD-GCN), which consists of two joint modules. i) Semantic Attention Module (SAM) for locating semantic regions and producing content-aware category representations for each image, and ii) Dynamic Graph Convolutional Network (D-GCN) for modeling the relation of content-aware category representations for final classification.
- Our ADD-GCN significantly outperforms recent state-of-the-art approaches on popular multi-label datasets: MS-COCO, VOC2007, and VOC2012. Specifically, our ADD-GCN achieves mAPs of 85.2% on MS-COCO, 96.0% on VOC2007, and 95.5% on VOC2012, respectively, which are new records on these benchmarks.

2 Related Work

Recent renaissance of deep neural network remarkably accelerates the progresses in single-label image recognition. Convolutional Neural Networks (CNNs) can learn powerful features from large scale image datasets such as MS-COCO [20], PASCAL VOC [7] and ImageNet [6], which greatly alleviates the difficulty of designing hand-crafted features. Recently, many CNN-based approaches have been proposed for multi-label image recognition as well [2,5,10,23,27,30,37], which can be roughly categorized into two main directions as following.

Region Based Methods. One direction aims to first coarsely localize multiple regions and then recognize each region with CNNs [5,10,23,37]. Wei *et al.* [31] propose a Hypotheses-CNN-Pooling (HCP) framework which generates a large number of proposals by objectness detection methods [5,37] and treats each proposal as a single-label image recognition problem. Yang *et al.* [32] formulate the task as a multi-class multi-instance learning problem. Specifically, they incorporate local information by generating a bag of instances for each image and enhance the discriminative features with label information. However, these object proposal based methods lead to numerous category-agnostic regions, which make the whole framework sophisticated and require massive computational cost. Moreover, these methods largely ignore the label dependencies and region relations, which are essential for multi-label image recognition.

Relation Based Methods. Another direction aims to exploit the label dependencies or region relations [2,4,17,27,29,30]. Wang *et al.* [27] propose CNN-RNN framework to predict the final scores and formulate label relation by utilizing Recurrent Neural Network (RNN). Wang *et al.* [30] attempt to discover such relations by iterative locating attention regions with spatial transformer [14] and LSTM [13]. Actually, these RNN/LSTM based methods explore the relation between labels or semantic regions in a sequential way, which cannot fully exploit the direct relations among them. Different from these sequential methods, some works resort to Graphical architectures. Li *et al.* [17] cope with such relations by image-dependent conditional label structures with Graphical Lasso framework. Li *et al.* [18] use a maximum spanning tree algorithm to create a tree-structured graph in the label space. Recently, the remarkable capacity of Graph Convolutional Networks (GCNs) has been proved in several vision tasks, Chen *et al.* [4] utilize GCN to propagate prior label representations (e.g. word embeddings) and generate a classifier, which replaces the last linear layer in a normal deep convolutional neural network such as ResNet [11]. With the help of label annotations, Chen *et al.* [2] compute a probabilistic matrix as the relation edge between each label in a graph. Our work is largely inspired by these GCN based methods for multi-label image recognition. However, instead of using external word embedding for category representations and label statistics for graph construction, our Attention-Driven Dynamic Graph Convolutional Network (ADD-GCN) directly decomposes the feature map extracted by a CNN backbone into content-aware category representations and optimizes the D-GCN, which consists of a static graph for capturing the global coarse category dependencies and a dynamic graph for exploiting content-dependent category relations, respectively.

3   Method

This section presents Attention-Driven Dynamic Graph Convolutional Network (ADD-GCN) for multi-label image recognition. We first give a brief overview of ADD-GCN, and then describe its key modules (Semantic Attention Module and Dynamic GCN module) in details.

3.1   Overview of ADD-GCN

As objects always co-occur in image, how to effectively capture the relations among them is important for multi-label recognition. Graph based representations

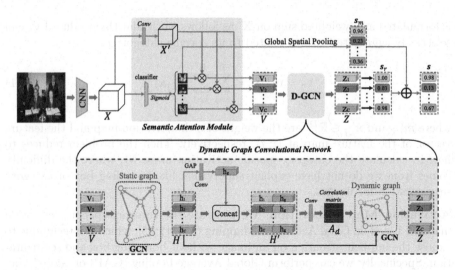

Fig. 2. Overall framework of our approach. Given an image, ADD-GCN first uses a CNN backbone to extract convolutional feature maps $\mathbf{X}$. Then, SAM decouples $\mathbf{X}$ to content-aware category representations $\mathbf{V}$, and D-GCN models global and local relations among $\mathbf{V}$ to generate the final robust representations $\mathbf{Z}$ that contains rich relation information with other categories.

provide a practical way to model label dependencies. We can use nodes $\mathbf{V} = [\mathbf{v}_1, \mathbf{v}_2, \ldots, \mathbf{v}_C]$ to represent labels and correlation matrix $\mathbf{A}$ to represent the label relations (edges). Recent studies [2,4] exploited Graph Convolutional Network (GCN) to improve the performance of multi-label image recognition with a clear margin. However, they construct correlation matrix $\mathbf{A}$ in a static way, which mainly accounts for the label co-occurrence in the training dataset, and is fixed for each input image. As a result, they fail to explicitly utilize the content of each specific input image.

To address this problem, this paper proposes ADD-GCN with two elaborately designed modules: We first introduce Semantic Attention Module (SAM) to estimate content-aware category representation $\mathbf{v}_c$ for each class c from the extracted feature map and the representations are input to another module, Dynamic-GCN, for final classification. We will detail them in the next part.

3.2 Semantic Attention Module

The objective of Semantic Attention Module (SAM) is to obtain a set of content-aware category representations, each of which describes the contents related to a specific label from input feature map $\mathbf{X} \in \mathbb{R}^{H \times W \times D}$. As shown in Fig. 2, SAM first calculates category-specific activation maps $\mathbf{M} = [\mathbf{m}_1, \mathbf{m}_2, \ldots, \mathbf{m}_C] \in \mathbb{R}^{H \times W \times C}$ and then they are used to convert the transformed feature map $\mathbf{X}' \in \mathbb{R}^{H \times W \times D'}$ into the content-aware category representations $\mathbf{V} = [\mathbf{v}_1, \mathbf{v}_2, \ldots, \mathbf{v}_C] \in \mathbb{R}^{C \times D}$. Specifically, each class representation $\mathbf{v}_c$

is formulated as a weighted sum on $\mathbf{X}'$ as follows, such that the produced $\mathbf{v}_c$ can selectively aggregate features related to its specific category c.

$$\mathbf{v}_c = \mathbf{m}_c^T \mathbf{X}' = \sum_{i=1}^{H} \sum_{j=1}^{W} m_{i,j}^c \mathbf{x}_{i,j}', \tag{1}$$

where $m_{i,j}^c$ and $\mathbf{x}_{i,j}' \in \mathbb{R}^{D'}$ are the weight of c th activation map and the feature vector of the feature map at (i,j), respectively. Then the problem reduces to how to calculate the category-specific activation maps $\mathbf{M}$, where the difficulty comes from we do not have explicit supervision like bounding box or category segmentation for images.

Activation Map Generation. We generate the category-specific activation maps $\mathbf{M}$ based on Class Activation Mapping (CAM) [35], which is a technique to expose the implicit attention on an image without bounding box and segmentation. Specifically, we can perform Global Average Pooling (GAP) or Global Max Pooling (GMP) on the feature map $\mathbf{X}$ and classify these pooled features with FC classifiers. Then these classifiers are used to identify the category-specific activation maps by convolving the weights of FC classifiers with feature map $\mathbf{X}$. Unlike CAM, we put a convolution layer as the classifier as well as a $Sigmoid(\cdot)$ to regularize $\mathbf{M}$ before the global spatial pooling, which has better performance in experiments. Ablation studies on these methods are presented in Table 6.

3.3 Dynamic GCN

With content-aware category representations $\mathbf{V}$ obtained in previous section, we introduce Dynamic GCN (D-GCN) to adaptively transform their coherent correlation for multi-label recognition. Recently Graph Convolutional Network (GCN) [16] has been widely proven to be effective in several computer vision tasks and is applied to model label dependencies for multi-label image recognition with a static graph [2,4]. Different from these works, we propose a novel D-GCN to fully exploit relations between content-aware category representations to generate discriminative vectors for final classification. Specifically, our D-GCN consists of two graph representations, static graph and dynamic graph, as shown in Fig. 2. We first revisit the traditional GCN and then detail our D-GCN.

Revisit GCN. Given a set of features $\mathbf{V} \in \mathbb{R}^{C \times D}$ as input nodes, GCN aims to utilize a correlation matrix $\mathbf{A} \in \mathbb{R}^{C \times C}$ and a state-update weight matrix $\mathbf{W} \in \mathbb{R}^{D \times D_u}$ to update the values of $\mathbf{V}$. Formally, The updated nodes $\mathbf{V}_u \in \mathbb{R}^{C \times D_u}$ can be formulated by a single-layer GCN as

$$\mathbf{V}_u = \delta(\mathbf{A}\mathbf{V}\mathbf{W}), \tag{2}$$

where $\mathbf{A}$ is usually pre-defined and $\mathbf{W}$ is learned during training. $\delta(\cdot)$ denotes an activation function, such as the $ReLU(\cdot)$ or $Sigmoid(\cdot)$, which makes the whole operation nonlinear. The correlation matrix $\mathbf{A}$ reflects the relations between the features of each node. During inference, the correlation matrix $\mathbf{A}$ first diffuse the correlated information among all nodes, then each node receives all necessary information and its state is updated through a linear transformation $\mathbf{W}$.

D-GCN. As shown in the bottom of Fig. 2, D-GCN takes the content-aware category representations $\mathbf{V}$ as input node features, and sequentially feeds them into a static GCN and a dynamic GCN. Specifically, the single-layer static GCN is simply defined as $\mathbf{H} = LReLU(\mathbf{A}_s\mathbf{V}\mathbf{W}_s)$, where $\mathbf{H} = [\mathbf{h}_1, \mathbf{h}_2, \dots, \mathbf{h}_C] \in \mathbb{R}^{C \times D_1}$, the activation function $LReLU(\cdot)$ is LeakyReLU, and the correlation matrix $\mathbf{A}_s$ and state-update weights $\mathbf{W}$ is randomly initialized and learned by gradient decent during training. Since $\mathbf{A}_s$ is shared for all images, it is expected that $\mathbf{A}_s$ can capture global coarse category dependencies.

In the next, we introduces dynamic GCN to transform $\mathbf{H}$, whose correlation matrix $\mathbf{A}_d$ is estimated adaptively from input features $\mathbf{H}$. Note this is different from static GCN whose correlation matrix is fixed and shared for all input samples after training, while our $\mathbf{A}_d$ is constructed dynamically dependent on input feature. Since every sample has different $\mathbf{A}_d$, it makes model increase its representative ability and decrease the over-fitting risk that static graph brings. Formally, the output $\mathbf{Z} \in \mathbb{R}^{C \times D_2}$ of the dynamic GCN can be defined as,

$$\mathbf{Z} = f(\mathbf{A}_d\mathbf{H}\mathbf{W}_d), \text{ where } \mathbf{A}_d = \delta(\mathbf{W}_A\mathbf{H}'), \tag{3}$$

where $f(\cdot)$ is the LeakyReLU activation function, $\delta(\cdot)$ is the Sigmoid activation function, $\mathbf{W}_d \in \mathbb{R}^{D_1 \times D_2}$ is the state-update weights, $\mathbf{W}_A \in \mathbb{R}^{C \times 2D_1}$ is the weights of a *conv* layer to formulate the dynamic correlation matrix $\mathbf{A}_d$, and $\mathbf{H}' \in \mathbb{R}^{2D_1 \times C}$ is obtained by concatenating $\mathbf{H}$ and its global representations $\mathbf{h}_g \in \mathbb{R}^{D_1}$, which is obtained by global average pooling and one *conv* layer, sequentially. Formally, $\mathbf{H}'$ is defined as,

$$\mathbf{H}' = [(\mathbf{h}_1; \mathbf{h}_g), (\mathbf{h}_2; \mathbf{h}_g), \dots, (\mathbf{h}_c; \mathbf{h}_g)]. \tag{4}$$

It is worth mentioning that the dynamic graph $\mathbf{A}_d$ is specific for each image which may capture content-dependent category dependencies. Overall, our D-GCN enhances the content-aware category representations from $\mathbf{V}$ to $\mathbf{Z}$ by the dataset-specific graph and the image-specific graph.

3.4   Final Classification and Loss

Final Classification. As shown in Fig. 2, the final category representation $\mathbf{Z} = [\mathbf{z}_1, \mathbf{z}_2, \dots, \mathbf{z}_C]$ is used for final classification. Due to each vector $\mathbf{z}_i$ is aligned with its specific class and contains rich relation information with others, we simply put each category vector into a binary classifier to predict its category score. In particular, we concatenate the score for each category to generate the final score vector $\mathbf{s}_r = [s_r^1, s_r^2, \dots, s_r^C]$. In addition, we can also get another confident scores $\mathbf{s}_m = [s_m^1, s_m^2, \dots, s_m^C]$ through global spatial pooling on the category-specific activation map $\mathbf{M}$ estimated by SAM in Section 3.2. Thus, we can aggregate the two score vectors to predict more reliable results. Here we simply average them to produce the final scores $\mathbf{s} = [s^1, s^2, \dots, s^C]$.

Training Loss. We supervise the final score **s** and train the whole ADD-GCN with the traditional multi-label classification loss as follows,

$$L(\mathbf{y}, \mathbf{s}) = \sum_{c=1}^{C} y^c \log(\sigma(s^c)) + (1 - y^c) \log(1 - \sigma(s^c)), \tag{5}$$

where $\sigma(\cdot)$ is $Sigmoid(\cdot)$ function.

4 Experiments

In this section, we first introduce the evaluation metrics and our implementation details. And then, we compare our ADD-GCN with other existing state-of-the-art methods on three public multi-label image recognition dataset, i.e., MS-COCO [20], Pascal VOC 2007 [7], and Pascal VOC 2012 [7]. Finally, we conduct extensive ablation studies and present some visualization results of the category-specific activation maps and the dynamic graphs.

4.1 Evaluation Metrics

To compare with other existing methods in a fair way, we follow previous works [2,4,36] to adopt the average of overall/per-class precision (OP/CP), overall/per-class recall (OR/CR), overall/per-class F1-score (OF1/CF1) and the mean Average Precision (mAP) as evaluation metrics. When measuring precision/recall/F1-score, the label is considered as positive if its confident score is great than 0.5. Besides, top-3 results of precision/recall/F1-score are also reported. Generally, the OF1, CF1 and mAP are more important than other metrics.

4.2 Implementation Details

For the whole ADD-GCN framework, we use ResNet-101 [11] as our backbone. The channel of **V** is 1024 and the nonlinear activation function LeakyReLU with negative slop of 0.2 is adopted in our SAM and D-GCN. During training, we adopt the data augmentation suggested in [4] to avoid over-fitting: the input image is random cropped and resized to 448×448 with random horizontal flips for data augmentation. To make our model converge quickly, we follow [2] to choose the model that trained on COCO as the pre-train model for Pascal VOC. We choose SGD as our optimizer, with momentum of 0.9 and weight decay of 10^{-4}. The batch size of each GPU is 18. The initial learning rate is set to 0.5 for SAM/D-GCN and 0.05 for backbone CNN. We train our model for 50 epoch in total and the learning rate is reduced by a factor of 0.1 at 30 and 40 epoch, respectively. During testing, we simply resize the input image to 512×512 for evaluation. All experiments are implemented based on PyTorch [22].

Table 1. Comparison of our ADD-GCN and other state-of-the-art methods on MS-COCO dataset. The best results are marked as bold.

Method	All							Top-3					
	mAP	CP	CR	CF1	OP	OR	OF1	CP	CR	CF1	OP	OR	OF1
RARL [1]	–	–	–	–	–	–	–	78.8	57.2	66.2	84.0	61.6	71.1
RDAR [30]	–	–	–	–	–	–	–	79.1	58.7	67.4	84.0	63.0	72.0
Multi-Evidence [8]	-	80.4	70.2	74.9	85.2	72.5	78.4	84.5	62.2	70.6	89.1	64.3	74.7
ResNet-101 [11]	79.7	82.7	67.4	74.3	86.4	71.8	78.4	85.9	60.5	71.0	90.2	64.2	75.0
DecoupleNet [21]	82.2	83.1	71.6	76.3	84.7	74.8	79.5	–	–	–	–	–	–
ML-GCN [4]	83.0	85.1	72.0	78.0	85.8	75.4	80.3	89.2	64.1	74.6	90.5	66.5	76.7
SSGRL [2]	83.8	**89.9**	68.5	76.8	**91.3**	70.8	79.7	**91.9**	62.5	72.7	**93.8**	64.1	76.2
Ours	**85.2**	84.7	**75.9**	**80.1**	84.9	**79.4**	**82.0**	88.8	**66.2**	**75.8**	90.3	**68.5**	**77.9**

4.3 Comparison with State of the Arts

To demonstrate the scalability and effectiveness of our proposed ADD-GCN, extensive experiments are conducted on three widely used benchmarks, i.e., MS-COCO [20], Pascal VOC 2007 [7], and Pascal VOC 2012 [7].

MS-COCO. Microsoft COCO [20] is primarily built for object segmentation and detection, and it is also widely used for multi-label recognition recently. It is composed of a training set with 82081 images, a validation set with 40137 images. The dataset covers 80 common object categories with about 2.9 object labels per image. The number of labels for each image varies considerably, rendering MS-COCO more challenging. Since the labels of the test set are not given, we compare our performance to other previous methods on the validation set.

Table 1 shows the comparison between our ADD-GCN and other state-of-the-art methods. In particular, we compare with RARL [1], RDAR [30], Multi-Evidence [8], ResNet-101 [11], DecoupleNet [21], ML-GCN [4], and SSGRL [2]. Our ADD-GCN consistently outperforms the other state-of-the-art approaches in terms of OF1, CF1, and mAP, as well as some other less important metrics. In particular, both ML-GCN and SSGRL also construct graphs for multi-label classification, our ADD-GCN respectively outperforms ML-GCN by 2.2% and SSGRL by 1.4% in terms of mAP. In addition, our ADD-GCN improves the baseline by 5.5%. This demonstrates the superiority of our approach.

VOC 2007. Pascal VOC 2007 [7] is widely used multi-label dataset, which contains 9963 images from 20 common object categories. It is divided into a train set, a validation set, and a test set. For fair comparisons, following previous works [2,4], we train our model on the trainval set (5011 images) and evaluate on the test set (4952 images). The evaluation metrics are the Average Precision (AP) and the mean of Average Precision (mAP).

The comparison between our ADD-GCN and other methods is presented in Table 2. Our method consistently outperforms these methods with a clear margin, and improves our baseline from 90.8% to 96.0%. Particularly, compared with other two current state-of-the-art methods ML-GCN and SSGRL [2], the gain of overall mAP is 2.0% and 1.0%, respectively.

Table 2. Comparison of our ADD-GCN and other state-of-the-art methods on Pascal VOC 2007 dataset. The best results are marked as bold.

Method	aero	bike	bird	boat	bottle	bus	car	cat	chair	cow	table	dog	horse	mbike	person	plant	sheep	sofa	train	tv	mAP
CNN-RNN [27]	96.7	83.1	94.2	92.8	61.2	82.1	89.1	94.2	64.2	83.6	70.0	92.4	91.7	84.2	93.7	59.8	93.2	75.3	99.7	78.6	84.0
RMIC [12]	97.1	91.3	94.2	57.1	86.7	90.7	93.1	63.3	83.3	76.4	92.8	94.4	91.6	95.1	92.3	59.7	86.0	69.5	96.4	79.0	84.5
RLSD [34]	96.4	92.7	93.8	94.1	71.2	92.5	94.2	95.7	74.3	90.0	74.2	95.4	96.2	92.1	97.9	66.9	93.5	73.7	97.5	87.6	88.5
VeryDeep [24]	98.9	95.0	96.8	95.4	69.7	90.4	93.5	96.0	74.2	86.6	87.8	96.0	96.3	93.1	97.2	70.0	92.1	80.3	98.1	87.0	89.7
ResNet-101 [11]	99.1	97.3	96.2	94.7	68.3	92.9	95.9	94.6	77.9	89.9	85.1	94.7	96.8	94.3	98.1	80.8	93.1	79.1	98.2	91.1	90.8
HCP [31]	98.6	97.1	98.0	95.6	75.3	94.7	95.8	97.3	73.1	90.2	80.0	97.3	96.1	94.9	96.3	78.3	94.7	76.2	97.9	91.5	90.9
RDAR [30]	98.6	97.4	96.3	96.2	75.2	92.4	96.5	97.1	76.5	92.0	87.7	96.8	97.5	93.8	98.5	81.6	93.7	82.8	98.6	89.3	91.9
FeV+LV [32]	98.2	96.9	97.1	95.8	74.3	94.2	96.7	96.7	76.7	90.5	88.0	96.9	97.7	95.9	98.6	78.5	93.6	82.4	98.4	90.4	92.0
RARL [1]	98.6	97.1	97.1	95.5	75.6	92.8	96.8	97.3	78.3	92.2	87.6	96.9	96.5	93.6	98.5	81.6	93.1	83.2	98.5	89.3	92.0
RCP [28]	99.3	97.6	98.0	96.4	79.3	93.8	96.6	97.1	78.0	88.7	87.1	97.1	96.3	95.4	99.1	82.1	93.6	82.2	98.4	92.8	92.5
ML-GCN [4]	99.5	98.5	**98.6**	98.1	80.8	94.6	97.2	98.2	82.3	95.7	86.4	98.2	98.4	96.7	99.0	84.7	96.7	84.3	98.9	93.7	94.0
SSGRL [2]	99.7	98.4	98.0	97.6	85.7	96.2	98.2	**98.8**	82.0	98.1	**89.7**	**98.8**	98.7	97.0	99.0	86.9	98.1	85.8	99.0	93.7	95.0
Ours	**99.8**	**99.0**	98.4	**99.0**	**86.7**	**98.1**	**98.5**	98.3	**85.8**	**98.3**	88.9	**98.8**	**99.0**	**97.4**	**99.2**	**88.3**	**98.7**	**90.7**	**99.5**	**97.0**	**96.0**

Table 3. Comparison of our ADD-GCN and other state-of-the-art methods on Pascal VOC 2012 dataset. The best results are marked as bold.

Methods	aero	bike	bird	boat	bottle	bus	car	cat	chair	cow	table	dog	horse	mbike	person	plant	sheep	sofa	train	tv	mAP
RMIC [12]	98.0	85.5	92.6	88.7	64.0	86.8	82.0	94.9	72.7	83.1	73.4	95.2	91.7	90.8	95.5	58.3	87.6	70.6	93.8	83.0	84.4
VeryDeep [24]	99.1	88.7	95.7	93.9	73.1	92.1	84.8	97.7	79.1	90.7	83.2	97.3	96.2	94.3	96.9	63.4	93.2	74.6	97.3	87.9	89.0
HCP [31]	99.1	92.8	97.4	94.4	79.9	93.6	89.8	98.2	78.2	94.9	79.8	97.8	97.0	93.8	96.4	74.3	94.7	71.9	96.7	88.6	90.5
FeV+LV [32]	98.4	92.8	93.4	90.7	74.9	93.2	90.2	96.1	78.2	89.8	80.6	95.7	96.1	95.3	97.5	73.1	91.2	75.4	97.0	88.2	89.4
RCP [28]	99.3	92.2	97.5	94.9	82.3	94.1	92.4	98.5	83.8	93.5	83.1	98.1	97.3	96.0	98.8	77.7	95.1	79.4	97.7	92.4	92.2
SSGRL [2]	99.7	96.1	97.7	96.5	86.9	95.8	95.0	98.9	88.3	97.6	87.4	99.1	**99.2**	97.3	99.0	84.8	98.3	85.8	99.2	94.1	94.8
Ours	**99.8**	**97.1**	**98.6**	**96.8**	**89.4**	**97.1**	**96.5**	**99.3**	**89.0**	**97.7**	**87.5**	**99.2**	99.1	**97.7**	**99.1**	**86.3**	**98.8**	**87.0**	**99.3**	**95.4**	**95.5**

VOC 2012. Pascal VOC 2012 [7] is the dataset that is widely used for multi-label image recognition task, which consists of 11540 images as trainval set and 10991 as test set from 20 common object categories. For fair comparisons with previous state-of-the-art methods, we train our model on the trainval set and evaluate our results on test set.

We present the AP of each category and mAP over all categories of VOC 2012 in Table 3. Our ADD-GCN also achieves the best performance compared with other state-of-the-art methods. Concretely, the proposed ADD-GCN obtains 95.5% mAP, which outperforms another state-of-the-art SSGRL by 0.7%. And the AP of each category is higher than other methods except "horse". The results demonstrate the effectiveness of our framework.

4.4 Ablation Studies

In this section, we conduct ablation experiments on MS-COCO and VOC 2007.

Evaluation of SAM and D-GCN. To investigate the contribution of each module in ADD-GCN, we separately apply SAM and D-GCN with certain adaptions upon a standard ResNet backbone. We evaluate the effectiveness of SAM by removing D-GCN and adding binary classifiers upon the output of SAM (**V**) directly, while for evaluating the effectiveness of D-GCN, we simply replace SAM with a *Conv-LReLU* block. The results are shown in Fig. 3. As can be seen, on both MS-COCO and VOC-2007, SAM and D-GCN individually improve the baseline with large margins. Compared to the baseline that directly learn classifier upon global-pooled features, SAM first decomposes the feature map into

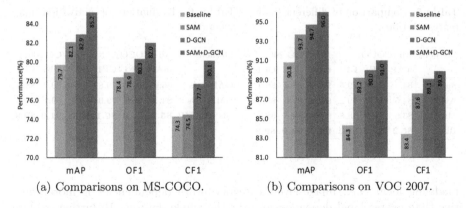

(a) Comparisons on MS-COCO.   (b) Comparisons on VOC 2007.

Fig. 3. Evaluation of SAM and D-GCN on MS-COCO and VOC 2007.

Table 4. The performance of different combinations of static and dynamic graph. "S": static graph and "D": dynamic graph. "P": we propagate information through the static and dynamic graph in a parallel way, and fuse them by either addition (add) or element-wise multiplication (mul) or concatenation (cat).

Methods	All (COCO)			All (VOC 2007)		
	mAP	OF1	CF1	mAP	OF1	CF1
ResNet-101	79.7	78.4	74.3	90.8	84.3	83.4
S	82.9	78.3	74.7	94.5	89.3	88.3
D	83.7	79.4	76.6	94.9	89.9	88.7
P (add)	84.0	79.4	76.9	94.6	88.8	88.2
P (mul)	83.7	80.8	78.5	94.6	89.6	88.5
P (cat)	83.3	80.0	76.9	94.9	89.7	88.8
D→S	84.5	81.4	79.3	95.0	90.1	88.8
S→D	**85.2**	**82.0**	**80.1**	**96.0**	**91.0**	**89.9**

content-aware category representations and train classifiers upon them. The improvement from SAM shows that the decomposed representations are more discriminative. We also find that D-GCN is able to enhance the discriminative ability of features from the results compared with the baseline. Combining SAM and D-GCN further boosts performance as we expect, since they focus on different aspects. Specifically, the gain of the mAP, OF1 and CF1 over the baseline is 5.5%, 3.6% and 5.8% on MS-COCO, while 5.2%, 6.7% and 6.5% on VOC 2007.

Static graph vs Dynamic graph. We investigate the effects of static graph and dynamic graph in D-GCN. Results are shown in Table 4. Firstly, we study the case with only one graph. Both static and dynamic graph can achieve better performance compared with baseline ResNet-101, and the dynamic graph performs better on both MS-COCO and VOC 2007. The results show that

Table 5. Comparison of different final representations.

Methods	All (COCO)			All (VOC 2007)		
	mAP	OF1	CF1	mAP	OF1	CF1
Sum	84.5	81.5	79.5	94.7	89.4	88.4
Avg	84.5	81.5	79.2	94.8	89.6	88.6
Max	83.9	81.2	78.8	94.7	89.6	88.8
Bi	**85.2**	**82.0**	**80.1**	**96.0**	**91.0**	**89.9**

Table 6. Evaluation of activation map generation.

Methods	All (COCO)			All (VOC 2007)		
	mAP	OF1	CF1	mAP	OF1	CF1
GAP→cls	85.0	**82.0**	79.8	94.8	89.7	88.5
GMP→cls	84.1	80.9	79.0	93.9	89.1	87.7
Cls → GMP	**85.2**	**82.0**	**80.1**	**96.0**	**91.0**	**89.9**

modeling local (*i.e.*, image-level) category-aware dependencies is more effective than coarse label dependencies over the whole dataset. To further explore whether the static graph is complementary with the dynamic graph, we attempt to combine them in different ways as shown in Table 4. "S" stands for static graph, "D" donates dynamic graph, "P" denotes that we propagate information through the static graph and dynamic graph in a parallel way, and fuse them by either addition (add) or element-wise multiplication (mul) or concatenation (cat). From the results, "S→D" achieves the best performance among all settings.

Final Representations. To demonstrate the effectiveness and rationality of category-specific feature representations, we compare it with image-level feature representations by aggregating the category-specific feature representations to an image feature vector. For aggregation, Sumation (Sum), Average (Avg) and Maximum (Max) are adopted to fuse category-specific feature representations Z, which are the output of D-GCN for obtaining image-level feature representations. "Bi" means that we utilize binary classifier for each category-specific feature representation to decide whether this class exists or not. Table 5 shows the results that the category-specific feature representations outperforms other aggregated representations on all metrics. Thus, we can believe it is an effective way to represent an input image by decomposing the feature map to category-specific representations for multi-label recognition.

Evaluation of Activation Map Generation. As mentioned in Section 3.2, we first adopt the standard CAM as baseline. Here we compare the final performance of ADD-GCN between our method and the standard CAM. Specifically, CAM can be donated as "GAP→cls" or "GMP→cls", and ours is "cls→GMP". "GAP→cls" equals to "cls→GAP" since the classifier is linear operator. The results are shown in Table 6. Comparing the results of GAP (*i.e.*, GAP→cls) and GMP (*i.e.*, GMP→cls), we believe that GMP loses lots of information as GMP only identify one discriminative part. However, our adaption "cls→GMP" outperforms "GAP→cls", which indicates that the modified GMP(cls→GMP) may compensate for the disadvantages that "GMP→cls" brings.

Fig. 4. Visualization results of category-specific activation maps on MS-COCO.

4.5   Visualization

In this section, we visualize some examples of category-specific activation maps and dynamic correlation matrix $\mathbf{A}_d$ to illustrate whether SAM can locate semantic targets and what relations dynamic graph has learned, respectively.

Visualization of Category-Specific Activation Maps. We visualize original images with their corresponding category-specific activation maps to illustrate the capability of capturing the semantic region of each category appeared in the image with our SAM module. Some examples are shown in Fig. 4, each row presents the original image, corresponding category-specific activation maps and the final score of each category. For the categories appeared in image, we observe that our model can locate their semantic regions accurately. In contrast, the activation map has low activation of categories that the image does not contain. For example, the second row has labels of "person", "tie" and "chair", our ADD-GCN can accurately highlight related semantic regions of the three classes. Besides, the final scores demonstrate that the category-aware representations are discrminative enough, and can be accurately recognized by our method.

Visualization of Dynamic Graph. As shown in Fig. 5, we visualize an original image with its corresponding dynamic correlation matrix $\mathbf{A}_d$ to illustrate what relations D-GCN has learned. For the input image in Fig. 5(a), its ground truths are "car", "dog" and "person". Figure 5(b) is the visualization of the $\mathbf{A}_d$ of the input image. We can find that $\mathbf{A}_d^{car;dog}$ and $\mathbf{A}_d^{car;person}$ rank top (about top

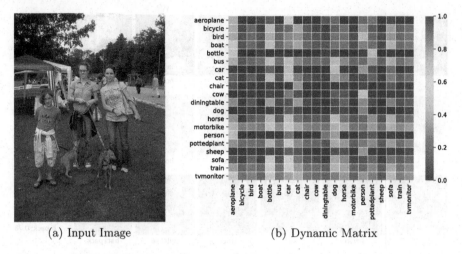

(a) Input Image   (b) Dynamic Matrix

Fig. 5. Visualization of an example and what its dynamic correlation matrix $\mathbf{A}_d$ looks like on Pascal-VOC 2007.

10%) in the row of "car". It means that "dog" and "person" are more relevant for "car" in the image. Similar results can also be found in the rows of "dog" and "person". From the observation of the dynamic graph's visualization, we can believe that D-GCN has capacity to capture such semantic relations for a specific input image.

5   Conclusion

In this work, we propose an Attention-Driven Dynamic Graph Convolutional Network (ADD-GCN) for multi-label image recognition. ADD-GCN first decomposes the input feature map into category-aware representations by the Semantic Attention Module (SAM), and then models the relations of these representations for final recognition by a novel dynamic GCN which captures content-aware category relations for each image. Extensive experiments on public benchmarks (MS-COCO, Pascal VOC 2007, and Pascal VOC 2012) demonstrate the effectiveness and rationality of our ADD-GCN.

Acknowledgements. This work is partially supported by National Natural Science Foundation of China (U1813218, U1713208), Science and Technology Service Network Initiative of Chinese Academy of Sciences (KFJ-STS-QYZX-092), Guangdong Special Support Program (2016TX03X276), and Shenzhen Basic Research Program (JSGG20180507182100698, CXB201104220032A), Shenzhen Institute of Artificial Intelligence and Robotics for Society. We also appreciate Xiaoping Lai and Hao Xing from VIPShop Inc. who cooperate this project with us and provide validation Fashion data.

References

1. Chen, T., Wang, Z., Li, G., Lin, L.: Recurrent attentional reinforcement learning for multi-label image recognition. In: Thirty-Second AAAI Conference on Artificial Intelligence (2018)
2. Chen, T., Xu, M., Hui, X., Wu, H., Lin, L.: Learning semantic-specific graph representation for multi-label image recognition. arXiv preprint arXiv:1908.07325 (2019)
3. Chen, Y., Rohrbach, M., Yan, Z., Shuicheng, Y., Feng, J., Kalantidis, Y.: Graph-based global reasoning networks. In: Proceedings of the IEEE Conference on Computer Vision and Pattern Recognition, pp. 433–442 (2019)
4. Chen, Z.M., Wei, X.S., Wang, P., Guo, Y.: Multi-label image recognition with graph convolutional networks. In: Proceedings of the IEEE Conference on Computer Vision and Pattern Recognition, pp. 5177–5186 (2019)
5. Cheng, M.M., Zhang, Z., Lin, W.Y., Torr, P.: Bing: binarized normed gradients for objectness estimation at 300fps. In: Proceedings of the IEEE Conference on Computer Vision and Pattern Recognition, pp. 3286–3293 (2014)
6. Deng, J., Dong, W., Socher, R., Li, L.J., Li, K., Fei-Fei, L.: Imagenet: a large-scale hierarchical image database. In: 2009 IEEE Conference on Computer Vision and Pattern Recognition, pp. 248–255. IEEE (2009)
7. Everingham, M., Van Gool, L., Williams, C.K., Winn, J., Zisserman, A.: The pascal visual object classes (VOC) challenge. Int. J. Comput. Vision **88**(2), 303–338 (2010). https://doi.org/10.1007/s11263-009-0275-4
8. Ge, W., Yang, S., Yu, Y.: Multi-evidence filtering and fusion for multi-label classification, object detection and semantic segmentation based on weakly supervised learning. In: Proceedings of the IEEE Conference on Computer Vision and Pattern Recognition, pp. 1277–1286 (2018)
9. Ge, Z., Mahapatra, D., Sedai, S., Garnavi, R., Chakravorty, R.: Chest x-rays classification: a multi-label and fine-grained problem. arXiv preprint arXiv:1807.07247 (2018)
10. Girshick, R.: Fast R-CNN. In: Proceedings of the IEEE international conference on computer vision, pp. 1440–1448 (2015)
11. He, K., Zhang, X., Ren, S., Sun, J.: Deep residual learning for image recognition. In: Proceedings of the IEEE Conference on Computer Vision and Pattern Recognition, pp. 770–778 (2016)
12. He, S., Xu, C., Guo, T., Xu, C., Tao, D.: Reinforced multi-label image classification by exploring curriculum. In: Thirty-Second AAAI Conference on Artificial Intelligence (2018)
13. Hochreiter, S., Schmidhuber, J.: Long short-term memory. Neural Comput. **9**(8), 1735–1780 (1997)
14. Jaderberg, M., Simonyan, K., Zisserman, A., et al.: Spatial transformer networks. In: Advances in Neural Information Processing Systems, pp. 2017–2025 (2015)
15. Jain, H., Prabhu, Y., Varma, M.: Extreme multi-label loss functions for recommendation, tagging, ranking & other missing label applications. In: Proceedings of the 22nd ACM SIGKDD International Conference on Knowledge Discovery and Data Mining, pp. 935–944. ACM (2016)
16. Kipf, T.N., Welling, M.: Semi-supervised classification with graph convolutional networks. In: International Conference on Learning Representations (ICLR) (2017)
17. Li, Q., Qiao, M., Bian, W., Tao, D.: Conditional graphical lasso for multi-label image classification. In: Proceedings of the IEEE Conference on Computer Vision and Pattern Recognition, pp. 2977–2986 (2016)

18. Li, X., Zhao, F., Guo, Y.: Multi-label image classification with a probabilistic label enhancement model. In: Proceedings of the Thirtieth Conference on Uncertainty in Artificial Intelligence, UAI 2014, AUAI Press, Arlington, Virginia, United States, pp. 430–439 (2014). http://dl.acm.org/citation.cfm?id=3020751.3020796
19. Li, Y., Huang, C., Loy, C.C., Tang, X.: Human attribute recognition by deep hierarchical contexts. In: Leibe, B., Matas, J., Sebe, N., Welling, M. (eds.) ECCV 2016. LNCS, vol. 9910, pp. 684–700. Springer, Cham (2016). https://doi.org/10.1007/978-3-319-46466-4_41
20. Fleet, D., Pajdla, T., Schiele, B., Tuytelaars, T. (eds.): Microsoft COCO: common objects in context. ECCV 2014. LNCS, vol. 8693, pp. 740–755. Springer, Cham (2014). https://doi.org/10.1007/978-3-319-10602-1_48
21. Liu, L., Guo, S., Huang, W., Scott, M.: Decoupling category-wise independence and relevance with self-attention for multi-label image classification. In: ICASSP 2019, pp. 1682–1686, May 2019. https://doi.org/10.1109/ICASSP.2019.8683665
22. Paszke, A., et al.: Automatic differentiation in pytorch. In: NIPS Workshop (2017)
23. Ren, S., He, K., Girshick, R., Sun, J.: Faster R-CNN: towards real-time object detection with region proposal networks. In: Advances in Neural Information Processing Systems, pp. 91–99 (2015)
24. Simonyan, K., Zisserman, A.: Very deep convolutional networks for large-scale image recognition. arXiv preprint arXiv:1409.1556 (2014)
25. Tommasi, T., Patricia, N., Caputo, B., Tuytelaars, T.: A deeper look at dataset bias. In: Csurka, G. (ed.) Domain Adaptation in Computer Vision Applications. ACVPR, pp. 37–55. Springer, Cham (2017). https://doi.org/10.1007/978-3-319-58347-1_2
26. Torralba, A., Efros, A.A.: Unbiased look at dataset bias. In: CVPR 2011, pp. 1521–1528. IEEE (2011)
27. Wang, J., Yang, Y., Mao, J., Huang, Z., Huang, C., Xu, W.: CNN-RNN: a unified framework for multi-label image classification. In: Proceedings of the IEEE Conference on Computer Vision and Pattern Recognition, pp. 2285–2294 (2016)
28. Wang, M., Luo, C., Hong, R., Tang, J., Feng, J.: Beyond object proposals: random crop pooling for multi-label image recognition. IEEE Trans. Image Process. 25(12), 5678–5688 (2016)
29. Wang, Y., et al.: Multi-label classification with label graph superimposing. arXiv preprint arXiv:1911.09243 (2019)
30. Wang, Z., Chen, T., Li, G., Xu, R., Lin, L.: Multi-label image recognition by recurrently discovering attentional regions. In: Proceedings of the IEEE International Conference on Computer Vision, pp. 464–472 (2017)
31. Wei, Y., et al.: HCP: a flexible CNN framework for multi-label image classification. IEEE Trans. Pattern Anal. Mach. Intell. 38(9), 1901–1907 (2015)
32. Yang, H., Tianyi Zhou, J., Zhang, Y., Gao, B.B., Wu, J., Cai, J.: Exploit bounding box annotations for multi-label object recognition. In: Proceedings of the IEEE Conference on Computer Vision and Pattern Recognition, pp. 280–288 (2016)
33. Yang, X., Li, Y., Luo, J.: Pinterest board recommendation for Twitter users. In: Proceedings of the 23rd ACM International Conference on Multimedia, pp. 963–966. ACM (2015)
34. Zhang, J., Wu, Q., Shen, C., Zhang, J., Lu, J.: Multilabel image classification with regional latent semantic dependencies. IEEE Trans. Multimedia 20(10), 2801–2813 (2018)
35. Zhou, B., Khosla, A., Lapedriza, A., Oliva, A., Torralba, A.: Learning deep features for discriminative localization. In: Computer Vision and Pattern Recognition (2016)

36. Zhu, F., Li, H., Ouyang, W., Yu, N., Wang, X.: Learning spatial regularization with image-level supervisions for multi-label image classification. In: Proceedings of the IEEE Conference on Computer Vision and Pattern Recognition, pp. 5513–5522 (2017)

37. Zitnick, C.L., Dollár, P.: Edge boxes: locating object proposals from edges. In: Fleet, D., Pajdla, T., Schiele, B., Tuytelaars, T. (eds.) ECCV 2014. LNCS, vol. 8693, pp. 391–405. Springer, Cham (2014). https://doi.org/10.1007/978-3-319-10602-1_26

Gen-LaneNet: A Generalized and Scalable Approach for 3D Lane Detection

Yuliang Guo$^{(\boxtimes)}$, Guang Chen, Peitao Zhao, Weide Zhang, Jinghao Miao, Jingao Wang, and Tae Eun Choe

Baidu Apollo, Sunnyvale, CA 94089, USA
`yuliang_guo@alumni.brown.edu`

Abstract. We present a generalized and scalable method, called Gen-LaneNet, to detect 3D lanes from a single image. The method, inspired by the latest state-of-the-art 3D-LaneNet, is a unified framework solving image encoding, spatial transform of features and 3D lane prediction in a single network. However, we propose unique designs for Gen-LaneNet in two folds. First, we introduce a new geometry-guided lane anchor representation in a new coordinate frame and apply a specific geometric transformation to directly calculate real 3D lane points from the network output. We demonstrate that aligning the lane points with the underlying top-view features in the new coordinate frame is critical towards a generalized method in handling unfamiliar scenes. Second, we present a scalable two-stage framework that decouples the learning of image segmentation subnetwork and geometry encoding subnetwork. Compared to 3D-LaneNet, the proposed Gen-LaneNet drastically reduces the amount of 3D lane labels required to achieve a robust solution in real-world applications. Moreover, we release a new synthetic dataset and its construction strategy to encourage the development and evaluation of 3D lane detection methods. In experiments, we conduct extensive ablation study to substantiate the proposed Gen-LaneNet significantly outperforms 3D-LaneNet in average precision (AP) and F-measure.

Keywords: 3D lane detection · Geometry-guided anchor · Two-stage framework · Monocular camera · Unified network

1 Introduction

Over the past few years, autonomous driving has drawn numerous attention from both academic and industry. To drive safely, one of the fundamental problems is to perceive the lane structure accurately in real-time. Robust detection on current lane and nearby lanes is not only crucial for lateral vehicle control and accurate localization [14], but also a powerful tool to build and validate high definition map [8] (Fig. 1).

Electronic supplementary material The online version of this chapter (https://doi.org/10.1007/978-3-030-58589-1_40) contains supplementary material, which is available to authorized users.

ⓒ Springer Nature Switzerland AG 2020
A. Vedaldi et al. (Eds.): ECCV 2020, LNCS 12366, pp. 666–681, 2020.
https://doi.org/10.1007/978-3-030-58589-1_40

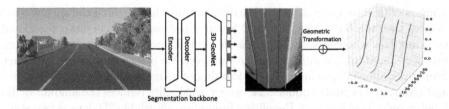

Fig. 1. Procedure of the Gen-LaneNet. A segmentation backbone(*image segmentation subnetwork*) first encodes an input image in deep features and decodes the features into a lane segmentation map. Given the segmentation as input, 3D-GeoNet(*geometry encoding subnetwork*) focuses on geometry encoding and predicts intermediate 3D lane points, specifically represented in top-view 2D coordinates and real heights. At last, the presented geometric transformation directly converts the network output to real-world 3D lane points.

The majority of image-based lane detection methods treat lane detection as a 2D task [1,4,21]. A typical 2D lane detection pipeline consists of three components: A semantic segmentation component, which assigns each pixel in an image with a class label to indicate whether it belongs to a lane or not; a spatial transform component to project image segmentation output to a flat ground plane; and a third component to extract lanes which usually involves lane model fitting with strong assumption,*e.g.*, fitting quadratic curves. By assuming the world is flat, a 2D lane represented in the flat ground plane might be an acceptable approximation for a 3D lane in the ego-vehicle coordinate system. However, this assumption could lead to unexpected problems, as well studied in [2,6]. For example, when an autonomous driving vehicle encounters a hilly road, an unexpected driving behavior is likely to occur since the 2D planar geometry provides incorrect perception of the 3D road.

To overcome the shortcomings associated with planar road assumption, the latest trend of methods [2,5,6,19] has started to focus on perceiving complex 3D lane structures. Specifically, the latest state-of-the-art 3D-LaneNet [6] has introduced an end-to-end framework unifying image encoding, spatial transform between image view and top view, and 3D curve extraction in a single network. 3D-LaneNet shows promising results to detect 3D lanes from a monocular camera. However, representing lane anchors in an inappropriate space makes 3D-LaneNet not generalizable to unobserved scenes, while the end-to-end learned framework makes it highly affected by visual variations.

In this paper, we present **Gen-LaneNet**[1], a generalized and scalable method to detect 3D lanes from a single image. We introduce a new design of geometry-guided lane anchor representation in a new coordinate frame and apply a specific geometric transformation to directly calculate real 3D lane points from the network output. In principle our anchor design is an intuitive extension to the anchors of 3D-LaneNet, yet representing the lane anchor in an appropriate coordinate

---

[1] https://github.com/yuliangguo/Pytorch_Generalized_3D_Lane_Detection.

frame is critical for generalization. We demonstrate that aligning the anchor coordinates with the underlying top-view features in essence breaks down the global encoding of a whole scene to local patch level. Thus it makes the method more robust in handling unfamiliar scenes. Moreover, we present a scalable two-stage framework allowing the independent learning of image segmentation subnetwork and geometry encoding subnetwork, which drastically reduces the amount of 3D labels required for learning. Benefiting from more affordable 2D data, a two-stage framework outperforms end-to-end learnt framework when expensive 3D labels are rather limited to certain visual variations. Besides, we present a highly realistic synthetic dataset of images with rich visual variation, which would serve the development and evaluation of 3D lane detection. Finally in experiments, we conduct extensive ablation study to substantiate that the proposed Gen-LaneNet significantly outperforms state-of-the-art [6] in AP and F-measure, as high as 13% in some test sets.

2   Related Work

Various techniques have been proposed to tackle the lane detection problem. Driven by the effectiveness of Convolutional Neural Network(CNN), lots of recent progress can be observed in improving the 2D lane detection. Some prior methods focus on improving the accuracy of lane segmentation [7,9,10,13,17,18,21,25,26] while others try to improve segmentation and curve extraction in a unified network [16,22]. More delicate network architectures are further developed to unify 2D lane detection and the following projection to planar road plane into an end-to-end learned network architecture [3,7,12,15,20]. However as discussed in Sect. 1, all these 2D lane detectors suffer from the specific planar world assumption. Indeed, even perfect 2D lanes are far from sufficient to imply accurate lane positions in 3D space.

As a better alternative, 3D lane detection methods assume no planar road and thus provide more reliable road perception. However, 3D lane detection is more challenging, because 3D information is generally unrecoverable from a single image. Consequently, existing methods are rather limited and usually based on multi-sensor or multi-view camera setups [2,5,19] rather than monocular camera. [2] takes advantage of both LiDAR and camera sensors to detect lanes in real world. But the high cost and high data sparsity of LiDAR limits its practical usage(e.g., effective detection range is 48 m in [2]). [5,19] apply more affordable stereo cameras to perform the 3D lane detection, but they also suffer from low accuracy of 3D information in the distance.

The current state of the art, 3D-LaneNet [6], predicts 3D lanes from a single image. It has made a first attempt to solve 3D lane detection in a single network unifying image encoding, spatial transform of features and 3D curve extraction. It is realized in an end-to-end learning-based method with a network processing information in two pathways: The *image-view pathway* processes and preserves information from the image while the *top-view pathway* processes features in top-view to output the 3D lane estimations. Image-view pathway features are passed to the top-view pathway through four projective transformation layers which

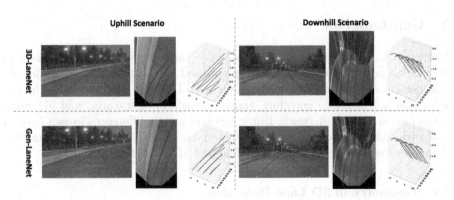

Fig. 2. Comparisons between 3D-LaneNet [6] **and Gen-LaneNet** in two typical scenes with ground height change. We have color-coded ground-truth lanes in blue and predicted lanes in red. Observed from **top-views** in each row, 3D-LaneNet represents anchor points in a coordinate frame not aligned with the underlying visual features(white lane marks). While the proposed Gen-LaneNet resolves this issue.

are conceptually built upon the spatial transform network [11]. Finally, top-view pathway features are fed into a *lane prediction head* to predict 3D lane points. Specifically, anchor representation of lanes has been developed to enable the *lane prediction head* to estimate 3D lanes in the form of polylines. 3D-LaneNet shows promising results in recovering 3D structure of lanes in frequently observed scenes and common imaging conditions, however, its practicality is questionable due to two major drawbacks.

First, 3D-LaneNet uses an inappropriate coordinate frame to represent lane points in anchor representation, in which the ground truth of lane points is misaligned with visual features. This is most evident in the hilly road scenario, where the parallel lanes projected to the virtual top-view appear nonparallel, as observed in the top row of Fig. 2. However the ground-truth lanes (blue lines) are prepared in coordinates not aligned with the underlying visual features (white lane marks). Learning a model against such "corrupt" ground-truth could force the model sort to global encoding of the whole scene. This global encoding behavior in turn may cause the model not generalizable to a new scene partially different from an existing one of the training data.

Second, the end-to-end learning-based framework indeed makes geometric encoding unavoidably affected by the change of image appearance, because it closely couples 3D geometry reasoning with image encoding. As a result, 3D-LaneNet might require exponentially increased amount of training data, in order to reason the same 3D geometry in the presence of occlusion by other traffic participants, varying lighting conditions in day and night, or different weather conditions. Unfortunately labeling 3D lanes is much more expensive than labeling 2D lanes in images. It often requires high-definition map built upon expensive multiple sensors (LiDAR, camera, etc.), accurate localization and online calibration, and even the expensive sensor data manual alignment in 3D space to produce correct ground truth. This further prevents the 3D-LaneNet from being practical in real world.

3   Gen-LaneNet

Motivated by the success of 3D-LaneNet [6] and its drawbacks discussed in Sect. 2, we propose Gen-LaneNet, a generalized and scalable framework for 3D lane detection. Compared to 3D-LaneNet, Gen-LaneNet is still a unified framework that solves image encoding, spatial transform of features, and 3D curve extraction in a single network. But it involves major differences in two folds: a geometric extension to lane anchor design and a scalable two-stage network that decouples the learning of image encoding and 3D geometry reasoning.

3.1   Geometry in 3D Lane Detection

We begin by reviewing the geometry to establish the theory motivating our method. In a common vehicle camera setup as illustrated in Fig. 3(a), 3D lanes are represented in the ego-vehicle coordinate frame defined by x, y, z axes and origin O. Specifically O defines the perpendicular projection of camera center on the road. Following a simple setup, only camera height h and pitch angle θ are considered to represent camera pose which leads to camera coordinate frame defined by x_c, y_c, z_c axes and origin C. A *virtual top-view* can be generated by first projecting a 3D scene to the image plane through a projective transformation and then projecting the captured image to the flat road-plane via a planer homography. Because camera parameters are involved, points in the virtual top-view in principle have different x, y values compared to their corresponding 3D points in the ego-vehicle system. In this paper, we formally considers the virtual top-view as a unique coordinate frame defined by axes $\bar{x}, \bar{y}, z$ and original O. The geometric transformation between virtual top-view coordinate frame and ego-vehicle coordinate frame is derived next.

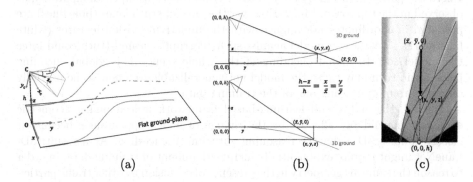

Fig. 3. Geometry in 3D lane detection. (a) Camera setup. **(b)** The co-linear relationship between a 3D lane point (x, y, z), its projection on the virtual top-view $(\bar{x}, \bar{y}, 0)$ and camera center $(0, 0, h)$ holds, no matter $z > 0$ (top) or $z < 0$ (bottom). **(c)** In the virtual top-view, estimating lane height z is conceptually equivalent to estimating the vector field(black arrows) moving top-view lane points (red curves) to their destination positions such that they can form parallel curves (blue curves). (Color figure online)

For a projective camera, a 3D point (x, y, z), its projection on the image plane, and the camera optical center $(0, 0, h)$ should lie on a single ray. Similarly, if a point $(\bar{x}, \bar{y}, 0)$ from the virtual top-view is projected to the same image pixel, it must be on the same ray. Accordingly, camera center $(0, 0, h)$, a 3D point (x, y, z) and its corresponding virtual top-view point $(\bar{x}, \bar{y}, 0)$ appear to be co-linear, as shown in Fig. 3(b) and (c). Formally, the relationship between these three points can be written as:

$$\frac{h - z}{h} = \frac{x}{\bar{x}} = \frac{y}{\bar{y}}. \tag{1}$$

Specifically, as illustrated in Fig. 3(b), this relationship holds no matter z is positive or negative. Thus we derive the geometric transformation from virtual top-view coordinate frame to 3D ego-vehicle coordinate frame as:

$$x = \bar{x} \cdot (1 - \frac{z}{h})$$
$$y = \bar{y} \cdot (1 - \frac{z}{h}), \tag{2}$$

It is worth mentioning that the obtained transformation describes a general relationship without assuming zero yaw and roll angles in camera orientation.

3.2 Geometry-Guided Anchor Representation

Following the presented geometry, we solve 3D lane detection in two steps: A network is first applied to encode the image, transform the features to the virtual top-view, and predict lane points represented in virtual top-view; afterwards the presented geometric transformation is adopted to calculate 3D lane points in ego-vehicle coordinate frame, as shown in Fig. 5. Equation 2 in principle guarantees the feasibility of this approach because the transformation is shown to be independent from camera orientations. This is an important fact to ensures the approach not affected by the inaccurate camera pose estimation.

Anchor representation is the core of a network realization unifying boundary detection and contour grouping in a structured scene because it effectively constrains the search space to a tractable level. Similar to 3D-LaneNet [6], we develop an anchor representation such that a network can directly predict 3D lanes in the form of polylines. Formally, as shown in Fig. 4, lane anchors are defined as N equally spaced vertical lines in x-positions $\{X_A^i\}_{i=1}^N$. Given a set of pre-defined fixed y-positions $\{y_i\}_{j=1}^K$, each anchor X_A^i defines a 3D lane line in $3 \cdot K$ attributes $\{(\bar{x}_j^i, z_j^i, v_j^i)\}_{j=1}^K$ or equivalently in three vectors as $(\mathbf{x}^i, \mathbf{z}^i, \mathbf{v}^i)$, where the values $\bar{x}_j^i$ are horizontal offsets relative to the anchor position and the attribute v_j^i indicates the visibility of every lane point. Denoting lane centerline type with c and lane-line type with l, each anchor can be written as $X_A^i = \{(\mathbf{x}_t^i, \mathbf{z}_t^i, \mathbf{v}_t^i, p_t^i)\}_{t \in \{c, l\}}$, where p_t^i indicates the existence probability of a lane. Based on this anchor representation, our network outputs 3D lane lines in

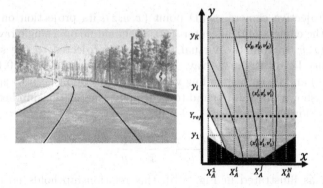

Fig. 4. Anchor representation. Lane anchors are defined as N equally spaced vertical lines in x-positions $\{X_A^i\}_{i=1}^N$. Given a set of pre-defined fixed y-positions $\{y_i\}_{j=1}^K$, a 3D lane can be represented with an anchor X_A^i composed of $3 \cdot K$ attributes $\{(\bar{x}_j^i, z_j^i, v_j^i)\}_{j=1}^K$. Specifically, $\bar{x}_j^i$ indicates x-position in the virtual top-view. A ground-truth lane is associated with its closest anchor based on x-value at Y_{ref}.

the virtual top-view. The derived transformation is applied afterwards to calculate their corresponding 3D lane points. Given predicted visibility probability per lane point, only those visible lane points will be kept in the final output.

Our anchor representation involves two major changes compared to 3D-LaneNet. First, lane point positions are represented in a different coordinate frame, the virtual top-view. This change guarantees the target lane position to align with projected image features, as shown in the bottom row of Fig. 2. Compared to the global encoding of the whole scene as in 3D-LaneNet, establishing the correlation at local patch-level is more robust to novel or unobserved scenes. Even a new scene's overall structure has not been observed from training, those local patches more likely have been. Second, additional attributes are introduced to the representation to indicate the visibility of each anchor point. As a result, our method is more stable in handling partially visible lanes starting or ending in halfway, as observed in Fig. 2.

3.3   Two-Stage Framework with Decoupled Image Encoding and Geometry Reasoning

Instead of adopting an end-to-end learned network, we propose a two-stage framework which decouples the learning of image encoding and 3D geometry reasoning. Basically, the two-stage framework relieves the dependence of 3D geometry on image appearance via introducing an intermediate representation in the form of 2D lane segmentation. As shown in Fig. 5, the first subnetwork focuses on lane segmentation in image domain; the second predicts 3D lane structure from the segmentation outputs of the first subnetwork. The two-stage framework is well motivated by an important fact that the encoding of 3D geometry is rather independent from image features. As observed from Fig. 3(b), ground height z

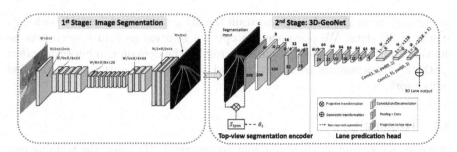

Fig. 5. The two-stage network. An input image is first fed into the image segmentation subnetwork to generate a lane segmentation map with the same resolution. The intermediate segmentation map then goes through 3D-GeoNet, which is composed of the top-view segmentation encoder and the lane prediction head. The output 3D lanes are represented in the virtual top-view. At last, the presented geometric transformation is applied to calculate 3D lane points in the ego-vehicle system.

is mostly correlated to the displacement vector from the position $(\bar{x}, \bar{y})$ to position (x, y). Therefore, estimating the ground heights is conceptually equivalent to estimating a vector field such that all the points corresponding to lanes in the top-view are moved to positions overall in parallelism. It can be anticipated that the geometric information carried by 2D lane segmentation suffices for the 3D lane prediction.

There are a bunch of off-the-shelves candidates [9,21,23,24] to perform 2D lane segmentation in image, any of which could be effortlessly integrated to the first stage of our framework. Although contemporary methods achieve higher performance, we choose ERFNet [24] for its simplicity hence to emphasize the raw power of the two-stage framework. For 3D lane prediction, we introduce **3D-GeoNet**, as shown in Fig. 5, to estimate 3D lanes from image segmentation. The segmentation map is first projected to the top-view and encoded into a top-view feature map through the *top-view segmentation encoder*. Then the *lane prediction head* recovers 3D lane attributes based on the proposed anchor representation. 3D Lane points produced by *lane prediction head* are represented in top-view positions, while 3D lane points in ego-vehicle coordinate frame are calculated afterwards through the introduced geometric transformation.

Decoupling the learning of image encoding and geometry reasoning makes the two-stage framework more cost-effective and scalable. As discussed in Sect. 2, an end-to-end learned framework like [6] is closely keen to image appearance. Consequently, it depends on huge amount of very expensive real-world 3D data for the leaning. On contrary, the two-stage pipeline drastically reduces the cost as it no longer requires to collect redundant real 3D lane labels in the same area under different weathers, day times, and occlusion cases. Moreover, the two-stage framework could leverage on more sufficient 2D real data, *e.g.*, [1,4,21], to train a more reliable 2D lane segmentation subnetwork. With extremely robust segmentation as input, 3D lane prediction would in turn perform better. In an optimal situation, the two-stage framework could train the image segmentation subnetwork from 2D real data and train the 3D geometry subnetwork with only synthetic 3D data. We postpone the optimal solution as future work because domain transfer technique is required to resolve the domain gap between perfect synthetic segmentation ground truth and segmentation output from the first subnetwork.

3.4 Training

Given an image and its corresponding ground-truth 3D lanes, the training proceeds as follows. Each ground-truth lane curve is projected to the virtual top-view, and is associated with the closest anchor at Y_{ref}. The ground-truth anchor attributes are calculated based on the ground-truth values at the pre-defined y-positions $\{y_i\}_{j=1}^K$. Given pairs of predicted anchor X_A^i and corresponding ground-truth $\hat{X}_A^i = \{(\hat{\mathbf{x}}_t^i, \hat{\mathbf{z}}_t^i, \hat{\mathbf{v}}_t^i, \hat{p}_t^i)\}_{t\in\{c,l\}}$, the loss function can be written as:

$$
\begin{aligned}
\ell = & -\sum_{t\in\{c,l\}}\sum_{i=1}^N (\hat{p}_t^i \log p_t^i + (1-\hat{p}_t^i)\log(1-p_t^i)) \\
& + \sum_{t\in\{c,l\}}\sum_{i=1}^N \hat{p}_t^i \cdot (\|\hat{\mathbf{v}}_t^i \cdot (\mathbf{x}_t^i - \hat{\mathbf{x}}_t^i)\|_1 + \|\hat{\mathbf{v}}_t^i \cdot (\mathbf{z}_t^i - \hat{\mathbf{z}}_t^i)\|_1) \\
& + \sum_{t\in\{c,l\}}\sum_{i=1}^N \hat{p}_t^i \cdot \|\mathbf{v}_t^i - \hat{\mathbf{v}}_t^i\|_1
\end{aligned}
\tag{3}
$$

There are three changes compared to the loss function introduced in 3D-LaneNet [6]. First, both $\mathbf{x}_t^i$ and $\hat{\mathbf{x}}_t^i$ are represented in virtual top-view coordinate frame rather than the ego-vehicle coordinate frame. Second, additional cost terms are added to measure the difference between predicted visibility vector and ground-truth visibility vector. Third, cost terms measuring $\bar{x}$ and z distances are multiplied by its corresponding visibility probability v such that those invisible points do not contribute to the loss.

4   Synthetic Dataset and Construction Strategy

Due to lack of 3D lane detection benchmark, we construct a synthetic dataset to develop and validate 3D lane detection methods. Our dataset[2] simulates abundant visual elements and specifically focuses on evaluating a method's generalization capability to rarely observed scenarios. We use Unity game engine to build highly diverse 3D worlds with realistic background elements and render images with diversified scene structure and visual appearance.

The synthetic dataset is rendered from three world maps with diverse terrain information. All the maps are based on real regions within the silicon valley in the United States. Lane lines and center lines involve adequate ground height

---

[2] https://github.com/yuliangguo/3D_Lane_Synthetic_Dataset.

Fig. 6. Examples of synthetic data. Images in the dataset are rendered from different world maps with diverse day-times respectively. In each image, lane lines and center lines are drawn in green and blue separately. Those black-colored segments of lanes in the distance are discarded in a post-process, as background-occluded segments are generally not desired from a lane detection method.

variation and turnings, as shown in Fig. 6. Images are sparsely rendered at different locations and different day-times(morning, noon, evening), under two levels of lane-marker degradation, random camera-height within 1.4–1.8 m and random pitch angles within 0°–10°. We used fixed intrinsic parameters during data rendering and placed a decent amount of agent vehicles driving in the simulation environment, such that the rendered images include realistic occlusions of lanes. In summary, a total of 6000 samples from virtual high-way map, 1500 samples from urban map, and 3000 samples from residential area, along with corresponding depth map, semantic segmentation map, and 3D lane lines information are provided. 3D lane labels are truncated at 200 m distance to the camera, and at the border of the rendered image.

So far, essential information about occlusion is still missing for developing reliable 3D lane detectors. In general, a lane detector is expected to recover the foreground-occluded portion but discard the background-occluded portion of lanes, which in turn requires accurate labeling of the occlusion type per lane point. In our dataset, we use ground-truth depth maps and semantic segmentation maps to deduce the occlusion type of lane points. First, a lane point is considered occluded when its y position is deviated from the value at the corresponding pixel in the depth map. Second, its occlusion type is further determined based on semantic segmentation map. The final dataset keeps the portion of lanes occluded by foreground but discard the portion occluded by background, as the black segments in the distance shown in Fig. 6.

5   Experiments

In the section, we first describe the experimental setups, including dataset splits, baselines, algorithm implementation details, and evaluation metrics. Then we conduct experiments to demonstrate our contributions in ablation. Finally, we design and conduct experiments to substantiate the advantages of our method, compared with prior state of the art [6].

5.1   Experimental Setup

Dataset Setup: In order to evaluate algorithms from different perspectives, we design three different rules to split the synthetic dataset:

(1) *Balanced Scenes:* the training and testing set follow a standard five-fold split of the whole dataset, to benchmark algorithms with massive, unbiased data.

(2) *Rarely Observed Scenes:* This dataset split contains the same training data as *balanced scenes*, but uses only a subset of the testing data, captured from the complex urban map. This dataset split is designed to examine a method's capability of generalization to the test data rarely observed from training. Because the testing images are sparsely rendered at different locations involving drastic elevation change and sharp turnings, the scenes in testing data are rarely observed from the training data.

(3) *Scenes with Visual Variations:* This split of dataset evaluates methods under the change of illumination, assuming more affordable 2D data compared to expensive 3D data is available to cover the illumination change for the same region. Specifically, the same training set as *balanced scenes* is used to train image segmentation subnetwork in the first stage of our Gen-LaneNet. However 3D examples from a certain day time, namely before dawn, are excluded from the training of 3D geometry subnetwork of our method(3D-GeoNet) and 3D-LaneNet [6]. In testing, on contrary, only examples corresponding to the excluded day time are used.

Baselines and Parameters: Gen-LaneNet is compared to two other methods: Prior state-of-the-art 3D-LaneNet [6] is considered as a major baseline; To honestly study the upper bound of our two-stage framework, we treats 3D-GeoNet subnetwork as a stand-alone method which is fed with ground-truth 2D lane segmentation. To conduct fair comparison, all the methods resize the original image into size 360×480 and use the same spatial resolution 208×108 for the first top-view layer to represent a flat-ground region with the range of $[-10, 10] \times [1, 101]$ meters along x and y axes respectively. For the anchor representation, we use y-positions $\{3, 5, 10, 15, 20, 30, 40, 50, 65, 80, 100\}$, where the intervals are gradually increasing due to the fact that visual information in the distance gets sparser in top-view. In label preparation, we set $Y_{ref} = 5$ to associate each lane label with its closest anchor. In training, all the networks are randomly initialized with normal distribution and trained from scratch with Adam optimization and with an initial learning rate $5 \cdot 10^{-4}$. We set batch size 8 and complete training in 30 epochs. For training ERFNet, we follows the same procedure described in [24], but with modified input image size and output segmentation maps sizes. To rule out the error caused by inaccurate camera parameters, we conduct all the experiments given perfect camera intrinsic and extrinsic parameters provided by the synthetic dataset.

Evaluation Metrics: We formulate the evaluation of 3D lane detection as a bipartite matching problem between predicted lanes and ground-truth lanes. The global best matching is sought via minimum-cost flow. Our evaluation method is so far the most strict compared to one-to-many matching in [1] or greedy search bipartite matching in [6].

To handle partial matching properly, we define a new pairwise cost between lanes in euclidean distance. Specifically, lanes are represented in $X^j = \{x_i^j, z_i^j, v_i^j\}_{i=1}^n$ at n pre-determined y-positions, where v^i indicates whether the y-position is covered by a given lane. Denser y-positions compared to the anchor points are used here, which are equally placed from 0 to 100 meters with 2 meter interval. Formally, the lane-to-lane cost between X^j and X^k is calculated as the square root of the squared sum of point-wise distances over all y-positions, written as $cost_{jk} = \sqrt{\sum_i^n d_i^{jk}}$, where

$$
d_i^{jk} = \begin{cases} (x_i^j - x_i^k)^2 + (z_i^j - z_i^k)^2, & \text{if } v_i^j = 1 \text{ and } v_i^k = 1 \\ 0, & \text{if } v_i^j = 0 \text{ and } v_i^k = 0 \\ d_{max}, & \text{otherwise.} \end{cases}
$$

Specifically, point-wise euclidean distance is calculated when a y-position is covered by both lanes. When a y-position is only covered by one lane, the point-wise distance is assigned to a max-allowed distance $d_{max} = 1.5$m. While a y-position is not covered by any of the lanes, the point-wise distance is set to zero. Following such metric, a pair of lanes covering different ranges of y-positions can still be matched, but at an additional cost proportional to the number of *edited* points. This defined cost is inspired by the concept of *edit distance* in string matching. After enumerating all pairwise costs between two sets, we adopt the solver included in Google OR-tools to solve the minimum-cost flow problem. Per lane from each set, we consider it matched when 75% of its covered y-positions have point-wise distance less than the max-allowed distance (1.5 m).

At last, the percentage of matched ground-truth lanes is reported as recall and the percentage of matched predicted lanes is reported as precision. We report both the Average Precision(AP) as a comprehensive evaluation and the maximum F-score as an evaluation of the best operation point in application.

5.2 Anchor Effect

We first demonstrate the superiority of the presented geometry-guided anchor representation compared to [6]. For each candidate method, we keep the architecture exact the same, except the anchor representation integrated. As reported in Table 1, all the three methods, no matter end-to-end 3D-LaneNet [6], "theoretical existing" 3D-GeoNet, or our two-stage Gen-LaneNet, benefit significantly from the new anchor design. Both AP and F-score achieve 3% to 10% improvements, across all splits of dataset.

5.3 The Upper Bound of the Two-Stage Framework

Experiments are designed to substantiate that a two-stage method potentially gains higher accuracy when more robust image segmentation is integrated, and

Table 1. The comparison in anchor representations. "w/o" represents the integration with anchor design in [6], while "w" represents the integration with our anchor design. On three dataset splits, we show the performance gain by integrating our anchor design.

		Balanced scenes			Rarely observed scenes			Scenes with visual variations		
		w/o	w/	gain	w/o	w/	gain	w/o	w/	gain
3D-LaneNet	**F-score**	86.4	90.0	+3.6	72.0	80.9	+8.9	72.5	82.7	+10.5
	AP	89.3	92.0	+2.7	74.6	82.0	+7.4	74.9	84.8	+9.9
3D-GeoNet	**F-score**	88.5	91.8	+3.3	75.4	84.7	+9.3	83.8	90.2	+6.4
	AP	91.3	93.8	+2.5	79.0	86.6	+7.6	86.3	92.3	+6.0
Gen-LaneNet	**F-score**	85.1	88.1	+3.0	70.0	78.0	+8.0	80.9	85.3	+4.4
	AP	87.6	90.1	+2.5	73.0	79.0	+6.0	83.8	87.2	+3.4

meanwhile to localize the upper bound of Gen-LaneNet when perfect image segmentation subnetwork is provided. As shown in Table 2, 3D-GeoNet consistently outperforms Gen-LaneNet and 3D-LaneNet across all three experimental setups. We notice that on *balanced scenes*, the improvement over Gen-LaneNet is pretty obvious, around 3% better, while on *rarely observed scenes* and *scenes with visual variations*, the improvement is significant from 5% to 7%. This observation is rather encouraging because the 3D geometry from hard cases(*e.g.,*new scenes or images with dramatic visual variations) can still be reasoned well from the abstract ground-truth segmentation or from the output of image segmentation subnetwork. Besides, Table 2 also shows promising upper bound of our method, as the 3D-GeoNet outperforms 3D-LaneNet [6] by a large margin, from 5% to 18% in F-score and AP.

Table 2. The upper bound of the two-stage framework, represented by 3D-GeoNet, outperforms the 3D-LaneNet [6] significantly on all the three dataset splits.

		Balanced scenes	Rarely observed scenes	Scenes with visual variations
3D-LaneNet	**F-score**	86.4	72.0	72.5
	AP	89.3	74.6	74.9
3D-GeoNet	**F-score**	91.8	84.7	90.2
	AP	93.8	86.6	92.3
Gen-LaneNet	**F-score**	88.1	78.0	85.3
	AP	90.1	79.0	87.2

5.4   Whole System Evaluation

We conclude our experiments with the whole system comparison between our two-stage Gen-LaneNet with prior state-of-the-art 3D-LaneNet [6]. The apple-to-apple comparisons have been taken on all the three splits of dataset, as shown in Table 3. On the *balanced scenes* the 3D-LaneNet works well, but our Gen-LaneNet still achieves 0.8% AP and 1.7% F-score improvement. Considering this data split is well balanced between training and testing data and covers various scenes, it means the proposed Gen-LaneNet have better generalization on various scenes; On the *rarely observed scenes*, both AP and F-score are improved 6% and 4.4% respectively by our method, demonstrating the superior robustness of our method when it meets uncommon test scenes; Finally on the *scenes with visual variations*, our method significantly surpasses the 3D-LaneNet by around 13% in F-score and AP, which shows that our two-stage algorithm successfully benefits from the decoupled learning of the image encoding and 3D geometry reasoning. For any specific scene, we could annotate more cost-effective 2D lanes in image, to learn a general segmentation subnetwork while label a limited number of expensive 3D lanes to learn the 3D lane geometry. This makes our method a more scalable solution in real-world application. Qualitative comparisons are presented in the supplemental material.

Besides F-score and AP, errors (euclidean distance) in meters over those matched lanes are respectively reported for **near range** (0–40) and **far range** (40–100). This is a complementary evaluation focusing on the quality of the detected portion. As observed, Gen-LaneNet maintains the error lower or on par with 3D-LaneNet, even more matched lanes are involved[3].

Running Time Analysis: The average running speed of Gen-LaneNet is 60 FPS on a single NVIDIA RTX 2080 GPU, compared to 53 FPS of 3D-LaneNet.

Table 3. Whole system comparison between 3D-LaneNet [6] and Gen-LaneNet.

Dataset splits	Method	F-score	AP	x error near (m)	x error far (m)	z error near (m)	z error far (m)
Balanced scenes	3D-LaneNet	86.4	89.3	0.068	0.477	0.015	0.202
	Gen-LaneNet	88.1	90.1	0.061	0.496	0.012	0.214
Rarely observed scenes	3D-LaneNet	72.0	74.6	0.166	0.855	0.039	0.521
	Gen-LaneNet	78.0	79.0	0.139	0.903	0.030	0.539
Scenes with visual variations	3D-LaneNet	72.5	74.9	0.115	0.601	0.032	0.230
	Gen-LaneNet	85.3	87.2	0.074	0.538	0.015	0.232

6 Conclusion

We present a generalized and scalable 3D lane detection method, Gen-LaneNet. A geometry-guided anchor representation has been introduced together with a two-stage framework decoupling the learning of image segmentation and 3D lane prediction. Moreover, we present a new strategy to construct synthetic dataset for 3D lane detection. We experimentally substantiate that our method surpasses 3D-LaneNet significantly in both AP and in F-score from various perspectives.

Acknowledgements. This work was supported by Apollo autonomous driving solution, Baidu USA.

---

[3] A method with higher F-score calculates the errors from more matched lane points.

References

1. Tusimple (2018). https://github.com/TuSimple/tusimple-benchmark
2. Bai, M., Máttyus, G., Homayounfar, N., Wang, S., Lakshmikanth, S.K., Urtasun, R.: Deep multi-sensor lane detection. CoRR (2019)
3. Brabandere, B.D., Gansbeke, W.V., Neven, D., Proesmans, M., Gool, L.V.: End-to-end lane detection through differentiable least-squares fitting. CoRR abs/1902.00293 (2019)
4. Cordts, M., et al.: The cityscapes dataset for semantic urban scene understanding. In: CVPR, pp. 3213–3223, June 2016
5. Coulombeau, P., Laurgeau, C.: Vehicle yaw, pitch, roll and 3D lane shape recovery by vision. In: Intelligent Vehicle Symposium, 2002. IEEE (2002)
6. Garnett, N., Cohen, R., Pe'er, T., Lahav, R., Levi, D.: 3D-lanenet: end-to-end 3D multiple lane detection. In: IEEE International Conference on Computer Vision, ICCV (2019)
7. He, B., Ai, R., Yan, Y., Lang, X.: Accurate and robust lane detection based on dual-view convolutional neutral network. In: Intelligent Vehicles Symposium (2016)
8. Homayounfar, N., Ma, W., Lakshmikanth, S.K., Urtasun, R.: Hierarchical recurrent attention networks for structured online maps. In: 2018 IEEE Conference on Computer Vision and Pattern Recognition, CVPR 2018, Salt Lake City, UT, USA, 18–22 June 2018, pp. 3417–3426 (2018)
9. Hou, Y., Ma, Z., Liu, C., Loy, C.C.: Learning lightweight lane detection CNNs by self attention distillation. In: IEEE International Conference on Computer Vision, ICCV (2019)
10. Huval, B., et al.: An empirical evaluation of deep learning on highway driving. CoRR (2015)
11. Jaderberg, M., Simonyan, K., Zisserman, A., Kavukcuoglu, K.: Spatial transformer networks. In: Advances in Neural Information Processing Systems 28: Annual Conference on Neural Information Processing Systems 2015, Montreal, Quebec, Canada, 7–12 December 2015, pp. 2017–2025 (2015)
12. Kheyrollahi, A., Breckon, T.P.: Automatic real-time road marking recognition using a feature driven approach. Mach. Vis. Appl. **23**, 123–133 (2012). https://doi.org/10.1007/s00138-010-0289-5
13. Kim, J., Lee, M.: Robust lane detection based on convolutional neural network and random sample consensus. In: Loo, C.K., Yap, K.S., Wong, K.W., Teoh, A., Huang, K. (eds.) ICONIP 2014. LNCS, vol. 8834, pp. 454–461. Springer, Cham (2014). https://doi.org/10.1007/978-3-319-12637-1_57
14. Kogan, V., Shimshoni, I., Levi, D.: Lane-level positioning with sparse visual cues. In: 2016 IEEE Intelligent Vehicles Symposium, IV 2016, Gotenburg, Sweden, 19–22 June 2016, pp. 889–895 (2016)
15. Lee, S., et al.: Vpgnet: vanishing point guided network for lane and road marking detection and recognition. In: IEEE International Conference on Computer Vision, ICCV 2017, Venice, Italy, 22–29 October 2017, pp. 1965–1973 (2017)
16. Li, J., Mei, X., Prokhorov, D.: Deep neural network for structural prediction and lane detection in traffic scene. IEEE Trans. Neural Netw. Learn. Syst. **28**, 690–703 (2016)
17. Meyer, A., Salscheider, N., Orzechowski, P., Stiller, C.: Deep semantic lane segmentation for mapless driving. In: IROS (2018). https://doi.org/10.1109/IROS.2018.8594450

18. Măriut, F., Foşalău, C., Petrisor, D.: Lane mark detection using Hough transform. In: 2012 International Conference and Exposition on Electrical and Power Engineering, pp. 871–875 (2012)
19. Nedevschi, S., et al.: 3D lane detection system based on stereovision. In: Proceedings the 7th International IEEE Conference on Intelligent Transportation Systems (IEEE Cat. No.04TH8749), pp. 161–166 (2004)
20. Neven, D., Brabandere, B.D., Georgoulis, S., Proesmans, M., Gool, L.V.: Towards end-to-end lane detection: an instance segmentation approach. In: 2018 IEEE Intelligent Vehicles Symposium, IV 2018, Changshu, Suzhou, China, 26–30 June 2018, pp. 286–291 (2018)
21. Pan, X., Shi, J., Luo, P., Wang, X., Tang, X.: Spatial as deep: Spatial CNN for traffic scene understanding. In: Proceedings of the Thirty-Second AAAI Conference on Artificial Intelligence, (AAAI 2018), New Orleans, Louisiana, USA, 2–7 February 2018, pp. 7276–7283 (2018)
22. Philion, J.: Fastdraw: addressing the long tail of lane detection by adapting a sequential prediction network. In: IEEE Conference on Computer Vision and Pattern Recognition, CVPR 2019, Long Beach, CA, USA, 16–20 June 2019, pp. 11582–11591 (2019)
23. Poudel, R.P.K., Bonde, U., Liwicki, S., Zach, C.: Contextnet: exploring context and detail for semantic segmentation in real-time. BMVC (2018)
24. Romera, E., Alvarez, J.M., Bergasa, L.M., Arroyo, R.: Erfnet: efficient residual factorized convnet for real-time semantic segmentation. IEEE Trans. Intell. Transp. Syst. **19**, 263–272 (2018)
25. Yuan, C., Chen, H., Liu, J., Zhu, D., Xu, Y.: Robust lane detection for complicated road environment based on normal map. IEEE Access **6**, 49679–49689 (2018)
26. Zhang, W., Mahale, T.: End to end video segmentation for driving: lane detection for autonomous car. CoRR (2018)

Sparse-to-Dense Depth Completion Revisited: Sampling Strategy and Graph Construction

Xin Xiong[1], Haipeng Xiong[1], Ke Xian[1], Chen Zhao[1], Zhiguo Cao[1(✉)], and Xin Li[2]

[1] School of AIA, Huazhong University of Science and Technology, Wuhan, China
{xiong_xin,hpxiong,kexian,hust_zhao,zgcao}@hust.edu.cn
[2] Lane Department of CSEE, West Virginia University, Virginia, USA
xin.li@mail.wvu.edu

Abstract. Depth completion is a widely studied problem of predicting a dense depth map from a sparse set of measurements and a single RGB image. In this work, we approach this problem by addressing two issues that have been under-researched in the open literature: sampling strategy (data term) and graph construction (prior term). First, instead of the popular random sampling strategy, we suggest that Poisson disk sampling is a much more effective solution to create sparse depth map from a dense version. We experimentally compare a class of quasi-random sampling strategies and demonstrate that an optimized sampling strategy can significantly improve the performance of depth completion for the same number of sparse samples. Second, instead of the traditional square kernel, we suggest that dynamic construction of local neighborhood is a better choice for interpolating the missing values. More specifically, we proposed an end-to-end network with a graph convolution module. Since the neighborhood relationship of 3D points is more effectively exploited by our novel graph convolution module, our approach has achieved not only state-of-the-art results for depth completion of indoor scenes but also better generalization ability than other competing methods.

Keywords: Depth completion · Graph neural network · Poisson disk sampling · Sparse-to-dense

X-Xiong, H-Xiong Contributed equally.
This work was supported in part by the National Key R&D Program of China (No. 2018YFB1305504) and the National Natural Science Foundation of China (Grant No. U1913602). Xin Li's work is partially supported by the DoJ/NIJ under grant NIJ 2018-75-CX-0032, NSF under grant OAC-1839909, IIS-1951504 and the WV Higher Education Policy Commission Grant (HEPC.dsr.18.5).

Electronic supplementary material The online version of this chapter (https://doi.org/10.1007/978-3-030-58589-1_41) contains supplementary material, which is available to authorized users.

A. Vedaldi et al. (Eds.): ECCV 2020, LNCS 12366, pp. 682–699, 2020.
https://doi.org/10.1007/978-3-030-58589-1_41

1   Introduction

Depth sensing and measurement has become an essential tool supporting a wide range of engineering applications from robotics and autonomous vehicles to 3D vision and augmented reality. Despite decades of research, existing depth sensors still have various limitations. More specifically LiDAen many ways to generalizeR-based (e.g., Velodyne) sensors are expensive and only provide sparse measurements for objects at a distance. Structure-light-based sensors (e.g., Kinect) are sensitive to sunlight and suffer from a short ranging distance (i.e., only suitable for indoor scenes). Stereo-camera based approaches often require a large baseline and careful calibration; and their computational complexity is demanding and performance around featureless regions is often poor. Most recently, 3D from a single image [15,20,39,61–63] has received increasingly more attention because it might lead to a low-cost and energy-efficient solution to 3D/depth sensing.

In view of the limitations of existing depth sensors, it has been suggested in [42] that sparse-to-dense depth completion is a promising remedy in practice. Sparse depth measurements are often readily available from low-cost LiDARs or computed from the output of Simultaneous Localization and Mapping (SLAM). The task of sparse-to-dense is to fill in the missing data and approximate the dense depth map as accurate as possible. Previous works [8,17,42,64] have all assumed that sparse samples are acquired in a random fashion - i.e., observing a Bernoulli distribution spatially [42]. However, the optimality of random-sampling strategy is often questionable. In fact, the sampling locations given by practical LiDAR sensors are seldom randomly distributed, but distributed uniformly in the sampling space due to mechanical spinning [37]. To the best of our knowledge, the issue of sampling strategy has not been studied for depth completion in the open literature.

The other important motivation behind this work is the definition of neighborhood and the construction of network for sparse-to-dense depth completion. Almost all existing works [7,8,12,17,18,42,44,47,65] have adopted a standard spatially-invariant square kernel (refer to Fig. 3b). Attempts to work with spatially varying kernels such as guided-image filter [24] have been made in recent works (e.g., GuideNet [55]) but requires a special implementation of guided convolution module. To manage the prohibitive computational complexity, GuideNet [55] has to count on a factorization strategy to speed up the implementation. How to seamlessly integrate adaptive selection of local neighborhood with network design has remained an under-researched topic.

Based on the above observations, we conduct a systematic study of sampling strategy and neighborhood construction in this paper. On one hand, we propose to leverage the idea of Poisson disk sampling [2] and low-discrepancy sequence [46] to generate a class of deterministic yet quasi-random sampling patterns. When compared with randomly sampled patterns, patterns generated by Poisson disk sampling appear more spatially uniformly distributed (please refer to Fig. 1). On the other hand, inspired by the latest advances in graph neural networks (GNN) [31,58,60,67] we propose a novel GNN-based implementation with the desirable spatially-varying kernels. More specifically, we first construct a k-Nearest-Neighbor(kNN) based neighborhood and compute pointwise features as the inputs for GNN; then the task of depth completion is done by a multi-layer perceptron (MLP) based propagation process on the constructed GNN. Overall, the main technical contributions of our work are summarized as follows:

- For the first time, we demonstrate that optimizing the sampling strategy can significantly improve the performance of sparse-to-dense depth completion. We have systematically compared a class of four competing quasirandom sampling strategies and found that Golden sequence based sampling represents the best approximation of ideal Poisson disk sampling.
- We propose to develop a GNN-based sparse-to-dense depth completion algorithm. Constructed on a kNN-based local neighborhood, our solution is capable of exploiting spatially varying kernel which elegantly fits the graph convolutional module of GNN. We have also developed a MLP-based propagation process for depth completion on the constructed GNN.
- On NYUDv2 benchmark dataset, our GNN-based method with Golden sampling outperforms previous state-of-the-art methods DeepLiDAR [49] and depth-normal constraints [64] by over 25% in terms of RMSE values. On Matterport3D test set, ours outperforms previous state-of-the-art [26] by over 20% in terms of RMSE values.

2   Related Work

Depth Completion. Traditional depth completion methods usually employ hand-craft features or specific kernels to predict missing values [12,18,44,47,65]. While these algorithms might be suitable for a specific task or scene, their performance and generalization capability is often unsatisfactory. Recently, deep learning based methods have shown promising performance on depth completion task - e.g., sparsity-invariant CNN [56], multi-scale sparsity-invariant network [27], confidence propagation CNN [17], color-guided encoder-decoder network [42], self-supervised depth completion from sparse LiDAR data [41], and global/local information fusion [57]. In addition to color information, other methods utilize the surface normal or object boundary to facilitate the task of depth completion - e.g., [26,49,66]. Among them, [26,66] recover depth from coarse depth map with missing values taken by structured-light scanners in indoor scenes. There also exist other competing methods exploiting the correlation among sparse data in the depth map - e.g., [6,8,64].

Spatial Sampling Strategy. Spatial sampling of an unknown function is a widely studied problem in spatial statistics [28,50], computer graphics [2,45], image processing [16], and remote sensing [14]. Unlike regular sampling on integral lattice, irregular sampling such as Poisson disk sampling [2] often demonstrates better coverage of the space in terms of uniformity (e.g., absent from cluster of points within a certain region). Despite the theoretical appeal, practical implementation of Poisson disk sampling is a nontrivial task especially when computational complexity is considered. To address the problem of large computational consumption, Monte Carlo methods [4] have been developed to generate quasi-random sequences and sampling patterns. The quasi-random sequences are also known as low-discrepancy sequences [46] which is constructed in a deterministic manner while ensuring that the whole sampling space is uniformly covered.

Graph Neural Network. Graph Neural Networks (GNNs) are a class of neural networks for modeling graph-based data. Some GNN methods [3,11], apply Convolution Neural Networks(CNNs) to a graph in the spectral domain characterized by the graph Laplacian. Other methods [21,52], aim at recurrently applying neural networks to every node of the graph. When GNNs contain an iterative process, they propagate the node states until reaching the equilibrium and produce an output for each node based on its state in the process. This idea was adopted and improved by [38], which used gated recurrent units [10] in the propagation step. The learning process of these methods can be achieved by the back-propagation through time (BPTT) algorithm [59].

3   Spatial Sampling Strategy.

In this section, we study and compare different quasi-random sampling strategies based on deterministic 1D low-discrepancy sequences. To facilitate the objective comparison, we propose a minimum-radius based criterion to evaluate the effectiveness of different sampling strategies for the depth completion task.

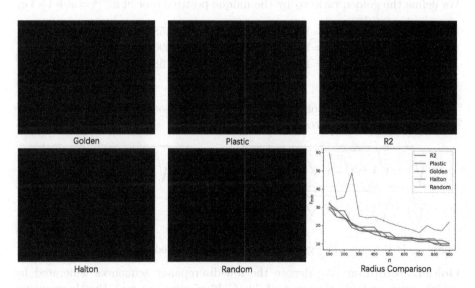

Fig. 1. Visualization of 5 different spatial sampling strategies ($N = 500$) and their performance comparison in terms of radius metric (r_{min}).

3.1   Low-Discrepancy Sequences and Quasi-Random Sampling

The problem of generating deterministic 1D low-discrepancy sequences has been well studied in the literature [1,32]. We opt to work on Van der Corput sequences, a series of low-discrepancy sequences defined on a unit interval [35]. Each Van Der Corput sequence is defined by a unique hyper-parameter b. Note that any natural number n can be represented using b as a radix: $n = \sum_{k=0}^{L-1} d_k(n)b^k$, where $d_k(n)$ is the k-th coefficient and L is the length. Then the n-th number in Van Der Corput reverse-radix sequence is given by $g_b(n) = \sum_{k=0}^{L-1} d_k(n)b^{-k-1}$. For example, the first few terms for $b = 2, b = 3$ are: $V_2 : \{\frac{1}{2}, \frac{1}{4}, \frac{3}{4}, \frac{1}{8}, \frac{5}{8}, \frac{3}{8}, \frac{7}{8}, \frac{1}{16}, \frac{9}{16}, ...\}$, $V_3 : \{\frac{1}{3}, \frac{2}{3}, \frac{1}{9}, \frac{4}{9}, \frac{7}{9}, \frac{2}{9}, \frac{5}{9}, \frac{8}{9}, \frac{1}{27}, ...\}$.

Halton Sampling: Since we want to sample a depth map, 1D low-discrepancy sequences need to be extended into 2D. A simple yet effective solution is to simply consider the Cartesian product of multiple 1D sequences. For instance, Halton sequence in 2D is constructed by using different 1D van de Corput sequences whose bases b's are co-prime. For instance, taking Halton(2,3) as an example, the coordinates of the k-th point in a Halton(2,3) sequence can be written by $Halton : \{(V_2[k], V_3[k])\}_{k=1,2,...}$.

It is also possible to construct low-discrepancy sequence which dose not require choosing any basis parameters. These methods are based on golden ratio [13] and there have been many ways to generalize the golden ratio (e.g., [33]). We define the golden ratio ϕ_d by the unique positive root of $x^{d+1} = x + 1$ - i.e.,

$$
\begin{aligned}
d &= 1, \quad \phi_1 = 1.61803398874989484820458683436563... \\
d &= 2, \quad \phi_2 = 1.32471795724474602596090885447809... \\
d &= 3, \quad \phi_3 = 1.22074408460575947536168534910883... \\
&\quad ...
\end{aligned}
\tag{1}
$$

This special sequence of constants, ϕ_d is called Harmonious numbers [43]. These special values can be expressed very elegantly as:

$$
\phi_1 = \sqrt{1 + \sqrt{1 + \sqrt{1 + \sqrt{1 + \sqrt{1 + ...}}}}}, \phi_2 = \sqrt[3]{1 + \sqrt[3]{1 + \sqrt[3]{1 + \sqrt[3]{1 + \sqrt[3]{1 + ...}}}}}
$$

$$
\phi_3 = \sqrt[4]{1 + \sqrt[4]{1 + \sqrt[4]{1 + \sqrt[4]{1 + \sqrt[4]{1 + ...}}}}}, \qquad ...
\tag{2}
$$

where ϕ_1 is the canonical golden ratio and ϕ_2 is called the plastic constant.

Golden Sampling: We denote the low-discrepancy sequences generated by golden ratio and plastic constant by Golden sequence and Plastic sequence respectively. Then Golden sampling is constructed by:

$$
Golden : \{(\frac{k+m}{N+n}, \frac{k}{\phi_1} \bmod 1)\}_{k=1,2,...}
\tag{3}
$$

where N is the total number of points, m and n are two hyper parameters, and mod denotes the modulo operator. In the classical packing problem, the objective is to maximize the smallest neighboring distance among N disks: $d_N = \min_{i \neq j} |x_i - x_j|$. We have chosen the pair of $m = 6, n = 11$ producing a large d_N in our implementation of Golden sampling.

Plastic Sampling: Similar to Golden sampling, Plastic sampling is defined by:

$$Plastic : \{(\frac{k}{\phi_2} \ mod \ 1, \frac{k\phi_2}{\phi_2 + 1} \ mod \ 1)\}_{k=1,2,...} \qquad (4)$$

R2 Sampling: Another variation is based on plastic number and the coordinates of the k-th point in R2 sampling is given by:

$$R2 : \{((0.5 + \frac{k+1}{\phi_2}) \ mod \ 1, (0.5 + \frac{\phi_2(k+1)}{\phi_2 + 1}) \ mod \ 1)\}_{k=1,2,...} \qquad (5)$$

Figure 1 shows the distribution of five different sampling strategies. It can be seen that ad-hoc random sampling exhibits clustering of points, which does not spread the sampling location uniformly in the 2D space. By contrast, quasi-random sampling more uniformly cover the whole 2D space, which is more desirable for signal reconstruction.

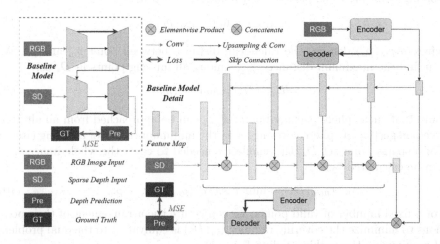

Fig. 2. The architecture of our baseline model for depth completion. Specifically, we substitute the normal convolution in the encoder part of baseline model with the basic residual block structure proposed by [25].

3.2   Quasi-Random Sampling Pattern Comparison and Criterion

How do we compare different quasi-random sampling patterns for sparse-to-dense completion? For low-discrepancy sequences, the definition of *discrepancy* has been articulated in in [46]. Let P denotes a point set consisting of $X_1, ...X_N \in$

I^d, where $I^d = [0,1]^d$ is the normalized integration domain (i.e., a closed d-dimensional unit cube). For an arbitrary subset $B \in I^d$, we can define

$$A(B; P) = \sum_{n=1}^{N} c_B(x_n) \tag{6}$$

where c_B is the characteristic function of B - i.e., $c_B = 1$ if $x_n \in B$; otherwise $c_B = 0$. Then $A(B; P)$ can be deemed as a counting function indicating the number of points falling into B. If $\mathcal{B}$ is a nonempty family of Lebesgue-measurable subsets of I^d, the discrepancy of a point set P is given by:

$$D_N(\mathcal{B}; P) = \sup_{B \in \mathcal{B}} \left| \frac{A(B; P)}{N} - \lambda_d(B) \right| \in [0,1] \tag{7}$$

where λ_d is a d-dimensional Lebesgue measure. However, such definition in the continuous space can not be directly used for discrete data such as depth maps.

Let D_s denotes the *sparse* depth map which requires the completion, and M_s the binary mask of D_s (1/0 denotes valid/missing depth values respectively). For each missing point, we need to gather information from its neighbouring valid points to evaluate the discrepancy. It turns out that the maximum distance to the closest valid points (d^v) can be a plausible approximation of Eq. (7), which is correlated with the overall difficulty of depth completion. The distance of a missing point p_m to its closest valid point is defined by:

$$d^v(p_m) = \min_{p_v} d(p_m, p_v) \tag{8}$$

where $d(p_m, p_v)$ denotes the Euclidean distance between point p_m and p_v. We can find the maximum distance d_{max}^v among all missing points in D_s as:

$$R_{max}(D_s) = \max_{p_m} d^v(p_m) \tag{9}$$

Note that such sphere packing bound d_{max}^v can be approached from an alternative perspective of sphere covering - i.e., the minimum radius of covering circles. For a sparse point set D_s, the smallest r needed to cover the whole image r_{min} can be calculated by:

$$r_{min}(D_s) = \min \ r \ s.t. \ d^v(p_m) \leq r, \ \forall \ p_m \tag{10}$$

For a fixed number of valid points, we conjecture the arrangement of their positions to minimize the covering radius $r_{min}(D_s)$ is equivalent to the dual problem of maximizing the packing radius $R_{max}(D_s)$.

In summary, we suggest that $R_{max} = r_{min}$ can serve as an objective metric for quantifying the performance of sparse sampling pattern D_s on the depth completion task - the smaller this value, the better performance we can achieve (at least in theory). Using this radius as a criterion for evaluating the sampling strategy, we have compared the five sampling strategies as shown in Fig. 1. It can be observed that: 1) all four quasirandom sampling strategies have much better performance than random sampling; 2) all four quasirandom sampling strategies have shown comparable performance. For $n \leq 350$, $R2$ shown slightly better performance; while *Golden* shows better performance for $n \in [350, 750]$.

4   Graph Construction for GNN-Based Depth Completion

To better motivate our network design, we start from a baseline model implementing spatially varying convolution kernel and then proceed with the full model incorporating 3D graph with GNN module. The construction of 3D graph using kNN-based neighborhood is the soul of our full model, which distinguishes our approach from the existing ones.

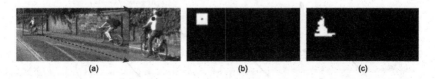

Fig. 3. Comparison of neighborhood in 2D and 3D space. (a) the color image (the red point in the red square is the point of interest); (b) the 2D neighborhood in the imaging plane; (c) the 3D neighborhood considering the pin-hole camera model (we first use model-inverted 3D coordinates to calculate the neighborhood of the red point in the physical world, then project those points to the 2D imaging plane).

4.1   Spatially-Variant Filter and Neighborhood Consideration

The concept of spatially-variant filter dated back to guided image filtering [24] which was originally proposed as an edge-preserving smoothing operator for images. Under the context of color-guided depth completion, guided image filter is still useful to capture rapid changes of depth values around a sharp depth discontinuity or an object boundary, which leads to recently developed GuideNet [55]. This concept was later extended in Dynamic Filter Networks [30] where filters are generated dynamically from the input. For different inputs, a special filter generating network will produce different filters which can be adaptively applied to the output by the dynamical filtering layer in the network. Inspired by both [30,55], we propose a multi-scale extension of dynamic filter networks to support sparse-to-dense depth completion in this work.

Our baseline model consists of two fully-convolutional subnetworks which have similar shapes as U-net [51]. One subnetwork (filter-generation) takes the color guide image as the input and generates feature maps at different scales that will be used as spatially-variant filters. Another subnetwork (depth-completion) takes sparse depth measurements as the input and generates a dense depth map as the final result. Both networks have almost identical encoder-decoder configurations but do not share the weights. As shown in Fig. 2, we multiply the feature maps generated by two subnetworks (filter generation and depth completion). Note that GuideNet [55] utilizes the generated filters as the weight of convolution kernels; however in view of the prohibitive complexity with convolution, we simply use element-wise product in our approach.

To further exploit spatially-variant mechanism, we propose a 3D extension that is specially tailored for depth completion tasks. Note that unlike color

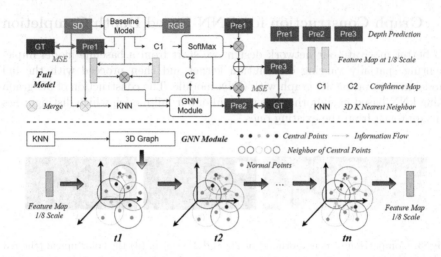

Fig. 4. The architecture of our full model including the graph convolution module.

images, depth maps indeed convey important 3D information about the physical world. If we consider the abstraction of projection onto imaging plane by a pin-hole camera model, the 2D neighborhood in the projected image plane is different from the actual 3D neighborhood in the physical world as shown in Fig. 3. For the definition of neighborhood in 3D space, it is intrinsically connected with the object which the point of interest belongs to. While for 2D, we might not have the luxury of accessing to the complete neighborhood information (e.g., due to occlusion). Such observation motivates us to construct a graph-based representation for spatially-variant neighborhood as we will elaborate next.

4.2 Graph Construction and Network Propagation

The input data to Graph Neural Networks (GNNs) [21,52] are often represented by a graph $G = \{V, E\}$, where V and E denote the nodes and edges. For each node $v \in V$, we use U_v, f_v and h_v^t to represent its neighborhood, input feature and state for time t respectively. Note that the hidden dynamics of all nodes in a GNN will evolve over time. Such time-evolving process can be modeled by [48]:

$$\begin{cases} m_v^t = \mathcal{M}(\{h_u^t\}), & u \in U_v \\ h_v^{t+1} = \mathcal{F}(h^t, m_v^t) \end{cases} \tag{11}$$

where $\mathcal{M}$ is a function to fuse information from the neighborhood of node v, $\mathcal{F}$ is a function to update its hidden state, m_v^t is a feature vector generated by $\mathcal{M}$ containing information from the node's neighborhood. Similar to a recurrent neural network (RNN) [29,48], $\mathcal{M}$ and $\mathcal{F}$ share weights at different time steps.

Since the baseline model computes features for each point, we directly use these features as the input to our GNN. Specifically, we use the feature map at 1/8 of original scale; we have used median pooling to downsample the initial dense depth map D_N to 1/8 scale. Moreover, we propose to merge the sparse depth input D_I and baseline output D_B as follows:

$$D_N^i = \begin{cases} D_I^i & D_I^i \neq 0 \\ D_I^i & D_I^i = 0 \end{cases} \tag{12}$$

where D_N^i, D_I^i and D_B^i are the depth values of point i in D_N, D_I and D_B, respectively. Then we construct the graph using D_N. Let $[u, v]$ be the coordinates of a point in the imaging plane and $[x, y, z]$ be the 3D coordinates of this point in the camera coordinate system. We can easily convert between 2D and 3D coordinates using the standard pinhole camera model [23].

To design a GNN for depth completion task, we need to identify the 3D neighborhood of each point in the color image first. Similar problem has been considered in a recent work 3DGNN [48] but for a different application (semantic segmentation) and with a more favorable assumption (access to the dense instead of sparse depth maps). Inspired by 3DGNN [48], we make each point in the image as a node and find out its k-Nearest-Neighbor(kNN) in 3D space (k is set to 64 in all experiments). Note that our constructed graph is a directed graph. This way all directed edges convey information about how each node obtains information from others. As shown in Fig. 4, we can create the new dense depth map D_N by searching 3D neighbors and propagating the information on GNN.

The propagation process on GNN ($\mathcal{M}$ and $\mathcal{F}$ in Eq. (11)) is implemented by a multi-layer perceptron (MLP) with ReLu [34]. The propagation process can be written as:

$$\begin{cases} m_v^t = \frac{1}{|U_v|} \sum_{v \in U_v} g(h_u^t) \\ h_v^{t+1} = \mathcal{F}(h^t, m_v^t) \end{cases} \tag{13}$$

where g denotes MLP. During the propagation, each node in the graph first collects the information from its neighbors; then features will be computed by g and m_v^t is the average feature over the whole neighborhood. Finally every node updates its state based on the previous information. We use MLP as the update function $\mathcal{F}$ because it is commonly used in RNN (e.g., [48]).

5   Experimental Results

5.1   Datasets

KITTI Dataset. The KITTI depth completion dataset [19] provides 86898 training samples, 1000 validating samples and 1000 testing samples taken in outdoor scenes. The ground truth depth is not 100% dense and only about 30% of points in the ground truth depth map have valid depth value. Since these points with valid depth values are at the bottom, following [57] we opt to crop the bottom portion of input images to $256 * 1216$ for training and testing.

NYUDv2 Dataset. The NYUDv2 dataset [54] provides RGB images and dense depth maps captured by Microsoft Kinect from 464 indoor scenes. Its raw data contains more than $100k$ samples and following [8,42,49], we use about 46k samples as training data. Besides, NYUDv2 provides an officially labeled subset containing 1449 samples (654 for testing and evaluation). To fill missing values, the depth values are in-painted using the official toolbox, which adopts the colorization scheme [36]. Following [42,64], we have down-sampled the origin images to half-resolution and center-cropped to the dimension of 320×256 with additional paddings.

Matterport3D. Matterport3D [5] is an indoor large-scale RGB-D dataset with $10.8k$ real panoramic views and 90 real indoor scenes. We use the same training and testing split as Zhang [66]. There are about 104k samples for training and 474 samples for testing. The ground truth depth map of Matterport3D is generated from Zhang [66] using multi-view reconstruction method.

Evaluation Metrics and Loss Function. In all of our experiments, we use Mean Square Error(MSE) as the loss function. For evaluation metric, we take root mean squared error (RMSE) as a prime index to evaluated our models. Besides, mean absolute error (MAE) and mean absolute relative error (REL) are used. For indoor scenes, percentage index δ_i which means the percentage of predicted pixels where the relative error is less a threshold i are used. Specifically, i is chosen as $1.25, 1.25^2, 1.25^3$ (Details of experiments are shown in supplement).

Table 1. Ablation studies on the test set of the NYUDv2 dataset. The effectiveness of GNN module and various sampling strategies are evaluated, respectively.

Method	Sample	Rel↓	RMSE↓	δ_1↑	δ_2↑	δ_3↑
Ours baseline	Random	0.020	0.112	99.4	**99.9**	100
	R2	0.017	0.101	99.5	**99.9**	100
	Plastic	**0.016**	0.095	99.5	**99.9**	100
	Golden	**0.016**	**0.094**	99.6	**99.9**	100
	Halton2,3	0.017	0.098	99.5	**99.9**	100
Ours GNN	Random	0.016	0.106	99.5	**99.9**	100
	R2	0.015	0.092	**99.6**	**99.9**	100
	Plastic	0.014	0.091	**99.6**	**99.9**	100
	Golden	**0.013**	**0.087**	**99.6**	**99.9**	100
	Halton2,3	0.015	0.093	99.5	**99.9**	100

5.2   Ablation Study

Neighborhood Construction via GNN. We show the effectiveness of neighborhood construction via GNN by comparing against its baseline without GNN. Quantitative results are listed in Table 1. Consistent improvement can be observed for different sampling strategies when comparing GNN to baseline models, which suggest the effectiveness of neighborhood construction via GNN.

Evaluation of Various Sampling Strategies. We have compared four quasi random sampling strategies against random sampling as shown in Table 1. We can observe that the RMSE values of quasirandom sampling are much smaller than that of random sampling, and Golden is the best one among four quasirandom strategies. This finding is consistent with our analysis in Sect. 3.2 and Fig. 1 - i.e., r_{min} metric of the Golden sampling is the smallest.

5.3   Comparison with State-of-the-Art

We compare our method against several state-of-the-art methods on two indoor datasets (NYUDv2 and Matterport3D) and one outdoor dataset (KITTI). Tables 2, 3 and 4 show the quantitative results. It can be observed that our GNN achieves much better results than previous methods on indoor datasets and highly competitive result on outdoor datasets. Especially for NYUDv2 dataset, our GNN with Golden sampling has dramatically outperformed all existing approaches in the open literature. Figures 5 and 6 include the visual comparison of reconstructed depth maps among several competing methods. It can be

Table 2. Comparison with state-of-the-art methods on the test set of the NYUDv2. For fair comparison, we add the random sampling strategy of our GNN.

Method	Sample	Rel↓	RMSE↓	δ_1↑	δ_2↑	δ_3↑
Bilateral [54]	Random	0.084	0.479	92.4	97.6	98.9
Ma et al. [42]	Random	0.044	0.230	97.1	99.4	99.8
Eldesokey et al. [17]	Random	0.018	0.129	99.0	99.8	**100**
CSPN [9]	Random	0.016	0.117	99.2	**99.9**	100
DeepLiDAR [49]	Random	0.022	0.115	99.3	**99.9**	100
Xu et al. [64]	Random	0.018	0.112	99.5	**99.9**	100
Ours baseline	Random	0.020	0.112	99.4	**99.9**	100
Ours GNN	Random	0.016	0.106	99.5	**99.9**	100
Ours GNN	Golden	**0.013**	**0.087**	**99.6**	**99.9**	100

Table 3. Comparison with other methods on the test set of the Matterport3D.

Method	RMSE↓	MAE↓
Huang et al. [26]	1.092	**0.342**
Zhang et al. [66]	1.316	0.461
MRF [22]	1.675	0.618
AD [40]	1.653	0.610
Ours GNN	**0.860**	0.462

Table 4. Comparison with other methods on the KITTI validation set.

Method	RMSE↓	MAE↓	Rel↓
Dfusenet [53]	1240	429	0.022
Ours baseline	883	280	0.016
Ma et al. [41]	858	311	0.019
Ours GNN	**831**	**247**	**0.013**

observed that our result are consistently the closest to the ground-truth on both datasets.

5.4   Cross-Dataset Evaluation

We further conduct cross-dataset experiments to show the transferability of our method. Two transfer scenes (outdoor→indoor and indoor→indoor) are considered, and the experimental results are shown in Tables 5 and 6. Our GNN achieves significant improvements in both scenarios. First, our approach (baseline model without GNN) shows much better transfer results than [57]. Second, our approach with quasi-random sampling achieved much better transfer results than random sampling. Besides the good transferability of our model, the feature is more transferable. We show the transferability by fixing the feature extractor and only fine-tuning the last layer (please refer to the supplemental material).

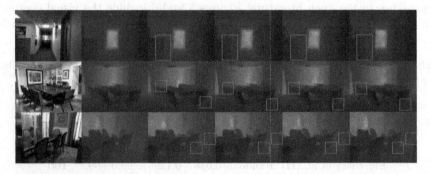

Fig. 5. Results on NYUDv2 dataset. From left to right is RGB image, the ground truth, result from [9], results of our baseline model, results of our full model and results of our full model with golden sampling strategy. Except the rightest one, other methods are tested using randomly sampling strategy.

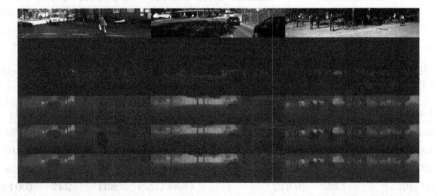

Fig. 6. Results on KITTI dataset. From top to bottom is RGB image, the sparse depth input, the ground truth, results of our baseline model, result from [41], results of our full model.

Similar improvements can be observed as transferring the whole model, and transferring the feature achieves better cross-dataset results.

It is worth mentioning that initial sparse depth maps in the Matterport3D dataset are generated differently from those in the NYUDv2 dataset, where the depth values of local areas are completely missing (i.e., noticeable holes in the depth map). To evaluate the generalization property of our work between different datasets, we have compared two transfer settings: $R2R$ using initial sparse depth maps (without filling the holes) and $Q2Q$ using a modified sparse depth maps (resampled by quasi-random sampling). As shown in Table 6, when adopting the original sparse depth maps as the initial input, our trained network on NYUDv2 with random sampling performs better than quasi-random sampling, which justifies that the original sparse depth maps are indeed more similar to random sampling. By contrast, when adopting the modified sparse depth maps as the input, our trained network with quasi-random sampling performs better than random sampling. As expected, the RMSE of $Q2Q$ setting is much smaller than that of $R2R$ setting. These findings suggest that our GNN model with quasi-random sampling enjoys good generalization properties.

Table 5. Result of models transferred from KITTI dataset to NYUDv2 dataset using different sampling strategies. All models are pre-trained on KITTI dataset. For method, Ours GNN is our full model with graph neural network. Mode "D" means directly evaluating the model on NYUDv2 test set with 654 samples. For methods with *, we have used the pre-trained model provided by their author.

Method	Mode	Sample	Rel↓	RMSE↓	δ_1↑	δ_2↑	δ_3↑
*Van et al. [57]	D	Random	0.090	0.487	85.7	93.6	97.2
Ours Baseline	D	Random	0.087	0.365	90.6	97.7	99.3
Ours GNN	D	Random	0.061	0.310	94.0	98.4	99.5
Ours Baseline	D	R2	0.060	0.261	95.8	99.0	**99.8**
Ours Baseline	D	Plastic	0.060	0.262	95.7	99.0	99.7
Ours Baseline	D	Golden	0.058	0.256	95.7	98.7	99.7
Ours Baseline	D	Halton2,3	0.070	0.294	94.6	98.7	99.7
Ours GNN	D	R2	**0.044**	**0.238**	**96.6**	**99.2**	**99.8**
Ours GNN	D	Plastic	**0.044**	**0.238**	**96.6**	**99.2**	**99.8**
Ours GNN	D	Golden	0.045	0.240	96.4	99.1	99.7
Ours GNN	D	Halton2,3	0.048	0.254	96.1	99.1	99.7

6   Conclusion

In this paper, we have studied different sampling strategies approximating Poisson disk sampling and their impact on sparse-to-dense depth completion. We

Table 6. RMSE results of our GNN transferred from NYUDv2 to Matterport3D in two different transfer settings (*R2R* and *Q2Q*) are evaluated.

Setting	Mode	Random	Plastic	Golden	R2	Halton
R2R	D	**1.831**	1.859	1.913	1.855	1.895
Q2Q	D	0.501	0.483	0.489	0.487	**0.460**

show that random sampling is far from being optimal and quasi-random sampling constructed by low-discrepancy sequences can significantly outperforms random sampling. We have also proposed to construct a 3D graph based on kNN neighborhood information and develop a novel propagation-driven depth completion algorithm based on Graph Neural Network. Our proposed depth completion method outperforms the state-of-the-art on indoor scenes and has good generalization property for outdoor scenes.

References

1. Bratley, P., Fox, B.L., Niederreiter, H.: Implementation and tests of low-discrepancy sequences. ACM Trans. Model. Comput. Simul. (TOMACS) **2**(3), 195–213 (1992)
2. Bridson, R.: Fast poisson disk sampling in arbitrary dimensions. SIGGRAPH sketches **10**, 1278780–1278807 (2007)
3. Bruna, J., Zaremba, W., Szlam, A., LeCun, Y.: Spectral networks and locally connected networks on graphs. arXiv preprint arXiv:1312.6203 (2013)
4. Caflisch, R.E.: Monte carlo and quasi-monte carlo methods. Acta numerica **7**, 1–49 (1998)
5. Chang, A., et al.: Matterport3d: Learning from rgb-d data in indoor environments. In: International Conference on 3D Vision (3DV) (2017)
6. Chen, Y., Yang, B., Liang, M., Urtasun, R.: Learning joint 2d–3d representations for depth completion. In: Proceedings of the IEEE International Conference on Computer Vision. pp. 10023–10032 (2019)
7. Chen, Z., Badrinarayanan, V., Drozdov, G., Rabinovich, A.: Estimating depth from rgb and sparse sensing. In: Proceedings of the European Conference on Computer Vision (ECCV). pp. 167–182 (2018)
8. Cheng, X., Wang, P., Yang, R.: Depth estimation via affinity learned with convolutional spatial propagation network. In: Proceedings of the European Conference on Computer Vision (ECCV). pp. 103–119 (2018)
9. Cheng, X., Wang, P., Yang, R.: Learning depth with convolutional spatial propagation network. arXiv preprint arXiv:1810.02695 (2018)
10. Cho, K., et al.: Learning phrase representations using rnn encoder-decoder for statistical machine translation. arXiv preprint arXiv:1406.1078 (2014)
11. Defferrard, M., Bresson, X., Vandergheynst, P.: Convolutional neural networks on graphs with fast localized spectral filtering. In: Advances in Neural Information Processing Systems. pp. 3844–3852 (2016)
12. Doria, D., Radke, R.J.: Filling large holes in lidar data by inpainting depth gradients. In: 2012 IEEE Computer Society Conference on Computer Vision and Pattern Recognition Workshops. pp. 65–72. IEEE (2012)
13. Dunlap, R.A.: The golden ratio and Fibonacci numbers. World Scientific (1997)

14. Early, D.S., Long, D.G.: Image reconstruction and enhanced resolution imaging from irregular samples. IEEE Trans. Geosci. Remote Sens. **39**(2), 291–302 (2001)
15. Eigen, D., Fergus, R.: Predicting depth, surface normals and semantic labels with a common multi-scale convolutional architecture. In: Proceedings of the IEEE International Conference on Computer Vision. pp. 2650–2658 (2015)
16. Eldar, Y., Lindenbaum, M., Porat, M., Zeevi, Y.Y.: The farthest point strategy for progressive image sampling. IEEE Trans. Image Process. **6**(9), 1305–1315 (1997)
17. Eldesokey, A., Felsberg, M., Khan, F.S.: Confidence propagation through cnns for guided sparse depth regression. IEEE Trans. Pattern Anal. Mach. Intell. (2019)
18. Ferstl, D., Reinbacher, C., Ranftl, R., Rüther, M., Bischof, H.: Image guided depth upsampling using anisotropic total generalized variation. In: Proceedings of the IEEE International Conference on Computer Vision. pp. 993–1000 (2013)
19. Geiger, A., Lenz, P., Urtasun, R.: Are we ready for autonomous driving? the kitti vision benchmark suite. In: 2012 IEEE Conference on Computer Vision and Pattern Recognition. pp. 3354–3361. IEEE (2012)
20. Godard, C., Mac Aodha, O., Brostow, G.J.: Unsupervised monocular depth estimation with left-right consistency. In: Proceedings of the IEEE Conference on Computer Vision and Pattern Recognition. pp. 270–279 (2017)
21. Gori, M., Monfardini, G., Scarselli, F.: A new model for learning in graph domains. In: Proceedings 2005 IEEE International Joint Conference on Neural Networks, 2005. vol. 2, pp. 729–734. IEEE (2005)
22. Harrison A., Newman P.: Image and Sparse Laser Fusion for Dense Scene Reconstruction. In: Howard A., Iagnemma K., Kelly A. (eds). Field and Service Robotics. Springer Tracts in Advanced Robotics, vol 62. Springer, Berlin, Heidelberg (2010) https://doi.org/10.1007/978-3-642-13408-1_20
23. Hartley, R., Zisserman, A.: Multiple view geometry in computer vision. Cambridge University Press, Chennai (2003)
24. He, K., Sun, J., Tang, X.: Guided image filtering. In: Daniilidis, K., Maragos, P., Paragios, N. (eds.) ECCV 2010. LNCS, vol. 6311, pp. 1–14. Springer, Heidelberg (2010). https://doi.org/10.1007/978-3-642-15549-9_1
25. He, K., Zhang, X., Ren, S., Sun, J.: Deep residual learning for image recognition. In: Proceedings of the IEEE Conference on Computer Vision and Pattern Recognition. pp. 770–778 (2016)
26. Huang, Y.K., Wu, T.H., Liu, Y.C., Hsu, W.H.: Indoor depth completion with boundary consistency and self-attention. In: Proceedings of the IEEE International Conference on Computer Vision Workshops (2019)
27. Huang, Z., Fan, J., Cheng, S., Yi, S., Wang, X., Li, H.: Hms-net: Hierarchical multi-scale sparsity-invariant network for sparse depth completion. IEEE Trans. Image Process. **29**, 3429–3441 (2019)
28. Illian, J., Penttinen, A., Stoyan, H., Stoyan, D.: Statistical analysis and modelling of spatial point patterns. John Wiley & Sons, New Jersey (2008)
29. Jaeger, H.: Tutorial on training recurrent neural networks, covering BPPT, RTRL, EKF and the" echo state network" approach. GMD-Forschungszentrum Informationstechnik Bonn, Bonn (2002)
30. Jia, X., De Brabandere, B., Tuytelaars, T., Gool, L.V.: Dynamic filter networks. In: Advances in Neural Information Processing Systems. pp. 667–675 (2016)
31. Kipf, T.N., Welling, M.: Semi-supervised classification with graph convolutional networks. arXiv preprint arXiv:1609.02907 (2016)
32. Kocis, L., Whiten, W.J.: Computational investigations of low-discrepancy sequences. ACM Trans. Math. Software (TOMS) **23**(2), 266–294 (1997)

33. Krcadinac, V.: A new generalization of the golden ratio. Fibonacci Quart. **44**(4), 335 (2006)
34. Krizhevsky, A., Sutskever, I., Hinton, G.E.: Imagenet classification with deep convolutional neural networks. In: Advances in Neural Information Processing Systems. pp. 1097–1105 (2012)
35. Kuipers, L., Niederreiter, H.: Uniform distribution of sequences. Courier Corporation (2012)
36. Levin, A., Lischinski, D., Weiss, Y.: Colorization using optimization. In: ACM SIGGRAPH 2004 Papers, pp. 689–694 (2004)
37. Li, Y., Ibanez-Guzman, J.: Lidar for autonomous driving: The principles, challenges, and trends for automotive lidar and perception systems. IEEE Signal Process. Mag.(2020)
38. Li, Y., Tarlow, D., Brockschmidt, M., Zemel, R.: Gated graph sequence neural networks. arXiv preprint arXiv:1511.05493 (2015)
39. Liu, F., Shen, C., Lin, G., Reid, I.: Learning depth from single monocular images using deep convolutional neural fields. IEEE Trans. Pattern Anal. Mach. Intell. **38**(10), 2024–2039 (2015)
40. Liu, J., Gong, X.: Guided depth enhancement via anisotropic diffusion. In: Huet, B., Ngo, C.-W., Tang, J., Zhou, Z.-H., Hauptmann, A.G., Yan, S. (eds.) PCM 2013. LNCS, vol. 8294, pp. 408–417. Springer, Cham (2013). https://doi.org/10.1007/978-3-319-03731-8_38
41. Ma, F., Cavalheiro, G.V., Karaman, S.: Self-supervised sparse-to-dense: Self-supervised depth completion from lidar and monocular camera. In: 2019 International Conference on Robotics and Automation (ICRA). pp. 3288–3295. IEEE (2019)
42. Mal, F., Karaman, S.: Sparse-to-dense: Depth prediction from sparse depth samples and a single image. In: 2018 IEEE International Conference on Robotics and Automation (ICRA). pp. 1–8. IEEE (2018)
43. Marohnić, L., Strmečki, T.: Plastic number: Construction and applications. In: International Virtual Conference ARSA (Advanced Researsch in Scientific Area) (2013)
44. Matsuo, K., Aoki, Y.: Depth image enhancement using local tangent plane approximations. In: Proceedings of the IEEE Conference on Computer Vision and Pattern Recognition. pp. 3574–3583 (2015)
45. McCool, M., Fiume, E.: Hierarchical poisson disk sampling distributions. In: Proceedings of the Conference on Graphics Interface. vol. 92, pp. 94–105 (1992)
46. Niederreiter, H.: Random number generation and quasi-Monte Carlo methods. Siam, Thailand (1992)
47. Park, J., Kim, H., Tai, Y.W., Brown, M.S., Kweon, I.S.: High-quality depth map upsampling and completion for rgb-d cameras. IEEE Trans. Image Process. **23**(12), 5559–5572 (2014)
48. Qi, X., Liao, R., Jia, J., Fidler, S., Urtasun, R.: 3d graph neural networks for rgbd semantic segmentation. In: Proceedings of the IEEE International Conference on Computer Vision. pp. 5199–5208 (2017)
49. Qiu, J., et al.: Deeplidar: Deep surface normal guided depth prediction for outdoor scene from sparse lidar data and single color image. In: Proceedings of the IEEE Conference on Computer Vision and Pattern Recognition. pp. 3313–3322 (2019)
50. Ripley, B.D.: Spatial statistics. John Wiley & Sons, New Jersey (2005)
51. Ronneberger, O., Fischer, P., Brox, T.: U-Net: convolutional networks for biomedical image segmentation. In: Navab, N., Hornegger, J., Wells, W.M., Frangi, A.F.

(eds.) MICCAI 2015. LNCS, vol. 9351, pp. 234–241. Springer, Cham (2015). https://doi.org/10.1007/978-3-319-24574-4_28

52. Scarselli, F., Gori, M., Tsoi, A.C., Hagenbuchner, M., Monfardini, G.: The graph neural network model. IEEE Trans. Neural Netw. **20**(1), 61–80 (2008)

53. Shivakumar, S.S., Nguyen, T., Miller, I.D., Chen, S.W., Kumar, V., Taylor, C.J.: Dfusenet: Deep fusion of rgb and sparse depth information for image guided dense depth completion. In: 2019 IEEE Intelligent Transportation Systems Conference (ITSC). pp. 13–20. IEEE (2019)

54. Silberman, N., Hoiem, D., Kohli, P., Fergus, R.: Indoor segmentation and support inference from rgbd images. In: Fitzgibbon, A., Lazebnik, S., Perona, P., Sato, Y., Schmid, C. (eds.) ECCV 2012. LNCS, vol. 7576, pp. 746–760. Springer, Heidelberg (2012). https://doi.org/10.1007/978-3-642-33715-4_54

55. Tang, J., Tian, F.P., Feng, W., Li, J., Tan, P.: Learning guided convolutional network for depth completion. arXiv preprint arXiv:1908.01238 (2019)

56. Uhrig, J., Schneider, N., Schneider, L., Franke, U., Brox, T., Geiger, A.: Sparsity invariant cnns. In: 2017 International Conference on 3D Vision (3DV). pp. 11–20. IEEE (2017)

57. Van Gansbeke, W., Neven, D., De Brabandere, B., Van Gool, L.: Sparse and noisy lidar completion with rgb guidance and uncertainty. In: 2019 16th International Conference on Machine Vision Applications (MVA). pp. 1–6. IEEE (2019)

58. Veličković, P., Cucurull, G., Casanova, A., Romero, A., Lio, P., Bengio, Y.: Graph attention networks. arXiv preprint arXiv:1710.10903 (2017)

59. Werbos, P.J.: Backpropagation through time: what it does and how to do it. Proc. IEEE **78**(10), 1550–1560 (1990)

60. Wu, Z., Pan, S., Chen, F., Long, G., Zhang, C., Yu, P.S.: A comprehensive survey on graph neural networks. arXiv preprint arXiv:1901.00596 (2019)

61. Xian, K., et al.: Monocular relative depth perception with web stereo data supervision. In: The IEEE Conference on Computer Vision and Pattern Recognition (CVPR) (2018)

62. Xian, K., Zhang, J., Wang, O., Mai, L., Lin, Z., Cao, Z.: Structure-guided ranking loss for single image depth prediction. In: The IEEE/CVF Conference on Computer Vision and Pattern Recognition (CVPR) (2020)

63. Xiong, X., Cao, Z., Zhang, C., Xian, K., Zou, H.: Binoboost: Boosting self-supervised monocular depth prediction with binocular guidance. In: 2019 IEEE International Conference on Image Processing (ICIP). pp. 1770–1774. IEEE (2019)

64. Xu, Y., Zhu, X., Shi, J., Zhang, G., Bao, H., Li, H.: Depth completion from sparse lidar data with depth-normal constraints. In: Proceedings of the IEEE International Conference on Computer Vision. pp. 2811–2820 (2019)

65. Xue, H., Zhang, S., Cai, D.: Depth image inpainting: Improving low rank matrix completion with low gradient regularization. IEEE Trans. Image Process. **26**(9), 4311–4320 (2017)

66. Zhang, Y., Funkhouser, T.: Deep depth completion of a single rgb-d image. In: Proceedings of the IEEE Conference on Computer Vision and Pattern Recognition. pp. 175–185 (2018)

67. Zhou, J., et al.: Graph neural networks: A review of methods and applications. arXiv preprint arXiv:1812.08434 (2018)

MEAD: A Large-Scale Audio-Visual Dataset for Emotional Talking-Face Generation

Kaisiyuan Wang[1], Qianyi Wu[1], Linsen Song[1,3,4], Zhuoqian Yang[2],
Wayne Wu[1(✉)], Chen Qian[1], Ran He[3,4], Yu Qiao[5],
and Chen Change Loy[6]

[1] SenseTime Research, Beijing, China
wuwenyan@sensetime.com
[2] Robotics Institute, Carnegie Mellon University, Pittsburgh, USA
[3] Center for Research on Intelligent Perception and Computing, CASIA,
Beijing, China
[4] University of Chinese Academy of Sciences, Beijing, China
[5] Shenzhen Institutes of Advanced Technology, Chinese Academy of Science,
Shenzhen, China
[6] Nanyang Technological University, Singapore, Singapore

Abstract. The synthesis of natural emotional reactions is an essential criterion in vivid talking-face video generation. This criterion is nevertheless seldom taken into consideration in previous works due to the absence of a large-scale, high-quality emotional audio-visual dataset. To address this issue, we build the **M**ulti-view **E**motional **A**udio-visual **D**ataset (MEAD), a talking-face video corpus featuring 60 actors and actresses talking with eight different emotions at three different intensity levels. High-quality audio-visual clips are captured at seven different view angles in a strictly-controlled environment. Together with the dataset, we release an emotional talking-face generation baseline that enables the manipulation of both emotion and its intensity. Our dataset could benefit a number of different research fields including conditional generation, cross-modal understanding and expression recognition. Code, model and data are publicly available on our project page [‡‡]https://wywu.github. io/projects/MEAD/MEAD.html.

Keywords: Video generation · Generative adversarial networks · Representation disentanglement

1 Introduction

Talking face generation is the task of synthesizing a video of a talking face conditioned on both the identity of the speaker (given by a single still image) and the

K. Wang, Q. Wu, L. Song—Equal contribution.

Electronic supplementary material The online version of this chapter (https:// doi.org/10.1007/978-3-030-58589-1_42) contains supplementary material, which is available to authorized users.

© Springer Nature Switzerland AG 2020
A. Vedaldi et al. (Eds.): ECCV 2020, LNCS 12366, pp. 700–717, 2020.
https://doi.org/10.1007/978-3-030-58589-1_42

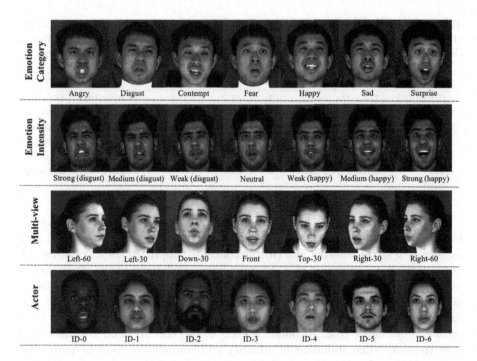

Fig. 1. MEAD overview. MEAD is a large-scale, high-quality audio-visual dataset with rich affective data, diversified speakers and multiple perspectives.

content of the speech (provided as an audio track). A major challenge in this task is constituted by the fact that natural human speech is often accompanied by several nonverbal characteristics, e.g. intonations, eye contact and facial expressions, which reflect the emotion of the speaker [11,54]. State-of-the-art methods are able to generate lip movements in perfect synchronization with the audio speech [6,58], but the faces in such videos are often emotionless and sometimes even impassive. Either the faces are devoid of any emotion or there is a distinct mismatch between the facial expression and the content of the audio speech.

A considerable number of recent advancements in the task of talking-face generation are deep learning based methods [6,52,57,58], where data has a significant influence on performance. We argue that the absence of a large-scale, high-quality emotional audio-visual dataset is the main obstacle to achieve vivid talking-face generation. As shown in Table 1, the available datasets are very limited in the diversity of the speakers, the duration of the videos, the number of the view-angles and the richness of the emotions. To address this issue, we build the **M**ulti-view **E**motional **A**udio-visual **D**ataset (MEAD), a talking-face video corpus featuring 60 actors talking with eight different emotions at three different intensity levels (except for neutral). The videos are simultaneously recorded at seven different perspectives in a strictly-controlled environment to provide high-quality details of facial expressions. About 40 hours of audio-visual clips are recorded for each person and view.

Table 1. Comparison of datasets. We compare with recent high-quality audio-visual datasets that are built in controlled environments. The symbol "#" indicates the number. The "Avg Dura" means average audio/video duration time per speaker.

Dataset	Avg Dura	#Actor	#Emo	#View	#Intensity	Resolution	#Clips
SAVESS [23]	7 min 21 s	4	7	1	1	1280 × 1024	480
RAVDESS [32]	3 min 42 s	24	8	1	2	1920 × 1080	7,356
GRID [9]	18 min 54 s	34	-	1	-	720 × 576	34,000
Lombard [1]	4 min 1 s	54	-	2	-	854 × 480	10,800
CREMA-D [4]	N/A	91	6	1	3(only 1/12)	1280 × 720	7442
Ours	**38 min 57 s**	**60**	**8**	**7**	**3**	**1920 × 1080**	**281,400**

A fundamental design choice for emotional talking-face corpus is the choice between (i) the in-the-wild approach [7,39], i.e., annotating videos gathered from sources such as the Internet and (ii) the controlled approach [29,33], i.e., recording coordinated performers in a constant, controlled environment. It is easy to scale up with the in-the-wild approach, but the data suffer from inconsistency in both the quality of the audio/video and the annotation of emotions [35]. The controlled approach, on the other hand, ensures the quality of the data but takes considerably higher costs to build. *MEAD* is an effort to build a dataset that is at the same time abundant in quantity and consistently good in quality. As far as we know, our dataset is the largest controlled dataset in terms of the number of video clips. In order to ensure the naturalness of the performed emotions, our data acquisition pipeline is carefully designed from the selection of actors/actresses to the performance feedback and correction. A team led by a professional actor guide the participants to speak in natural and accurate emotional status. To ensure the quality of the audio, we carefully select emotionally consistent speech texts that cover different phonemes.

Together with the dataset, we propose an emotional talking face generation baseline that enables the manipulation of the emotion and its intensity. A two-branch architecture is designed to process the audio and emotional conditions separately. Specifically, one of the branches is responsible for mapping audio to lip movements and the other branch is responsible for synthesizing the desired emotion on the target face. Finally, the intermediate representations are fused in a refinement network to render the emotional talking-face video.

In summary, our contributions are twofold:

- We build a large-scale, high-quality emotional audio-visual dataset *MEAD*, which is the largest emotional audio-visual corpus in terms of the number of video clips and viewpoints.
- Together with the dataset, we propose an emotional talking face generation baseline that enables the manipulation of the emotion and its intensity. Extensive experiments measure video generation and emotion manipulation performance for future reference.

2 Related Work

Talking-Face Generation. Talking-face generation is a long-standing problem [14,31,36] which is gaining attention [15,49,51]. Researchers' current focus is mainly on improving the realisticness of the generated videos. Chung et al.[8] proposes the Speech2Vid model to animate static target faces. Zhou et al. [58] adopted a representation disentanglement framework to drive different identities to utter the same speech contents. Chen et al. [6] used a cascade GAN approach to improve the temporal continuity of the generation. Song et al. [48] propose an audio-driven talking face generation method to solve head pose and identity challenges by utilizing 3D face model. However, how to manipulate the *emotion* in generated talking-face is still an open question.

Emotion Conditioned Generation. Emotion conditioned image generation has been advancing under the inspiration of recent progress in unsupervised image translation [22,30,43,55,60]. These frameworks are able to transfer expression according to specified emotion categories. However, obvious artifacts are frequently observed in dynamic changing areas in the results. Pumarola et al. [42] propose the GANimation model, which is based on an unsupervised framework to describe expressions in a continuous rather than discrete way, representing facial expressions as action units activations. Ding et al. [12] designed a novel controller module in an encoder-decoder network to adjust expression intensity continuously, however, the method is not explicit enough for more fine-grained control. Several works have also studied the generation of emotional talking sequences as well. Najmeh et al. [46] introduces a conditional sequential GAN model to learn the relationship between emotion and speech content, and generate expressive and naturalistic lip movements. Konstantinos [52] uses three discriminators to enhance details, synchronization, and realistic expressions like blinks and eyebrow movements. Both of the methods could basically capture the facial movements related to emotion categories, however, the emotion manipulation is completely determined by the speech audio, and cannot be more delicate to achieve manipulation in different intensities. Although some works [5,13,25,59] have proposed thought-provoking solutions towards this problem from a 3D facial animation perspective, the lack of suitable emotional audio-visual dataset still hinders further progress.

Emotional Audio-visual Dataset. There are some high-quality in-the-lab audio-visual datasets [1,3,9], but none of these take emotion information into consideration in design. The SAVEE [23] dataset is one of the datasets that considers emotion in speech. But only 4 actors are featured to read the designed TIMIT corpus [16]. Some datasets annotated not just emotion categories but also the intensities. AffectNet [38] and SEWA [28] included continuous intensity annotations based on dimensional model valence and arousal circumflex [45]. There are also datasets with discrete emotion intensity annotations. CREMA-D dataset [4] contains affective data with three intensity levels. Actors are required to express each emotion in two intensities when collecting data for RAVESS dataset [32]. However, the limited number of recorded sentences makes it hard for networks trained on it to generalize to real-life applications.

3   MEAD

In order to ensure the naturalness of the performed emotions, our data acquisition pipeline is carefully designed from the selection of actor/actresses to the performance feedback and correction. A team led by a professional actor guide the participants to speak in natural and accurate emotional status. To ensure the quality of the audio, we carefully select emotionally consistent speech texts that cover different phonemes.

3.1   Design Criteria

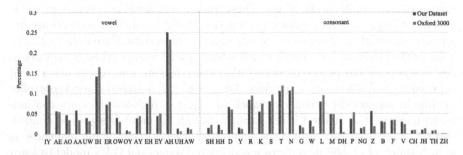

Fig. 2. Vowel and consonant distribution. Although we design different corpora for different emotions, for simplicity, we demonstrate the distribution for all emotion categories. The distribution of our corpus is basically consistent with that of frequently used 3000 words in [10].

Definition of Emotion Categories and Intensities. We use eight emotion categories following [32] and three levels of emotion intensity which is intuitive to human understanding. More intensity levels would be hard to distinguish and cause confusion and inconsistency in data acquisition. The first level is defined as *weak*, which describes delicate but detectable facial movements. The second level *medium* is the normal state of the emotion, which stands for the typical expression of the emotion. The third level is defined as *strong*, which describes the most exaggerated expressions of this emotion, requiring intense movements in related facial areas.

Design of the Speech Corpus. For audio speech content, we follow the phonetically-diverse speech corpus TIMIT [16], which is originally designed in automatic speech recognition [19,40,50]. We carefully select the sentences covering all phonemes in each emotion category, and the sentences in each emotion category is divided into three parts: 3 common sentences, 7 emotion-related sentences, and 20 generic sentences. We provide more details of the speech corpus in supplementary materials.

3.2 Data Acquisition

In the data acquisition process we mainly consider the following two aspects. First, the captured emotion should be natural and stable during talking. Second, the three levels of emotion intensities should be distinguishable. A guidance team led by a professional actor oversees the process.

Speaker Selection. We recruit fluent English speakers aged from 20 to 35 with previous acting experience. To evaluate the skills of the actors, video samples of each emotion in different intensities performed by the professional actor is recorded and the candidates are required to imitate the expressions in the videos. The guidance team evaluated the performance of each speaker according to the imitation result to ensure the main features of emotions are expressed correctly and naturally.

Recording. Before the recording begins, training courses are offered to help the speaker get into desired emotional status. We require the speakers to express different emotions spontaneously in their mother tongue to help them release tension. Then, an emotion-arousal session helps the speakers raise their emotional status so that the speakers can manage extreme expressions in level 3. During the recording, the guidance team arranges recording order of the different intensities to suit each speaker. Most of the speakers are recorded in the order of *weak*, *strong* and *medium*, as it will be easier to master medium intensity when the speaker is aware of the two extremes of one emotion.

Supervision and Feedback. During the recording, the guidance team would provide supervision from both emotion and speaking perspectives. In terms of emotion, the expression must cover all the features of the corresponding category, and the expressiveness should be consistent with the given intensity. Meanwhile, the speaker is requested to read the whole sentence with no pause and no pronunciation error. The guidance team would make the final judgment of whether the clip is qualified or not. In general, it would take the speaker two or three times to finish a qualified clip when first switching to another emotion category.

3.3 Analysis and Comparison

In this section, we show statistics of MEAD and compare with related datasets.

Analysis of the Speech Corpus. We keep track of the number of occurrences of different phonemes, including 15 vowels and 24 consonants based on the ARPAbet symbol set [27]. The distributions of vowels and consonants are shown as Fig. 2. Our corpus fully covers all vowels and consonants, and their occurrence frequency intuitively accords with the frequency of daily usage [10]. In RAVDESS [32], GRID [9], Lombard [1] and CREMA-D [4], the speech corpora have been greatly simplified, *e.g.* RAVDESS [32] uses only two sentences, GRID [9] and Lombard [1] use fixed sentence patterns, and CREMA-D [4] provides only 12 sentences for each emotion. These corpora are much less diverse than the TIMIT [16] corpus used by SAVEE [23] and our dataset. In our dataset,

30 sentences are used for each of the 7 basic emotion categories and 40 for the neutral category. Please refer to the supplementary materials for more details.

Analysis of the Audio-Visual Dataset. We demonstrate the quantitative comparison in Table 1. To enable the manipulation of emotion and its intensity in a more fine-grained way, our dataset is designed to contain neutral and 7 basic emotions including 3 intensity levels. The emotion categories are designed following [32], but our dataset provides 3 intensity levels for each emotion additionally. Thus, compared to recent datasets [23,32], our dataset provides richer emotion information. Another feature of MEAD is the inclusion of multi-view data. We place 7 cameras at different viewpoints to capture our portrait videos simultaneously, the detailed set-up is shown in supplementary. The viewpoint number is the largest among recent audio-visual datasets. The AV Digits [41] database is recorded from three angles (front, 45°, and profile), and the Ouluvs2 [2] dataset extends the view setting of the former to five for more fine-grained coverage. The Lombard [1] dataset and TCD-TIMIT [17] dataset only provide the front and side views. [9,23,32] provide data captured from the front view only. In terms of resolution, our dataset provides videos in the resolution of 1920 × 1080, which can be used in high fidelity portrait video generation. Similar to recent datasets [1,9,23,32], our audio sample rate is 48 kHz and video frame rate is 30. Note that SAVEE provides video data of fps 60, which is higher than that of us. However, many video feature extraction networks [53] usually downsample video frames of 30 fps. Thus, we adopt 30 fps which is sufficient for many video tasks like emotion recognition and portrait video generation.

3.4 Evaluation

We design a user study to evaluate the quality of our dataset, specifically, to examine if (i) the emotion performed by actors can be correctly and accurately recognized and (ii) the three levels of emotion intensity can be correctly distinguished. A hundred volunteers are gathered from universities for this experiment. The age range of the participants is from 18 to 40. We randomly selected six actors' data from MEAD for the user study, the testing data includes four males and two females.

Two types of experiments are conducted, namely emotion classification and intensity classification. For each type of experiment, two kinds of evaluations are performed – one on normal videos and the other on silenced videos. For emotion classification, we prepare test videos with varying emotion intensities in random order and the user needs to select one of 8 emotion categories. This evaluation is conducted 144 times for each user. The results of the silent video experiment are shown in Table 2, where the "Generated" stands for the user study results of the videos generated by our proposed baseline. It demonstrates that most of the testing video convey correct emotion to users. In emotions such as angry, happy and sad, we get an accuracy rate of over 0.90, while the results in neutral is much less satisfying, as the neutral expression is much easier to be in confusion with delicate emotional expressions in level 1. Considering that the intensity of

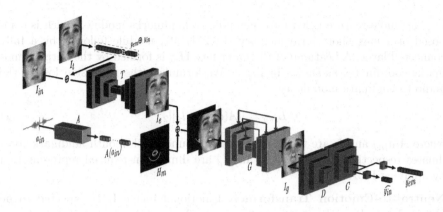

Fig. 3. Architecture of our baseline method. The overview of our emotional controllable talking-face approach. Our method includes three modules to drive the input neutral portrait image, audio clip and controllable emotion condition vector to obtain the output speech video.

emotion could affect the perception of the category of the emotion, we design the intensity classification experiment as follows: three videos of different intensities of one emotion is displayed in random order, and the participants are required to give the correct order from weak to strong. This experiment is conducted 42 times for each participant. From the results shown in the upper part of Table 3, we can observe that progressive emotion intensities are well distinguishable. Level 1 and level 2 are sometimes confused on some difficult emotions like disgust and contempt. More details about the user study can be found in the supplementary materials.

4   Emotional Talking-Face Baseline

Based on MEAD, we propose an emotional talking-face generation baseline that is able to manipulate emotion category and emotion intensity in talking-faces. The overview of our generation approach is depicted in Fig. 3. A two-branch architecture is designed to process the audio and emotional conditions separately. Specifically, the audio-to-landmarks module maps the audio to lip movements and neutral-to-emotion transformer synthesizes the desired emotion on the upper face of the target. Finally, the intermediate representations are fused in a refinement network to render the emotional talking-face video. The training phase requires three inputs, namely the input audio feature a_{in}, the identity-specifying image $\mathbf{I}_{in}$ and a target emotional image $\mathbf{I}_t$. Note that only the first two inputs are required in the testing phase.

Audio-to-Landmarks Module. We extract the D-dim Mel-Frequency Cepstral Coefficients (MFCC) feature from the input audio. A one-second temporal sliding window is used to pair audio features with video frames. The sample rate for the audio feature is set as 30, same as the video frame rate.

The audio feature is fed into the audio-to-landmarks module, which is composed of a long short term memory (LSTM) [21] module followed by a fully connected layer. A heatmap of the lower face $\mathbf{H}_m$ is formed by the output landmarks coordinates as shown in Fig. 3. We formulate the regression loss of the audio-to-landmark module as:

$$L_{reg} = ||A(a_{in}) - l_p||_2, \tag{1}$$

where $A(a_{in})$ and l_p are the predicted and ground truth mouth landmark coordinates respectively. Both $A(a_{in})$ and l_p are dimension-reduced representation by applying PCA.

Neutral-to-Emotion Transformer. Emotional image $\mathbf{I}_t$ is expected to be obtained based on the input neutral image $\mathbf{I}_{in}$ and emotion status vector y, specifically, $y = y_{em} \oplus y_{in}$ is the concatenation of two one-hot vectors: emotion category y_{em} and emotion intensity y_{in}. The module is composed of an encoder-decoder architecture where the encoder and decoder are constructed by symmetric 6-layer residual blocks [18] and 4 convolutional layers. We expand the emotion status vector y to the width and height of neutral face image $\mathbf{I}_{in}$ and concatenate them along the color channel as the input of our neutral-to-emotion transformer [42].

We supervise the neutral-to-emotion transformer with the reconstruction loss L_{rec} and perceptual loss [47] L_{con1} as follows:

$$\begin{aligned} L_{rec} &= \lambda_{rec}||\mathbf{I}_t - T(\mathbf{I}_{in}|y)||_1 \\ L_{con1} &= ||VGG_i(T(\mathbf{I}_{in}|y)) - VGG_i(\mathbf{I}_t)||_1, \end{aligned} \tag{2}$$

where the $T(\mathbf{I}_{in}|y)$ is the transformed emotional image, VGG_i is the activation layer for specific layer i on pre-trained VGG_16 [47] model. We can transform the neutral face into a face with input emotion category and intensity to achieve the emotion manipulation on the upper face.

Refinement Network. A refinement network is used to produce the final high-resolution face image $\mathbf{I}_g$ conditioned on the input lower face heatmap $\mathbf{H}_m$ and the generated emotional upper face image $T(\mathbf{I}_{in}|y)$. A U-Net [44] structure is adopted as the generator in this module. To generate realistic talking sequences with natural emotion, we first reconstruct the mouth region with the supervision of the target emotional image $\mathbf{I}_t$, and then constrain a content loss between the output image and $\mathbf{I}_t$ on the whole face. The mouth reconstruction loss and the content loss are defined as:

$$\begin{aligned} L_{mou} &= \lambda_{rec}||M(\mathbf{I}_t) - M(G(T(\mathbf{I}_{in}|y), \mathbf{H}_m))||_1 \\ L_{con2} &= ||VGG_i(G(T(\mathbf{I}_{in}|y), \mathbf{H}_m)) - VGG_i(\mathbf{I}_t)||_1, \end{aligned} \tag{3}$$

where M stands for a mouth area crop function and G is the generation network. We adopt LSGAN [34] scheme to train this module with the following adversarial loss:

$$L_{adv} = \frac{1}{2}E(D(\mathbf{I}_t)^2) + \frac{1}{2}E((1 - D(G(T(\mathbf{I}_{in}|y), \mathbf{H}_m)))^2). \tag{4}$$

Besides, we also use the total variation loss [24] to reduce the spiky artifacts and make the output image smooth:

$$L_{TV} = \lambda_{TV} \sum_{i,j}^{H,W} ||(\mathbf{I}_{g(i+1,j)} - \mathbf{I}_{g(i,j)})^2 + (\mathbf{I}_{g(i,j+1)} - \mathbf{I}_{g(i,j)})^2||, \qquad (5)$$

where $\mathbf{I}_{g(i,j)}$ means the (i,j) pixel of $\mathbf{I}_g$. To further improve the generation quality, two pre-trained classifier models, both trained by cross-entropy loss from given labels, are used for emotion and intensity monitoring. We add two classification losses $L_{c_{em}}$ and $L_{c_{in}}$ to improve the performance of our generation network. Therefore, the final loss function should be formulated as:

$$L_{total} = L_{reg} + L_{rec} + L_{mou} + \lambda_{con}L_{con} + L_{adv} + L_{TV} + L_{c_{em}} + L_{c_{in}}, \qquad (6)$$

where L_{con} is the summation of L_{con1} and L_{con2}. We empirically set all the coefficients of loss terms as 1, except 1e-5 for λ_{TV}.

5   Experiments and Results

5.1   Experiment Setup

Pre-processing. We evaluate our baseline method on our proposed dataset and set aside 20% of all collected data as the test set. For the videos, we crop and align the face in each frame using facial landmarks detected with an open source face alignment tool [56]. For the audios, we extract the 28-dim MFCC features. The size of the audio features of a one-second clip is set as 30×28 in accordance with the frame rate of the video.

Implementation Details. We train our network with Adam [26] optimizer using a learning rate of 0.001, $\beta_1 = 0.5$ and $\beta_2 = 0.999$. We determine the weight coefficients of different loss functions through empirical validation. It takes nearly 4 hours to train the audio-to-landmark module, 24 hours to train the emotion transformer module and another 36 hours to continue training the result refinement module. Note that the transformer module can be trained together with the result refinement as well, but separate training is more stable according to experiment results. The training and testing phases are conducted on a single Nvidia GTX1080Ti GPU with a batch size of 1.

5.2   Baseline Comparison

We compare our emotion controllable talking-face generation approach with the following three methods.

CycleGAN [60] is an unsupervised image translation framework designed by Zhu et al.

ADAVR [58] is an talking-face generation method based on adversarially disentangled audio-visual representation proposed by Zhou et al.

Table 2. User study on emotion category discrimination. The accuracy rate of generated videos is nearly 10% lower compared to that of the captured videos, however, the accuracy distributions of generated and captured videos are basically consistent.

Emotion	angry	disgust	contempt	fear	happy	sad	surprise	neutral	mean
Captured	0.91	0.86	0.82	0.87	0.90	0.93	0.86	0.65	0.85
Generated	0.75	0.76	0.67	0.79	0.81	0.85	0.78	0.52	0.74

Table 3. User study on emotion intensity discrimination. The result is not satisfying enough, illustrating the intensity manipulation strategy still needs further improvement.

Groups	wrong types	angry	disgust	contempt	fear	happy	sad	surprised	mean
Captured	weak medium	0.13	0.17	0.19	0.11	0.12	0.13	0.08	0.13
	medium strong	0.07	0.05	0.09	0.13	0.07	0.10	0.07	0.08
	weak strong	0.03	0.03	0.04	0.04	0.02	0.04	0.03	0.03
	all wrong	0.02	0.01	0.02	0.01	0.01	0.02	0.00	0.01
Generated	weak medium	0.40	0.32	0.37	0.34	0.41	0.32	0.29	0.35
	medium strong	0.31	0.36	0.28	0.31	0.32	0.41	0.28	0.32
	weak strong	0.10	0.08	0.09	0.12	0.07	0.09	0.09	0.09
	all wrong	0.04	0.03	0.06	0.07	0.06	0.06	0.03	0.05

ATVGnet [6] is a hierarchical cross-modal method for talking-face generation.

Emotion Category Manipulation. This experiment attempts to generate talking-face videos with desired emotion category. Note that all the three baselines mentioned above are not capable of directly controlling facial expressions through conditional manipulation as our proposed method is. Therefore, we train several emotion-specific models for the audio-driven methods ATVGnet [6] and ADAVR [58], i.e. they are a set of models, each trained to generate emotional talking-face videos of only one kind of emotion. Similarly, the CycleGANs are each trained to translate a neutral face to a face of one specific emotion category. In Fig. 4, our method is able to generate diversified expressions from the input emotion categories while the audio-driven methods ATVGnet and ADAVR do not produce convincing emotional expressions. The results produced by CycleGAN contain obvious artifacts around areas where intense modification is needed, such as teeth, lips and eyes regions. The unpaired training style of CycleGAN also inherently impairs the temporal continuity of the produced talking-face videos.

We also conduct an compound emotion generation experiment by first generating an intermediate image with one kind of emotion and then modifying the mouth area with another kind of emotion in accordance with the emotion of the input audio. Please see the supplementary materials for the interesting results.

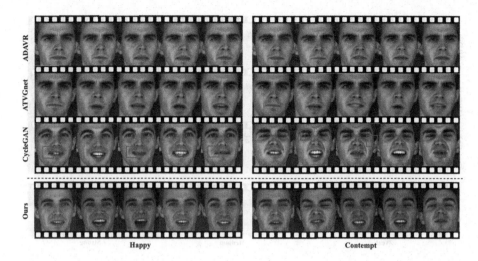

Fig. 4. **Emotion category manipulation.** Comparative results of emotion category manipulation on our dataset: three baseline methods against our method. The face regions in the red rectangles contain obvious artifacts caused by rapid dynamic variation.

Emotion Intensity Manipulation. This experiment attempts to generate talking-face videos with desired emotion category and intensity. Similar to the Emotion Catagory Manipulation experiment, ATVGnet [6], ADAVR [58] and CycleGAN [60] are trained as emotion-intensity-specific models. Figure 5 shows the intensity manipulation results. Specifically, the results from audio-driven methods ATVGnet and ADAVR do not exhibit observable intensity distinction, and the results of the CycleGAN exhibit slight differences between intensity levels while being prone to generating artifacts.

5.3 Evaluation Results for Our Baseline

Generation Quality. To quantitatively evaluate the quality of our generated portrait videos, we generate 48 portrait video clips of 8 different emotions (6 clips each emotion) of each actor for evaluation. We first adopt an emotion-video classification network [37] which achieves state-of-the-art performance on CK+ [33] to evaluate the emotion categories generation accuracy. Note that, the emotion categories in our dataset are consistent with CK+, and we retrained the network with the video data in our dataset and got the best result of accuracy 86.29%. Then we use this model to test on our generated videos of different emotions and got an accuracy of 86.26%, which reflects the genuineness and correctness of the emotion videos we generated. Then, we compare FID score [20] between baseline methods based on videos generated by same audios or videos. As shown in Tab. 4, neutral and surprise videos get the best quality and similarity compared to the training data. Since these baseline methods include no specified module to generate or manipulate emotions, the generated videos either are always recognized

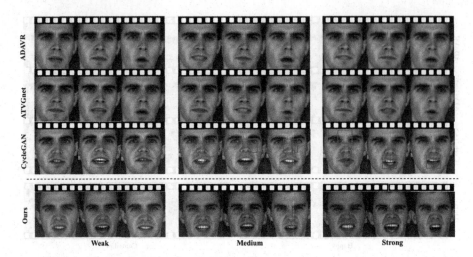

Fig. 5. Emotion intensity manipulation. Comparative results of emotion intensity manipulation on disgust: three baseline methods against our method. The face regions in the red rectangles contain unnatural skin texture and blurred artifacts

Table 4. Evaluation on generation quality. FID scores of different emotion categories. The results indicates that the generated videos can clearly express discrepancy between categories.

emotion	angry	disgust	contempt	fear	happy	sad	surprise	neutral	mean
FID	36.14	36.99	43.02	35.06	32.81	32.64	25.97	28.06	33.84

as neutral or contain strong artifacts which influence the evaluation seriously, resulting in the final FID scores over 100. We also provide quantitative and qualitative results of detailed ablation study in our supplementary materials, when different loss term in Eq. 6 is removed.

Emotion Generation Accuracy. Consistent with the dataset evaluation in Sect. 3.4, we also conduct a user study with 100 participants on our generated videos. The study includes two aspects: the accuracy of the generated facial emotion category and the accuracy of the generated emotion intensity. The emotion category accuracy is shown in Table 2, and the user study results on the ground truth video clips can be used as a reference. In general, the user study accuracy of generated facial emotion has decreased by nearly 10% on average, however, the accuracy distribution still mostly remains the same with the reference. Since some emotions are hard to distinguish, *e.g. fear* and *surprise, disgust* and *contempt*, even nearly 15% of the ground truth video clips are not rated to contain the correct emotion. We get the overall accuracy of 0.74 in all categories, indicating that our method performs well in manipulating facial emotion in portrait videos. As shown in Table 4, neutral and surprise videos get the best quality and similarity with the training data.

Similar as the intensity experiments of the dataset, the participants are given a sample with 3 video clips concatenated that respectively contain 3 different intensities of the same emotion. Then, the participants are asked to rank the emotion intensities and the accuracy is shown in Table 3. In Table 3, we note that the different levels of "sad" is best distinguished while the levels of "fear" and "surprised" are worst distinguished. Only about 20% of participants agree with the comparative emotional intensities generated by our method, which proves that our method cannot provide intensity manipulation results that are convincing enough in all categories.

6 Limitations and Future Work

Although our approach shows the ability of manipulating emotion for talking-face generation, there still exists some limitations. First, our method cannot disentangle emotion from input audio signal. The emotion in the lip is totally up to the input audio, we have not achieved emotion manipulation in the lips. The next step is to edit the whole emotional talking-face driven by neutral audio features. Second, according to the user study results, the generation results still need improvement in image quality and discrimination in emotion intensities. How to measure discrimination in emotion intensities is a challenging problem. One direction towards it is to depend on more explicit auxiliary annotations like FACS, which will be further implemented in our dataset. Furthermore, based on our three levels of intensity, we are getting close to achieving more accurate emotion intensity manipulation on talking-face task. However, the inadequacy of intensity levels in the recent datasets still forms the biggest obstacle for more fine-grained emotion generation. Collecting more fine-grained emotion data may requires smarter design and more complex procedure.

7 Conclusion

The generation of emotion in talking-face generation task is often neglected in previous works due to the absence of suitable emotional audio-visual dataset. We contribute a large-scale high-quality emotional audio-visual dataset, MEAD, providing rich and accurate affective visual and audio information with great detail. The emotional talking head generation baseline trained on MEAD achieves the manipulation of emotion and its intensity with favorable performance compared with current methods. We believe MEAD would benefit the community of talking-face generation and other research fields such as conditional generation, cross-modal understanding and expression recognition.

Acknowledgement. This work is supported by the SenseTime-NTU Collaboration Project, Singapore MOE AcRF Tier 1 (2018-T1-002-056), NTU SUG, and NTU NAP.

References

1. Alghamdi, N., Maddock, S., Marxer, R., Barker, J., Brown, G.J.: A corpus of audio-visual Lombard speech with frontal and profile views. J. Acoust. Soc. Am. **143**(6), EL523 (2018)
2. Anina, I., Zhou, Z., Zhao, G., Pietikäinen, M.: Ouluvs2: a multi-view audiovisual database for non-rigid mouth motion analysis. In: 2015 11th IEEE International Conference and Workshops on Automatic Face and Gesture Recognition (FG), vol. 1, pp. 1–5. IEEE (2015)
3. Burkhardt, F., Paeschke, A., Rolfes, M., Sendlmeier, W.F., Weiss, B.: A database of German emotional speech. In: Ninth European Conference on Speech Communication and Technology (2005)
4. Cao, H., Cooper, D.G., Keutmann, M.K., Gur, R.C., Nenkova, A., Verma, R.: Crema-d: crowd-sourced emotional multimodal actors dataset. IEEE Trans. Affect. Comput. **5**(4), 377–390 (2014)
5. Cao, Y., Tien, W.C., Faloutsos, P., Pighin, F.: Expressive speech-driven facial animation. ACM Trans. Graph. (TOG) **24**(4), 1283–1302 (2005)
6. Chen, L., Maddox, R.K., Duan, Z., Xu, C.: Hierarchical cross-modal talking face generation with dynamic pixel-wise loss. In: Proceedings of the IEEE Conference on Computer Vision and Pattern Recognition, pp. 7832–7841 (2019)
7. Chung, J.S., Nagrani, A., Zisserman, A.: Voxceleb2: deep speaker recognition. In: INTERSPEECH (2018)
8. Chung, J.S., Jamaludin, A., Zisserman, A.: You said that? In: British Machine Vision Conference (2017)
9. Cooke, M., Cunningham, S., Shao, X.: An audio-visual corpus for speech perception and automatic speech recognition. J. Acoust. Soc. Am. **120**, 2421 (2006)
10. Cowie, A.P., Gimson, A.: Oxford Advanced Learner's Dictionary of Current English. Oxford University Press, Oxford (1992)
11. Cowie, R., et al.: Emotion recognition in human-computer interaction. IEEE Signal Process. Mag. **18**(1), 32–80 (2001)
12. Ding, H., Sricharan, K., Chellappa, R.: Exprgan: facial expression editing with controllable expression intensity. In: Thirty-Second AAAI Conference on Artificial Intelligence (2018)
13. Edwards, P., Landreth, C., Fiume, E., Singh, K.: Jali: an animator-centric viseme model for expressive lip synchronization. ACM Trans. Graph. (TOG) **35**(4), 127 (2016)
14. Ezzat, T., Geiger, G., Poggio, T.: Trainable videorealistic speech animation, vol. 21. ACM (2002)
15. Fried, O., et al.: Text-based editing of talking-head video. ACM Trans. Graph. (TOG) **38**, 1–14 (2019)
16. Garofolo, J.S.: Timit acoustic phonetic continuous speech corpus. Linguistic Data Consortium (1993)
17. Harte, N., Gillen, E.: Tcd-timit: an audio-visual corpus of continuous speech. IEEE Trans. Multimedia **17**(5), 603–615 (2015)
18. He, K., Zhang, X., Ren, S., Sun, J.: Deep residual learning for image recognition. In: Proceedings of the IEEE conference on computer vision and pattern recognition, pp. 770–778 (2016)

19. Healy, E.W., Yoho, S.E., Wang, Y., Wang, D.: An algorithm to improve speech recognition in noise for hearing-impaired listeners. J. Acoust. Soc. Am. **134**(4), 3029–3038 (2013)

20. Heusel, M., Ramsauer, H., Unterthiner, T., Nessler, B., Hochreiter, S.: Gans trained by a two time-scale update rule converge to a local nash equilibrium. In: Advances in Neural Information Processing Systems, pp. 6626–6637 (2017)

21. Hochreiter, S., Schmidhuber, J.: Long short-term memory. Neural Comput. **9**(8), 1735–1780 (1997)

22. Huang, X., Liu, M.Y., Belongie, S., Kautz, J.: Multimodal unsupervised image-to-image translation. In: Proceedings of the European Conference on Computer Vision (ECCV), pp. 172–189 (2018)

23. Jackson, P., Haq, S.: Surrey audio-visual expressed emotion (savee) database. http://kahlan.eps.surrey.ac.uk

24. Johnson, J., Alahi, A., Fei-Fei, L.: Perceptual losses for real-time style transfer and super-resolution. In: ECCV (2016)

25. Karras, T., Aila, T., Laine, S., Herva, A., Lehtinen, J.: Audio-driven facial animation by joint end-to-end learning of pose and emotion. ACM Trans. Graph. (TOG) **36**(4), 94 (2017)

26. Kingma, D.P., Ba, J.: Adam: A method for stochastic optimization. arXiv preprint arXiv:1412.6980 (2014)

27. Klautau, A.: Arpabet and the timit alphabet (2001)

28. Kossaifi, J., et al.: Sewa db: A rich database for audio-visual emotion and sentiment research in the wild. arXiv preprint arXiv:1901.02839 (2019)

29. Langner, O., Dotsch, R., Bijlstra, G., Wigboldus, D.H., Hawk, S.T., Van Knippenberg, A.: Presentation and validation of the radboud faces database. Cogn. Emot. **24**(8), 1377–1388 (2010)

30. Lee, H.Y., Tseng, H.Y., Huang, J.B., Singh, M., Yang, M.H.: Diverse image-to-image translation via disentangled representations. In: Proceedings of the European Conference on Computer Vision (ECCV), pp. 35–51 (2018)

31. Lewis, J.: Automated lip-sync: background and techniques. J. Vis. Comput. Animation **2**(4), 118–122 (1991)

32. Livingstone, S.T., Russo, F.A.: The ryerson audio-visual database of emotional speech and song (ravdess): a dynamic, multimodal set of facial and vocal expressions in north American English. PLoS One **13**, e0196391 (2018)

33. Lucey, P., Cohn, J.F., Kanade, T., Saragih, J., Ambadar, Z., Matthews, I.: The extended cohn-kanade dataset (ck+): a complete dataset for action unit and emotion-specified expression. In: 2010 IEEE Computer Society Conference on Computer Vision and Pattern Recognition-Workshops, pp. 94–101. IEEE (2010)

34. Mao, X., Li, Q., Xie, H., Lau, R.Y., Wang, Z., Paul Smolley, S.: Least squares generative adversarial networks. In: Proceedings of the IEEE International Conference on Computer Vision, pp. 2794–2802 (2017)

35. Mariooryad, S., Busso, C.: Analysis and compensation of the reaction lag of evaluators in continuous emotional annotations. In: 2013 Humaine Association Conference on Affective Computing and Intelligent Interaction, pp. 85–90. IEEE (2013)

36. Mattheyses, W., Verhelst, W.: Audiovisual speech synthesis: an overview of the state-of-the-art. Speech Commun. **66**, 182–217 (2015)

37. Meng, D., Peng, X., Wang, K., Qiao, Y.: frame attention networks for facial expression recognition in videos. In: 2019 IEEE International Conference on Image Processing (ICIP), pp. 3866–3870. IEEE (2019). https://github.com/Open-Debin/Emotion-FAN

38. Mollahosseini, A., Hasani, B., Mahoor, M.H.: Affectnet: a database for facial expression, valence, and arousal computing in the wild. IEEE Trans. Affect. Comput. **10**(1), 18–31 (2017)
39. Nagrani, A., Chung, J.S., Zisserman, A.: Voxceleb: a large-scale speaker identification dataset. In: INTERSPEECH (2017)
40. Narayanan, A., Wang, D.: Ideal ratio mask estimation using deep neural networks for robust speech recognition. In: 2013 IEEE International Conference on Acoustics, Speech and Signal Processing, pp. 7092–7096. IEEE (2013)
41. Petridis, S., Shen, J., Cetin, D., Pantic, M.: Visual-only recognition of normal, whispered and silent speech. In: 2018 IEEE International Conference on Acoustics, Speech and Signal Processing (ICASSP), pp. 6219–6223. IEEE (2018)
42. Pumarola, A., Agudo, A., Martinez, A.M., Sanfeliu, A., Moreno-Noguer, F.: Ganimation: anatomically-aware facial animation from a single image. In: Proceedings of the European Conference on Computer Vision (ECCV), pp. 818–833 (2018)
43. Qian, S., et al.: Make a face: towards arbitrary high fidelity face manipulation. In: ICCV (2019)
44. Ronneberger, O., Fischer, P., Brox, T.: U-Net: convolutional networks for biomedical image segmentation. In: Navab, N., Hornegger, J., Wells, W.M., Frangi, A.F. (eds.) MICCAI 2015. LNCS, vol. 9351, pp. 234–241. Springer, Cham (2015). https://doi.org/10.1007/978-3-319-24574-4_28
45. Russell, J.A.: A circumplex model of affect. J. Pers. Soc. Psychol. **39**(6), 1161 (1980)
46. Sadoughi, N., Busso, C.: Speech-driven expressive talking lips with conditional sequential generative adversarial networks. IEEE Trans. Affect. Comput. (2019)
47. Simonyan, K., Zisserman, A.: Very deep convolutional networks for large-scale image recognition. arXiv preprint arXiv:1409.1556 (2014)
48. Song, L., Wu, W., Qian, C., Qian, C., Loy, C.C.: Everybody's talkin': let me talk as you want. arXiv preprint arXiv:2001.05201 (2020)
49. Song, Y., Zhu, J., Wang, X., Qi, H.: Talking face generation by conditional recurrent adversarial network. arXiv preprint arXiv:1804.04786 (2018)
50. Srinivasan, S., Roman, N., Wang, D.: Binary and ratio time-frequency masks for robust speech recognition. Speech Commun. **48**(11), 1486–1501 (2006)
51. Suwajanakorn, S., Seitz, S.M., Kemelmacher-Shlizerman, I.: Synthesizing obama: learning lip sync from audio. ACM Trans. Graph. (TOG) **36**(4), 95 (2017)
52. Vougioukas, K., Petridis, S., Pantic, M.: Realistic speech-driven facial animation with gans. Int. J. Comput. Vis., 1–16 (2019)
53. Wang, T.C., et al.: Video-to-video synthesis. In: NeurIPS (2018)
54. Williams, C.E., Stevens, K.N.: Emotions and speech: some acoustical correlates. J. Acoust. Soc. Am. **52**(4B), 1238–1250 (1972)
55. Wu, W., Cao, K., Li, C., Qian, C., Loy, C.C.: Transgaga: geometry-aware unsupervised image-to-image translation. In: CVPR (2019)
56. Wu, W., Qian, C., Yang, S., Wang, Q., Cai, Y., Zhou, Q.: Look at boundary: a boundary-aware face alignment algorithm. In: Proceedings of the IEEE Conference on Computer Vision and Pattern Recognition, pp. 2129–2138 (2018)
57. Zakharov, E., Shysheya, A., Burkov, E., Lempitsky, V.: Few-shot adversarial learning of realistic neural talking head models. In: Proceedings of the IEEE International Conference on Computer Vision, pp. 9459–9468 (2019)
58. Zhou, H., Liu, Y., Liu, Z., Luo, P., Wang, X.: Talking face generation by adversarially disentangled audio-visual representation. In: The Association for the Advancement of Artificial Intelligence Conference (2019)

59. Zhou, Y., Xu, Z., Landreth, C., Kalogerakis, E., Maji, S., Singh, K.: Visemenet: audio-driven animator-centric speech animation. ACM Trans. Graph. (TOG) **37**(4), 161 (2018)
60. Zhu, J.Y., Park, T., Isola, P., Efros, A.A.: Unpaired image-to-image translation using cycle-consistent adversarial networks. In: Proceedings of the IEEE International Conference on Computer Vision, pp. 2223–2232 (2017)

Detecting Human-Object Interactions with Action Co-occurrence Priors

Dong-Jin Kim[1], Xiao Sun[2], Jinsoo Choi[1], Stephen Lin[2], and In So Kweon[1(✉)]

[1] KAIST, Daejeon, South Korea
{djnjusa,jinsc37,iskweon77}@kaist.ac.kr
[2] Microsoft Research, Beijing, China
{xias,stevelin}@microsoft.com
https://github.com/Dong-JinKim/ActionCooccurrencePriors/

Abstract. A common problem in human-object interaction (HOI) detection task is that numerous HOI classes have only a small number of labeled examples, resulting in training sets with a long-tailed distribution. The lack of positive labels can lead to low classification accuracy for these classes. Towards addressing this issue, we observe that there exist natural correlations and anti-correlations among human-object interactions. In this paper, we model the correlations as *action co-occurrence matrices* and present techniques to learn these priors and leverage them for more effective training, especially on rare classes. The utility of our approach is demonstrated experimentally, where the performance of our approach exceeds the state-of-the-art methods on both of the two leading HOI detection benchmark datasets, HICO-Det and V-COCO.

1 Introduction

Human-object interaction (HOI) detection aims to localize humans and objects in an image and infer the relationships between them. An HOI is typically represented as a human-action-object triplet with the corresponding bounding boxes and classes. Detecting these interactions is a fundamental challenge in visual recognition that requires both an understanding of object information and high-level knowledge of interactions.

A major issue that exists in HOI detection is that its datasets suffer from long-tailed distributions, in which many HOI triplets have few labeled instances. Similar to datasets for general visual relationship detection (VRD) [37], a reason for this is missing labels, where the annotation covers only a subset of the interactions present in an image. For the widely-used HICO-Det dataset [3], 462 out of the 600 HOI classes have fewer than 10 training samples. For such classes, the lack of positive labels can lead to inadequate training and low classification performance. How to alleviate the performance degradation on rare classes is thus a key issue in HOI detection.

Electronic supplementary material The online version of this chapter (https://doi.org/10.1007/978-3-030-58589-1_43) contains supplementary material, which is available to authorized users.

© Springer Nature Switzerland AG 2020
A. Vedaldi et al. (Eds.): ECCV 2020, LNCS 12366, pp. 718–736, 2020.
https://doi.org/10.1007/978-3-030-58589-1_43

P(operate-hair_drier) = 0.001   P(blow-cake) = 0.007

P(operate-hair_drier | hold-hair_drier) = 0.958   P(blow-cake | cut-cake) = 0.005

Fig. 1. Marginal/conditional probability values are computed from the distribution of the training label. Intuitively, detection of HOIs (operate-hair dryer) can be facilitated by detection of commonly co-occurring HOIs (hold-hair dryer). Also, non-detection of HOIs (blow-cake) can be aided by detection of incompatible HOIs (cut-cake). We leverage this intuition as a prior to learn an HOI detector on long-tailed datasets.

To address the problem of long-tailed distributions, we propose to take advantage of natural *co-occurrences* in human actions. For example, the HOI of 'operate-hair dryer' is rarely labeled and consequently hard to detect in the left image of Fig. 1. However, 'operate-hair dryer' often occurs when the more commonly labeled HOI of 'hold-hair dryer' is present. As a result, detection of 'operate-hair dryer' can be facilitated by detection of 'hold-hair dryer' in an image. On the other hand, the detection of an HOI may preclude other incompatible HOIs, such as for 'cut-cake' and 'blow-cake' in the right image of Fig. 1.

In this paper, we introduce the new concept of utilizing co-occurring actions as prior knowledge, termed as action co-occurrence priors (ACPs), to train an HOI detector. In contrast to the language-based prior knowledge which requires external data sources [27,37,61], the co-occurrence priors can be easily obtained from the statistics of the target dataset. We also propose two novel ways to exploit them. First, we design a neural network with hierarchical structure where the classification is initially performed with respect to *action groups*. Each action group is defined by one anchor action, where the anchor actions are mutually exclusive according to the co-occurrence prior. Then, our model predicts the fine-grained HOI class within the action group. Second, we present a technique that employs knowledge distillation [20] to expand HOI labels so they can have more positive labels for potentially co-occurring actions. During training, the predictions are regularized by the refined objectives to improve robustness, especially for classes in the data distribution tail. To the best of our knowledge, we are the first to leverage the label co-occurrence in HOI detection to alleviate long-tailed distribution.

The main contributions of this work can be summarized as: (1) The novel concept of explicitly leveraging correlations among HOI labels to address the problem of long-tailed distributions in HOI detection; (2) Two orthogonal ways to leverage action co-occurrence priors, namely through a proposed hierarchical architecture and HOI label expansion via knowledge distillation. The resulting model is shown to be consistently advantageous in relation to state-of-the-art techniques on both the HICO-Det [3] and V-COCO [16] benchmark datasets.

2   Related Work

Human-Object Interaction. Human-Object Interaction was originally studied in the context of recognizing the function or 'affordance' of objects [6,12,14, 15,46]. Early works focus on learning more discriminative features combined with variants of SVM classifiers [7,8,57], and leverage the relationship with human poses for better representation [7,8,59] or mutual context modeling [58].

Recently, a completely data-driven approach based on convolutional neural networks (CNNs) has brought dramatic progress to HOI. Many of the pioneering works created large scale image datasets [3,4,16,68] to set new benchmarks in this field. Henceforth, significant progress have been seen in using CNNs for this problem [1,3,11,13,18,25,32,33,35,43,45,49,50,52–54]. Most of these works follow a two-step scheme of CNN feature extraction and multi-information fusion, where the multiple information may include human and object appearance [3,11,13,43]; box relation (either box configuration or spatial map) [1,3,13,18,54]; object category [1,18,41]; human pose [18,32,33]; and particularly, linguistic prior knowledge [25,41]. More recent works tend to combine these various cues [18,33,50,53]. These works differ from one another mainly in their techniques for exploiting external knowledge priors. Kato *et al.* [25] incorporate information from WordNet [39] using a Graph Convolutional Network (GCN) [29] and learn to compose new HOIs. Xu *et al.* [54] also use a GCN to model the general dependencies among actions and object categories by leveraging a VRD dataset [37]. Li *et al.* [33] utilize interactiveness knowledge learned across multiple HOI datasets. Peyre *et al.* [41] transfer knowledge from triplets seen at training to new unseen triplets at test time by analogy reasoning.

Different from the previous works, our approach is to reformulate the target action label space and corresponding loss function in a manner that leverages *co-occurrence* relationships among action classes for HOI detection. In principle, the proposed technique is complementary to all of the previous works and can be combined with any of them. For our experiments, we implemented our approach on a baseline presented in [18], with details given in Sect. 3.3.

Visual Relationship Detection. The closest problem to HOI detection is Visual Relationship Detection (VRD) [5,31,42,56,60–62,64,65,67], which deals with general visual relationships between two arbitrary objects. In the VRD datasets [30,37], the types of visual relationships that are modeled include verb (action), preposition, spatial and comparative phrase. The two tasks share common challenges such as long-tail distributions or even zero-shot problems [37]. Our work focuses on HOI detection, as co-occurrences of human-object interactions are often strong, but the proposed technique could be extended to model the general co-occurrences that exist in visual relationships.

Label Hierarchy in Multi-label Learning. The hierarchical structure of label categories has long been exploited for multi-label learning in various vision tasks, *e.g.*, image/object classification [9,55], detection [23,38], and human pose estimation [47,48]. In contrast, label hierarchy has rarely been considered in

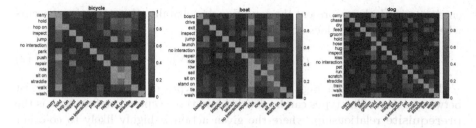

Fig. 2. Examples of co-occurrence matrices constructed for several objects (bicycle, boat, dog). Along the Y-axis is the given action, and the X-axis enumerates conditional actions. Each element represents the conditional probability that an action occurs when another action is happening.

HOI detection. Inspired by previous work [9] that uses Hierarchy and Exclusion (HEX) graphs to encode flexible relations between object labels, we present the first method to take advantage of an action label hierarchy for HOI recognition. While label hierarchies have commonly been used, our method is different in that it is defined by co-occurrences (rather than semantics or a taxonomy [9]). This co-occurrence based hierarchy can be determined statistically, without direct human supervision.

3   Proposed Method

Our method for utilizing the co-occurrence information of HOI labels consists of three key components: (1) establishing action co-occurrence priors (Sect. 3.1), (2) hierarchical learning including anchor action selection (Sect. 3.2) and devising the hierarchical architecture (Sect. 3.3), and (3) ACP projection for knowledge distillation (Sect. 3.4).

3.1   Action Co-occurrence Priors

Here, we formalize the action co-occurrence priors. The priors for the actions are modeled by a co-occurrence matrix $C \in \mathbb{R}^{N \times N}$ where an entry c_{ij} in C represents the conditional probability that action j occurs when action i is happening:

$$c_{ij} = p(j|i), i, j \in [0, N), \tag{1}$$

where N denotes the total number of actions classes and i, j are indices of two actions. C is constructed from the target HOI detection dataset by counting the image-level statistics of its training labels. Examples of co-occurrence matrices constructed for single object are visualized in Fig. 2.

Meanwhile, we also consider the complementary event of action i (*i.e.*, where the i-th action does not occur) and denote it as i', such that $p(i') + p(i) = 1$. The complementary action co-occurrence matrix $C' \in \mathbb{R}^{N \times N}$ can thus be defined by entries c'_{ij} in C' that represent the conditional probability that an action j occurs when another action i does not occur:

$$c'_{ij} = p(j|i'), i, j \in [0, N). \tag{2}$$

It can be seen from Fig. 2, that different types of relationships can exist between actions. The types can be divided into three types. The first type is the **prerequisite** relationship, where the given action is highly likely to co-occur with the conditional action. For example, the HOI 'sit on-bicycle' is a prerequisite of the HOI 'ride-bicycle'. In this case, p(sit on-bicycle|ride-bicycle) is close to 1. Next, **exclusion**, where the given action is highly unlikely to co-occur with the conditional action. An example is that the HOI 'wash-bicycle' and HOI 'ride-bicycle' are unlikely to happen together. As a result, p(wash-bicycle|ride bicycle) is close to 0. Finally, **overlapping**, where the given action and conditional action may possibly co-occur, for example HOI 'hold-bicycle' and HOI 'inspect-bicycle', such that p(hold-bicycle|inspect-bicycle) is in between 0 and 1.

The strong relationships that may exist between action labels can provide strong priors on the presence or absence of an HOI in an image. In contrast to previous works where models may implicitly learn label co-occurrence via relational architectures [2,63], we explicitly exploit these relationships between action labels as priors, to effectively train our model especially for rare HOIs.

3.2 Anchor Action Selection via Non-exclusive Action Suppression

From a co-occurrence matrix for an object, it can be seen that some actions are close in semantics or commonly co-occur while others are not. Intuitively, closely related actions (*e.g.*, 'sit on-bicycle' and 'straddle-bicycle') tend to be harder to distinguish from each other. If the positive labels for these actions are rare, then they become even more difficult to distinguish. Such cases require fine-grained recognition [10] and demand more dedicated classifiers. This motivates us to learn HOI classes in a coarse-to-fine manner. Specifically, we first identify a set of mutually exclusive action classes, called *anchor actions*, which tend to be distinguishable from one another. The anchor actions will be used to partition the entire action label space into fine-grained sub-spaces. The other action classes will be attributed to one or more sub-spaces and recognized in the context of a specific anchor action. In summary, unlike previous HOI detection works which predict action probabilities independently of one another, we divide the whole action label set into two sets, one for anchor actions and one for regular actions, which are modeled in different ways as explained in detail in Sect. 3.3.

In selecting anchor actions, we seek a set of action classes that are exclusive of one another. Toward this end, we define the exclusiveness of an action class as counting the number of actions that will never occur if action i is happening ($e_i = \sum_j (1$ if $(c_{ij} = 0)$, else $0))$. e_i will have a high value if few other actions can occur when i does. Based on exclusiveness values, the anchor action label set $\mathcal{D}$ is generated through ***non-exclusive suppression (NES)***. It iteratively finds the most exclusive action class as an anchor action and removes remaining action classes that are not exclusive to the selected anchor actions. The anchors in the list are action classes that never occur together in the training labels. For example, if an action such as 'toast' is on the anchor list, then actions like 'stand' and 'sit' cannot be on the list because they may co-occur with 'toast', while actions such as 'hunt' or 'hop on' can potentially be on the list. While there may exist other ways the anchor action selection could be done, we empirically found this approach to be simple, effective (detection accuracy), and efficient (less than 0.01 s).

The anchor action label set acts as a finite partition of the action label space (a set of pairwise disjoint events whose union is the entire action label space). To form a complete action label space, we add an *'other'* anchor action, denoted as $\mathcal{O}$, for when an action class does not belong to $\mathcal{D}$. Finally, we get $|\mathcal{D}| + 1$ anchor actions including $\mathcal{D}$ and the 'other' action class $\mathcal{O}$.

There are several benefits to having this anchor action label set. First, only one anchor action can happen at one time between a given human and object. Thus, we can use the relative (one hot) probability representation with *softmax* activation, whose features were shown to compare well against distance metric learning-based features [21]. Second, anchor actions tend to be easier to distinguish from one another since they generally have prominent differences in an image. Third, it decomposes the action label space into several sub-spaces, which facilitates a coarse-to-fine solution. Each sub-task will have a much smaller solution space, which can improve learning. Finally, each sub-task will use a standalone sub-network which focuses on image features specific to the sub-task, which is an effective strategy for fine-grained recognition [10].

After selecting anchor actions, the entire action label set $\mathcal{A}$ is divided into the anchor action label set $\mathcal{D}$ and the remaining set of 'regular' action classes $\mathcal{R}$, so that $\mathcal{A} = \{\mathcal{D}, \mathcal{R}\}$. Each of the regular action classes is then associated with one or more anchor actions to form $|\mathcal{D}| + 1$ action groups $\mathbf{G} = \{\mathcal{G}_i; i \in \mathcal{D} \cup \mathcal{O}\}$, one for each anchor action. A regular action class $j \in \mathcal{R}$ will be assigned to the group of anchor action i ($\mathcal{G}_i$) if action j is able to co-occur with the anchor action i,

$$j \in \mathcal{G}_i, \text{if } c_{ij} > 0 \quad (i \in \mathcal{D} \cup \mathcal{O}, j \in \mathcal{R}). \tag{3}$$

Note that the anchor actions themselves are not included in the action groups and a regular action j can be assigned to multiple action groups since it may co-occur with multiple anchor actions.

3.3 Hierarchical Architecture

We implemented our hierarchical approach upon the 'No-Frills' (NFs) baseline presented in [18] on account of its simplicity, effectiveness, and code availability [17]. Here, we give a brief review of the NFs architecture.

Baseline Network. NFs follows the common scheme of CNN feature extraction followed by multi-information fusion. It uses the off-the-shelf Faster R-CNN [44] object detector with ResNet152 [19] backbone network to detect human and object bounding boxes. As illustrated in Fig. 3, the multiple information used in [18] (denoted as X) are fed through four separate network streams to generate fixed dimension features. Then, all the features are added together and sent through a *sigmoid* activation to get the action probability prediction $\hat{A}$:

$$\hat{A} = sigmoid(F(X)) \in \mathbb{R}^N, \tag{4}$$

where $\hat{A}(a) = p(a|X)$ represents the probability prediction for action class a.

To eliminate training-inference mismatch, NFs directly optimizes the HOI class probabilities instead of separating the detection and interaction losses as done in [11,13]. The final HOI prediction is a joint probability distribution over M number of HOI classes computed from the probabilities $\hat{H}$, $\hat{O}$, and $\hat{A}$ for human, object, and action, respectively:

$$\hat{Y} = joint(\hat{H}, \hat{O}, \hat{A}) \in \mathbb{R}^M. \tag{5}$$

Specifically, for a HOI class (h, o, a),

$$\hat{Y}(h, o, a) = \hat{H}(h) * \hat{O}(o) * \hat{A}(a) = p(h|I) * p(o|I) * p(a|X), \tag{6}$$

where $\hat{H}(h) = p(h|I)$ and $\hat{O}(o) = p(o|I)$ are the probability of a candidate box pair being a human h and object o, provided by the object detector [44]. Finally, the binary cross-entropy loss $\mathcal{L}(\hat{Y}, Y^{gt})$ is directly computed from the HOI prediction. This 'No-Frills' baseline network is referred to as **Baseline**.

Modified Baseline Network. For a stronger baseline comparison, we make two simple but very effective modifications on the baseline network. (1) Replace the one-hot representation with the Glove word2vec [40] representation for the object category. (2) Instead of directly adding up the multiple information, we average them and forward this through another *action prediction module* to obtain the final action probability prediction. For a naive approach (the **Modified Baseline**), we simply use a sub-network f_{sub} of a few FC layers as the *action prediction module*. Then Eq. (4) is modified to

$$\hat{A} = sigmoid(f_{sub}(F(X))). \tag{7}$$

Our hierarchical architecture further modifies the *action prediction module* by explicitly exploiting ACP information which is described in the next paragraph.

Proposed Hierarchical Architecture. Now we introduce the *action prediction module* for our hierarchical architecture (illustrated in Fig. 3) that better exploits the inherent co-occurrence among actions. While the baseline network predicts all the action probabilities directly from $F(\cdot)$ with a single feed-forward sub-network f_{sub}, we instead use $|\mathcal{D}| + 2$ sub-networks where one $(f_{anchor}(\cdot))$ is first applied to predict the anchor action set and then one of the $|\mathcal{D}| + 1$ other sub-networks $(f_{g_i}(\cdot))$ which corresponds to the predicted anchor action is used to estimate the specific action within the action group. Because of the mutually exclusive property of anchor actions, we use the *softmax* activation for anchor action predictions, while employing the *sigmoid* activation for regular action predictions conditional to the action groups:

$$\hat{A}_{anchor} = softmax(f_{anchor}(F(X))) \in \mathbb{R}^{|\mathcal{D}|+1} \tag{8}$$

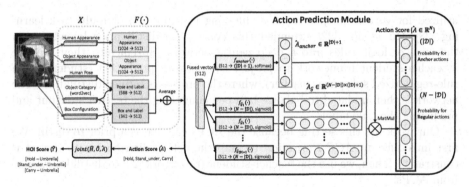

Fig. 3. Illustration of our overall network architecture. Our work differs from the baseline [18] by the addition of a hierarchical *action prediction module*. For our hierarchical architecture, anchor action probability is directly generated by a *softmax* sub-network. Regular action probability is generated by a matrix multiplication of the anchor probability and the output from a few *sigmoid* based conditional sub-networks.

$$\hat{A}_{\mathcal{G}_i} = sigmoid(f_{\mathcal{G}_i}(F(X))) \in \mathbb{R}^{N-|\mathcal{D}|}, \text{where } i \in \mathcal{D} \cup \mathcal{O}, \tag{9}$$

where $\hat{A}_{anchor}(i)$ is directly used as the final probability predictions for the anchor actions $(p(i|X) = \hat{A}_{anchor}(i), i \in \mathcal{D})$. We let $\hat{A}_{\mathcal{G}_i}(j)$ represent the learned conditional probability that action j occurs when action i is happening $(p(j|i, X) = \hat{A}_{\mathcal{G}_i}(j))$. Since the anchor action set is a finite partition of the entire action label space, the probability of a regular action j can be predicted according to the law of total probability:

$$\hat{A}_{regular}(j) = p(j|X) = \sum_{i \in \mathcal{D} \cup \mathcal{O}} p(i|X) * p(j|i, X) = \sum_{i \in \mathcal{D} \cup \mathcal{O}} \hat{A}_{anchor}(i) * \hat{A}_{\mathcal{G}_i}(j), \tag{10}$$

where $j \in \mathcal{R}$. Thus, instead of Eq. (7), we obtain the final action probability predictions for our hierarchical architecture $\hat{A}(a) = p(a|X)$ as

$$\hat{A}(a) = \begin{cases} \hat{A}_{anchor}(a), & \text{if } a \in \mathcal{D} \\ \sum_{i \in \mathcal{D} \cup \mathcal{O}} \hat{A}_{anchor}(i) * \hat{A}_{\mathcal{G}_i}(a), & \text{otherwise.} \end{cases} \tag{11}$$

We use the same method as in Eq. (6) and cross-entropy loss to compute the final HOI probability prediction $\hat{Y}$ and the corresponding loss $\mathcal{L}$.

To demonstrate the effectiveness of the hierarchical learning, we introduce another two baselines, *MultiTask* and *TwoStream*, that lie between the *Modified Baseline* and our hierarchical learning. *MultiTask* only uses the anchor action classification as an additional multi-task element to the *Modified Baseline*. *TwoStream* separately predicts the anchor and the regular actions but without using the hierarchical modeling between anchor and regular actions.

3.4 ACP Projection for Knowledge Distillation

Knowledge distillation [20] was originally proposed to transfer knowledge from a large network to a smaller one. Recently, knowledge distillation has been utilized for various purpose such as life-long learning [34] or multi-task learning [28]. Hu *et al.* [22] extended this concept to distill prior knowledge in the form of logic rules into a deep neural network. Specifically, they propose a teacher-student framework to project the network prediction (student) to a rule-regularized subspace (teacher), where the process is termed distillation. The network is then updated to balance between emulating the teacher's output and predicting the true labels.

Our work fits this setting because the ACPs can act as a prior to distill. We first introduce an *ACP Projection* to map the action distribution to the ACP constraints. Then, we use the teacher-student framework [22] to distill knowledge from ACPs.

ACP Projection. In ACP Projection, an arbitrary action distribution $A = \{p(i), i \in [0, N)\} \in \mathbb{R}^N$ is projected into the ACP-constrained probability space:

$$A^* = project(A, C, C') \in \mathbb{R}^N, \tag{12}$$

where A^* is the projected action prediction. The projected probability for the j-th action $A^*(j) = p(j^*)$ is generated using the law of total probability:

$$p(j^*) = \frac{1}{N} \sum_{i=1}^{N} (p(i)*p(j|i) + p(i')*p(j|i')) = \frac{1}{N} (\sum_{i=1}^{N} p(i)*c_{ij} + \sum_{i=1}^{N} (1-p(i))*c'_{ij}). \tag{13}$$

In matrix form, the ACP projection is expressed as

$$project(A, C, C') = \frac{AC + (1 - A)C'}{N}. \tag{14}$$

In practice, we use the object-based action co-occurrence matrices $C_o \in \mathbb{R}^{N \times N}$ and $C'_o \in \mathbb{R}^{N \times N}$ which only count actions related to a specific object o. Figure 2 shows examples of C_o with respect to object classes. Also, we give different weights α and β as hyper-parameters to the action co-occurrence matrix C_o and its complementary matrix C'_o, with the weights subject to $\alpha + \beta = 2, \alpha > \beta$. The projection function is then modified as

$$project(A, C_o, C'_o) = \frac{\alpha A C_o + \beta(1 - A)C'_o}{N}. \tag{15}$$

This is done because we empirically found the co-occurrence relationships in C_o to generally be much stronger then the complementary actions in C'_o.

Teacher-Student Framework. Now we can distill knowledge from the ACPs using ACP Projection in both the training and inference phases. There are three ways ACP Projection can be used: (1) Directly project the action prediction $\hat{A}$ into the ACP-constrained probability space at the testing phase to obtain the final action output (denoted as **PostProcess**). (2) Project the action prediction $\hat{A}$ in the training phase and use the projected action as an additional learning target [22,61]. (3) Project the ground truth label H^{gt}, O^{gt}, and $A^{gt\,1}$ to the ACP space in the training phase and use the projected action $project(A^{gt}, C_{O^{gt}}, C'_{O^{gt}})$ as an additional learning target. The second and third items are incorporated into the teacher-student framework as terms in a new objective function (denoted as **Distillation**):

$$\mathcal{L}_{total} = \lambda_1 \mathcal{L}(\hat{Y}, Y^{gt}) + \lambda_2 \mathcal{L}(\hat{Y}, \hat{Y}_{projO}) + \lambda_3 \mathcal{L}(\hat{Y}, Y^{gt}_{projO}), \qquad (16)$$

where

$$\hat{Y}_{projO} = joint(\hat{H}, \hat{O}, project(\hat{A}, C_{\hat{O}}, C'_{\hat{O}})) \in \mathbb{R}^M, \qquad (17)$$

$$Y^{gt}_{projO} = joint(H^{gt}, O^{gt}, project(A^{gt}, C_{O^{gt}}, C'_{O^{gt}})) \in \mathbb{R}^M. \qquad (18)$$

$\lambda_1, \lambda_2, \lambda_3$ are balancing weights among the ground truth HOI term and the teacher objectives. The object type can be easily determined from the object probability predictions $\hat{O}$ or the ground truth label O^{gt}.

4   Experiments

The goal of the experiments is to show the effectiveness and generalizability of our method. In particular, we show that our method can consistently alleviate the long-tailed distribution problem in various setups by improving performance especially for rare HOI classes. In this section, we describe the experimental setups, competing methods and provide performance evaluations of HOI detection.

4.1   Datasets and Metrics

We evaluate the performance of our model on the two popular HOI detection benchmark datasets, **HICO-Det** [3] and **V-COCO** [16]. HICO-Det [3] extends the HICO (Humans Interacting with Common Objects) dataset [4] which contains 600 human-object interaction categories for 80 objects. Different from the HICO dataset, HICO-Det additionally contains bounding box annotations for humans and objects of each HOI category. The vocabulary of objects matches the 80 categories of MS COCO [36], and there are 117 different verb (action) categories. The number of all possible triplets is 117×80, but the dataset contains positive examples for only 600 triplets. The training set of HICO-Det contains 38,118 images and 117,871 HOI annotations for 600 HOI classes. The test set has 9,658 images and 33,405 HOI instances.

For evaluation, HICO-Det uses the mean average precision (mAP) metric. Here, an HOI detection is counted as a true positive if the minimum of the human overlap IOU and object overlap IOU with the ground truth is greater than 0.5. Following [3], HOI detection performance is reported for three different HOI category sets: (1) all 600 HOI categories (Full), (2) 138 categories with fewer

---

[1] The triplet ground truth labels H^{gt}, O^{gt}, and A^{gt} are straightforward to determine from the HOI ground truth label Y^{gt}.

Table 1. Ablation study on the HICO-Det dataset. Our final model that includes both hierarchical architecture and distillation followed by post processing (Ours, ACP) shows the best performance among the baselines.

	Full	Rare	Non-rare
Baseline	17.56	13.23	18.85
Modified Baseline	19.09	13.09	20.89
+Hierarchical only	20.03	14.52	21.67
+Distillation only	19.98	13.67	21.86
+Hierarchical+Distillation	20.25	15.33	21.72
+Hierarchical+Distillation +Post (Ours, ACP)	**20.59**	**15.92**	**21.98**

Table 2. Performance of our models with different architectures for action prediction module. Our model ((D) Hierarchical) shows the best performance among the design choices.

	Full	Rare	Non-rare
(A) Modified Baseline	19.09	13.09	20.89
(B) MultiTask	19.54	13.93	21.22
(C) TwoStream	19.63	13.67	21.41
(D) Hierarchical	**20.03**	**14.52**	**21.67**

than 10 training samples (Rare), and (3) the remaining 462 categories with more than 10 training samples (Non-rare).

V-COCO (Verbs in COCO) is a subset of MS-COCO [36], which consists of 10,346 images (2,533, 2,867, 4,946 for training, validation and test, respectively) and 16,199 human instances. Each person is annotated with binary labels for 26 action classes. For the evaluation metric, same as for evaluation on HICO-Det, we use the AP score.

4.2 Quantitative Results

Ablation Study. In the ablations, the 'No-Frills' baseline network [18] we used is denoted as the *Baseline*. We first evaluate the effectiveness of the core design components in the proposed method including (1) our simple modification to *Baseline* in Sect. 3.3, denoted as *Modified Baseline*; (2) the hierarchical learning technique introduced in Sect. 3.3, denoted as *Hierarchical*; and (3) the knowledge distillation technique presented in Eq. (16) of Sect. 3.4, denoted as *Distillation*. Table 1 gives a comprehensive evaluation for each component. We draw conclusions from it one-by-one.

First, our baseline network is strong. Our *Modified Baseline* achieves 19.09 mAP and surpasses the 'No-Frills' *Baseline* by 1.51 mAP (a relative 8.7% improvement), which is already competitive to the state-of-the-art result [41] and serves as a strong baseline.

Second, both hierarchical learning and knowledge distillation are effective. This is concluded by adding *Hierarchical* and *Distillation* to the *Modified Baseline*, respectively. Specifically, *+Hierarchical* improves the modified baseline by 0.94 mAP (a relative 4.9% improvement), and *+Distillation* (training with Eq. (16)) improves the modified baseline by 0.89 mAP (a relative 4.7% improvement). Including both obtains 1.16 mAP improvement (relatively better by 6.1%).

Table 3. Results on the HICO-Det dataset compared to the previous state-of-the-art methods. Our model shows favorable performance against the current state-of-the-art models on all the metrics.

	Full	Rare	Non-rare
Shen et al. [45]	6.46	4.24	7.12
HO-RCNN [3]	7.81	5.37	8.54
Gupta et al. [16] impl. by [13]	9.09	7.02	9.71
InteractNet [13]	9.94	7.16	10.77
GPNN [43]	13.11	9.34	14.23
iCAN [11]	14.84	10.45	16.15
iHOI [53]	13.39	9.51	14.55
with Knowledge [54]	14.70	13.26	15.13
Interactiveness Prior [33]	17.22	13.51	18.32
Contextual Attention [51]	16.24	11.16	17.75
No-Frills [18]	17.18	12.17	18.68
RPNN [66]	17.35	12.78	18.71
PMFNet [50]	17.46	15.65	18.00
Peyre et al. [41]	19.40	15.40	20.75
Our baseline	17.56	13.23	18.85
ACP (Ours)	**20.59**	**15.92**	**21.98**

Table 4. Results on the V-COCO dataset. For our method, we show results both for constructing the ACP from V-COCO and for using the ACP constructed from HICO-Det instead. Both of these models show favorable performance against the current state-of-the-art models.

	AP_{role}
Gupta et al. [16] impl. by [13]	31.8
InteractNet [13]	40.0
GPNN [43]	44.0
ICAN [11]	45.3
iHOI [53]	45.79
with Knowledge [54]	45.9
Interactiveness Prior [33]	48.7
Contextual Attention [51]	47.3
RPNN [66]	47.53
PMFNet [50]	52.0
Our baseline	48.91
ACP (Ours, V-COCO)	**52.98**
ACP (Ours, HICO-Det)	**53.23**

Third, the proposed ACP method achieves a new state-of-the-art. Our final result is generated by further using the *PostProcess* step (introduced in Sect. 3.4) that projects the final action prediction into the ACP constrained space. Our method achieves 20.59 mAP (relative 7.9% improvement) for Full HOI categories, 15.92 mAP (relative 21.6% improvement) for Rare HOI categories, and 21.98 mAP (relative 5.2% improvement) for Non-rare HOI categories. Note that our method made especially significant improvements for Rare classes, which supports the claim that the proposed method can alleviate the long-tailed distribution problem of HOI detection datasets. This result sets the new state-of-the-art on both the HICO-Det and V-COCO datasets as shown in Table 3 and Table 4.

In addition, the *MultiTask*, *TwoStream*, and our *Hierarchical* architecture are compared in Table 2. From *MultiTask*, it can be seen that the *softmax* based anchor action classification already brings benefits to the *Modified Baseline* when used only in a multi-task learning manner. From *TwoStream*, separately modeling the anchor and the regular classes leads to a slight more improvements compared to *MultiTask*. Moreover, our *Hierarchical* architecture improves upon *TwoStream* by explicitly modeling the hierarchy between anchor and regular action predictions.

Comparison with the State-of-the-Art. We compare our method with the previous state-of-the-art techniques in Table 3. Among the methods included in this comparison are the benchmark model of the HICO-Det dataset [3], the baseline model that we modified from [18], and the current published state-of-the-art method [41]. As shown in Table 3, our final model (Ours) shows significant improvements over our baseline model on all metrics, and shows favorable performance against the current state-of-the-art model in terms of all the metrics. In particular, our model surpasses the current state-of-the-art model [41] by 1.19 mAP.

Results on V-COCO Dataset. To show the generalizability of our method, we also evaluate our method on the V-COCO dataset. Note that the exact same method is directly applied to both HICO-Det and V-COCO, including the co-occurrence matrix, anchor action selection, and the architecture design. We also constructed a co-occurrence matrix from V-COCO, but the matrix was sparse. Thus, to better take advantage of our idea, we instead use the co-occurrence matrix collected from the HICO-Det dataset to train on the V-COCO. Table 4 shows the performance of our model (Ours, HICO-Det) compared to the recent state-of-the-art HOI detectors on the V-COCO dataset. In addition, we show results of our model with the co-occurrence matrix constructed from the V-COCO dataset (Ours, V-COCO). Both of these models show favorable performance on the V-COCO dataset against the previous state-of-the-art model [50].

Table 5. Results on the zero-shot triplets of the HICO-Det dataset. Our final model shows better performance than Peyre *et al.* by a large margin. Note that our ACP model under the zero-shot setting even outperforms the supervised setting of our baseline.

	mAP
Peyre *et al.* [41] Supervised	33.7
Peyre *et al.* [41] Zero-Shot	24.1
Peyre *et al.* [41] Zero-Shot with Aggregation	28.6
Ours Supervised (Modified Baseline)	33.27
Ours Zero-Shot (Modified Baseline)	20.34
Ours Zero-Shot (ACP)	**34.95**

Results of the Zero-Shot Setup on the HICO-Det Dataset. The *zero-shot setting* on the HICO-Det dataset is defined by Peyre *et al.* [41]. Specifically, we select a set of 25 HOI classes that we treat as unseen classes and exclude them and their labels in the training phase. However, we still let the model predict those 25 unseen classes in the test phase, which is known as the zero-shot problem. These HOI classes are randomly selected among the set of non-rare HOI classes. Since Peyre *et al.* did not provide which specific HOI classes they selected, we select the unseen HOI classes such that the performance (mAP) for these classes in our *Modified Baseline* model (introduced in Sect. 3.3) is similar to the corresponding *Supervised* baseline in [41]. In Table 5, we show results of our final model (ACP) and our modified baseline model compared to the corresponding setting reported in [41]. Our final model shows better performance (35.0 mAP) than Peyre *et al.* (28.6 mAP) by a large margin (relative 22.4% improvement). This result is remarkable in that our ACP model under the zero-shot setting even *outperforms* the supervised setting of our baseline model, indicating the power of the proposed ACP method to effectively leverage prior knowledge on action co-occurrences. Furthermore, the analogy transfer method proposed by Peyre *et al.* (denoted as aggregation) requires large-scale linguistic knowledge to train a word representation, whereas our model only requires the co-occurrence information of the labels in the dataset, which is much easier to obtain. We conclude that the proposed method is effective for the zero-shot problem while being easy to implement.

4.3 Additional Analysis

Score Improvement after ACP Projection. We also show the HOI probability change from before to after applying the projection function $project(\cdot)$ on our model's HOI prediction (*i.e.*, the effect of **PostProcess** introduced in Sect. 3.4) in Fig. 4. Leveraging co-occurrence matrix C can not only increase the score for true classes (top) but also reduce the score for false classes (bottom). Note that this change can be achieved *without any optimization* process.

Performance on Various Sets with Different Number of Training Samples. Finally, in Fig. 5, we show the relative mAP score improvements of our model compared to the baseline model by computing mAP on various sets of HOI classes that have different number of training samples. Our method shows positive performance improvements for all numbers of training samples. Also, there

watch-bird (62.41 → 97.90) hold-potted_plant (49.98 → 90.07)

walk-dog (60.04 → 24.53) hold-horse (75.16 → 42.95)

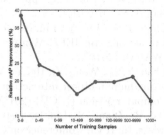

Fig. 4. The HOI probability before and after applying the projection function $project(\cdot)$ on our model's HOI prediction (**PostProcess**). Note that **PostProcess** can be done without any optimization.

Fig. 5. The relative mAP score improvements for various HOI sets with the different number of training samples. Our method is able to improve the performance especially when the number of training samples is small (38.24% improvement for 0–9 samples).

is a trend that HOI classes with a small number of training samples mostly show larger performance improvements. In particular, for HOI classes with the number of training samples between 0 and 9, our model achieves 38.24% improvement compared to the baseline model. These results indicate that the proposed method is able to improve the performance of an HOI detector, especially for classes with few training samples.

5   Conclusion

We introduced a novel method to effectively train an HOI detector by leveraging prior knowledge on action co-occurrences in two different ways, via the architecture and via the loss function. Our proposed method consistently achieves favorable performance compared to the current state-of-the-art methods in various setups. Co-occurrence information not only is helpful for alleviating the long-tailed distribution problem but also can be easily obtained. A direction for future work is to construct and utilize co-occurrence priors for other relationship-based vision tasks [24,26,37].

Acknowledgements. This work was supported by the Institute for Information & Communications Technology Promotion (2017-0-01772) grant funded by the Korea government.

References

1. Bansal, A., Rambhatla, S.S., Shrivastava, A., Chellappa, R.: Detecting human-object interactions via functional generalization. In: AAAI Conference on Artificial Intelligence (AAAI) (2020)
2. Baradel, F., Neverova, N., Wolf, C., Mille, J., Mori, G.: Object level visual reasoning in videos. In: Ferrari, V., Hebert, M., Sminchisescu, C., Weiss, Y. (eds.) ECCV 2018. LNCS, vol. 11217, pp. 106–122. Springer, Cham (2018). https://doi.org/10.1007/978-3-030-01261-8_7
3. Chao, Y.W., Liu, Y., Liu, X., Zeng, H., Deng, J.: Learning to detect human-object interactions. In: IEEE Winter Conference on Applications of Computer Vision (WACV) (2018)
4. Chao, Y.W., Wang, Z., He, Y., Wang, J., Deng, J.: HICO: a benchmark for recognizing human-object interactions in images. In: IEEE International Conference on Computer Vision (ICCV) (2015)
5. Dai, B., Zhang, Y., Lin, D.: Detecting visual relationships with deep relational networks. In: IEEE Conference on Computer Vision and Pattern Recognition (CVPR) (2017)
6. Delaitre, V., Fouhey, D.F., Laptev, I., Sivic, J., Gupta, A., Efros, A.A.: Scene semantics from long-term observation of people. In: Fitzgibbon, A., Lazebnik, S., Perona, P., Sato, Y., Schmid, C. (eds.) ECCV 2012. LNCS, vol. 7577, pp. 284–298. Springer, Heidelberg (2012). https://doi.org/10.1007/978-3-642-33783-3_21
7. Delaitre, V., Laptev, I., Sivic, J.: Recognizing human actions in still images: a study of bag-of-features and part-based representations. In: British Machine Vision Conference (BMVC) (2010)

8. Delaitre, V., Sivic, J., Laptev, I.: Learning person-object interactions for action recognition in still images. In: Advances in Neural Information Processing Systems (NIPS) (2011)
9. Deng, J., et al.: Large-scale object classification using label relation graphs. In: Fleet, D., Pajdla, T., Schiele, B., Tuytelaars, T. (eds.) ECCV 2014. LNCS, vol. 8689, pp. 48–64. Springer, Cham (2014). https://doi.org/10.1007/978-3-319-10590-1_4
10. Fu, J., Zheng, H., Mei, T.: Look closer to see better: recurrent attention convolutional neural network for fine-grained image recognition. In: IEEE Conference on Computer Vision and Pattern Recognition (CVPR) (2017)
11. Gao, C., Zou, Y., Huang, J.B.: iCAN: instance-centric attention network for human-object interaction detection. In: British Machine Vision Conference (BMVC) (2018)
12. Gibson, J.J.: The Ecological Approach to Visual Perception, Classic edn. Psychology Press, New York (2014)
13. Gkioxari, G., Girshick, R., Dollár, P., He, K.: Detecting and recognizing human-object interactions. In: IEEE Conference on Computer Vision and Pattern Recognition (CVPR) (2018)
14. Grabner, H., Gall, J., Van Gool, L.: What makes a chair a chair? In: IEEE Conference on Computer Vision and Pattern Recognition (CVPR) (2011)
15. Gupta, A., Davis, L.S.: Objects in action: an approach for combining action understanding and object perception. In: IEEE Conference on Computer Vision and Pattern Recognition (CVPR) (2007)
16. Gupta, S., Malik, J.: Visual semantic role labeling. arXiv preprint arXiv:1505.04474 (2015)
17. Gupta, T., Schwing, A., Hoiem, D.: No-Frills Pytorch Github. https://github.com/BigRedT/no_frills_hoi_det
18. Gupta, T., Schwing, A., Hoiem, D.: No-frills human-object interaction detection: factorization, layout encodings, and training techniques. In: IEEE International Conference on Computer Vision (ICCV) (2019)
19. He, K., Zhang, X., Ren, S., Sun, J.: Deep residual learning for image recognition. In: IEEE Conference on Computer Vision and Pattern Recognition (CVPR) (2016)
20. Hinton, G., Vinyals, O., Dean, J.: Distilling the knowledge in a neural network. arXiv preprint arXiv:1503.02531 (2015)
21. Horiguchi, S., Ikami, D., Aizawa, K.: Significance of softmax-based features in comparison to distance metric learning-based features. IEEE Trans. Pattern Anal. Mach. Intell. (TPAMI) **42**, 1279–1285 (2019)
22. Hu, Z., Ma, X., Liu, Z., Hovy, E., Xing, E.: Harnessing deep neural networks with logic rules. In: Annual Meeting of the Association for Computational Linguistics (ACL) (2016)
23. Hwang, S.J., Sha, F., Grauman, K.: Sharing features between objects and their attributes. In: IEEE Conference on Computer Vision and Pattern Recognition (CVPR) (2011)
24. Johnson, J., Hariharan, B., van der Maaten, L., Fei-Fei, L., Lawrence Zitnick, C., Girshick, R.: CLEVR: a diagnostic dataset for compositional language and elementary visual reasoning. In: IEEE Conference on Computer Vision and Pattern Recognition (CVPR) (2017)
25. Kato, K., Li, Y., Gupta, A.: Compositional learning for human object interaction. In: Ferrari, V., Hebert, M., Sminchisescu, C., Weiss, Y. (eds.) Computer Vision – ECCV 2018. LNCS, vol. 11218, pp. 247–264. Springer, Cham (2018). https://doi.org/10.1007/978-3-030-01264-9_15

26. Kim, D.J., Choi, J., Oh, T.H., Kweon, I.S.: Dense relational captioning: triple-stream networks for relationship-based captioning. In: IEEE Conference on Computer Vision and Pattern Recognition (CVPR) (2019)
27. Kim, D.J., Choi, J., Oh, T.H., Kweon, I.S.: Image captioning with very scarce supervised data: adversarial semi-supervised learning approach. In: Proceedings of the 2019 Conference on Empirical Methods in Natural Language Processing and the 9th International Joint Conference on Natural Language Processing (EMNLP-IJCNLP) (2019)
28. Kim, D.J., Choi, J., Oh, T.H., Yoon, Y., Kweon, I.S.: Disjoint multi-task learning between heterogeneous human-centric tasks. In: IEEE Winter Conference on Applications of Computer Vision (WACV) (2018)
29. Kipf, T.N., Welling, M.: Semi-supervised classification with graph convolutional networks. In: International Conference on Learning Representations (ICLR) (2017)
30. Krishna, R., et al.: Visual genome: connecting language and vision using crowd-sourced dense image annotations. Int. J. Comput. Vis. (IJCV) **123**(1), 32–73 (2017)
31. Li, Y., Ouyang, W., Wang, X.: VIP-CNN: a visual phrase reasoning convolutional neural network for visual relationship detection. In: IEEE Conference on Computer Vision and Pattern Recognition (CVPR) (2017)
32. Li, Y.L., et al.: Detailed 2D–3D joint representation for human-object interaction. In: IEEE Conference on Computer Vision and Pattern Recognition (CVPR) (2020)
33. Li, Y.L., et al.: Transferable interactiveness knowledge for human-object interaction detection. In: IEEE Conference on Computer Vision and Pattern Recognition (CVPR) (2019)
34. Li, Z., Hoiem, D.: Learning without forgetting. In: Leibe, B., Matas, J., Sebe, N., Welling, M. (eds.) ECCV 2016. LNCS, vol. 9908, pp. 614–629. Springer, Cham (2016). https://doi.org/10.1007/978-3-319-46493-0_37
35. Liao, Y., Liu, S., Wang, F., Chen, Y., Feng, J.: PPDM: parallel point detection and matching for real-time human-object interaction detection. In: IEEE Conference on Computer Vision and Pattern Recognition (CVPR) (2020)
36. Lin, T.Y., et al.: Microsoft COCO: common objects in context. In: Fleet, D., Pajdla, T., Schiele, B., Tuytelaars, T. (eds.) ECCV 2014. LNCS, vol. 8693, pp. 740–755. Springer, Cham (2014). https://doi.org/10.1007/978-3-319-10602-1_48
37. Lu, C., Krishna, R., Bernstein, M., Fei-Fei, L.: Visual relationship detection with language priors. In: Leibe, B., Matas, J., Sebe, N., Welling, M. (eds.) ECCV 2016. LNCS, vol. 9905, pp. 852–869. Springer, Cham (2016). https://doi.org/10.1007/978-3-319-46448-0_51
38. Marszalek, M., Schmid, C.: Semantic hierarchies for visual object recognition. In: IEEE Conference on Computer Vision and Pattern Recognition (CVPR) (2007)
39. Miller, G.A.: WordNet: a lexical database for English. Commun. ACM **38**(11), 39–41 (1995)
40. Pennington, J., Socher, R., Manning, C.: Glove: global vectors for word representation. In: Conference on Empirical Methods in Natural Language Processing (EMNLP) (2014)
41. Peyre, J., Laptev, I., Schmid, C., Sivic, J.: Detecting unseen visual relations using analogies. In: IEEE International Conference on Computer Vision (ICCV) (2019)
42. Plummer, B.A., Mallya, A., Cervantes, C.M., Hockenmaier, J., Lazebnik, S.: Phrase localization and visual relationship detection with comprehensive linguistic cues (2017)

43. Qi, S., Wang, W., Jia, B., Shen, J., Zhu, S.-C.: Learning human-object interactions by graph parsing neural networks. In: Ferrari, V., Hebert, M., Sminchisescu, C., Weiss, Y. (eds.) ECCV 2018. LNCS, vol. 11213, pp. 407–423. Springer, Cham (2018). https://doi.org/10.1007/978-3-030-01240-3_25

44. Ren, S., He, K., Girshick, R., Sun, J.: Faster R-CNN: towards real-time object detection with region proposal networks. In: Advances in Neural Information Processing Systems (NIPS) (2015)

45. Shen, L., Yeung, S., Hoffman, J., Mori, G., Fei-Fei, L.: Scaling human-object interaction recognition through zero-shot learning. In: IEEE Winter Conference on Applications of Computer Vision (WACV) (2018)

46. Stark, L., Bowyer, K.: Achieving generalized object recognition through reasoning about association of function to structure. IEEE Trans. Pattern Anal. Mach. Intell. (TPAMI) **13**(10), 1097–1104 (1991)

47. Sun, X., Li, C., Lin, S.: Explicit spatiotemporal joint relation learning for tracking human pose. In: Proceedings of the IEEE International Conference on Computer Vision Workshops (2019)

48. Sun, X., Wei, Y., Liang, S., Tang, X., Sun, J.: Cascaded hand pose regression. In: IEEE Conference on Computer Vision and Pattern Recognition (CVPR) (2015)

49. Ulutan, O., Iftekhar, A., Manjunath, B.: VSGNet: spatial attention network for detecting human object interactions using graph convolutions. In: IEEE Conference on Computer Vision and Pattern Recognition (CVPR) (2020)

50. Wan, B., Zhou, D., Liu, Y., Li, R., He, X.: Pose-aware multi-level feature network for human object interaction detection. In: IEEE International Conference on Computer Vision (ICCV) (2019)

51. Wang, T., et al.: Deep contextual attention for human-object interaction detection. In: IEEE International Conference on Computer Vision (ICCV) (2019)

52. Wang, T., Yang, T., Danelljan, M., Khan, F.S., Zhang, X., Sun, J.: Learning human-object interaction detection using interaction points. In: IEEE Conference on Computer Vision and Pattern Recognition (CVPR) (2020)

53. Xu, B., Li, J., Wong, Y., Zhao, Q., Kankanhalli, M.S.: Interact as you intend: intention-driven human-object interaction detection. IEEE Trans. Multimed. **22**, 1423–1432 (2019)

54. Xu, B., Wong, Y., Li, J., Zhao, Q., Kankanhalli, M.S.: Learning to detect human-object interactions with knowledge. In: IEEE Conference on Computer Vision and Pattern Recognition (CVPR) (2019)

55. Yan, Z., et al.: HD-CNN: hierarchical deep convolutional neural networks for large scale visual recognition. In: IEEE International Conference on Computer Vision (ICCV) (2015)

56. Yang, X., Zhang, H., Cai, J.: Shuffle-then-assemble: learning object-agnostic visual relationship features. In: Ferrari, V., Hebert, M., Sminchisescu, C., Weiss, Y. (eds.) ECCV 2018. LNCS, vol. 11216, pp. 38–54. Springer, Cham (2018). https://doi.org/10.1007/978-3-030-01258-8_3

57. Yao, B., Fei-Fei, L.: GroupLet: a structured image representation for recognizing human and object interactions. In: IEEE Conference on Computer Vision and Pattern Recognition (CVPR) (2010)

58. Yao, B., Fei-Fei, L.: Modeling mutual context of object and human pose in human-object interaction activities. In: IEEE Conference on Computer Vision and Pattern Recognition (CVPR) (2010)

59. Yao, B., Jiang, X., Khosla, A., Lin, A.L., Guibas, L., Fei-Fei, L.: Human action recognition by learning bases of action attributes and parts. In: IEEE International Conference on Computer Vision (ICCV) (2011)

60. Yin, G., et al.: Zoom-Net: mining deep feature interactions for visual relationship recognition. In: Ferrari, V., Hebert, M., Sminchisescu, C., Weiss, Y. (eds.) ECCV 2018. LNCS, vol. 11207, pp. 330–347. Springer, Cham (2018). https://doi.org/10.1007/978-3-030-01219-9_20

61. Yu, R., Li, A., Morariu, V.I., Davis, L.S.: Visual relationship detection with internal and external linguistic knowledge distillation. In: IEEE International Conference on Computer Vision (ICCV) (2017)

62. Zhan, Y., Yu, J., Yu, T., Tao, D.: On exploring undetermined relationships for visual relationship detection. In: IEEE Conference on Computer Vision and Pattern Recognition (CVPR) (2019)

63. Zhang, H., Zhang, H., Wang, C., Xie, J.: Co-occurrent features in semantic segmentation. In: IEEE Conference on Computer Vision and Pattern Recognition (CVPR) (2019)

64. Zhang, H., Kyaw, Z., Chang, S.F., Chua, T.S.: Visual translation embedding network for visual relation detection. In: IEEE Conference on Computer Vision and Pattern Recognition (CVPR) (2017)

65. Zhang, J., Elhoseiny, M., Cohen, S., Chang, W., Elgammal, A.: Relationship proposal networks. In: IEEE Conference on Computer Vision and Pattern Recognition (CVPR) (2017)

66. Zhou, P., Chi, M.: Relation parsing neural network for human-object interaction detection. In: IEEE International Conference on Computer Vision (ICCV) (2019)

67. Zhuang, B., Liu, L., Shen, C., Reid, I.: Towards context-aware interaction recognition for visual relationship detection. In: IEEE Conference on Computer Vision and Pattern Recognition (CVPR) (2017)

68. Zhuang, B., Wu, Q., Shen, C., Reid, I., van den Hengel, A.: HCVRD: a benchmark for large-scale human-centered visual relationship detection. In: AAAI Conference on Artificial Intelligence (AAAI) (2018)

Learning Connectivity of Neural Networks from a Topological Perspective

Kun Yuan$^{(\boxtimes)}$, Quanquan Li , Jing Shao, and Junjie Yan

SenseTime Research Institute, Hong Kong, China
{yuankun,liquanquan,shaojing,yanjunjie}@sensetime.com

Abstract. Seeking effective neural networks is a critical and practical field in deep learning. Besides designing the depth, type of convolution, normalization, and nonlinearities, the topological connectivity of neural networks is also important. Previous principles of rule-based modular design simplify the difficulty of building an effective architecture, but constrain the possible topologies in limited spaces. In this paper, we attempt to optimize the connectivity in neural networks. We propose a topological perspective to represent a network into a complete graph for analysis, where nodes carry out aggregation and transformation of features, and edges determine the flow of information. By assigning learnable parameters to the edges which reflect the magnitude of connections, the learning process can be performed in a differentiable manner. We further attach auxiliary sparsity constraint to the distribution of connectedness, which promotes the learned topology focus on critical connections. This learning process is compatible with existing networks and owns adaptability to larger search spaces and different tasks. Quantitative results of experiments reflect the learned connectivity is superior to traditional rule-based ones, such as random, residual and complete. In addition, it obtains significant improvements in image classification and object detection without introducing excessive computation burden.

Keywords: Learning connectivity · Topological perspective

1 Introduction

Deep learning successfully transits the feature engineering from manual to automatic design. It marks the mapping function from sample to feature can be optimized accordingly. As a tendency, seeking effective neural networks gradually becomes an important and practical direction. But the design of architecture is still a challenging and time-consuming effort. Part of the research focuses on how depth [7,17,17,26], type of convolution [3,13], normalization [23,43] and nonlinearities [24,27] affect the performance. In addition to these endeavors, another group of work also attempted to simplify the architecture design through stacking blocks/modules and wiring topological connections.

This strategy was demonstrably first popularized by the VGGNet [32] that is directly stacked by a series of convolution layers with plain topology. Due to the

A. Vedaldi et al. (Eds.): ECCV 2020, LNCS 12366, pp. 737–753, 2020.
https://doi.org/10.1007/978-3-030-58589-1_44

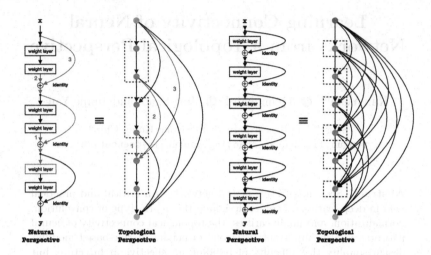

Fig. 1. From a natural perspective to the topological perspective for networks with residual connectivity. Two types of networks with 1/2 interval are given. Red node denotes the input **x**, and green one means the output feature **y**. Red arrows give an example of this mapping for a node with in-degree of 3.

problems of gradient vanishing and exploding, extending the network to a deeper level for better representation is nearly difficult. To better adapt the optimization process of gradient descent, GoogleNet [36] adopted parallel modules, and Highway networks [33] utilized gating units to regulate the flow of information, resulting in elastic topologies. Driven by the significance of depth, the residual block consisted of residual mapping and shortcut was raised in ResNet [10]. Topological changes in neural networks successfully scaled up neural networks to hundreds or even thousands of layers. The proposed residual connectivity was widely approved and applied in the following works, e.g. MobileNet [12,31] and ShuffleNet [45]. Divergent from aforementioned relative sparse topologies, DenseNet [14] wired densely among blocks to reuse features fully. Recent advances in computer vision also explore neural architecture search (NAS) methods [21,37,46] to search convolutional blocks. To trade-off efficiency and performance, most of them used hand-designed stacked patterns, and constrained the search space in limited ones. These trends reflect the great impact of topology on the optimization of neural networks. To a certain degree, previous principles of modular design simplify the difficulty of building an effective architecture. But how to aggregate and distribute these blocks is still an open question. Echoing this perspective, we wonder: can connectivity in neural networks be learned? What is the suitable route to do this?

To answer these questions, we propose a topological perspective to represent neural networks, resulting in a directed acyclic graph as shown in Fig. 1. Under this perspective, transformations (e.g. convolution, normalization and activation) are mapped into a node, and connections between layers are projected to edges which indicate the flow of information. We *first* unfold the residual connections to be a complete graph. This gives another way to explain the effectiveness of the residual topology, and inspires us to define the search space using a complete graph. Instead of choosing a predefined rule-based topology, we assign learnable parameters which determine the importance of corresponding connections to the edges. To adequately promote generalization and concentrate on critical connections, we attach auxiliary sparsity constraint on the weights of edges. Particularly, we propose two updating methods to optimize the weights of topology. One is a *uniform* type that regulates different edges uniformly. The other is an *adaptive* type that is logarithmically related to the in-degree of a node. Then the connectivity is learned simultaneously with the weights of the network by optimizing the loss function according to the task using a modified version of gradient descent.

We evaluate our optimization method on classical networks, such as ResNets and MobileNet. It demonstrates the compatibility with existing networks and adaptability to larger search spaces. To exhibit the benefits of connectivity learning, we construct a larger search space in which different topologies can be compared strictly. We also evaluate our method on various tasks and dataset, concretely, image classification on CIFAR-100 and ImageNet, object detection on COCO. Our contributions are as follows:

- The proposed topological perspective can be used to represent most existing neural networks. For the residual topology, we reveal for the first time the properties of its dense connections, which can be used for the search space.
- The proposed optimization method is compatible with existing networks. Without introducing much additional computing burden, we achieve 2.23% improvement using ResNet-110 on CIFAR-100, and 0.75% using deepened MobileNet on ImageNet.
- We design an architecture called TopoNet for larger search spaces and restrict comparison. Quantitative results prove the learned connectivity is superior to random, residual and complete ones, and surpasses ResNet in the similar computation cost by 2.10% on ImageNet.
- This method owns good generalization. The optimized topology with learned connectivity surpasses the best rule-based one by 0.95% in AP on COCO. To the equal-sized backbone of ResNet, the improvement is 5.27%. We also explore the properties of the optimized topology for future work.

2   Related Work

We briefly review related works in the aspects of neural network structure design and relevant optimization methods.

Neural network design is widely studied in previous literature. From shallow to deep, the shortcut connection plays an important role. Before ResNet, an early practice [41] also added linear layer connected from input to the output to train multi-layer perceptrons. Besides, "Inception" layer was proposed in [36] that is composed of a shortcut branch and a few deeper branches. Except on large networks, shortcut also proved effective in small networks, e.g. MobileNet [31], ShuffleNet [45] and MnasNet [37]. The existence of shortcut eases vanishing/exploding gradients [10,33]. In this paper, we explain from a *topological perspective* that shortcuts offer dense connections and benefit optimization. On the macrostructure, there also exist many networks with dense connections. DenseNet [14] contacted all preceding layers and passed on the feature maps to all subsequent layers in a block. HRNet [34] benefits from dense high-to-low connections for fine representations. Densely connected networks promote the specific task of localization [39]. Differently, we optimize the desired network from the complete graph in a differentiable way. And it is different from MaskConnect [1] which is constrained by K discrete in-degree and owns binary connections. This also provides an extension to [44] where random graphs generated by different generators are employed to form a network.

For the learning process, our method is consistent with DARTS [21] which is differentiable. In contrast to DARTS, we do not adopt alternative optimization strategies for weights and architecture. Joint training can replace the transferring step from one task to another, and obtain task-related topology. Different from sample-based optimization methods [29], the connectivity is learned simultaneously with the weights of the network using our modified version of gradient descent. [2,8] also explored this type and utilized weight-sharing across models to amortize the cost of training. Searching from the full space is evaluated in object detection by NAS-FPN [5], in which the feature pyramid is sought in all cross-scale connections. In the aspect of semantic segmentation, Auto-DeepLab [20] formed a hierarchical architecture to enlarge search spaces. The sparsity constraint can be observed in other applications, e.g. path selection for a multi-branch network [15], and pruning unimportant channels for fast inference [9].

3   Methodology

3.1   Topological Perspective of Neural Networks

We represent the neural network using a directed acyclic graph (DAG) in topology. Specifically, we map both combining (e.g., addition) and transformation (e.g., convolution, normalization and activation) to a node. And connections between layers are represented as edges, which determine the flow of information. Then we can get a new representation of the architecture $\mathcal{G} = (\mathcal{N}, \mathcal{E})$, where $\mathcal{N}$ is the set of nodes, and $\mathcal{E}$ denotes the set of edges.

In the graph, each node $n_i \in \mathcal{N}$ performs a transformation operation o_i, parametrized by $\mathbf{w}_i$, where i stands for the topological ordering of the node. While the edge $e_{ji} = (j, i, \alpha_{ji}) \in \mathcal{E}$ means the flow of features from node j to node i, and the importance of the connection is determined by the weight of α_{ji}. During forward computation, each node aggregates inputs from preorder nodes where connections exist. Then it performs a feature transformation to get an output tensor $\mathbf{x}_i$. And $\mathbf{x}_i$ is sent out to the postorder nodes through the output

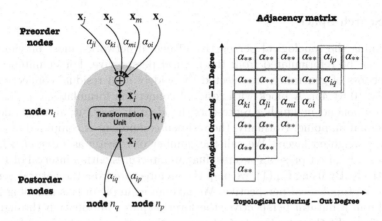

Fig. 2. Details of node operations and the adjacency matrix. For each node, features generated by preorder nodes are aggregated through weights of edges. Then a transformation unit which consists of convolutional layers, batch normalization and the activation function is used to transform features. Next, features are allocated to postorder nodes where connections exist. For a stage, weights of edges can be represented in an adjacency matrix, in which rows denote the weights of input edges, and columns stand for the weights of output edges.

edges. It can be seen in the left of Fig. 2. It can be formulated as follows:

$$\mathbf{x}_i = o_i(\mathbf{x}'_i; \mathbf{w}_i), \text{ where } \mathbf{x}'_i = \sum_{(j<i) \wedge (e_{ji} \in \mathcal{E})} \alpha_{ji} \cdot \mathbf{x}_j. \tag{1}$$

In each graph, the first node in topological ordering is the input one, which only performs the distribution of features. The last node is the output one, which only generates final output of the graph by gathering preorder inputs. We also propose an adjacency matrix as the memory space to store weights of edges. As shown in the right of Fig. 2, each row denotes the weights of input edges, and each column is the outputs. For nodes where there are no edges attached, the corresponding α is 0. The dimension of row (with $\alpha \neq 0$) is called in-degree for a node, and the dimension of column (with $\alpha \neq 0$) is named as out-degree.

For a network with k stages, k DAGs are initialized and connected in series. Each graph is linked to its preceding or succeeding stage by output or input node. We rewrite the weights of nodes as $\mathbf{w}_i^k$ and the weights of edges as α_{ji}^k. For the k-th stage, $\mathcal{T}^k(\cdot)$ denotes the mapping function established by $\mathcal{G}^k$ with parameters of $\mathbf{W}^k$ and α^k, where $\mathbf{W}^k$ is the set of $\{\mathbf{w}_i^k\}$, α^k is the set of $\{\alpha_{ji}^k\}$. Given an input $\mathbf{x}$ and corresponding label $\mathbf{y}$, the mapping function from the sample to the feature representation can be written as:

$$\mathcal{F}(\mathbf{x}) = \mathcal{T}^k(\cdots \mathcal{T}^2(\mathcal{T}^1(\mathbf{x}; \alpha^1, \mathbf{W}^1); \alpha^2, \mathbf{W}^2) \cdots; \alpha^k, \mathbf{W}^k), \tag{2}$$

3.2   Search Space

By defining the topological perspective of neural networks, most previous networks can be reformulated from the natural perspective. For definiteness and without loss of generality, we selected the widely used residual connections for analysis [10,31,45]. A block with residual connection formulates $x + \varphi(x)$ as a basic component, in which x represents the identity shortcut, and $\varphi(x)$ denotes the residual mapping. Normally, the residual component is composed of several repeated weighted layers. We call the number of repeats as *interval*, which is noted as l. Figure 1 presents two residual architectures with a interval of 1 and 2 respectively. By using Eq.(1), we map the architecture from the natural perspective to the topological perspective. We give an example of this mapping in red lines. From a natural perspective, the layer acquires information through skip connections. In the new topological perspective, the node obtains information by corresponding edges. It should be pointed out that these two perspectives are completely equivalent in results. It also can be seen that the residual connections are rather denser than the original view, and perform multiple feed-forward paths instead of a single deep network. Our topological view explains the reason why residual connectivity is effective from a new aspect different from [40].

If the interval degrades to 1, as shown in the right of Fig. 2, its topology evolves into a complete graph. Structurally, all nodes are directly connected to the input and output, resulting in indirect access to the gradients and the original input. Different from stacking blocks using predefined connectivity, the complete graph provides all possible connections and is suitable to be the search space. For a complete graph with N nodes, the search space contains $2^{N(N-1)/2}$ possible topological structures. For a network with k stages, the total search space can be noted as $\prod_k 2^{N^k(N^k-1)/2}$. And it is much wider than cell-based or block-based approaches [8,12,37]. By assigning learnable parameters which reflect the magnitude of connections to edges, it changes to a weighted graph. Within the search space, the connectivity can be optimized by learning continuous weights of edges.

3.3   Optimization of Topological Connectivity

We put forward a differentiable type to optimize the topological connectivity by learning a set of continuous weights of edges α. And they are learned simultaneously with all the other weights in the network via the loss generated by the concurrent task, noted as $\mathcal{L}_t(\cdot)$. Different from [1], we do not transform the weights of α into binary. This allows us to assign discriminating weights to different feature inputs. Different from the selection of node type [21], we do not select the maximum input edge using arg max operation. Instead, the continuous weights guarantees the consistency between training and testing. The optimization objective can be viewed as:

$$\min_{\mathbf{W},\alpha} \mathcal{L}_t(\mathcal{F}(\mathbf{x}; \mathbf{W}, \alpha), \mathbf{y}) \tag{3}$$

Algorithm 1: Optimization of Topological Connectivity.

Input: Network represented with k graphs $\mathcal{G}(\mathcal{N}, \mathcal{E}; \mathbf{W}, \boldsymbol{\alpha})$, training set $\{(\mathbf{x}, \mathbf{y})^{(s)}\}$, *Type* of sparsity, λ and η.

Output: Optimized weights $\mathbf{W}$ and a weighted graph defined by topology $\boldsymbol{\alpha}$.

for $(\mathbf{x}^s, \mathbf{y}^s)$ *in training set* **do**

    for $\mathcal{G}^k$ *in network* **do**

        Set *rev_adj_dict* $= \{\}$ to save directed edges;

        Set *memory* $= \{\}$ to save features generated by nodes;

        for e_{ji} *in* $\mathcal{E}^k$ **do**

            | *rev_adj_dict*[i].*append*(j);

        end

        for n_i *in* $\mathcal{N}^k$ **do**

            Obtain indices of input nodes of n_i from *rev_adj_dict*[i] and fetch corresponding input features from *memory*;

            Aggregate inputs by weights of $\boldsymbol{\alpha}^k$ and transform features using Eq. (1) and store transformed features *memory*[i]$= \mathbf{x}_i$;

        end

        Fetch the features generated by the output node for $\mathcal{G}^{k+1}$;

    end

    Obtain final representation and compute the loss w.r.t $\mathbf{y}^s$ by Eq.(6);

    Compute gradients w.r.t $\mathbf{w_i}$ in $\mathbf{W}$ by Eq.(4)and update weights;

    if *Type* **is** *uniform* **then**

        | Compute gradients w.r.t α_{ji} in $\boldsymbol{\alpha}$ by Eq.(7) and update weights;

    else if *Type* **is** *adaptive* **then**

        | Compute gradients w.r.t α_{ji} in $\boldsymbol{\alpha}$ by Eq.(8) and update weights;

end

Set $\frac{\partial \mathcal{L}_t}{\partial \mathbf{w}_i}$ be the gradients that the network flows backwards to $\mathbf{w}_i$. And let $\frac{\partial \mathcal{L}_t}{\partial \mathbf{x}_i}$ be the gradients to $\mathbf{x}_i$. Then the gradients update to $\mathbf{w}_i$ and α_{ji} are of the form:

$$\mathbf{w}_i \leftarrow \mathbf{w}_i + \eta \frac{\partial \mathcal{L}_t}{\partial \mathbf{w}_i} \tag{4}$$

$$\alpha_{ji} \leftarrow \alpha_{ji} + \eta \sum \frac{\partial \mathcal{L}_t}{\partial \mathbf{x}_i} \odot \frac{\partial o_i}{\partial \mathbf{x}_i'} \odot \mathbf{x}_j, \tag{5}$$

where η is the learning rate, and $\odot$ indicates entrywise product.

Since the features generated by different layers exhibit different semantic representations, they contribute differently to subsequent layers, resulting in diversities of the importance of connections. Much as the mammalian brain [28] in biology, where synapses are created in the first few months of a child's development, followed by gradual re-weighting through postnatal knowledge, growing into a typical adult with relative sparse connections.

To facilitate this process appropriately, we raise to attach sparsity constraint as a regularization on the distribution of weights of edges. Similar thought also has been verified in hashing representation [42] that sparsity can bring effective gain through minimizing a hash collision. We choose L1 regularization, denoted as $\mathcal{L}_1(\cdot)$, to penalize non-zero parameters of edges resulting in more parameters near zero. This sparsity constraint promotes attention to more critical connections. Then the loss function of our proposed method can be reformulated as:

$$\mathcal{L} = \mathcal{L}_t + \lambda \mathcal{L}_1 = \mathcal{L}_t(\mathcal{F}(\mathbf{x}; \mathbf{W}, \boldsymbol{\alpha}), \mathbf{y}) + \lambda \cdot \|\boldsymbol{\alpha}\|_1, \tag{6}$$

and λ is a hyper-parameter to balance the sparse level. Due to the properties of a complete graph, we propose two types to update α_{ji}. The first one is *uniform sparsity* that attaches constraint on all weights of edges uniformly. And let $\frac{\partial \mathcal{L}_1}{\partial \mathbf{x}_i}$ be the gradients to $\mathbf{x}_i$, we rewrite the Eq. (5) as:

$$\alpha_{ji} \leftarrow \alpha_{ji} + \eta \sum \left(\frac{\partial \mathcal{L}_t}{\partial \mathbf{x}_i} + \lambda \frac{\partial \mathcal{L}_1}{\partial \mathbf{x}_i} \right) \odot \frac{\partial o_i}{\partial \mathbf{x}_i'} \odot \mathbf{x}_j, \tag{7}$$

The second one is *adaptive sparsity* which is logarithmically related to the in-degree δ_i of a node n_i. It performs larger constraints on dense input and smaller on sparse input. For the nodes with fewer input edges, this can ensure the smooth flow of information and avoid being blocked. In this type, the α_{ji} is updated by:

$$\alpha_{ji} \leftarrow \alpha_{ji} + \eta \sum \left(\frac{\partial \mathcal{L}_t}{\partial \mathbf{x}_i} + \lambda \log(\delta_i) \frac{\partial \mathcal{L}_1}{\partial \mathbf{x}_i} \right) \odot \frac{\partial o_i}{\partial \mathbf{x}_i'} \odot \mathbf{x}_j. \tag{8}$$

These two types will be further discussed in the experiments section. Algorithm 1 summarizes the optimization procedure detailedly.

4   Experiments and Analysis

4.1   Connectivity Optimization for Classical Networks

Our optimization method is compatible with classical networks. To investigate the applicability, we select ResNet-CIFAR [10] consisted of 3 × 3 conv and MobileNetV2-1.0 [31] consisted of Inverted Bottleneck. For the optimization of ResNets, we rewire the interval of 2 in the BasicBlock to 1 to form the complete graph. For MobileNetV2-1.0, each node involves a residual connection and can be viewed as a complete graph naturally. In the case that MobileNet owns fewer layers in each stage, we also increase the depth by increasing the node in each stage, resulting in larger search spaces. It is also a common skill to expand networks [38]. Through assigning learnable parameters to their edges, the topologies can be optimized using Algorithm 1. It should be mentioned that the additional computations and parameters introduced by the edges are negligible compared with convolution.

First, we evaluate the optimization of the connectivity with ResNets on CIFAR-100 [16]. The experiments are trained using 2 GPUs[1] with batchsize

---

[1] All of our experiments were performed using NVIDIA Tesla V100 GPUs with our implementation in PyTorch [25].

Table 1. Optimization Top-1 Accuracy of ResNets on CIFAR-100.

Network	Params(M)	FLOPs(G)	Original	Optimized	Gain
ResNet-20	0.28	0.04	69.01	**69.91** ± 0.12	**0.90**
ResNet-32	0.47	0.07	72.07	**73.34** ± 0.09	**1.37**
ResNet-44	0.67	0.10	73.73	**75.60** ± 0.14	**1.87**
ResNet-56	0.86	0.13	75.22	**76.90** ± 0.03	**1.68**
ResNet-110	1.74	0.25	76.31	**78.54** ± 0.15	**2.23**

128 and weight decay 5e–4. We follow the hyperparameter settings in paper [4], which initializes $\eta = 0.1$ and divides by 5 times at 60th, 120th, 160th epochs. The training and test size is 32×32. We report classification accuracy on the validation set by 5 repeat runs. The results are shown in Table 1. Under similar Params and FLOPs, the optimization brings 2.22% improvement on Top-1 accuracy for ResNet-110, which reflects larger search spaces lead to more improvements.

Next, we extend our method to ImageNet dataset [30] using MobileNets. We train MobileNetV2 using 16 GPUs for 200 epochs with a batch size of 1024. The initial learning rate is 0.4 and cosine shaped learning rate decay [22] is adopted. Following [31], we use a weight decay of 4e–5 and dropout [11] of 0.2. Nesterov momentum of 0.9 without dampening is also used. The training and test size is 224×224. The network with 2 times of layers is denoted as 2N. Under the mobile-setting, we achieve 76.4% Top-1 accuracy. Under the larger optimization space of 6N, the optimization brings a 0.75% improvement. This further demonstrates the benefits of topology optimization for different networks.

Table 2. Optimization Top-1 Accuracy of Scaled MobileNets on ImageNet

Network	Params(M)	FLOPs(G)	Original	Optimized	Gain
MobileNetV2-1.0	3.51	0.30	72.60	**72.86** ± 0.13	**0.24**
MobileNetV2-1.0-2N	6.43	0.60	75.93	**76.40** ± 0.05	**0.47**
MobileNetV2-1.0-4N	9.62	1.10	77.33	**77.87** ± 0.09	**0.54**
MobileNetV2-1.0-6N	12.00	2.06	77.61	**78.36** ± 0.14	**0.75**

4.2 Expanding to Larger Search Spaces by TopoNet

Due to restricted optional topologies of classical networks, the topology can be only optimized in small search spaces, which limits the representation ability of topology. These may limit the influence caused by topological changes and affect the search for optimal topology. In this section, we propose a larger search space, and fully illustrate the improvement brought by topology optimization. The properties of edges and nodes in the optimized topology are also analyzed.

Table 3. Architectures of TopoNets for ImageNet.

Layers	Output Size	Component and channels	Edges of different topologies		
			Random, p	Residual, l	Complete
Head	112×112	7×7 conv, C	–	–	–
Stage 1	56×56	N_1 nodes, $2 \times C$	$p \cdot \binom{N_1}{2}$	$\frac{N_1-2}{l} + \binom{\frac{N_1-2}{l}+2}{2}$	$\binom{N_1}{2}$
Stage 2	28×28	N_2 nodes, $4 \times C$	$p \cdot \binom{N_2}{2}$	$\frac{N_2-2}{l} + \binom{\frac{N_2-2}{l}+2}{2}$	$\binom{N_2}{2}$
Stage 3	14×14	N_3 nodes, $8 \times C$	$p \cdot \binom{N_3}{2}$	$\frac{N_3-2}{l} + \binom{\frac{N_3-2}{l}+2}{2}$	$\binom{N_3}{2}$
Stage 4	7×7	N_4 nodes, $16 \times C$	$p \cdot \binom{N_4}{2}$	$\frac{N_4-2}{l} + \binom{\frac{N_4-2}{l}+2}{2}$	$\binom{N_4}{2}$
Classifier	1×1	GAP, 1k-d fc, softmax	–	–	–

We design a series of architectures named as TopoNets that can flexibly adjust search space, types of topology and node. As shown in Table 3, it consists of four stages with number of nodes of $\{N_1, N_2, N_3, N_4\}$. The topology in each stage is defined by a graph, whose type can be chosen from {complete, random, residual}. The complete graph is used for the optimization of topology. For a more strict comparison, we also take the other two types as baselines. The residual one is a well-designed topology. In the random one, an edge between two nodes is linked with probability p, independent of all other nodes and edges. The higher the probability, the denser it is. We follow two simple design rules used in [10], (i) in each stage, the nodes have the same number of filters C; (ii) and if the feature map size is halved, the number of filters is doubled. The change of filters is implied by the first calculation node in each graph. For the head of the network, we use a single convolutional layer for simplicity. The network ends with a classifier composed of a global average pooling (GAP), a 1000-dimensional fully-connected layer and softmax function.

Setup for TopoNet. To demonstrate the optimization capability in the larger search space and to compare with existing network, we designed a set with similar computation cost as ResNet-50. We select the separable depthwise convolution that includes a 3×3 depthwise convolution followed by a 1×1 pointwise convolution, and build a triplet unit ReLU-conv-BN as the node. The number of nodes in each stage is $\{14, 20, 26, 14\}$. In this setting, the number of possible discrete topologies is 6×10^{209}. The weights of α are initialized to be 1. And C is set to be 64, resulting in Params of $23.23\,M$ and FLOPs of $3.95\,G$ (e.g ResNet-50 with Params of $25.57\,M$ and FLOPs of $4.08\,G$).

Strict Comparisons. To demonstrate the effectiveness of our optimization method, we select graphs with random, residual connectivity as baselines. For comparison, we also reproduce Erdös-Rényi (ER), Barabási-Albert (BA) and Watts-Strogatz WS graphs [44] using NetworkX[2]. Since original paper does not

---

[2] https://networkx.github.io.

release codes, we compare the best configurations of their method. We use these graphs to build networks under the same setup in TopoNet. Two types of sparsity constraints are also demonstrated. For a fair comparison, all experiments are conducted on ImageNet with training 100 epochs. We use a weight decay of 1e–4 and a Nesterov momentum of 0.9 without dampening. Dropout is not used. Label smoothing regularization [35] with a coefficient of 0.1 is also used.

Table 4. Comparision with Different Topologies on ImageNet.

Network	Top-1 Acc(%)	Top-5 Acc(%)
ResNet-50	76.50	93.10
random, $p = 0.8$	77.56 ± 0.22	93.69 ± 0.32
random, $p = 0.6$	77.84 ± 0.19	93.65 ± 0.43
random, $p = 0.4$	77.90 ± 0.27	93.76 ± 0.37
residual, $l = 4$	77.72 ± 0.13	93.57 ± 0.20
residual, $l = 3$	78.10 ± 0.07	93.83 ± 0.17
residual, $l = 2$	78.26 ± 0.14	93.78 ± 0.22
ER, $p = 0.2$	77.76 ± 0.33	93.20 ± 0.41
BA, $m = 5$	78.08 ± 0.17	93.46 ± 0.34
WS, $k = 4, p = 0.75$	78.19 ± 0.25	93.78 ± 0.24
complete	77.24 ± 0.12	93.40 ± 0.23
complete, α	78.22 ± 0.13	93.80 ± 0.15
complete, α, *uniform*	$\mathbf{78.46 \pm 0.14}$	93.81 ± 0.32
complete, α, *adaptive*	$\mathbf{78.60 \pm 0.16}$	93.92 ± 0.11

The validation results are shown in Table 4. Some conclusions can be drawn from the results. (i) Topological connectivity of network largely affects the performance of representation. (ii) The performance is related to the density of connections according to different p and l. (iii) For the complete graphs, direct optimization with α can yield 0.98% improvement on Top-1. (iv) Through assigning sparsity constraints, performances have been further improved. The complete graph with *adaptive* sparsity constraint gets the best Top-1 of 78.60%. This proves the benefits of sparseness for the connectivity. (v) The connectivity can be optimized in neural networks, and is superior to rule-based designed ones, such as random, residual, BA and WS.

In order to intuitively understand the optimization effect of sparsity constraints on dense topological connections, we give the distributions of the learned α Fig. 3. Sparsity constraints push more parameters near zero, resulting in focusing on critical connections. With the enhancement of constraints, more connections disappear. Excessive constraints will damage the representation of features, so we set the weight of balance λ to be e−4 in all experiments. And it is also robust in the range from e−5 to e−4, resulting in similar effects. In the right of

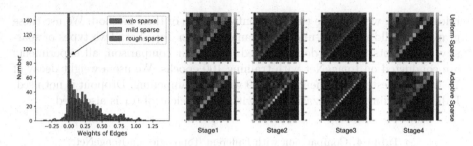

Fig. 3. The effect of sparsity constraint on distributions of α. The histogram on the left indicates that sparsity drives most of the weights near zero. Adjacency matrices on the right shows the difference between *uniform* and *adaptive* one, whose rows correspond to the input edges for a particular node and columns represent the output ones. Colors indicate the weights of edges.

Fig. 3, we give the optimized results with two types of constraints. The *adaptive* one penalizes denser connections a lot and keeps the relative sparser but critical connections, resulting in better performance.

4.3 Transferability on Different Tasks

To evaluate the generalization and transferability for both optimization method and TopoNets, we also conduct experiments on COCO object detection task [19]. We adopt FPN [18] as the object detection method. The backbone is replaced with corresponding pretrained one in Table 4, and is fine-tuned on COCO train2017 dataset. We test using the COCO val2017 dataset. Our fine-tuning is based on 1× setting of the publicly available Detectron [6]. The training configurations of different models are consistent. Test performances are given in Table 5. To comparable ResNet-50, TopoNets obtain significant promotions in AP with lower computation costs. Contrast with elegant residual topology, our optimization method can also achieve increase by 0.95%. These results indicate the effectiveness of the proposed network and the optimization method.

Table 5. Transferability Results on COCO object detection.

Backbone	AP	AP_{50}	AP_{75}	AP_S	AP_M	AP_L
ResNet-50	36.42	58.66	38.90	21.93	39.84	46.74
Residual, $l = 2$	40.74(+ 4.32)	63.22	44.62	25.01	44.18	52.74
Complete, α	41.35(+ 4.93)	63.32	45.08	25.63	44.99	53.47
Complete, α, *uniform*	**41.46**(+ 5.04)	63.83	44.91	25.07	45.31	53.52
Complete, α, *adaptive*	**41.69**(+ 5.27)	63.86	45.45	25.58	45.52	53.69

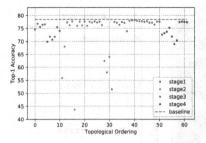 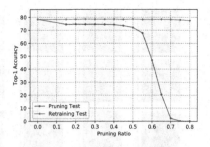

Fig. 4. Impact of node (left) and edge (right) removal for the optimized topology.

4.4 Exploring Topological Properties by Graph Damage

We further explore the properties of the optimized topology. First, we remove individual nodes according to its topological ordering in the graph and evaluate it without extra training. We expect them to break because dropping any layer drastically changes the input distribution of all subsequent layers. Surprisingly, most removals do not lead to a noticeable change as shown in Fig. 4 (left). It can be explained that the available paths is reduced from $(n-1)!$ to $(n-2)!$, leaving sufficient paths. This suggests that each node in the complete graph do not strongly rely on others, although it is trained jointly. Direct links with input/output nodes make each node contribute to the final feature representation, and benefits the optimization process. Another observation is that nodes in the front of topological orderings contribute more. This can be explained that for a node with the ordering of i, the generated x_i can be only received by node j (where $j > i$). This causes the feature generated by the front nodes to participate in aggregation as a downstream input. It makes the front nodes contribute more, which can be used to reallocate calculation resources in future work.

Second, we consider the impact of edge removal. All edges with α below a threshold are pruned from the graph, only remaining the important connections. Accuracies before and after retraining are given in Fig. 4 (right). Without retraining, accuracy decreases as the degree of pruning deepens. It is interesting to see that we have the "free lunch" of reducing less than 40% without losing much accuracy. If we fix α of remaining edges and retrain the weights, it can maintain accuracy with 80% of the nodes removed. This proves that the optimization process has found indeed important connections. After pruning edges, nodes with zero in-degree or zero out-degree maybe safely removed. It can be used to reduce the parameters and accelerate inference in practical applications.

4.5 Visualization of the Optimization Process

We visualize the optimization process in Fig. 5. During the initial phase, there are strong connections between all nodes. As the optimization process progresses, the connections become sparse, leaving the critical ones. We sample topologies with different connectivity during the process and retrain them from scratch

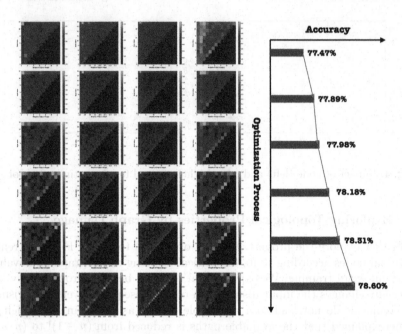

Fig. 5. The changes of the connectivity and corresponding accuracies after retraining. It proves that the representation capability of topological connectivity improves along with the process, and demonstrates the effectiveness of optimization.

with α froze. This allows us to compare the change of topology capabilities during optimization. Validation accuracies are given in the right of the figure. It can be seen the representation ability of connectivity increases with the training process, not just the weights of networks.

5 Conclusion and Future Work

In this work, we proposed a feasible way for the learning of topological connectivity in neural networks. Motivated by our *topological perspective*, the optimization space is defined as a complete graph. By assigning learnable continuous weights which reflect the importance of connections, the optimization process is transformed into a differentiable type with less extra cost. The sparsity constraint further improve the generalization and performance. This method is compatible with existing networks, and the optimized connectivity is superior to rule-based designed ones. Experiments on different tasks proved the effectiveness and transferability. Moreover, the observed properties of topology can be used for future work and practical applications. Our work has a wide application and is complementary to existing neural architecture search methods. We will consider verifying NAS-inspired networks in the future work.

References

1. Ahmed, K., Torresani, L.: Maskconnect: connectivity learning by gradient descent. In: Proceedings of the European Conference on Computer Vision (ECCV), pp. 349–365 (2018)
2. Bender, G., Kindermans, P.J., Zoph, B., Vasudevan, V., Le, Q.: Understanding and simplifying one-shot architecture search. In: International Conference on Machine Learning, pp. 550–559 (2018)
3. Dai, J., et al.: Deformable convolutional networks. In: Proceedings of the IEEE International Conference on Computer Vision, pp. 764–773 (2017)
4. DeVries, T., Taylor, G.W.: Improved regularization of convolutional neural networks with cutout. arXiv preprint arXiv:1708.04552 (2017)
5. Ghiasi, G., Lin, T.Y., Le, Q.V.: NAS-FPN: Learning scalable feature pyramid architecture for object detection. In: Proceedings of the IEEE Conference on Computer Vision and Pattern Recognition, pp. 7036–7045 (2019)
6. Girshick, R., Radosavovic, I., Gkioxari, G., Dollár, P., He, K.: Detectron (2018)
7. Glorot, X., Bordes, A., Bengio, Y.: Deep sparse rectifier neural networks. In: Proceedings of the Fourteenth International Conference on Artificial Intelligence and Statistics, pp. 315–323 (2011)
8. Guo, Z., et al.: Single path one-shot neural architecture search with uniform sampling. arXiv preprint arXiv:1904.00420 (2019)
9. Han, S., Pool, J., Tran, J., Dally, W.: Learning both weights and connections for efficient neural network. In: Advances in Neural Information Processing Systems, pp. 1135–1143 (2015)
10. He, K., Zhang, X., Ren, S., Sun, J.: Deep residual learning for image recognition. In: Proceedings of the IEEE Conference on Computer Vision and Pattern Recognition, pp. 770–778 (2016)
11. Hinton, G.E., Srivastava, N., Krizhevsky, A., Sutskever, I., Salakhutdinov, R.R.: Improving neural networks by preventing co-adaptation of feature detectors. arXiv preprint arXiv:1207.0580 (2012)
12. Howard, A., et al.: Searching for mobilenetv3. arXiv preprint arXiv:1905.02244 (2019)
13. Howard, A.G., et al.: Mobilenets: efficient convolutional neural networks for mobile vision applications. arXiv preprint arXiv:1704.04861 (2017)
14. Huang, G., Liu, Z., Van Der Maaten, L., Weinberger, K.Q.: Densely connected convolutional networks. In: Proceedings of the IEEE Conference on Computer Vision and Pattern Recognition, pp. 4700–4708 (2017)
15. Huang, Z., Wang, N.: Data-driven sparse structure selection for deep neural networks. In: Proceedings of the European Conference on Computer Vision (ECCV), pp. 304–320 (2018)
16. Krizhevsky, A., Hinton, G., et al.: Learning multiple layers of features from tiny images. Technical report, Citeseer (2009)
17. Krizhevsky, A., Sutskever, I., Hinton, G.E.: Imagenet classification with deep convolutional neural networks. In: Advances in Neural Information Processing Systems, pp. 1097–1105 (2012)
18. Lin, T.Y., Dollár, P., Girshick, R., He, K., Hariharan, B., Belongie, S.: Feature pyramid networks for object detection. In: Proceedings of the IEEE Conference on Computer Vision and Pattern Recognition, pp. 2117–2125 (2017)
19. Lin, T.Y., et al.: Microsoft COCO: common objects in context. In: Fleet, D., Pajdla, T., Schiele, B., Tuytelaars, T. (eds.) ECCV 2014. LNCS, vol. 8693, pp. 740–755. Springer, Cham (2014). https://doi.org/10.1007/978-3-319-10602-1_48

20. Liu, C., et al.: Auto-deeplab: Hierarchical neural architecture search for semantic image segmentation. In: Proceedings of the IEEE Conference on Computer Vision and Pattern Recognition, pp. 82–92 (2019)
21. Liu, H., Simonyan, K., Yang, Y.: Darts: differentiable architecture search. arXiv preprint arXiv:1806.09055 (2018)
22. Loshchilov, I., Hutter, F.: SGDR: stochastic gradient descent with warm restarts. arXiv preprint arXiv:1608.03983 (2016)
23. Luo, P., Ren, J., Peng, Z., Zhang, R., Li, J.: Differentiable learning-to-normalize via switchable normalization. arXiv preprint arXiv:1806.10779 (2018)
24. Maas, A.L., Hannun, A.Y., Ng, A.Y.: Rectifier nonlinearities improve neural network acoustic models. In: Proc. ICML, vol. 30, p. 3 (2013)
25. Paszke, A., et al.: Automatic differentiation in pytorch (2017)
26. Pérez-Rúa, J.M., Baccouche, M., Pateux, S.: Efficient progressive neural architecture search. arXiv preprint arXiv:1808.00391 (2018)
27. Ramachandran, P., Zoph, B., Le, Q.V.: Searching for activation functions. arXiv preprint arXiv:1710.05941 (2017)
28. Rauschecker, J.: Neuronal mechanisms of developmental plasticity in the cat's visual system. Hum. Neurobiol. **3**(2), 109–114 (1984)
29. Real, E., Aggarwal, A., Huang, Y., Le, Q.V.: Regularized evolution for image classifier architecture search. Proceedings of the AAAI Conference on Artificial Intelligence, vol. 33, pp. 4780–4789 (2019)
30. Russakovsky, O., et al.: Imagenet large scale visual recognition challenge. Int. J. Comput. Vis. **115**(3), 211–252 (2015)
31. Sandler, M., Howard, A., Zhu, M., Zhmoginov, A., Chen, L.C.: Mobilenetv 2: Inverted residuals and linear bottlenecks. In: Proceedings of the IEEE Conference on Computer Vision and Pattern Recognition, pp. 4510–4520 (2018)
32. Simonyan, K., Zisserman, A.: Very deep convolutional networks for large-scale image recognition. arXiv preprint arXiv:1409.1556 (2014)
33. Srivastava, R.K., Greff, K., Schmidhuber, J.: Highway networks. arXiv preprint arXiv:1505.00387 (2015)
34. Sun, K., et al.: High-resolution representations for labeling pixels and regions. arXiv preprint arXiv:1904.04514 (2019)
35. Szegedy, C., Ioffe, S., Vanhoucke, V., Alemi, A.A.: Inception-v4, inception-resnet and the impact of residual connections on learning. In: Thirty-first AAAI conference on artificial intelligence (2017)
36. Szegedy, C., et al.: Going deeper with convolutions. In: Proceedings of the IEEE Conference on Computer Vision and Pattern Recognition, pp. 1–9 (2015)
37. Tan, M., et al.: Mnasnet: platform-aware neural architecture search for mobile. In: Proceedings of the IEEE Conference on Computer Vision and Pattern Recognition, pp. 2820–2828 (2019)
38. Tan, M., Le, Q.V.: Efficientnet: rethinking model scaling for convolutional neural networks. arXiv preprint arXiv:1905.11946 (2019)
39. Tang, Z., Peng, X., Geng, S., Wu, L., Zhang, S., Metaxas, D.: Quantized densely connected u-nets for efficient landmark localization. In: Proceedings of the European Conference on Computer Vision (ECCV), pp. 339–354 (2018)
40. Veit, A., Wilber, M.J., Belongie, S.J.: Residual networks behave like ensembles of relatively shallow networks. In: NIPS, pp. 550–558 (2016)
41. Venables, W.N., Ripley, B.D.: Modern Applied Statistics with S-PLUS. Springer Science & Business Media, Berlin (2013)

42. Weinberger, K., Dasgupta, A., Langford, J., Smola, A., Attenberg, J.: Feature hashing for large scale multitask learning. In: Proceedings of the 26th annual international conference on machine learning. pp. 1113–1120 (2009)
43. Wu, Y., He, K.: Group normalization. In: Proceedings of the European Conference on Computer Vision (ECCV), pp. 3–19 (2018)
44. Xie, S., Kirillov, A., Girshick, R., He, K.: Exploring randomly wired neural networks for image recognition. arXiv preprint arXiv:1904.01569 (2019)
45. Zhang, X., Zhou, X., Lin, M., Sun, J.: Shufflenet: an extremely efficient convolutional neural network for mobile devices. In: Proceedings of the IEEE Conference on Computer Vision and Pattern Recognition, pp. 6848–6856 (2018)
46. Zoph, B., Vasudevan, V., Shlens, J., Le, Q.V.: Learning transferable architectures for scalable image recognition. In: Proceedings of the IEEE Conference on Computer Vision and Pattern Recognition, pp. 8697–8710 (2018)

JSTASR: Joint Size and Transparency-Aware Snow Removal Algorithm Based on Modified Partial Convolution and Veiling Effect Removal

Wei-Ting Chen[1], Hao-Yu Fang[2], Jian-Jiun Ding[2], Cheng-Che Tsai[1], and Sy-Yen Kuo[2(✉)]

[1] Graduate Institute of Electronics Engineering, National Taiwan University, Taipei 106, Taiwan
jimmy3505090@gmail.com, danielfang60609@gmail.com
[2] Department of Electrical Engineering, National Taiwan University, Taipei 106, Taiwan
{r08943148,jjding,sykuo}@ntu.edu.tw
https://github.com/weitingchen83/JSTASR-DesnowNet-ECCV-2020

Abstract. Snow removal usually affects the performance of computer vision. Comparing with other atmospheric phenomenon (e.g., haze and rain), snow is more complicated due to its transparency, various size, and accumulation of veiling effect, which make single image de-snowing more challenging. In this paper, first, we reformulate the snow model. Different from that in the previous works, in the proposed snow model, the veiling effect is included. Second, a novel joint size and transparency-aware snow removal algorithm called JSTASR is proposed. It can classify snow particles according to their sizes and conduct snow removal in different scales. Moreover, to remove the snow with different transparency, the transparency-aware snow removal is developed. It can address both transparent and non-transparent snow particles by applying the modified partial convolution. Experiments show that the proposed method achieves significant improvement on both synthetic and real-world datasets and is very helpful for object detection on snow images.

Keywords: Transparency and size-aware snow removal · Snow detection · Differentiable dark channel prior

1 Introduction

Snow is an atmospheric phenomenon which usually obstructs the visibility and degrades the image quality severely. Therefore, snow removal is significant for several computer vison missions. Compared with other atmospheric

---

W.-T. Chen—Equal contribution.

Electronic supplementary material The online version of this chapter (https://doi.org/10.1007/978-3-030-58589-1_45) contains supplementary material, which is available to authorized users.

A. Vedaldi et al. (Eds.): ECCV 2020, LNCS 12366, pp. 754–770, 2020.
https://doi.org/10.1007/978-3-030-58589-1_45

| Input | Eigen [6] | Zheng [35] | DesnowNet [17] | Ours |

Fig. 1. Comparison of the proposed method with state-of-the-art methods [6,17,35]. The proposed method can achieve superior performance in both snow removal and veiling effect recovery. The red bounding box demonstrates the complicated snow scenario. (i.e. the snow particles with different sizes, transparencies, and the veiling effect)

phenomenon, due to the non-transparency property, snow removal is even more challenging because it may occlude the background, which leads to the loss of information. In the open literature, many snow removal algorithms [17,19,27,28,32,35] have been developed for decades. Xu *et al.* [28] proposed a snow flake removal algorithm using the guided filter without any pixel-based statistical information. Zheng *et al.* [35] proposed to use the multi-guided filter to separate the snow part from background. Pei *et al.* [19] utilized the features based on the color information for snow removal. In [17], a learning-based snow removal algorithm was proposed and the formation of snow was modeled by

$$K(x) = J(x)(1 - Z(x)) + C(x)Z(x) \tag{1}$$

where K is snow image acquired by the camera, J is the clean image, Z is the snow mask, C denotes the chromatic aberration map for snow images. Although these algorithms have good performance in some scenarios, they may have several problems which limit the performance. First, for the conventional snow removal strategies, the over-smoothness on the snow-free region may be produced because the snow region is not identified in prior. Second, owing to the complicated formation of the snow image, the conventional snow removal algorithms fail to address the crucial factor in real snow images, such as the veiling effect caused from haze veil and the accumulation of tiny snow particles. Third, unlike haze or rain, which are transparent, snow particles may be non-transparent and the background information is lost. Last, snow images usually consist of various size of particles, however, the existing methods do not address this problem, which leads to the limited recovered performance.

To address these limitations, we develop a novel snow removal model which can cope with the complex snow scenes. The main contributions of this paper are summarized as follows:

1. First, a novel de-snowing model called the **J**oint **S**ize and **T**ransparency-**A**ware **S**now **R**emoval (**JSTASR**) is proposed. It applies three stages: (i) veiling effect removal, (ii) the size-aware snow identifier, and (iii) transparency-aware snow removal. The proposed architecture can retrieve snow information (i.e., size, location and transparency), remove snow, and remove the veiling effect jointly.

2. A new snow formation model is proposed based on the observation of snow images in real world. The snow region mask and the veiling effect are included in this model. In order to train our network in this scenario, a large scale snow dataset called **S**now **R**emoval in **R**ealistic **S**cenario **(SRRS)** is proposed.
3. The size-aware snow identifier, discriminator and loss function is developed to address the complicated snow scenarios according to the snow size. With these modules, the performance of snow removal can be much improved.
4. In order to address particles with different transparency (i.e. transparent and non-transparent snow), the modified partial convolution operation and the transparency-aware snow removal algorithm are proposed.
5. To deal with the veiling effect and increase the visual quality of the recovered result, a differentiable dark channel prior layer and an atmospheric light prediction module are proposed and embedded into the network for fully end-to-end learning in the de-snowing pipeline.

In Fig. 1, the de-snowing results recovered by the proposed JSTASR are compared with other state-of-the-art algorithms. One can see that the proposed method can achieve more desirable results compared with other methods. More experimental results will be shown in Sect. 4.

2   Related Work

2.1   Single Image Snow Removal Algorithm

For snow removal in the single image [17,19,27,28,32,35], Zheng *et al.* [35] investigated the difference between snow streaks and clear background edges. By this statistical information, the multi-guided filter is applied to remove the snow flake. Wang *et al.* [27] proposed a three-layer hierarchical scheme which combines image decomposition and dictionary learning. Voronin *et al.* [26] developed the anisotropic gradient in Hamiltonian quaternions to remove rain and snow. Li *et al.* [15] applied the generative adversarial network (GAN) for snowflake removal. Liu *et al.* [17] proposed a learning-based snow removal architecture called the DesnowNet.

2.2   Single Image Haze Removal Algorithm

Although the proposed algorithm is related to snow removal, since it also adopts the dark channel prior (DCP), which is a dehazing technique, several existing image dehazing algorithms are also reviewed here. For the single image dehazing, several haze and smoke removal algorithms have been proposed for a decade [1–5,10,13,22,25,36]. He *et al.* [10] investigated haze-free images in nature and proposed the dark channel prior to compute the transmission value. Berman *et al.* [1] proposed a non-local image dehazing algorithm. For learning-based algorithms, Cai *et al.* [2] computed the transmission map by a learning architecture called DehazeNet. Chen *et al.* [3] proposed a new feature called the patch map to select the patch size adaptively and addressed the color distortion problem of the DCP. Ren *et al.* [22] proposed the multi-scale CNN to predict the transmission map.

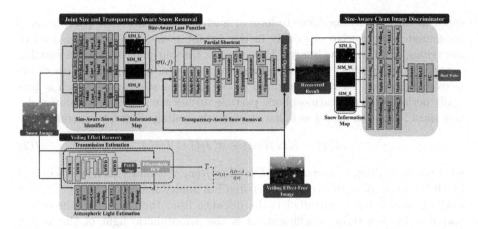

Fig. 2. The overall flow chart of the proposed de-snowing architecture. The proposed method consists three parts: snow removal, veiling effect removal, and clean image discriminator.

3   Proposed Method

The flowchart of the proposed snow removal algorithm is depicted in Fig. 2. Based on the observation of snow images in real world (see Fig. 1) and the limitations mentioned in Sect. 1, several mechanisms are designed. First, the snow formation model is reformulated. Second, the **J**oint **S**ize and **T**ransparency-**A**ware **S**now **R**emoval (**JSTASR**) module is designed to remove both transparent and non-transparent snows. The proposed JSTASR consists of two parts: (i) the size-aware snow identifier and (ii) transparency-aware snow removal. The size-aware snow identifier will generate the snow information map according to different snow scales. Based on this information, the transparency-aware snow removal module can deal with the snow in different scales and transparencies separately. Moreover, to address non-transparent snow, the modified partial convolution is proposed. Third, the differentiable dark channel prior layer is proposed and embedded in the network to remove the veiling effect.

3.1   Snow Model Formulation

In this section, the snow model is reformulated due to the distribution and the veiling effect of snow images. First, unlike the global-distributed atmospheric phenomenon such as haze, mist, or rain, the snow may distribute locally. However, there is no region information in (1) and it may cause error in the snow removal process for snow-free pixels. By using existing snow removal algorithms, the recovered results tend to be over-smoothed and the detail may vanish in those pixels. Second, in snow images, the veiling effect, which is similar to haze and mist, usually occurs. There are two reasons that can explain this phenomenon. First, when snow starts to fall, the temperature on the ground decreases and the vapor in the air may condense to haze or mist. Second, tiny snow particles may accumulate and overlap especially in the distant scene. This may also lead to the veiling effect in snow images. However, for the conventional snow formation model, the veiling effect is not considered and it may limit the performance of the recovered process. Therefore, inspired by the Koschmieder model, which is generally performed on the atmospheric particle turbulence problem, we proposed a new snow formation model as follows to take the veiling effect into consideration

$$I(x) = K(x)T(x) + A(x)(1 - T(x)) \tag{2}$$

where $K(x) = J(x)(1-Z(x)R(x))+C(x)Z(x)R(x)$, I denotes the image acquired by the camera, K is the veiling effect-free but snowy image. $T(x) = e^{-\beta d(x)}$ is media transmission where $d(x)$ is the distance from the camera to the object and β is the scattering coefficient. A is the atmospheric light of the veiling effect. J is the clean image. C and Z are the same definition as (1). R is a binary mask to denote the snow location information. By this reformulation, the veiling effect problem in snow scenario can be well addressed. Moreover, different from previous snow removal algorithms, with the introduction of R, the local reconstruction is performed on the snow pixels. Therefore, with the veiling effect recovery and the local reconstruction, the recovered image can achieve sharper result containing more fruitful edge comparing with other methods.

3.2 Joint Size and Transparency-Aware Snow Removal

In this section, we illustrate the proposed JSTASR module in detail. It can address the aforementioned snow removal problem via the fully end-to-end learning system. As discussed before, the conventional snow removal algorithms may tend to over or under reconstruction because they do not consider the variety of snow particles. Compared with haze and rain, snow is more complicated because snow particles have different sizes and transparencies and are distributed locally. Thus, in the proposed JSTASR module, the information of (i) size, (ii) transparency, and (iii) localization of snow particles are incorporated to the network design. The flowchart of JSTASR module is shown in upper side of Fig. 2.

Size-Aware Snow Identifier: In the proposed architecture, the size-aware snow identifier is designed. That is, given a snow image, one can acquire three snow information maps in different scales (i.e., large, medium, and small). These maps are gray-scale image consisting of the location and the intensity information. For the design of the proposed snow identifier, three different networks are designed to predict large, medium, and small snow particles and the multi-scale convolution and deconvolution layers are adopted. In these layers, several convolution and deconvolution filters with different scales are connected in parallel. (i.e., $L = L_1 \| L_2 \| L_3$ for large snow particles, $M = M_1 \| M_2 \| M_3$ for medium snow particles, $S = S_1 \| S_2 \| S_3$ for small snow particles, where $\|$ denotes the parallel connection and L_n, M_n, and S_n present the convolution kernels with different kernel sizes). In general, the kernel size in L is largest and that in S is smallest. With this architecture, the size and the location information can be

acquired. Moreover, the shape information of snow particles can also be estimated by this operation. Then, the accurate snow information map (SIM) can
be computed and it will be used in the snow removal process in next subsection.

Transparency-Aware Snow Removal: For snow particles, unlike haze and
rain, its transparency may be more divergent. From Fig. 1, one can observe
that, in a snow image, some snow particles are non-transparent while some
are transparent. Recovering transparent particles may be easier because it has
been developed for a while in [30,33]. However, recovering non-transparent particles may be challenging because the background information is totally occluded
by invisible snow, which means that these regions can be only estimated by
the neighboring area. Thus, inspired by other computer vision applications
[16,18,31], the technique of image inpainting is adopted to mitigate this problem. Image inpainting aims to fill the holes (broken pixels) in the image, which
is similar to the problem in non-transparent snow removal. In this paper, the
modified partial convolution which can perform the transparency-aware snow
removal is proposed. This method is inspired by the inpainting strategy in [16]
which applies the partial convolution to inpaint irregular area. The proposed
snow removal operation can be expressed as

$$
\mathbf{y'} = \begin{cases} H^T\,(S \odot M)\,\frac{sum(1)}{sum(U)} + b, & \text{if } sum(U) \geq 0 \\ 0, & \text{otherwise} \end{cases} \tag{3}
$$

where y' denotes the output of inpainting, H presents the convolution filter
weight; b is the corresponding bias, S is the input feature map of the snow
image for the current convolutional kernel, U is the corresponding binary mask
which records the useful (transparent or clean) information for the inpainting
process, $\odot$ means element-wise multiplication, and sum(1) presents summation
of the matrix which has the same dimension as U but all values are one. Sum(1)
/sum(U) is a scaling factor to adjust the varying amount of unmasked inputs and
prevent the feature map from vanishing. After the operation in (3) is performed
for one round, the elements of the original snow mask in U will be updated to
U' by (4)

$$
U'(m,n) = \begin{cases} 1, & \text{if } \underset{(p,q)\in\Omega_{m,n}}{sum}\,(U(p,q)) \geq 0 \\ 0, & \text{otherwise} \end{cases} \tag{4}
$$

where $\Omega_{m,n}$ is a patch centered at $(m,\,n)$. By (4), the pixel value at $(m,\,n)$ will
be updated as a useful information if there is at least one useful pixel (transparent or clear pixels) within the patch $\Omega_{m,n}$. Finally, after the inpainting process
repeated for sufficient rounds, the whole snow mask U will be set to 1, which
means that all non-transparent snow particles are filled and transparent pixels
are also recovered to clear pixels. To better remove the snow with different transparencies, different from the original partial convolution, the transparency-aware
mechanism is proposed to initialize the snow mask U in (3) according to the snow
information maps predicted by the size-aware snow identifier. This operation is
important because it can determine whether this information is useful in the

inpainting process. The transparency-aware mechanism in this architecture can help us to select proper reference pixels for inpainting. The operation can be expressed as

$$\sigma(i,j) = \begin{cases} 1, & \text{if } N(i,j) \geq \gamma \\ 0, & \text{otherwise} \end{cases} \tag{5}$$

where $\sigma(i,j)$ denotes the transparency-aware snow region map (i.e., initial U in (3)), $N(i,j)$ presents the summation of snow information maps in three scales generated from the size-aware snow identifier, and γ is the transparent threshold which determines whether the pixel is visible or not. With this operation, snow pixels can be recovered according to their transparencies, that is, non-transparent pixels are recovered only based on the neighboring pixels while the transparent pixels are recovered partially by itself and partially by neighbor pixels.

The proposed transparency-aware snow removal architecture is inspired by U-Net [23]. For the encoder part, different from the traditional partial convolution which applies single convolution kernel size, the multi-scale architecture is adopted. With it, the encoder can extract features in various levels adaptively according to the different sizes of snow. For the decoder part, different from the original partial convolution which applies up-sampling and convolution in the decoder, we adopt both multi-scale deconvolution and up-sampling operations to prevent the recovered feature from blurring. Moreover, the partial shortcut as follows is applied to improve the quality of the reconstructed images :

$$k_i = \rho\left[(y_i' \odot U_i) \cup k_{i-1}\right] \tag{6}$$

where k_i denotes the output of the decoder in the layer i, ρ is a combination of the traditional convolution operation, the global convolutional network (GCN) [20] module, and the boundary refinement (BR) [20]. y_i' and U_i indicate the output of the partial convolution and the snow mask in the layer i, which has been presented in (3), and $\cup$ presents the concatenate operation. The proposed idea is that, for the original design, the decoded information usually tends to be blurring. Inspired by the U-Net [23], the shortcut of original information is applied to increase the recovered resolution. However, instead of passing the original information directly, we pass the partial information to the decoder because the background information in non-transparent region is occluded in the original data and this information may damage the recovered process. Therefore, we discard the information on non-transparent pixels and only adopt useful information. Moreover, the adopted GCN and BR modules are helpful for enhancing the boundary and semantic information. With these operations, high-quality and high-resolution de-snowing results can be acquired.

Size-Aware Loss Function: In order to enhance the performance of the generator, we leverage the snow information maps which are predicted by the snow identifier and proposed size-aware loss function in (7).

$$L_{size-aware} = (w_s + 0.5w_m)\left[\lambda_1 L_{pixel} + \lambda_2 L_{TV}\right] + (w_l + 0.5w_m)\left[\lambda_3 L_S + \lambda_4 L_P\right] \tag{7}$$

where λ_1, λ_2, λ_3, λ_4 are constants, w_s, w_m and w_l are the composition percentages of snow pixels in the small, medium, and large snow information maps, respectively. L_{pixel} [16], L_{TV} [16], L_S [16] and L_P [11] denote the pixel loss, the total variation loss, the style loss and perceptual loss, respectively. The idea is based on that, for the region recovered from small snow particles, the total variation loss and the pixel loss should have more weight because the results are generally suffered from noise and the under-smoothed problem. On the other hand, the area damaged by the large snow should put more emphasis on the style loss and the perceptual loss because the recovered region may lose the texture and global information [11,16]. Moreover, for the pixels damaged by medium snow, since the recovered results are between large and small snow, the weights should be shared with w_s and w_l. By designing the size-aware loss function, the error can be calculated according to different snow scenarios.

Size-Aware Discriminator: To improve the performance of the transparency-aware snow removal module, the size-aware discriminator which can identify whether the recovered result looks like a real image is proposed. Inspired from [21], which applied the attention map to improve the performance of the discriminator in rain drop removal. Based on the oberservation of snow above, to further improve the recovered process in the snow scenario, three snow information maps based on different sizes (i.e, L, M, S) are applied in the proposed network. The features are extracted with snow information map in different scales by multi-pooling as follows

$$l = \underset{e \in s}{\psi} M_e^k (x). \tag{8}$$

where x represents the features determined from the previous layer, M_e^k denotes the stride convolution operation with the kernel size e and the dilated level k, ψ denotes the concatenate operation, and s is the scale range for the stride convolution where $s \in \{2, 3, 5\}$. In this work, k is set to 2. By the operation of multi pooling, the extracted features can be well preserved in different sizes and scales. Moreover, the concatenating operation is performed instead of addition because we hope to preserve the properties at each scale. The idea of size-aware discriminator is based on that, compared to rain drop, snow is more complicated. It may be non-transparent and the size variation is large. By extracting the features from different scales and focusing on different transparencies, the performance of the discriminative process will be improved effectively. Then, these features will be concatenated together at the end of network and the fully connected operation is performed to identify the recovered result.

Overall Loss Function: The overall loss function of the proposed JSTASR module can be presented as

$$L_{JSTASR} = L_{size-aware} + \lambda_5 L_{Identifier} + \lambda_6 L_{Adv} \tag{9}$$

$$L_{Adv} = \underset{G}{\min}\underset{D}{\max} \, E_x[\log D(x)] + E_x[\log(1 - D(G(x')))] \tag{10}$$

where λ_5 and λ_6 are constants. x and x' are snow-free image and snow image. $L_{Identifier}$ presents the MSE between the predicted snow information maps and their corresponding ground truths. L_{JSTASR} and L_{Adv} are the generative loss and the adversarial loss, respectively.

3.3   Veiling Effect Removal

In veiling effect removal, we aim to address the inverse problem in (2), that is, estimating the accurate transmission and atmospheric light values. For the transmission value, we apply the differentiable DCP layer with the patch map [3]. The flowchart of veiling effect removal is shown in the bottom of Fig. 2.

Differentiable Dark Channel Prior: To cope with the veiling effect problem in the snow removal process, in the proposed method, we apply the DCP [10] to estimate the transmission because it has been proved to be an effective way to deal with the veiling effect in other works about atmospheric particle removal [19,34]. However, this method has been indicated several limitations such as the color degradation in the white and bright scene [3,29]. Although in PMS-Net [3], the patch map can address the limitation of the dark channel prior effectively, sometimes the performance is limited in color fidelity and recovered quality. The reason is that the PMS-Net cannot train the whole system including the atmospheric light estimation and refined network because the patch-map based dark channel cannot be embedded into this method. Moreover, due to the complicated scenario in our proposed method, end-to-end training and optimization is necessary to achieve better performance. Therefore, in this paper, we develop the patch-map based differentiable dark channel prior layer to improve the performance of this method and further embed into the proposed snow removal process to achieve fully end-to-end learning. This layer is extended with a learnable patch map. The inputs are the image with veil, the predicted patch map, and the predicted atmospheric light. The output is the estimated transmission map. First, the minimum operation in the local patch is performed in each color channel. The operation can be expressed as follows

$$H(x,y,z,p) = \min_{\alpha,\beta \in [1,p]} [I(x+\alpha, y+\beta, z)] \tag{11}$$

where I is the input image. x and y are the indexes of pixel location. z indicates the different color channel. p is the index of the patch size axis, and $H(x, y, z. p)$ is the result of the minimum operation. Then, $H'(x, y, p)$ is formulated by applying the minimum filter along the color channel axis z on H (x, y, z, p). The patch map is then projected to the patch map box $PMB(x, y, p)$:

$$PMB(x,y,p) = \begin{cases} 1, & \text{if } P(x,y) = p \\ 0, & \text{otherwise} \end{cases} \tag{12}$$

where P (x, y) is the predicted patch map. Then, $H'(x, y, p)$ will be multiplied with $PMB(x, y, p)$ and the coarse transmission map will be computed by summing the value along the patch size axis (i.e., p-axis). With this process, the

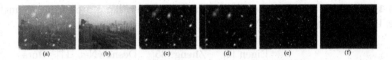

Fig. 3. The example in SRRS: (a) Snow image; (b) ground truth; (c) combined snow mask (d) large snow mask; (e) medium snow mask; (f) small snow mask. Note that the snow masks are gray-scale images which contain the information of snow location, intensity, and size.

fully differentiable DCP layer can be achieved in the end-to-end veiling effect removal network and the performance of recovered quality and color distortion can be improved effectively.

Learning Atmospheric Light: To estimate the global atmospheric light A from input images, we adopt the learning architecture to predict the air light intensity because we want to achieve end-to-end learning with the differentiable DCP layer. For this sub-network, we apply the VGG-16 architecture [24] as the backbone, but each convolution operation is replaced by the multi-level pooling operation mentioned in Subsect. 3.2.

Loss Function: The loss functions of the veiling effect removal module is formulated as

$$L_{Veil} = L_{PatchMap} + \lambda_7 L_A \tag{13}$$

where L_{Veil} is the overall loss of veiling effect removal $L_{PatchMap}$ is the loss of the patch map selection network proposed in [3]. L_A is the atmospheric light loss, which is defined as the MSE between the predicted atmospheric light and the ground truth.

The overall loss function of the proposed whole network can be expressed as

$$L_{Overall} = L_{JSTASR} + L_{Veil} \tag{14}$$

4   Experimental Result

4.1   Dataset Generation

Although there is one large-scale synthetic dataset [17] for image de-snowing, it does not contain the veiling effect. In this work, in order to address various snow scenarios, a novel dataset called the Snow Removal in Realistic Scenario (SRRS)[1] is proposed. It contains 15000 synthesized snow images and 1000 snow images in real scenarios downloaded from the Internet. We synthesize the SRRS by two steps. First, we apply the popular haze benchmark dataset called RESIDE dataset [14] to synthesize the image with veiling effect based on the procedure in their paper. We set $\beta \in [0.4, 1.6]$ and $A \in [0.5, 1]$. Then, for each snow image, various types of snow are synthesized by Photoshop and the corresponding snow

---

[1] The dataset can be downloaded from our project page.

Table 1. Quantitative evaluation for comparison with other state-of-the-art methods on the Test A and Test B. (PSNR/SSIM)

		Eigen [6]	Zheng [35]	DesnowNet [17]	Ours
Test A	w/o MSCNN [22]	14.51/0.58	15.48/0.729	16.32/0.80	**25.62/0.89**
	w MSCNN [22]	17.36/0.57	18.02/0.76	18.42/0.82	–
Test B	–	18.57/0.43	23.72/0.83	25.58/0.85	**25.69/0.86**

Table 2. The evaluation of sharpness in terms of JNBM [7] and e [8] on 1000 real-world images. Note: Larger values of JNBM and e means sharper and more newly visible edges on recovered image

	Zheng [35]	Eigen [6]	DesnowNet [17]	Ours
JNBM/e	2.19/−0.08	3.04/−0.31	2.95/−0.09	**3.92/0.25**

Table 3. Run time comparison with state-of-the-art snow removal methods.

	Zheng (C) [35]	Eigen (G) [6]	DesnowNet (G) [17]	Ours (G)
Time (s)	1.4	2.81	1.38	**0.36**

Table 4. Comparing the performances of object detection when using the proposed algorithm and other state-of-the-art methods for de-snowing in prior.

	Baseline	Zheng+MSCNN	Eigen+MSCNN	DesnowNet+MSCNN	Ours
Accuracy (IoU)	0.23	0.43	0.33	0.50	**0.72**

information (i.e., transparency, size, and location) is labeled. The example of SRRS is shown in Fig. 3. For the training procedure, we randomly pick 2500 snow images as the training dataset. For the test dataset, 1000 snow images are picked randomly from the proposed SRRS dataset and we call it as **Test A**. Moreover, in order to test the generalization of the proposed network, the test dataset proposed by [17] is applied. We call this dataset as **Test B** which contains 1000 snow images only with snow particles but without veiling effect.

4.2 Training Detail

The proposed de-snowed network mainly consists of two sub-networks: the JSTASR module and veiling effect removal. First, we train the veiling effect removal network with 2500 hazy images based on the RESIDE [14] dataset. Second, for JSTASR, initially, we train the size-aware snow identifier to predict the snow information. Then, the identifier is fixed to train the transparency-aware snow removal module. These two processes are pre-trained with the fixed veiling effect removal network. After the pre-trained process, two sub-networks are trained together in the fine-tuned state. All these modules and the fine-tuned

Table 5. Quantitative evaluation for the ablation study in the proposed modules of snow removal on Test A (PSNR/SSIM)

Module	J	J+P	J+P+M	J+P+M+T	All w/o EVER	All w EVER
PSNR/SSIM	24.01/0.81	24.86/0.87	23.02/0.88	24.27/0.89	20.32/0.77	**25.85/0.89**

Table 6. Quantitative evaluation for the ablation study of the proposed differential dark channel prior layer on the indoor dataset in SOTS [14]. The lower value of CIEDE 2000 means less color distortion.

	DCP [10]	PMS-Net [3]	Ours
PSNR/SSIM/CIEDE 2000	18.2/0.83/9.42	21.14/0.88/5.88	**24.68/0.90/4.25**

process are trained based on the training set of *SRRS* dataset. The learning rate is set to e^{-4} and the Adam [12] optimizer is adopted in both JSTASR and veiling effect removal. In this network, the pre-trained process is adopted. For the parameter setting in the proposed network, we set $\lambda_1 = \lambda_2 = \lambda_3 = \lambda_4 = \lambda_5 = 1$, $\lambda_6 = 0.5$, and $\lambda_7 = 10^{-4}$. We set the threshold value $\gamma = 0.1$. For each epoch, we cut 15% of the training data as the validation set. Finally, two pre-trained sub-networks will be combined and trained together to fine-tune the entire network for several thousand iterations. The proposed method is implemented on a workstation with a 3.5 GHz CPU, 64 G RAM, and Nvidia GTX 1080 Ti GPU. The number of the parameter in the proposed network is 6.5×10^7. Moreover, the whole training process of this network takes five days.

4.3 Comparison with State-of-the-art Methods

Analysis in Synthetic Snow Dataset. In Table 1, the proposed de-snowed algorithm is evaluated on the synthetic snow dataset. We apply two metrics: the peak-to-peak signal to noise ratio (PSNR) and the structural similarity (SSIM). We apply state-of-the-art snow removal algorithms (i.e., Zheng [35], and DesnowNet [17]) and atmospheric particle removal methods (i.e., Eigen [6]) for comparison. To compare with these methods fairly, we apply the conventional haze removal method (i.e., MSCNN [22]) as the veiling effect removal strategy because the images in Test A contain the veiling effect. The quantitative comparison is presented in Table 1. We can find that the proposed method outperforms state-of-the-art methods on both Test A and B in all metrics no matter the veiling removal strategy is applied or not. Note that, the proposed method is trained on our proposed dataset but still can achieve better performance on that proposed by the DesnowNet [17]. These results can prove that our proposed method for snow removal is guaranteed and effective.

Analysis in Real-World Snow Dataset. The qualitative comparison of the proposed method with other state-of-the-art de-snowing algorithms is performed. In Fig. 4, some examples of our real-world dataset are shown. One can notice that, the results recovered by Eigen [6] and Zheng [35] may still contain snow

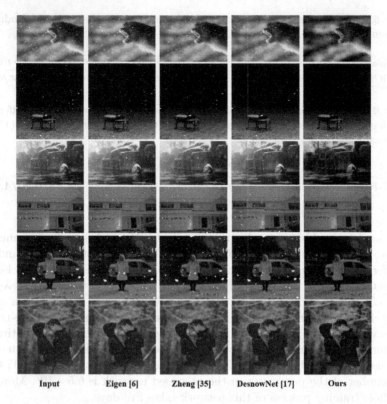

Input   Eigen [6]   Zheng [35]   DesnowNet [17]   Ours

Fig. 4. A qualitative comparison of the proposed method with state-of-the-art algorithms on real-world images. (Zoom-in to view in detail.)

particles and those recovered by the later one may have severe blurring problem. Although the results recovered by DesnowNet [17] can achieve better performance comparing with Eigen [6] and Zheng [35], the large and non-transparent snow particles cannot be removed clearly (see the 2nd, 3rd and 4th rows). Moreover, all other methods cannot address the veiling effect problem (see the 1st, 3rd and 6th and the Fig. 1). By contrast, the proposed algorithm can well remove snow particle with different sizes and transparencies, and veiling effect because we propose the size-aware snow identifier, transparency-aware snow removal and differentiable dark channel prior. Furthermore, with the snow information maps predicted by the size-aware snow identifier, the proposed method can prevent the recovered images from blurring effectively because the snow location is considered during the snow removal procedure. To evaluate the effectiveness of this mechanism, we apply the just noticeable blur metric (JNBM) [7] and the newly visible edge ratio (e) in VLD [8] to evaluate the performance of detail and sharpness preservation. The results shown in Table 2 demonstrate that the proposed method achieve the best performance on retaining both the sharpness and edge comparing with other methods. It can prove the effectiveness of the proposed method in the real-world images.

Run Time Comparison. We further compare the runtime in Table 3. The results show that the proposed snow removal method can achieve much better performance in computational time[2]. It proves that the proposed method has superior performance comparing to state-of-the-art methods in terms of both the run time and the recovered quality.

Improvement in High-Level Vision Application of Object Detection. In Table 4, we verify that the proposed snow removal algorithm can benefit to the high-level vision tasks such as object detection in the snow scenario. We apply the RTTS dataset in RESIDE [14], which contains the hazy images with annotation and conduct the snow synthesis process in Subsect. 4.1 to generate snow images with annotation. We compare the proposed algorithm with three desnowing methods. The object detection algorithm is fixed to the Mask R-CNN [9]. From Table 4, one can see that, with the utility of the proposed de-snowing algorithm, the accuracy of object detection can be improved by 44% and 210% comparing to the DesnowNet and baseline. It proves that the proposed approach is helpful for achieving better performance for the object detection.

4.4   Ablation Study

To verify the effectiveness of the proposed modules, the ablation study is performed. The evaluation consists of two part: the proposed JSTASR module and the differentiable DCP. For the former one, six component combinations are constructed, that is, 1) JSTASR with traditional partial convolution (**J**); 2) the J module with partial shortcut (**J+P**); 3) the J+PS module with multi-convolution (**J+P+M**); 4) the J+P+M module with the transparency-aware mechanism (**J+P+M+T**); 5) the J+P+M+T module with the size-aware mechanism (i.e., the size-aware loss and the size-aware discriminator) but without end-to-end veiling effect removal (**All** w/o EVER); 6) the JSTASR with end-to-end veiling effect removal (**(All)** w EVER); The result is shown in Table 5. One can see that all the proposed modules are effective for the snow removal process. Furthermore, for the veiling effect removal, the results illustrate that the image quality achieves great improvement with the proposed differentiable DCP. To prove the effectiveness of the proposed differentiable dark channel prior, we adopt the ablation study on the famous haze image benchmark called SOTS dataset without the snow removal process in Table 6. In this experiment, we only apply indoor dataset to evaluate the improvement. The results show that applying the differentiable DCP layer can achieve higher performance on dehazing comparing to the PMS-Net [3] and DCP [10].

---

[2] The input size of DesnowNet in this experiment is &480 × 480&.

5 Conclusion

In this paper, a novel de-snowed algorithm based on joint size and transparency aware filters and veiling effect removal is proposed. The differentiable DCP is proposed to remove the veiling effect. In the snow removal procedure, first, the size-aware snow identifier is proposed to identify snow particles according to their sizes. In each snow information map, the intensity, location, and size information are recorded to perform adaptive snow removal. Then, the transparency-aware snow removal process based on the modified partial convolution is developed to address various snow types. Also, a binary mask is applied to select the useful information for snow pixel recovery. Moreover, to further improve the performance of the modified partial convolution, the partial shortcut and the multi-scale encoder and decoder are proposed. Last, to better optimize the snow recovered results in different sizes of snow particles, size-aware loss functions and the snow-free discriminator are designed. Experimental results showed that the proposed method can achieve better performance even in the complicated snow scenarios compared to other methods.

References

1. Berman, D., Avidan, S., et al.: Non-local image dehazing. In: Proceedings of the IEEE Conference on Computer Vision and Pattern Recognition, pp. 1674–1682 (2016)
2. Cai, B., Xu, X., Jia, K., Qing, C., Tao, D.: Dehazenet: an end-to-end system for single image haze removal. IEEE Trans. Image Process. **25**(11), 5187–5198 (2016)
3. Chen, W.T., Ding, J.J., Kuo, S.Y.: PMS-Net: robust haze removal based on patch map for single images. In: Proceedings of the IEEE Conference on Computer Vision and Pattern Recognition, pp. 11681–11689 (2019)
4. Chen, W.T., Fang, H.Y., Ding, J.J., Kuo, S.Y.: PMHLD: Patch map based hybrid learning dehazenet for single image haze removal. IEEE Trans. Image Process. (2020)
5. Chen, W.T., Yuan, S.Y., Tsai, G.C., Wang, H.C., Kuo, S.Y.: Color channel-based smoke removal algorithm using machine learning for static images. In: 2018 25th IEEE International Conference on Image Processing (ICIP), pp. 2855–2859. IEEE (2018)
6. Eigen, D., Krishnan, D., Fergus, R.: Restoring an image taken through a window covered with dirt or rain. In: Proceedings of the IEEE International Conference on Computer Vision, pp. 633–640 (2013)
7. Ferzli, R., Karam, L.J.: A no-reference objective image sharpness metric based on the notion of just noticeable blur (JNB). IEEE Trans. Image Process. **18**(4), 717–728 (2009)
8. Hautière, N., Tarel, J.P., Aubert, D., Dumont, E.: Blind contrast enhancement assessment by gradient ratioing at visible edges. Image Anal. Stereology **27**(2), 87–95 (2008)
9. He, K., Gkioxari, G., Dollár, P., Girshick, R.: Mask R-CNN. In: Proceedings of the IEEE International Conference on Computer Vision, pp. 2961–2969 (2017)
10. He, K., Sun, J., Tang, X.: Single image haze removal using dark channel prior. IEEE Trans. Pattern Anal. Mach. Intell. **33**(12), 2341–2353 (2010)

11. Johnson, J., Alahi, A., Fei-Fei, L.: Perceptual losses for real-time style transfer and super-resolution. In: European Conference on Computer Vision, pp. 694–711 (2016)
12. Kingma, D.P., Ba, J.: Adam: a method for stochastic optimization. arXiv preprint arXiv:1412.6980 (2014)
13. Li, B., Peng, X., Wang, Z., Xu, J., Feng, D.: AOD-Net: All-in-one dehazing network. In: Proceedings of the IEEE International Conference on Computer Vision, pp. 4770–4778 (2017)
14. Li, B., et al.: Benchmarking single-image dehazing and beyond. IEEE Trans. Image Process. **28**(1), 492–505 (2018)
15. Li, Z., et al.: Single image snow removal via composition generative adversarial networks. IEEE Access **7**, 25016–25025 (2019)
16. Liu, G., Reda, F.A., Shih, K.J., Wang, T.C., Tao, A., Catanzaro, B.: Image inpainting for irregular holes using partial convolutions. In: Proceedings of the European Conference on Computer Vision (ECCV), pp. 85–100 (2018)
17. Liu, Y.F., Jaw, D.W., Huang, S.C., Hwang, J.N.: DesnowNet: context-aware deep network for snow removal. IEEE Trans. Image Process. **27**(6), 3064–3073 (2018)
18. Pathak, D., Krahenbuhl, P., Donahue, J., Darrell, T., Efros, A.A.: Context encoders: feature learning by inpainting. In: Proceedings of the IEEE Conference on Computer Vision and Pattern Recognition, pp. 2536–2544 (2016)
19. Pei, S.C., Tsai, Y.T., Lee, C.Y.: Removing rain and snow in a single image using saturation and visibility features. In: 2014 IEEE International Conference on Multimedia and Expo Workshops (ICMEW), pp. 1–6 (2014)
20. Peng, C., Zhang, X., Yu, G., Luo, G., Sun, J.: Large kernel matters-improve semantic segmentation by global convolutional network. In: Proceedings of the IEEE Conference on Computer Vision and Pattern Recognition, pp. 4353–4361 (2017)
21. Qian, R., Tan, R.T., Yang, W., Su, J., Liu, J.: Attentive generative adversarial network for raindrop removal from a single image. In: Proceedings of the IEEE Conference on Computer Vision and Pattern Recognition, pp. 2482–2491 (2018)
22. Ren, W., Liu, S., Zhang, H., Pan, J., Cao, X., Yang, M.-H.: Single image dehazing via multi-scale convolutional neural networks. In: Leibe, B., Matas, J., Sebe, N., Welling, M. (eds.) ECCV 2016. LNCS, vol. 9906, pp. 154–169. Springer, Cham (2016). https://doi.org/10.1007/978-3-319-46475-6_10
23. Ronneberger, O., Fischer, P., Brox, T.: U-Net: convolutional networks for biomedical image segmentation. In: Navab, N., Hornegger, J., Wells, W.M., Frangi, A.F. (eds.) MICCAI 2015. LNCS, vol. 9351, pp. 234–241. Springer, Cham (2015). https://doi.org/10.1007/978-3-319-24574-4_28
24. Simonyan, K., Zisserman, A.: Very deep convolutional networks for large-scale image recognition. arXiv preprint arXiv:1409.1556 (2014)
25. Tarel, J.P., Hautiere, N.: Fast visibility restoration from a single color or gray level image. In: 2009 IEEE 12th International Conference on Computer Vision, pp. 2201–2208 (2009)
26. Voronin, V., Semenishchev, E., Zhdanova, M., Sizyakin, R., Zelenskii, A.: Rain and snow removal using multi-guided filter and anisotropic gradient in the quaternion framework. In: Artificial Intelligence and Machine Learning in Defense Applications, vol. 11169, p. 111690 (2019)
27. Wang, Y., Liu, S., Chen, C., Zeng, B.: A hierarchical approach for rain or snow removing in a single color image. IEEE Trans. Image Process. **26**(8), 3936–3950 (2017)
28. Xu, J., Zhao, W., Liu, P., Tang, X.: An improved guidance image based method to remove rain and snow in a single image. Comput. Inf. Sci. **5**(3), 49 (2012)

29. Yang, D., Sun, J.: Proximal dehaze-net: a prior learning-based deep network for single image dehazing. In: Proceedings of the European Conference on Computer Vision (ECCV), pp. 702–717 (2018)
30. Yang, W., Tan, R.T., Feng, J., Liu, J., Guo, Z., Yan, S.: Deep joint rain detection and removal from a single image. In: Proceedings of the IEEE Conference on Computer Vision and Pattern Recognition, pp. 1357–1366 (2017)
31. Yu, J., Lin, Z., Yang, J., Shen, X., Lu, X., Huang, T.S.: Generative image in painting with contextual attention. In: Proceedings of the IEEE Conference on Computer Vision and Pattern Recognition, pp. 5505–5514 (2018)
32. Yu, S., et al.: Content-adaptive rain and snow removal algorithms for single image. In: International Symposium on Neural Networks, pp. 439–448 (2014)
33. Zhang, H., Patel, V.M.: Densely connected pyramid dehazing network. In: Proceedings of the IEEE Conference on Computer Vision and Pattern Recognition, pp. 3194–3203 (2018)
34. Zhang, Z., Liu, W., Ma, H., Liu, X.: Going clear from misty rain in dark channel guided network
35. Zheng, X., Liao, Y., Guo, W., Fu, X., Ding, X.: Single-image-based rain and snow removal using multi-guided filter. In: International Conference on Neural Information Processing, pp. 258–265 (2013)
36. Zhu, Q., Mai, J., Shao, L.: A fast single image haze removal algorithm using color attenuation prior. IEEE Trans. Image Process. 24(11), 3522–3533 (2015)

Ocean: Object-Aware Anchor-Free Tracking

Zhipeng Zhang[1,2], Houwen Peng[2(✉)], Jianlong Fu[2], Bing Li[1], and Weiming Hu[1]

[1] NLPR, CASIA and AI School, UCAS and CEBSIT, Beijing, China
{zhipeng.zhang,bli,wmhu}@nlpr.ia.ac.cn
[2] Microsoft Research, Beijing, China
{houwen.peng,jianf}@microsoft.com

Abstract. Anchor-based Siamese trackers have achieved remarkable advancements in accuracy, yet the further improvement is restricted by the lagged tracking robustness. We find the underlying reason is that the regression network in anchor-based methods is only trained on the positive anchor boxes (*i.e.*, $IoU \geq 0.6$). This mechanism makes it difficult to refine the anchors whose overlap with the target objects are small. In this paper, we propose a novel object-aware anchor-free network to address this issue. First, instead of refining the reference anchor boxes, we directly predict the position and scale of target objects in an anchor-free fashion. Since each pixel in groundtruth boxes is well trained, the tracker is capable of rectifying inexact predictions of target objects during inference. Second, we introduce a feature alignment module to learn an object-aware feature from predicted bounding boxes. The object-aware feature can further contribute to the classification of target objects and background. Moreover, we present a novel tracking framework based on the anchor-free model. The experiments show that our anchor-free tracker achieves state-of-the-art performance on five benchmarks, including VOT-2018, VOT-2019, OTB-100, GOT-10k and LaSOT. The source code is available at https://github.com/researchmm/TracKit.

Keywords: Visual tracking · Anchor-free · Object-aware

1 Introduction

Object tracking is a fundamental vision task. It aims to infer the location of an arbitrary target in a video sequence, given only its location in the first frame.

Z. Zhang—Work performed when Zhipeng was an intern of Microsoft Research.
Z. Zhang, B. Li, W. Hu are with the Institution of Automation, Chinese Academy of Sciences (CASIA) and School of Artificial Intelligence, University of Chinese Academy of Sciences (UCAS) and CAS Center for Excellence in Brain Science and Intelligence Technology (CEBSIT).

Electronic supplementary material The online version of this chapter (https://doi.org/10.1007/978-3-030-58589-1_46) contains supplementary material, which is available to authorized users.

© Springer Nature Switzerland AG 2020
A. Vedaldi et al. (Eds.): ECCV 2020, LNCS 12366, pp. 771–787, 2020.
https://doi.org/10.1007/978-3-030-58589-1_46

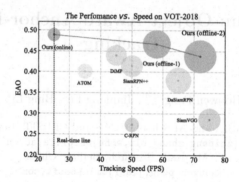

Fig. 1. A comparison of the performance and speed of state-of-the-art tracking methods on VOT-2018. We visualize the Expected Average Overlap (EAO) with respect to the Frames-Per-Seconds (FPS). *Offline-1* and *Offline-2* indicate the proposed offline trackers with and without feature alignment module, respectively.

The main challenge of tracking lies in that the target objects may undergo heavy occlusions, large deformation and illumination variations [44,49]. Tracking at real-time speeds has a variety of applications, such as surveillance, robotics, autonomous driving and human-computer interaction [16,25,33].

In recent years, Siamese tracker has drawn great attention because of its balanced speed and accuracy. The seminal works, *i.e.*, SINT [35] and SiamFC [1], employ Siamese networks to learn a similarity metric between the object target and candidate image patches, thus modeling the tracking as a search problem of the target over the entire image. A large amount of follow-up Siamese trackers have been proposed and achieved promising performances [9,11,21,22,50]. Among them, the Siamese region proposal networks, dubbed SiamRPN [22], is representative. It introduces region proposal networks [31], which consist of a classification network for foreground-background estimation and a regression network for anchor-box refinement, *i.e.*, learning 2D offsets to the predefined anchor boxes. This anchor-based trackers have shown tremendous potential in tracking accuracy. However, since the regression network is only trained on the positive anchor boxes (*i.e.*, $IoU \geq 0.6$), it is difficult to refine the anchors whose overlap with the target objects are small. This will cause tracking failures especially when the classification results are not reliable. For instance, due to the error accumulation in tracking, the predictions of target positions may become unreliable, *e.g.*, $IoU < 0.3$. The regression network is incapable of rectifying this weak prediction because it is previously unseen in the training set. As a consequence, the tracker gradually drifts in subsequent frames.

It is natural to throw a question: *can we design a bounding-box regressor with the capability of rectifying inaccurate predictions?* In this work, we show the answer is affirmative by proposing a novel object-aware anchor-free tracker. Instead of predicting the small offsets of anchor boxes, our object-aware anchor-free tracker directly regresses the positions of target objects in a video frame. More specifically, the proposed tracker consists of two components: an object-aware classification network and a bounding-box regression network. The classification is in charge of determining whether a region belongs to foreground or background, while the regression aims to predict the distances from each pixel within the target objects to the four sides of the groundtruth bounding boxes. Since each pixel in the groundtruth box is well trained, the regression network is able to localize the target object even when only a small region is identified as the foreground. Eventually, during inference, the tracker is capable of rectifying the weak predictions whose overlap with the target objects are small.

When the regression network predicts a more accurate bounding box (*e.g.*, rectifying weak predictions), the corresponding features can in turn help the classification of foreground and background. We use the predicted bounding box as a reference to learn an object-aware feature for classification. More concretely, we introduce a feature alignment module, which contains a 2D spatial transformation to align the feature sampling locations with predicted bounding boxes (*i.e.*, regions of candidate objects). This module guarantees the sampling is specified within the predicted regions, accommodating to the changes of object scale and position. Consequently, the learned features are more discriminative and reliable for classification.

The effectiveness of the proposed framework is verified on five benchmarks: VOT-2018 [17], VOT-2019 [18], OTB-100 [44], GOT-10k [14] and LaSOT [8]. Our approach achieves state-of-the-art performance (an EAO of 0.467) on VOT-2018 [17], while running at 58 *fps*, as shown in Fig. 1. It obtains up to 92.2 % and 12.8 % relative improvements over the anchor-based methods, *i.e.*, SiamRPN [22] and SiamRPN++ [21], respectively. On other datasets, the performance of our tracker is also competitive, compared with recent state-of-the-arts. In addition, we further equip our anchor-free tracker with a plug-in online update module, and enable it to capture the appearance changes of objects during inference. The online module further enhances the tracking performance, which shows the scalability of the proposed anchor-free tracking approach.

The main contributions of this work are two-fold. 1) We propose an object-aware anchor-free network based on the observation that the anchor-based method is difficult to refine the anchors whose overlap with the target object is small. The proposed algorithm can not only rectify the imprecise bounding-box predictons, but also learn an object-aware feature to enhance the matching accuracy. 2) We design a novel tracking framework by combining the proposed anchor-free network with an efficient feature combination module. The proposed tracking model achieves state-of-the-art performance on five benchmarks while running in real-time speeds.

2   Related Work

In this section, we review the related work on anchor-free mechanism and feature alignment in both tracking and detection, as well as briefly review recent Siamese trackers.

Siamese Trackers. The pioneering works, *i.e.*, SINT [35] and SiamFC [1], employ Siamese networks to offline train a similarity metric between the object target and candidate image patches. SiamRPN [22] improves it with a region proposal network, which amounts to a target-specific anchor-based detector. With the predefined anchor boxes, SiamRPN [22] can capture the scale changes of objects effectively. The follow-up studies mainly fall into two camps: designing more powerful backbone networks [21,50] or proposing more effective proposal networks [9]. Although these offline Siamese trackers have achieved very promising results, their tracking robustness is still inferior to the recent state-of-the-art online trackers, such as ATOM [4] and DiMP [2].

Anchor-Free Mechanism. Anchor-free approaches recently became popular in object detection tasks, because of their simplicity in architectures and superiority in performance [7,19,36]. Different from anchor-based methods which estimate the offsets of anchor boxes, anchor-free mechanisms predict the location of objects in a direct way. The early anchor-free work [47] predicts the intersection over union with objects, while recent works focus on estimating the keypoints of objects, *e.g.*, the object center [7] and corners [19]. Another branch of anchor-free detectors [30,36] predicts the object bounding box at each pixel, without using any references, *e.g.*, anchors or keypoints. The anchor-free mechanism in our method is inspired by, but different from that in the recent detection algorithm [36]. We will discuss the key differences in Sect. 3.4.

Feature Alignment. The alignment between visual features and reference ROIs (Regions of Interests) is vital for localization tasks, such as detection and tracking [40]. For example, ROIAlign [12] are commonly recruited in object detection to align the features with the reference anchor boxes, leading to remarkable improvements on localization precision. In visual tracking, there are also several approaches [15,41] considering the correspondence between visual features and candidate bounding boxes. However, these approaches only take account of the bounding boxes with high classification scores. If the high scores indicate the background regions, then the corresponding features will mislead the detection of target objects. To address this, we propose a novel feature alignment method, in which the alignment is independent of the classification results. We sample the visual features from the predicted bounding boxes directly, without considering the classification score, generating object-aware features. This object-aware features, in turn, help the classification of foreground and background.

3   Object-Aware Anchor-Free Networks

This section proposes the **O**bject-aware **a**nchor-free **n**etworks (Ocean) for visual tracking. The network architecture consists of two components: an object-aware classification network for foreground-background probability prediction and a regression network for target scale estimation. The input features to these two networks are generated by a shared backbone network (elaborated in Sect. 4.1). We introduce the regression network first, followed by the classification branch,

(a) Regression   (b) Regular-region Classification (c) Object-aware Classification

Fig. 2. (a) Regression: the pixels in groundtruth box, *i.e.* the red region, are labeled as the positive samples in training. (b) Regular-region classification: the pixels closing to the target's center, *i.e.* the red region, are labeled as the positive samples. The purple points indicate the sampled positions of a location in the score map. (c) Object-aware classification: the *IoU* of predicted box and groundtruth box, *i.e.*, the region with red slash lines, is used as the label during training. The cyan points represent the sampling positions for extracting object-aware features. The yellow arrows indicate the offsets induced by spatial transformation. Best viewed in color. (Color figure online)

because the regression branch provides object scale information to enhance the classification of the target object and background.

3.1 Anchor-Free Regression Network

Revisiting recent anchor-based trackers [21,22], we observed that the trackers drift speedily when the predicted bounding box becomes unreliable. The underlying reason is that, during training, these approaches only consider the anchor boxes whose *IoU* with groundtruth are larger than a high threshold, *i.e.*, $IoU \geq 0.6$. Hence, these approaches lack the competence to amend the weak predictions, *e.g.*, the boxes whose overlap with the target are small.

To remedy this issue, we introduce a novel anchor-free regression for visual tracking. It considers all the pixels in the groundtruth bounding box as the training samples. The core idea is to estimate the distances from each pixel within the target object to the four sides of the groundtruth bounding box. Specifically, let $B = (x_0, y_0, x_1, y_1) \in \mathbb{R}^4$ denote the top-left and bottom-right corners of the groundtruth bounding box of a target object. A pixel is considered as the regression sample if its coordinates (x, y) fall into the groundtruth box B. Hence, the labels $T^* = (l^*, t^*, r^*, b^*)$ of training samples are calculated as

$$l^* = x - x_0, t^* = y - y_0,$$
$$r^* = x_1 - x, b^* = y_1 - y, \tag{1}$$

which represent the distances from the location (x, y) to the four sides of the bounding box B, as shown in Fig. 2(a). The learning of the regression network is through four 3×3 convolution layers with channel number of 256, followed by one 3×3 layer with channel number of 4 for predicting the distances. As shown in Fig. 3, the upper "Conv" block indicates the regression network.

This anchor-free regression allows for all the pixels in the groundtruth box during training, thus it can predict the scale of target objects even when only a small region is identified as foreground. Consequently, the tracker is capable of rectifying weak predictions during inference to some extent.

3.2   Object-Aware Classification Network

In prior Siamese tracking approaches [1,21,22], the classification confidence is estimated by the feature sampled from a fixed regular region in the feature map, e.g., the purple points in Fig. 2(b). This sampled feature depicts a fixed local region of the image, and it is not scalable to the change of object scale. As a result, the classification confidence is not reliable in distinguishing the target object from complex background.

To address this issue, we propose a feature alignment module to learn an object-aware feature for classification. The alignment module transforms the fixed sampling positions of a convolution kernel to align with the predicted bounding box. Specifically, for each location (d_x, d_y) in the classification map, it has a corresponding object bounding box $M = (m_x, m_y, m_w, m_h)$ predicted by the regression network, where m_x and m_y denote the box center while m_w and m_h represent its width and height. Our goal is to estimate the classification confidence for each location (d_x, d_y) by sampling features from the corresponding candidate region M. The standard 2D convolution with kernel size of $k \times k$ samples features using a fixed regular grid $\mathcal{G} = \{(- \lfloor k/2 \rfloor, - \lfloor k/2 \rfloor), ..., (\lfloor k/2 \rfloor, \lfloor k/2 \rfloor)\}$, where $\lfloor \cdot \rfloor$ denotes the floor function. The regular grid $\mathcal{G}$ cannot guarantee the sampled features cover the whole content of region M.

Therefore, we propose to equip the regular sampling grid $\mathcal{G}$ with a spatial transformation $\mathcal{T}$ to convert the sampling positions from the fixed region to the predicted region M. As shown in Fig. 2(c), the transformation $\mathcal{T}$ (the dashed yellow arrows) is obtained by measuring the relative direction and distance from the sampling positions in $\mathcal{G}$ (the purple points) to the positions aligned with the predicted bounding box (the cyan points). With the new sampling positions, the object-aware feature is extracted by the feature alignment module, which is formulated as

$$\mathbf{f}[u] = \sum_{g \in \mathcal{G}, \Delta t \in \mathcal{T}} \mathbf{w}[g] \cdot \mathbf{x}[u + g + \Delta t], \qquad (2)$$

where $\mathbf{x}$ represents the input feature map, $\mathbf{w}$ denotes the learned convolution weight, u indicates a location on the feature map, and $\mathbf{f}$ represents the output object-aware feature map. The spatial transformation $\Delta t \in \mathcal{T}$ represents the distance vector from the original regular sampling points to the new points aligned with the predicted bounding box. The transformation is defined as

$$\mathcal{T} = \{(m_x, m_y) + \mathcal{B}\} - \{(d_x, d_y) + \mathcal{G}\}, \qquad (3)$$

where $\{(m_x, m_y) + \mathcal{B}\}$ represents the sampling positions aligned with M, e.g., the cyan points in Fig. 2(c), $\{(d_x, d_y) + \mathcal{G}\}$ indicates the regular sampling positions used in standard convolution, e.g., the purple points in Fig. 2(c), and

$\mathcal{B} = \{(-m_w/2, -m_h/2), ..., (m_w/2, m_h/2)\}$ denotes the coordinates of the new sampling positions (*e.g.*, the cyan points in Fig. 2(c)) relative to the box center (*e.g.*, (m_x, m_y)). It is worth noting that when the transformation $\Delta t \in \mathcal{T}$ is set to 0 in Eq. (2), the feature sampling mechanism is degenerated to the fixed sampling on regular points, generating the regular-region feature. The transformations of the sampling positions are adaptive to the variations of the predicted bounding boxes in video frames. Thus, the extracted object-aware feature is robust to the changes of object scale, which is beneficial for feature matching during tracking. Moreover, the object-aware feature provides a global description of the candidate targets, which enables the distinguish of the object and background to be more reliable.

We exploit both the object-aware feature and the regular-region feature to predict whether a region belongs to target object or image background. For the classification based upon the object-aware feature, we apply a standard convolution with kernel size of 3 × 3 over $\mathbf{f}$ to predict the confidence p_o (visualized as the "OA.Conv" block of the classification network in Fig. 3). For the classification based on the regular-region feature, four 3 × 3 standard convolution layers with channel number of 256, followed by one standard 3 × 3 layer with channel number of one are performed over the regular-region feature $\mathbf{f}'$ to predict the confidence p_r (visualized as the "Conv" block of the classification network in Fig. 3). Calculating the summation of the confidence p_o and p_r obtains the final classification score. The object-aware feature provides a global description of the target, thus enhancing the matching accuracy of candiate regions. Meanwhile, the regular-region feature concentrates on local parts of images, which is robust to localize the center of target objects. The combination of the two features improves the reliability of the classification network.

3.3   Loss Function

To optimize the proposed anchor-free networks, we employ IoU loss [47] and binary cross-entropy (BCE) loss [6] to train the regression and classification networks jointly. In regression, the loss is defined as

$$\mathcal{L}_{reg} = -\sum_i ln(IoU(p_{reg}, T^*)), \tag{4}$$

where p_{reg} denotes the prediction, and i indexes the training samples. In classification, the loss $\mathcal{L}_o$ based upon the object-aware feature $\mathbf{f}$ is formulated as

$$\mathcal{L}_o = -\sum_j p_o^* log(p_o) + (1 - p_o^*)log(1 - p_o), \tag{5}$$

while the loss $\mathcal{L}_r$ based upon the regular-region feature $\mathbf{f}'$ is formulated as

$$\mathcal{L}_r = -\sum_j p_r^* log(p_r) + (1 - p_r^*)log(1 - p_r), \tag{6}$$

where p_o and p_r are the classification score maps computed over the object-aware feature and regular-region feature respectively, j indexes the training samples for classification, and p_o^* and p_r^* denote the groundtruth labels. More concretely, p_o^* is a probabilistic label, in which each value indicates the IoU between the predicted bounding box and groundtruth, $i.e.$, the region with red slash lines in Fig. 2(c). p_r^* is a binary label, where the pixels closing to the center of the target are labeled as 1, $i.e.$, the red region in Fig. 2(b), which is formulated as

$$p_r^*[v] = \begin{cases} 1, & if\,||v-c|| \leq R, \\ 0, & otherwise. \end{cases} \tag{7}$$

The joint training of the entire object-aware anchor-free networks is to optimize the following objective function:

$$\mathcal{L} = \mathcal{L}_{reg} + \lambda_1 \mathcal{L}_o + \lambda_2 \mathcal{L}_r, \tag{8}$$

where λ_1 and λ_2 are the tradeoff hyperparameters.

3.4   Relation to Prior Anchor-Free Work

Our anchor-free mechanism shares similar spirit with recent detection methods [7,19,36] (discussed in Sect. 2). In this section, we further discuss the differences to the most related work, $i.e.$, FCOS [36]. Both FCOS and our method predict the object locations directly on the image plane at pixel level. However, our work differs from FCOS [36] in two fundamental ways. 1) In FCOS [36], the training samples for the classification and regression networks are identical. Both are sampled from the positions within the groundtruth boxes. Differently, in our method, the data sampling strategies for classification and regression are asymmetric which is tailored for tracking tasks. More specifically, the classification network only considers the pixels closing to the target as positive samples ($i.e.$, $R \leq 16$ pixels), while the regression network considers all the pixels in the ground-truth box as training samples. This fine-grained sampling strategy guarantees the classification network can learn a robust similarity metric for region matching, which is important for tracking. 2) In FCOS [36], the objectness score is calculated with the feature extracted from a fixed regular-region, similar to the purple points in Fig. 2(b). By contrast, our method additionally introduce an object-aware feature, which captures the global appearance of target objects. The object-aware feature aligns the sampling regions with the predicted bounding box ($e.g.$, cyan points in Fig. 2(c)), thus it is adaptive to the scale change of objects. The combination of the regular-region feature and the object-aware feature allows the classification to be more reliable, as verified in Sect. 5.3.

4   Object-Aware Anchor-Free Tracking

This section depicts the tracking algorithm building upon the proposed object-aware anchor-free networks (Ocean). It contains two parts: an offline anchor-free model and an online update model, as illustrated in Fig. 3.

4.1   Framework

The offline tracking is built on the object-aware anchor-free networks, consisting of three steps: feature extraction, combination and target localization.

Feature Extraction. Following the architecture of Siamese tracker [1], our approach takes an image pair as input, *i.e.*, an exemplar image and a candidate search image. The exemplar image represents the object of interest, *i.e.*, an image patch centered on the target object in the first frame, while the search image is typically larger and represents the search area in subsequent video frames. Both inputs are processed by a modified ResNet-50 [13] backbone and then yield two feature maps. More specifically, we cut off the last stage of the standard ResNet-50 [13], and only retain the first fourth stages as the backbone. The first three stages share the same structure as the original ResNet-50. In the fourth stage, the convolution stride of down-sampling unit [13] is modified from 2 to 1 to increase the spatial size of feature maps, meanwhile, all the 3×3 convolutions are augmented with a dilation with stride of 2 to increase the receptive fields. These modifications increase the resolution of output features, thus improving the feature capability on object localization [3,21].

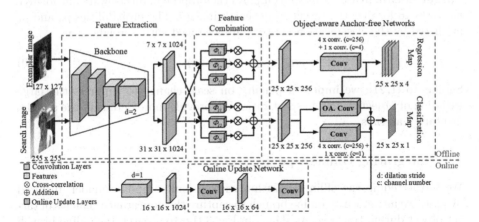

Fig. 3. Overview of the proposed tracking framework, consisting of an offline anchor-free part (top) and an online model update part (bottom). The offline tracking includes feature extraction, feature combination and target localization with object-aware anchor-free networks, as elaborated in Sect. 4.1. The plug-in online update network models the appearance changes of target objects, as detailed in Sect. 4.2. Φ_{ab} indicates a 3×3 convolution layer with dilation stride of a along the X-axis and b along the Y-axis.

Feature Combination. This step exploits a depth-wise cross-correlation operation [21] to combine the extracted features of the exemplar and search images, and generates the corresponding similarity features for the subsequent target localization. Different from the previous works performing the cross-correlation on multi-scale features [21], our method only performs over a single scale, *i.e.*, the last stage of the backbone. We pass the single-scale features through three parallel dilated convolution layers [48], and then fuse the correlation features through point-wise summation, as presented in Fig. 3 (feature combination).

For concreteness, the feature combination process can be formulated as

$$S = \sum_{ab} \Phi_{ab}(\mathbf{f}_e) * \Phi_{ab}(\mathbf{f}_s) \tag{9}$$

where $\mathbf{f}_e$ and $\mathbf{f}_s$ represent the features of the exemplar and search images respectively, Φ_{ab} indicates a single dilated convolution layer, and $*$ denotes the cross-correlation operation [1]. The kernel size of the dilated convolution Φ_{ab} is set to 3×3, while the dilation strides are set to a along the X-axis and b along the Y-axis. Φ_{ab} also reduces the feature channels from 1024 to 256 to save computation cost. In experiments, we found that increasing the diversity of dilations can improve the representability of features, thereby we empirically choose three different dilations, whose strides are set to $(a, b) \in \{(1, 1), (1, 2), (2, 1)\}$. The convolutions with different dilations can capture the features of regions with different scales, improving the scale invariance of the final combined features.

Target Localization. This step employs the proposed object-aware anchor-free networks to localize the target from search images. The probabilities p_o and p_r predicted by the classification network are averaged with a weight ω as

$$p_{cls} = \omega p_o + (1 - \omega)p_r. \tag{10}$$

Similar to [1, 21], we impose a penalty on scale change to suppress the large variation of object size and aspect ratio. We provide more details in the supplementary materials.

4.2 Integrating Online Update

We further equip the offline algorithm with an online update model. Inspired by [2, 4], we introduce an online branch to capture the appearance changes of target object during tracking. As shown in Fig. 3 (bottom part), the online branch inherits the structure and parameters from the first three stages of the backbone network, *i.e.*, modified ResNet-50 [13]. The fourth stage keep the same structure as the backbone, but its initial parameters are obtained through the pretraining strategy proposed in [2]. For model update, we employ the fast conjugate gadient algorithm [2] to train online branch during inference. The foreground score maps estimated by the online branch and the classification branch are weighted as

$$p = \omega' p_{onl} + (1 - \omega')\hat{p}_{cls}, \tag{11}$$

where ω' represents the weights between the classification score $\hat{p}_{cls}$ and the online estimation score p_{onl}. Note that the IoUNet in [2, 4] is not used in our model. We refer readers to [2, 4] for more details.

5   Experiments

This section presents the results of our Ocean tracker on five tracking benchmark datasets, with comparisons to the state-of-the-art algorithms. Experimental analysis is provided to evaluate the effects of each component in our model.

5.1   Implementation Details

Training. The backbone network is initialized with the parameters pretrained on ImageNet [32]. The proposed trackers are trained on the datasets of Youtube-BB [29], ImageNet VID [32], ImageNet DET [32], GOT-10 k [14] and COCO [26]. The size of input exemplar image is 127×127 pixels, while the search image is 255×255 pixels. We use synchronized SGD [20] on 8 GPUs, with each GPU hosting 32 images, hence the mini-batch size is 256 images per iteration. There are 50 epochs in total. Each epoch uses 6×10^5 training pairs. For the first 5 epochs, we start with a warmup learning rate of 10^{-3} to train the object-aware anchor-free networks, while freezing the parameters of the backbone. For the remaining epochs, the backbone network is unfrozen, and the whole network is trained end-to-end with a learning rate exponentially decayed from 5×10^{-3} to 10^{-5}. The weight decay and momentum are set to 10^{-3} and 0.9, respectively. The threshold R of the classification label in Eq. (7) is set to 16 pixels. The weight parameters λ_1 and λ_2 in Eq. (8) are set to 1 and 1.2, respectively.

We noticed that the training settings (data selection, iterations, *etc.*) are often different in recent trackers, *e.g.*, SiamRPN [22], SiamRPN++[21], ATOM [4] and DiMP [4]. It is difficult to compare different models under a unified training schedule. But for a fair comparison, we additionally evaluate our method and SiamRPN++ [21] under the same training setting, as discussed in Sect. 5.3.

Testing. For the offline model, tracking follows the same protocols as in [1, 22]. The feature of the target object is computed once at the first frame, and then is continuously matched to subsequent search images. The fusion weight ω of the object-aware classification score in Eq. (10) is set to 0.07, while the weight ω' in Eq. (11) is set to 0.5. These hyper-parameters in testing are selected with the tracking toolkit [50], which contains an automated parameter tuning algorithm. Our trackers are implemented using Python 3.6 and PyTorch 1.1.0. The experiments are conducted on a server with 8 Tesla V100 GPUs and a Xeon E5-2690 2.60 GHz CPU. Note that we run the proposed tracker three times, the standard deviation of the performance is ±0.5%, demonstrating the stability of our model. We report the average performance of the three-time runs in the following comparisons.

Evaluation Datasets and Metrics. We use five benchmark datasets including VOT-2018 [17], VOT-2019 [18], OTB-100 [44], GOT-10k [14] and LaSOT [8] for tracking performance evaluation. In particular, VOT-2018 [17] contains 60 sequences. VOT-2019 [18] is developed by replacing the 20% least challenging videos in VOT-2018 [17]. We adopt the Expected Average Overlap (EAO) [18] which takes both accuracy (A) and robustness (R) into account to evaluate

Table 1. Performance comparisons on VOT-2018 benchmark. Red, green and blue fonts indicate the top-3 trackers. "Ocean" denotes our propose model.

	CRPN [9]	ECO [5]	STRCF [23]	LADCF [45]	ATOM [4]	SRCNN [39]	SiamRPN++ [21]	DiMP [2]	Ocean offline	Ocean online
EAO ↑	0.273	0.280	0.345	0.389	0.401	0.408	0.414	**0.440**	0.467	0.489
A ↑	0.550	0.484	0.523	0.503	0.590	0.609	0.600	0.597	**0.598**	0.592
R ↓	0.320	0.276	0.215	0.159	0.204	0.220	0.234	**0.153**	0.169	0.117

Table 2. Performance comparisons on VOT-2019. The "DiMP^r" and "DiMP^b" indicate realtime and baseline performances of DiMP, as reported in [18].

	MemDTC [46]	SiamMASK [43]	SiamRPN++ [21]	ATOM [4]	STN [37]	DiMP^r [2]	DiMP^b [2]	Ocean offline	Ocean online
EAO ↑	0.228	0.287	0.292	0.301	0.314	0.321	0.379	**0.327**	0.350
A ↑	0.485	0.594	0.580	0.603	0.589	0.582	0.594	0.590	0.594
R ↓	0.587	0.461	0.446	0.411	**0.349**	0.371	0.278	0.376	0.316

Table 3. Performance comparisons on GOT-10k test set.

	CFNet [38]	MDNet [27]	SiamFC [1]	ECO [5]	DSiam [11]	SiamRPN++ [21]	ATOM [4]	DiMP [2]	Ocean offline	Ocean online
AO ↑	0.261	0.299	0.392	0.395	0.417	0.518	0.556	0.611	0.592	0.611
SR$_{0.5}$ ↑	0.243	0.303	0.406	0.407	0.461	0.618	0.634	0.712	**0.695**	0.721

overall performance. The standardized OTB-100 [44] benchmark consists of 100 videos. Two metrics, *i.e.*, precision (Prec.) and area under curve (AUC) are used to rank the trackers. GOT-10k [14] is a large-scale dataset containing over 10 thousand videos. The trackers are evaluated using an online server on a test set of 180 videos. It employs the widely used average overlap (AO) and success rate (SR) as performance indicators. Compared to these benchmark datasets, LaSOT [8] has longer sequences, with an average of 2,500 frames per sequence. Success (SUC) and precision (Prec.) are used to evaluate tracking performance.

5.2 State-of-the-art Comparison

To extensively evaluate the proposed method, we compare it with 22 state-of-the-art trackers, which cover most of current representative methods. There are 9 anchor-based Siamese framework based methods (SiamFC [1], GradNet [24], DSiam [11], MemDTC [46], SiamRPN [22], C-RPN [9], SiamMASK [43], SiamRPN++ [21] and SiamRCNN [39]), 8 discriminative correlation filter based methods (CFNet [38], ECO [5], STRCF [23], LADCF [45], UDT [42], STN [37], ATOM [4] and DiMP [2]), 3 multi-domain learning based methods (MDNet [27], RT-MDNet [15] and VITAL [34]), 1 graph network based method (GCT [10]) and 1 meta-learning based tracker (MetaCREST [28]). The results are summarized in Table. 1 - 3 and Fig. 4.

VOT-2018. The evaluation on VOT-2018 is performed by the official toolkit [17]. As shown in Table. 1, Our offline Ocean tracker outperforms the champion method of VOT-2018, *i.e.*, (LADCF [45]), by 7.8 points. Compared to the state-of-the-art offline tracker SiamRPN++ [21], our offline model achieves EAO improvements of 5.3 points, while running faster, as shown in Fig. 1. It is worth noting that the improvements mainly come from the robustness score, which obtains 27.8% relative increases over SiamRPN++. Moreover, our offline model is superior to the recent online trackers ATOM [4] and DiMP [2]. The online augmented model further improves our tracker by 2.2 points in terms of EAO.

VOT-2019. Table. 2 reports the evaluation results with the comparisons to recent prevailing trackers on VOT-2019. We can see that the recent proposed DiMP [2] achieves the best performance, while our method ranks second. However, in real-time testing scenarios, our offline Ocean tracker achieves the best performance, surpassing DiMP^r by 0.6 points in terms of EAO. Moreover, the EAO of our offline model surpasses SiamRPN++ [21] by 3.5 points.

GOT-10k. The evaluation on GOT-10k follows the protocols in [14]. The proposed offline Ocean tracker model achieves the state-of-the-art AO score of 0.592, outperforming SiamRPN++ [21], as shown in Table. 3. Our online model improves the AO by 4.5 points over ATOM [4], while outperforming DiMP [2] by 0.9 points in terms of success rate.

OTB-100. The last evaluation in short-term tracking is performed on the classical OTB-100 benchmark. As reported in Fig. 4, among the compared methods, our online tracker achieves the best precision score of 0.920, while DiMP [21] achieves best AUC score of 0.686.

LaSOT. To further evaluate the proposed models, we report the results on LaSOT, which is larger and more challenging than previous benchmarks. The results on the 280 videos test set is presented in Fig. 4. Our offline Ocean tracker achieves SUC score of 0.527, outperforming SiamRPN++ with score of 0.496. Com-

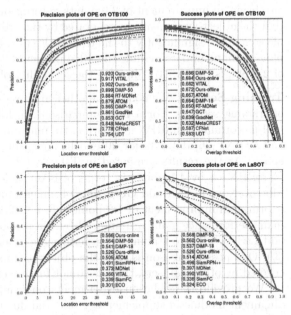

Fig. 4. Success and precision plots on OTB-100 [44] (top) and LaSOT [8] (bottom).

pared to ATOM [4], our online tracker improves the SUC score by 4.6 points, giving comparable results to top-ranked tracker DiMP-50 [2]. Moreover, the proposed online tracker achieves the best precision score of 0.566.

5.3 Analysis of the Proposed Method

Component-Wise Analysis. To verify the efficacy of the proposed method, we perform a component-wise analysis on the VOT-2018 benchmark, as presented in Table. 4. The baseline model consists of a backbone network (detailed in Sect. 4.1), an anchor-free regression network (detailed in Sect. 3.1) and a classification network using regular-region feature (detailed in Sect. 3.2). In the training of baseline model, all pixels in the groundtruth box are considered as positive samples. The baseline model obtains an EAO of 0.358. In ②, the "centralized sampling" indicates that we only consider the pixels closing to the target's center as positive samples in the training of classification (formulated as Eq. (7)). It brings significant gains, *i.e.*, 3.8 points on EAO (② *vs.* ①). This verifies that the sampling helps to learn a robust similarity metric for region matching. Adding the feature combination module (detailed in Sect. 4.1) can bring a large improvement of 4.2 points in terms of EAO (③ *vs.* ②). This demonstrates the effectiveness of the proposed irregular dilated convolution module. It introduces a multi-scale modeling of target objects, without increasing much computation overhead. Furthermore, the object-aware classification (detailed in Sect. 3.2) can also bring an improvement of 2.9 points in terms of EAO (④ *vs.* ③). This shows that the object-aware features generated by the proposed feature alignment module contribute significantly to the tracker. Finally, the tracker equipped with the plug-in online update module (detailed in Sect. 4.2) yields another improvement of 2.2 points (⑤ *vs.* ④), showing the scalability of the proposed framework.

Training Setting. We conduct another ablation study to evaluate the impact of training settings. For a fair comparison, we follow the same setting as the well-performing SiamRPN++[21], *i.e.* training on YTB [29], VID [32], DET [32] and COCO [26] datasets for 20 epochs and using 6×10^5 image pairs in each epoch. As the results presented in Table. 5 (② *v.s.* ①), our model surpasses SiamRPN++ by 4.1 points in terms of EAO under the same training settings. Moreover, we further add GOT-10k [14] images into training, and observe that it brings an improvement of 1.2 points (③ *v.s.* ②). This demonstrates that the main performance gains are induced by the proposed model. If we continue to add LaSOT [8] into training, the performance cannot improve further (④ *v.s.* ③). One possible reason is that the object categories in LaSOT [8] have been covered by other datasets, thus it cannot further elevate model capacities.

We provide the ablation experiments on feature combination module and alignment module in supplementary material due to space limit.

Table 4. Component-wise analysis.

#Num	Components	EAO
①	baseline	0.358
②	+ centralized sampling	0.396
③	+ feature combination	0.438
④	+ object-aware classification	0.467
⑤	+ online update	0.489

Table 5. Analysis of training settings on VOT2018.

#Num	Settings	EAO
①	SiamRPN++ [21]	0.414
②	Ocean tracker (ours)	0.455
③	+ GOT-10k in training	0.467
④	+ LaSOT in training	0.462

6 Conclusion

In this work, we propose a novel object-aware anchor-free tracking framework (Ocean) based upon the observation that the anchor-based method is difficult to refine the anchors whose overlap with the target objects are small. Our model directly regresses the positions of target objects in a video frame instead of predicting offsets for predefined anchors. Moreover, the learned object-aware feature by the alignment module provides a global description of the target, contributing to the reliable matching of objects. The experiments demonstrate that the proposed tracker achieves state-of-the-art performance on five benchmark datasets. In the future work, we will study applying our framework to other online video tasks, *e.g.*, video object detection and segmentation.

Acknowledgement. This work is supported by the Natural Science Foundation of China (Grant No. U1636218, 61751212, 61721004, 61902401, 61972071), Natural Science Foundation of Beijing (Grant No.L172051), the CAS Key Research Program of Frontier Sciences (Grant No. QYZDJ-SSW-JSC040), and the Natural Science Foundation of Guangdong (No. 2018B030311046).

References

1. Bertinetto, L., Valmadre, J., Henriques, J.F., Vedaldi, A., Torr, P.H.S.: Fully-convolutional siamese networks for object tracking. In: Hua, G., Jégou, H. (eds.) ECCV 2016. LNCS, vol. 9914, pp. 850–865. Springer, Cham (2016). https://doi.org/10.1007/978-3-319-48881-3_56

2. Bhat, G., Danelljan, M., Gool, L.V., Timofte, R.: Learning discriminative model prediction for tracking. In: ICCV, pp. 6182–6191 (2019)

3. Chen, L.C., et al.: DeepLab: Semantic image segmentation with deep convolutional nets, atrous convolution, and fully connected CRFS. TPAMI **40**(4), 834–848 (2017)

4. Danelljan, M., Bhat, G., Khan, F.S., Felsberg, M.: Atom: accurate tracking by overlap maximization. In: CVPR, pp. 4660–4669 (2019)

5. Danelljan, M., Bhat, G., Shahbaz Khan, F., Felsberg, M.: ECO: efficient convolution operators for tracking. In: CVPR, pp. 6931–6939 (2017)

6. De Boer, P.T., Kroese, D.P., Mannor, S., Rubinstein, R.Y.: A tutorial on the cross-entropy method. Ann. Oper. Res. **134**(1), 19–67 (2005)

7. Duan, K., Bai, S., Xie, L., Qi, H., Huang, Q., Tian, Q.: Centernet: keypoint triplets for object detection. In: ICCV, pp. 6569–6578 (2019)

8. Fan, H., Lin, L., Yang, F., et al.: LaSOT: A high-quality benchmark for large-scale single object tracking. In: CVPR, pp. 5374–5383 (2019)
9. Fan, H., Ling, H.: Siamese cascaded region proposal networks for real-time visual tracking. In: CVPR. pp, 7952–7961 (2019)
10. Gao, J., et al.: Graph convolutional tracking. In: CVPR, pp. 4649–4659 (2019)
11. Guo, Q., Feng, W., Zhou, C., Huang, R., Wan, L., Wang, S.: Learning dynamic siamese network for visual object tracking. In: ICCV, pp. 1763–1771 (2017)
12. He, K., Gkioxari, G., et al.: Mask R-CNN. In: ICCV, pp. 2961–2969 (2017)
13. He, K., Zhang, X., Ren, S., Sun, J.: Deep residual learning for image recognition. In: CVPR, pp. 770–778 (2016)
14. Huang, L., Zhao, X., Huang, K.: Got-10k: a large high-diversity benchmark for generic object tracking in the wild. TPAMI 1(1), 1–17 (2019)
15. Jung, I., Son, J., et al.: Real-Time MDNet. In: ECCV, pp. 83–98 (2018)
16. Kiani Galoogahi, H., Fagg, A., Lucey, S.: Learning background-aware correlation filters for visual tracking. In: ICCV, pp. 1135–1143 (2017)
17. Kristan, M., Leonardis, A., Matas, et al.: The sixth visual object tracking vot2018 challenge results. In: ECCVW, pp. 1–52 (2018)
18. Kristan, M., Matas, et al.: The seventh visual object tracking vot2019 challenge results. In: ICCVW, pp. 1–36 (2019)
19. Law, H., Deng, J.: Cornernet: detecting objects as paired keypoints. In: ECCV, pp. 734–750 (2018)
20. LeCun, Y., Boser, B., et al.: Backpropagation applied to handwritten zip code recognition. Neural Comput. 1(4), 541–551 (1989)
21. Li, B., Wu, W., Wang, Q., Zhang, F., Xing, J., Yan, J.: SiamRPN++: evolution of siamese visual tracking with very deep networks. In: CVPR, pp. 4282–4291 (2019)
22. Li, B., Yan, J., Wu, W., Zhu, Z., Hu, X.: High performance visual tracking with siamese region proposal network. In: CVPR, pp. 8971–8980 (2018)
23. Li, F., Tian, C., Zuo, W., Zhang, L., Yang, M.H.: Learning spatial-temporal regularized correlation filters for visual tracking. In: CVPR, pp. 4904–4913 (2018)
24. Li, P., Chen, B., Ouyang, W., Wang, D., Yang, X., Lu, H.: GradNet: gradient-guided network for visual object tracking. In: ICCV, pp. 6162–6171 (2019)
25. Li, X., Hu, W., Shen, C., Zhang, Z., Dick, A., Hengel, A.V.D.: A survey of appearance models in visual object tracking. TIST 4(4), 1–48 (2013)
26. Lin, T.S., et al.: Microsoft COCO: common objects in context. In: Fleet, D., Pajdla, T., Schiele, B., Tuytelaars, T. (eds.) ECCV 2014. LNCS, vol. 8693, pp. 740–755. Springer, Cham (2014). https://doi.org/10.1007/978-3-319-10602-1_48
27. Nam, H., Han, B.: Learning multi-domain convolutional neural networks for visual tracking. In: CVPR, pp. 4293–4302 (2016)
28. Park, E., Berg, A.C.: Meta-tracker: fast and robust online adaptation for visual object trackers. In: ECCV, pp. 569–585 (2018)
29. Real, E., Shlens, J., et al.: Youtube-boundingboxes: a large high-precision human-annotated data set for object detection in video. In: CVPR, pp. 7464–7473 (2017)
30. Redmon, J., Divvala, S., Girshick, R., Farhadi, A.: You only look once: unified, real-time object detection. In: CVPR, pp. 779–788 (2016)
31. Ren, S., He, K., Girshick, R., Sun, J.: Faster R-CNN: towards real-time object detection with region proposal networks. In: NIPS, pp. 91–99 (2015)
32. Russakovsky, O., et al.: Imagenet large scale visual recognition challenge. Int. J. Comput. Vis. 115(3), 211–252 (2015). https://doi.org/10.1007/s11263-015-0816-y
33. Smeulders, A.W., Chu, D.M., Cucchiara, R., Calderara, S., Dehghan, A., Shah, M.: Visual tracking: an experimental survey. TPAMI 36(7), 1442–1468 (2013)

34. Song, Y., et al.: VITAL: visual tracking via adversarial learning. In: CVPR, pp. 8990–8999 (2018)
35. Tao, R., Gavves, E., Smeulders, A.W.: Siamese instance search for tracking. In: CVPR, pp. 1420–1429 (2016)
36. Tian, Z., Shen, C., Chen, H., He, T.: FCOS: fully convolutional one-stage object detection. In: ICCV, pp. 9627–9636 (2019)
37. Tripathi, A.S., Danelljan, M., Van Gool, L., Timofte, R.: Tracking the known and the unknown by leveraging semantic information. In: BMVC, pp. 1192–1198 (2019)
38. Valmadre, J., Bertinetto, L., Henriques, et al.: End-to-end representation learning for correlation filter based tracking. In: CVPR, pp. 2805–2813 (2017)
39. Voigtlaender, P., Luiten, J., Torr, P.H., Leibe, B.: Siam R-CNN: visual tracking by re-detection. arXiv preprint arXiv:1911.12836 (2019)
40. Vu, T., Jang, H., Pham, T.X., Yoo, C.: Cascade RPN: delving into high-quality region proposal network with adaptive convolution. In: NIPS, pp. 1430–1440 (2019)
41. Wang, G., et al.: SPM-tracker: series-parallel matching for real-time visual object tracking. In: CVPR, pp. 3643–3652 (2019)
42. Wang, N., Song, Y., Ma, C., Zhou, W., Liu, W., Li, H.: Unsupervised deep tracking. In: CVPR, pp. 1308–1317 (2019)
43. Wang, Q., Zhang, L., Bertinetto, L., Hu, W., Torr, P.H.: Fast online object tracking and segmentation: a unifying approach. In: CVPR, pp. 1328–1338 (2019)
44. Wu, Y., Lim, et al.: Object tracking benchmark. TPAMI, 37(9), 1834–1848 (2015)
45. Xu, T., Feng, Z.H., Wu, X.J., Kittler, J.: Learning adaptive discriminative correlation filters via temporal consistency preserving spatial feature selection for robust visual object tracking. TIP 28(11), 5596–5609 (2019)
46. Yang, T., Chan, A.B.: Learning dynamic memory networks for object tracking. In: ECCV, pp. 152–167 (2018)
47. Yu, J., Jiang, Y., Wang, Z., Cao, Z., Huang, T.: Unitbox: an advanced object detection network. In: ACM MM, pp. 516–520 (2016)
48. Zhang, H., et al.: Context encoding for semantic segmentation. In: CVPR, pp. 7151–7160 (2018)
49. Zhang, K., Zhang, L., Yang, M.H.: Fast compressive tracking. TPAMI 36(10), 2002–2015 (2014)
50. Zhang, Z., Peng, H.: Deeper and wider siamese networks for real-time visual tracking. In: CVPR, pp. 4591–4600 (2019)

Author Index

Agrawal, Susmit 497
Ak, Kenan E. 18
AlRegib, Ghassan 206
Ando, Takahiro 514
Ang Jr., Marcelo H. 530
Anguelov, Dragomir 279
Ashwath, Balraj 497

Babu, R. Venkatesh 497
Badrinarayanan, Vijay 104
Bai, Chunyan 279
Banerjee, Pratyay 379
Bansal, Ankan 51
Bansal, Mohit 447
Bao, Wentao 581
Baral, Chitta 379
Bartolozzi, James 104
Beker, Deniz 514
Berg, Tamara L. 447
Braunegg, A. 35

Caine, Benjamin 279
Cao, Zhiguo 682
Chakraborty, Amartya 35
Chan, Stanley H. 122
Chang, Xiaojun 431
Chaudhuri, Syomantak 312
Chellappa, Rama 51
Chen, Guang 666
Chen, Junwen 581
Chen, Lei 397
Chen, Qi 68
Chen, Qifeng 598, 615
Chen, Wei-Ting 754
Chen, Xuanlin 190
Chen, Yun 156
Chen, Yuxuan 598
Chen, Zailiang 190
Chen, Zhao-Min 633
Cheng, Shuyang 279
Chi, Yiheng 122
Choe, Tae Eun 666

Choi, Jinsoo 718
Cord, Matthieu 173
Cubuk, Ekin Dogus 279

Darrell, Trevor 360
Ding, Jian-Jiun 754
Ding, Wenchao 598
Duan, Lixin 547

Fang, Hao-Yu 754
Fu, Jianlong 771

Gaidon, Adrien 514
Gidaris, Spyros 173
Gnanasambandam, Abhiram 122
Gokhale, Tejas 379
Guo, Yanwen 633
Guo, Yuliang 666
Guo, Zhengkui 85

Han, Mingfei 431
He, Jiawei 397
He, Junjun 649
He, Ran 700
Hu, Han 227
Hu, Weiming 771
Hu, Xuefeng 312
Huang, Huaiyi 85
Huang, Qingqiu 85

Jain, Himalaya 173
Jia, Kui 68
Jiang, Jianmin 329
Jiang, Qinhong 547
Jiang, Zhenye 312
Jimeno Yepes, Antonio 564
Jin, Jiongchao 295
Jin, Xin 633
Jin, Ying 464

Kato, Hiroharu 514
Kehl, Wadim 514

Kim, Dong-Jin 718
Koltun, Vladlen 122
Komodakis, Nikos 173
Kong, Yu 581
Krishna Reddy, Mahesh Kumar 261
Krumdick, Michael 35
Kuo, Sy-Yen 754
Kweon, In So 718
Kwon, Gukyeong 206

Lape, Nicole 35
Le, Quoc V. 279
Leary, Sara 35
Lee, Gim Hee 530
Lei, Jie 447
Leng, Zhaoqi 279
Li, Bing 771
Li, Congcong 279
Li, Duo 615
Li, Quanquan 737
Li, Shuo 190
Li, Wei 1
Li, Xin 682
Liao, Miao 1
Liao, Renjie 156
Lin, Dahua 85
Lin, Guosheng 414
Lin, Stephen 227, 718
Lin, Zhe 18
Lindell, David B. 139
Liu, Fayao 414
Liu, Ziwei 85
Long, Mingsheng 464
Loy, Chen Change 700
Lu, Feixiang 1
Luo, Ping 547
Lv, Fengmao 414

Manville, Keith 35
Matsuoka, Toru 514
Merkhofer, Elizabeth 35
Metzler, Christopher 139
Miao, Jinghao 666
Morariu, Mihai Adrian 514
Mori, Greg 397
Murez, Zak 104

Nagahara, Hajime 345
Nawhal, Megha 397

Nevatia, Ram 312
Ngiam, Jiquan 279
Nishimura, Mark 139

Park, Jae Sung 360
Patil, Akshay Gadi 295
Peng, Houwen 771
Peng, Xiaojiang 649
Pérez, Patrick 173
Prabhakar, K. Ram 497
Prabhushankar, Mohit 206

Qian, Chen 700
Qiao, Yu 431, 649, 700

Rabinovich, Andrew 104
Rochan, Mrigank 261
Rohrbach, Anna 360

Sasagawa, Yukihiro 345
ShafieiBavani, Elaheh 564
Shao, Jing 737
Shen, Shaojie 598
Shlens, Jonathon 279
Singh, Durgesh Kumar 497
Sinha, Ayan 104
Song, Haoran 598
Song, Linsen 700
Song, Yang 279
Strickhart, Laura 35
Sun, Lin 68
Sun, Xiao 718

Tang, Ming 481
Tao, Chaofan 547
Tao, Dacheng 245
Temel, Dogancan 206
Tian, Meng 530
Tsai, Cheng-Che 754
Tung, Frederick 397

Urtasun, Raquel 156, 227

Vasudevan, Vijay 279

Walmer, Matthew 35
Wang, Jianmin 464
Wang, Jingao 666
Wang, Jinghua 329

Wang, Jinqiao 481
Wang, Kaisiyuan 700
Wang, Liwei 227
Wang, Michael Yu 598
Wang, Shenlong 156
Wang, Tong 481
Wang, Ximei 464
Wang, Yali 431
Wang, Yang 261
Wang, Yaowei 481
Wang, Yilin 18
Wang, Zhixin 68
Wei, Xiu-Shen 633
Wetzstein, Gordon 139
Wu, Qianyi 700
Wu, Wayne 700
Wu, Wenhao 649

Xian, Ke 682
Xiong, Haipeng 682
Xiong, Xin 682
Xiong, Zhang 295
Xu, Chang 245
Xu, Ning 18
Xu, Yinghao 227
Xue, Han 227

Yan, Junjie 737
Yang, Bin 156
Yang, Ruigang 1
Yang, Xiaofeng 414
Yang, Yezhou 379

Yang, Ze 227
Yang, Zhenheng 312
Yang, Zhuoqian 700
Yao, Anbang 615
Ye, Jin 649
Ye, Linwei 261
Yu, Licheng 447
Yuan, Kun 737
Yuille, Alan 68

Zeng, Wei 481
Zeng, Wenyuan 156
Zhai, Mengyao 397
Zhang, Hao 295
Zhang, Sibo 1
Zhang, Weide 666
Zhang, Xikun 245
Zhang, Yuqi 85
Zhang, Yuting 51
Zhang, Zheng 227
Zhang, Zhihan 312
Zhang, Zhipeng 771
Zhao, Borui 633
Zhao, Chaoyang 481
Zhao, Chen 682
Zhao, Peitao 666
Zhao, Rongchang 190
Zhong, Xu 564
Zhou, Dingfu 1
Zhu, Yousong 481
Zoph, Barret 279

nted in the United States
kmasters